SUSTAINABLE HYDROGEN
PRODUCTION

SUSTAINABLE HYDROGEN PRODUCTION

IBRAHIM DINCER

CALIN ZAMFIRESCU

ELSEVIER

AMSTERDAM · BOSTON · HEIDELBERG · LONDON · NEW YORK · OXFORD
PARIS · SAN DIEGO · SAN FRANCISCO · SINGAPORE · SYDNEY · TOKYO

Elsevier
Radarweg 29, PO Box 211, 1000 AE Amsterdam, Netherlands
The Boulevard, Langford Lane, Kidlington, Oxford OX5 1GB, United Kingdom
50 Hampshire Street, 5th Floor, Cambridge, MA 02139, United States

Notices
Knowledge and best practice in this field are constantly changing. As new research and experience broaden our understanding, changes in research methods, professional practices, or medical treatment may become necessary.

Practitioners and researchers must always rely on their own experience and knowledge in evaluating and using any information, methods, compounds, or experiments described herein. In using such information or methods they should be mindful of their own safety and the safety of others, including parties for whom they have a professional responsibility.

To the fullest extent of the law, neither the Publisher nor the authors, contributors, or editors, assume any liability for any injury and/or damage to persons or property as a matter of products liability, negligence or otherwise, or from any use or operation of any methods, products, instructions, or ideas contained in the material herein.

Library of Congress Cataloging-in-Publication Data
A catalog record for this book is available from the Library of Congress

British Library Cataloguing-in-Publication Data
A catalogue record for this book is available from the British Library

ISBN: 978-0-12-801563-6

For information on all Elsevier publications
visit our website at https://www.elsevier.com/

Working together
to grow libraries in
developing countries

www.elsevier.com • www.bookaid.org

Publisher: Joe Hayton
Acquisition Editor: Raquel Zanol
Developmental Editor: Mariana Kühl Leme
Production Project Manager: Sruthi Satheesh
Designer: Greg Harris

Typeset by SPi Global, India

Contents

Preface

Increases in population and living standards in growing economies have put an unprecedented pressure on the production infrastructure of worldwide economies. There is a rapidly increasing need for synthetic materials, fuels, power, and thermal energy. The hydrogen economy is a key solution in this context because hydrogen is at the same time a constituent of a large palette of essential materials, chemicals, and fuels, and it is a source of power and heat.

Therefore, this book on sustainable methods of hydrogen production will benefit scientists and engineers working on methods of efficient and clean hydrogen generation from renewable (sustainable) energies. It stresses the scientific understanding, analysis, and assessment of sustainable hydrogen production, and the evaluation and improvement of these processes and methods. This approach is designed to present an intellectually rich and interesting text that is also practical. This is accomplished by introducing the fundamental methods of hydrogen production categorized by the type of energy source used: electrical, thermal, photonic, and biochemical. Where appropriate, the historical context of these methods is introduced. Thermodynamic concepts, through energy and exergy analysis methods, are used as examples and case studies to solve concrete power engineering problems. The book also serves as a unique source of information on advanced hydrogen generation systems and application, including integrated, hybrid, and multi-generation. It links advanced and clean technologies to environmental impact issues and sustainable development, and provides methodologies, models, and analysis techniques, to help achieve a more efficient and cost-effective use of resources and to improve energy security and sustainability. Finally, its content may also make it a distinguished book for senior undergraduate and graduate students and practicing engineers.

Chapter 1 introduces the fundamental aspects of thermodynamics, as well as process and system assessment methods. The laws of thermodynamics and related analysis methods with energy and exergy are presented in detail, particularly through balance equations. Both energy and exergy efficiencies for various hydrogen production systems and applications are defined for use. The use of exergy methods for economic, environmental, and sustainability analysis are presented. Two case studies are given to illustrate the application of fundamental concepts of system assessment for renewable energy and hydrogen.

In Chapter 2, hydrogen and its production methods are classified, and the sustainable development and use of hydrogen as an energy storage medium and carrier to harvest renewable energy sources is discussed. A comparative evaluation of hydrogen and other fuel options is also presented, as well as an introduction to the main material resources from which hydrogen can be extracted, and to the main methods and pathways of hydrogen production.

Chapter 3 primarily focuses on hydrogen production methods driven by electrical energy by covering the fundamentals of electrochemical hydrogen production, and most importantly thermodynamic parameters and electrochemical reactions. Further, the main types of water electrolyzers, including alkaline, proton exchange membrane, and solid oxide, are analyzed. The efficiency of hydrogen production through electrochemical methods is formulated in terms of energy, exergy, and Faraday efficiency. Beyond water or steam electrolysis, some other electrochemical methods of hydrogen production are included, such as the chloralkali process, and both hydrogen sulphide, and ammoniated water, electrolysis methods.

In Chapter 4, hydrogen production methods driven by thermal energy are introduced, analyzed, and discussed through fundamentals of thermochemical hydrogen production, thermodynamic parameters, thermochemical reactions, and the main types of thermochemical processes, such as water thermolysis, thermochemical cycles and the gasification and reforming of fuels. The chapter ends with a presentation of some of the integrated systems that produce green hydrogen from thermal energy. These systems are based on nuclear energy, solar thermal energy, biomass energy, and clean coal technology.

Chapter 5 dwells on hydrogen production methods driven by photonic energy. It focuses on the photochemical reactions potentially available for hydrogen production. It also presents some of the integrated systems that produce green hydrogen from photonic energy. Such systems do not emit harmful chemicals or greenhouse gases (GHG) during operation.

Chapter 6 is about how to produce hydrogen through biochemical energy and bio-photochemical energy. It discusses the basic pathways of biochemical hydrogen production, and some of the integrated systems that produce sustainable hydrogen from biochemical energy: either light assisted, or dark fermentation.

Chapter 7 focuses on some hybrid water splitting cycles, such as photothermochemical, photoelectro-thermochemical and radiothermochemical. It presents nuclear-based hydrogasification of coal and nuclear-based natural gas reforming as two potential methods for very large-scale hydrogen production. Solar fuel reforming for hydrogen production is also covered, and has a potential application in small and intermediate scales of production. Some fuels envisaged for this process are natural gas, or synthetic natural gas, and methanol. Another less common fuel is urea, which is also a fertilizer. Urea is a source of hydrogen as well, as it can be converted thermochemically. Water electrolysis in molten alkali hydroxides is studied as a method in conducting electrolysis at high temperatures. This chapter ends with a focus on the hydrogen-ammonia paradigm, discussing ammonia synthesis, storage, and its conversion to hydrogen.

Chapter 8 discusses the recent progress made on some novel and newly developed hydrogen production processes, methods, techniques, technologies, etc. in order to reflect potential options and future directions on hydrogen production technologies. The topics covered in this chapter involve numerous hydrogen production methods, ranging from biological to photoelectrochemical options. Various pathways, through which the four kinds of energies (electrical, thermal, photonic and biochemical) drive hydrogen production, are discussed. Note that electrical and thermal energy can be derived from renewable energies, such as solar, wind, geothermal, tidal, wave, ocean thermal, hydro, and biomass; from nuclear energy; or from recovered energy.

References are included to direct the reader to sources where more details can be found, and to assist the reader who is simply curious to learn more. These references can also help identify information on topics not covered comprehensively in this book.

In this book, green methods for hydrogen production are presented comprehensively. Conventional, but environmentally benign, methods as well as potential methods are introduced, discussed, assessed and compared for various applications, along with many illustrative examples and case studies using various sources of renewables. This approach is aimed to make this book a unique source on hydrogen production technologies and to satisfy readers.

The selection of the topic for this book was designed to provide the reader with an introduction to the language, concepts and technologies used in all major hydrogen production methods that are expected to contribute to the 21st century.

We hope this book will open a new window to carbon free solutions through sustainable hydrogen production.

Ibrahim Dincer and Calin Zamfirescu
January 2016

Acknowledgments

The illustrations provided by several past and current graduate students of Prof. Dincer are gratefully acknowledged, including Canan Acar, Seyedali Aghahosseini, Ehsan Baniasadi, Camilo Cassallas, Sayantan Gosh, Hasan Ozcan, Musharaf Rabbani, Tahir Ratlamwala, Ron Roberts, and Rafay Shamim.

In addition, Yusuf Bicer's help in materializing the literature review for Chapter 8 is acknowledged.

Last but not least, we warmly thank our wives, Gulsen Dincer and Iuliana Maria Zamfirescu, and our children Meliha, Miray, Ibrahim Eren, Zeynep, and Ibrahim Emir Dincer, and Ioana, Cosmin, and Paulina Zamfirescu. They have been a great source of support and motivation.

Ibrahim Dincer and Calin Zamfirescu
January 2016

CHAPTER

1

Fundamental Aspects

1.1 INTRODUCTION

Hydrogen becomes increasingly important to society, as it is considered one of the key solutions to sustainability. Since the global population and its demand for services, materials, transportation, commodities, etc., are rapidly increasing, many concerns arise related to the supply of energy and the environmental impact related to its conversion/transformation, distribution and use. Thence, there is a clear motivation worldwide to develop cleaner processes and more efficient production methods in all sectors, which will become able to supplement and replace the conventional power generation and production methods by incorporating hydrogen and sustainable energies, including renewables and nuclear.

Hydrogen can be produced from a wealth of materials existent in abundance on earth and easily accessible. Water is the most abundant source of hydrogen. Biomass, anthropogenic wastes and even fossil fuels, when processed through clean methods, represent sources for sustainable hydrogen production. As reviewed in Dincer and Zamfirescu (2012), five general methods can be identified for hydrogen production, namely: electrochemical, photochemical, biochemical, thermochemical and radiochemical. In addition, hybrid methods may be possible such as photo-electrochemical, photo-biochemical, etc.

Developing, designing, optimizing, and assessing the methods for hydrogen production, including hydrogen storage and distribution, requires an interdisciplinary effort. Thermodynamics as a tool and its applications to electrochemical, thermochemical, photochemical, biochemical processes, and catalysis are treated as the masterpiece for system design, analysis, assessment, and optimization in view of making hydrogen production more efficient, more cost-effective, more environmentally benign and more sustainable.

In this chapter, fundamental aspects related to hydrogen production are reviewed with a focus on thermodynamics. Of particular attention is thermodynamics application in the study of chemical reactions, especially those relevant to hydrogen production. Some general aspects chemical thermodynamics are introduced and complemented with key methods on chemical kinetics. Of particular interest is the use of the concept of exergy and chemical exergy for process assessment in accordance with the second law of thermodynamics (SLT) and by accounting for the influence of the surroundings environment. The chapter is expanded with the enhanced use of exergoeconomic, exergoenvironmental and exergosustainability assessments.

1.2 PHYSICAL QUANTITIES AND UNIT SYSTEMS

The scientific method in natural sciences appeals to empirical evidences in order to construct and validate models to predict and understand nature's phenomena. Therefore, one needs to relate to quantities that are measurable by instruments, which are known as "physical quantities." Physical quantities are of two types, namely:

- Extensive quantity: which by definition is the sum of quantities in all system constituents (e.g., mass, volume, electric charge, energy)
- Intensive quantity: which is independent of the extent of the system (e.g., temperature, pressure, specific volume)

Seven physical quantities—denoted as fundamental quantities—have been chosen to represent the International System of Units (ISU) according to ISU (2006). These are length, mass, time, electric current, thermodynamic temperature, amount of substance (molar) and luminous intensity (measured in candela). The standard definition of these quantities is given in Table 1.1. Note that all other physical quantities can be derived from the fundamental ones based on constituency relationships.

Each of the physical quantity types has an associated unit, forming thus the ISU. Table 1.2 gives the definitions of the fundamental units of measure according to ISU formulations. Adoption of a system of units is an important step in the analyses. There are two main systems of units: the ISU, which is usually referred to as SI units, and the English System of Units (sometimes referred as Imperial). The SI units are used most widely throughout the world, although the English System is traditional in the United States. SI units are primarily employed throughout this book; however, relevant unit

TABLE 1.1 Definitions and Standard Symbols for the Fundamental Physical Quantities

Quantity	Symbol	Quantity definition
Length	l	Geometric distance between two points in space
Mass	m	Quantitative measurement of inertia that is the resistance to acceleration
Time	t	A measurable period of the progress of observable or nonobservable events
Electric current	I	Rate of flow of electric charge
Thermodynamic temperature	T	A measure of kinetic energy stored in a substance that has a minimum of zero (no kinetic "internal") energy
Amount of substance	n	The number of unambiguously specified entities of a substance such as electrons, atoms, molecules, etc.
Luminous intensity	I_v	Luminous flux of a light source per direction and solid angle, for a standard model of human-eye sensitivity

From: ISU, 2006. The International System of Units (SI), eighth ed.

TABLE 1.2 Definition of the Fundamental Units According to International System of Units

Quantity	Dimensional symbol	Unit	Symbol	Measurement unit definition
Length	L	Meter	m	The length of the path traveled by light in a vacuum during a time interval of 1/299,792,458 of a second
Mass	M	Kilogram	kg	The weight of the International Prototype of the kilogram made in platinum-iridium alloy
Time	T	Second	s	The duration of 9,192,631,770 periods of the radiation corresponding to the transition between the two hyperfine levels of the ground state of the cesium 133 atom
Electric current	I	Ampere	A	The constant current which, if maintained in two straight parallel conductors of infinite length, of negligible circular cross-section, and placed 1 m apart in a vacuum, would produce between these conductors a force equal to 2×10^{-7} N/m of length
Thermodynamic temperature	Θ	Kelvin	K	The fraction 1/273.16 of the thermodynamic temperature of the triple point of water (having the isotopic composition defined exactly by the following amount of substance ratios: 0.00015576 mole of ^{2}H per mole of ^{1}H, 0.0003799 mole of ^{17}O per mole of ^{16}O, and 0.0020052 mole of ^{18}O per mole of ^{16}O)
Amount of substance	N	Mole	mol	The amount of substance of a system that contains as many elementary entities as there are atoms in 0.012 kg of carbon 12
Luminous intensity	J	Candela	cd	The luminous intensity, in a given direction, of a source that emits monochromatic radiation of frequency 540 THz and that has a radiant intensity in that direction of 1/683 W per steradian

From: ISU, 2006. The International System of Units (SI), eighth ed.

conversions and relationships between the SI and English unit systems for fundamental properties and quantities are listed in Appendix A.

Any other physical quantity besides those given in Table 1.1 can be derived from the fundamental quantities using constitutive relationships. The units of measure of any derived physical quantity can be determined based on the units of fundamental quantities and the constitutive relationships according to the dimensional analysis method. In this respect, the dimensional symbols for the fundamental quantity must be used to express the dimension for the derived quantities. The list of dimension symbols for the fundamental physical quantities is given in Table 1.2.

Now, we will review the main derived physical quantities which are the most relevant within this book. First, the thermodynamic properties pressure, specific volume and temperature are introduced, and their interconnections and interpretations are provided. As known, the thermodynamic state of a system is generally specified when its pressure, temperature and specific volume are known.

To begin, let us introduce the specific volume, defined as the ratio between the volume and the mass of a system. Here, mass is a fundamental quantity but volume is not. Thence, the notion of volume must be first defined. Volume is a physical quantity, representing a three-dimensional extent of a system. This is an extensive property, having the dimension equal to length to the power 3, noted L^3. The common symbol for volume is V, therefore one writes

$$V = L^3 \left[\text{m}^3\right]$$

Here, the notation between brackets is the unit of volume, namely cubic meter. The specific volume is thus defined as follows

$$v = \frac{V}{m} \left[\text{m}^3/\text{kg}\right]$$

In dimensional analysis notation, one writes that the dimension of the specific volume is given by L^3M^{-1}, where L is the dimension for length and M is the dimension for mass. Specific volume is in fact the reciprocal of density, $\rho = v^{-1}$. Furthermore, one notes that specific volume can be expressed with respect to the amount of substance in moles (molar-specific volume) by the following equation

$$v = \frac{V}{n} \left[\frac{\text{m}^3}{\text{mol}}\right]$$

Here, the dimension of the molar-specific volume is, therefore, L^3N^{-1}. Pressure is defined as the force (action) per unit of surface area. Since both the force and surface are not fundamental physical quantities, definition relationships must first be introduced for those. Area, denoted with symbol A, represents the two-dimensional extent of a system, having thus the dimension L^2 and the unit m².

Force represents a measure of action or interaction between systems. Forces do exist in any region as they manifest their presence in the form of force fields (e.g., electromagnetic or gravitational fields). The force acting between two systems can produce acceleration, meaning a change of velocity. If an object changes its velocity, then it necessarily behaves so because it is under the action of a net force. Here we define velocity as a vector (having thus direction and magnitude) and representing the speed of a system toward a defined target. Since speed is defined as the displacement per unit of time, one notes that the dimension for velocity is LT^{-1} with the unit of SI (ISU) m/s, and the definition relationship follows

$$v = \frac{dl}{dt}[\text{m/s}]$$

The acceleration magnitude is the derivative of speed with respect to time and whence will have the dimension of LT^{-2} and the SI unit of m/s²; the definition acceleration is written as

$$a = \frac{dv}{dt} = \frac{d^2l}{dt^2}\left[\text{m/s}^2\right]$$

Force is a vector quantity having the same direction as the acceleration that it produces. Force is defined as the product of mass and acceleration. That is, if a system of mass m is accelerated with the acceleration a, then a net force acts upon it as follows

$$\vec{F} - m\,\vec{a}$$

The dimension for force is MLT^{-2}, meaning that force is measured in SI in kg m/s², a unit that is shorthanded as Newton (symbol N); thence 1 N = 1 kg m/s². In British System of Units, force is measured in pound-force denoted as lb$_f$. Pressure is defined now as the ratio between a force and the normal area of the surface over which force is exerted uniformly (here the term "normal" stands for a direction perpendicular on force vector direction). Therefore, one writes

$$P = \frac{F}{A}[\text{Pa}]$$

Note that although force is a vector, pressure is defined as a scalar. Furthermore, pressure is an intensive quantity. The dimension of pressure is equal to the dimension of force divided by the dimension of area, that is, $ML^{-1}T^{-2}$, having the SI unit of kg/s²m, which is the same as N/m² and shorthanded as Pa (Pascal). One says that 1 Pascal represents the force of 1 N (Newton) exerted on 1 m² surface. Pressure is a very important parameter for the thermodynamics of gases and liquids.

The atmosphere that surrounds the earth acts as a reservoir of low-pressure air. Its weight exerts a pressure that varies with temperature, humidity, and altitude. Atmospheric pressure also varies temporarily at a given geographic location, due to the movement of weather patterns. The standard value of the atmospheric pressure (or the pressure of standard atmosphere) is 101,325 Pa or 760 mmHg.

The atmospheric pressure is typically measured with an instrument called barometer; from here it comes the name of *barometric pressure*. While the changes in barometric pressure are usually as minor as <12.5 mm of mercury, they need to be taken into account only when precise measurements are required.

Gauge pressure is any pressure for which the base for measurement is the atmospheric pressure. The gauge pressure is expressed as kPa gauge. Atmospheric pressure serves as a reference level for other types of pressure measurements, for example, gauge pressure. As shown in Fig. 1.1, the gauge pressure is either positive or negative, depending on its level above or below atmospheric level. At the level of atmospheric pressure, the gauge pressure becomes zero.

A different reference level is utilized to obtain a value for absolute pressure. The absolute pressure can be any pressure for which the base for measurement is a complete vacuum, expressed in kPa (absolute). Absolute pressure is composed of the sum of the gauge pressure (positive or negative) and the atmospheric pressure as follows

$$\text{Pressure (gauge)} + \text{Atmospheric pressure} = \text{Pressure (absolute)} \tag{1.1}$$

For example, to obtain the absolute pressure, we simply add the value of atmospheric pressure. Absolute pressure is the most common type used in thermodynamic calculations, despite having the pressure difference between the absolute pressure and the atmospheric pressure existing in the gauge being what is read by most pressure gauges and indicators.

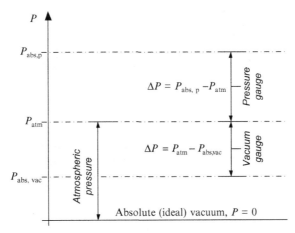

FIG. 1.1 Illustration of relationships among pressures.

A vacuum is a pressure lower than atmospheric and occurs only in closed systems, except in outer space. It is also called *negative gauge pressure*. A vacuum is usually divided into four levels: (i) low vacuum, representing pressures above 133 Pa ($\sim$1 Torr) absolute; (ii) medium vacuum, varying between 1 and 10^{-3} 0.1333 to 133 Pa absolute; (iii) high vacuum, ranging between 10^{-3} and 10^{-6} Torr absolute (1 Torr $=$ 133.3 Pa); and (iv) very high vacuum, representing absolute pressure below 10^{-6} Torr. The *ideal vacuum* is characterized by the lack of any form of matter. Note that in physics, *matter* is defined as particles with rest mass.

As known, the particles possessing rest mass include *quarks* and *leptons*, which are able to combine and form *protons*, *neutrons* and *electrons*. Among them, protons and neutrons combine to form nuclei, having a positive electric charge. Nuclei combine with electrons of negative electric charge to form atoms. Therefore, atoms become neutral with respect to the electric charge. In total, there are 116 known kinds of atoms corresponding to 116 chemical elements from the periodic table of elements.

Besides matter, in the universe force fields manifest of which of particular importance is the electromagnetic field. The electromagnetic field is said to propagate via photons moving at the speed of light. Photons have mass, but no rest; which means that photons have no rest mass. Being matter with no rest, photons do propagate in space regions characterized by absolute vacuum. Gravitational fields also propagate through vacuum.

It is generally understood that bulk matter (or matter with rest mass) and fields (photons, gravity, etc.) constitute the substance of the universe. Since the thermodynamic temperature characterizes the kinetic energy stored in substance, it means that this physical quantity is not restricted to bulk matter, but also may refer to force fields (which do not have the property of rest mass). In further chapters of this book, an interpretation of the temperature of electromagnetic radiation (blackbody, nonblackbody, monochromatic) will be given. One essential aspect related to thermodynamic temperature comes from the fact that the kinetic energy at the microscale (where many particles manifest) cannot be entirely retrieved at the macroscale due to probabilistic reasons, as discussed in the next section of this chapter where ideal-gas theory is introduced. The ideal-gas theory explains at a basic level the correlation between pressure, temperature and specific volume of an ideal gas.

The force action resulting on a displacement of matter is quantified by a physical quantity referred to as work (also called "the work of a force"). Consider a situation as in Fig. 1.2 where a force acts on a system obliquely at angle α and produces a displacement l. Then the dot product of force and displacement represent the work of the force. Provided that the force and displacement have the same direction ($\alpha=0$), the work becomes $W=Fl$.

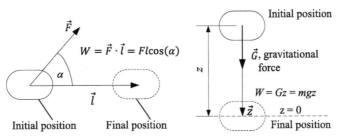

FIG. 1.2 Illustration of the concept of work.

Being defined as a dot product, work is a scalar. It is said that work represents the amount of energy spent to perform a displacement action. If there is no displacement, then there is no work. If there is no force, then there is no work produced. Fig. 1.2 shows the work of gravity, displacing a system in the gravitational field with the distance z. Thence, force and displacement are both vertical, $W=mgz$, $g=9.81$ m/s^2. Generally, the magnitude of work is given by

$$W = Fl\cos(\alpha) \tag{1.2}$$

Here, the work has dimension of force multiplied by length, which is ML^2T^{-2} with the unit kg m^2/s^2 or N m (Newton-meter); this unit is known as Joule, J, where 1 J $= 1$ N m. The capacity of a system to do work (or receive work) defines the physical quantity known as energy. Energy has the same dimension and unit as work. A rigorous definition of energy requires the introduction of the notion of thermodynamic system.

By definition, a *thermodynamic system* is part of the universe, delimited by a real or imaginary boundary that separates the system from the rest of the universe, whereas the rest of the universe is denoted as the *surroundings*. If a thermodynamic system exchanges energy but not matter with its surroundings, it is said to be a *closed thermodynamic system* or *control mass*; see the representation from Fig. 1.3a. On the other hand, an *open system* or *control volume* is a thermodynamic system that, as represented in Fig. 1.3b, can interact with its surroundings both by mass and energy transfer. A closed thermodynamic system that does not exchange energy in any form with its surroundings is denoted as an isolated system. Fig. 1.4 shows an isolated thermodynamic system.

Energy can be exchanged by a closed system with its surrounding in only two forms: by work transfer or by heat transfer. The symbol for energy is E and any energy change of a system is denoted with ΔE. If a closed thermodynamic system has a deformable boundary that is a perfect thermal insulator, then the system is capable of exchanging work with its surroundings in an adiabatic manner. The system is adiabatic. Therefore, the change of energy equals the work transfer under adiabatic conditions as follows

$$\Delta E = W \quad \text{(so-called : work transfer)}$$

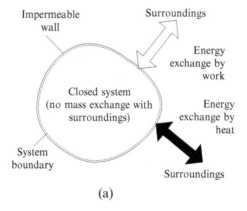

(a)

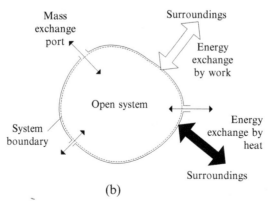

(b)

FIG. 1.3   Illustrating the concepts of thermodynamic systems as (a) closed and (b) open.

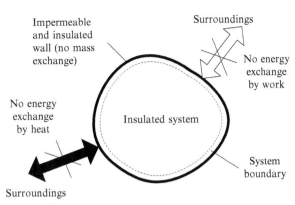

FIG. 1.4 Representation of an insulated thermodynamic system.

If a system has a rigid boundary, but is permeable to heat transfer and does not exchange work with its surroundings, then the only way to transfer energy is said to be by heat. Any change of system energy equals a heat transfer, denoted with symbol Q; therefore, one notes

$$\Delta E = Q, \quad \text{(so-called : heat transfer)}$$

In order for a transfer of energy by heat to be possible, a temperature difference must exist between the system and its surrounding. Therefore, heat represents energy transferred due to a temperature difference; however, one remarks that although a temperature difference may exist between an insulated system and its surroundings, there will not be heat transfer due to the perfect insulation property of the system boundary.

Internal energy represents a summation of all microscopic forms of energy including vibrational, chemical, electrical, magnetic, surface, and thermal. Internal energy is an extensive quantity. By definition, an extensive thermodynamic quantity depends on the amount of matter. When internal energy U is divided by the mass of the control volume m, then an intensive property is obtained $u = U/m$, the specific internal energy. Intensive properties are independent of the amount of matter (e.g., intensive properties are pressure, temperature, specific volume, etc.). Using the specific internal energy, the following expression for a system's internal energy is obtained for a closed system

$$\Delta U = m\,(u_2 - u_1)$$

For many thermodynamic processes in closed systems, the only significant energy changes are of internal energy, and the significant work done by the system in the absence of friction is the work of pressure-volume expansion, such as in a piston-cylinder mechanism. Other important energy changes are occurring due to liquid-vapor-phase change. The vapor quality changes during phase-change processes such as boiling, condensation, absorption or desorption. Thence, the specific internal energy of the two-phase liquid-vapor mixture-changes according to the change Δx in vapor quality. In general, the specific internal energy of a mixture of liquid and vapor at equilibrium can be written as follows

$$u = (1 - x)u' + x\,u'' \tag{1.3}$$

where u' is the specific internal energy of the saturated liquid, and whereas u'' is the specific internal energy of saturated vapor. *Enthalpy* is another state function. Specific enthalpy, usually expressed in kJ/kg, is defined based on internal energy, pressure and specific volume. According to its definition, the specific enthalpy is given by

$$h = u + P v \tag{1.4}$$

Entropy is another important state function defined by ratio of the infinitezimal heat added to a substance to the absolute temperature at which it was added, and is a measure of the molecular disorder of a substance at a given state. Entropy quantifies the molecular (microscale) random motion within a thermodynamic system and is related to the *thermodynamic probability* (p) of possible microscopic states as indicated by Boltzmann equation $S = k_B \ln p$. Entropy is an extensive property, whereas specific enthalpy (s) is an intensive property.

It is known that all substances can "hold" a certain amount of heat; that property is their thermal capacity. When a liquid is heated, the temperature of the liquid rises to the boiling point. This is the highest temperature that the liquid can reach at the measured pressure. The heat absorbed by the liquid in raising the temperature to the boiling point is called *sensible heat*. The thermodynamic quantity known as *specific heat* is a parameter that can quantify the state change of a system that performs a process with sensible heat exchange. Specific heat is defined based on internal energy or enthalpy, depending on the nature of the thermodynamic process. The specific heat at constant volume is equal to the change in internal energy with temperature at constant volume as defined below

$$C_v \equiv \left(\frac{\partial u}{\partial T}\right)_v \tag{1.5}$$

The *specific heat at constant pressure* C_p represents the amount of heat required to increase the temperature of a system evolving at constant pressure with 1 K. The specific heat is the change of enthalpy with temperature at constant pressure defined according to

$$C_p \equiv \left(\frac{\partial h}{\partial T}\right)_p \tag{1.6}$$

For example, the specific heat at constant pressure and the specific heat at constant volume of any incompressible substance (e.g., solid and liquid) are equal. The heat required for converting liquid to vapor at the same temperature and pressure is called *latent heat*. This is the change in enthalpy during a state change (the amount of heat absorbed or rejected at constant temperature at any pressure, or the difference in enthalpies of a pure condensable fluid between its dry saturated state and its saturated-liquid state at the same pressure).

Fusion is associated with the melting and freezing of a material. For most pure substances, there is a specific melting/freezing temperature, relatively independent of the pressure. For example, ice begins to melt at 0°C. The amount of heat required to melt 1 kg of ice at 0°C to 1 kg of water at 0°C is called the latent heat of fusion of water, and equals 334.92 kJ/kg. The removal of the same amount of heat from 1 kg of water at 0°C changes liquid water back to ice.

A number of fundamental physical constants are very relevant for thermodynamics; examples are the universal gas constant, Boltzmann constant, Faraday constant and elementary electric charge. In addition, some standard parameters, such as standard atmospheric pressure and temperature, standard molar volume and solar constant are very important for thermodynamic analysis. Table 1.3 presents fundamental physical constants and standard parameters. The table lists the constant name, its symbol, the value and units and a brief definition.

TABLE 1.3 Fundamental Constants and Standard Parameters

Constant/ parameter	Value	Definition
Speed of light in vacuum	$c = 299{,}792{,}458 \, \text{m/s}$	Maximum speed at which matter and information can be transported in the known cosmos
Elementary charge	$e = 1.60218 \times 10^{-19} \, \text{C}$	Electrical charge carried by a single proton
Faraday's constant	$F = 96{,}485 \, \text{C/mol}$	Electric charge of one mole of electrons
Gravitational acceleration	$g = 9.80665 \, \text{m/s}^2$	Gravitational force (G) per unit of mass as $g = G/m$
Planck's constant	$h = 6.626 \times 10^{-37} \, \text{kJs}$	Magnitude of energy of a quanta (particle) that expresses the proportionality between frequency of a photon and its energy according to $E = h \cdot \nu$
Boltzmann constant	$k_B = 1.3806 \times 10^{-23} \, \text{J/K}.$ $k_B = \mathcal{R}/N_A$	A measure of kinetic energy of one molecule of ideal gas
Number of Avogadro	$N_A = 6.023 \times 10^{26} \, \text{molecules/kmol}$	Ratio of constituent entities of a bulk substance to the amount of substance. $N_A = N/n$
Standard atmospheric pressure	$P_0 = 101.325 \, \text{kPa}$	Pressure of the terrestrial atmosphere at sea level in standard conditions
Universal gas constant	$\mathcal{R} = 8.314 \, \text{J/mol.K}$ $\mathcal{R} = Pv/T$	A measure of kinetic energy of one mole of an ideal gas at molecular level
Stefan-Boltzmann constant	$\sigma = 5.670373 \times 10^{-8} \, \text{W/m}^2\text{K}^4$	A constant in Stefan-Boltzmann law expressing the proportionality between forth power of temperature and black body's emissive power

1.3 IDEAL-GAS THEORY

Ideal-gas theory is very important for analysis of processes because in most of the practical situations gases behave as rarefied matter (the interaction between gas molecules can be neglected). An ideal gas can be described in terms of three parameters: the volume that it occupies, the pressure that it exerts on boundaries, and its temperature. According to the definition, ideal gas represents a special state of matter that can be delimited by a system boundary. Ideal gas is formed by a number N of freely moving molecules or atoms with perfect elastic behavior at collisions among each other and with the system boundary. It is assumed that:

- All particles have rest mass ($m > 0$; the particles are not photons)
- The number of particles with respect to the system volume is small
- The total volume of particles is negligible with respect to system volume
- The collisions of particles with each other are much less probable than the collisions with the system boundary

The practical advantage of treating real gases as ideal is that a simple equation of state with only one constant can be applied in the following form

$$P V = m R T \text{ or } P v = R T \text{ or } P \mathit{v} = \mathcal{R} T \tag{1.7}$$

where P is the pressure in Pa, V is the gas volume in m^3, m is mass of gas in kg, T is gas temperature in K, R is known as gas constant and is given in J/kg K, v is mass-specific volume in m^3/kg, v is molar-specific volume in m^3/kmol, $\mathcal{R}$ is the universal gas constant of 8.134 J/mol K (Table 1.3). Observe that the gas constant is specific to each particular gas and depends on the universal gas constant and the molecular mass (M) of the gas according to

$$R = \frac{\mathcal{R}}{M} \tag{1.8}$$

Note that Eq. (1.7) is named "the thermal equation of state" of the ideal gas because it expresses the relationship between pressure, specific volume and temperature. It is possible to express the ideal-gas equation in terms of internal energy, specific volume and temperature. In this case, the equation of state is called caloric equation of state. In particular, for ideal gas only, the internal energy depends on temperature only. The caloric equation of state for a monoatomic ideal gas is $u = 1.5 \mathcal{R} T$, where u is the molar-specific internal energy. Since $h = u + P v$, it follows that the enthalpy of monoatomic ideal gas is given by $h = 2.5 \mathcal{R} T$. Based on the specific heats definitions from Eqs. (1.5) and (1.6), the well-known Robert Meyer equation for ideal gas can be derived

$$C_p [\text{J/kg K}] = C_v + R \tag{1.9}$$

Note that, specific heats for ideal monoatomic gas are $C_v = 1.5 R$ and $C_p = 2.5 R$. Thence, the ratio of specific heat at constant pressure and constant volume, known as the *adiabatic exponent*, namely

$$\gamma = \frac{C_p}{C_v}$$

has the following values for ideal gas: monoatomic gas $5/3 = 1.67$, whereas for diatomic gas it is $7/5 = 1.4$.

There are some special cases if one of P, v and T is constant. At a fixed temperature, the volume of a given quantity of ideal gas varies inversely with the pressure exerted on it (this is known as Boyle's law), describing gas compression or expansion as follows

$$P_1 V_1 = P_2 V_2 \tag{1.10}$$

where the subscripts refer to the initial and final states. This equation is employed by analysts in a variety of situations: when selecting an air compressor, for calculating the consumption of compressed air in reciprocating air cylinders, and for determining the length of time required for storing air, for selecting air motors (or expanders) and determining their air consumption etc. If the process is at constant pressure or at constant volume, then Charles laws apply

$$\frac{V_1}{T_1} = \frac{V_2}{T_2} \text{ and } \frac{P_1}{T_1} = \frac{P_2}{T_2} \tag{1.11}$$

TABLE 1.4 Simple Thermodynamic Processes and Corresponding Equations for Ideal-Gas Model

Process	Definition	Equation	Specific work expression
Isothermal	$T = \text{const.}$	$P_1 v_1 = P_2 v_2$	$w_{1-2} = P_1 v_2 \ln(v_2/v_1)$
Isochoric	$v = \text{const.}$	$\dfrac{P_1}{T_1} = \dfrac{P_2}{T_2}$	$w_{1-2} = 0$
Isobaric	$P = \text{const.}$	$\dfrac{v_1}{T_1} = \dfrac{v_2}{T_2}$	$w_{1-2} = P(v_2 - v_1)$
Polytropic	$Pv^n = \text{const.}$	$\dfrac{P_2}{P_1} = \left(\dfrac{V_1}{V_2}\right)^n$ $= \left(\dfrac{T_2}{T_1}\right)^{n/(n-1)}$	$w_{1-2} = \dfrac{1}{n-1}(P_2 v_2 - P_1 v_1)$
General	P, v, T vary constant mass	$\dfrac{P_1 v_1}{T_1} = \dfrac{P_2 v_2}{T_2}$	$w_{1-2} = \displaystyle\int_1^2 P dv$

If the number of moles of ideal gas does not change in an enclosed volume, then the combined ideal equation of state is

$$\frac{P_1 V_1}{T_1} = \frac{P_2 V_2}{T_2} \tag{1.12}$$

If there is no heat exchange with the exterior ($dq = 0$), then the process is called adiabatic. Also, if a process is neither adiabatic nor isothermal, it could be modeled as polytropic. Table 1.4 gives the principal features of simple processes for ideal gas. Fig. 1.5 shows representation in $P - v$ diagram for four simple processes with ideal air, modeled as ideal gas, by using the Engineering Equation Solver (EES) software.

The entropy change of an ideal gas with constant specific heats is given by the following equations, depending on the type of process (at constant pressure or constant volume)

$$s_2 - s_1 = C_{p0} \ln\left(\frac{T_2}{T_1}\right) + R \ln\left(\frac{v_2}{v_1}\right) \quad \text{and} \quad s_2 - s_1 = C_{v0} \ln\left(\frac{T_2}{T_1}\right) - R \ln\left(\frac{P_2}{P_1}\right) \tag{1.13}$$

The shape of the thermodynamic system can be arbitrary, but for simplicity, we assume here a cubical volume as indicated in Fig. 1.6. The average particle velocity is denoted with v and the cube edge is l; it follows that the time

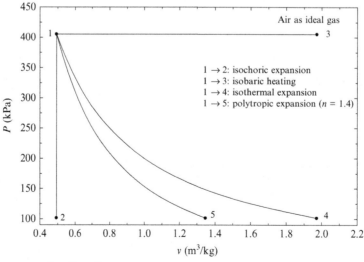

FIG. 1.5 Ideal-gas processes represented on $P - v$ diagram.

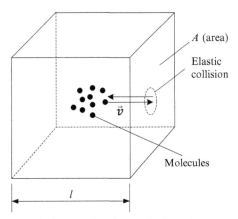

FIG. 1.6 A thermodynamic system formed by the ideal gas enclosed in a cube boundary.

between two collisions, which is approximated with the time needed for the particle to travel between two opposite walls is

$$\Delta t = \frac{2l}{v}$$

Based on momentum conservation law, the force exerted by one particle on the wall, during collision, is

$$F = \frac{m\,v - m\,(-v)}{\Delta t} = \frac{m v^2}{l} \tag{1.14}$$

One also assumes that there is a uniform distribution of particle collisions for the three Cartesian directions; thus only 1/3 of particles exert force on a wall; therefore the pressure expression is determined by divided the force—given by Eq. (1.14) with wall area A

$$P = \frac{1}{3} N m \frac{v^2}{lA} = \frac{Nm}{3V} = \frac{1}{3} \rho\, v^2 \tag{1.15}$$

where $V = lA$ is the volume of thermodynamic system and A is the surface area of the face, and

$$\rho = \frac{Nm}{V}$$

which is the gas density (total mass of particles divided to the volume of the enclosure). The kinetic energy of a single gas particle can be expressed based on the average particle velocity; in this respect, Eq. (1.15) is solved for v^2 and it results in

$$KE = \frac{1}{2} m\, v^2 = \frac{3}{2} \frac{PV}{N} \tag{1.16}$$

The *degree of freedom* of monoatomic gas molecules is DOF = 3, because there are only three possible translation movements along Cartesian axes. According to its thermodynamic definition, *temperature* (T) is a measure of the average kinetic energy of molecules per degree of freedom. The quantitative relationship between temperature and kinetic energy of one single molecule is

$$\frac{KE}{DOF} \equiv \frac{1}{2} k_B T \tag{1.17}$$

where k_B is Boltzmann constant, defined in Table 1.2. Solving Eq. (1.17) for T it results the thermodynamic expression for temperature as follows

$$T \equiv \frac{2\,KE}{k_B\,DOF} \tag{1.18}$$

Kinetic energy expression from Eq. (1.16) can be introduced in Eq. (1.18) and the following it is obtained

$$T = \frac{PV}{k_B N} = \frac{PV}{k_B\,(N/N_A)\,N_A} = \frac{PV}{k_B\,n\,N_A} = \frac{Pv}{k_B N_A}$$

TABLE 1.5 Amagat and Dalton models for ideal gas mixtures Models

Definition	Dalton model	Amagat model
Assumptions	T and V are constant $P_{\text{tot}} = P_1 + P_2 + \ldots + P_N$	T and P are constant $V_{\text{tot}} = V_1 + V_2 + \ldots + V_N$
Equations for the components	$P_i V = n_i \mathcal{R} T$	$P V_i = n_i \mathcal{R} T$
Equation for the mixture	$P_{\text{tot}} V = \left(\sum n \right) \mathcal{R} T$	$P V_{\text{tot}} = \left(\sum n \right) \mathcal{R} T$

TABLE 1.6 Relevant Parameters of Mixtures of Ideal-Gas

Parameter	Equation(s)
Total mass of a mixture of N components	$m_{\text{tot}} = \sum m_i$
Total number of moles of a mixture of N components	$n_{\text{tot}} = \sum n_i$
Mass fraction for each component	$c_i = m_i / m_{\text{tot}}$
Mole fraction for each component	$y_i = \dfrac{n_i}{n_{\text{tot}}} = \left(\dfrac{P_i}{P_{\text{tot}}} \right)_{\text{Dalton model}} = \left(\dfrac{V_i}{V_{\text{tot}}} \right)_{\text{Amagat model}}$
Molecular weight of the mixture	$M_{\text{mix}} = \dfrac{m_{\text{tot}}}{n_{\text{tot}}} = \dfrac{\sum (n_i M_i)}{n_{\text{tot}}} = \sum (y_i M_i)$
Internal energy of the mixture	$U_{\text{mix}} = \sum (n_i U_i)$
Enthalpy of the mixture	$H_{\text{mix}} = \sum (n_i H_i)$
Entropy of the mixture	$S_{\text{mix}} = \sum (n_i S_i)$
Entropy difference for the mixture	$S_2 - S_1 = -\mathcal{R} \sum (n_i \ln y_i)$

where v is the *molar-specific volume* and n is the amount of substance (number of moles). Therefore, the following thermodynamic definition of temperature is obtained, which is equivalent with that given in Eq. (1.18)

$$T \equiv \frac{P v}{k_B N_A} \equiv \frac{P v}{\mathcal{R}} \tag{1.19}$$

where $\mathcal{R}$ the *universal gas constant* given in Table 1.2 as follows

$$\mathcal{R} = k_B N_A$$

In many practical situations, mixtures of real gases can be approximated as mixtures of ideal gases. There are two ideal-gas models for gas mixtures: the Dalton model and Amagat model. For both models, it is assumed that each gas is unaffected by the presence of other gases. The Dalton model assumes that the mixture is at constant temperature and volume, whereas the Amagat volume considers the case when temperature and pressure are constant.

Table 1.5 compares models of Amagat and Dalton for ideal-gas mixtures. The equations relating the thermodynamic parameters of the component gases with the parameters of the mixture are given in Table 1.6.

1.4 EQUATIONS OF STATE

Equations of state describe the thermodynamic behavior of bulk matter (which is formed by groups of atoms, molecules and clusters of them) generally in terms of temperature, pressure and specific volume. Other state variables such as specific internal energy, specific entropy and specific volume can be also used to formulate equations of state. There are four *forms of aggregation* of substances, denoted also as phases or states, namely *solid, liquid, gas*, and *plasma*. Each of the properties of a substance in a given state has only one definite value, regardless of how the substance reaches the state. Temperature and specific volume represent a set of thermodynamic properties that defines completely the thermodynamic state and the state of aggregation of a substance.

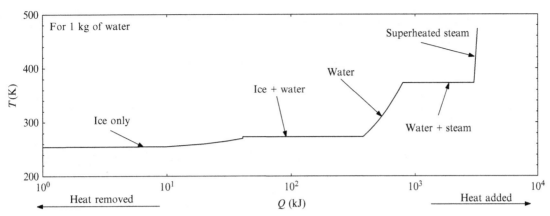

FIG. 1.7 Representation of phase diagram of water.

The thermodynamic state of a system can be modified via various interactions, among which heat transfer is one. Heat can be added or removed from a system. When sufficient heat is added or removed at a certain condition, most substances undergo a state change. For pure substances, the temperature remains constant until the state change is complete. The state transition can transform from solid to liquid, liquid to vapor, or vice versa.

Fig. 1.7 shows the phase diagram of water, calculated using EES (see EES, 2013), which is a typical example of temperature nonvariation during latent heat exchange as in melting and boiling. Ice reaches its melting point at 273.15 K. During the melting process, an ice-water mixture is formed. Due to phase change, the temperature remains constant (see Fig. 1.7), although heat is continuously added. At the end of the melting process, all water is in a liquid state of aggregation. Water is further heated and its temperature increases until it reaches the boiling point at 373.15 K. Additional heating produces boiling, which evolves at constant temperature, while a water+steam mixture is formed. The boiling process is completed when all liquid water is transformed in steam. Further heating leads to temperature increase and generation of superheated steam.

A representation of solid, liquid, and vapor phases of a pure substance is qualitatively exhibited also on a temperature-volume (T-v) diagram in Fig. 1.8. In this diagram, "T" is the triple point of the pure substance. The triple point represents that thermodynamic state where solid, liquid and vapor can coexist. For example, the triple point of water occurs at 273.16 K, 6.117 mbar, and specific volume is 1.091 dm^3/kg for ice, 1 dm^3/kg for liquid water, and 206 m^3/kg for vapor. Below the triple-point isobar, there is no liquid phase. A sublimation or desublimation process occurs, which represents phase transition between solid and vapor.

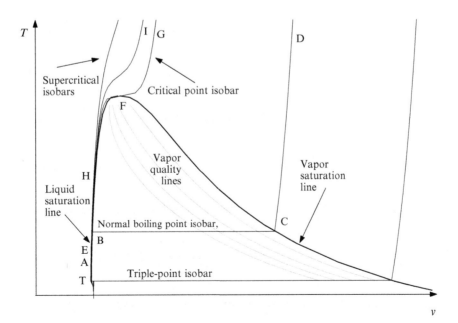

FIG. 1.8 Illustration of the temperature-volume diagram features for a pure substance.

Between triple-point isobar and critical-point isobar, three phases do exist: solid, liquid, and vapor. In addition, there are defined thermodynamic regions of subcooled liquid, two-phase and superheated vapor. Subcooled liquid regions exist between the critical isobar and liquid saturation line (see Fig. 1.8). The two-phase region is delimited by the liquid saturation line at the left, vapor saturation line at the right and triple-point isobar at the bottom. Superheated vapor exists above the vapor saturation line and below the critical isobar. At temperatures higher than the temperature of critical point and above the critical isobar, there is a thermodynamic region denoted as *supercritical fluid region* where the substance is neither liquid nor gas, but has some common properties with gases and liquid; supercritical fluids will be discussed in detail in other chapters of this book.

The specific volume along the boiling pallier can be expressed based on *vapor quality*, which is defined based v, v', v'', the specific volumes of mixture, saturated liquid and saturated vapor, respectively, according to

$$x = \frac{v - v'}{v'' - v'} \tag{1.20}$$

On the diagram from Fig. 1.8, the constant vapor quality lines are suggested and state points A–I are indicated. These state points are representative for various processes as follows:

- A-B-C-D: Represents a constant-pressure process
- A-B: Represents the process where the substance in liquid phase is heated from the initial temperature to the saturation temperature (liquid) at constant pressure; at point B, a fully saturated liquid with a quality $x = 0$ is has formed
- B-C: Represents a constant-temperature boiling process in which there is phase change from a saturated liquid to a saturated vapor; as this process proceeds, the vapor quality varies from 0% to 100%; within this zone, the substance is a mixture of liquid and vapor; at point C, we have a completely saturated vapor and the quality is 100%.
- C-D: Represents the constant-pressure process in which the saturated vapor is superheated with increasing temperature
- E-F-G: Represents a constant-pressure heating process evolving at critical pressure. Point F is called the critical point where the saturated-liquid and saturated-vapor states are identical; the thermodynamic properties at this point are called critical thermodynamic properties, for example, critical temperature, critical pressure and critical specific volume. State G represents a thermodynamic state point along the critical point isobar, at a temperature higher than the critical temperature
- H-I: Represents a constant-pressure heating process in which there is no change from one phase to another (because the pressure is set at a super-critical value.); however, there is a continuous change in density during this process

Another important state diagram is the pressure versus volume diagram for pure substances. The pressure versus specific volume diagram of pure water is presented in Fig. 1.9, which has been constructed using the EES software. This plot indicates the saturation lines where liquid and vapor reach the saturation temperature at a given pressure.

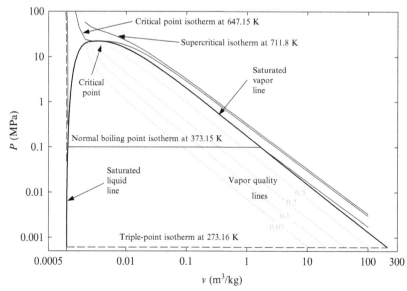

FIG. 1.9 The pressure-volume diagram for pure water.

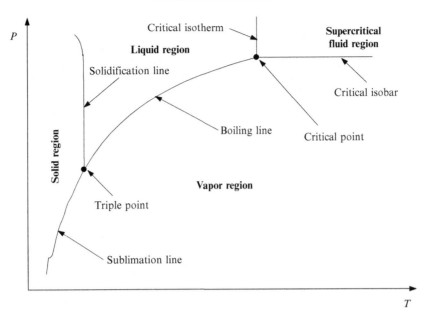

FIG. 1.10 Pressure versus temperature diagram of water. *Data from Haynes, W.M., Lide, D.R., 2012. CRC Handbook of Chemistry and Physics, 92nd Ed. Internet version. CRC Press, New York.*

One may observe that the specific volume of saturated vapors is ~1000 times higher than the volume of liquid for a process evolving along the normal boiling point isotherm. The normal boiling point isotherm corresponds to a temperature of 373.15 K (and 1 atm pressure).

The *pressure versus temperature diagram* is also an important tool that shows phase transitions of any substance. Fig. 1.10 qualitatively shows the *P-T* diagram of a pure substance. There are four regions delimited in the diagrams: solid, vapor, liquid, and supercritical fluid. The phase transition lines are sublimation, solidification, boiling, critical isotherm, and critical isobar; the last two lines are represented only for supercritical region (at pressure and temperature higher than critical).

The ideal-gas equation of state ($Pv = nRT$) is not applicable at the liquid-vapor transition region, liquid and solid region. Therefore, more accurate models are required to predict the behavior of real substances for a larger extent of the thermodynamic parameters. The ideal gas is applicable only for rarefied gases (low pressures and high temperature, far from the liquid-vapor saturation line).

In this respect, the *compressibility factor* (Z) is introduced to measure the deviation of a real substance from the ideal-gas equation of state. The compressibility factor is defined by the following relation

$$Z = \frac{Pv}{RT} \tag{1.21}$$

where specific volume is expressed on mass basis.

The order of magnitude is about 0.2 for many fluids. For accurate thermodynamic calculations, compressibility charts can be used, which express compressibility factor as a function of pressure and temperature. In this way, an equation of state is obtained based on compressibility factor by the following

$$Pv = ZRT$$

where the compressibility factor is a function of pressure and temperature.

According to the so-called *principle of corresponding states*, compressibility factor has a quantitative similarity for all gases when it is plotted against reduced pressure and reduced temperature. The reduced pressure is defined by the actual pressure divided by the pressure of the critical point

$$P_r = \frac{P}{P_c}$$

where subscript c refers to critical properties and subscript r to reduced properties. Analogously, the reduced temperature is defined by

$$T_r = \frac{T}{T_c}$$

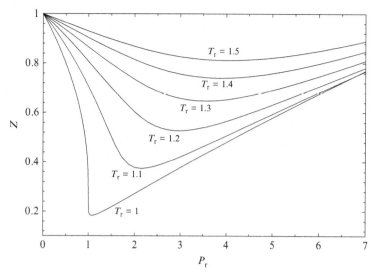

FIG. 1.11   Generalized compressibility chart averaged for water, oxygen, nitrogen, carbon dioxide, carbon monoxide, methane, ethane, propane, *n*-butane, isopentane, cyclohexane, *n*-heptane.

TABLE 1.7   Description of the Van der Waals Equation of State

Item	Equation
Reduced pressure, temperature, and specific volume	$P_r = \dfrac{P}{P_c};\quad T_r = \dfrac{T}{T_c};\quad v_r = \dfrac{v}{v_c}$
Reduced internal energy	$u_r = \dfrac{u}{(P_c v_c)}$
Thermal equation of state	$P_r = 8\,T_r/(3\,v_r - 1) - 3/v_r^2$
Caloric equation of state	$u_r = 4\,\mathcal{R}\,T_r$

The compressibility charts showing the dependence of the compressibility factor on reduced pressure and temperature can be obtained from accurate P, v, T data for fluids. These data are obtained primarily based on measurements. Accurate equations of state exist for many fluids; these equations are normally fitted to the experimental data to maximize the prediction accuracy. A generalized compressibility chart $Z = f(P_r, T_r)$ is presented in Fig. 1.11. As seen in the figure, at all temperatures, Z tends to 1 as P_r tends to 0. This means that the behavior of the actual gas closely approaches ideal-gas behavior as the pressure approaches zero.

In the literature, there are also several equations of state for accurately representing the *P-v-T* behavior of a gas over the entire superheated vapor region, for example, the Benedict–Webb–Rubin equation, the van der Waals equation, and the Redlich and Kwong equation; However, some of these equations of state are complicated, due to the number of empirical constants they contain, and are more conveniently used with computer software to obtain results. The most basic equation of state is that of *Van der Waals*, which is capable to predict the vapor and liquid saturation line and a qualitatively correct fluid behavior in the vicinity of the critical point. This equation is described as given in Table 1.7.

1.5 THE LAWS OF THERMODYNAMICS

There are three laws of thermodynamics. The *zeroth law of thermodynamics* is a statement about thermodynamic equilibrium expressed as follows: "if two thermodynamic systems are in thermal equilibrium with a third, they are also in thermal equilibrium with each other." A system at internal equilibrium has a uniform pressure, temperature and chemical potential throughout its volume.

Note that two thermodynamic systems are said to be in *thermal equilibrium* if they cannot exchange heat, or in other words, they have the same temperature. Two thermodynamic systems are in *mechanical equilibrium* if they cannot exchange energy in the form of work. Two thermodynamic systems are in *chemical equilibrium* if they are not able

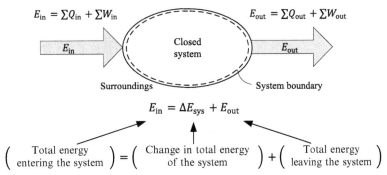

FIG. 1.12 Illustrating the first law of thermodynamics for a closed system.

to change their chemical composition. An insulated thermodynamic system is said to be in *thermodynamic equilibrium* when no mass, heat, work, chemical energy, etc., are exchanged between any parts within the system.

The *first law of thermodynamics* (FLT) postulates the energy conservation principle: "energy can be neither created nor destroyed." The FLT can be phrased as "you can't get something from nothing." If one denotes E the energy (in KJ) and ΔE_{sys} the change of energy of the system, then the FLT for a closed system undergoing any kind of process is written in the manner illustrated in Fig. 1.12. There are three mathematical forms for FLT, namely on an amount basis on rate basis and on mass specific basis. These mathematical formulations are written according to the following three equations:

$$E_{in} = E_{out} + \Delta E_{sys}, \text{ on amount basis} \tag{1.22a}$$

$$\dot{E}_{in} = \dot{E}_{out} + dE_{sys}/dt, \text{ on rate basis} \tag{1.22b}$$

$$e_{in} = e_{out} + \Delta e_{sys}, \text{ on mass specific basis}, e = E/m \tag{1.22c}$$

Energy can be transferred to or from a thermodynamic system in three basic forms, namely as work, heat and through energy associated with mass-crossing the system boundary. In classical thermodynamics, there is, however, a sign convention for work and heat transfer, which is the following:

- The heat is positive when given to the system, that is, $Q = \sum Q_{in} - \sum Q_{out}$ is positive when there is net heat provided to the system
- The net output work, $W = \sum W_{out} - \sum W_{in}$, is positive when work is generated by the system

Using the sign convention, the FLT for closed systems becomes

$$Q - W = \Delta E_{sys} = m(e_2 - e_1) \tag{1.23}$$

where e is the specific total energy of the system comprising internal energy, kinetic energy and potential energy, and expressed as follows

$$e = u + \frac{1}{2}v^2 + g z \tag{1.24}$$

The FLT can be expressed in differential form. For a closed system here is the expression of FLT in differential form

$$de = dq - dw = dq - P\,dv$$

If it is assumed that there is no kinetic and potential energy change, the FLT for closed system becomes

$$du = dq - P\,dv \tag{1.25}$$

If the system is a control volume, then the energy term will comprise the additional term of flow work. In this case, the total specific energy of a flowing matter is

$$\theta = u + P\,v + 0.5\,v^2 + g z = h + 0.5\,v^2 + g z \tag{1.26}$$

Using enthalpy formulation, the FLT for a control volume that has neither velocity nor elevation becomes

$$dh = dq + v\,dP \tag{1.27}$$

The FLT for control volume, using the sign convention for heat and work is formulated mathematically, in rate form, in the following way

$$\dot{Q} + \sum_{\text{in}} \dot{m}\,\theta = \dot{W} + \sum_{\text{out}} \dot{m}\,\theta + \frac{d(m\,e)}{dt} \tag{1.28}$$

Because $u = u(T, v)$ and $h = h(T, P)$, the following two relationships can be obtained from FLT

$$u = u(T, v) \rightarrow du = \left(\frac{\partial u}{\partial T}\right)_v dT + \left(\frac{\partial u}{\partial v}\right)_T dv = C_v\,dT - P\,dv$$

$$h = h(T, P) \rightarrow dh = \left(\frac{\partial h}{\partial T}\right)_P dT + \left(\frac{\partial h}{\partial P}\right)_T dP = C_p dT + v\,dP$$

From the above two expressions, the pressure and specific volume can be obtained from the specific internal energy and specific enthalpy, respectively as follows

$$P \equiv -\left(\frac{\partial u}{\partial u}\right)_T \tag{1.29a}$$

$$v \equiv \left(\frac{\partial h}{\partial P}\right)_T \tag{1.29b}$$

The first law of thermodynamics is practically complemented by the second law of thermodynamics (SLT) which provides a means to predict the direction of any process in time, to establish conditions of equilibrium, to determine the maximum attainable performance of machines and processes, to assess quantitatively the irreversibilities and determine their magnitude for the purpose of identifying ways of improvement of processes and engineered systems. The SLT is related to the concepts of reversibility and irreversibility. One says that a thermodynamic process is reversible if during a transformation both the thermodynamic system and its surroundings can be returned to their initial states. Reversible processes are of three kinds as follows:

- *externally reversible*: with no associated irreversibilities outside the system boundary
- *internally reversible*: with no irreversibilities within the boundary of the system during the process
- *totally reversible*: with no irreversibilities within the system and surroundings

There are two classical statements of SLT, which state that heat cannot be completely converted into work although the opposite is possible:

- *The Kelvin–Plank statement*: It is impossible to construct a device, operating in a cycle (e.g., heat engine), that accomplishes only the extraction of heat energy from some source and its complete conversion to work. This simply shows the impossibility of having a heat engine with a thermal efficiency of 100%.
- *The Clausius statement*: It is impossible to construct a device, operating in a cycle (e.g., refrigerator and heat pump), that transfers heat from the low-temperature side (cooler) to the high-temperature side (hotter). This simply shows the imposibility of having a heat pump or refrigerator working spontaneously. The Clausius inequality provides a mathematical statement of the SLT, namely:

$$\oint \frac{dQ}{T} \leq 0 \tag{1.30}$$

where the circular integral indicates that the process must be cyclical.

At the limit when the inequality becomes zero, then the processes are reversible (ideal situation). A useful mathematical artifice is to attribute to the integral from Eq. (1.30) a new physical quantity. Any real process must have generated entropy. The following cases may thus occur:

(i) $S_{\text{gen}} > 0$, real, irreversible process;
(ii) $S_{\text{gen}} = 0$, ideal, reversible process;
(iii) $S_{\text{gen}} < 0$, impossible process.

Generated entropy of a system during a process is a superposition of entropy change of the thermodynamic system and the entropy change of the surroundings. This will define entropy generated by the system S_{gen}

$$S_{\text{gen}} = -\oint \frac{dQ}{T} = \Delta S_{\text{sys}} + \Delta S_{\text{surr}} \geq 0 \rightarrow \Delta S_{\text{sys}} > \Delta S_{\text{surr}} = \left(\frac{Q}{T}\right)_{\text{surr}} \tag{1.31}$$

Since for a reversible process $S_{gen} = 0$, it results that entropy change of the system is the opposite of the entropy change of the surroundings

$$\Delta S_{rev} = -\Delta S_{surr} = \left(\frac{Q}{T}\right)_{rev}$$

Although the change in entropy of the system and its surroundings may individually increase, decrease or remain constant, the total entropy change (the sum of entropy change of the system and the surroundings or the total entropy generation) cannot be less than zero for any process. Note that entropy change along a process 1–2 results from the integration of the following equation

$$dQ = T\,dS \tag{1.32}$$

hence

$$S_{1-2} = \int_1^2 \frac{dQ}{T}$$

The SLT is a useful tool in predicting the limits of a system to produce work while generating irreversibilities to various imperfections of energy conversion or transport processes. The most fundamental device with cyclical operation with which thermodynamic operates is the heat engine; or the other important device is a heat pump. These devices operate between a heat source and a heat sink. A heat sink represents the thermal reservoir capable of absorbing heat from other systems. A heat source represents a thermal reservoir capable of providing thermal energy to other systems. A heat engine operates cyclically by transferring heat from a heat source to a heat sink. While receiving more heat from the source (Q_H) and rejecting less to the sink (Q_C), a heat engine can generate work (W). As stated by the FLT, energy is conserved, thus $Q_H = Q_C + W$.

A typical "black box" representation of a heat engine is presented in Fig. 1.13a. According to the SLT, the work generated must be strictly smaller than the heat input, $W < Q_H$. The thermal efficiency of a heat engine—also known as energy efficiency—is defined as the net work generated by the total heat input. Using notations from Fig. 1.13a, energy efficiency of a heat engine is expressed (by definition) with

$$\eta = \frac{W}{Q_H} = 1 - \frac{Q_C}{Q_H} \tag{1.33}$$

If a thermodynamic cycle operates as a refrigerator or heat pump, then its performance can be assessed by the *coefficient of performance* (COP) defined as useful heat generated per work consumed. As observed in Fig. 1.13b, the energy balance equation (EBE) for a heat pump is written as $Q_C + W = Q_H$. According to SLT, $Q_H \geq W$ (this means that work can be integrally converted into heat). Based on its definition, the COP is

$$COP = \frac{Q_H}{W} = \frac{Q_H}{Q_H + Q_C} \tag{1.34}$$

The Carnot cycle is a fundamental model in thermodynamics, representing a heat engine (or heat pump) that operates between a heat source and a heat sink, both of them being at constant temperature. This cycle is a conceptual

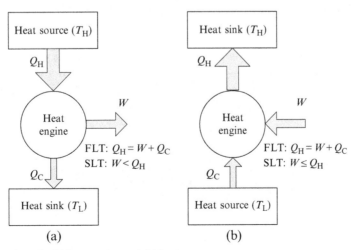

FIG. 1.13 Conceptual representation of a (a) heat engine and (b) heat pump.

(theoretical) cycle and was proposed by Sadi Carnot in 1824. The cycle comprises fully reversible processes, namely two adiabatic and two isothermal processes. The efficiency of the Carnot cycle is independent of working fluid, which performs the processes cyclically. Based on the definition of the Carnot cycle, it results that $s_2 = s_3$ and $s_4 = s_1$ (for heat engine). The heat transferred at the source and sink are $Q_H = T_H(s_3 - s_4) = T_H(s_2 - s_1)$ and $Q_L = T_L(s_2 - s_1)$, respectively. Therefore, the energy efficiency of a reversible Carnot heat engine is defined as

$$\eta = 1 - \frac{T_L}{T_H} \tag{1.35}$$

so the COP of the reversible Carnot heat pump becomes

$$COP = \frac{T_H}{(T_H - T_L)} \tag{1.36}$$

thus the COP of the reversible Carnot refrigerator becomes

$$COP = \frac{T_L}{(T_H - T_L)} \tag{1.37}$$

using the temperature scale as $(Q_H/Q_L)_{rev} = (T_H/T_L)$.

In summary, the above given Carnot efficiency and Carnot COPs are useful criteria to assess practical heat engines, refrigerators, heat pumps or other energy conversion systems with respect to the idealized case of reversible devices. Accordingly, energy efficiency (η) and the COP of a reversible thermodynamic cycle (Carnot) is the highest possible and any actual (irreversible) cycle has smaller efficiency ($\eta_{rev} > \eta_{irrev}$) and ($COP_{rev} > COP_{irrev}$).

1.6 EXERGY

Exergy represents the maximum work that can be produced by a thermodynamic system when it comes into equilibrium with its surrounding environment. This statement assumes that at an initial state, there is a thermodynamic system that is not in equilibrium with the environment. In addition, it is assumed that, at least potentially, mechanisms of energy (and mass) transfer between the system and the environment must exist, such that eventually the system can evolve such equilibrium condition will eventually occur.

During the assumed process, system must exchange work with the environment. By definition, exergy assumes the existence of a reference environment. The system under the analysis will interact only with that environment. Exergy analysis is a method appertaining to engineering thermodynamics and can be used to determine the alleviation of manmade and natural systems from the ideal case. Here, by an ideal system, one understands a reversible system.

In many practical problems, the reference environment is assumed to be the earth's atmosphere, characterized by its average temperature and pressure. Often standard pressure and temperature are used for the reference environment: $P_0 = 101.325 \, kPa$, $T_0 = 298.15 \, K$.

In some classes of the problems when reacting systems are present, the chemical potential of the reference environment must be specified. In such cases, thermodynamic equilibrium will refer to all types of interactions, including chemical reactions. One can say that a system is in thermodynamic equilibrium with the environment if it shares the same temperature (thermal equilibrium), the same pressure (mechanical equilibrium) and the same chemical potential. Therefore, exergy includes at least two components, one being thermomechanical and one chemical.

Exergy cannot be conserved. Any real process destroys exergy as, similarly, it generates entropy. Exergy is destroyed and entropy is generated due to irreversibilities. According to Dincer and Rosen (2012), the exergy of a closed (nonflow) thermodynamic system comprises four terms, namely physical (or thermomechanical), chemical, kinetic, and potential. In brief, total exergy of a nonflow system is

$$Ex_{nf} = Ex_{ph} + Ex_{ch} + Ex_{ke} + Ex_{pe} \tag{1.38}$$

The exergy of a flowing stream of matter Ex_f represents the sum of the nonflow exergy and the exergy associated with the flow work of the stream $(P - P_0)V$, therefore

$$Ex_f = Ex_{nf} + (P - P_0)V \tag{1.39}$$

The physical exergy for a nonflow system is defined by

$$Ex_{ph} = (U - U_0) + P_0 (V - V_0) - T_0 (S - S_0) \qquad (1.40)$$

where U is internal energy, V volume and S entropy of a closed system that is in nonequilibrium with the environment, T_0 is the reference temperature of the surroundings environment, and index 0 refers to the values of the parameters when the system is in thermomechanical equilibrium with the environment.

The kinetic and potential exergies of the system equal the kinetic and potential energy respectively, which are given by known formulas, namely

$$Ex_{ke} = \frac{1}{2} m \, v^2$$

$$Ex_{pe} = m g (z - z_0)$$

where m is the system mass, and v is its (macroscopic) velocity, z is the system elevation and z_0 is a reference elevation of the environment (e.g., ground level).

Consider a system that is in thermomechanical equilibrium with the reference environment (it has the same temperature and pressure as the environment—T_0, P_0), but it is not in chemical equilibrium with the reference environment, because it has another chemical composition.

Chemical exergy represents the maximum work that can be extracted during a process when the system, being initially in thermo-mechanical equilibrium with the reference environment (both have the same temperature and pressure), it still chemically reacts with the environment until the largest extent, such that a chemical equilibrium is achieved. There are two main components of chemical exergy: (i) exergy due to chemical reaction; (ii) exergy due to concentration difference. When a chemical compound is let to interact with the environment, chemical reactions may occur, involving unstable species. Eventually, more stable species are formed and further reaction is not possible.

Typical reference systems for exergy on earth are the atmosphere, hydrosphere and lithosphere (sea water, oceans). If a substance is not present in the atmosphere, then the reference for zero chemical exergy is the most stable state of that substance in seawater.

There are developed tables of chemical exergy of elements in past literature data. A recent source for tabulated data of standard chemical exergy of elements is Rivero and Grafias (2006). Table 1.8 tabulates the chemical exergies of some of the most encountered chemical elements in industrial processes. The standard chemical exergy of elements is useful for the calculation of chemical exergy of chemical compounds provided that their Gibbs energy of formation is known.

Moreover, if system compounds have another concentration or phase as that corresponding to the environment, then various processes such as dilution or concentration may occur until there is no difference in the concentration between system components and the environment.

The chemical exergy depends on the difference between the chemical potential of system components $\left(\sum n_i \mu_i^0 \right)$ being in thermomechanical equilibrium but not in chemical equilibrium with reference environment (μ_i^0), and the chemical potential of system components (μ_i^{00}) occurring when the chemical species are brought into chemical equilibrium with the reference environment, $\left(\sum n_i \mu_i^{00} \right)$. Therefore, the chemical exergy of the system is defined as

$$Ex_{ch} = \sum n_i \left(\mu_i^0 - \mu_i^{00} \right) \qquad (1.41)$$

Let us analyze the chemical exergy due to the concentration difference between the system and the surrounding environment. Let us assume the thermodynamic system at state 1 in nonequilibrium with the environment. If mass transfer is permitted with the environment, a dilution process occurs until the moment when the system components are fully diluted and there is no concentration gradient; this state is denoted with a 2. The maximum work extractable from process 1–2 represents the exergy due to concentration difference and is given by

$$\Delta Ex_{conc}^{ch} = Ex_1 - Ex_2 = (U_1 - U_2) + (P_1 V_1 - P_2 V_2) - T_0(S_1 - S_2) = T_0(S_2 - S_1) \qquad (1.42)$$

TABLE 1.8   Standard Chemical Exergy of Some Elements

Element	B	C	Ca	Cl₂	Cu	F₂	Fe	H₂	I₂	K	Mg
ex^{ch}, kJ/mol	628.1	410.27	729.1	123.7	132.6	505.8	374.3	236.12	175.7	336.7	626.9
Element	Mo	N₂	Na	Ni	O₂	P	Pb	Pt	Pu	Si	Ti
ex^{ch}, kJ/mol	731.3	0.67	336.7	242.6	3.92	861.3	249.2	141.2	1100	855.0	907.2

Data from Rivero, R., Grafias, M., 2006. Standard chemical exergy of elements updated. Energy 31, 3310–3326.

Here, one accounts that the process of diffusion is isothermal and one assumes that the gases involved are ideal gas $U_2 = U_1$ and $P_2 V_2 = P_1 V_1$. Furthermore, according to the FLT, $T \, dS = dU + P \, dV$; therefore, for an isothermal process of ideal gas for which $dU = 0$ and $d(PV) = 0$, one has $T \, dS = d(PV) - v \, dP$, or $T \, dS = -v \, dP$. Consequently, the chemical exergy due to difference in concentration of the gas component i having molar fraction y_i is given as follows

$$Ex_{conc,i}^{ch} = -\mathcal{R} \, T_0 \, \ln{(y_i)} \tag{1.43}$$

The notion of *Gibbs free energy* is introduced as a *state function* defined by $g = h - Ts$; this function can be used to determine the maximum work related to chemical processes. Rivero and Grafias (2006) give a general equation for the chemical exergy calculation; any chemical compound can be derived in a similar manner as illustrated above for water chemical exergy. In order to determine the chemical exergy of a compound, it is required to know its standard Gibbs free energy of formation, $\Delta^f g^0$. Then, using $\Delta^f g^0$ and the standard exergy of the elements, the following formula must be used to determine the chemical exergy of the compound as

$$ex^{ch} = \Delta^f g^0 + \sum_{element} \left(\nu \, ex^{ch} \right)_{element} \tag{1.44}$$

where ν is the stoichiometric factor representing the number of moles of element per one mole of chemical compound.

Because the intensive properties (pressure, temperature, chemical potential) of the natural environment varying temporally and spatially, the natural environment is far from a thermodynamic equilibrium. Thenceforth, being departed from equilibrium, the natural environment possesses work potential. This means that the natural environment has nonzero exergy. However, in most of the cases, the environment changes slowly as chemical reactions are not activated due to a reduced rate of the involved transport processes. Fossil fuel supplies reserves or forests do not burn spontaneously since no activation energy is provided.

Therefore, a compromise can be made between the theoretical requirements of the reference environment and the actual behavior of the natural environment. That is, a standard reference model can be adopted that is useful for exergy and environmental impact calculations. The reference environment is in stable equilibrium, acting as an infinite sink and source for heat and materials, experiencing only internally reversible processes with unaltered intensive states.

Natural environment models attempt to realistically simulate subsystems of the natural environment. Some relevant chemical component for the reference environment are water (H_2O), gypsum ($CaSO_4 \cdot 2H_2O$), and limestone ($CaCO_3$). The stable configurations of C, O, and N, respectively, may be taken to be those of CO_2, O_2, and N_2, as they exist in air saturated with liquid water at T_0 and P_0 (the temperature and pressure for the reference environment). Hydrogen reaches the equilibrium with the environment after reaction with oxygen and forming liquid water at T_0 and P_0. Calcium reaches thermodynamic equilibrium with the environment after reacting with either sulfur or carbon dioxide and forming, respectively $CaSO_4 \cdot 2H_2O$ and $CaCO_3$ at T_0 and P_0.

Equilibrium and constrained-equilibrium models were also formulated in which all the materials present in the atmosphere, oceans and a layer of the crust of the earth are pooled together and an equilibrium composition is calculated for a given temperature. The selection of the thickness of crust considered as reference environment is subjective as it is intended to include all materials accessible to thermal processes. Thicknesses varying from 1 to 1000 m, and a temperature of 25°C were considered. Exergy values obtained using these environments are significantly dependent upon the thickness of crust considered, and represent the absolute maximum amount of work obtainable from a material.

Assuming that the environment is a large reservoir that, when a particular substance interacts with it, the following type of processes occur after a sufficiently long time: (i) mechanical equilibration (the substance pressure will equal to that of the surroundings); (ii) thermal equilibrium (meaning that the substance temperature becomes equal to that of the environment); (iii) chemical reaction equilibrium (meaning that the substance enters into a series of spontaneous chemical reactions with the environment such that it decomposes and eventually forms only chemical species present in the environment); (iv) concentration equilibrium (the chemical species resulting from substance reaction with the environment dilute or concentrate such that they reach the concentration in the environment). The processes (i) and (ii) refer to the thermomechanical equilibrium. The departure of the substance stream from the temperature and pressure of the environment is a measure of the thermomechanical exergy. Moreover, as mentioned previously, the processes (iii) and (iv) represent chemical equilibrium, which is associated with chemical exergy.

Rivero and Grafias (2006) proposed a model for the standard environment, considering that the relative humidity in the atmosphere is 70%, carbon dioxide concentration is 345 ppm in volume and the salinity of sea water is 35‰. The list of stable components in the atmosphere and their concentration in the reference environment is given in

TABLE 1.9 Reference Environment Described in Rivero and Grafias (2006)

Atmosphere	E:	Ar	CO_2	D_2O	H_2O	He			
	y:	9.13E$-$3	3.37E$-$4	3.37E$-$6	2.17E$-$2	4.89E$-$6			
	E:	Kr	N_2	Ne	O_2	Xe			
	y:	9.87E$-$7	0.7634	1.76E$-$5	0.2054	8.81E$-$8			
Hydrosphere	E:	$HAsO_4^{2-}$	$B(OH)_3$	BiO^+	Br^-	Cl^-	Cs^+	IO_3^-	K^+	Li^+
	x:	3.87E$-$8	3.42E$-$4	9.92E$-$11	8.73E$-$4	0.5658	2.34E$-$9	5.23E$-$7	1.04E$-$2	2.54E$-$5
	E:	MoO_4^{2-}	Na^+	HPO_4^{2-}	Rb^+	SO_4^{2-}	SeO_4^{2-}	WO_4^{2-}		
	x:	1.08E$-$7	0.4739	4.86E$-$7	1.46E$-$6	1.24E$-$2	1.18E$-$9	5.64E$-$10		
Lithosphere	E:	AgCl	Al_2SiO_5	Au	$BaSO_4$	Be_2SiO_4	$CaCO_3$	$CdCO_3$	CeO2	$CoFe_2O_4$
	y:	1E$-$9	2.07E$-$3	1.36E$-$9	4.2E$-$6	2.1E$-$7	1.4E$-$4	1.22E$-$8	1.17E$-$6	22.85E$-$7
	E:	$K_2Cr_2O_7$	$CuCO_3$	$Dy(OH)_3$	$Er(OH)_3$	$Eu(OH)_3$	$CaF_2 \cdot 3Ca_3(PO_4)_2$		Fe_2O_3	Ga_2O_3
	y:	1.35E$-$6	5.89E$-$6	4.88E$-$8	4.61E$-$8	2.14E$-$8	2.24E$-$4		6.78E$-$3	2.89E$-$7
	E:	$Gd(OH)_3$	GeO_2	HfO2	$HgCl_2$	$Ho(OH)_3$	In_2O_3	IrO_2	$La(OH)_3$	$Lu(OH)_3$
	y:	9.21E$-$8	9.49E$-$8	1.15E$-$7	5.42E$-$10	1.95E$-$8	2.95E$-$9	3.59E$-$12	5.96E$-$7	7.86E$-$9
	E:	$Mg_3Si_4O_{10}(OH)_2$		MnO_2	Nb_2O_3	$Nb(OH)_3$	NiO	OsO_4	$PbCO_3$	PdO
	y:	8.67E$-$4		2.3E$-$5	1.49E$-$7	5.15E$-$7	1.76E$-$6	3.39E$-$13	1.04E$-$7	6.37E$-$11
	E:	$Pr(OH)_3$	PtO_2	PuO_2	$RaSO_4$	Re_2O_7	Rh_2O_3	RuO_2	Sb_2O_5	Sc_2O_3
	y:	1.57E$-$7	1.76E$-$11	8.4E$-$20	2.98E$-$14	3.66E$-$12	3.29E$-$12	6.78E$-$13	1.08E$-$10	3.73E$-$7
	E:	SiO_2	$Sm(OH)_3$	SnO_2	$SrCO_3$	Ta_2O_5	$Tb(OH)_3$	TeO_2	ThO_2	TiO_2
	y:	0.407	1.08E$-$7	4.61E$-$7	2.91E$-$5	7.45E$-$9	1.71E$-$8	9.48E$-$12	2.71E$-$7	1.63E$-$4
	E:	Tl_2O_4	$Tm(OH)_3$	$UO_3 \cdot H_2O$		V_2O_5	$Y(OH)_3$	$Yb(OH)_3$	$ZnCO_3$	$ZrSiO_4$
	y:	1.49E$-$9	7.59E$-$9	1.48E$-$8		1.83E$-$6	1E$-$6	4.61E$-$8	7.45E$-$6	2.44E$-$5

Here, x is the mass fraction in hydrosphere; y is the molar fraction used for atmosphere and lithosphere, "E" is abbreviation for "elements".

Table 1.9. The chemical elements in the atmosphere for this model are Ar, C, D_2, H_2, He, Kr, N_2, Ne, O_2, and Xe. The molar fraction of the species in the atmosphere is related to the standard chemical exergy under the assumption of ideal-gas behavior as follows

$$y = \exp\left(-\frac{ex^{ch}}{\mathcal{R}T_0}\right) \tag{1.45}$$

The standard reference model for the environment is taken as a base for calculating the chemical exergy of the chemical elements for which an extract is given in Table 1.8. Once the exergy of chemical elements is known, the chemical exergy of any chemical compound can be determined based on Eq. (1.44).

EXAMPLE 1.1

An example of chemical exergy calculation is given here for methane. The formation reaction of methane is $C + 2H_2 \rightarrow CH_4$, while the standard formation Gibbs energy is $\Delta^f g^0 = -50.53$ kJ/mol. The standard chemical exergy of hydrogen is 236.12 kJ/mol, while that of carbon (graphite) is 410.27 kJ/mol.

Solution

The chemical exergy of methane is computed with Eq. (1.44) as follows

$$ex^{ch}_{CH_4} = -50.53 + 410.27 + 2 \times 236.12 = 832 \text{ kJ/mol}$$

1.7 THERMODYNAMIC ANALYSIS THROUGH ENERGY AND EXERGY

Thermodynamic analysis is generally based on four types of balance equations, which will be presented here in detail. These are mass balance equation (MBE), energy balance equation (EBE), entropy balance equation (EnBE), and exergy balance equation (ExBE). Thermodynamic analysis using balance equations is documented in detail in Dincer and Rosen (2012). Subsequently, a brief introduction on this method is presented.

1.7.1 Mass Balance Equation

The effect of mass addition or extraction on the energy balance of control volume is proportional with the *mass flow rate* defined as the amount of mass flowing through a cross-section of a flow stream per unit of time. For a control volume, according to the *conservation of mass principle*, the net mass transferred to the system is equal to the net change in mass within the system plus the net mass leaving the system.

Assume that a number of streams with total mass flow rate $\sum \dot{m}_{in}$ enter the system, while a number of streams of total mass flow rate $\sum \dot{m}_{out}$ leave the system. Consequently, the mass of the control volume will change with differential amount dm_{cv}. The mass balance equation for a general control volume (see Fig. 1.14) can be written for nonsteady state system as in Eq. (1.46a) or for a steady state system as in Eq. (1.46b)

$$\text{MBE}: \quad \sum \dot{m}_{in} = \sum \dot{m}_{out} + \frac{dm_{cv}}{dt} \tag{1.46a}$$

$$\text{MBE}_{\text{steady flow}}: \quad \sum \dot{m}_{in} = \sum \dot{m}_{out} \tag{1.46b}$$

1.7.2 Energy Balance Equation

The EBE is an expression of the FLT with a sign convention relaxed (see Fig. 1.12). Therefore, the variation of system energy of a closed between states 1 and 2 is written as

$$\Delta E_{sys} = m \Delta e_{sys} = m \left[\left(u_2 + \frac{1}{2} v_2^2 + g z_2 \right) - \left(u_1 + \frac{1}{2} v_1^2 + g z_1 \right) \right] \tag{1.47}$$

For a closed system, the EBE is written with the help of the total specific energy of a nonflowing thermodynamic system $e = u + 0.5 v^2 + g z$, namely

$$\text{EBE}_{\text{closed system}}: \quad \sum \dot{q}_{in} + \sum \dot{w}_{in} = \sum \dot{q}_{out} + \sum \dot{w}_{out} + \frac{de}{dt} \tag{1.48}$$

The EBE for control volumes must account for the existence of flow work and boundary work and for the rate of change of total energy $[d(me)/dt]$; thence it can be formulated as follows

$$\text{EBE}_{\text{open system}}: \quad \sum_{in} \dot{m}\theta + \sum \dot{Q}_{in} + \sum \dot{W}_{in} = \sum_{out} \dot{m}\theta + \sum \dot{Q}_{out} + \sum \dot{W}_{out} + \left[\frac{d(me)}{dt} \right]_{sys} \tag{1.49}$$

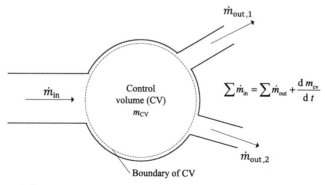

FIG. 1.14   Illustrative sketch for mass balance equation.

where θ is the total *energy of a flowing matter*, which represents the sum of internal energy, flow work, kinetic energy and potential energy defined by

$$\theta = u + P\,v + \frac{1}{2}\,v^2 + g\,z = h + \frac{1}{2}\,v^2 + g\,z \qquad (1.50)$$

In a steady flow system, mass flow rate, pressure, temperature, etc., do not change in time, thence the integration of the following equations $d(mh) = (\dot{m}h)dt$, $dQ = \dot{Q}d$, and $dW = \dot{W}dt$ between initial state 1 and a latter state 2 of the open system is straightforward. In steady flow regime, the EBE can be written in rate form

$$\text{EBE}_{\text{steady flow}}: \quad \dot{m}_1 e_1 + \dot{Q}_{\text{in}} + \dot{W}_{\text{in}} + \sum_{\text{in}}(\dot{m}h) = \dot{m}_2 e_2 + \dot{Q}_{\text{out}} + \dot{W}_{\text{out}} + \sum_{\text{out}}(\dot{m}h) \qquad (1.51)$$

1.7.3 Entropy Balance Equation

The SLT can be expressed in the form of an EnBE that states for a thermodynamic system, that entropy input plus generated entropy is equal to entropy output plus change of entropy within the system. In other words, the EnBE postulates that the entropy change of a thermodynamic system is equal to entropy generated within the system plus the net entropy transferred to the system across its boundary (ie, the entropy entering minus the entropy leaving). Entropy can be transferred as heat, but it cannot be transferred as work. Fig. 1.15 illustrates schematically the EnBE, which is written mathematically according to

$$\text{EnBE}: \quad \sum \dot{S}_{\text{in}} + \dot{S}_{\text{gen}} = \sum \dot{S}_{\text{out}} + \frac{dS_{\text{sys}}}{dt} \qquad (1.52)$$

The entropy transferred across the system boundary or along a process 1–2 is $S_{1-2} = \int_1^2 \frac{dQ}{T}$. The general EnBE takes a special form for closed systems. For a closed system, there is no mass transfer at the system boundary. Therefore, entropy can be transferred only by heat. If the closed system is also adiabatic, then there is neither entropy transfer due to mass nor due to heat transfer, henceforth $\dot{S}_{\text{sys}} = \dot{S}_{\text{gen}}$.

If the closed system is not adiabatic, then the EnBE becomes

$$\text{EnBE}_{\text{closed system}}: \quad \sum_{\text{in}}\left(\int \frac{d\dot{Q}}{T}\right) + \dot{S}_{\text{gen}} = \frac{dS_{\text{sys}}}{dt} + \sum_{\text{out}}\left(\int \frac{d\dot{Q}}{T}\right) \qquad (1.53)$$

The EnBE for an open system (control volume, cv) has the following expression in rate form

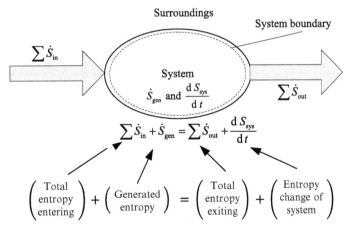

FIG. 1.15   Explanatory sketch for the entropy balance equation, a statement of SLT.

TABLE 1.10   Entropy Transfer Across a Wall Boundary

Case 1: No wall effect considered	Case 2: Wall ΔT considered	Case 3: Wall and boundary layer
$S_{\text{gen},1} = \dfrac{Q}{T_0} - \dfrac{Q}{T_{\text{sys}}}$	$S_{\text{gen},2} > \dfrac{Q}{T_0} - \dfrac{Q}{T_{\text{sys}}}$	$S_{\text{gen},3} > S_{\text{gen},2} > \dfrac{Q}{T_0} - \dfrac{Q}{T_{\text{sys}}}$

Modified from Cengel, Y.A., Boles, M.A., 2011. Fundamentals of Thermodynamics. McGraw Hill, New York.

$$\text{EnBE}_{\text{cv}}: \quad \sum_{\text{in}}\left(\int \frac{\mathrm{d}\dot{Q}}{T}\right) + \sum_{\text{in}}\dot{m}s + \dot{S}_{\text{gen}} = \frac{\mathrm{d}S_{\text{CV}}}{\mathrm{d}t} + \sum_{\text{out}}\left(\int \frac{\mathrm{d}\dot{Q}}{T}\right) + \sum_{\text{out}}\dot{m}s \tag{1.54}$$

The EnBE for a steady flow through a control volume must account for the fact that there are no temporal variations of parameters; thence, the mass enclosed in the control volume and specific entropy of the control volume remains constant in time; consequently,

$$\text{EnBE}_{\text{steady state}}: \quad \sum_{\text{in}}\left(\int \frac{\mathrm{d}\dot{Q}}{T}\right) + \sum_{\text{in}}\dot{m}s + \dot{S}_{\text{gen}} = \sum_{\text{out}}\left(\int \frac{\mathrm{d}\dot{Q}}{T}\right) + \sum_{\text{out}}\dot{m}s \tag{1.55}$$

In the case when $\dot{m}_{\text{out}} = \dot{m}_{\text{in}} = \dot{m}$ applies, the EnBE simplifies to

$$\text{EnBE}: \quad \sum_{\text{in}}\left(\int \frac{\mathrm{d}\dot{Q}}{T}\right) + m(s_{\text{in}} - s_{\text{out}}) + \dot{S}_{\text{gen}} = \sum_{\text{out}}\left(\int \frac{\mathrm{d}\dot{Q}}{T}\right)$$

If in addition to these where the process is adiabatic, there is no heat transfer across the system boundary; therefore, the EnBE simplifies to EnBE: $ms_{\text{in}} + \dot{S}_{\text{gen}} = ms_{\text{out}}$.

The generated entropy is the sum of entropy change of the system and of its surroundings. In regard to determination of entropy generation, there are three relevant cases that can be assumed for heat transfer across the system boundary as given subsequently. Consider first a thermodynamic system that has a diabatic boundary. As illustrated in Table 1.10, the EnBE for this system is given by the difference between Q/T_0 and Q/T_{sys}. It is assumed in this case that there is no wall with finite thickness at the system boundary. Therefore, the temperature profile has a sharp change. A more accurate assumption is to assume the existence of a wall at the boundary. In this case, there will be a variation of temperature across the wall. In Case 2 represented in Table 1.10, once considers a wall attached to the system boundary. In this case the entropy generation has to be calculated by integration accounting of the temperature profile. In the third case, in addition to a wall, one considers the existence of boundary layers at the inner and outer sides of the wall. Therefore, the entropy generation will be the highest in assumption for Case 3.

1.7.4 Exergy Balance Equation

The ExBE introduces the term *exergy destroyed*, which represents the maximum work potential that cannot be recovered for useful purpose due to irreversibilities. For a reversible system, there is no exergy destruction, since all work generated by the system can be made useful. The exergy destruction and entropy generation are related by the following expression: $Ex_{\text{d}} = T_0 \Delta S_{\text{gen}}$, where T_0 is the reference temperature. If $Ex_{\text{d}} > 0$, then the process is irreversible; if $Ex_{\text{d}} = 0$, then the process is reversible; if $Ex_{\text{d}} < 0$, the process is impossible.

The total exergy entering a thermodynamic system must be balanced by the total exergy leaving the system, plus the change of exergy content of the system plus, the exergy destruction. Fig. 1.16 shows an explanatory sketch for the ExBE. Exergy can be transferred to or from a system by three means: work, heat, and mass. Therefore, the ExBE can be expressed generally in rate form as follows

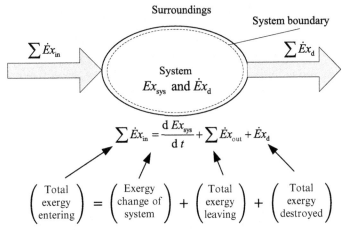

FIG. 1.16 Explanatory sketch for the exergy balance equation.

$$\text{ExBE}: \quad \sum_{\text{in}}\left[\dot{W} + \dot{m}\,\varphi + \left(1 - \frac{T_0}{T}\right)\dot{Q}\right] = \frac{dEx}{dt} + \sum_{\text{out}}\left[\dot{W} + \dot{m}\,\varphi + \left(1 - \frac{T_0}{T}\right)\dot{Q}\right] - P_0\frac{dV_{\text{CV}}}{dt} + \dot{Ex}_d \tag{1.56}$$

where the *total specific exergy* is defined with

$$\varphi = (h - h_0) + T_0(s - s_0) + \frac{1}{2}v^2 + g(z - z_0) + ex^{\text{ch}}.$$

Exergy transfer between the system and surrounding environment can be done by work, mass transfer or heat transfer. The exergy due to work transfer (Ex^W) is by definition equal to the work: $Ex^W = W$. However, if the system impinges against a moving boundary, then the exergy must be diminished accordingly, thence in that case $Ex^W = W - P_0(V - V_0)$. The exergy associated to mass transfer (Ex^m) is $Ex^m = m\varphi$. The exergy due to heat transfer can be expressed based on the Carnot factor (or temperature factor) according to

$$Ex^Q = \int_{\text{system boundary}} \left(1 - \frac{T_0}{T}\right) dQ$$

For a thermodynamic system at steady state, the ExBE simplifies to:

$$\text{ExBE}_{\text{steady state}}: \quad \sum_{\text{in}}\left[\dot{W} + \dot{m}\,\varphi + \left(1 - \frac{T_0}{T}\right)\dot{Q}\right] = \sum_{\text{out}}\left[\dot{W} + \dot{m}\,\varphi + \left(1 - \frac{T_0}{T}\right)\dot{Q}\right] + \dot{Ex}_d \tag{1.57}$$

If in Eq. (1.57) denotes one $\dot{W} = \sum\dot{W}_{\text{out}} - \sum\dot{W}_{\text{in}}$ and $\dot{Q} = \sum\dot{Q}_{\text{in}} - \sum\dot{Q}_{\text{out}}$ and one assumes that there is no exergy destroyed then, the reversible work can be obtained as follows

$$\dot{W}_{\text{rev}} = \dot{m}(\varphi_1 - \varphi_2) + \sum\left(1 - \frac{T_0}{T}\right)\dot{Q} \tag{1.58}$$

1.7.5 Formulations for System Efficiency

The term efficiency originates mainly from thermodynamics when the attempt of assessing the heat conversion into work led to its initial formulation as the "work generated per total heat energy input". However, efficiency as assessment criterion can be applied widely for any system and process. A general efficiency expression of a system, as a measure of its performance and effectiveness, is represented by the ratio of useful output per required input. Here, energy efficiency it is recognized as a figure of merit based on FLT. If the system is an energy system, then its input and output (delivered energy) must be forms of energy. Therefore, for an energy system, the energy efficiency is written as

$$\eta = \frac{\dot{E}_{\text{deliv}}}{\dot{E}_{\text{cons}}} = 1 - \frac{\dot{E}_{\text{loss}}}{\dot{E}_{\text{cons}}} \tag{1.59}$$

Any source of energy is characterized by an associated exergy. By analogy with energy efficiency, the exergy efficiency is defined as the ratio between exergy associated to the useful output (exergy delivered to the user) and the exergy associated to the consumed input, namely

$$\psi = \frac{\dot{Ex}_{\text{deliv}}}{\dot{Ex}_{\text{cons}}} = 1 - \frac{\dot{Ex}_{\text{d}}}{\dot{Ex}_{\text{cons}}} \tag{1.60}$$

In Table 1.11, the efficiency formulations for main devices used in process engineering are given. Turbine is the first device analyzed in the table. A high enthalpy flow enters the turbine; work is produced, and a lower enthalpy flow exits the turbine. The turbine efficiency quantifies various losses such as the isentropic losses, the heat loses from turbine shell, the friction losses. Isentropic efficiency is one of the most used assessment parameter for turbines. Isentropic efficiency (η_{s}) is defined by the ratio of actual power generated to the power generated during an isentropic expansion. For an isentropic expansion, there is no entropy generation; the turbine operation is reversible. Therefore, isentropic efficiency is a relative measure of alleviation from thermodynamic ideality.

The expansion process 1–2 is the actual process, while the process 1–2s is the reversible process (isentropic). Exergy efficiency of a turbine is defined as the ratio of generated power and rate of exergy consumed. A compressor is a device used to increase the pressure of a fluid under the expense of work consumption. Compressors are typically assessed by the isentropic efficiency that, for the case of compressors, is the ratio of isentropic work and actual work.

Pumps are organs used to increase the pressure of liquids on the expense of work input. The liquid is incompressible and, therefore, the power required for pumping the liquid for a reversible process is $\dot{W}_{\text{s}} = \dot{m}v(P_2 - P_1)$. Hydraulic turbines are devices that generate work from potential energy of a liquid. Nozzles and diffusers are adiabatic devices used to accelerate or decelerate a fluid, respectively. The exergy efficiency of nozzle is nil because it produces no work, although the expanded flow has work potential. A heat exchanger is a device that facilitates heat transfer between two fluids without mixing. It is known that heat exchangers are assessed by their effectiveness parameter, which represents the ratio between the actual amount of heat transfer and the maximum amount of heat possible to transfer.

For a heat exchanger, the exergy source is derived from the hot fluid that, during the process, reduces its exergy. The exergy of the cold fluid represents the delivered exergy as the useful product of a heat exchanger. Regarding the energy exchange, ideally if there are no energy losses, all energy from the hot fluid is transferred to the cold fluid; however, some loses are unavoidable in practical systems, therefore, one can define an energy efficiency of heat exchangers as the ratio between energy delivered and energy consumed. Regarding the exergy efficiency of a heat exchanger, this is given by exergy retrieved from the cold fluid divided by exergy provided by the hot fluid.

Many practical devices are used to mix streams. Mixing chambers accept multiple stream inputs and have one single output. Mixers can be isothermal or one can mix a hot fluid with a cold fluid, etc. A combustion chamber or a reaction chamber can be modeled from a thermodynamic point of view as a mixing device, whereas mixing is accompanied by chemical reaction. Very similar to mixers, there are also stream separators. In this case, an input stream is separated into two (or more) different output streams.

1.7.6 Cost Accounting of Exergy

We delve now into an exergy-based theory of value, on which exergoeconomic analysis can be based. In economics, the theory of value attempts to explain the correlation between value and price of traded goods and services. The trades are typically made using the monetary values as a standardized currency for payments of goods, services and debts. The monetary price must reflect the value of the trade, which is related to costs and profitability. The theory of value offers an ideological basis for quantification of the benefit from a traded good or service. This helps assigning a monetary price to a value.

Three theories of value received much attention. The first is the power theory of value, which states that political power and the economy (which is constrained by laws of trade) are so highly interlaced that prices are established based on an internal hierarchy of values of the society rather than on a production and demand balance. The second is the labor theory of value, which states that the value is determined by the labor developed to produce the good or the service, including the labor spent to accumulate any required capital for the production process.

TABLE 1.11 Energy and Exergy Efficiencies of Devices

Device	Equations
1. Turbine 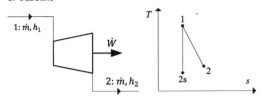	*Balance equations* MBE: $\dot{m}_1 = \dot{m}_2 = \dot{m}$ EBE: $\dot{m}_1 h_1 = \dot{W} + \dot{m}_2 h_2$ EnBE: $\dot{m}_1 s_1 + \dot{S}_{gen} = \dot{m}_2 s_2$ ExBE: $\dot{m}[(h_1 - h_2) - T_0(s_1 - s_2)] = \dot{W} + \dot{E}x_d$ *Efficiency equations* $\eta = \dfrac{\dot{W}}{\dot{W}_s} = \dfrac{\dot{m}(h_1 - h_2)}{\dot{m}(h_1 - h_{2s})}$ $\psi = \dfrac{\dot{W}}{\dot{E}x_{cons}} = \dfrac{\dot{m}(h_1 - h_2)}{\dot{m}[h_1 - h_2 - T_0(s_1 - s_2)]}$
2. Compressor 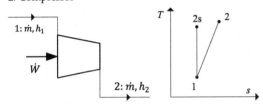	*Balance equations* MBE: $\dot{m}_1 = \dot{m}_2 = \dot{m}$ EBE: $\dot{m}_1 h_1 + \dot{W} = \dot{m}_2 h_2$ EnBE: $\dot{m}_1 s_1 + \dot{S}_{gen} = \dot{m}_2 s_2$ ExBE: $\dot{W} = \dot{m}[h_2 - h_1 - T_0(s_2 - s_1)] + \dot{E}x_d$ *Efficiency equations* $\eta = \dfrac{\dot{W}_s}{\dot{W}} = \dfrac{\dot{m}(h_{2s} - h1)}{\dot{m}(h_2 - h_1)}$ $\psi = \dfrac{\dot{W}_s}{\dot{E}x_{cons}} = \dfrac{\dot{m}(h_{2s} - h_1)}{\dot{m}[h_1 - h_2 - T_0(s_1 - s_2)]}$
3. Pump 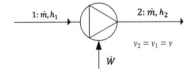	*Balance equations* MBE: $\dot{m}_1 = \dot{m}_2 = \dot{m}$ EBE: $\dot{m}_1 h_1 + \dot{W} = \dot{m}_2 h_2$ EnBE: $\dot{m}_1 s_1 + \dot{S}_{gen} = \dot{m}_2 s_2$ ExBE: $\dot{W} = \dot{m}[h_2 - h_1 - T_0(s_2 - s_1)] + \dot{E}x_d$ *Efficiency equations* $\eta = \dfrac{\dot{W}_s}{\dot{W}_{cons}} = \dfrac{\dot{m}v(P_2 - P_1)}{\dot{m}(h_2 - h_1)}$ $\psi = \dfrac{\dot{W}_s}{\dot{E}x_{cons}} = \dfrac{\dot{m}[h_2 - h_1 - T_0(s_2 - s_1)]}{\dot{m}(h_2 - h_1)}$
4. Hydraulic turbine 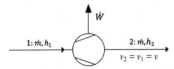	*Balance equations* MBE: $\dot{m}_1 = \dot{m}_2 = \dot{m}$ EBE: $\dot{m}_1 h_1 = \dot{W} + \dot{m}_2 h_2$ EnBE: $\dot{m}_1 s_1 + \dot{S}_{gen} = \dot{m}_2 s_2$ ExBE: $\dot{m}[(h_1 - h_2) - T_0(s_1 - s_2)] = \dot{W} + \dot{E}x_d$ *Efficiency equations* $\eta = \dfrac{\dot{W}}{\dot{W}_s} = \dfrac{\dot{m}(h_1 - h_2)}{\dot{m}v(P_1 - P_2)}$ $\psi = \dfrac{\dot{W}}{\dot{E}x_{cons}} = \dfrac{\dot{m}(h_1 - h_2)}{\dot{m}[h_1 - h_2 - T_0(s_1 - s_2)]}$

Continued

TABLE 1.11 Energy and Exergy Efficiencies of Devices—cont'd

Device	Equations
5. Diffuser and nozzle Diffuser: 1: $\dot{m}, h_1$　　2: $\dot{m}, h_2$ Nozzle: 1: $\dot{m}, h_1$　　2: $\dot{m}, h_2$	*Balance equation* MBE: $\dot{m}_1 = \dot{m}_2 = \dot{m}$ EBE: $\dot{m}_1 h_1 = \dot{m}_2 h_2$ EnBE: $\dot{m}_1 s_1 + \dot{S}_{\text{gen}} = \dot{m}_2 s_2$ ExBE: $\dot{m}[(h_1 - h_2) - T_0(s_1 - s_2)] = \dot{E}x_{\text{d}}$ *Efficiency equations* $\eta = \dfrac{h_1 - h_2}{h_1 - h_{2\text{s}}}$ and $\psi = 0$
6. Heat exchanger 1h: $\dot{m}_{\text{h}} h_{1\text{h}}$ 1c: $\dot{m}_{\text{c}} h_{1\text{c}}$　　2c: $\dot{m}_{\text{c}} h_{2\text{c}}$ 2h: $\dot{m}_{\text{h}} h_{2\text{h}}$	*Balance equations* MBE: $\dot{m}_{1\text{h}} = \dot{m}_{2\text{h}} = \dot{m}_{\text{h}}$ and $\dot{m}_{1\text{c}} = \dot{m}_{2\text{c}} = \dot{m}_{\text{c}}$ EBE: $\dot{m}_{\text{h}}(h_{1\text{h}} - h_{2\text{h}}) = \dot{m}_{\text{c}}(h_{2\text{c}} - h_{1\text{c}})$ EnBE: $\dot{S}_{\text{gen}} = \dot{m}_{\text{c}}(h_{2\text{c}} - h_{1\text{c}}) + \dot{m}_{\text{h}}(h_{2\text{h}} - h_{1\text{h}})$ ExBE: $\dot{m}_{\text{h}}[(h_{1\text{h}} - h_{2\text{h}}) - T_0(s_{1\text{h}} - s_{2\text{h}})] + \dot{m}_{\text{c}}[(h_{1\text{c}} - h_{2\text{c}}) - T_0(s_{1\text{c}} - s_{2\text{c}})] = \dot{E}x_{\text{d}}$ *Efficiency equations* $\varepsilon = \dfrac{\dot{Q}_{\text{cold}}}{\dot{Q}_{\text{max}}} = \dfrac{(\dot{m}C_{\text{p}}\Delta T)_{\text{cold}}}{(\dot{m}C_{\text{p}})_{\text{min}}(T_{1\text{h}} - T_{2\text{h}})}$ $\eta = \dfrac{\dot{Q}_{\text{cold}}}{\dot{Q}_{\text{hot}}}; \quad \psi = \dfrac{(\dot{E}x_{2\text{c}} - \dot{E}x_{1\text{c}})_{\text{cold}}}{(\dot{E}x_{1\text{h}} - \dot{E}x_{2\text{h}})_{\text{hot}}}$ $\psi = \dfrac{\dot{m}[h_{2\text{c}} - h_{1\text{c}} - T_0(s_{2\text{c}} - s_{1\text{c}})]}{\dot{m}_{\text{ht}}[h_{2\text{h}} - h_{1\text{h}} - T_0(s_{2\text{h}} - s_{1\text{h}})]}$
7. Mixing chamber or chemical reactor 3: $\dot{m}_3, h_3$ $\dot{E}_{\text{in}}$ and $\dot{E}x_{\text{in}}$ 1: $\dot{m}_1, h_1$　　2: $\dot{m}_2, h_2$	*Balance equations* MBE: $\dot{m}_1 + \dot{m}_2 = \dot{m}_3$ EBE: $\dot{m}_1 h_1 + \dot{m}_2 h_2 + \dot{E}_{\text{in}} = \dot{m}_3 h_3$; assume adiabatic EnBE: $\dot{m}_1 s_1 + \dot{m}_2 s_2 + \dot{S}_{\text{gen}} = \dot{m}_3 s_3$; assume adiabatic ExBE: $\dot{m}_1 ex_1 + \dot{m}_2 ex_2 + \dot{E}x_{\text{in}} = \dot{m}_3 ex_3 + \dot{E}x_{\text{d}}$ *Efficiency equations* $\eta = \dfrac{\dot{m}_3 h_3}{\dot{m}_1 h_1 + \dot{m}_2 h_2 + \dot{E}_{\text{in}}}$ $\psi = \dfrac{\dot{m}_3 ex_3}{\dot{m}_1 ex_1 + \dot{m}_2 ex_2 + \dot{E}x_{\text{in}}}$
8. Separation device 3: $\dot{m}_3, h_3$ $\dot{E}_{\text{in}}, \dot{E}x_{\text{in}}$ and $\dot{Q}_{\text{in}}$ 1: $\dot{m}_1, h_1$　　2: $\dot{m}_2, h_2$	*Balance equations* MBE: $\dot{m}_1 = \dot{m}_2 + \dot{m}_3$ EBE: $\dot{m}_1 h_1 + \dot{Q}_{\text{in}} + \dot{E}_{\text{in}} = \dot{m}_2 h_2 + \dot{m}_3 h_3$ EnBE: $\dot{m}_1 s_1 + \dfrac{\dot{Q}_{\text{in}}}{T_{\text{in}}} + \dot{S}_{\text{gen}} = \dot{m}_2 s_2 + \dot{m}_3 s_3$ ExBE: $\dot{m}_1 ex_1 + \dot{Q}_{\text{in}}\left(1 - \dfrac{T_0}{T_{\text{in}}}\right) + \dot{E}x_{\text{in}} = \dot{m}_2 ex_2 + \dot{m}_3 ex_3 + \dot{E}x_{\text{d}}$ *Efficiency equations* $\eta = \dfrac{\dot{m}_2 h_2 + \dot{m}_3 h_3}{\dot{m}_1 h_1 + \dot{Q}_{\text{in}} + \dot{E}_{\text{in}}}$ $\psi = \dfrac{\dot{m}_2 ex_2 + \dot{m}_3 ex_3}{\dot{m}_1 ex_1 + \dot{Q}_{\text{in}}\left(1 - \dfrac{T_0}{T_{\text{in}}}\right) + \dot{E}x_{\text{in}}}$

The third is the utility theory of value, which quantifies the value of tradable goods and services based on their utility. There is no direct way of measuring the utility as a representation of the preferences for trading of various services or goods, because this depends on subjective factors of human individuals, such as wishes and wants. However, the utility can be observed indirectly through the price that is established by trading activity. The price is determined by the balance between marginal utility and marginal cost. Here, the term marginal means an infinitesimal change, or derivative with respect to quantity (amount). Let us assume that a quantity Q of products are traded; if one denotes U a quantified utility of the product, then the utility marginality is the derivative dU/dQ. If the production cost is C, then the marginal cost would be dC/dQ. According to the utility theory of value the monetary price is spontaneously established by the trading market such that

$$\frac{d\mathcal{U}}{dQ} = \frac{d\mathcal{C}}{dQ} \tag{1.61}$$

Eq. (1.61) explains why the price of gold is much higher than that of water. The marginality of gold cost (the change in cost for an infinitesimal change of quantity) is obviously much higher than the marginality of water cost. The utility of water is much higher than the utility of gold, but the scarcity of gold is high, whereas the availability of water is high. This makes the marginality of water utility much smaller than the marginality of gold utility, and therefore the marginalities of the costs of water and gold must be on the same relationship.

If a theory quantifies value through a conservable quantity, then it falls into fallacy. For example, if when applying the theory of labor, one assumes that any type of labor can be valued through the amount of mechanical work (or energy) deployed to do it, then this leads to misappropriations (which are in fact common), because energy conserves and value does not. Value decreases or increases.

Another example: if when applying the theory of utility one assumes that the utility is expressed in term of mass (amounts, quantities) of a precious metal (gold), then this is a fallacy because although finite, mass of gold is conserved.

The true values that humans and also any other living organisms appreciate are the sources of low entropy or high exergy. These quantities do not conserve. When used, a low entropy source is converted into a high entropy waste by a living organism, which in the meantime performs its activity. Equivalently, one says that the high exergy sources are degraded by earth systems (living species, natural cycles, etc.); exergy degrades from the source to waste. The net exergy absorbed by the earth, consequently, is gradually destroyed, but during this destruction, it manages to drive the earth's water, wind and other natural systems, as well as life on earth.

Because it does not conserve, a source of low entropy can be used only once and never reused. The same stands for exergy; once destroyed, it cannot be reused. Exergy is also related to the surrounding environment as it accounts for its temperature, pressure and species concentration. Therefore, due to these attributes, exergy can be used in establishing a theory of value. In fact, exergy represents the part of energy that is useful to society and therefore it has economic value.

Furthermore, once the economic value of exergy is expressed in terms of currency, then it can effectively be used for exergoeconomic and exergoenvironmental analyses. Various methods can be approached to price exergy for an analyse purpose. It is important to determine sound methods to set the prices and costs in relation to exergy content. This, in fact, requires formulation of a theory of costs based on exergy. It has been suggested that when analyzing a thermal system, is reasonable to distribute costs in relation to outputs and accumulations of exergy. With regard to the prices of physical resources (fuels, materials), these also must be set in a tight relation with the resource exergy content to foster resource savings and effective technology.

Let us do a simple attempt to quantify exergy in terms of monetary currency. In doing this, one considers the main fuels in a society. The energy content is taken as the lower heating value (LHV) of the fuel, and the exergy content is the chemical exergy of each fuel. Table 1.12 gives the specific energy and exergy content of fuels considered in this brief analysis. The price per unit of mass of each fuel is also given. When the price is divided to chemical exergy, then the exergy-specific price C_{ex} is obtained. Table 1.12 is constructed for Canada, however, the methodology presented subsequently is general.

In our approach, a country or region must be considered first. Then, the primary energy sources are inventoried. For Canada, the following primary energy sources can be considered: coals, refined natural gas, natural gas liquids, crude oil derivate, hydro, nuclear and biomass derivate (here, wind and solar are neglected as not highly represented). Further, the method for exergy price estimation goes as follows:

- Cost of each fuel type is obtained from the market and expressed in $/kg. For the case of hydro and nuclear, this step is skipped.
- Based on the available statistics, the consumed energy fraction (CEF) for each type of fuel is determined. In Table 1.12, the CEF is obtained from the previous work by Dincer and Zamfirescu (2011), Chapter 17. The CEF represents the fraction of specified primary energy source from the total energy consumed from primary sources.

TABLE 1.12　Calculation Table for Exergy Price Based on Primary Energy Sources for Canada (data for year 2008)

Fuel		Data input				Calculated item							
		LHV (MJ/kg)	ex^{ch} (MJ/kg)	C_f ($/kg)	CEF (%)	$\overline{LHV}$ (MJ/kg)	$\overline{ex}^{ch}$ (MJ/kg)	$\overline{C_f}$ ($/kg)	γ	C_{ex} ($/GJ)	CEF·γ	CExF (%)	CExF·C_{ex} ($/GJ)
Coals		24.0	15.0	0.15	8.3	24.0	15.0	0.15	0.62	10.0	0.052	5.1	0.51
Refined natural gas		50.7	52.4	0.18	30.8	50.7	52.4	0.18	1.03	3.43	0.318	31.3	1.07
Natural gas liquids	LPG	46.0	54.9	1.43	2.9	31.6	35.6	0.93	1.13	26.1	0.033	3.3	0.85
	Methanol	19.9	22.4	0.47									
	Ethanol	28.8	29.5	0.88									
Crude oil derivate	Gasoline	43.5	47.7	1.74	45.1	42.4	45.5	1.37	1.07	30.1	0.484	47.7	14.3
	Diesel	42.8	44.2	1.78									
	Kerosene	43.1	49.1	0.94									
	Fuel oil	40.1	41.1	1.03									
Hydro		N/A	N/A	N/A	7.4	N/A	N/A	N/A	1	34.7	0.074	7.3	0.16
Nuclear		N/A	N/A	N/A	5.0	N/A	N/A	N/A	1	89.6	0.050	4.9	6.6
Biomass derivate	Whole tree	19.7	22.1	0.2	0.5	15.6	17.8	0.05	1.14	2.8	0.005	0.4	0.14
	Wood pellets	14.6	18.5	0.2									
	Wood chips	10.0	11.0	0.2									
	Pine wood	18.9	24.8	0.5									
	Sawdust	8.0	8.5	0.2									
	Straw	14.5	16.5	0.3									
	Rice straw	14.1	15.9	0.8									
	Waste paper	17.7	20.1	0.1									
	Biogas	22.5	23.2	2.0									
Equations used		$\gamma = \dfrac{\overline{ex}^{ch}}{\overline{LHV}}$		$C_{ex} = \dfrac{\overline{C_f}}{\overline{ex}^{ch}}$				$CExF = \dfrac{CEF\gamma}{\sum(CEF\gamma)}$	Averaged price of exergy				8.4 ¢/kWh
									Averaged price of electricity				11.0 ¢/kWh

N/A, not applicable.

- The LHV and specific chemical exergy for categories of fuels are averaged. For example, the mean LHV for natural gas liquids is an average of the LHVs for LPG, methanol and ethanol.
- The specific price of fuel is averaged for each fuel category. For example, the mean fuel price $\overline{C_f}$ for fuels obtained from crude oil (crude oil derivate) is the average of prices of gasoline, diesel, kerosene, and fuel oil.
- The quality factor for each category of fuel is determined as indicated in the table, that is, by the ratio between averaged specific chemical exergy and LHV. The quality factor for hydropower and nuclear energy is one.
- The exergetic price C_{ex} for each fuel category is determined as shown in the table, by the ratio between the averaged fuel cost and specific chemical exergy. The exergetic price of hydropower results from the specific price of electric power divided by 0.8, as it is fair to assume that the exergy efficiency of the hydro power plant is 80%. The exergetic price of nuclear energy is determined from the price of electric power divided by 0.31 based on the fact that, as shown in Dincer and Zamfirescu (2014), the exergy efficiency of CANDU power plants is 31.3%, on average. The cost of electric power in Canada is taken at an average of 11 ¢/kWh.
- The consumed exergy factor for each fuel category is calculated with $CExF_i = CEF_i\gamma_i/\sum(CEF_i\gamma_i)$, where i is an index representing each type of fuel. The factor γ_i represents the ratio between chemical exergy and lower heating value of a fuel and it is used as a weighting factor.
- The average price of exergy results as a weighted average $C_{ex} = \sum CExF_i \times C_{ex,i}$.

- The results show that the exergy price for Canada, based on the exergy of the primary resources, is 8.4 ¢/kWh. This compares with the average electricity price of 11 ¢/kWh.

Once a price of exergy is determined, further models can be created to establish a costing scheme for other items of interest in an exergoeconomic analysis. Nonenergetic costs such as labor, material supply, environment remediation expenditure, incidental expenditures, etc., can be priced using exergy content as a basis for cost accounting. The economic value of system outputs can be also allocated based on exergy.

1.8 EXERGOECONOMIC ANALYSIS

In the real world, the economic factor has one of the most important influences in selecting any technical design. Ultimately, a business case must be presented in terms of investment cost and profitability, and this guides the system development. Noteworthy is that the cost is a very volatile factor; therefore, any technical-economic analysis is not absolute, but subject to regional and temporal economic constraints.

The common technical-economic analysis of thermal systems is referred to as thermo-economic analysis, which combines the method of energy analysis with economic analysis. Basically, in thermo-economics, the ratio between energy loss and capital cost is determined and then guides the path toward the best design. Although the thermo-economics (energy/cost analysis) gives a valuable answer for the design and selection of processes and systems in engineering, this method is much connected to the economy fluctuation.

Exergoeconomic analysis instead, provides a fairer comparative assessment of competing designs in a market, because it is based on exergy, which, as known, represents the maximum potential of a system to do work with respect to a reference environment. The exergoeconomic analysis is a combination of exergy and economic analyses in which a unique cost for exergy is rationally defined.

1.8.1 EXCEM Method

Among other methods of exergoeconomic analysis, the so-called exergy-cost-energy-mass (EXCEM) method is distinguished by the fact that it represents an extension of the typical exergy analysis, which is based on four balance equations: mass, energy, entropy, and exergy. In EXCEM, there is one more balance equation to be written, namely the cost balance equation. Note that in EXCEM, the balance equations are written for quantities that do not conserve, such as entropy, exergy, and costs. For such quantities, the generation and consumption terms must be considered in order to obtain a balance equation. The EXCEM method originated from the work of Rosen (1986) and has been further developed in the subsequent works by Rosen and Dincer (2003) and Dincer and Rosen (2012).

The basic rationale underlying an EXCEM analysis is that an understanding of the performance of a system requires an examination of the flows of each of the quantities represented by EXCEM into, out of and at all points within a system. Fig. 1.17 illustrates the application of the EXCEM method. We give the general thermodynamic balances here, in terms of mass, energy, entropy, and exergy balance, respectively, as follows

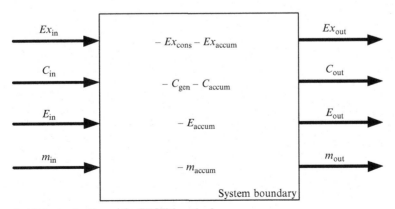

FIG. 1.17 System modeling sketch to application of the EXCEM method.

$$\sum \dot{m}_{in} - \sum \dot{m}_{out} = \frac{dm}{dt} \tag{1.62a}$$

$$\sum \dot{E}_{in} - \sum \dot{E}_{out} = \frac{dE}{dt} \tag{1.62b}$$

$$\sum \dot{S}_{in} - \sum \dot{S}_{out} + \dot{S}_{gen} = \frac{dS}{dt} \tag{1.62c}$$

$$\sum \dot{E}x_{in} - \sum \dot{E}x_{out} - \sum \dot{E}x_{d} = \frac{d\dot{E}x}{dt} \tag{1.62d}$$

Note that in Eq. (1.62d), the rate of exergy destruction $\dot{E}x_d$ represents a consumption of exergy due to internal irreversibilities. The exergy output includes the exergy in products and the exergy lost at system interaction with the surroundings: $\sum \dot{E}x_{out} = \sum \dot{E}x_{prod} + \dot{E}x_{lost}$. It is remarked from Eqs. (1.62a)–(1.62d) that the general form of a balance equation in rate format is as follows

$$\text{Input rate} + \text{Generation rate} - \text{Output rate} - \text{Consumption rate} = \text{Accumulation} \tag{1.63}$$

The balance equations are sometimes written in terms of amounts, as follows

$$\text{Input amount} + \text{Generated amount} - \text{Output amount} - \text{Consumed amount} = \text{Amount accumulated} \tag{1.64}$$

The cost balance is essentially different than the thermodynamic balances, although it has the same general form given by Eqs. (1.63) and (1.64). In a cost balance equation, only the cost input and cost generation are defined, whereas in thermodynamic balances, all terms are defined with the help of scientific relationships. For cost accumulation and cost outputs, there is no strict formulation; rather, these types of costs are subjectively allocated, depending on the type and purpose of the system and other economic considerations. For example, costs may be distributed proportionally to all outputs and accumulations of a quantity (such as mass, energy or exergy), or all nonwaste outputs and accumulations of a quantity. The cost balance equation in rate and amount forms is written, respectively, as follows:

$$\dot{\mathcal{C}}_{in} + \dot{\mathcal{C}}_{gen} - \dot{\mathcal{C}}_{out} = \frac{d\mathcal{C}}{dt} \tag{1.65a}$$

$$\mathcal{C}_{in} + \mathcal{C}_{gen} - \mathcal{C}_{out} = \Delta\mathcal{C} \tag{1.65b}$$

The cost inputs are generally defined by the knowledge of the prices for each input and the rate of supply as determined for a specified market. For example, the fuel rate supplied to the system is generally known; also known is the specific cost of the fuel. Thus, determination of the cost rate due to fuel input is just obtained by multiplication of the rate of fuel consumption and the specific cost of fuel. Cost generation corresponds to the appropriate capital and other costs associated with the creation and maintenance of a system. That is

$$\mathcal{C}_{gen} = \mathcal{C}_{cap} + \mathcal{C}_{o\&m} + \mathcal{C}_{occ} \tag{1.66}$$

where index gen refers to the capital costs, o&m refers to operation and maintenance cost, and occ refers to other cost creation due to various reasons.

Capital costs are often the most significant component of the total cost generation. Hence, the consideration of only capital costs closely approximates the results when cost generation is considered. The other reasons for cost creations may be, depending on the case, interest, insurance, etc. The "cost generation rate" represents the total cost generation levelized over the operating life time of the system. The cost term denoting the "amount of cost generated" in an integral cost balance representing the portion of the total cost generation accounted for in the time interval under consideration. Cost generation components other than capital costs often are proportional to capital costs. Hence, the trends described are in qualitative agreement with those identified when the entire cost generation term is considered.

The system assessment in EXCEM is made based on indicators expressing the thermodynamic loss versus the capital cost. The energetic loss ratio is defined therefore as

TABLE 1.13 A List of Parameters Required for an Economic Analysis

Parameter	Symbol	Description
System lifetime	n	Number of years (or discount periods) of the system under investigation
Real discount rate	r	Rate of return for the discounted cash flow to determine value
Inflation rate	i	A subunitary factor accounting for inflation rate
Tax on credit due to renewable energy	t_c	A tax credit that may apply to renewable energy, $t_c \in [0, 1)$
Tax on income	t_i	Tax applied in any income of money coming from output costs, $t_i \in [0, 1)$
Tax on property	t_p	Tax applied to the property value, $t_p \in [0, 1)$
Tax on salvage	t_s	Tax applied on the salvage value of the system, $t_s \in [0, 1)$
Capital salvage factor	CSF	The recoverable end-of-lifetime fraction CSF of the invested capital
Levelized product price	LPP	The price in dollars with which the unit of product is sold (in present worth)
Annual production	AP	Number of units of product for one year (or one discount period)
Levelized cost of consumables	LCC	The sum of all input costs (fuels, supplies) levelized for present worth
Annual consumption	AC	The annual consumption of consumables
Operation and maintenance cost factor	OMCF	A subunitary fraction of either the net income or the invested capital
Invested capital	$\mathcal{C}_{cap}$	A monetary value (in $) paid upfront (in equity) to initiate the business

$$R_{en} = \frac{E_{loss}}{\mathcal{C}_{cap}} \tag{1.67}$$

where $E_{loss} = \sum E_{in} - \sum E_{prod} - \Delta E_{accum}$ and $\sum E_{prod}$ represents the energy retrieved in form of products; this is a part of the energy output: $\sum E_{out} = \sum E_{prod} + E_{loss}$.

Eq. (1.67) is written in general in terms of amounts, where E_{loss} represents the total energy lost during the system lifetime and $\mathcal{C}_{cap}$ is the present worth of capital discounted for the entire lifetime. Eq. (1.67) can be also given in terms or rates of lost energy and rate of capital cost. The exergetic loss ratio is defined as

$$R_{ex} = \frac{Ex_{loss}}{\mathcal{C}_{cap}} \tag{1.68}$$

where $Ex_{loss} = \sum Ex_{in} - \sum Ex_{prod} - \sum Ex_d - \Delta Ex_{accum}$ represents the exergy wasted into the surroundings.

Eq. (1.68) is written in general in terms of amounts, where Ex_{loss} represents the total energy lost during the system lifetime. If analysis requires, Eq. (1.68) can be given in terms or rates of lost exergy divided by the rate of capital cost. In this case, the discounted capital in present worth is levelized for the entire lifetime.

Although there may be some variations in the economic engineering analysis of the system, here we describe a simple approach that leads to determination of the discounted capital in present worth and does not consider loans, but only invested capital. The economic analysis requires first the establishment of a set of parameters as given in Table 1.13.

The cost generation and cost outputs can be determined based on a set of economic calculations for the cash flow, as given in Table 1.14. Note that the accumulated cost represents a generated capital. Assume that an equity capital $\mathcal{C}_{cap}$ is invested at the beginning of the system lifetime. Due to the generated capital at the end of the system lifetime, the total available capital is increased to GC + $\mathcal{C}_{cap}$, where GC is the generated capital. Therefore, a capital productivity can be defined as shown in the table, as the ratio of total available capital at the end of the lifetime to the invested capital at the beginning of the lifetime.

The capital productivity factor, as well as the exergetic loss ratio or energetic loss ratio, may help in selecting a design option of the system among many possibilities. In general, a higher invested capital may lead to better systems with less exergetic and energetic losses; however, the loss ratio can sometimes show a maximum.

If a maximum of loss ratio exits, then this is an indication for the best system selection or the optimal capital investment. In addition, the capital productivity factor may be affected by the invested capital, which, in some

TABLE 1.14 A List of Equations for Economic Analysis

Parameter	Equation	Description
Market discount rate	$r_m = (r+1)(i+1) - 1$	The discount rate (taking into account the inflation rate)
Present value factor	$PVF = (1 + r_m)^{-N}$	A factor used to determine the present value of cash flow
Capital recovery factor	$CRF = \dfrac{r_m}{1 - PVF}$	The ratio of a constant annuity to the present value of receiving that annuity for a given number of discount periods
Present worth factor	$PWF = CRF^{-1}$	A factor used to determine the present value of a series of values
Present worth income ($)	$PWI = LPP\,AP\,PWF$	The income from a business expressed in the present worth of money, equal to cost output from product selling
Present worth costs ($)	$C_{in} = LCC\,AC\,PWF$	A cost input due to the cost of fuels and supplies
Cost of O & M	$C_{o\&m} = PWI\,OMCF$	The O & M costs taken as a fraction of income from product sales
Net income ($)	$NI = PWI - C_{o\&m} - C_{in}$	The income from a business expressed in present worth
Tax credit deduction ($)	$TCD = t_c\,C_{cap}$	The tax deduction due to investment in renewable energy
Taxable income ($)	$TI = NI - TCD$	The amount of income subjected to taxation
Tax on income ($)	$TOI = t_l\,TI$	The tax applied to the income
Tax on property($)	$TOP = t_p\,C_{cap}(1 - t_i)$	The tax applied on the property
Other cost creation	$C_{occ} = TOI + TOP$	The cost creation to support taxation on income and property
Generated costs ($)	$C_{gen} = C_{cap} + C_{o\&m} + C_{occ}$	The total costs generated by the system during its lifetime
Salvage value ($)	$SV = CSF\,C_{cap}\,PVF(1 - t_s)$	The salvage value expressed in present worth
Generated capital (M$)	$GC = NI + SV - C_{cap} - TOI - TOP$	The amount equal to the accumulated cost within the system, according to the cost balance Eq. (1.65b)
Capital productivity	$CP = (C_{cap} + GC)/C_{cap}$	A factor that compares the amount of available cash at the end of the lifetime $(C_{cap} + GC)$ to the invested equity capital

conditions may be a maximum capital productivity for a given investment cost. However, the system design showing the maximum capital productivity is generally different than the design showing the lowest exergetic loss ratio, and also different than the system showing the lowest energetic loss ratio. These facts may lead to tradeoff selection problems for the best system design.

1.8.2 SPECO Method

Another method to perform exergoeconomic analysis is the SPECO method (specific exergy cost). This method has been applied to assess and improve various devices, such as power plants. The SPECO method has been described by various authors (Tsatsaronis and Moran, 1997; Bejan, 1996). Cost accounting with SPECO methods uses cost balances, and is generally concerned with determining the actual cost of goods or services, providing a rational basis for pricing them. This method simultaneously provides a means for allocating and controlling expenditures, and determines useful assessment parameters to assist in creating and evaluating design and operating decisions.

For a system operating at a steady state, there may be a number of entering and exiting material streams, as well as heat and work interactions with the surroundings. Since exergy measures the true thermodynamic value of such effects and cost should only be assigned to commodities of value, it is meaningful to use exergy as a basis for assigning costs in energy systems. Such "exergy costing" provides a rational basis for assigning costs to the interactions that a thermal system experiences with its surroundings and to the sources of inefficiencies within it. In this method, the costs are determined based on the exergy content. Therefore, a monetary value of exergy can be use as determined as also mentioned above.

In a SPECO economic analysis, a cost balance is usually formulated in a rate form (cost rates $\dot{C}$ given in $/h) and applied to the overall system operating at steady state. Let us consider a general system. Exergy is applied to the system as input, together with other types of inputs, including an invested capital. Each physical (heat, work, material) stream entering and exiting the system is expressed through its exergy content. To each stream i, a specific exergy cost is allocated $C_{ex,i}$. Fig. 1.18 shows a system representation for SPECO method application. The following streams can be remarked

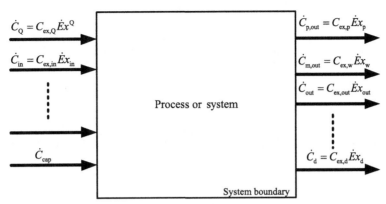

FIG. 1.18   SPECO method application to a system showing the input and output cost stream.

- $\dot{C}_Q = C_{ex,Q}\dot{E}x^Q$, represents the rate of cost input due to the supplied heat; $C_{ex,Q}$, is a specific cost for exergy provided in form of heat; $\dot{E}x^Q$ is the exergy rate provided to the system in form of heat
- $\dot{C}_{in} = C_{ex,in}\dot{E}x_{in}$, represents the cost rate in physical inputs; all types of inputs are considered (energy inputs, materials inputs); for each input, the exergy rate is given $\dot{E}x_{in}$, and the specific cost of the respective type of exergy
- $\dot{C}_{cap}$, represents the rate of capital cost, which will be explained later
- $\dot{C}_{p,out} = C_{ex,p}\dot{E}x_p$, represents the cost rate in dried product, where $\dot{E}x_p$ is the exergy rate in dried product and $C_{ex,p}$ is the allocated cost to the exergy of the dried product
- $\dot{C}_{m,out} = C_{ex,w}\dot{E}x_w$, represents the cost of moisture generated by the system with
- $\dot{E}x_w$ being the exergy rate of moisture and $C_{ex,w}$ the specific cost of exergy allocated for the moisture
- $\dot{C}_{out} = C_{out}\dot{E}x_{out}$, represents the cost rate of any other type of output
- $\dot{C}_d = C_{ex,d}\dot{E}x_d$, represents the cost rate of exergy destruction, which is related to the allocated specific cost of exergy destruction $C_{ex,d}$ and the rate of exergy destruction $\dot{E}x_d$

The cost balance for the system, is written as

$$\dot{C}_Q + \sum \dot{C}_{in} + \dot{C}_{cap} = \dot{C}_{p,out} + \dot{C}_{m,out} + \sum \dot{C}_{out} \qquad (1.69)$$

The cost balance Eq. (1.69) states that the total cost of the exiting exergy streams (including the cost in exergy destruction) equals the total expenditure to obtain them, namely the cost of the entering exergy streams plus the capital and other costs. The capital investment rate $\dot{C}_{cap}$ expressed in \$/h is determined based on the invested capital expressed in present worth value C_{cap}, the capital recovery factor (*CRF*) (defined according to equation given in Table 1.14) and operation and maintenance cost factor (*OMCF*), which is here defined with respect to the invested capital. A typical value of around 1.06 can be assumed for *OMCF* for SPECO analysis, if not specified otherwise. Therefore, the capital investment rate becomes

$$\dot{C}_{cap} = \frac{C_{cap}\ CRF\ OMCF}{n_{hour}} \qquad (1.70)$$

where n_{hour} represents the number of operating hours per year.

All cost data used in an economic analysis must be brought to the same reference year, the year used as a basis for the cost calculations. For the cost data based on conditions at a different time, a normalization is performed with the aid of an appropriate cost index.

Cost balances as given by Eq. (1.69) can be written for each of the system components and for the overall system. When the cost balances are solved simultaneously, the cost rate of exergy destructions can be estimated. Exergoeconomic factors can be devised to account for the contribution of nonexergy-related costs. The system assessment can be done with respect to the cost rate of exergy destruction divided by the cost rate for the capital, or as the ratio of the difference between the product and the fuel cost.

1.9 EXERGOENVIRONMENTAL ANALYSIS

It is important to enhance the use of exergy analysis with cost, environmental impact and even sustainability quantities. As shown in Dincer and Rosen (2012), when considering exergy in environmental impact analysis, the inventory analysis phase has to account more carefully for mass and energy flows into, out of and through all the stages of the life cycle; next, the energy flow is associated with an exergy flow; eventually, environmental impact indicators (such as the emissions of greenhouse gases) based on exergy can be developed and studied.

When the environmental impact is accounted for, one must consider the impact of the system effluents to the environment, but also the pollution and resource depletion related to the system construction. Therefore, the system boundary for the analysis is set at larger limits than that of the actual system. In fact, one can model a system construction process in terms of required exergy for construction and the environmental impact. Also, one can elaborate a model for the environmental impact due to system effluents. Therefore, the system itself is expanded with two subsystems: upstream (the system construction process), and downstream (effluent-environmental and decommissioning processes). All three subsystems, upstream process, actual system and its process(es), and downstream process, must be supplied with exergy; otherwise, no subsystem can run. For the system itself, the exergy supply in the form of any kind of fuels is referred to as direct exergy. For the upstream and downstream subsystems, the supplied exergy is referred to as the indirect exergy. Thus, the total exergy supply to be considered in an extended exergoenvironmental analysis is

$$\dot{E}x_{\text{supply}} = \dot{E}x_{\text{direct}} + \dot{E}x_{\text{indirect}}^{\text{upstream}} + \dot{E}x_{\text{indirect}}^{\text{downstream}} \tag{1.71}$$

A similar equation can also be written for the environmental impact, namely

$$EI_{\text{total}} = EI_{\text{direct}} + EI_{\text{indirect}} \tag{1.72}$$

where EI is a general environmental impact indicator; EI_{direct} stands for the downstream environmental impact, that is, the environmental impact caused by the system due to its operation, and EI_{indirect} stands for the environmental impact associated with system construction.

In many practical situations, the energy which drives a process is electricity or a petrochemical fuel generated elsewhere (power plant or refinery, respectively). It is generally known in a certain economy what is the environmental impact and exergy efficiency for electric power or consumer fuels production. Once the rate of fuel consumption is specified, the indirect exergy consumption and environmental impact associated with fuel supply can be easily determined.

Other types of environmental impacts may be considered in the analysis. There are several types of impact categories, such as: depletion of resources (abiotic or biotic), land use (competition, biodiversity loss, life support function loss), climate change, desiccation, stratospheric ozone depletion, human- and eco-toxicity, acidification, photo-oxidant formation, eutrophication, waste heat generation, noise and odor (in water and air), and ionizing radiation. The toxicity includes aquatic (marine, freshwater) and terrestrial ecosystems. The eutrophication shows the over-increase of chemical nutrients in the ecosystem, which causes productivity augmentation, which in turn creates an imbalance in the ecosystem, because other biological species may experience an increase in number and thus deplete other species.

A primary objective of exergoenvironmental analysis is the identification and quantification of direct and/or indirect environmental impact in correlation with exergy destroyed by the system. Another objective is to minimize the environmental impact. Often this is done by finding the means of increasing the energy efficiency of the system. Another approach to environmental impact minimization is the reduction of the polluting effluents.

In recent years, particular emphasis has been placed on releases of carbon dioxide, since it is the main greenhouse gas, and optimization of thermal systems based on this parameter has received much attention. A focus of many studies is to consider emissions of the following types of atmospheric pollutants: CO, NO_x, SO_2, CO_2, CH_4. In the case of systems that are fueled by a combustion process, the atmospheric pollution through flue gas emissions is very relevant.

The inventory analysis is the core of the environmental impact analysis, as it determines the streams of materials and their impact on the environment. All flows of matter exchanged with the environment and their interrelationship within the analyzed system must be inventoried. The impact categories are then determined according to the definitions given in Table 1.15.

The SPECO and EXCEM analyses can be extended for exergoenvironmental analysis. Regardless of the actual method used to conduct an exergoenvironmental analysis, the environmental impact by the system and all its components must first be determined. This is a purely environmental impact analysis phase.

For each exergy stream entering or exiting the system, a compounded environmental impact rate can be assigned. As an example, let us consider a combustion process. The stream of flue gases injects into the surrounding atmosphere an exergy at a rate that can be easily determined from the exergy analysis. The same flow impacts the environment at a rate that can be determined based on specific method of life-cycle environmental impact assessment. The ratio of the rate of environmental impact to the rate of exergy can be then determined for any exergy stream, including for streams of exergy destruction. Therefore, a balance equation for the environmental impact of a system can be formulated by

TABLE 1.15 A List of Quantifying Parameters of Environmental Impact

Impact category	Unit	Definition
Depletion of abiotic resources	kg antimony equivalent	Abiotic resources include all nonliving resources (coal, oil, iron ore, renewable energies, etc.)
Climate change	GDP, (CO_2 equivalent)	kg of CO_2 equivalent (See other chapters of this book for a more detailed definition)
Ozone depletion	ODP, (CFC11 equivalent)	ODP = ozone depletion potential is a time-dependent parameter characterizing the potential of a substance to deplete the stratospheric ozone layer; represents the amount of a substance that depletes the ozone layer relative to CFC11 (trichlorfluoromethane)
Photo-oxidant formation	kg of ethylene equivalent	Depends on photochemical ozone creation potential expressed with respect to the reference substance, ethylene
Toxicity	kg DCB equivalent	Human toxicity and ecotoxicity produced in air, freshwater, seawater or terrestrial; quantified with respect to the toxicity of DCB
Acidification	kg SO_2 equivalent	Expressed in SO_2 emitted in Switzerland equivalent;.three acidification factors exist: ammonia, NOx and SO_2
Eutrophication	kg PO_4 equivalent	Covers all potential impacts of high levels of nutrients, the most important of which are nitrogen (N) and phosphorus (P)

DCB, 1,4-dichlorobenzene.

$$\dot{E}I_Q + \sum \dot{E}I_{in} + \dot{E}I_{sys} = \dot{E}I_{p,out} + \dot{E}I_{m,out} + \sum \dot{E}I_{out} \tag{1.73}$$

where $\dot{E}I_Q$ is the environmental impact rate due to the heating source for the process, $\sum \dot{E}I_{in}$ is related to any input stream of energy or matter, $\dot{E}I_{sys}$ represents the environmental impact of the system (or its subsystems when the component are analyzed), $\dot{E}I_{p,out}$ is due to the release in the environment of the dried product themselves, $\dot{E}I_{m,out}$ refers to moisture release, and $\sum \dot{E}I_{out}$ refers to other outputs and exergy destruction.

Carbon dioxide emission from fuel combustion can be relatively easily correlated with a fuel combustion rate based on general combustion parameters such as stoichiometry, excess air, etc. More difficult to determine through modeling are the rates of CO and NOx, since detailed kinetics modeling for the combustion process would be required. Some simplified models can be used as an alternative, such those by Toffolo and Lazzaretto (2004), which relate the mass flow rate of CO and NOx emissions with the fuel (natural gas) consumption

$$\frac{\dot{m}_{CO}}{\dot{m}_f} = 179,000 \frac{\exp\left(\frac{7800}{T}\right)}{P^2 \tau \left(\frac{\Delta P}{P}\right)^{0.5}} \tag{1.74a}$$

$$\frac{\dot{m}_{NOx}}{\dot{m}_f} = 150,000 \frac{\tau^{0.5} \exp\left(-\frac{71,100}{T}\right)}{P^{0.05} \tau \left(\frac{\Delta P}{P}\right)^{0.5}} \tag{1.74b}$$

where T is the adiabatic flame temperature, P is combustion pressure, ΔP is the pressure drop across the combustion chamber, and τ is the residence time in the combustion zone (~2 ms).

1.10 EXERGOSUSTAINABILITY ASSESSMENT

At the beginning of the 21st century, a new branch of science came out, namely the sustainability science, having an interdisciplinary character involving the following main disciplines: physics, chemistry, biology, medicine, social and economic sciences and engineering. One of the goals of sustainability science is to model the complex interactions between society, the economy and the environment, accounting also for resource depletion. One key parameter aspect is the provision of theoretical foundations and tools for sustainability assessment.

Often in policy planning, indicators are used to assess a country or geopolitical region from various points of view: economic, social, etc. For example, the GDP (gross domestic product) is an economic indicator that indicates the

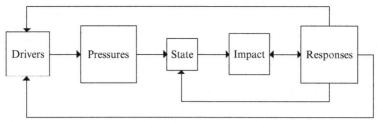

FIG. 1.19 The DPSIR model for sustainability assessment.

wellbeing of a society. The purpose of sustainability assessment indicators is to confer a basis for decision-making and policy elaboration toward sustainable development, considering the integrated nature-society systems for a certain temporal perspective. In general, sustainability assessment indicators quantify in an integrative manner the economic, social, environmental, and institutional development of a country or region.

The sustainability indicators (or indices) are generally extensions of the environmental assessment indicators. Therefore, they are based on a model such as the one shown in Fig. 1.19, and denoted with the acronym DPSIR, meaning drivers-pressures-state-impact-responses. In this mechanism, the drivers are the developments in society, economy and the environment. These developments (or changes) exert pressure on the sustainability and as a consequence, the sustainability state changes in some direction.

This eventually leads to a foreseeable impact on sustainable development; it improves or degrades the sustainability. The society can respond to this by taking actions to change the impacts directly (see the double arrow in the figure). In addition, the society can send feedback to drivers to impose educated changes in the society, economy, environment. The responses can also act directly on the pressures and the state of sustainability by applying adequate measures, if possible. The responses are expected be effective, because their effectiveness directly affects sustainable development. For sustainability assessment, this model simplifies to DSR (drivers-state-responses) or PSR (pressure-state-response).

Table 1.16 gives some significant indicators for sustainability assessment. Certainly, the indicators have some degree of subjectivity, because it is not actually possible to objectively determine and fully understand the interrelationships between social change and environment and economic development. The independent formulation of environmental, economic and social indicators is more facile.

Once independent sustainability indicators covering various aspects are determined, aggregate indicators can be constructed using weighting and scaling methods. The aggregated sustainability indicators (ASI) are generally known

TABLE 1.16 Categories and Types of Indicators Influencing the Sustainability Assessment

	Kind		
Category	**Drivers indicators**	**State indicators**	**Responses indicators**
Social	- Unemployment rate - Population growth rate - Adult literacy rate - Motor fuel consumption per capita - Loss rate due to natural disasters	- Poverty gap index - Income inequality index - School life expectancy - House price per income - Floor area per person	- GDP on education - Childhood immunization - Infrastructure expenditure per capita - Health expenditure per capita - Hazardous chemicals in foods
Economic	- Per capita GDP - Investment share of GDP - Annual energy consumption - Natural resource consumption - Capital goods imports	- Proven mineral reserves - Fossil fuel reserves - Lifetime of energy reserves - Share of renewable energy - Manufacturing added value	- Environmental protection expenditure - Funding rate on sustainable development - Amount of funds on sustainability - Percentage of new funds on sustainability - Funding grants on technology
Environmental	- Water consumption per capita - Generation of wastes - ODP substances emission - Emission of GHG, SO_2, NOx - Energy use in agriculture	- Ground water reserves - Monthly rainfall index - Desertification rate - Pollutants concentration - Acute poisoning	- Waste water treatment expenditures - Natural resource management - Air pollution mitigation expenditures - Waste management expenditures - Number of restricted chemicals
Institutional		- Scientists number per capita - Engineers number per capita - Internet access per capita - Telephone line access - Other information channels	- Policy on sustainable development - Environment protection programs - GDP share of R&D expenditures - Number of R&D personnel per capita - Ratification/implementation of agreements

From: UN, 2007. Indicators of Sustainable Development: Guidelines and Methodology third ed. United Nations, New York.

as sustainability assessment indices. The United Nations developed more than 130 indicators for sustainability assessment, of which 30% are social indicators, 17% are economic, 41% are environmental and the remaining 12% represent institutional indicators. Approximately 33% of sustainability indicators are of the driving force (drivers) kind; 39% are state indicators; and 18% are response indicators.

Exergy analysis offers a basis for sustainability assessment, because exergy is a measure of abatement of the system subjected to the analysis from the environment. Because sustainable development includes a component of energy security, the exergy-based assessment of sustainability becomes very relevant. Through exergy analysis and exergy-based thermodynamic optimization, increased efficiency is obtained and systems with reduced exergy destruction are developed. Less exergy destruction implicitly leads to reduced environmental impact.

The conceptual representation of exergy at the confluence of energy, environment and sustainable development is shown in Fig. 1.20. Here it is also suggested that the energy and material flow balance methods connect energy analysis with the environment. Indeed, in energy analysis material and energy flows, balances are inventoried as they must conserve. These balances ultimately allow us to find the relation between system performance and system discharges into the environment. In addition, the energy efficiency is at the frontier between energy and sustainable environment. Better energy efficiency means less resource consumption and less environmental impact. Furthermore, at the frontier between the environment and sustainable development, the environmental impact is interlaced, which suggests that less environmental impact leads to better system sustainability. The unique position at the center, in the figure, shows that exergy unites the three fields: energy, environment and sustainable development. In addition, exergy is connected with efficiency, environmental impact and energy, and material inventories.

As shown in Dincer (2011), exergy efficiency can be correlated with sustainability index (SI) and with the environmental impact index (EI). It can also be correlated with the lifecycle sustainability index (LCSI). Here, the difference between SI and LCSI consists of the fact that SI refers to the system utilization phase only, while LCSI considers the system lifecycle, including the construction and scraping phases.

By using exergy destruction as a measure to quantify environmental effects associated with emissions and resource depletion, the environmental impact assessment is made based on purely physical principles. An understanding of the relationship between exergy and the environment may reveal the underlying fundamental patterns and forces affecting changes in the environment, and help researchers to deal better with environmental damage. The discharged wastes in the environment and their impact can be quantified by accounting for exergy destructions. It is known that the exergy destroyed (and wasted) by the system is of two kinds:

- internal exergy destruction, which represents the lost opportunity to perform work; the environmental impact and rejected wastes due to all upstream processes (e.g., power generation) can be related to the internal exergy destruction.

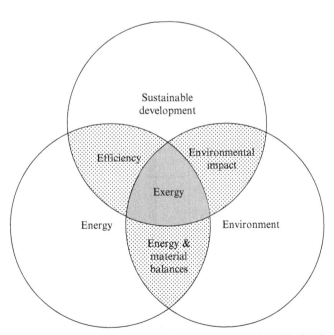

FIG. 1.20 Representation of the exergy at the confluence of energy, environment and sustainable development. *Modified from Dincer, I., 2011. Exergy as a tool for sustainable drying systems. Sustain. Cities Soc. 1, 91–96.*

- lost exergy, or exergy destruction at system interaction with its surroundings, which is related to the discharged wastes by the process itself. In principle, all discharged wastes by the system can be recovered to use their exergy and reduce the environmental impact; however, this action may be very expensive and generally is not undertaken in practice, except when it is justified economically or enforced by sustainability policies.

The practical connection between wasted exergy and environmental impact can be discovered by correlating recorded emission data at the regional level or globally with the chemical exergy of emitted pollutants. Here, some exergy destruction versus emission data correlation is presented for Ontario. The general aspects of environmental policy in Ontario can be described as follows:

- In Ontario, an Environmental Protection Act by the Ministry of Environment exists, giving the legislation on environmental quality and air pollution limits, which are conceived such that human health and the ecosystem are not endangered
- The potential of a substance to impact the environment is evaluated using a set of ten parameters:
  - transport
  - persistence
  - bioaccumulation
  - acute lethality
  - sublethal effects on mammals
  - sublethal effects on plants
  - sublethal effects on nonmammalian animals
  - teratogenicity
  - mutagenicity/genotoxicity
  - carcinogenicity
- An aggregated indicator is determined based on the ten impact parameters (above), referred as the Point of Impingement (PoI), which is determined based on the best known available pollution control technology
- The methodology denoted as removal pollution costs (RPCs) is applied to correlate the exergy of the waste stream with the cost of removing pollutants from the waste stream prior to discharge into the surroundings. The cost for waste emissions is evaluated as the total fuel cost per unit of fuel exergy multiplied by the chemical exergy per unit of fuel exergy, and then divided by the exergy efficiency of the pollution removal process.
- In Canada, the environmental pollution costs (EPCs) are estimated based on qualitative and quantitative evaluations of the pollution cost to the society for compensation and correction of the environmental damage and to prevent harmful discharges into the environment. Table 1.17 gives the EPCs of Ontario pollutants.
- The average composition of VOCs emissions in Ontario are approximated, as given in Table 1.18.
- The average composition of particulate matter (PM) emissions in Ontario are given in Table 1.19.
- The fuel cost for three types of fossil fuels, namely, coal, no. 6 fuel oil and natural gas, are in average value of CN$2013 as follows:

TABLE 1.17 Estimations of Exergetic Cost of Atmospheric Pollutants for Ontario

Pollutant	EPC ($/kg)	M (kg/kmol)	MEPC ($/kmol)	ex^{ch} (MJ/kmol)	ex^{PC} ($/MJ)	PoI ($\mu g/m_{air}^3$)
CO_2	0.0402	44	1.77	19.60	0.090	56,764
CH_4	1.2998	16	20.80	831.66	0.025	N/A
NOx	3.8944	38	148.18	72.4	2.047	500
SO_2	3.551	64	227.26	310.99	0.731	830
CO	5.5074	28	154.21	275.00	0.561	6000
VOCs	0.603	44	26.53	1233.00	0.021	N/A
PM	5.5074	68	374.50	436.00	0.860	N/A

Note: Costs are in $ for 2013 based on the Canadian consumer price index.
Data from: Carpenter, S., 1990. The environmental cost of energy in Canada. In: Sustainable Energy Choices for the 90's. Proceedings of the 16th Annual Conference of the Solar Energy Society of Canada, Halifax, NS, pp. 337–342.; De Gouw, J.A., Warneke, C., Stohl, A., Wollny, A.G., Brock, C.A., Cooper, O.R., Holloway, J.S., Trainer, M., Fehsenfeld, F.C., Atlas, E.L., Donnelly, S.G., Stroud, V., Lueb, A., 2006. Volatile organic compounds composition of merged and aged forest fire plumes from Alaska and western Canada. J. Geophys. Res. 111, D10303; Pellizzari, E.D., Clayton, C.A., Rodes, C.E., Mason, R.E., Piper, L.L., Forth, B., Pfeifer, G., Lynam, D., 1999. Particulate matter and manganese exposures in Toronto, Canada. Atmos. Environ. 33, 721–734; Ontario Regulation 419/05, 2013. Air pollution: local air quality. Internet source: http://www.ontario.ca/laws/regulation/050419.

TABLE 1.18 Approximated Average VOCs Composition and Characteristics in Ontario

VOCs	Formula	M (kg/kmol)	y (kmol/kmol)	ex^{ch} (MJ/kmol)	PoI ($\mu g/m^3_{air}$)
Methanol	CH_3OH	32	0.020	612	12,000
Acetonitrile	CH_3CN	41	0.823	1169	180
Acetaldehyde	CH_3CHO	44	0.001	1063	500
Acetone	CH_3COCH	58	0.111	1636	48,000
Acetic acid	CH_3COOH	60	0.025	780	2500
Butanone	$C_2H_5COCH_3$	72	0.015	2755	250
Toluene	$C_6H_5CH_3$	92	0.005	3771	2000

Data from: De Gouw, J.A., Warneke, C., Stohl, A., Wollny, A.G., Brock, C.A., Cooper, O.R., Holloway, J.S., Trainer, M., Fehsenfeld, F.C., Atlas, E.L., Donnelly, S.G., Stroud, V., Lueb, A., 2006. Volatile organic compounds composition of merged and aged forest fire plumes from Alaska and western Canada. J. Geophys. Res. 111, D10303; Ontario Regulation 419/05, 2013. Air pollution: local air quality. Internet source: http://www.ontario.ca/laws/regulation/050419.

TABLE 1.19 Approximated Average PM Composition and Characteristics in Ontario

PM	Formula	M (kg/kmol)	y (kmol/kmol)	ex^{ch} (MJ/kmol)	PoI ($\mu g/m^3_{air}$)
Lead	Pb	207	0.053	249.2	10
Cadmium	Cd	112	0.097	298.4	5
Nickel	Ni	59	0.185	242.6	5
Chromium	Cr	52	0.210	584.4	5
Copper	Cu	63	0.173	132.6	100
Manganese	Mn	55	3.115E-7	487.7	7.5
Vanadium	V	51	0.214	721.3	5
Aluminum	Al	27	0.006	795.7	26
Calcium	Ca	40	0.054	729.1	14
Magnesium	Mg	24	0.008	626.9	60

Data from: Pellizzari, E.D., Clayton, C.A., Rodes, C.E., Mason, R.E., Piper, L.L., Forth, B., Pfeifer, G., Lynam, D., 1999. Particulate matter and manganese exposures in Toronto, Canada. Atmos. Environ. 33, 721–734; Ontario Regulation 419/05, 2013. Air pollution: local air quality. Internet source: http://www.ontario.ca/laws/regulation/050419.

- Average coal: $1.411/GJ LHV
- Average no. 6 fuel oil: $1.864/GJ LHV
- Average natural gas: $3.899/GJ LHV
- (volatile organic compounds) Table 1.20 gives the EPC and RPC for the main fossil fuels in Ontario.

A simplified method to estimate the cost of pollutant removal from the waste stream is based on the exergy efficiency of pollutants removal. According to Rosen and Dincer (1999) this exergy efficiency is in the range of 1–5%.

TABLE 1.20 Environmental Pollution Costs (EPCs) and Removal Pollution Costs (RPC) for Fossil Fuels in Ontario

Pollutant	EPC ($/GJ_{fuel exergy})			RPC ($/GJ_{fuel exergy})		
	Coal	No. 6 fuel oil	Natural gas	Coal	No. 6 fuel oil	Natural gas
CO_2	5.2662	5.7352	7.7184	3.35	2.7604	1.7822
CH_4	5.1724	1.1524	0	0.8174	0.1608	0
NOx	0.0469	0.03082	0.03082	0.938	0.469	0.3886
CO	2.1038	0.03082	0.3484	7.5442	0.0938	0.4556
SO_2	0.4958	0.402	0	2.5594	1.5544	0

Note: Monetary values are in CN$ 2013.
Data from: Rosen, M.A., Dincer, I., 1999. Exergy analysis of waste emissions. Int. J. Energy Res. 23, 1153–1163.

Therefore, once the exergy destroyed due to pollutant discharge is known, the required exergy to remove the pollutant from the waste stream can be calculated. Furthermore, in Chapter 8, it is shown that the average price of exergy can be estimated for any geopolitical region; for Canada, it is approximated $C_{ex} = 8.4\,¢/kWh = 2.3\,¢/MJ$. When the exergy required to remove the pollutants is multiplied with exergy price, the removal pollutant cost is obtained. Therefore, one has

$$RPC = C_{ex}\psi_{pr}Ex_{d,pw} \tag{1.75}$$

where C_{ex} is the exergy price, ψ_{pr} is the exergy efficiency of pollutant removal from the waste stream, $Ex_{d,pw}$ is the exergy destroyed due to pollutant waste in the environment.

Assuming an average exergy efficiency of pollutant removal from waste stream of 3%, the RPC can be estimated as \$0.7/GJ. The RPC of power generation can be roughly estimated based on statistical data that allow for the estimation of the exergy destructions. Table 1.21 gives the rough estimate of RPC associated with Canadian power generation. The cost of pollution associated with the construction of power generation facilities, reparations and maintenance is not included in the results shown in Table 10.12. As given, the total RPC for power generation is ~\$1870 Mill. and the overall exergy efficiency of the power generation sector is 51%. The RPC becomes \$2.5/MWh generated power.

Lifecycle pollutant emission from power generation technologies are determined in Table 1.22 based on multiple literature data sources as follows: Carpenter (1990), De Gouw et al. (2006), Pellizzari et al. (1999), and Rosen and Dincer (1999). The amounts of atmospheric pollutant is given with respect to the gigajoule (GJ) of source exergy. Using the data from Table 1.21, weighted average pollutants emissions are obtained for the Canadian power generation mix. The averages are given in kilograms (kg) of pollutant per megawatt hour (MWh) of power generated; in order to convert from source exergy basis to generated power basis, the average Canadian exergy efficiency of power generation is used. Then the exergy-based EPC for power generation is calculated for each pollutant in \$/MWh; the Canadian average of EPC$_{ex}$ is \$17.8/MWh, therefore the cost of pollutant removal from the waste stream is much lower to the society than the cost of pollutant emission.

Materials used for system construction bring associated embodied energy and pollution. Table 1.23 gives the embodied energy and pollution amount and cost with various construction materials. Concrete, copper and fiberglass bring the highest among the EPCs of the listed materials. The highest embodied energy is due to aluminum fabrication, which is known to require an energy-intensive electrochemical process.

Using the concepts introduced here, the exergy destruction can be utilized to determine the environmental impact of a system in various manners as follows:

- The removal of pollution cost can be approximated by multiplying the exergy destruction with the exergy price estimated for a specific region or country.
- The rate of exergy input into the system can be used to determine the system's physical size.
- From the system's physical size, the mass amounts of materials required for system construction are determined.

TABLE 1.21   Pollution Removal Cost at Power Generation in Canada

Primary energy	Exergy input (PJ)	Power output (PJ)	Exergy destruction (PJ)	RPC (Mill.\$)	ψ (%)
Nuclear	1073	339	734	514	32
Hydro	1698	1358	340	238	80
Coal	630	378	252	176	60
Fuel oil	62.5	23	39.5	28	37
Natural gas	254	112	142	99	44
Natural gas liquids	33.6	8	25.6	20	24
Diesel fuel	8.3	3.2	5.1	4	38
Biomass	66	19	47	33	29
Secondary sources	1629	543	1086	760	33
Overall	5454	2783	2671	1870	51

Data from: Dincer, I., Zamfirescu, C., 2011. Sustainable Energy Systems and Applications. Springer, New York, (Chapter 17).

TABLE 1.22 Lifecycle Emissions Into the Atmosphere for Power Generation Technologies (kg/GJ)

Technology	kg per GJ fuel exergy						
	CO$_2$	CH$_4$	NOx	SO$_2$	CO	VOCs	PM
Coal-fired power plants	274	73	0.180	0.400	1.374	0.251	8.463E − 6
Fuel oil-fired power plants	176	47	0.115	0.200	0.876	0.160	5.439E − 6
Natural gas-fired power plants	112	30	0.073	0.150	0.558	0.102	2.557E − 6
PV power generation	44	12	0.045	0.090	0.038	0.003	1.336E − 6
Wind power generation	33	9	0.033	0.070	0.028	0.003	1.024E − 6
Hydro power	9	2	0.006	0.015	0.044	0.008	0.268E − 6
Nuclear power generation	48	13	0.032	0.065	0.241	0.044	1.488E − 6

	Canada Averages							
	CO$_2$	CH$_4$	NOx	SO$_2$	CO	VOCs	PM	Total ($/MWh)
kg pollutant per MWh power	42	11	0.029	0.060	0.194	0.035	1.288E − 6	
EPC$_{ex}$ ($ per MWh power)	1.7	14.7	0.1	0.2	1.1	0.02	7.1E − 6	17.8

EPC$_{ex}$, exergetic environmental pollution cost.
Data from: Carpenter, S., 1990. The environmental cost of energy in Canada. In: Sustainable Energy Choices for the 90's. Proceedings of the 16th Annual Conference of the Solar Energy Society of Canada, Halifax, NS, pp. 337–342.; De Gouw, J.A., Warneke, C., Stohl, A., Wollny, A.G., Brock, C.A., Cooper, O.R., Holloway, J.S., Trainer, M., Fehsenfeld, F.C., Atlas, E.L., Donnelly, S.G., Stroud, V., Lueb, A., 2006. Volatile organic compounds composition of merged and aged forest fire plumes from Alaska and western Canada. J. Geophys. Res. 111, D10303; Pellizzari, E.D., Clayton, C.A., Rodes, C.E., Mason, R.E., Piper, L.L., Forth, B., Pfeifer, G., Lynam, D., 1999. Particulate matter and manganese exposures in Toronto, Canada. Atmos. Environ. 33, 721–734; Rosen, M.A., Dincer, I., 1999. Exergy analysis of waste emissions. Int. J. Energy Res. 23, 1153–1163.

TABLE 1.23 Embodied Energy, Pollution, and Environmental Pollution Cost in Construction Materials

Material	EE (GJ/t)	SE (kgCO$_2$/GJ)	EPC$_{ex}$ ($_{2013}$/GJ)
Concrete	1.4	24	70.5
Iron	23.5	11	29.3
Steel	34.4	11	29.3
Stainless steel	53	62	29.3
Aluminum	201.4	10	29.9
Copper	131	57	60.1
Fiberglass	13	62	66.0

EE, embodied energy; SE, specific GHG emissions; EPC$_{ex}$, exergetic environmental pollution cost.
Data from: Rosen, M.A., Dincer, I., 1999. Exergy analysis of waste emissions. Int. J. Energy Res. 23, 1153–1163.

- The amount of each construction material correlates with embedded energy required for its extraction and with GHG emissions and EPC (see Table 1.23).
- The lifecycle total exergy input into the system is equal to the exergy required for system construction, system operation and system salvage.
- Based on the lifecycle total exergy input and the exergy efficiency of power and heat generation system, the exergy destruction can be determined at power and heat generation.
- The emissions and EPC due to lifecycle total exergy supply are determined based on the exergy destruction and exergetic EPC of the power and heat generation subsystem.
- The exergy destruction of the system itself allows for determination of pollutant wastes and EPC for operation during the entire lifetime.
- If a percent of materials recycling is provided, then the wasted energy and emissions of scrapped system can be determined.
- The total pollutant emissions and EPC result from the summation of the terms associated to power and heat generation from primary sources, system manufacturing, system operation and system scraping.

In many cases, it is useful to compare the environmental performance of the studied system with a reference system. If the environmental impact is smaller with respect to the reference, the studied system is "greener." The concept of greenization has been introduced by Dincer and Zamfirescu (2012). When applied to a system, the greenization concept can be illustrated as shown in Fig. 1.21.

As can be observed from the figure, the general idea is to reduce the environmental impact of a system by relating in an increased manner to renewable energy sources and greener processes with improved exergy efficiency and reduced exergy destructions. It is noteworthy that the societal reaction to pollution can be of two kinds: adaptation and mitigation.

Adaptation involves taking measures aimed at reducing the causes leading to society's vulnerability to various pollution effects, such as climate change. In this respect, the system must be made more effective to reduce the polluting fuels consumption and thence to reduce the GHG emissions. Furthermore, the share of renewable must be increased gradually in the energy supply.

The mitigation refers to a set of measures aimed to eliminate the effects of environmental impact. Here also the GHGs are the most important target of mitigation. It appears that a good policy to encourage carbon mitigation entails imposing a carbon tax to encourage sustainable energy development. The carbon tax seems likely to be applied in most of the developed countries. By 2050, it is projected that developed countries will implement a carbon tax in the range of $25–250 per ton of carbon dioxide equivalent emitted. Some other environmental impact mitigation measures of anthropogenic environmental impact are as follows:

- Applying energy efficiency and energy conservation measures
- Encouraging development of nuclear power generation systems
- Switching fuels from coal to natural gas
- Making use of renewable thermal energy from geothermal, biomass or solar radiation
- Making use of renewable electricity from solar radiation, hydro-energy, wind, biomass, geothermal, tidal and ocean thermal (via ocean thermal energy conversion)
- Using combined heat and power systems and multigeneration systems
- Applying carbon dioxide capture and storage
- Applying waste heat recovery and energy extraction from waste materials

In view of the greenization of systems (and other thermal systems), a greenization factor can be developed such that the system is compared in relative terms with a reference system. A greenization factor of zero indicates that the system

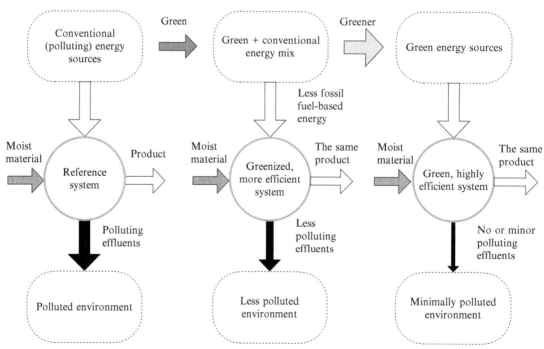

FIG. 1.21 Illustrating the concept of greenization applied to systems.

is not greenized with respect to the reference system. If the system is fully greenized, then the greenization factor is 1. Fully greenized systems depend on sustainable energy sources that have zero or minimal environmental impact during the utilization stage (although some environmental impact is associated with system construction). In greenization factor definition equation, the environmental impact factor must be specified for two cases: the reference system and the greenized system.

Depending on the specific problem analyzed, various types of environmental impact factors may be formulated. An effective and comprehensive way to define the greenization factor is through exergy destructions as follows:

$$GF = \frac{Ex_{d,ref} - Ex_d}{Ex_{d,ref}} \quad (1.76)$$

where Ex_d is the lifecycle exergy destruction of the studied system and $Ex_{d,ref}$ is the exergy destruction of the reference system.

Both the system and the reference system process generate the same amount of product. But, the energy, exergy, environmental and cost parameters of the two systems differ. Observe from Eq. (1.76) that if $Ex_d \rightarrow 0$ then $GF \rightarrow 1$, meaning that the system tends to be greener. If $Ex_d = Ex_{d,ref}$, no greenization effect is observed, and both the reference and the studied system having similar environmental impacts.

Noteworthy that the greenization factor can be defined based on other environmental and sustainability parameters. The aggregated sustainability index (ASI, mentioned above) can be used for determining the greenization factor. In this case, Eq. (1.76) becomes

$$GF = \frac{ASI - ASI_{ref}}{ASI} \quad (1.77)$$

The aggregated sustainability index accounts for thermodynamic, technical, environmental, economic and societal impacts of the technology. Thenceforth, this greenization factor formulation allows for a multidimensional comparison of systems.

1.11 CASE STUDY 1: EXERGOSUSTAINABILITY ASSESSMENT OF A CONCENTRATED PHOTOVOLTAIC-THERMAL SYSTEM FOR RESIDENTIAL COGENERATION

In this case study, a novel concentrated photovoltaic-thermal system that combines PV power generation with a special Rankine engine is to be assessed, as an exemplary case. Fig. 1.22 shows a system description. The system has several compound parabolic concentrators (CPCs) of through type installed on the southward face (assumed equatorial side) of a residence (house). The CPC concentrates light both on a vapor generator and small-area PV module. The vapor generators produce high-pressure cyclohexane vapors in a calandria (thermosiphon) loop, which makes part of an organic Rankine cycle. The system generates power through concentrated PV and the ORC (organic Rankine cycle) jointly, and heat for water heating through the ORC condenser.

The following emerging energy technologies are integrated in this system: concentrated PV with CPC as hybridized with an organic vapor generator, and cyclohexane ORC with thermo-mechanical solar energy storage and cogeneration.

The hybridized CPC system is detailed in Fig. 1.23. It captures sunlight under a half acceptance angle ($\theta_c = 60°$) and an aperture A_a. All light captured within this angle will reach the receiver surface of aperture A_r and never escape back. Part of the light is absorbed directly by a copper tube coated in black. The tube is placed in a glass shell, which is vacuumed. Inside the tube, saturated cyclohexane vapors are generated in form of bubbles. A part of the light falls on the PV modules installed at the back. The modules are covered with low-band reflection coating.

The PV coating is of dielectric type and reflects all light with wavelength longer than 900 nm. This radiation is eventually absorbed by the vapor generator (mainly) after repeated reflection on the CPC, as it cannot escape back through the aperture. The vapor accumulator is well insulated thermally as it keeps (all day including overnight) hot vapors at 120°C under pressure. The pressurized vapor in #3 (see Fig. 1.22) generate work by passing through an expander. A part of the expanded vapors are extracted at an intermediate pressure in #5, cooled to 58°C as in #7 and condensed in a coil 7–8 immersed in a water tank that stores water at ∼45°C for service (kitchen, bathroom). A part of the expanded vapors is extracted to low pressure in #4, and cooled at 30°C in #6 and then condensed at a condenser temperature of ∼20°C. The condenser, rejects the heat in a ground-coil, buried at a depth of ∼2 m such that the temperature remains constant throughout the year.

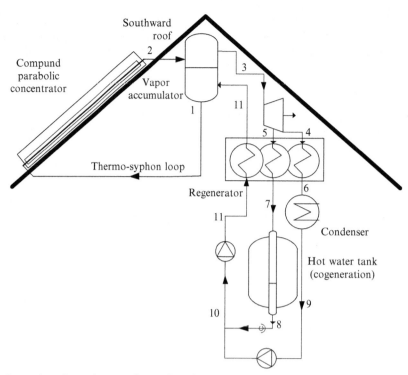

FIG. 1.22   Concentrated photovoltaic-thermal system for residential cogeneration.

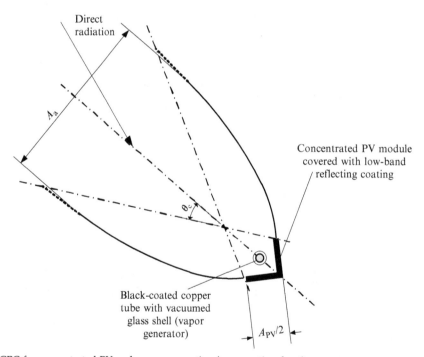

FIG. 1.23   Hybridized CPC for concentrated PV and vapor generation (cross sectional cut).

1.11.1 Assumptions

The following assumptions are given for this case study:

- For determination of exergy input $\dot{E}x_1$
    - total area concentrator exposed to $A_a = 100\,\mathrm{m}^2$
    - total area of PV arrays $A_{PV} = 1\,\mathrm{m}^2$

- the total annual irradiance is of $E_{light} = 1.8\,\mathrm{MWh/m^2}$
- the total number of sufficient sunshine is $t_{year} = 4000\,\mathrm{h/year}$
- the sunlight spectrum can be assimilated to that of a blackbody at $T_{sun} = 6000\,\mathrm{K}$
- the reference temperature is taken as $T_0 = 298\,\mathrm{K}$
- For calculating the photocurrent:
  - the transmittance of the PV coating $\mathcal{T}_\lambda$ is zero for $\lambda < 200\,\mathrm{nm}$ and for $\lambda > 900\,\mathrm{nm}$
  - the average quantum efficiency of the PV cell is $\Phi_e = 0.8$
  - the optical efficiency of concentrator is $\eta_{opt} = 0.8$
  - the band gap temperature $T_g = 11{,}000\,\mathrm{K}$
  - the diode nonideality factor $n_i = 1.5$
  - the temperature of the PV cell is $T_{PV} = 90°\mathrm{C}$
  - the resistance of PV array is $R_s = 1\mathrm{E} - 5\,\Omega$
- For calculation of the ORC
  - thermal efficiency of solar concentrator $\eta_{th} = 0.7$
  - energy efficiency of the ORC $\eta_{ORC} = 0.25$
  - the temperature of heated water $T_4 = 50°\mathrm{C}$
  - supply temperature of water $T_w = 15°\mathrm{C}$
  - amount of hot water produced per day $m_w = 2000\,\mathrm{kg}$
  - specific heat of water is $c_p = 4285\,\mathrm{J/kgK}$
- For the economic analysis
  - specific exergy cost $\mathrm{SExC} = \$23/\mathrm{GJ}$
  - fraction of capital bonds (CB) from total capital cost, $r = 0.3$
  - the operation and maintenance fraction $f_{om} = 1\mathrm{E} - 9$
  - the payback period $\mathrm{PBP} = 5\,\mathrm{years}$
- For the exergosustainability analysis
  - the system lifetime, $\mathrm{LT} = 20\,\mathrm{years}$

1.11.2 Thermodynamic Analysis

Here, the thermodynamic analysis aims ultimately to quantify the irreversibilities to determine the exergy destruction. ExBE applied for the overall system is described as shown in Fig. 1.24. The only input is the exergy from the sunlight while the outputs delivered are the power (produced by the ORC engine and PV) plus the exergy associated with the heating of water. The ExBE for the overall system states that exergy input is equal to the exergy delivered to the user plus exergy destroyed. The exergy balance is

$$\dot{E}x_1 = \dot{E}x_5 + \dot{E}x_d \tag{1.78}$$

where the exergy delivered to the user is $\dot{E}x_5 = \dot{W}_2 + \dot{W}_3 + \dot{E}x_4$, with $\dot{W}_2$ being the power generated by PV, $\dot{W}_3$ the power generated by ORC and $\dot{E}x_4$ is the exergy associated to the water heating process.

Therefore, the exergy rate of the light falling on the heliostat mirrors, input $\dot{E}x_1$ is

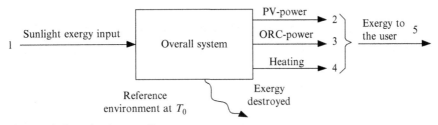

FIG. 1.24 Description of exergy balance for the overall energy system.

$$\dot{Ex}_1 = \frac{E_{light}}{t_{year}}\left(1 - \frac{T_0}{T_{sun}}\right)A_a \tag{1.79}$$

For the calculation of $\dot{W}_2$, one accounts for the fact that when the concentrated sunlight of partial spectrum falls on the photocathode, a photonic current is generated as follows:

$$J_{ph} = \eta_{optic}\left(\frac{e}{hc}\right)\frac{E_{light}A_r}{t_{year}A_a\left(\sigma T_{sun}^4\right)}\Phi_e\int_0^\infty \lambda T_\lambda I_{\lambda,b}\,d\lambda$$

where σT_{sun}^4 approximates the irradiance of sun, Φ_e is the average quantum efficiency, T_λ is the transmittance of the PV coating, η_{optic} is the optical efficiency of the concentrator, and $I_{\lambda,b}$ is the spectral irradiance of blackbody at temperature T_{sun}. The power developed by the PV array is given by

$$\dot{W}_2 = FF\,V_{oc}J_{ph} \tag{1.80}$$

The following must be calculated for the PV array:

Saturation current density: $\qquad J_0 = (1.5E-9)\exp\left(-\frac{T_g}{T_{PV}}\right)$

Dimensionless open-circuit voltage: $\qquad v_{oc} = n_i\ln\left(1 + \frac{J_{ph}}{J_0}\right)$

Open-circuit voltage: $\qquad V_{oc} = \frac{k_B T_{PV}}{e}v_{oc}$

Filling factor: $\qquad FF = \frac{v_{oc}-\ln(v_{oc}-0.72)}{v_{oc}+1}\left(1 - \frac{R_s J_{ph}A_{PV}}{V_{oc}}\right)$

The concentrated light irradiance incident on the PV surface: $\qquad I_{PV} = \eta_{optic}\frac{E_{light}A_r}{t_{year}A_a\left(\sigma T_{sun}^4\right)}\int_0^\infty T_\lambda I_{\lambda,b}\,d\lambda$

The following EES code will calculate the values of J_{ph}, v_{oc}, I_{PV}:

```
$units J K
//Given
E_light=1.8E6 "Wh/m2"
t_year=4000"h"
T_sun=6000
A_a=100"m2"
A_PV=1
T_g=11000
n_i=1.5
T_PV=90+T_zero#
R_s=1E-5
eta_optic=0.8
Phi=0.8
Tau=0.9
integr=integral(lambda/1e6*Phi*Tau*Eb(T_sun,lambda),lambda,0.2,0.9)
J_ph=eta_field*e#/h#/c#*A_a/A_PV*E_light/t_year/sigma#/
T_sun^4*integr
J_0=(1.5E9)*exp(-T_g/T_PV)
v.oc=n_i*ln(1+J_ph/J_0)
I_PV=integral(Phi*Tau*Eb(T_sun,lambda),lambda,0.2,0.9)*E_light/
t_year/sigma#/T_sun^4*A_a/A_PV
```

The important results to be used by the student are given as follows:

$$J_{ph} = 7949\,(A/m^2)$$
$$v_{oc} = 27.21$$
$$I_{PV} = 21{,}976\,(W/m^2)$$

Note that:

$$k_B = 1.381E-23\,(J/K)$$
$$e = 1.602E-19\,C$$

For the ORC, the following parameters are defined:

- Thermal efficiency of solar concentrator being the ratio between heat absorbed $\dot{Q}_a$ and light energy on the absorber $\dot{E}_{light,abs}$:

$$\eta_{th} = \frac{\dot{Q}_{abs}}{\dot{E}_{light,abs}}$$

- Energy efficiency of the ORC being the ratio between the net work generated and heat absorbed by vapor generator $\dot{Q}_{abs}$:

$$\eta_{ORC} = \frac{\dot{W}_3}{\dot{Q}_{abs}}$$

The energy balance for the photonic radiation on the solar concentrator states that the light energy rate passing through the aperture must be equal to the light energy rate absorbed on the PV array plus the light energy rate on the thermal absorber (ie, the vapor generator); this statement is written as follows:

$$\frac{E_{light}}{t_{year}} A_a = I_{PV} A_{PV} + \dot{E}_{light,a}$$

Based on the thermal efficiency of the solar concentrator, the above equation becomes:

$$\frac{\dot{Q}_{abs}}{\eta_{th}} = \frac{E_{light}}{t_{year}} A_a - I_{PV} A_{PV}$$

Furthermore, based on the ORC efficiency, the above equation can be manipulated to determine the power generated by the ORC engine, as follows:

$$\dot{W}_3 = \eta_{ORC}\eta_{th}\left(\frac{E_{light}}{t_{year}} A_a - I_{PV} A_{PV}\right) \tag{1.81}$$

The following data is given for calculating $\dot{W}_3$, namely $\eta_{ORC} = 0.25$ and $\eta_{th} = 0.7$. For calculating the exergy associated with the water heating process, the following value is given for the temperature of heated water, namely $T_4 = 50°C$, and the amount of hot water produced per day is $m_w = 2000$ kg. The water is heated from its supply temperature of $T_w = 15°C$. The specific heat of water is $c_p = 4285$ J/kgK. The exergy associated with the water heating delivered to the user is given as follows:

$$\dot{E}x_4 = \frac{m_w c_p (T_4 - T_w)}{24 \times 3600}\left(1 - \frac{T_0}{T_4}\right) \tag{1.82}$$

From Eqs. (1.80)–(1.82), the total exergy delivered to the user becomes

$$\dot{E}x_5 = \frac{v_{oc} - \ln(v_{oc} - 0.72)}{v_{oc} + 1}\left(1 - \frac{eR_s J_{ph} A_{PV}}{k_B T_{PV} v_{oc}}\right)\frac{k_B T_{PV}}{e} v_{oc} J_{ph}$$
$$+ \eta_{ORC}\eta_{th}\left(\frac{E_{light}}{t_{year}} A_a - I_{PV} A_{PV}\right) + \frac{m_w c_p (T_4 - T_w)}{24 \times 3600}\left(1 - \frac{T_0}{T_4}\right) \tag{1.83}$$

1.11.3 Environmental Impact Analysis

Environmental impact analysis aims to determine system pollution as a prerequisite of environmental impact assessment. The analyzed system generates power and heat from solar energy. As solar energy is not harmful on earth, there will be no negative environmental impact of the system during its operation period; however, the system construction involves certain amounts of construction materials and energy spent for manufacturing. The construction phase and even the scrapping phase of the system lifetime are definitely affecting the environment.

Exergy inputs and/or exergy of products are used to determine the materials inventory. Similar to the capital cost, the materials expenditure is proportional to the system scale to the power of ~ 0.6. The system scale represents the production scale and here it is the same as the total exergy output $\dot{E}x_5$. Also, because $\dot{E}x_5 = \psi\dot{E}x_1$, where the exergy efficiency ψ may be considered an invariant, it came out that the following factor denoted as scaling factor (SF), namely:

$$SF = \frac{AM}{\dot{E}x_1^{0.6}} \tag{1.84}$$

which is a constant. The amount of material AM is given in tons and the exergy rate $\dot{E}x_1$ is given in W. Here, Table 1.24 gives the scaling factors for the materials used by the system, as well as the specific pollution due to fabrication processes using various materials.

TABLE 1.24 Materials Amount Correlation and Environmental Parameters

Material	Scaling factor SF	SE ($kg_{pollutant}/t_{material}$)						
		SE_{CO_2}	SE_{CH_4}	SE_{NO_x}	SE_{SO_2}	SE_{CO}	SE_{VOC}	SE_{PM}
Concrete	1E−5	2.1E+1	5.3E+0	1.4E−2	2.8E−2	9.2E−3	1.7E−2	6.4E−7
Iron	1.3E−4	5.8E−1	1.5E−1	4.0E−4	8.1E−4	2.7E−4	4.7E−4	1.8E−8
Steel	6.6E−4	3.9E−1	1.0E−1	2.7E−4	5.5E−4	1.8E−4	3.2E−4	1.2E−8
Stainless Steel	2.6E−4	1.4E+0	3.8E−1	1.0E−3	2.0E−3	6.7E−4	1.2E−3	4.4E−8
Aluminum	4.5E−4	6.1E−2	1.6E−2	4.2E−5	8.8E−5	2.8E−5	5.1E−5	1.9E−9
Copper	1.5E−4	5.4E−1	1.4E−1	3.7E−4	7.5E−4	2.5E−4	4.5E−4	1.7E−8
Fiberglass	3E−5	5.9E+0	1.5E+0	4.1E−3	8.2E−3	2.7E−3	4.9E−3	1.8E−7

1.11.4 Economic Analysis

The capital investment must be estimated as a first step of the economic analysis. Exergy is used as a base for the estimation of costs. For each material used, the so-called embedded exergy (EEx) is given, where the embedded exergy represents the exergy required to produce that material. The cost of the material (CM) is the product of the embedded exergy and the amount of material (AM) and the specific exergy cost ($SExC$); the specific exergy cost from Canada results from its sectorial exergy analysis and it is taken $SExC = \$23/GJ$.

$$CM = EEx \times AM \times SExC \qquad (1.85)$$

In addition, the cost associated to the environmental pollution can be specified for each material with the help of the indicator called EPC given in \$per ton of pollutant. Thenceforth, the cost of pollution is found with

$$CP = AM \times EPC \qquad (1.86)$$

Table 1.25 gives the embedded exergy and EPCs for each material assumed for the purpose of this case study. The total cost of material and total cost of pollution are added to determine the capital cost (CC, where $CC = CM + CP$). Assume that the money to cover capital cost is covered by CB offered by the government due to policy on alternative energy and invested capital (IC) sourced both from private funds and government. Denote r the fraction of CB from total capital cost (CC). Then the invested capital becomes

$$IC = (1-r)CC$$

The operation and maintenance cost must be a fraction ($f_{o\&m} < 1$) of the exergy destruction. Denote with PBP the payback period in years. Then the total o&m cost for the PBP is given as follows

$$C_{o\&m} = 3600 t_{year} \dot{E}x_d f_{o\&m} PBP$$

The total cost during the PBP, meaning the sum of invested capital and the $C_{o\&m}$, is

TABLE 1.25 Embedded Exergy (EEx) and the Environmental Pollution Costs (EPCs)

Material	EEx (GJ/t)	EPC (\$/t)
Concrete	1.3	102
Iron	21.1	687
Steel	31	1009
Stainless Steel	47.7	1553
Aluminum	181.3	6023
Copper	117.9	7873
Fiberglass	11.7	858

$$C_{\text{tot}} = IC + C_{\text{o\&m}} = (1-r)CC + 3600 t_{\text{year}} \dot{E} x_{\text{d}} f_{\text{o\&m}} PBP$$

The product of the system is represented by the exergy delivered to the user in form of power and heating. The total exergy delivered during the PBP will be

$$Ex_5[\text{GJ}] = \frac{3600\ PBP\ t_{\text{year}} \dot{E} x_5}{10^9}$$

We denote LExS as the levelized exergy savings measured in [\$/GJ]. The following economic balance states holds

$$Ex_5(SExC - LExS) = C_{\text{tot}}$$

Thenceforth, based on the above equations, one determines LExS as follows:

$$LExS[\$/\text{GJ}] = SExC - 10^9 \frac{(1-r)CC + 3600 t_{\text{year}} \dot{E} x_{\text{d}} f_{\text{o\&m}} PBP}{3600\ PBP\ t_{\text{year}} \dot{E} x_5} \tag{1.87}$$

If the *LExS* is positive, then the system saves on exergy cost, because it provides to the user exergy at a cost smaller than the market value of *SExC* (specific exergy cost).

1.11.5 Exergosustainability Analysis

We base the sustainability assessment on the exergy method and the exergetic lifecycle assessment method. The ExBE must be written for the whole lifetime of the system. As a rule of thumb, the lifetime (*LT*) is at least three times longer than *PBP*.

Fig. 1.25 shows the system representation for sustainability analysis, as based on the exergetic lifecycle assessment. If there is net exergy output on hydrogen in state 2, Fig. 1.25, then the system is sustainable because in that case, the system is able to produce sufficient exergy to construct itself and pay for its emissions. Otherwise, if net exergy in #5 is zero or negative, the system is not sustainable.

As can be deduced from the figure, the net exergy amount Ex_7 can be approximated with

$$Ex_7 = Ex_5 - Ex_6$$

The term Ex_6 represents the exergy amount embedded in construction materials plus the exergy required to remove the pollutants emitted during the construction process from the atmosphere. This exergy can be related to the capital cost with the help of the *SExC*. One obtains the following

$$Ex_6 = \frac{CC}{SExC}$$

FIG. 1.25 System model for exergosustainability analysis (Case Study 1).

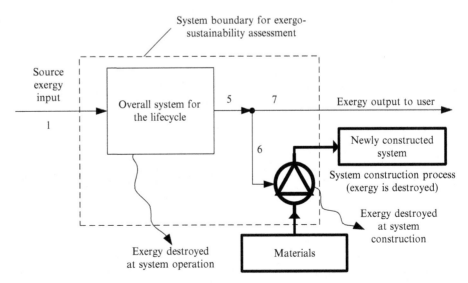

Therefore, the net exergy amount becomes

$$Ex_7 = 3600 LT\, t_{year} \dot{E}x_5 - \frac{CC}{SExC}$$

An exergetic sustainability index can be defined showing the proportion of exergy

$$ExSI = \frac{Ex_7}{Ex_5} = 1 - \frac{CC}{3600 LT\, t_{year} \dot{E}x_5 SExC} \tag{1.88}$$

For a system to be sustainable, the exergetic sustainability index must be positive and close to 1 (but subunitary).

1.11.6 Results

The following results are obtained based on the assumed data for the case study analyses presented above. The light exergy input becomes

$$\dot{E}x_1 = \frac{E_{light}}{t_{year}}\left(1 - \frac{T_0}{T_{sun}}\right)A_a = 42{,}765\,\text{W}$$

Total exergy output is calculated as follows

$$\dot{E}x_5 = \frac{v_{oc} - \ln(v_{oc} - 0.72)}{v_{oc} + 1}\left(1 - \frac{eR_s J_{ph} A_{PV}}{k_B T_{PV} v_{oc}}\right)\frac{k_B T_{PV}}{e} v_{oc} J_{ph} + \eta_{ORC}\eta_{th}\left(\frac{E_{light}}{t_{year}}A_a - I_{PV}A_{PV}\right) + \frac{m_w c_p (T_4 - T_w)}{24 \times 3600}\left(1 - \frac{T_0}{T_4}\right) = 9506\,\text{W}$$

The total exergy destroyed results as follows

$$\dot{E}x_d = \dot{E}x_1 - \dot{E}x_5 = 33{,}259\,\text{W}$$

The pollutant emissions are calculated for $\dot{E}x_1 = 42{,}765\,\text{W}$ as given in Table 1.26. Also given the assumptions: $r = 0.3$, $f_{o\&m} = 1E - 9$, $PBP = 5$ years, the levelized exergy savings by using the system become

$$LExS = SExC - 10^9 \frac{(1 - r)CC + 3600 t_{year}\dot{E}x_d f_{o\&m} PBP}{3600\, PBP\, t_{year}\dot{E}x_5} = 4.36[\$/\text{GJ}]$$

Given a system lifetime $LT = 20$ years, the exergetic sustainability index is written as follows

TABLE 1.26 The Results Regarding the Atmospheric Pollutants for Case Study 1

Material	Amount AM (t)	Pollutant emissions, $E_{pollutant}$ ($kg_{pollutant}$) = AM × $SE_{pollutant}$						
		E_{CO_2}	E_{CH_4}	E_{NO_x}	E_{SO_2}	E_{CO}	E_{VOC}	E_{PM}
Concrete	1.20E − 02	2.5E − 01	6.4E − 02	1.7E − 04	3.4E − 04	1.1E − 04	2.0E − 04	7.7E − 09
Iron	1.56E − 01	9.1E − 02	2.3E − 02	6.2E − 05	1.3E − 04	4.2E − 05	7.3E − 05	2.8E − 09
Steel	7.93E − 01	3.1E − 01	7.9E − 02	2.1E − 04	4.4E − 04	1.4E − 04	2.5E − 04	9.5E − 09
Stainless Steel	3.12E − 01	4.4E − 01	1.2E − 01	3.1E − 04	6.2E − 04	2.1E − 04	3.7E − 04	1.4E − 08
Aluminum	5.41E − 01	3.3E − 02	8.6E − 03	2.3E − 05	4.8E − 05	1.5E − 05	2.8E − 05	1.0E − 09
Copper	1.80E − 01	9.7E − 02	2.5E − 02	6.7E − 05	1.4E − 04	4.5E − 05	8.1E − 05	3.1E − 09
Fiberglass	3.60E − 02	2.1E − 01	5.4E − 02	1.5E − 04	3.0E − 04	9.7E − 05	1.8E − 04	6.5E − 09
Total (kg)		1.4E + 00	3.7E − 01	9.9E − 04	2.0E − 03	6.6E − 04	1.2E − 03	4.4E − 08

$$ExSI = 1 - \frac{CC}{3600 LT \, t_{year} \dot{E}x_5 \, SExC} = 0.76$$

Therefore, the system is sustainable since only $100 - 76 = 24\%$ of produced exergy can be used to build the system itself and to compensate pollution, whereas the rest of 76% is delivered to the user.

In order to encourage the use of renewable energy, a policy plan can be established as follows:

(a) To encourage research and development by providing focused grants aiming at reducing exergy destruction $\dot{E}x_d$. This will lead to an increase of savings on exergy and therefore to better sale of the system. Also, the exergetic sustainability index will improve. Assume that this measure aims at 5% reduction of exergy destruction. Then the expected effect of this measure is written as $\dot{E}x_d \rightarrow 0.95 \dot{E}x_d$.

(b) To improve the system's economic performance by increasing the CB (that applies a higher fraction r representing the ratio between CB and capital cost). This will again increase the exergy savings obtained with the system. Assume that one increases the CB with 10%; then $r \rightarrow 1.1r$.

(c) To apply higher carbon tax to the coal, petroleum and natural gas power generators such that the $SExC$ increases in the jurisdiction. This will lead to an improved exergetic sustainability index and better savings with the system. Consider the application of carbon tax to induce $SExC$ in the jurisdiction higher with 5%, then $SExC \rightarrow 1.05 SExC$.

With the considered policy measured, the new values for $LExS$ and $ExSI$, become, respectively:

$$LExS = 1.05 \, SExC - 10^9 \frac{(1 - 1.1r)CC + 3600 t_{year} 0.95 \dot{E}x_d f_{o\&m} PBP}{3600 \, PBP \, t_{year} \dot{E}x_5} = 6.34 \, [\$/GJ]$$

$$ExSI = 1 - \frac{CC}{3600 \, LT \, t_{year} \dot{E}x_5 1.05 \, SExC} = 0.78$$

1.11.7 Closing Remark

In conclusion, the policy measures are necessary to encourage the system sales, since the exergy savings increase as well the exergetic sustainability index.

1.12 CASE STUDY 2: EXERGOSUSTAINABILITY ASSESSMENT OF A HIGH-TEMPERATURE STEAM PHOTO-ELECTROLYSIS PLANT

In this case study, a novel solar hydrogen production system is considered in which a heliostat field concentrates the sunlight on the surface of a high-temperature photo-electrochemical cell for steam electrolysis, having a light-exposed photocathode. Liquid water is pumped atop the tower and H_2 and O_2 gases are generated. The solar radiation is concentrated on photo-electrochemical cells; those assembled in modules form arrays.

The cell has a transparent glass with a concave shape to increase the strength as the gas inside is at $\sim$20 atm. A mixture of 90% steam and 10% H_2 is fed, and, due to the photo-electrochemical reaction occurring there, a mixture of 90% H_2 and 10% steam is extracted at $\sim$1000°C. The permeable photocathode has a rough surface (3D, "volumetric" configuration) and it is doped with cheap metallic electrocatalysts and CuO, Cu_2O semiconductors, which operate as photosensitizers (excite electrons into the conduction band when photons are absorbed).

1.12.1 Assumptions

- The following data are assumed for determination of $\dot{E}x_1$:
  - total reflecting area heliostat mirrors $A_r = 915E3 \, m^2$
  - total aperture area $A_a = 725 \, m^2$
  - the total annual irradiance is of $E_{light} = 2.0 \, MWh/m^2$
  - the total number of sufficient sunshine hours is $t_{year} = 4000 \, h/year$

- the sunlight spectrum can be assimilated to that of a blackbody at $T_{sun} = 6000\,K$
- the reference temperature is taken as $T_0 = 298\,K$
- The following data for calculating the photocurrent are used:
  - the heliostats have an aluminum reflective surface. The spectral reflectance of aluminum is approximated as follows

$$R_\lambda = \begin{cases} 0.92 \text{ for } \lambda < 700\,nm \\ 0.88 \text{ for } \lambda < 700\,nm \text{ and } \lambda < 900\,nm \\ 0.98 \text{ for } \lambda > 900\,nm \end{cases}$$

  - the spectral quantum efficiency $\Phi_{e,\lambda} = 0.9$ for $\lambda \le 580\,nm$ and $\Phi_{e,\lambda} = 0.6$ for $\lambda \le 1033$ nm and $\Phi_{e,\lambda} = 0$ for $\lambda > 1033\,nm$
  - the field efficiency is given as $\eta_{field} = 0.7$
- The following data are assumed for calculation of the exergy content of produced hydrogen:
  - the chemical exergy of hydrogen $ex^{ch}_{H2} = 236\,MJ/kmol$
- The following data are assumed for economic analysis:
  - the specific exergy cost SExC $= \$23/GJ$
  - fraction of CB from total capital cost, $r = 0.3$
  - the operation and maintenance fraction $f_{om} = 1E-9$
  - the payback period PBP $= 20$ years
- For the sustainability analysis, the following data are assumed:
  - The system lifetime, LT $= 40$ years

1.12.2 Thermodynamic Analysis

Thermodynamic analysis aims to determine the exergy destruction. ExBE is applied for the overall system as shown in Fig. 1.26. The inputs are sunlight and water; the only useful output is hydrogen (carrying its chemical exergy). The exergy balance determines the exergy destroyed, which accounts for any irreversibility plus the release of oxygen in the atmosphere (as a waste from the process).

The ExBE for the overall system states that exergy input is equal to the exergy output plus exergy destroyed. The exergy input is:

$$\dot{E}x_{in} = \dot{E}x_3 + \dot{E}x_d$$

where the exergy input is $\dot{E}x_{in} = \dot{E}x_1 + \dot{E}x_2 \cong \dot{E}x_1$, since the exergy carried by water $(\dot{E}x_2)$ is negligible with respect to the exergy rate of light input: $(\dot{E}x_1)$. Therefore, the ExBE becomes

$$\dot{E}x_1 = \dot{E}x_3 + \dot{E}x_d \tag{1.89}$$

Therefore, the exergy rate of the light falling on the heliostat mirrors input $\dot{E}x_1$ is:

$$\dot{E}x_1 = \frac{E_{light}}{t_{year}}\left(1 - \frac{T_0}{T_{sun}}\right)A_r \tag{1.90}$$

When the concentrated sunlight falls on the photocathode, a photonic current is generated. The photonic current density is given by

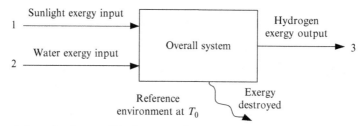

FIG. 1.26 Description of exergy balance for the overall hydrogen production system.

$$J_{ph} = \eta_{field} \left(\frac{e}{hc}\right) \frac{E_{light} A_r}{t_{year} A_a \left(\sigma T^4_{sun}\right)} \int_0^\infty \lambda \Phi_{e,\lambda} R_\lambda I_{\lambda,b} d\lambda$$

where σT^4_{sun} approximates the irradiance of sun, $\Phi_{e,\lambda}$ is the spectral quantum efficiency, R_λ is the reflectance of heliostat mirrors, η_{field} is the efficiency of the heliostat field, and $I_{\lambda,b}$ is the spectral irradiance of blackbody at temperature T_{sun}. The following sequence of code can be used to calculate the photonic current density in EES (conversion from microns to m for λ is required).

```
$units K J
A_r=915E3
A_a=725
E_light=2E6
t_year=4000
Int_a=0.92*0.6*integral(Lbd_3a/1e6*Eb(T_sun,Lbd_3a),
Lbd_3a,0.2,0.58)
Int_b=0.92*0.6*integral(Lbd_3b/1e6*Eb(T_sun,Lbd_3b),
Lbd_3b,0.58,0.7)
Int_c=0.88*0.3*integral(Lbd_3c/1e6*Eb(T_sun,Lbd_3c),
Lbd_3c,0.7,0.9)
Int_d=0.98*0.3*integral(Lbd_3d/1e6*Eb(T_sun,Lbd_3d),
Lbd_3d,0.7,0.9)
J_ph=(Int_a+Int_b+Int_c+Int_d)*eta_field*e#/h#/c#/sigma#/
T_sun^4*A_r/A_a*E_light/t_year
```

The important results to be used by the student are given as follows:

$J_{ph} = 75{,}061 \ (A/m^2)$

Note that:

$k_B = 1.381E - 23 \, J/K$

$e = 1.602E - 19 \, C$

The molar flow rate of produced hydrogen results from the photonic current

$$\dot{n}_{H_2} = \frac{J_{ph} A_a}{zF}$$

where $z = 2$, $F = 96{,}486{,}700 \, C/kmol$ (Faraday's constant).

The exergy of generated hydrogen becomes

$$\dot{Ex}_3 = \dot{n}_{H_2} ex^{ch}_{H_2} = \frac{J_{ph} A_a}{zF} ex^{ch}_{H_2} \tag{1.91}$$

where $ex^{ch}_{H_2} = 118 \, MJ/kmol$ is the chemical exergy of hydrogen.

1.12.3 Environmental Impact Analysis

Environmental impact analysis aims to determine system pollution as a prerequisite of environmental impact assessment. The analyzed system generates hydrogen from solar energy. As solar energy is not harmful on earth, there will be no negative environmental impact of the system during its operation period; however, the system construction involves certain amounts of construction materials and energy spent for manufacturing. The construction phase and even the scrapping phase of the system lifetime are definitely affecting the environment.

Similar to Case Study 1, the scaling factor is defined by Eq. (1.84) and the specific emissions are given in Table 1.24. The assumed values of the scaling factor for Case Study 2 are given in Table 1.27.

1.12.4 Economic Analysis

The economic analysis is similar to that presented for Case Study 1. Once the capital cost and total cost are determined, the total hydrogen produced in kg for the payback period results as follows

$$m_{H_2} = 7200 \, t_{year} \frac{J_{ph} A_a}{zF} PBP$$

Denote LH_2P as the levelized hydrogen price for selling. The following economic balance states that the revenues from hydrogen selling during the PBP balance the total cost

$$m_{H_2} LH_2P = C_{tot}$$

TABLE 1.27   Materials Amount Correlation and Environmental Parameters

Material	SF
Concrete	1381
Iron	24
Steel	236
Stainless steel	118
Aluminum	2
Copper	5
Fiberglass	1

Based on the above equations, one determines LH_2P as follows

$$LH_2P = \frac{(1-r)CC + 3600\, t_{year}\dot{E}x_d f_{o\&m}PBP}{7200\frac{J_{ph}A_a}{zF}t_{year}PBP} \tag{1.92}$$

1.12.5 Exergosustainability Analysis

Fig. 1.27 shows the system representation for sustainability analysis, as based on the exergetic lifecycle assessment. If there is net exergy output on hydrogen, in state 2, Fig. 1.27, then the system is sustainable, because in that case, the system is able to produce sufficient exergy to construct itself and pay for its emissions. Otherwise, if net exergy in #5 is zero or negative, the system is not sustainable.

As deduced from the figure, the net exergy amount Ex_5 can be approximated with

$$Ex_5 = Ex_3 - Ex_4$$

The term Ex_4 represents the exergy amount embedded in construction materials plus the exergy required to remove the pollutants emitted during the construction process from the atmosphere. This exergy can be related to the capital cost with the help of the $SExC$. One has:

$$Ex_4 = \frac{CC}{SExC}$$

Therefore, the net exergy amount becomes

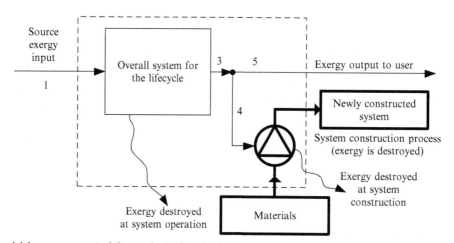

FIG. 1.27   System model for exergosustainability analysis (Case Study 2).

$$Ex_5 = 3600 LT\, t_{year} \dot{Ex}_3 - \frac{CC}{SExC}$$

An exergetic sustainability index can be defined showing the proportion of exergy

$$ExSI = \frac{Ex_5}{Ex_3} = 1 - \frac{CC}{3600\, LT\, t_{year} \dot{Ex}_5\, SExC} \tag{1.93}$$

For a system to be sustainable, the $ExSI$ must be positive and close to 1.

1.12.6 Results

The following results are obtained based on the assumed data for the case study analyses presented above. The light exergy input becomes

$$\dot{Ex}_1 = \frac{E_{light}}{t_{year}}\left(1 - \frac{T_0}{T_{sun}}\right)A_r = \frac{2E6}{4000}\left(1 - \frac{298}{6000}\right)915E3 = 434{,}777{,}500\,W$$

Given $J_{ph} = 113{,}107\,A/m^2$, one calculates the exergy carried by hydrogen as follows:

$$\dot{Ex}_3 = \frac{J_{ph}A_a}{zF}ex_{H_2}^{ch} = \frac{75{,}061 \times 725}{2 \times 96{,}486{,}700} \times 118E6 = 66{,}552{,}888\,W$$

The total exergy destroyed becomes:

$$\dot{Ex}_d = \dot{Ex}_1 - \dot{Ex}_3 = 434{,}77{,}500 - 30{,}521{,}184 = 368{,}224{,}612\,W$$

The pollutants are calculated for $\dot{Ex}_1 = 434{,}777{,}500\,W$ as given in Table 1.28. The capital cost calculations are given in Table 1.29. Given the assumptions made and the estimated capital cost, the levelized hydrogen price LH_2P is determined as follows:

$$LH_2P = \frac{(1-r)CC + 3600\,t_{year}\dot{Ex}_d\, f_{o\&m}PBP}{7200\dfrac{J_{ph}A_a}{zF}t_{year}PBP} = \$1.31/kg$$

TABLE 1.28 The Results Regarding the Atmospheric Pollutants for Case Study 2

Material	Amount AM (t)	Pollutant emissions, $E_{pollutant}$ ($kg_{pollutant}$) = AM × SE$_{pollutant}$						
		E_{CO_2}	E_{CH_4}	E_{NO_x}	E_{SO_2}	E_{CO}	E_{VOC}	E_{PM}
Concrete	210,452,416	4.3E+9	1.1E+9	3.0E+6	6.0E+6	1.9E+6	3.6E+6	1.3E+2
Iron	3,657,392	2.1E+6	5.5E+5	1.5E+3	2.9E+3	9.7E+2	1.7E+3	6.6E−2
Steel	35,964,352	1.4E+7	3.7E+6	9.7E+3	2.0E+4	6.5E+3	1.2E+4	4.4E−1
Stainless steel	17,982,176	2.6E+7	6.7E+6	1.8E+4	3.6E+4	1.2E+4	2.1E+4	7.9E−1
Aluminum	304,783	1.9E+4	4.9E+3	1.3E+1	2.7E+1	8.6E+0	1.5E+1	5.7E−4
Copper	761,957	4.1E+5	1.1E+5	2.8E+2	5.8E+2	1.9E+2	3.4E+2	1.3E−2
Fiberglass	152,391	9.0E+5	2.3E+5	6.2E+2	1.3E+3	4.2E+2	7.4E+2	2.7E−2
Total (kg)		4.4E+9	1.1E+9	3.0E+6	6.1E+6	2.0E+6	3.6E+6	1.4E+2

TABLE 1.29 Capital Cost Calculations for Case Study 2

Material	CM ($)	CP ($)	Total ($)
Concrete	6,292,527	21,466,146	27,758,674
Iron	1,774,932	2,512,628	4,287,560
Steel	25,642,583	36,288,031	61,930,614
Stainless steel	19,728,245	27,926,319	47,654,565
Aluminum	1,270,913	1,835,706	3,106,619
Copper	2,066,198	5,998,884	8,065,082
Fiberglass	41,009	130,752	171,760
Total	56,816,407	96,158,467	CC = 152,974,874

Given system lifetime $LT = 40$ years, the exergetic sustainability index becomes

$$ExSI = 1 - \frac{CC}{3600\,LT\,t_{year}\dot{E}x_3\,SExC} = 0.73$$

In order to encourage the use of hydrogen, a policy plan can be established (similarly as for the Case Study 1) as follows:

(a) Encourage research and development by providing focused grants aiming at reducing exergy destruction $\dot{E}x_d$. This will lead to a decrease of levelized hydrogen price and therefore to better revenue from sales. Also, the exergetic sustainability index will improve.

   Assumption: Encourage research aiming at 5% reduction of exergy destruction, then $\dot{E}x_d \rightarrow 0.95\dot{E}x_d$.

(b) Try to improve the system's economic performance by increasing the CB (that applies a higher fraction r representing the ratio between CB and capital cost). This will again increase the exergy savings obtained with the system.

   Assumption: Increase the CB with 10%, then $r \rightarrow 1.1r$.

(c) Apply higher carbon tax to the coal, petroleum and natural gas power generators such that the $SExC$ increases in the jurisdiction. This will lead to an improved exergetic sustainability index and better revenues with the system.

   Assumption: Apply carbon tax to induce $SExC$ in the jurisdiction higher with 5%, then $SExC \rightarrow 1.05\,SExC$.

Due to the considered policy measures, the new values for the levelized hydrogen price for sale LH_2P and the exergetic sustainability index ($ExSI$) become:

$$LH_2P = \frac{(1-1.1r)CC + 3600\,t_{year}0.95\dot{E}x_d f_{o\&m}PBP}{7200\frac{J_{ph}A_a}{zF}t_{year}PBP} = \$1.22/kg$$

$$ExSI = 1 - \frac{CC}{3600\,LT\,t_{year}\dot{E}x_3\,1.05\,SExC} = 0.743$$

1.12.7 Closing Remarks

In conclusion, based on the system assessment, the system is sustainable since only 27% of produced exergy is used to build itself and compensate pollution; however, for better economic success, a system improvement may be beneficial.

Once policy measures are applied by encouraging research through grants, increase the investment CB and application of carbon taxation to power generators, the levelized hydrogen price becomes compatible with the price of hydrogen produced by conventional methods and the exergetic sustainability index increases to 0.83.

1.13 CONCLUDING REMARKS

In this chapter, fundamental aspects of thermodynamics and process and system assessment methods are introduced. The fundamental properties and quantities are introduced first, and the main notions from thermodynamics, such as pressure, temperature, volume, work, heat, energy enthalpy, etc. are discussed. The state diagrams are presented to give a true idea about the processes. Ideal and real gas equations are given, as they are of high importance. Furthermore, the chapter introduces the ideal-gas law, van der Waals law and compressibility factor charts as well, as it discusses the properties of gas mixtures. The laws of thermodynamics and related analysis methods with energy and exergy concepts are presented in detail, particularly through the balance equations. Both energy and exergy efficiencies are defined for later use.

The use of exergy method for economic, environmental, and sustainability analysis is also presented by introducing the exergoeconomics, exergoenvironmental, and exergosustainability methods. Two case studies are given to illustrate the applications of fundamental concepts to system analysis, assessment and evaluation of hydrogen production systems and applications.

Nomenclature

a	acceleration, m/s^2
A	area, m^2
C	specific heat, kJ/kgK
COP	coefficient of performance
DOF	degree of freedom
e	specific total energy, kJ/kg
$\dot{E}$	energy rate, kW
ex	specific exergy, kJ/kg
$\dot{Ex}$	total specific exergy, kW
g	gravity acceleration, m/s^2
g	specific Gibbs free energy, kJ/kg
h	specific enthalpy, kJ/kg
k_B	Boltzmann constant
KE	kinetic energy, kJ
l	displacement, m
F	force, N
m	mass, kg
M	molecular mass, kg/kmol
n	number of moles, mol
N	number of particles
N_A	number of Avogadro, molecules/kmol
p	probability
P	pressure, Pa
PE	potential energy, kJ
q	mass-specific heat, kJ/kg
Q	heat, kJ
$\dot{Q}$	heat rate, kW
$\mathcal{R}$	universal gas constant, kJ/kmolK
R	gas constant, kJ/kgK
s	specific entropy, kJ/kgK
S	entropy, J/K
$\dot{S}$	entropy rate, W/K
t	time, s
T	temperature, K
v	specific volume, m^3/kg
$\mathcal{v}$	velocity, m/s
V	volume, m^3
u	specific internal energy, kJ/kg
U	internal energy, kJ
w	mass-specific work, kJ/kg
W	work, J
$\dot{W}$	work rate, kW
x	vapor quality
z	elevation, m
Z	compressibility factor

Greek Letters

θ	total specific nonflow energy, kJ/kg
ϕ	total specific flow energy, kJ/kg

η	energy efficiency
μ	chemical potential, kJ/kg
ψ	exergy efficiency
γ	adiabatic expansion coefficient
ω	humidity ratio

Subscripts

0	reference state
b	boundary work
C	cold
ch	chemical
conc	concentration
deliv	delivered
f	flow work
gen	generated
in	input
ke	kinetic energy
H	hot
loss	loss
L	low
nf	nonflow
out	output
p	constant pressure
pe	potential energy
ph	physical
r	reduced value
rev	reversible
surr	surroundings
sys	system
u	useful work
v	constant volume

Superscripts

'	saturated liquid
''	saturated vapor
ch	chemical

References

Bejan, A., 1996. Entropy Generation Minimization. CRC Press, Boca Raton, FL.

Carpenter, S., 1990. The environmental cost of energy in Canada. In: Sustainable Energy Choices for the 90's. Proceedings of the 16th Annual Conference of the Solar Energy Society of Canada, Halifax, NS, pp. 337–342.

De Gouw, J.A., Warneke, C., Stohl, A., Wollny, A.G., Brock, C.A., Cooper, O.R., Holloway, J.S., Trainer, M., Fehsenfeld, F.C., Atlas, E.L., Donnelly, S. G., Stroud, V., Lueb, A., 2006. Volatile organic compounds composition of merged and aged forest fire plumes from Alaska and western Canada. J. Geophys. Res. 111. D10303.

Dincer, I., 2011. Exergy as a tool for sustainable drying systems. Sustain. Cities Soc. 1, 91–96.

Dincer, I., Rosen, M.A., 2012. Exergy: Energy, Environment and Sustainable Development. Elsevier, Oxford, UK.

Dincer, I., Zamfirescu, C., 2011. Sustainable Energy Systems and Applications. Springer, New York, NY.

Dincer, I., Zamfirescu, C., 2012. Potential options to greenize energy systems. Energy 46, 5–15.

Dincer, I., Zamfirescu, C., 2014. Advanced Power Generation Systems. Elsevier, New York, NY.

EES, 2013. Engineering Equation Solver. F-Chart Software (Developed by S.A. Klein). EES, Middleton, WI.

ISU, 2006. The International System of Units (SI), eighth ed. The International Bureau of Weights and Measures. Paris, France.

Pellizzari, E.D., Clayton, C.A., Rodes, C.E., Mason, R.E., Piper, L.L., Forth, B., Pfeifer, G., Lynam, D., 1999. Particulate matter and manganese exposures in Toronto, Canada. Atmos. Environ. 33, 721–734.

Rivero, R., Grafias, M., 2006. Standard chemical exergy of elements updated. Energy 31, 3310–3326.

Rosen, M.A., 1986. The Development and Application of a Process Analysis Methodology and Code Based on Exergy, Cost, Energy and Mass. Ph.D. Thesis, Department of Mechanical Engineering, University of Toronto, Toronto.

Rosen, M.A., Dincer, I., 1999. Exergy analysis of waste emissions. Int. J. Energy Res. 23, 1153–1163.

Rosen, M.A., Dincer, I., 2003. Exergy-cost-energy-mass analysis of thermal systems and processes. Energy Convers. Manag. 44, 1633–1651.

Toffolo, A., Lazzaretto, A., 2004. Energy, economy and environment as objectives in multi-criteria optimization of thermal system design. Energy 29, 1139–1157.

Tsatsaronis, G., Moran, M.J., 1997. Exergy-aided cost minimization. Energy Convers. Manag. 38, 1535–1542.

STUDY PROBLEMS

1.1 Explain the difference between flow work and boundary work.

1.2 Explain the difference between closed and open thermodynamic systems.

1.3 What is the relationship between temperature and kinetic energy of molecules?

1.4 What is the difference between thermal and caloric equation of state?

1.5 Explain the principle of corresponding states. How much is the value of compressibility factor for ideal gas?

1.6 Derive the equation for mechanical work during an isothermal transformation of ideal gas.

1.7 Explain the difference between internal energy and enthalpy.

1.8 Define the notion of adiabatic process and explain the role of expansion coefficient for determining the work exchange during this type of process.

1.9 What is the difference between externally reversible and internally reversible processes?

1.10 What is the relationship between entropy change of a system, of its surroundings and of generated entropy?

1.11 Using thermodynamic tables or EES software, calculate the thermomechanical exergy of saturated steam at 16 MPa. Assume standard values for temperature and pressure of the environment.

1.12 A cistern vehicle transports 20 t of propane and runs with 80 km/h, being exactly in the top of a hill of 100 m altitude. Calculate the total exergy stored in the cistern.

1.13 Stack gas of a power plant comprises 15% carbon dioxide by volume and the rest is made of other gaseous combustion products. The pressure of stack gas is 1.5 atm. and the temperature 425 K. Calculate the amount of reversible work needed to separate carbon dioxide.

1.14 Calculate chemical exergy of octane based on standard exergy of the elements.

1.15 Write balance equations for a building wall assuming that exterior temperature is 5°C, interior temperature is 25°C, the wall thermal conductivity is 1 W/mK, and the heat transfer coefficient is 10 W/m^{2}K on both sides. Calculate the entropy generation and exergy destruction per unit of wall surface.

1.16 Does an exergy analysis replace an energy analysis? Describe any advantages of exergy analysis over energy analysis.

1.17 Noting that heat transfer does not occur without a temperature difference and that heat transfer across a finite temperature difference is irreversible, is there such a thing as reversible heat transfer? Explain.

1.18 Write mass, energy, entropy and exergy balances for the following devices: (a) an adiabatic steam turbine; (b) an air compressor with heat loss from the air to the surroundings; (c) an adiabatic nozzle; and (d) a diffuser with heat loss to the surroundings.

1.19 A turbine expands superheated steam at 10 MP having a superheating degree of 50 K. The expansion ratio is 2.5. The isentropic efficiency of the turbine is 0.8. Write balance equations and calculate exergy efficiency of the turbine.

1.20 In a combustion chamber, natural gas is supplied at 10 atm, while air comes in an excess ratio of 5. Assume a complete combustion of natural gas modeled as methane. Write a balance equation and calculate the first and second law efficiency and exergy destruction of the combustion chamber.

1.21 Consider adiabatic mixing of 10 g/s moist air at 25°C with relative humidity of 60% with 2 g/s of moist air at 10°C with relative humidity of 40%. Write a balance equation and determine the temperature and relative humidity of air for the mixed stream, entropy generation, exergy destruction and second law efficiency for the mixer. EES software or a Molier diagram can be used.

C H A P T E R

2

Hydrogen and Its Production

OUTLINE

2.1 INTRODUCTION

Hydrogen appears to be a potential energy carrier (and carbon-free fuel) for satisfying power demand of the world in the future because the conventional energy resources are being depleted, and there is need of an energy carrier that can effectively help harvest energy from nonconventional renewable sources. The current energy consumption per capita is ~2.5 kW worldwide average. Based on the data documented previously by the authors in Dincer and Zamfirescu (2014), if this per-capita consumption is maintained, with the prediction of 9.6 billion of world population by 2050, the energy demand will reach 24 TW; however, accounting for the world tendency to increase the standards of life, the per-capita power demand may increase to 6.5 kW, representing the average of Europe and North America predicted for 2050. If this happens, then the world power demand will be 50 TW by 2050.

It is evident that petroleum resources will become scarce and only coal will still be sufficiently abundant with ~10^{13} GJ in remaining resources by 2050. Nuclear power cannot be a major share of energy mix, because even in the extreme scenario that one installs one 1 GW nuclear power plant per year for 35 years, then nuclear may cover a maximum 30% of power demand of 2050. However, in a more realistic scenario, one can guess perhaps that only 30% of power demand will be covered jointly by coal and nuclear resource. This concludes that the rest of the demand must be sourced from renewables. As shown in Naterer et al. (2013) (Chapter 1, Fig. 1.1) about 60% of energy demand in 2050 will be satisfied from renewable sources.

Among the renewables, the solar resource is the most important. If 30% of the Sahara desert is occupied with solar collectors to run a system that produces hydrogen as an energy carrier and uses fuel cells for continuous power production with an estimated overall efficiency of 10%, then the power generated would be ~50 TW. Other potential renewables can be given as follows (see Dincer and Zamfirescu, 2014): hydropower through hydroelectric power stations, ~1.5 TW; geothermal energy via Rankine power generation stations, ~3 TW; biomass energy of all fertile land that could be destined to energy crops, ~10 TW; and wind energy harvested with inland installed wind turbines,

Sustainable Hydrogen Production
http://dx.doi.org/10.1016/B978-0-12-801563-6.00002-9

~2 TW. Considering the future expansion of ocean power through OTEC (ocean thermal energy conversion) platforms, ocean currents, ocean wave and tidal generators, one expects an even greater demand for conversion technologies to produce hydrogen from renewable sources and power and synthetic fuels from hydrogen.

Therefore, hydrogen as an energy carrier and intermediate chemical in energy conversion steps and synthesis of fuels will have a drastically increased role. Renewable energies have a fluctuating and intermittent nature. One needs to really hunt for renewable resources when and where they occur in their surrounding environment. Once harvested, such renewable energy must be converted into a storable form, and hydrogen is the most promising candidate for the task. The development of sustainable hydrogen and fuel cell technology is not an optional choice, but rather something that must happen! The role of sustainable hydrogen and fuel cells become very important today for research and development and in the near future as an energy solution on a planetary scale.

At present, the hydrogen production is around 50 Mt per year, worldwide. The hydrogen demand will increase even more because other sectors will be benefited. Hydrogen is used intensely in the fertilizer industry as a constituent of ammonia and urea. Furthermore, hydrogen is a key element for the petrochemical industry where it is used for specific petrochemical operations and heavy oil upgrading. Hydrogen has been named an ideal fuel for road, rail, sea and airway transport. By 2025, four major industrial sectors in Canada are predicted to demand hydrogen: crude oil upgrading (62% share), ammonia synthesis (25% share), methanol synthesis (6% share), and metallurgy (most of the rest of the industrial share of hydrogen demand); see Dincer and Zamfirescu (2011).

In this chapter, the role and importance of hydrogen toward a cleaner environment and a sustainable development is discussed. Furthermore, hydrogen properties, hydrogen production methods, hydrogen storage and distribution are introduced and classified. The chapter continues with a presentation of fuel cells and hydrogen utilization in various sectors for numerous applications.

2.2 HYDROGEN AND THE ENVIRONMENT

The world energy consumption is currently $\sim 445 \times 10^{18}$ J annually and will continue to increase as the population grows, especially as the standards of living increase. The style of living becomes an important element to consider when the energy consumption is analyzed. The highest energy consumption per capita in the residential sector occurs in European countries and North America, as shown in Fig. 2.1.

The per-capita consumption is predicted to decrease in North America; however, the predictions show that the residential energy per capita will increase in the non-Organisation for Economic Co-operation and Development (OECD) European countries. Overall, the world energy consumption per capita in the residential sector will increase, showing that humans tend to enhance their lifestyle toward higher energy consumption.

Energy demand increases at the same pace as population growth, as shown in Fig. 2.2. The energy demand depends on the population density and the economic development of the world region. The world predicted energy demand by 2040 can be appreciated for each of the major sectors of activity as shown in Fig. 2.3. The main energy consumers are in

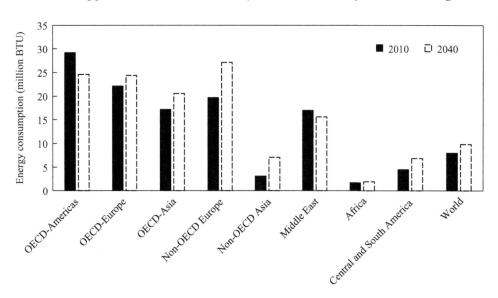

FIG. 2.1 Recorded (2010) and predicted (2040) residential energy consumption per capita in the main economic regions of the world.

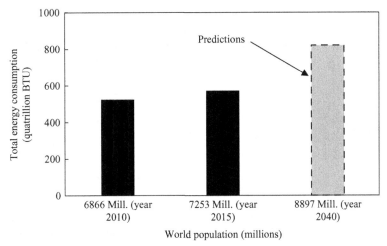

FIG. 2.2 Correlation between world population and energy consumption.

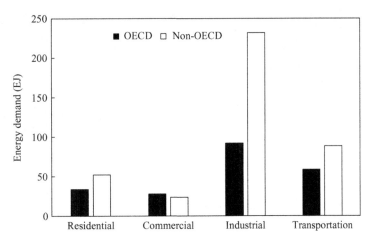

FIG. 2.3 Predicted energy demand of main activity sectors in 2040.

approx. equal shares the industrial, transportation and residential sectors. Since the electricity generation is overwhelmingly based on coal technology and transportation on petroleum, there are predicted increased amounts of atmospheric pollutants coming within the next few decades.

There is an alarming situation with respect to environmental impact and global warming owning to the increased fossil fuel consumption worldwide with more associated greenhouse gas (GHG) emissions among which CO_2 is the most relevant. The carbon dioxide emission records and predictions caused by combustion of the three major fossil fuels: liquid petrochemical fuels, natural gas, and coal is shown in Figure 2.4. All these types of emissions tend to increase as predicted, with coal being responsible for the majority of this type of atmospheric pollution.

The GHG emission by sectors of activity and by major GHG type is depicted as shown in Figure 2.5a. Since the industrial revolution anthropogenic activities, especially those related to fossil fuel combustion in industrial, commercial and residential settings and in the transportation sector, have dangerously affected the global environment. Among carbon dioxide sources, Fig. 2.5b shows that the combustion of fossil fuels is the most important with 57% of shares.

The transportation sector is one of the most important consumers of liquid fuels worldwide and a major contributor to air pollution (AP) through GHG, CO, SOx, NOx, VOCs (volatile organic compounds), and PM (particulate matter). The *air pollution indicator* quantifies the integrated impact of airborne pollutants from automobile exhaust gases other than GHG, which include CO, NOx, SOx, and VOCs. The effect of those pollutants is very diverse and has negative impacts on the environment, such as global warming and acid rains.

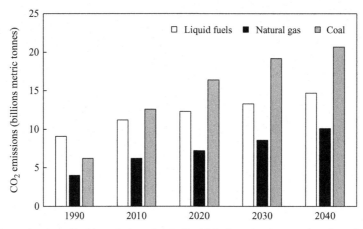

FIG. 2.4   Records and predictions of carbon dioxide emissions due to liquid fuels, natural gas and coal combustion.

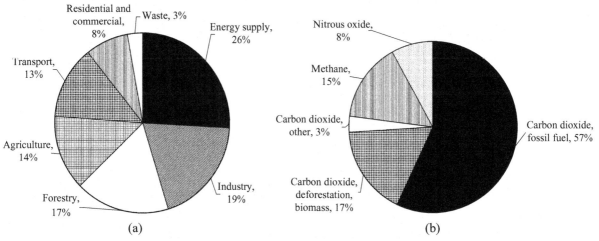

FIG. 2.5   Anthropogenic GHG emissions by sectors (a) and major GHG gases (b).

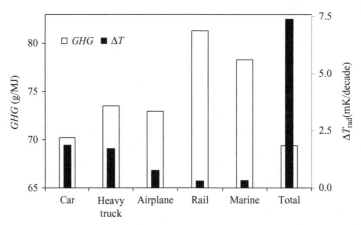

FIG. 2.6   Energy-specific GHG emissions of vehicles in Canadian and their contribution to earth's temperature increase per decade.

Fig. 2.6 illustrates the specific GHG emission of the main classes of vehicles in Canada, given in g CO_2 equivalent per MJ with respect to the lower heating value (LHV) of fuel. Based on the estimated GHG-specific emission and the number of vehicles in operation, the total GHG emissions can be determined. Further contribution of the Canadian transportation sector to the change in annual CO_2 concentration of earth's atmosphere can be calculated. As shown, these emissions lead to 7.5 mK in temperature increases per decade.

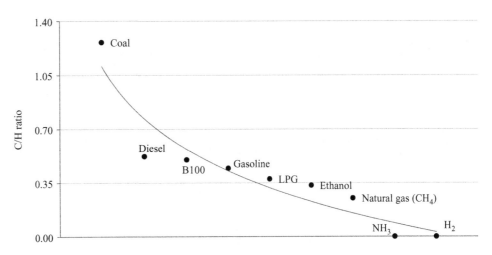

FIG. 2.7 Carbon-hydrogen ratio of common transportation fuels.

One of the most immediate factors to consider in comparing fuel alternatives is their potential reduction of high levels of GHG and other emissions hazardous to environment and health. The carbon-hydrogen ratio of various transportation fuels is shown in Fig. 2.7. Although biodiesel (B100) has a similar carbon content as petroleum diesel, this content is largely neutral in combustion depending on the feedstock. Compared to other fuels used in combustion applications, ammonia has the highest hydrogen energy density, higher even than pressurized and liquefied hydrogen fuel, based on current storage methods, and it contains no carbon, has a global warming potential of zero, and produces only nitrogen and water when combusted.

The plot in Fig. 2.7 shows a forecast toward the decarbonization of the world's energy supply, from one based mostly on carbon (molecular mass 12) to one that is based on hydrogen (molecular mass 2). This reflects a trend in the direction of "hydrogenation" of the world's energy supply, which is an evolution toward global use of hydrogen as an energy carrier.

In a hydrogen economy, sustainable energy sources like hydro, wind, solar, biomass, geothermal, ocean thermal, and nuclear are used to extract hydrogen from water (or other materials). Hydrogen is used as energy carrier (which by combustion with air reforms water) and as a chemical required in many crucial processes (manufacturing of plastics, foods, fertilizers, synthetic fuels, metallurgical procedures, etc.). Transportation can be made almost zero-polluting with hydrogen and hydrogen-based synthetic fuels, which are carbon-neutral.

The beneficial effect of hydrogen as a clean energy carrier can be observed from a comparative lifecycle assessment of conventional and hydrogen-fuel-cell vehicles. The lifecycle emissions from passenger vehicles of current technology are compared to those of fuel cell vehicles as shown in Fig. 2.8. The electric vehicle is characterized by AP during the utilization stage due to its battery recycling; however, if the GHGs of hydrogen production could be decreased, the fuel-cell vehicle has the best likelihood to have the lowest carbon footprint. This can definitely be achieved when hydrogen is produced exclusively from renewable sources.

FIG. 2.8 Lifecycle GHG and AP emissions of four categories of road vehicles, including the vehicle and fuel production and utilization phases. *Data from Dincer et al. (2010).*

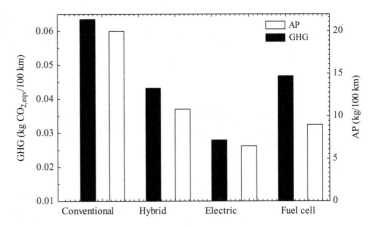

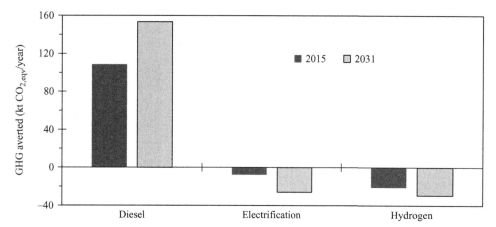

FIG. 2.9 Total equivalent GHGs averted by train transport. *Modified from Dincer, I., Hogerwaard, J., Zamfirescu, C., 2015. Clean Rail Transportation Options. Springer, New York.*

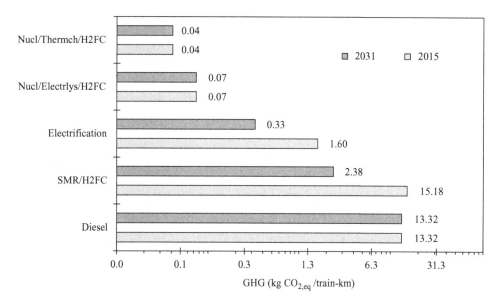

FIG. 2.10 Total GHG emissions from GO trains predicted to 2015 and 2031. *Modified from Dincer, I., Hogerwaard, J., Zamfirescu, C., 2015. Clean Rail Transportation Options. Springer, New York.*

Fig. 2.9 provides a comparison of averted emissions based on different power sources, against distance traveled by trains annually. On the basis of the source energy, whether nuclear, OPG (Ontario Power Generation) mix in Ontario, Canada, or hydrogen from steam methane reforming (SMR), the utilization of heat rejected from thermal power plants, combined with hydrogen generated through a thermochemical Cu-Cl cycle and proton exchange membrane fuel cell (PEMFC) trains results in the lowest level of GHG emissions, on the order of 0.10 kg/GJ. The higher efficiency of the PEMFC complements the performance of thermochemical hydrogen production, yielding the lowest GHG emissions at the prime mover, on the order of 0.04 kg/km (345 tonnes by 2015, and 488 tonnes by 2031). The actual impact of CO_2 emissions reduction is significant when compared to the diesel alternative. Further GHG emissions forecast from GO trains, depending on the propulsion option, are indicated in Fig. 2.10.

2.3 HYDROGEN AND SUSTAINABILITY

The energy crisis from the beginning of the 1970s influenced the consideration of a hydrogen economy as a potential global solution. In this view, hydrogen is produced using renewable energies or even fossil fuels, then hydrogen is stored and distributed at consumption points. Muradov and Veziroglu (2008) outline the potential green paths of hydrogen production from fossil fuels. A variety of uses of hydrogen is possible, namely direct use as fuel for fuel cells to generate power, use as fuel for transportation vehicles, production of a variety of synthetic fuels, and use as chemical feedstock for fertilizers, metallurgy, petrochemical fuels, other industries, etc.

A key event affecting the progress of the hydrogen economy concept was the THEME conference held in Miami, Florida in 1974. The International Association of Hydrogen Energy (IAHE) was established at that time, which became a world leader in promoting the hydrogen economy concept. Several other associations were formed since that time. One of the most known and prolific conferences, initiated and maintained by IAHE, is the World Hydrogen Energy Conference (WHEC), the first being held in Miami, 1976. By 2005, the Kyoto Protocol was signed, which represents an engagement of many countries to develop a cleaner energy system; hydrogen is an important part of this goal.

Hydrogen and fuel-cell systems are key components for sustainable power generation, sustainable transportation and sustainable economy. The compatibility of hydrogen with renewable energy systems is strictly required in the future for better energy security and economic growth facilitation. Hydrogen will satisfy a triple demand, namely that of power, heating and a large variety of chemicals and materials, all produced with high efficiency. Fuel cells are very efficient with better savings of primary energy. The variety of hydrogen and fuel cell technologies provides a flexible array of options for their use in various applications with reduced environmental impact and increased efficiency. Hydrogen favors the decentralization of power generation, thus providing more energy security. Even when fossil fuels are used to produce hydrogen, it is almost certain that hydrogen and fuel-cell systems are more effective than conventional systems, thus leading to reduced resource consumption. Therefore, hydrogen and fuel-cell systems are an important part of the overall solution for sustainable power generation at the global level.

Hydrogen is only a secondary energy source (a "storage carrier") and must be produced from a primary energy source. Because of physical reasons, there will always be losses from these conversion processes and, therefore, in any case, the costs of hydrogen must be higher than the costs of the energy used to produce hydrogen. This simple physical reason makes the decision on priorities and time scales for the introduction of hydrogen extremely complex. It is the same simple methodology leading to a higher CO_2-mitigation effect while using the input energy carrier for the production of hydrogen directly.

Some scientific and technical challenges for hydrogen economy may be given as follows:

- Lowering the cost of hydrogen production to a level comparable to the energy cost of petrol
- Development of a CO_2-free route for the mass production of sustainable hydrogen at a competitive cost
- Development of a safe and efficient national infrastructure for hydrogen delivery and distribution
- Development of viable hydrogen storage systems for both vehicular and stationary applications
- Dramatic reduction in costs and significant improvement in the durability of fuel-cell systems

The advantages of hydrogen versus fossil fuels can be listed as follows:

- Liquid hydrogen is the best transportation fuel when compared to liquid fuels such as gasoline, jet fuel and alcohols; and gaseous hydrogen in the best gaseous transportation fuel.
- While hydrogen can be converted to useful energy forms (thermal, mechanical, and electrical) at the user end through five different processes, fossil fuels can only be converted through one process, ie, flame combustion; therefore, hydrogen is the most versatile fuel.
- Hydrogen has the highest utilization efficiency when it comes to conversion to useful energy forms (thermal, mechanical, and electrical) at the user end. Overall, hydrogen is 39% more efficient than fossil fuels. In other words, hydrogen will save primary energy resources. It could also be termed as the most energy-conserving fuel.
- When fire hazards and toxicity are taken into account, hydrogen becomes the safest fuel.

Some examples of hydrogen uses in a projected hydrogen economy are given subsequently. The implementation of hydrogen and fuel cells to satisfy a significant portion of energy demands and for load-leveling at regional levels is illustrated schematically in Fig. 2.11. Two manifolds exist for this grid system. First, hydrogen is produced from primary energy sources through sustainable methods. If fossil fuels are used for hydrogen production, then ecological measures must be undertaken such as carbon capture and sequestration. Hydrogen is then stored at a large-scale facility and used on demand according to grid need through fuel-cell generators. The second manifold implies the use of excess power when the demand is low to store energy in the form of hydrogen. Again, when demand is high, hydrogen can be reconverted to power through fuel-cell power generators.

For an electrical grid, the hourly average electricity price (HOEP) changes in pace with the generation capacity reserve; the price is higher when the reserve is lower. In addition, the HOEP tends to increase during peak demand. Therefore, in order to reduce the HOEP, the capacity reserve must be increased. This can easily be done when energy storage is available, e.g., when a hydrogen and fuel-cell system is implemented.

As shown in Naterer et al. (2013), load leveling through hydrogen and fuel-cell systems can reduce the HOEP, especially when hydrogen production is done at large scale. The HOEP is directly influenced by the cost of the stored

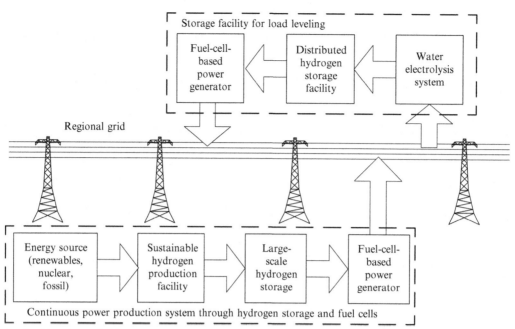

FIG. 2.11   Regional grid with coupled hydrogen and fuel-cell systems.

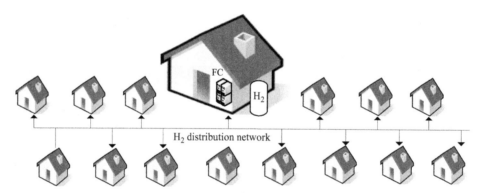

FIG. 2.12   Powering and heating residences through hydrogen and fuel cells.

hydrogen. As deduced from Naterer et al. (2013) (Fig. 2.12) if the hydrogen production increases from 10 tonnes/day to 100 tonnes/day, the HOEP is expected to decrease by 10%.

The residential sector can also benefit by the hydrogen and fuel-cell systems. In a residence, power can be generated locally using a fuel cell fed with hydrogen. In addition, hydrogen can be fed in a furnace and combusted to generate heat for water and space heating on demand. Hydrogen can be distributed either through the existent network for natural gas (with some modifications) or transported in cisterns and loaded in pressurized tanks in the residence. Fig. 2.12 illustrates the hydrogen and fuel-cell system for powering residences.

There are many advantages to using such a system, one of which is the important reduction of GHG emissions. Another advantage is the ability to generate local power with fuel cells, which will reduce the peak loads on regional grids. A third advantage is that delocalized energy harvested from renewable energy can be easily applied if hydrogen and fuel-cell systems are in place. For example, power can be generated locally at residences or quarters from biomass/ biofuels combustion, from wind energy or solar energy. The environmental benefit of hydrogen and fuel-cell systems in residences connected to a distribution network for hydrogen can be exemplified for the case of Canada, where in the vast majority, of cases the residences are heated with natural gas from distribution networks.

Deduced from Naterer et al. (2013) is that if Canadian residences are powered with hydrogen and fuel cells, then the reduction of GHG emission is more than 48%. Hydrogen generated from solar and wind energy is distributed through pipeline networks; the generation sites for hydrogen can be delocalized.

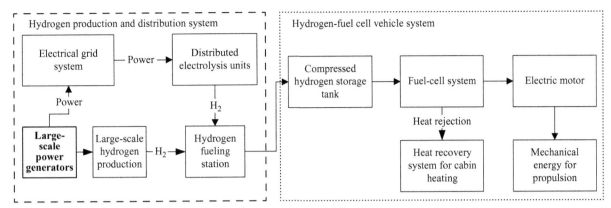

FIG. 2.13 Hydrogen and fuel-cell system for the transportation sector.

Once a hydrogen and fuel-cell system is put into practice for transportation, such global warming effects can be mitigated. In a hydrogen economy, the transportation sector can be represented as shown schematically in Fig. 2.13. Hydrogen is produced from two manifolds: large-scale concentrated generators and many small-scale distributed generators. The large-scale power generators are coupled with large-scale hydrogen production facilities. The distributed hydrogen generators consist of electrolysis units connected to the grid. Hydrogen thus produces is loaded on vehicles where it is converted to motive power and useful heating by fuel cells.

2.4 HYDROGEN PROPERTIES

Hydrogen is the simplest chemical element with the atomic number $Z = 1$, and it is the most abundant chemical element in the universe. As a consequence of its simplicity, hydrogen can easily lose valence electrons and therefore is very reactive. Because of this reason, hydrogen cannot be found as an individual element on earth, but rather is embedded within other materials.

Water is the most abundant resource of hydrogen on earth; also, hydrogen is part of most fossil fuels and biomass. In nature, hydrogen also can be found in the form of hydrogen sulfide (H_2S), which is abundant in some springs, geothermal sites and seas. Regarding its properties, hydrogen has a high calorific value, lowest molecular weight, highest thermal conductivity among all gases, and the lowest viscosity. Table 2.1 gives the main thermophysical properties of hydrogen. Table 2.2 gives a comparison of hydrogen's main properties compared to that of other conventional and alternative fuels.

TABLE 2.1 Thermophysical Properties of Hydrogen

Property	Description	Property	Description
Density	0.1 kg/m³ at 1 atm/298 K (gas)	Critical point	32. 97 K, 12.9 bar
	1.34 kg/m³ at 1 atm/20.3 K (vapor)	Normal boiling point	20.3 K (1 atm)
	70.79 kg/m³ at 1 atm and 20.3 K		99.8% parahydrogen + 0.02% orthohydrogen
	11.69 kg/m³ at 350 atm/298 K (gas)		
Composition	75% ortho/25% para at 298.15 K	Melting point	14.01 K (1 atm)
	Over 99.8% para in cryogenic H_2	Liquefaction	14.1 MJ/kg theoretical energy
LHV/HHV	119.9/141.8 MJ/kg	Diffusivity	In air, 0.61 cm/s
Exergy	118.05 MJ/kg (chemical)	Conductivity	0.177 W/mK
Autoignition temperature	850 K	Detonation	13–65% mix with air with 2 km/s, suppression of shock wave at 1.47 MPa
Flammability	4–18% and 59–75% mix with air	Flame	Flame speed 2.75 m/s

TABLE 2.2   Thermophysical and Combustion Properties of Fuel Options

Property	Gasoline	ULSD	CNG	LNG	B100	Hydrogen	Ammonia
Feedstock	Crude oil		NG reserves, biogas (SNG)		Soy, waste oil, fats	NG, H_2O electrolysis	NG
Molecule	C_8H_{18}	$C_{12}H_{23}$	CH_4	CH_4	$(C_{12}\text{–}C_{22})$	H_2	NH_3
Physical state	Liquid	Liquid	Comp. gas	Cryogenic liquid	Liquid	Comp. gas or liquid	Liquid
HHV (MJ/kg)	46.7	45.5	42.5	50	40.1	5.7	22.5
Storage density (kg/m³)	736	850	188	450	880	10 ($H_{2,g}$) 70.8 ($H_{2,\ell}$)	603
Storage pressure (bar)	1	1	20–25	~8	1	200–500	10
Cetane number	–	40–55	–	–	48–65	–	–
Octane number	84–93	–	120+	120+	–	130+	110–130
Flash point (°C)	−43	74	−184	−188	100–170	–	–
Autoignition temperature (°C)	257	225	540	540	150	565–585	651
Ignition limits (vol% in air)		0.6–5.5	5–15	5–15		4–75	16–25

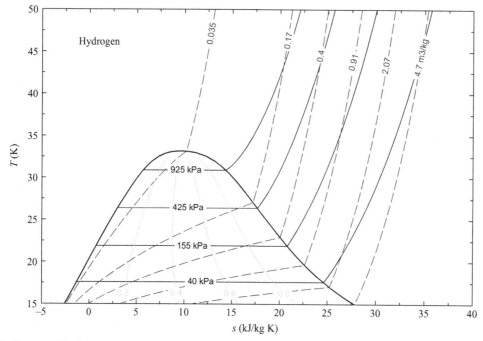

FIG. 2.14   The *T-s* diagram of hydrogen.

An important technical difficulty related to hydrogen storage is due to the existence of two isomers of molecular hydrogen. In orthohydrogen, the spin of the two protons of the hydrogen molecule are parallel. In the parahydrogen configuration, the spins are in opposite directions. At standard temperatures, the hydrogen gas composition is 25% ortho and the rest para. At cryogenic temperatures, there is practically no orthohydrogen (<0.2%).

The *T-s* diagram of hydrogen is shown in Fig. 2.14. Hydrogen has its normal boiling point at 20.3 K with the density of the saturated liquid of 70.77 kg/m³. The heating value of hydrogen is much higher than that of conventional fuels if taken as per unit of weight; however, hydrogen cannot be kept in a condensed phase with current technology. Mostly, hydrogen is stored in the form of compressed gas or as cryogenic liquid. In any of these storage conditions, the heating value of hydrogen per unit of volume is the lowest compared to conventional fuels. Nevertheless, the efficiency of power generation systems fueled with hydrogen is much higher than that obtained with conventional fuel, and this fact compensates for the low storage density problem of hydrogen. Furthermore, the hydrogen combustion is completely clean, producing only water vapor in the exhaust gas. Heat recovery can be applied intensively to hydrogen combustion to obtain liquid water, which may be recycled to produce back hydrogen via various methods for extracting hydrogen from water, including electrolysis and thermochemical splitting.

2.5 GREEN HYDROGEN SOURCES

In order to identify pathways for sustainable hydrogen production, one needs to inventory the natural resources of hydrogen, the available sources of energy that can be used to extract hydrogen from natural resources and the applicable methods of hydrogen production. These items are summarized in Fig. 2.15.

The hydrogen-containing natural resources are water, fossil hydrocarbons, biomass, hydrogen sulfide and anthropogenic wastes as indicated in Fig. 2.15a; municipal sewage waters containing urea, farming wastes like manure, crops residues, etc. (which are sources of biogas), other wastes that generate landfill gas, recycled plastic and cellulosic materials, etc., are all anthropogenic wastes from which hydrogen can be extracted.

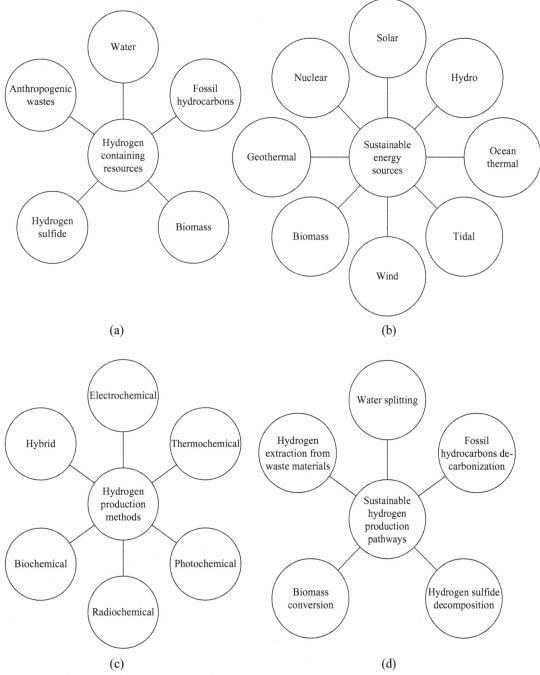

FIG. 2.15   (a) Natural hydrogen containing resources, (b) sustainable energy sources, (c) hydrogen production methods and (d) sustainable hydrogen production pathways for applications.

Although hydrogen is not viewed as a natural resource, but rather an energy carrier, here are some oil wells that yield a hydrogen-rich gas containing 60% hydrogen and 40% nitrogen plus traces of hydrocarbons. Other natural sources of hydrogen are water, fossil hydrocarbon, biomass, and hydrogen sulfide. The Black Sea contains a major natural resource of hydrogen sulfide, which occurs at a nearly constant concentration of 9.5 mg/L. In addition, there is an anthropogenic source of hydrogen, such as waste plastics, manure, landfill gas, biogas, urea containing wastes, and sewage sludge and other waste materials.

Petroleum cannot be neglected as a source of energy and hydrogen during the transition period toward hydrogen economy. In order to use fossil hydrocarbons in a cleaner way as energy supply, they must be converted to hydrogen with carbon dioxide sequestration. The sequestration of carbon dioxide results in cost penalties that can be balanced through a carbon taxation policy, which can be enforced by authorities. A hydrogen production system that cogenerates carbon black coproduct for better cost effectiveness was proposed by Muradov and Veziroglu (2008). In this vision, fossil hydrocarbons are used as a natural resource to extract hydrogen and carbon, where hydrogen is further used as a fuel or chemical in industry, and carbon is used mostly as a valuable material in the form of carbon black, carbon nanofibers, amorphous carbon, etc.

Sustainable energy is required to extract hydrogen from any resource in a clean, nonpolluting manner. Fig. 2.15b lists the main energy sources that can be considered sustainable: solar, hydro, ocean thermal, tidal (including ocean currents), wind, biomass, geothermal, and nuclear. The careful use of any of these sources can generate electricity and/or high-temperature heat and/or nuclear radiation with or without minor environmental impact. Such energy is used for hydrogen production via one of the methods, as listed in Fig. 2.15c. These methods are categorized in six classes, namely, electrochemical, thermochemical, photochemical, radiochemical, biochemical, and hybrid (where hybrid methods refer to integrated systems that use any kind of combination of the first five listed hydrogen production methods, e.g., electrophotochemical, photobiochemical, electrothermochemical, etc.).

Based on the above analysis, we identified five possible pathways to generate hydrogen in a sustainable manner. As indicated in Fig. 2.15d, these are: water splitting, fossil hydrocarbons decarbonization, hydrogen sulfide decomposition, biomass conversion to hydrogen and extraction of hydrogen from waste materials resulting from the anthropogenic activity. Each pathway corresponds to a natural resource (including the anthropogenic waste) from which hydrogen can be extracted. For any of the pathways, the use of a specific combination of sustainable energy (see Fig. 2.15b) and hydrogen production method (see Fig. 2.15c) is possible.

2.6 HYDROGEN PRODUCTION METHODS

The most common hydrogen source is water, from where hydrogen can be produced by a large variety of methods, according to the overall reaction: $2H_2O \rightarrow 2H_2 + O_2$. Fig. 2.16 depicts the reaction enthalpy ΔH for water-splitting reaction in function of the process temperature. The enthalpy of reaction is given according to $\Delta H = H_P - H_R$, where H is molar enthalpy and subscripts P and R represent products and reactants, respectively.

It is remarked from Fig. 2.16 that the total enthalpy of the reaction increases with the process temperature; however, when free Gibbs energy (ΔG) of the reaction is analyzed

$$\Delta G = \Delta H - T\Delta S \tag{2.1}$$

where $\Delta G = G_P - G_R$ and $\Delta S = S_P - S_R$ (the entropy of the reaction) and T is the process temperature, one observes in Fig. 2.16 a quasilinear decrease of ΔG.

The Gibbs energy is that form of organized energy that must be given to the reaction in as work, or electrical work, or work corresponding to electromagnetic or nuclear radiation, or exergy carried by a heat transfer process, etc. The entropic term, $T\Delta S$, is transmitted to the reaction in the form of heat at temperature T. There are some key cases to discuss:

- If water splitting is driven by electricity only (e.g., low-temperature water electrolysis), the entropic component of total enthalpy must be obtained by Joule effect heating (the electric current is dissipated as heat)
- If water splitting is driven by radiation, the entropic term is associated with all dissipative effects due to irreversibilities, which are converted in heat and transferred to the process; the entropic term can be determined based on the difference between the energy and exergy content of the radiation
- It water splitting is driven by heat only, the entropic term is transmitted to the process directly, while the generation of required Gibbs energy is obtained by heat to work conversion; this conversion is governed by the Carnot factor in a reversible process and depends on the reaction temperature

FIG. 2.16 Required energy to split one water molecule (reaction enthalpy ΔH, free Gibbs energy ΔG, total heat required Q_{tot}).

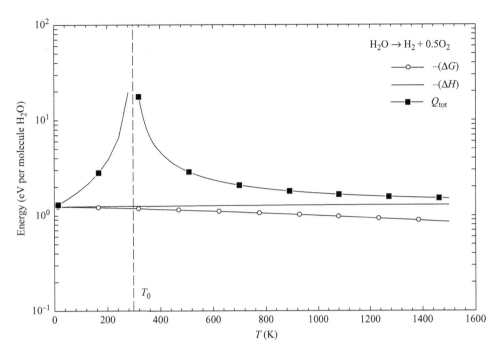

For the third case above, assume that the process of water splitting is driven by a heat Q_{tot} at the process temperature T. According to Eq. (2.1), the portion of the heat that must be converted in work form to supply the required Gibbs energy of the reaction is $(Q_{tot} - T\Delta S)$. Using the Carnot factor, one expresses the Gibbs energy as follows

$$\Delta G = (Q_{tot} - T\Delta S)(1 - T_0/T) \tag{2.2}$$

where T_0 is the reference temperature of the environment.

Eq. (2.2) can be solved for Q_{tot} for any given temperature T; the result is superimposed on the plot from Fig. 2.16. This shows that the process is favored either by very low temperatures or by very high temperatures, when the input source is of a thermal kind. In practical systems, a high temperature is generally obtained much more cost-effectively than a cryogenic temperature; therefore, it appears logical to conduct the water-splitting process at a high temperature. The challenge will remain in generating a high temperature in a nonpolluting manner. Hybridization is also a good, practical option, whereas the required energy is supplied partly electrically or by radiation and the remaining part as heat.

Fig. 2.17 shows energy conversion pathways for hydrogen production methods. The following four forms of energy are extracted from basic energy sources: thermal, electrical, biochemical, and photonic (or radiation).

As suggested in Fig. 2.17, two main energy conversion pathways do exist that lead to green or nongreen hydrogen production. Also, the hydrogen production methods can be classified based on production scale in three categories.

In Dincer and Zamfirescu (2012), the production scales are low ($\lesssim70\,kg/day$), medium (70–$70,000\,kg/day$) and large ($\gtrsim70,000\,kg/day$). Fig. 2.18 shows the classification of hydrogen production methods in function of the production capacity. In what follows, we will give some detail on main hydrogen production methods, namely: water electrolysis, thermochemical water splitting, gasification, hydrocarbon reforming, photocatalysis, and photobiochemical conversion.

Possible routes to generate hydrogen from water using nuclear energy shown in Fig. 2.19 with additional paths that assume processes other than water decomposition, e.g., coal gasification, natural gas reforming, petroleum naphtha reforming, or hydrogen sulfide cracking. In all such processes, high-temperature nuclear heat is used to conduct the required chemical reactions. Coal gasification and natural gas reforming methods include the water gas-shift reaction to generate additional hydrogen and reduce the carbon monoxide to CO_2. Nuclear-based reforming of fossil fuels is a promising method of high viability in the transition toward a fully implemented hydrogen economy, which will reduce the world dependence on fossil fuel energy.

Although fossil fuel reforming can be used to generate hydrogen, it is attractive to use nuclear heat to synthesize Fischer-Tropsch fuels or methanol. Such fuels can also be used for transportation vehicles.

In this respect, nuclear heat is used to generate synthesis gas, and then to generate either diesel or methanol. Another process of high interest is fertilizer production such as ammonia and urea with nuclear energy. A nuclear ammonia/urea production facility would be comprised of a nuclear reactor, a heat transfer system for high-temperature nuclear process heat, a nuclear power plant, a nuclear hydrogen generation unit, an air separation unit, H_2/CO_2 separators, a Haber-Bosch synthesis unit and a NH_3/CO_2 reactor to generate urea.

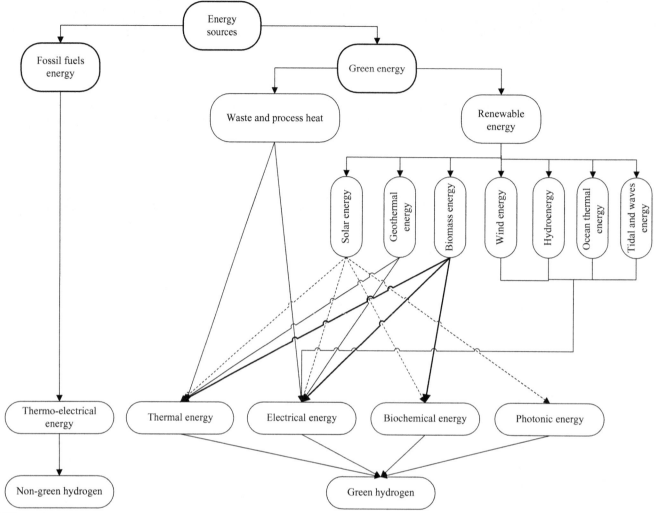

FIG. 2.17 Energy conversion paths for hydrogen production.

Nuclear energy is generated in fission reactors using uranium-based fuel (or thorium in the near future). The generated nuclear radiation is converted to high-temperature heat, which is transported by a heat transfer fluid for further use. The regular use of nuclear heat is power generation in large-scale power plants. It is possible to divert some of the high-temperature heat from nuclear reactors to supply chemical processes that eventually generate hydrogen from water splitting. Moreover, nuclear radiation present in nuclear reactors and during all phases of nuclear-fuel processing can be used directly to generate hydrogen from water.

Five general methods can be envisaged to generate hydrogen from nuclear energy through water decomposition: (i) radiolysis; (ii) electrolysis; (iii) high-temperature steam electrolysis; (iv) hybrid thermochemical water splitting; and (v) thermochemical water splitting. Method (i) uses nuclear radiation to directly split the molecule of water into hydrogen and oxygen; method (ii) uses electricity derived from nuclear energy to electrolyze water; methods (iii) and (iv) are called hybrid, because they use both electricity and high-temperature heat to split water; and method (v) directly uses high-temperature heat. All possible paths of water splitting using nuclear energy are represented in Fig. 2.20. As shown in the figure, nuclear energy is converted into nuclear radiation (either in the reactor, or during the fuel processing cycle). Nuclear radiation is converted further to high-temperature heat. Both the radiation and heat are used to generate hydrogen, as indicated.

The radiolytic (also known as chemonuclear) water-splitting method uses high-energy radiation or kinetic energy of fission products to excite water molecules and generate hydrogen and oxygen. Experiments of Carty et al. (1981) show that water can be split with fission fragments and steam can be split by alpha particle irradiation. An experiment of alpha irradiation of steam split six water molecules with 100 eV radiation, which corresponds to 15% efficiency; however, with current commercial reactors, a configuration for radiolytic water splitting is not practical because of

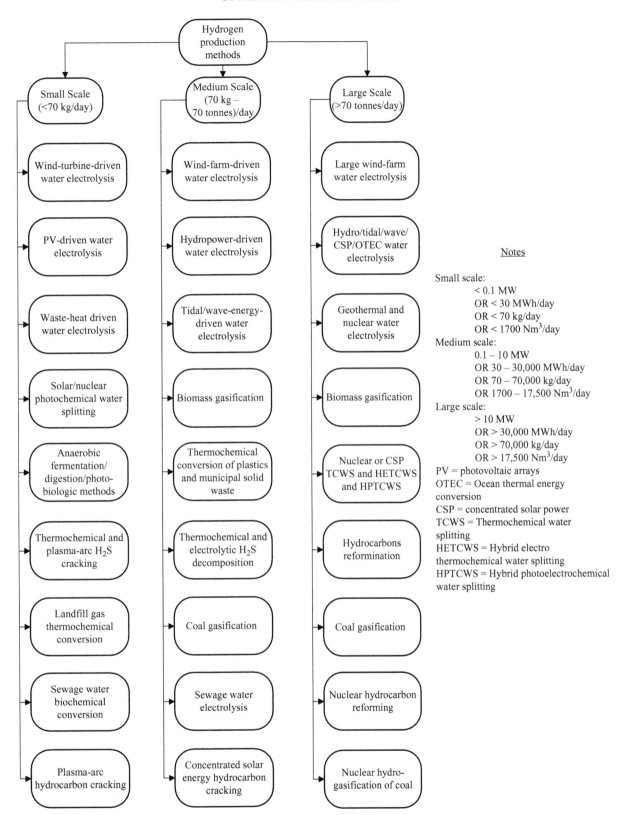

FIG. 2.18 Hydrogen production methods according to production scale.

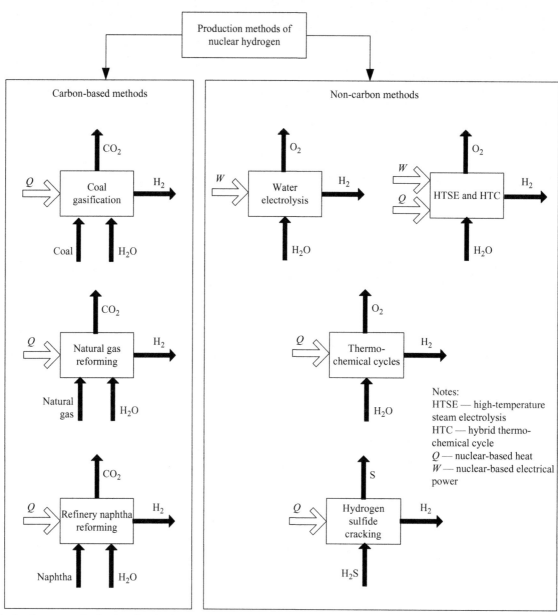

FIG. 2.19 Production methods of nuclear hydrogen that use heat and/or electricity inputs.

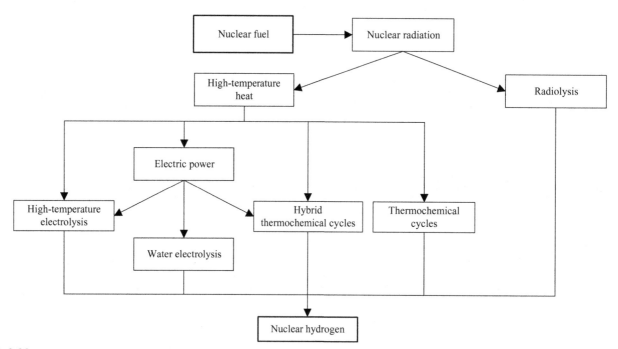

FIG. 2.20 Nuclear water-splitting pathways for hydrogen production.

containment requirements of the radioactive material, although radiolysis can be implemented at spent fuel pools of existing nuclear power plants.

Water-splitting pathways via sustainable methods are summarized in Table 2.3. The first option to consider for water splitting is water electrolysis. Various configurations are possible, depending on the kind of energy source used in conjunction with the capacity (size) of the hydrogen production facility.

TABLE 2.3 Water-Splitting Options for Sustainable Hydrogen Production

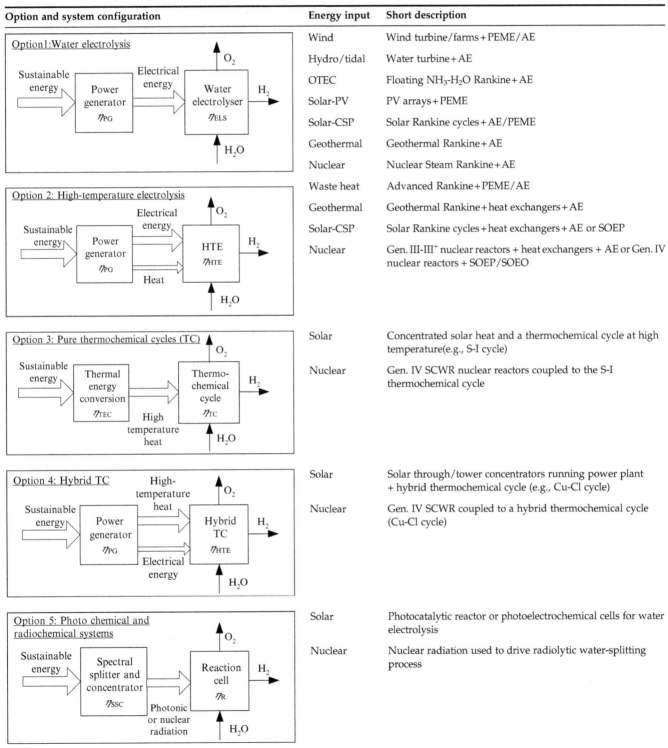

Option and system configuration	Energy input	Short description
Option 1: Water electrolysis	Wind	Wind turbine/farms + PEME/AE
	Hydro/tidal	Water turbine + AE
	OTEC	Floating NH_3-H_2O Rankine + AE
	Solar-PV	PV arrays + PEME
	Solar-CSP	Solar Rankine cycles + AE/PEME
	Geothermal	Geothermal Rankine + AE
	Nuclear	Nuclear Steam Rankine + AE
	Waste heat	Advanced Rankine + PEME/AE
Option 2: High-temperature electrolysis	Geothermal	Geothermal Rankine + heat exchangers + AE
	Solar-CSP	Solar Rankine cycles + heat exchangers + AE or SOEP
	Nuclear	Gen. III-III⁺ nuclear reactors + heat exchangers + AE or Gen. IV nuclear reactors + SOEP/SOEO
Option 3: Pure thermochemical cycles (TC)	Solar	Concentrated solar heat and a thermochemical cycle at high temperature(e.g., S-I cycle)
	Nuclear	Gen. IV SCWR nuclear reactors coupled to the S-I thermochemical cycle
Option 4: Hybrid TC	Solar	Solar through/tower concentrators running power plant + hybrid thermochemical cycle (e.g., Cu-Cl cycle)
	Nuclear	Gen. IV SCWR coupled to a hybrid thermochemical cycle (Cu-Cl cycle)
Option 5: Photo chemical and radiochemical systems	Solar	Photocatalytic reactor or photoelectrochemical cells for water electrolysis
	Nuclear	Nuclear radiation used to drive radiolytic water-splitting process

Notes: PEME = proton exchange membrane electrolyzer; AE = alkaline electrolyzer; PV = photovoltaic, SOEP = intermediate-temperature solid oxide electrolysis cell with proton conduction; SOEO = high-temperature solid oxide electrolysis cells with oxygen ion conduction; TC = thermochemical cycle.

As indicated in Table 2.3, proton exchange membrane electrolysis (PEME) cells are the logical choice for low-production capacity, whereas alkaline electrolyzers must be used in conjunction with medium production rates. For large production rates, groups of alkaline electrolyzer units can be employed. The renewable sources that can be selected for low-capacity hydrogen production are individual wind turbines of medium capacity, solar photovoltaic arrays and some applications involving waste heat recovery from various processes. Small-capacity heat engines, based on advanced Rankine cycles with organic fluids or other fluids such as ammonia-water or CO_2, can be used in conjunction with waste heat recovery systems to generate power (with a possible option to cogenerate low-grade heating for various applications). The generated power is further used to run a PEME and to produce hydrogen.

An interesting option is to use low-power solar dishes, which generate concentrated solar radiation to run heat engines with a power spectrum of the order of few kW. The power can be either directly used (e.g., in residences) or converted to hydrogen with the help of a PEME. Concentrated solar power at medium- to high-capacity range is possible with solar tower settings or solar through fields. In such cases, the facility should be coupled to an alkaline water electrolysis facility to generate hydrogen. Other sustainable energy sources available for "option 1" are wind farms, ocean thermal, geothermal, and nuclear. With wind farms, the generated power is in the medium-range spectrum of capacity, thus wind farms are suitable to be coupled with water electrolysis. OTEC systems comprise advanced power plants placed on floating platforms in warm seas (oceans). The platforms also include hydrogen production systems with alkaline water electrolyzers. Geothermal sites with enough energy content to run power plants with production in the range above 100 kW show economic promise; these facilities are suitable for hydrogen production with alkaline electrolyzers. In the vicinity of nuclear reactors, large electrolysis plants can be installed to generate hydrogen with limited electrical losses due to electricity transport. The logical choice is to use alkaline electrolyzer units in conjunction with nuclear power plants.

Table 2.3 also gives the options of water splitting via high-temperature water electrolysis. A variety of choices is possible in this case. The source of sustainable energy must provide both electricity and heat. One possibility is to use geothermal heat from a large-capacity geothermal heat generator that generates heat at 100–300°C; one part of the geothermal heat generator is used to generate power with a geothermal Rankine cycle-based plant; another part of the heat produced at the geothermal well is used to supply heat to an electrolysis cell. Alkaline electrolyzers can operate at temperatures up to 200 °C with heat supplied externally to reduce the electricity input requirements. Solar tower systems and solar through fields can be used to generate power and heat to drive (depending on the capacity and temperature level) either alkaline electrolyzers or intermediate-temperature solid oxide electrolysis cells with proton conduction (SOEP). Future nuclear reactors of generations III and III$^+$ can be designed with embedded facility to generate heat and power. For these generations of nuclear reactors, though, which will be available commercially in the next decade, the temperature level is limited to 300–350°C due to materials and safety constraints; they can be coupled with existent alkaline electrolyzers, which have the option to run at $\cong 200$°C with heat input for an improved efficiency. Future generations of reactors of IV generation can operate at high temperatures (500–1000°C); therefore, the use of solid oxide electrolysis cells with oxygen ion conduction (SOEO) or SOEP is applicable.

Pure thermochemical cycles run with sustainable high-temperature heat are indicated in Table 2.3 as "option 3." This cycle is closed chemical-processed that splits water into hydrogen and oxygen while recycling all intermediate chemicals. The sulfur-iodine (S-I) cycle is the most promising system for this hydrogen production option. The temperature level of high-temperature heat must be over 850°C. Only concentrated solar radiation and nuclear reactors of generation IV are capable to deliver this heat in a sustainable manner.

Hybrid thermochemical cycles that require heat at an intermediate temperature level ($\sim$500°C) and electricity to be driven are given as options in the same table. The promising candidate for this temperature level is the Cu-Cl thermochemical cycle. The input energy can be concentrated solar radiation from a solar tower or solar through system or nuclear heat from VI generation supercritical water-cooled reactors (SCWRs). The heat input is used in part for electricity generation and in part to supply the endothermic reaction within the thermochemical cycle.

Those methods to split water with high-energy radiation such as solar photonic radiation and nuclear radiation (e.g., gamma photons and alpha particles) are introduced in Table 2.3. The upper spectrum of solar radiation (above 450 nm, blue color) can be used directly for photocatalytic reactions, which evolve hydrogen from water. Some photoreactors are under development for this technology. In addition, photoelectrochemical cells were developed that are capable for a multiband capture of solar photons to split water in a hybrid manner; high-energy photons are directly used to split water photochemically, while low-energy photons are used to generate photovoltaic currents to split water electrochemically. Both methods are supposed work synergistically within a hybrid photoelectrochemical cell. Nuclear radiation from various fissionable materials can be used to split water through a radiochemical or radiocatalytic process. Spent fuel pools at nuclear power plants generate alpha particles and gamma photons, which can conveniently split water. In addition, nuclear reactors generate various fissionable materials from the range of actinides that can potentially be used with special radiochemical reactors to split water molecules and generate hydrogen.

Table 2.4 summarizes those options to generate hydrogen from other natural or anthropogenic resources besides water. Hydrogen can be extracted from biomass using thermochemical methods or methods based on biological processes such as anaerobic or aerobic digestion, depending on the feedstock. Water is used in the process to supplement the source of hydrogen. Solar energy can be used in conjunction with some algae and bacteria to generate hydrogen via photobiological processes.

TABLE 2.4 Hydrogen Extraction Options From Resources Other Than Water

Hydrogen resource and configuration	Energy input	Short description
	Biomass	Thermochemical conversion by gasification
		Aerobic fermentation
		Anaerobic digestion
	Solar + biomass	Photobiological processes (bacteria or algae process the biomass and generate hydrogen)
	Solar	Thermochemical $H_2S \rightarrow H_2 + S$
	Sustainable electricity[a]	Electrolytic of H_2S decomposition
		Plasma arc decomposition of H_2S
	Solar	Thermochemical conversion of plastics
	Biomass	Thermochemical conversion of biomass
		Biodigestion of municipal solid wastes
		Landfill gas generation
		Bacterial sewage waters treatment to generate hydrogen
	Sustainable electricity[a]	Sewage waters electrolysis
	Fossil fuels[b]	Coal gasification with CO_2 capture and sequestration
		Hydrocarbons reforming with CO_2 capture and sequestration
	Nuclear	Coal gasification with CO_2 sequestration
		Nuclear-assisted hydrocarbons cracking
	Solar	Solar-assisted hydrocarbon cracking
	Sustainable electricity[a]	Plasma arc cracking of hydrocarbons

[a]Electricity derived from all sustainable energy sources such as nuclear energy and renewable energies.
[b]Energy derived from fossil fuel combustion followed by CO_2 capture and sequestration.

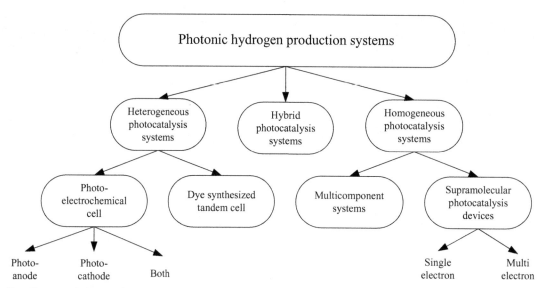

FIG. 2.21   Classification of photonic-based hydrogen production systems.

Pathways to extract hydrogen from hydrogen sulfide, which is a natural resource found at geothermal sites, volcanoes, and dissolved in some seas are also illustrated in Table 2.4. The basic methods are thermal cracking using high-temperature heat, plasma arc decomposition and electrolysis; all methods lead to generation of hydrogen with sulfur as a by-product.

Hydrogen can be extracted from anthropogenic wastes as indicated in sustainable options given in Table 2.4. Plastic material wastes resulting from human activity are an important source of hydrocarbons from which hydrogen can be reformed using thermochemical methods. Nuclear heat can be used to gasify coal and sequester carbon dioxide or to reform fossil hydrocarbons (some nuclear research programs worldwide consider this alternative as a transition period toward a hydrogen economy). Biogas and landfill gas are other examples of hydrogen-rich gas resulting from farming or municipal wastes. Sewage waters are an important source of hydrogen in towns and urban agglomerations. The sewage water contains important quantities of urea and other hydrogen-rich compounds. Hydrogen can be extracted from sewage water by microbial methods or electrochemical methods.

Conventional (coal, petroleum, natural gas) or nonconventional (oil shale, oil sands) fossil fuels are a source of hydrogen that cannot be neglected. Sustainable methods to extract hydrogen from those sources are given in the same table. The hydrogen content of some fossil hydrocarbon is poor (e.g., coal, oil sands) such that the extraction process may require water addition as a supplementary source of hydrogen. In order to make the processes sustainable, clean energy sources such as nuclear and renewables must be used to generate high-temperature heat and electricity. One class of processes involves hydrocarbon cracking to generate hydrogen and carbon black as a by-product. Another process is nuclear-assisted coal gasification with carbon dioxide sequestration and generation IV of nuclear reactors. In addition, autothermal processes can be applied in a sustainable manner, that is, with CO_2 capture and sequestration, to generate hydrogen, e.g., coal gasification, natural gas or petroleum reforming.

In order to use solar photons directly for water splitting, a photosensitizer capable of absorbing solar radiation must be dissolved in solution, because water itself is transparent to the visible spectrum. Photonic-based hydrogen production systems are categorized in Fig. 2.21. The following systems for water splitting with solar light are analyzed here, as they are found to be the most relevant processes in this category (photoelectrochemical): (i) photoelectrochemical cell; (ii) dye-sensitized electrolysis cell; and (iii) homogenous photocatalysis systems with supramolecular devices.

2.7 HYDROGEN STORAGE AND DISTRIBUTION

Before the realization of a hydrogen economy, there are challenges that require innovative solutions. The first, and most significant, is development of efficient storage methods; hydrogen is the lightest element and requires either large volumes, high pressure, low temperature, or advanced material storage techniques to hold sufficient fuel for practical operating range. The characteristics of the main storage options are given in Table 2.5.

Storage of hydrogen is very challenging because in normal atmospheric conditions, it is a gas with very low density, $40.8 \, g/m^3$. Comparison of the hydrogen energy densities of conventional and other alternative fuels is shown in

TABLE 2.5 Comparison of Hydrogen Storage Methods

Storage method	Energy intensity (MJ/kg-H₂)	wt%-H₂/tank	wt%-H₂/kg-system	g-H₂/tank	g-H₂/L-system
Compressed H₂ (35 MPa)	10.2	6	4–5	20	15
Liquid H₂	28–45	20	15	63	52
Low-temperature hydrides ($T < 100°C$)	10–12	2	1.8	105	70
High-temperature hydrides ($T > 300°C$)	20–25	7	5.5	90	55

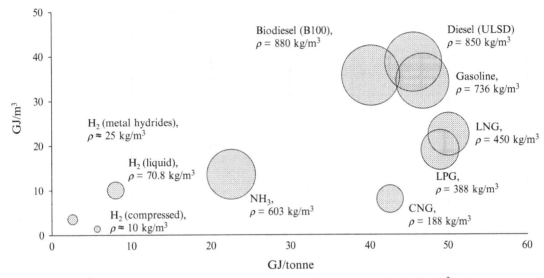

FIG. 2.22 Volumetric (GJ/m³) and gravimetric (GJ/tonne) energy density and mass storage density (kg/m³) of transportation fuels.

Fig. 2.22. Three methods of hydrogen storage are indicated, namely: compressed gas; cryogenic liquid; and hydrogen storage in metal hydrides.

As remarked, the energy density carried by hydrogen in all its physical forms of storage is essentially lower than that of other fuels. Based on higher heating values, 324 g of hydrogen have the same content of 1 kg of gasoline. The corresponding volume of 1 kg of gasoline is about 1.3 L; if 324 g of hydrogen is stored as gas under standard temperature and pressure conditions, the occupied volume is 3932 L. At normal boiling point (20.3 K), under atmospheric pressure, the density of the cryogenic liquid hydrogen is 70.77 kg/m³; the volume of 324 g of liquid hydrogen is 4.6 L, which is 3.5 times larger than that of gasoline with the same energy content. Under standard temperature (298.15 K) and 400 bar pressure, the hydrogen gas density is 25.98 kg/m³ and the gas volume becomes 12.5 L or 9.6 times more than gasoline volume with same energy content.

For compressed storage, the current technology target is to develop commercial systems capable of storing hydrogen at 700 bar, at which 5 kg of hydrogen occupy 125 L. Such systems were built and tested for vehicles within some proof-of-concept research programs, but the common storage pressure as of today is 300 bar. The adiabatic compression work can be easily calculated under the reasonable assumption of ideal gas behavior of hydrogen, according to

$$W|_{s=ct} = \frac{\gamma}{\gamma - 1} RT_1 \left[\left(\frac{P_2}{P_1} \right)^{\frac{\gamma}{\gamma - 1}} - 1 \right] \tag{2.3}$$

where $\gamma = C_p / C_v$.

The isentropic efficiency of the compression process depends on a compression ratio; typical values from practice are 70–80% for small compression ratios of the order of 3–4% and 50–60% for larger compression ratios as of 100. For compressing hydrogen at 350 bar, the expected energy consumed is around 10 MJ/kg.

Liquefaction of hydrogen implies its cooling to cryogenic temperature. This is a well documented process and there are mature technologies in industry for obtaining cryogenic hydrogen. The following are the main industrial processes to obtain liquefied hydrogen: (i) the Precooled Linde-Hampson process; (ii) the Claude process; and (iii) the helium-hydrogen cycle.

Some peculiar problems occur in the liquefaction process, namely: (i) hydrogen has a positive Joule-Thompson coefficient at a temperature higher than 190 K; (ii) there is an orthopara hydrogen conversion process that happens at 20 K and adds to the latent heat of 450 kJ/kg an amount of 530 kJ/kg; (iii) the latent heat of boiling of parahydrogen at 20 K is relatively very low; therefore, any heat small penetration can generate important amounts of hydrogen vapor.

At cryogenic conditions, all hydrogen converts in a para configuration. This process is slow, and for an acceptable rate of hydrogen production, it must be accelerated through catalysts. Catalysts (iron-oxide) are normally integrated in the heat exchanger that cools the hydrogen.

Liquefaction of hydrogen requires expensive materials (e.g., chrome-based metals), which are suitable for very low temperatures. All the equipment must be placed in a double-wall vacuum box, very well thermally insulated. In the first stage, liquid nitrogen is used to cool hydrogen down to 80 K. For further cooling, hydrogen itself is used as working fluid in a two-stage Brayton cycle cryogenic refrigerator, according to the Linde process. Due to the auxiliary systems and irreversibility, the energy required to liquefy hydrogen is ~50 MJ/kg, which represents about a third of the higher heating value.

The Linde process recycles the hydrogen evaporated from the cryogenic tank, which is drawn as shown in a simplified way in Fig. 2.23. After compression to about 20 bar, hydrogen is cooled down to 80 K. This cooling (not shown in the figure) is done in three stages: cooling with water to 300 K; cooling with a vapor compression refrigeration plant down to ~250 K; and cooling with liquid nitrogen down to 80 K. With further cooling, within the cryogenic cooler the main stream temperature is decreased to 21 K; the throttling process generates colder two-phase liquid-vapor hydrogen introduced in the cryogenic hydrogen tank.

Above the liquid level, hydrogen vapors are always generated because of heat penetrations. They are extracted from the tank at 20 K and 1 bar and used for cooling the main stream of hydrogen down to 21 K; further, the vapors are compressed in two stages with intercooling, first stage from 1 bar to about 8 bar, the second stage, from 8 to 50 bar. A part of the compressed hydrogen is expanded to intermediate pressure (8 bar) and reaches 50 K.

The cold stream at 50 K is used to cool the remaining stream to the second stage of expansion; before the second stage of expansion, the flow is divided into two streams. Note that the second stage of expansion occurs also from 50 to 8 bar, with the difference being that the flow is much colder prior to expansion. Consequently, the temperature at the end of the second stage of expansion reaches around 30 K. The remaining stream is cooled further, a little above 30 K prior to the final expansion process that occurs in a throttling valve, after which the temperature of the hydrogen stream reaches 30 K.

In Fig. 2.24, a modified Claude cycle, which includes two scroll expanders for work recovery, is shown. The first expander is positioned between states 3 and 4 in the diagram, and the second expander replaces the typical

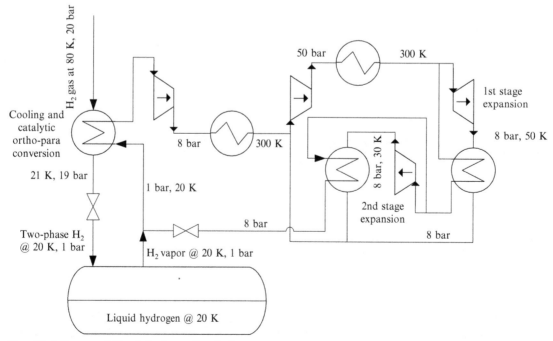

FIG. 2.23 Simplified diagram of Claude hydrogen liquefaction process.

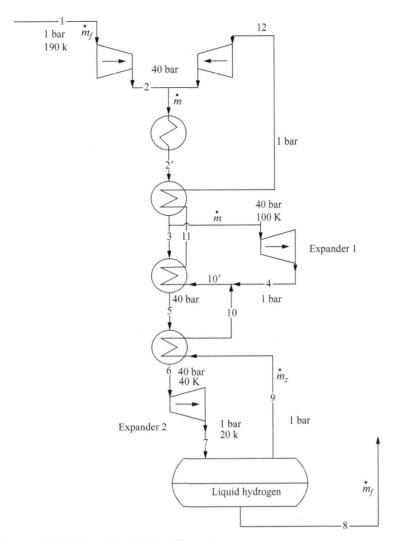

FIG. 2.24 Modified Claude process for hydrogen liquefaction with work recovery.

Joule-Thompson valve of the basic Claude. In this way, the expansion processes can recover maximum work, but has more chances to face two-phase expansion or, in extreme cases, wet expansion in the second expander.

The process starts with precooled hydrogen at 190 K (1 bar), which is obtained using the cooling effect of liquid nitrogen. As it results from the schematics, hydrogen is compressed to 40 bar in #2 and then cooled in subsequent processes $3 \rightarrow 5 \rightarrow 6 \rightarrow 7$ until it reaches 1 bar and 20 K and is in the two-phase region. The cryogenic liquid separates gravitationally (8), while the cold vapors are extracted and used to cool the liquid in processes $9 \rightarrow 10$.

Some vapors are extracted from high-pressure stream 3 at 100 K and expanded to 1 bar to generate cooling effect in 4 and power. The cold stream resulted from mixing 4 and 10 is further used for a cooling effect in process $10' \rightarrow 11 \rightarrow 12$. The vapors are recycled by recompressing and mixing with makeup stream 2. The temperature in state 7 must be around 20.24 K, while the temperature at state 3 is 100 K and the compression/expansion ratio is ~ 40.

Cryogenic hydrogen is normally stored in double-wall stainless steel vessels with vacuum between the walls, referred to as Dewar. The typical evaporation losses due to heat penetration with current storage tank technology varies between 0.3% and 5% per day. The energy associated with the liquefaction process is estimated to the average of 50 MJ/kg of hydrogen, which adds substantially to the hydrogen cost. It is therefore important to reduce the GHG emissions associated with hydrogen liquefaction.

Hydrogen storage on a large scale is difficult in hydrogen's pure form. Alternatively, hydrogen can be easily converted into ammonia or methanol using green processes, and the storage density is substantially increased. Furthermore, hydrogen can be converted in synthetic diesel to maximize the energy density and to obtain a reliable transportation fuel, provided that the manufacturing path is sufficiently "green." Before storage through any of these methods, hydrogen must be purified at an acceptable degree; for compressed storage, 4 ppm impurities are acceptable, while for cryogenic storage, 1 ppm is recommended.

Some chemical elements like sodium, lithium, magnesium, and boron can create hydrides (called "chemical hydrides") that store hydrogen more densely than it is possible in metal hydrides via physisorption and chemisorption. There are two processes that can be applied for storing hydrogen in a solid matrix: physisorption (absorption of molecular hydrogen by a solid structure) and chemisorption (dissociation of the hydrogen molecule and bonding of the protons with the metal lattice). Many metal-hydride-based methods were developed up to date for both physisorption and chemisorption of hydrogen. In general, they require some thermal energy at the discharging phase, and some cooling should be applied to enhance the efficiency of the hydrogen charging process. The ammonia-boron compound with chemical formula NH_4BH_4 appears to be the most "dense" hydrogen storage.

Hydrogen can be stored chemically in the form of ammonia, which can be decomposed thermocatalytically to release hydrogen according to the reaction $NH_3 \rightarrow 1.5H_2 + 0.5N_2$. One interesting possibility of storing hydrogen in solid form is by first producing ammonia from it, and then absorbing the ammonia-gas in metal amines. About six molecules of ammonia can be bounded weakly by magnesium chloride to form $Mg(NH_3)_6Cl_2$, which embeds hydrogen of 109 g/L and 92 g/kg of magnesium amine. Nevertheless, metal amines are an attractive solution for hydrogen storage, but one has to account for the diminishing of the calorific value due to hydrogen extraction, or alternatively, this method of storage should be applied where some combustion heat can be recovered to drive the ammonia-cracking reaction itself.

Storing hydrogen chemically in the form of urea is also an attractive solution, as urea is very stable, can be stored for very long terms and can be safely transported. Urea is a particulate material delivered in small-sized prills. It can be combusted having a low calorific value comparable to that of wood, that is, of about 10 MJ/kg. Urea can be synthesized from biomass or other renewable energy sources such that it results in a CO_2-free fuel. Urea, with chemical formula $CO(NH_2)_2$, is massively produced in industry and used as fertilizer. It is considered a nontoxic substance, and several car manufacturers use it on their passenger vehicles for NOx exhaust reduction.

Depending on the production method, the hydrogen costs vary from \$1/kg at coal gasification to 9.50 \$/kg using solar energy for electricity generation that in turn is used for water electrolysis. After production, hydrogen is stored at the manufacturer location for a certain period prior to delivery. Liquefaction adds at least 30% to the hydrogen price per kg, and on top of this, one must add the energy consumed to keep the storage tank at cryogenic temperatures during the storage time. The minimum cost penalty for hydrogen storage (when hydrogen is stored for 1–3 days) is CN\$0.3/kg for compressed H_2 and CN\$0.7/kg for liquefied H_2.

Hydrogen distribution to the consumption points is also expensive. If one assumes, for example, that the hydrogen transport is made in pressurized containers at 345 bars, the transported energy content is 8 GJ/m^3, ie, four times smaller than for gasoline case (32 GJ/m^3). The high explosion risk of hydrogen will raise the price even more because of the safety measures. Due to these factors, the estimated minimum cost of hydrogen regional distribution is ~\$1/kg H_2. Thus, if one considers the production, storage and distribution costs, the minimum expected hydrogen price at delivery point should be ~CN\$2.5/kg, if produced from coal, and respectively ~CN\$11/kg from solar energy-driven water electrolysis.

2.8 FUEL CELLS

A fuel cell performs the opposite process of an electrolysis cell. Water is formed from hydrogen and oxygen gases and electricity is generated. Fuel cells are generally constructed in a stack arrangement with planar electrodes and hydrogen fed at the anode where it loses electrons. An external circuit is attached to the anode to transfer electrons to the cathode, thus transferring power. The oxygen is fed at the cathode side where it receives electrons. Water is then formed from protons and oxygen ions either at the anode or the cathode side, depending on the fuel cell type.

There are eight main types of fuel cells as illustrated in Fig. 2.25. Most of the fuel cell types are fed with hydrogen fuel; there is also direct ammonia and direct methanol fuel cells. In addition, methane (natural gas) through reforming can be used as fuel in solid oxide fuel cells.

The applications of fuel-cell systems are mostly found in power generation and propulsion for transportation vehicles. In special applications, fuel cells can be used to generate pure water or pure oxygen. Fig. 2.26 illustrates a classification of fuel-cell applications. In order to generate extra pure oxygen, a larger capacity fuel cell is connected to a smaller capacity electrolyzer, which is fed with pure water produced by the fuel cell. The electrolyzer generates pure oxygen for use and hydrogen, which is fed back to the fuel cell.

The most important fuel-cell application is for stationary power generation at intermediate scale, either connected to or disconnected from the regional grid. Solid oxide and molten carbonate fuel cells are suitable for cogeneration of

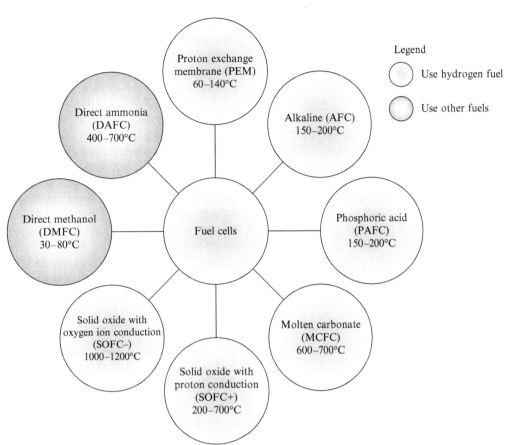

FIG. 2.25 Types of fuel cells and their operating temperature ranges.

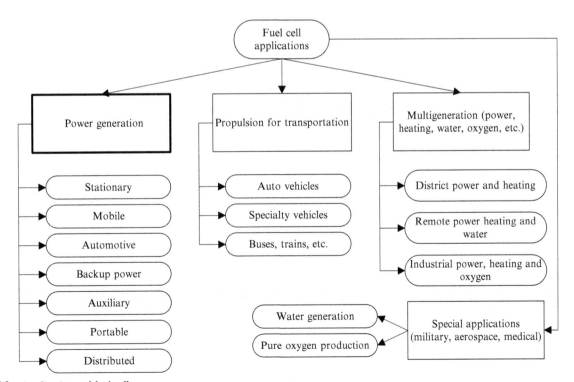

FIG. 2.26 Applications of fuel cells.

high-temperature heat and power, thus they can be coupled with processes that require high-grade heat. One of the processes that can be benefited by high-grade heat is hydrocarbon fuel reforming. Due to this fact, SOFC and MCFC can be fueled with natural gas or other fuels without the need of external reformers. Direct methanol and PEMFC can be used to cogenerate power and heating for residences (hot water and space heating).

The startup time is important in some applications. In vehicle propulsion and small-power applications such as residences and portable devices, PEM, PAFC, alkaline fuel cells (AFC), and DMFC are the best choices because they require very small startup time. The alkaline fuel cell is the preferred technology for space applications and is very attractive for road vehicles, although it requires a complicated system of electrolyte recirculation. For road vehicles, DMFC represents a very reliable choice, because it does not require any gas cleaning units, is simpler to implement and requires a fuel tank of the same size as gasoline vehicles. Nevertheless, PEMFC is the most developed for vehicular application, although it requires a compromise in driving range, since the fuel must be hydrogen. On the contrary, higher-temperature fuel cells such as SOFC, MCFC, DAFC are better suited to larger-power stationary applications. In addition, DAFC appears to be a good choice for large vehicles that have long running times in steady modes, such that railway locomotives, heavy-truck road transporters, and ships. Also, SOFC can be implemented on vehicles as an auxiliary power unit, which is in operation at steady running.

One of the most attractive fuel-cell applications today is for powering laptops, mobile cells and other small electronic devices. Here, PEM and DMFC play a major role due to their reduced operating temperatures.

2.8.1 Proton Exchange Membrane Fuel Cells

The simplified representation of a PEMFC is shown in Fig. 2.27. The particular aspect to this type of fuel cell is represented by the solid polymer electrolyte, usually denoted as a proton-conducting membrane. The electrolyte is very thin and allows for proton conduction through hydronium ions when the membrane is wetted. The half reactions at the fuel cell electrodes are as follows:

$$\begin{cases} \text{at anode}: H_2(g) \rightarrow 2H^+(aq) + 2e^- \\ \text{at cathode}: 0.5O_2(g) + 2H^+ + 2e^- \rightarrow H_2O(l) \end{cases} \tag{2.4}$$

For electrolyte, a polystyrene sulfonate polymer (PSP) can be used; however, the proprietary membrane from DuPont—Nafion®, which consists of a polytetrafluoroethylene (PTFE)—is proven to be more stable and has better conductivity than PSP membranes. Another proprietary membrane that is proven to be very effective is the perfluorocarbonic sulfonic acid Dow® from Dow Chemicals.

Since the protons are transported in hydronium form (H_3O^+), these acidic membranes must be always hydrated in order to operate. Therefore, water management in PEMFC is a major design issue. In order to assure good performance, with reduced ohmic losses, the PEM is constructed in a sandwich-like architecture with porous planar electrodes forming a so-called membrane-electrode assembly (MEA). Noble metal catalysts must be coated to the electrodes. The electrodes must be designed such that a three-phase boundary is formed between the solid catalytic center, the aqueous ionic species and the gaseous reactants. The catalytic center must be in direct contact with the electrode to ensure that electrons are supplied to or from the reaction site. Platinum is the preferred catalyst for the

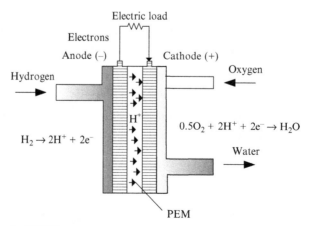

FIG. 2.27   Simplified representation of a PEMFC.

cathode, which is required to compensate for the slow kinetics. Also, for the anodic reaction, a platinum catalyst is required when pure hydrogen is used as fuel; however, if CO impurities are present in hydrogen, the platinum catalyst is immediately poisoned and becomes inactive. Due to the poisoning, complicated fuel purification systems may be required for PEMFC, depending on the case.

2.8.2 Phosphoric Acid Fuel Cells

The PAFC runs the same electrode half-reactions as PEMFC. Protons are the charge carriers through an acidic electrolyte, and water is formed at the cathode. The difference between PAFC and PEMFC is due to the electrolyte. In PEMFC, the electrolyte is a solid polymer, whereas for PAFC, a liquid acidic electrolyte is used. The electrolyte is almost pure pressurized phosphoric acid ($\sim$100% concentration), which has a very low volatility at the specific temperatures of operation of $\sim$175°C (the normal boiling point of the acid is 158°C). PAFCs are most developed at the commercial level for megawatt range (up to 25 MW); it is well tolerant to CO_2 presence in the air stream.

The liquid electrolyte is stabilized in a solid matrix of silicon carbide (SiC). The arrangement is similar to that of PEMFC, shown in Fig. 2.28. Due to the stabilization in the SiC, the loss of electrolytes due to evaporation are very much minimized such that no acid must be replenished during the system lifetime. The SiC matrix is comprised of particles of micrometer size that are able to assure reduced ohmic losses. The electrodes are of planar geometry made of polytetrafluoroethylene with bonded platinum-black of 46 μmol/cm^2 load.

2.8.3 Solid Oxide Fuel Cells

An emerging technology is that of solid oxide fuel cells with proton conduction. This is an acidic electrolyte, therefore the half reactions are as given in Eq. (2.4); however, the electrolyte is a proton-conducting metal oxide layer placed in between two porous, planar electrodes.

Fig. 2.29 shows a simplified representation of a proton-conducting solid oxide fuel cell. The SOFC+ (solid oxide fuel cell with proton conduction) normally use a barium oxide. The proton-conducting solid oxide membranes provide an important advantage of letting the protons migrate from anode to cathode. As a consequence, the water formation reaction occurs at the cathode. Complete hydrogen utilization is therefore possible (in principle) in SOFC+ with direct implication in increasing the system's simplicity and compactness by eliminating the need for the afterburner. Moreover, because all the hydrogen is reacted electrochemically at the fuel-cell cathode, practically no NOx is formed, thus the fuel-cell emission consists only of steam and nitrogen, ie, it is clean. At present, certain efforts are devoted worldwide to developing proton-conducting membranes. In this respect, barium cerate ($BaCeO_3$)-based materials were identified as excellent solid oxide electrolytes because of their high proton-conducting capability over a wide range of temperatures (300–1000°C).

The main problem with barium cerate results from the difficulty to sinter it in the form of a solid membrane. One option appears to be doping the barium cerate with Samarium (Sm), which allows for sintering thin membranes featuring a thickness as low as 50 μm and high-power densities in the range of 1300–3400 W/m^2.

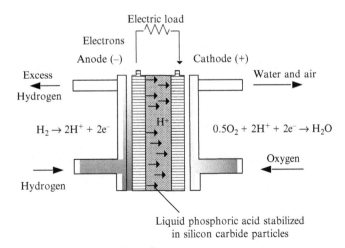

FIG. 2.28   Simplified representation of a phosphoric acid fuel cell.

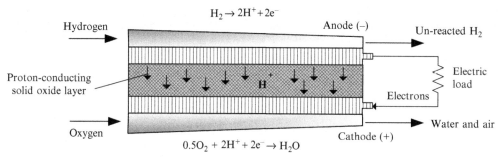

FIG. 2.29 Simplified representation of a proton-conducting solid oxide fuel cell.

Fig. 2.30 illustrates the common configuration of a SOFC – in bipolar-plate stack arrangement. SOFC – are also an established technology. Their electrolyte is a solid layer of yittria-stabilized zirconia, which operates at a high temperature (1000°C), a fact that leads to some important advantages for the SOFC –, namely:

- They are cheap and present a long lifetime, because no noble metal catalysts are needed for the electrodes
- Internal reforming of alternative fuels (e.g., methane, syngas, methanol, ammonia) to hydrogen is facilitated so they can use a reduced size fuel tank
- The exhaust gases have high exergy, which can be converted into additional power and low-temperature heating

The fact that there is no liquid present is an advantage for the design, since only two-phase solid-gas processes occur. During the operation, at the anode, the hydrogen is consumed and steam generated. Because of this fact, the hydrogen's partial pressure decreases. As a consequence of the low partial pressure of hydrogen, the reaction kinetics is degraded and the only solution to compensate for this effect is to supply hydrogen in excess. The excess hydrogen must then be consumed. This can be done in multiple ways. One common method is to combust the excess hydrogen in an afterburner and then release heat recovered or converted into work by a gas turbine. Thus, certain amounts of NOx are formed during the combustion of hydrogen with air.

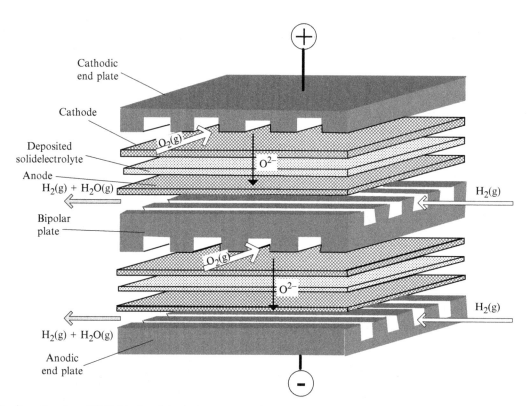

FIG. 2.30 Configuration for a SOFC- bipolar-plate stack arrangement.

2.8.4 Alkaline Fuel Cells

AFCs have slightly superior efficiency than the PEMFC with over 60% (potentially, they can reach 70%), taking advantage of the hydroxyl as an excellent charge carrier with faster kinetics for oxygen reduction reaction. These systems were used in NASA space programs with great success as they demonstrate an excellent stability. The catalyst requirements for AFC are less stringent than that for PEMFC; in principle, cheap nickel-based catalysts can be used. The electrolyte is a solution of potassium hydroxide (KOH, 30–40% by weight). The best configuration of AFC is with a circulating liquid electrolyte, although options exist to embed the electrolyte in a matrix. The half reactions at the electrodes are as follows:

$$\begin{cases} \text{at anode}: H_2(g) + 2OH^-(aq) \rightarrow 2H_2O(l) + 2e^- \\ \text{at cathode}: 0.5O_2(g) + H_2O(l) + 2e^- \rightarrow 2OH^-(aq) \end{cases} \tag{2.5}$$

One of the major problems with the KOH electrolyte is due to its spontaneous reaction with carbon dioxide from air. Air fed at the anode carries some amount of carbon dioxide; also, hydrogen fuel can be altered with CO_2. If carbon dioxide is present in the electrolyte, it reacts with hydroxyls and forms carbonate in an aqueous solution (CO_3^{2-}).

This fact degrades severely the electrolyte. Solutions exist to absorb CO_2, including that of swing absorption. Fig. 2.31 shows the system configuration for an AFC system. Hydrogen penetrates through the porous anode and water is generated at the electrode-electrolyte interface. The generated water is removed from the system with the electrolyte. The electrolyte is reconditioned in an external unit by eliminating the water; if a fresh electrolyte is needed, it may be added. Thereafter, the electrolyte is returned to the cell; the fresh electrolyte comes in contact with the cathode, where hydroxyls are generated.

2.8.5 Molten Carbonate Fuel Cells

MCFCs have efficiency higher than 60% while operating at high temperatures. Moreover, the MCFC can be used in cogeneration systems, in which case the fuel utilization efficiency may exceed 85%.

The electrodes do not need to be coated with expensive catalysts; simple carbon based, un-expensive electrodes operate very well at the high temperature specific to these cells. The electrolyte is a carbonate (e.g., LiK or LiNa carbonates) which is stabilized in an alumina-based porous matrix (e.g., $LiAlO_2$ with Al_2O_3 inserts). The cell efficiency can be over 55%; more efficiency can be obtained with integrated system that use the exergy of the expelled hot gases (CO_2 and steam). An advantage is that this fuel cell is insensitive to CO_2 or CO contamination. The durability of the fuel cell is low because the electrolyte is highly corrosive. The electrode reactions are given below:

$$\begin{cases} \text{at anode}: H_2(g) + CO_3^{2-} \rightarrow H_2O(g) + CO_2 + 2e^- \\ \text{at cathode}: 0.5O_2(g) + CO_2 + 2e^- \rightarrow CO_3^{2-} \end{cases} \tag{2.6}$$

MCFC require inexpensive nickel-based electrodes and can be constructed as a stack of bipolar plates. The system must include a system for recycling carbon dioxide in gaseous phase that includes a water separation process. The simplified configuration diagram for a molten carbonate fuel cell is given in Fig. 2.32.

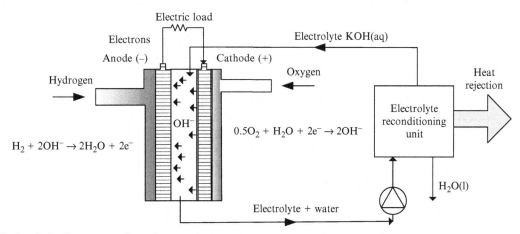

FIG. 2.31 Alkaline fuel-cell system configuration.

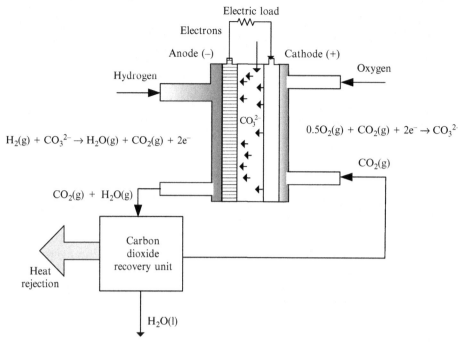

FIG. 2.32 Molten carbonate fuel-cell system configuration.

2.8.6 Direct Methanol Fuel Cells

DMFCs use a proton-conducting polymer electrolyte and operate with a 1:1 molar methanol-water mixture at low temperatures (average 50°C). These fuel cells are compact and store energy at high density; however, the rate of energy discharge (ie, the power density) is rather low. The electrode reactions for DMFC are

$$\begin{cases} \text{at anode}: CH_3OH(l) + H_2O(l) \rightarrow CO_2(g) + 6H^+ + 6e^- \\ \text{at cathode}: 1.5O_2(g) + 6H^+ + 6e^- \rightarrow 3H_2O(l) \end{cases} \tag{2.7}$$

The technology of PEMFC is fully applicable to DMFC; the only major difference is the fuel that comes in liquid phase. Because water is formed at the cathode and the fuel fed at the anode requires water, the water recycling is very important.

Fig. 2.33 illustrates a simplified diagram of a methanol fuel cell system. Besides the water recycling requirement, there are other challenges with methanol fuel cells, the main challenge being the crossover of methanol. Due to the fact that it has a dipole-like molecule similar to that of water, methanol crosses over relatively easily through NAFION® membranes. The reduction of crossover is obtained by using thicker membranes, modified membranes or composite membranes, which use several types of polymers. As a consequence of these types of issues, the efficiency of DMFC is rather low (~30%) as well as their power generation per unit of volume.

2.8.7 Direct Ammonia Fuel Cells

DAFCs are of the SOFC+ type with selected catalysts at the anode, where gaseous ammonia is fed as a source of hydrogen. The schematics of a DAFC are shown in Fig. 2.34. Ammonia fed as an anode decomposes thermocatalytically and generates protons that diffuse through the porous electrolyte. Water is formed at the cathode where protons encounter the oxygen. The achievable DAFC efficiency is of order of SOFC+ fueled with hydrogen, ie, over 55%. The system half reactions are as follows

$$\begin{cases} \text{at anode}: NH_3(g) \rightarrow N_2(g) + 3H^+ + 3e^- \\ \text{at cathode}: 1.5O_2(g) + 3H^+ + 3e^- \rightarrow 1.5H_2O(g) \end{cases} \tag{2.8}$$

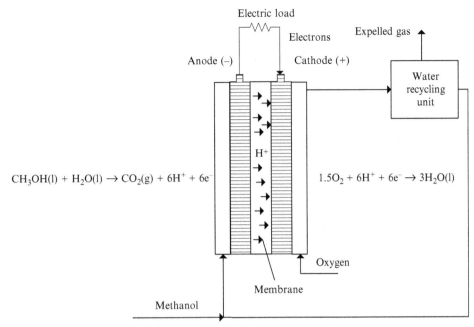

FIG. 2.33 Direct methanol fuel-cell system configuration.

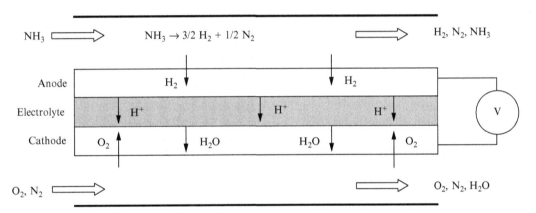

FIG. 2.34 Process description for a direct ammonia fuel cell.

2.9 HYDROGEN APPLICATIONS

Hydrogen is a basic chemical that is used in many industrial processes in chemical and petrochemical industry, fertilizer production, glass production, oil refining and metallurgical processes. The projected hydrogen economy is based on hydrogen use as a fuel and energy carrier medium. In industry, mostly the use of hydrogen is as

- a reactant for ammonia production
- a reactant for methanol production and petroleum processing
- a reactant for silicon tetrachloride reduction to silicon that is needed by semiconductor processing in the electronics industry
- an oxygen remover in annealing, sintering and furnace brazing in the metallurgical industry
- an oxygen scavenger and impedance against the possible corrosion that it can produce, used in nuclear power reactors when water dissociates under neutrons flux

Significant use of hydrogen as fuel is remarked in the aerospace industry where liquid hydrogen and liquid oxygen propel most rockets. Very promising applications of hydrogen are the fuel cells, which can produce clean electricity

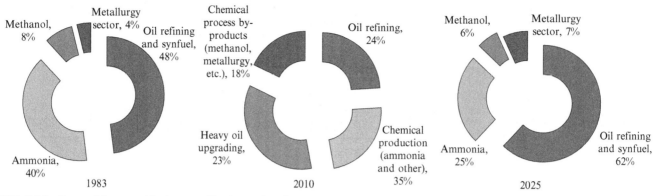

FIG. 2.35 Past and forecasted hydrogen utilization in Canada.

without GHG emission on the utilization side. Moreover, hydrogen is proposed as fuel for internal combustion engines for improved efficiency and lowered emissions.

The present worldwide production of hydrogen is around 50 million tonnes per year. In Canada, which is very active in hydrogen development, hydrogen is utilized as a feedstock for various chemical processes in industries and refineries. Canada produces more than 3 megatonnes of hydrogen per year (see Dincer and Zamfirescu, 2011). Presently there is a significant use of hydrogen in Canada for upgrading heavy oil, particularly for the oil sand sites in Alberta.

In the past, hydrogen has been used in Canada for oil refining, ammonia production, methanol production and processing gas in metallurgical sectors. The growth of hydrogen utilization trends in Canada is reported in Fig. 2.35. In 1983, the oil refining sector and ammonia production accounted in 1983 for 40% each, while at present it is observed that the oil refining sector increased its share of utilization to 46%. Based on forecasts, by 2025, the oil refining share for hydrogen demand will reach 62%.

One major hydrogen application is the nitrogen fixation process represented by ammonia synthesis from nitrogen and hydrogen, where nitrogen is derived from air and hydrogen from water or fossil hydrocarbons. Ammonia is a major fertilizer worldwide and a very well traded feedstock. In a subsequent process, ammonia is converted to urea, which is also a major fertilizer.

Other major uses of hydrogen include synthesis of amines (R—CH_2—NH_2) starting also for ammonia feedstocks; these include amine aniline (R—RH_2). Other products are alcohols, with methanol being an important product highly synthesized from hydrogen and carbon dioxide. More examples are cyclohexanol, hydrogen peroxide, butandiol and a large palette of pharmaceuticals.

One major use of hydrogen is for petrochemical fuels upgrade. The world dependence on crude oil and petroleum products will not be eliminated in the current century. The crude oil consumption in recent years increased from 11 million barrels per day (mbd) in 1950, to 57 mbd in 1970, to around 85 mbd at present. Crude oil (or petroleum) is a naturally occurring hydrocarbon-based liquid or solid (e.g., bitumen in oil shale, oil sands), found in underground rock formations. The main hydrocarbons included in petroleum are alkanes, cycloalkanes, aromatics, asphaltics, naphthalenes, and paraffins. From the total world crude oil reserves, 30% is conventional oil, 15% heavy oil, 25% extra heavy oil, and 30% of petroleum is found in forms of bitumen in oil shale and oil sands.

2.10 CONCLUDING REMARKS

In this chapter, hydrogen and its production methods are introduced in a classified manner for better discussion and illustration. Hydrogen represents a key element for sustainable development since it can be used as an energy storage medium and carrier to harvest renewable energy sources. Hydrogen's conversion into numerous types of synthetic fuels and chemicals is also discussed for an enhanced use. A comparative assessment and evaluation of hydrogen and other fuel options is also presented. The main material resources from which hydrogen can be extracted and the main methods and pathways of hydrogen production are introduced.

Nomenclature

C_p specific heat and constant pressure (kJ/kg K)
C_v specific heat at constant volume (kJ/kg K)
G molar Gibbs free energy (kJ/mol)
H molar enthalpy (kJ/mol)

P   pressure (kPa)
R   universal gas constant (kJ/kmol K)
S   molar entropy (kJ/mol K)
T   temperature (K)

Greek Letter

γ   adiabatic exponent

Subscripts

0   reference state
s   constant entropy
P   products
R   reactants
tot total

References

Carty, R.H., Mazumder, M.M., Schreiber, J.D., Pangborn, J.B., 1981. Thermo-chemical hydrogen production. Final report IGT Project 30517, Institute of Gas Technology, USA.

Dincer, M.A.Rosen, Zamfirescu, C., 2010. Economic and environmental comparison of conventional, hybrid, electric and hydrogen fuel cell vehicles. In: Pistoia, G. (Ed.), Electric and hybrid vehicles: power sources, models, sustainability, infrastructure and market. Elsevier.

Dincer, I., Zamfirescu, C., 2011. Sustainable Energy Systems and Applications. Springer, New York.

Dincer, I., Zamfirescu, C., 2012. Sustainable hydrogen production and the role of IAHE. Int. J. Hydrog. Energy 37, 16266–16286.

Dincer, I., Zamfirescu, C., 2014. Advanced Power Generation Systems. Elsevier, New York.

Dincer, I., Hogerwaard, J., Zamfirescu, C., 2015. Clean Rail Transportation Options. Springer, New York.

Muradov, N., Veziroglu, T.N., 2008. "Green" path from fossil-based to hydrogen economy: an overview of carbon-neutral technologies. Int. J. Hydrog. Energy 33, 6804–6839.

Naterer, G.F., Dincer, I., Zamfirescu, C., 2013. Hydrogen Production from Nuclear Energy. Springer, New York.

STUDY PROBLEMS

2.1 Describe the idea of hydrogen economy.

2.2 What characteristics of hydrogen make it attractive for the world economy?

2.3 In what way can a hydrogen economy influence the power generation sector?

2.4 Categorize the methods for hydrogen production.

2.5 Calculate the reaction enthalpy and the Gibbs energy of water decomposition reaction at 25°C, 1000°C, and 2500°C and then compare the results.

2.6 Calculate the work needed to compress 1 kg of hydrogen from 1 bar pressure to 800 bar and compare it with hydrogen's higher heating value.

2.7 What represents ortho and para hydrogen?

2.8 Calculate the simplified Claude cycle under reasonable assumptions.

2.9 Comment on the storage density of various hydrogen storage methods.

2.10 Describe the operation principle for fuel cells with a proton-conducting electrolyte.

2.11 What is the major advantage of an alkaline fuel cell?

2.12 What is the major advantage of a solid oxide fuel cell?

3

Hydrogen Production by Electrical Energy

3.1 INTRODUCTION

Hydrogen can be produced via electrochemical methods in which electrical energy is required to promote the reactions. Provided that electrical energy is obtained through environmentally benign processes, green hydrogen can be generated electrochemically. A variety of electrochemical methods are available for generating hydrogen, among which water electrolysis is best known. Commercial large-scale alkaline electrolyzers units do exist since long time with a mature technology. More recently, proton exchange electrolyzers were made commercially available at small scale of hydrogen production. Solid oxide electrolyzers of high-temperature steam are currently under research, with very high promises of green hydrogen generation, especially when coupled with nuclear derived process heat and power. Other electrochemical processes are possible, to generate hydrogen. However, an important aspect for green hydrogen production by electrical energy regards the clean technology through which electrical energy is produced.

The discovery of electrolysis around 1800 by Nicholson was one of the major landmarks in electrochemistry. Another landmark is due to Faraday who in 1834 stated that the mass of reacted matter at an electrode is proportional to the circulated electric charge. The well-known Faraday laws of electrochemistry were thus formulated.

Electrolysis achieved an industrial scale by the late 1800s because of a high demand of hydrogen for thermocatalytic ammonia synthesis in the newly developed fertilizer industry. The interest in developing higher-scale production of electrolysis diminished sharply in the first half of the 20th century because coal gasification and steam methane reforming became the most cost-competitive technologies. Water electrolysis was used in niche areas such as the pharmaceutical and food industries where a high-purity hydrogen is needed or the chloralkali industry where alkaline electrolysis is typically applied to extract chlorine and caustic soda from saline water, having hydrogen as by-product.

In the military marine sector, nuclear-powered water electrolyzers were used to generate oxygen from water for deep-sea subsistence. The energy crisis in the 1970s brought again more interest in electrolysis by scientists, this time as a means to produce hydrogen from renewable resources. By 1980, seven worldwide major manufacturers produced ammonia using hydrogen generated with alkaline electrolysis plants that were operating at capacities from 535 to 30,000 standard cubic meters of hydrogen per hour (with an equivalent electric power consumption in the of range 2–100 MW). By the end of the 20th century, proton exchange membrane electrolyzers (PEMEs) were commercially available with production capacities of about 100 kW. More recently, the research and development of solid oxide electrolysis cells have become one of the most promising technologies for generating electrolytic hydrogen on a large scale.

In this chapter, electrochemical hydrogen production is considered with a focus on water electrolysis and renewable sources for electrical energy generation as well as with coupling them for sustainable hydrogen production. The fundamentals of electrolysis are reviewed, and a classification of processes with respect to the state of the electrolyte (solid, liquid) and type of electrolyte (acid, alkaline, or ceramic) is provided. The calculations of reversible potentials and overpotentials are significantly treated in the design and modeling of electrolyzers. Some technological aspects of four types of electrolyzers are discussed: alkaline, proton exchange membrane, and solid oxide electrolyzers with oxygen ion and with proton conduction. Some special types of electrolyzers are discussed. The integration of electrolyzers with various types of renewable electricity generators is presented with the focus on green hydrogen production.

3.2 FUNDAMENTALS OF ELECTROCHEMICAL HYDROGEN PRODUCTION

In an electrochemical reaction, a process is conducted in which chemical reactions occur under the influence of an electrical field produced by electrodes. Usually, the electrodes are solid metals or semiconductors. In addition to the electrode, two types of electric charge conductors are required to form an external circuit (electronic conductor) and an internal circuit (ionic species conductor). The medium that is conductive to ionic species is denoted as electrolyte. Thenceforth, in electrochemistry, the electric energy is the driver that provides at least the free energy required for the desired reaction to occur.

Some electrochemical reactions are reversible, such as those in rechargeable batteries and in reversible fuel cells/electrolyzer cells. The electrons are directly transferred from electrodes to or from molecules and atoms involved into reactions. This is why at electrodes oxidation-reduction (or redox) reactions occur. In a complete electrochemical cell, a complete reaction occurs consisting of two half-reactions that take place at separate locations. There will be an oxidizing reaction and a reduction reaction. The ionic species are exchanged between the two reactions through the medium of electrolyte, whereas the electrons are transferred through the external circuit.

Electrolysis is a nonspontaneous electrochemical reaction that requires the consumption of an electric energy to occur. The reverse of the electrolysis—that is, the fuel cell process—is spontaneous and will produce electricity, consuming fuels (usually hydrogen).

The water (or steam) electrolysis process can be viewed as a superposition of concurrent or sequential electrochemical reactions occurring in the vicinity of electrodes (half reactions) with the overall effect of splitting the water molecule and separating the gaseous products (hydrogen and oxygen). The overall water-splitting reaction is written as

$$H_2O \rightarrow H_2(g) + 1/2 O_2(g) \tag{3.1}$$

In order to obtain 1 mole of hydrogen, 1 mole of water is required. The negative electric charge that passes though the external circuit (electrons) must be equal to the positive electric charge of ionic specifies passing through the electrolyte. The charge comprised in 1 mole of elementary charges is denoted with F and represents the Faraday constant $F = N_A e$, where $N_A = 6.022 \times 10^{23}$ is Avogadro's number and $e = 1.602 \times 10^{-19}$ C is the elementary charge; thus $F = 96{,}490$ C/mol.

Fig. 3.1 shows the general layout of an electrolysis cell which (anode and cathode) immersed an electrolyte. The electrodes are connected to a direct current (DC) power supply. The electrode connected at the positive pole of the power source is denoted as an anode, while the negative pole is the cathode. In a typical water electrolysis process, oxygen evolves at the anode while hydrogen is produced at the cathode.

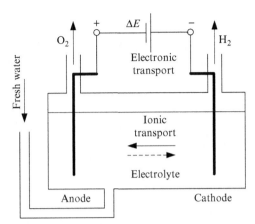

FIG. 3.1 A general layout of an electrolysis system.

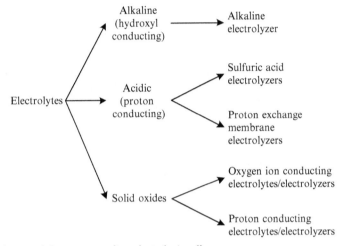

FIG. 3.2 Classification of electrolytes and the corresponding electrolysis cells.

The analysis of an electrochemical process must consider various aspects. The most fundamental is the thermodynamic analysis that determines the driving forces of the process based on an application of conservation laws such as energy conservation (first law of thermodynamics), mass conservation, electric charge conservation, thermodynamic equilibrium principles and other specific methods. The thermodynamic analysis must be accompanied by a kinetic analysis that studies the reaction rates. Moreover, the study of transport phenomena within electrolyzers is important and cannot be neglected. Some additional aspects can be derived from materials analysis, economic analysis, and environmental impact analysis.

A classification of electrolytes and water electrolyzes cells is shown in Fig. 3.2. Electrolytes allow conduction of ions and do not conduct electrons. Under the influence of the electric field, positive ions (or cations) move from the anode to cathode. The anions (or negatively charged ions) move through the electrolyte from the cathode to anode where they "consume" while releasing electrons. Depending on the nature of the electrolyte, the ions that are most mobile may be anions or cations. In acidic electrolytes, cations are those that are transported while in alkaline electrolytes; anions are the charge carriers. The current status, operating conditions and conversion efficiencies of the various electrolysis technologies are presented in Table 3.1.

TABLE 3.1 The Current Status of Different Electrolysis Technologies

Technology		Development status	T (°C)	P (bar)	SEC (kWh/kg H_2)
Alkaline	Large-scale	Commercial	70–90	1–25	48–60
	High-pressure	Commercial	70–90	Up to 690	56–60
	Advanced	Pre-commercial	80–140	Up to 120	42–48
PEM		Commercial	80–150	Up to 400	40–60
SOEC		Prototype	900–1000	Up to 30	28–39

3.2.1 Thermodynamic Analysis of Electrochemical Reactions

The main focus of the thermodynamic analysis of an electrochemical reaction referred to the is the of the electrochemical potential. In addition, the thermodynamic analysis determines the reversible potentials for the half reactions (or electrode reaction) and for the full cell. Any electrochemical reaction is defined by three parameters, such as temperature, pressure, and electrochemical potential. Note that the electrochemical potential is defined as a superposition of chemical potential and the potential energy caused by the presence of the electric field (electric potential). The chemical potential is defined by $\Delta G = zFE$, where E represents the electric potential difference between the two electrodes, ΔG, and z is the number of electric charges.

Once an electrochemical cell receives a potential greater than or equal to that corresponding to the Gibbs free energy of the reaction, then one expects that the reaction starts to proceed forward. This minimum potential is denoted as the reversible cell potential, E_{rev}, defined as follows:

$$E_{rev} = -\frac{\Delta G}{zF} \tag{3.2}$$

Note that the reversible potential is an intensive property. Once a potential $E_{cell} > E_{rev}$ is applied to the cell, the reaction products start to be generated at a rate equal to the current intensity. Since the current intensity (Ampere) is defined as the number of electric charges (Coulomb) per unit of time (s), the following law, known as the Faraday law is valid, namely

$$\int_0^t I(t)dt = zFn \tag{3.3}$$

where z represents the valence of ionic species of interest (the number of electrons per ion), F is the Faraday number and n is the number of moles of released product, $I(t)$ is the intensity of electric current through the exterior circuit (in amperes) and t is time. The electric current intensity may vary in time, or it can be stable; depending on the electrolyzer's regime of operation. For a hydrogen evolving reaction, $z = 2$, while for an oxygen evolving reaction, $z = 4$.

As stated by the Faraday's law, the electric charge is conserved in an electrochemical process. This means that the electric charge consumed by an ionic species which is reduced (or oxidized) in an electrolyzer equals to the electric charge of electrons flowing through the exterior circuit.

Table 3.2 gives the main electrode reactions relevant to water electrolysis and their associated values of reaction entropy, enthalpy, and the Gibbs free energy. The Gibbs free energy, is defined with $G = H - TS$, where H and S represent the molar enthalpy and entropy, respectively, while T is the temperature in K. The reaction enthalpy and entropy are calculated as the difference between the product and reactant enthalpy and entropy, respectively, giving ΔH and ΔS. Therefore, the Gibbs free energy of the reaction is calculated with $\Delta G = \Delta H - T\Delta S$.

TABLE 3.2   Thermodynamic Properties of Some Relevant Electrochemical Reactions[a]

Reaction Type	#	Chemical equation (half-cell)	ΔH(eV)	ΔS (meV/K)	ΔG (eV)	E(V)	E_{th}(V)
H_2 evolving ($z = 2$)	1	$H_2O(g) + 2e^- \rightarrow H_2(g) + O^{2-}$ @ 1000K	1.901	0.949	0.952	0.476	0.951
	2	$2H_2O(l) + 2e^- \rightarrow H_2(g) + 2OH^-(aq)$	2.914	3.413	1.896	0.948	1.457
	3	$2H^+ + 2e^- \rightarrow H_2(g)$	−31.840	−1.339	−31.44	−15.72[b]	−15.92
O_2 evolving ($z = 4$)	4	$2H_2O(l) \rightarrow O_2(g) + 4H^+ + 4e^-$	69.610	6.062	67.800	16.95[b]	17.40
	5	$2H_2O(g) \rightarrow O_2(g) + 4H^+ + 4e^-$ @ 775 K	69.320	4.852	65.560	16.39	17.33
	6	$2O^{2-} \rightarrow O_2(g) + 4e^-$ @ 1000 K	1.335	−0.753	2.087	0.522	0.334
	7	$4OH^-(aq) \rightarrow 2H_2O(l) + O_2(g) + 4e^-$	0.097	−2.707	0.904	0.226	0.024
Overall reaction	8	$H_2O(l) \rightarrow H_2(g) + 0.5O_2(g)$	2.962	1.691	2.458	1.229	1.481
	9	$H_2O(g) \rightarrow H_2(g) + 0.5O_2(g)$ @1000 K	2.569	0.571	1.998	0.999	1.284

[a]If not otherwise specified, the reaction temperature is 298.15 K, and the partial pressure for each species is assumed to be $P = P_0 = 101.325$ kPa.
[b]Here, the Gibbs free energy of protons is calculated as per vacuum conditions.

The energy associated to electrochemical reactions is sometimes expressed in electron-Volt (eV) units. By definition, an electron-Volt represents the energy needed to displace an elementary electrical charge in an electrical field, across a potential difference of 1 V. In practical situations, the species involved in an electrochemical reaction may be at a different partial pressure than P_0 (standard pressure) or at a thermodynamic activity different than 1 (see subsequently a definition of the "activity"), and this influences the value of the Gibbs free energy of the reaction. In a subsequent paragraph the notion of "activity" is introduced formally (see Eq. 3.7).

The equation $2H^+(aq) + 2e^- \rightarrow H_2(g)$ given in Table 3.2 assumes that the proton evolves from vacuum. The Gibbs energy of $2H^+ + 2e^- \rightarrow H_2(g)$ when a proton evolves from vacuum is -3033.79 kJ/mol. The electron-volt an electron-Volt is defined as the energy needed to displace an elementary electrical charge in an electrical field, across a potential difference of 1 V. The removal into a vacuum of protons from water according to the equation $2H^+(aq) \rightarrow 2H^+$ requires a Gibbs free energy of 2176 k/mol. Therefore, the Gibbs energy of the reaction $2H^+(aq) + 2e^- \rightarrow H_2(g)$ becomes as follows: $\Delta G = -858.24$ kJ/mol. Consequently, the absolute difference of potential for reaction $2H^+(aq) + 2e^- \rightarrow H_2(g)$ becomes $E = -4.44$ V.

In a mixture of chemically reacting species (e.g., ions such as OH^-, H^+, O^{2-} or molecules such as H_2O, H_2, O_2 etc.), the Gibbs energy variation can be calculated using chemical potentials μ, defined for the component "i" of the system according to the equation

$$dG = \sum \mu_i dn_i, \text{ where } \mu_i = \frac{\partial G}{\partial n_i} \tag{3.4}$$

If a mixture of gases obeys the ideal gas law, then the variation of chemical potential with pressure is expressed by the equation

$$\mu_i(P_i) - \mu_i(P_0) = RT \ln\left(\frac{P_i}{P_0}\right) \tag{3.5}$$

where P_0 is a reference pressure while P_i is the partial pressure of component "i."

A similar equation does exist for ideal solutions, where partial pressure is replaced by molar concentration, namely

$$\mu_i(c_i) - \mu_i(c_0) = RT \ln\left(\frac{c_i}{c_0}\right) \tag{3.6}$$

where c_0 represents the concentration at reference conditions.

For real chemical systems, which do obey neither with ideal gas law nor with ideal solution model, an equation similar to Eqs. (3.5) and (3.6) is introduced, using the notion of "thermodynamic activity" (also known as chemical activity). The thermodynamic activity "a_i" is defined for a species "i" by the following equation:

$$\mu_i(a_i) - \mu_i(a_i = 1) = RT \ln(a_i) \tag{3.7}$$

In Eq. (3.7) potential for $a_i = 1$, $\mu_i(a_i = 1)$ must be defined. This chemical potential, and it is known as the standard state thermodynamic potential, denoted by $\mu_i^0 = \mu_i(a_i = 1)$.

The thermodynamic activity has the unit of measure similar to the unit of concentration. Thermodynamic activity is often expressed through a dimensionless activity coefficient f that quantifies the abatement from ideal conditions of a component (species) of a chemical system. The activity coefficient is given through the equation

$$a_i = c_i f_i \tag{3.8}$$

According to the definitions for chemical potential and thermodynamic activity given in Eqs. (3.7), and (3.8), the Gibbs energy of a chemical reaction has the general expression given as follows:

$$\Delta G = \sum \left(n_i \mu_i^0\right) + RT \ln\left(\Pi(cf)_i^{n_i}\right) \tag{3.9}$$

where n_i is the stoichiometric number that is positive for the products (P) and negative for the reactants (R) the term $\sum(n_i\mu_i^0)$ can be expressed as the sum of products minus the sum of reactants, where reactants are the chemical species that are consumed and the products represent the species that are generated.

standard Gibbs free energy of an electrochemical reaction (ΔG^0) is given as follows

$$\sum \left(n_i\mu_i^0\right) = \sum_P \left(n_i\mu_i^0\right) - \sum_R \left(n_i\mu_i^0\right) = \Delta G^0. \tag{3.10}$$

From Eqs. (3.9) and (3.10) correlates the Gibbs free energy with the standard Gibbs free energy as follows::

$$\Delta G = \Delta G^0 + RT \ln\left(K_{eq}\right) \tag{3.11}$$

where

$$K_{eq} = \Pi(cf)_i^{n_i} \tag{3.12}$$

is known as the equilibrium constant of the electrochemical reaction.

The enthalpy of an electrochemical reaction relates to the Gibbs free energy according to the known equation, $\Delta H = \Delta G + T\Delta S$. Using this in Eq. (3.2) one obtains

$$E_{rev} = -\frac{\Delta H - T\Delta S}{zF} = -E_{th} + \frac{T\Delta S}{zF} \tag{3.13}$$

Equation (3.13) introduces the electric thermoneutral potential E_{th} that is defined with

$$E_{th} = -\frac{\Delta H}{zF} \tag{3.14}$$

When an electrochemical reaction is supplied with a potential equal to E_{th}, the reaction becomes adiabatic, in which the provided electrical energy compensates for all enthalpy of reaction: that is, work component (or ΔG) and heat (entropic) component (or $T\Delta S$). Depending on the operating parameters, in an electrochemical reaction, one can have either $\Delta H = \Delta G + T\Delta S \geq \Delta G$ (case in which the process may be endothermic) or $\Delta H \leq \Delta G$ (case in which the process rejects heat—it is exothermic). In an endothermic reaction, the required heat may be provided via a Joule effect (electric heating of the electrolyte due to the ionic current passing through it).

Furthermore, if $\Delta G < 0$, then the reaction is spontaneous and from Eq. (3.2) the reversible cell potential is positive, meaning that energy is delivered to the exterior, that is, the operation is in galvanic mode (or fuel cell mode). If $\Delta G > 0$, then the reaction is not spontaneous, requiring energy input (negative cell voltage), and the operation is in electrolytic cell mode (or electrolysis cell).

From Eqs. (3.2) and (3.11) one obtains the well-known Nernst equation for the reversible potential of an electrochemical reaction

$$E_{rev} = E^0 + \frac{RT}{zF} \ln\left(K_{eq}\right) \tag{3.15}$$

where E^0 is the standard reversible cell potential, introduced as follows:

$$E_0 = \frac{\Delta G_0}{zF} \tag{3.16}$$

An electrochemical reaction of major importance is the reaction of hydrogen oxidation:

$$H_2(g) \rightarrow 2H^+(aq) + 2e^- \tag{3.17}$$

The reverse of reaction given by Eq. (3.17) defines by convention the potential of the hydrogen electrode, which is adopted as the standard for electrochemical potentials. The hydrogen electrode reaction reads as follows:

$$2H^+(aq) + 2e^- \rightarrow H_2(g) \tag{3.18}$$

The absolute value of a standard hydrogen electrode is $+4.44\,V$. However, by convention, the standard hydrogen electrode is assigned to zero for all temperatures, such that the electrode potential tables can be calculated for any electrodes. The standard hydrogen electrode potential is denoted with $E^0(H/H^+) = 0V$.

The reversible potentials of all other electrochemical reactions can be expressed with reference to the standard hydrogen electrode. Generally defined, the standard electrode potential (of any electrode) represents the voltage potential of a reversible electrode at standard conditions.

More exactly, the standard electrode potential is defined for standard temperature, namely $(T_0 = 25°C)$ and standard pressure $(P_0 = 10^5\,Pa)$, having the solutes at a concentration of $1\,mol/dm^3$ and gas phase at 1 atm and a chemical activity of 1 for involved pure liquids, pure solids, and water solvent. The standard potential of electrodes are given as reduction potential, which indicates the gain of electrons for the chemical species involved.

Oxidation, in which the chemical species give electrons, have potentials that are the negatives of the reduction potentials. Values of some electrode potentials of basic half reaction are given in Table 3.3. In the table, "reductant" represents a reduction agent or species. The standard reversible potential of the whole cell, namely E^0, is calculated by subtracting the half reaction potentials of the reduction and the oxidation reactions.

Many of the half reactions from Table 3.2 are relevant in various ways to water electrolysis. The proton reduction reaction shown in bold is the key reaction for the table. Any reaction listed below can pair with the proton reduction reaction to form a full cell reaction. One possible way is to pair the reaction with the oxygen evolving one having a

TABLE 3.3 Standard Electrode Potentials for Some Basic Half Reactions

Reductant	Reduction reaction	$E^0(V)^a$
Ca	$Ca^+ + e^- \leftrightarrows Ca$	−3.8
$NH_3(aq)$	$3N_2(g) + 2H^+ + 2e^- \leftrightarrows 2NH_3(aq)$	−3.09
Li(s)	$Li^+ + e^- \leftrightarrows Li(s)$	−3.04
$Ca + 2OH^-(aq)$	$Ca(OH)_2 + 2e^- \leftrightarrows Ca + 2OH^-(aq)$	−3.02
$Ba + 2OH^-(aq)$	$Ba(OH)_2 + 2e^- \leftrightarrows Ba + 2OH^-(aq)$	−2.99
K(s)	$K^+ + e^- \leftrightarrows K(s)$	−2.931
Sr(s)	$Sr^{2+} + 2e^- \leftrightarrows Sr(s)$	−2.899
Ca(s)	$Ca^{2+} + 2e^- \leftrightarrows Ca(s)$	−2.868
Na(s)	$Na^+ + e^- \leftrightarrows Na(s)$	−2.71
Mg	$Mg^+ + e^- \leftrightarrows Mg$	−2.70
Y(s)	$Y^{3+} + 3e^- \leftrightarrows Y(s)$	−2.372
Mg(s)	$Mg^{2+} + 2e^- \leftrightarrows Mg(s)$	−2.372
Ce	$Ce^{3+} + 3e^- \leftrightarrows Ce$	−2.33
$Al + 4OH^-(aq)$	$AlH_2O_3^- + H_2O + 3e^- \leftrightarrows Al + 4OH^-(aq)$	−2.33
$Al(s) + 3OH^-(aq)$	$Al(OH)_3(s) + 3e^- \leftrightarrows Al(s) + 3OH^-(aq)$	−2.31
H^-	$H_2 + 2e^- \leftrightarrows 2H^-$	−2.23
U	$U^{3+} + 3e^- \leftrightarrows U$	−1.798
Al(s)	$Al^{3+} + 3e^- \leftrightarrows Al(s)$	−1.662
Ti(s)	$Ti^{2+} + 2e^- \leftrightarrows Ti(s)$	−1.630
Ti(s)	$Ti^{3+} + 3e^- \leftrightarrows Ti(s)$	−1.370
$Ti(s) + H_2O$	$TiO(s) + 2H^+ + 2e^- \leftrightarrows Ti(s) + H_2O$	−1.310
$Fe(s) + 2OH^-(aq)$	$Fe(OH)_2(s) + 2e^- \leftrightarrows Fe(s) + 2OH^-(aq)$	−0.890
$Ti(s) + H_2O$	$TiO^{2+} + 2H^+ + 4e^- \leftrightarrows Ti(s) + H_2O$	−0.860
$H_2(g) + 2OH^-(aq)$	$2H_2O + 2e^- \leftrightarrows H_2(g) + 2OH^-(aq)$	−0.828
Zn(s)	$Zn^{2+} + 2e^- \leftrightarrows Zn(s)$	−0.762
$Pb(s) + 2OH^-(aq)$	$PbO(s) + H_2O + 2e^- \leftrightarrows Pb(s) + 2OH^-(aq)$	−0.580
Fe(s)	$Fe^{2+} + 2e^- \leftrightarrows Fe(s)$	−0.440
Ni(s)	$Ni^{2+} + 2e^- \leftrightarrows Ni(s)$	−0.250
Fe(s)	$Fe^{3+} + 3e^- \leftrightarrows Fe(s)$	−0.040
$H_2(g)$	$2H^+ + 2e^- \leftrightarrows H_2(g)$	0.000
$3Fe(s) + 4H_2O$	$Fe_3O_4(s) + 8H^+ + 8e^- \leftrightarrows 3Fe(s) + 4H_2O$	+0.085
$NH_4OH(aq)$	$N_2(g) + 2H_2O + 6H^+ + 6e^- \leftrightarrows NH_4OH(aq)$	+0.092
$Hg(l) + 2OH^-(aq)$	$HgO(s) + H_2O + 2e^- \leftrightarrows Hg(l) + 2OH^-(aq)$	+0.098
$Cu(NH_3)_2^+ + 2NH_3$	$Cu(NH_3)_4^{2+} + e^- \leftrightarrows Cu(NH_3)_2^+ + 2NH_3$	+0.100
$CH_4(g)$	$C(s) + 4H^+ + 4e^- \leftrightarrows CH_4(g)$	+0.130
$CH_3OH(aq)$	$HCHO(aq) + 2H^+ + 2e^- \leftrightarrows CH_3OH(aq)$	+0.130
$H_2S(g)$	$S(s) + 2H^+ + 2e^- \leftrightarrows H_2S(g)$	+0.140

Continued

TABLE 3.3   Standard Electrode Potentials for Some Basic Half Reactions—cont'd

Reductant	Reduction reaction	$E^0(V)$
Cu^+	$Cu^{2+} + e^- \leftrightharpoons Cu^+$	$+0.159$
$SO_2(aq) + 2H_2O$	$HSO_4^- + 3H^+ + 2e^- \leftrightharpoons SO_2(aq) + 2H_2O$	$+0.160$
$SO_2(aq) + 2H_2O$	$SO_4^{2-} + 4H^+ + 2e^- \leftrightharpoons SO_2(aq) + 2H_2O$	$+0.170$
$OH^-(aq)$	$O_2(g) + H_2O + 4e^- \leftrightharpoons 4OH^-(aq)$	$+0.401$
$Cu(s)$	$Cu^+ + e^- \leftrightharpoons Cu(s)$	$+0.520$
$C(s) + H_2O$	$CO(g) + 2H^+ + 2e^- \leftrightharpoons C(s) + H_2O$	$+0.520$
I^-	$I_3^- + 2e^- \leftrightharpoons 3I^-$	$+0.530$
I^-	$I_2(s) + 2e^- \leftrightharpoons 2I^-$	$+0.540$
$Ag(s)$	$Ag^+ + e^- \leftrightharpoons Ag(s)$	$+0.780$
H_2O	$O_2(g) + 4H^+ + 4e^- \leftrightharpoons 2H_2O$	$+1.229$
Cl^-	$Cl_2(g) + 2e^- \leftrightharpoons 2Cl^-$	$+1.360$
$Au(s)$	$Au^+ + e^- \leftrightharpoons Au(s)$	$+1.830$

^{a}Given for one electron transfer only.

standard reversible half-cell potential of $+1.229$ V. The full reaction then becomes the water-splitting reaction, and the electrolytic cell potential results as follows: $E^0(H/H^+) - E^0(H_2O/O) = 0 - 1.229 = -1.229\,V$.

For the hydrogen electrode, the activity of hydrogen gas—modeled as an ideal gas—is equal to the concentration; therefore, the Nernst equation takes the following particular form:

$$E_{rev,H^+} = \frac{RT}{2F} \ln\left[\frac{(fc)^2_{H^+}}{c_{H_2}}\right] \tag{3.19}$$

The generic equation for the reversible potential of a reduction half reaction (described by the equation "red $\leftrightharpoons$ oxi $+ ze^-$") is given as follows:

$$E_{rev,red} = E^0 + \frac{RT}{zF} \ln\left(\frac{a_{oxi}}{a_{red}}\right) \tag{3.20}$$

The hydroxyl half-cell—specific to alkaline electrolyzers—has the Nernst equation taking the specific form

$$E_{rev,OH^-} = E^0_{OH^-} + \frac{RT}{2F} \ln\left(\frac{a^2_{H_2O}}{c_{H_2}a^2_{OH^-}}\right) \tag{3.21}$$

where, in standard conditions, the activity of water is 1, the activity of hydroxyls is $a_{OH^-} = 1$, and the hydrogen concentration is $c_{H_2} = 1$. Since Eqs. (3.20) and (3.21) are essentially equivalent, one can obtain the standard potential for basic calculations:

$$E^0_{OH^-} = \frac{RT}{2F} \ln\left(a^2_{H^+}\right) \tag{3.22}$$

For the overall cell, the reversible potential can then be calculated with the Nernst equation for the overall reaction,

$$E_{rev} = E^0 + \frac{RT}{2F} \ln\left(\frac{c_{H_2}c^{0.5}_{O_2}}{a_{H_2O}}\right) \tag{3.23}$$

where the standard potential at 25°C is $E^0 = 1.229$ V.

In many practical situations, in Eq. (3.23), the activity of water is 1, the hydrogen concentration is $c_{H_2} = P_{H_2}/P_0$ and the oxygen concentration is $c_{O_2} = P_{O_2}/P_0$.

A scale for hydrogen potential - denoted as pH scale - is established in chemistry, pH is defined as which is the extend of the scale covering the range from 0 to 14. The potential of hydrogen $(pH) = -\log_{10}(a_{H^+})$. Water with pH $= 0$ has a protonic activity of $a_{H^+} = 1$ mol/l. The equilibrium constant of water dissociation in solution $H_2O(l) \rightarrow H^+(aq) + OH^-(aq)$ is given by the constant $K_{eq,w} = a_{H^+}a_{OH^-}$, which has a value of 1.27×10^{-14} (mol/l)^2 at 25°C. For

$a_{OH^-} = 1$, it results in $a_{H^+} = 1.27 \times 10^{-14}$; and from Eq. (3.22), one has $E^0_{OH^-} = -0.822$ V. Determining the reversible potential of half-cell reactions is essential in calculations for reaction kinetics at each electrode.

Here it can be observed that at 25°C for pH = 0 when proton activity is $a_{H^+} = (fc)_{H^+} = 1$, the electrode potential is zero. For pH = 7, the electrode potential is −0.414 V; and for pH = 14, it is −0.828 V. However, the hydrogen electrode potential is a half-cell potential in acidic electrolyte, specific to PEM electrolyzers (pH < 7). If the electrolyte is basic (pH > 7), hydroxyl ions are formed instead of protons, according to reaction #2 in Table 3.2.

3.2.2 Kinetics and Transport Process Analyses

The kinetic and transport process analyses are related and focus on determination of the relationship between reaction rate (or the rate of products generation) and operating conditions as well as between the current and the voltage consumed by the electrolysis cell. The kinetic analysis determines the relationship between the rate of reaction at the electrodes and the electrode overpotential and the relation between the current of ionic species across the electrolyte and the electric potential over it. When the cell assembly is more complex, comprising membranes or diaphragms or anolyte/catholyte, a kinetic analysis accounts for the overpotentials across them and the currents of matter and electric charges through them.

For an electrolytic cell to operate, a voltage higher than the reversible DC voltage must be applied to the electrodes. The overvoltage is necessary to compensate for the activation energy of the reaction and to act as a driving force for the involved transport processes. In this case, reactant species must reach the active electrode surface, and the products leave the active sites. As well, the reaction kinetics impose a residence time of the chemicals at the active sites where the reactions occur. Denote the actual voltage with E_{tot} and the current consumed with I. The energy rate balance on the electrolysis cell requires

$$E_{tot} I = E_{rev} I + \dot{W}_{loss} \tag{3.24}$$

where E_{rev} is the reversible cell voltage given by Eq. (3.23), and $\dot{W}_{loss}$ represents the energy dissipated due to various losses and I is the electric current.

An equation for the electric current can be obtained by differentiating from the Faraday law

$$I = zF\dot{n} \tag{3.25}$$

where $\dot{n}(t) = dn/dt$ is the reaction rate, and n is the number of moles of generated products.

The presence of the electric field in the vicinity of electrodes enhances the reaction rate. Reduction reactions—that is, Eqs. (3.21) and (3.23)—can occur only at the electrode in the vicinity of which the potential is more negative than the reversible potential of the reaction, whereas oxidation reactions require an electrode with a potential more positive than it.

In this context, the reaction kinetics process can be described with the help of a potential surface diagram as shown in Fig. 3.3 for a redox process. There must be two potential surfaces, one corresponding to reactants and another for products. Moving along the reaction coordinate, the potential energy of the system decreases and reaches a minimum in #2. An energy barrier 2–3 of magnitude ΔG_{act} has to be overcome in order for the system to move on the product's potential surface, toward the right. If an electric field is applied by polarization of the electrode (anode for the case shown in the figure), the whole potential surface characterizing the thermodynamic system of the reaction products displaces at lower energy with a magnitude proportional to the polarization potential, namely $F\Delta E_{act}$, where ΔE_{act} represents the activation overpotential. Consequently, the activation energy of the reaction under the influence of polarization decreases as it evolves along the path $1 - 2 - \overline{3} - \overline{4}$.

The activation energy of the forward reaction in the presence of the electric field $\overline{\Delta G}_{act}$ is reduced with respect to the activation energy in the absence of polarization ΔG_{act}. As shown in Fig. 3.3 the activation energy reduces with a quantity that is proportional to the intensity of the electric field. Mathematically, one can write

$$\overline{(\Delta G)}_{act} = \Delta G_{act} - \alpha F \Delta E_{act} \tag{3.26}$$

where $\alpha \in [0, 1]$ is a parameter known as "transfer coefficient."

Note that the backward reaction can also occur (along path 4–3–2 in the absence of the electric field, or $\overline{4} - \overline{3} - 2$ in its presence). For the situation depicted in the figure, the backward reaction is disfavored by the presence of the electric field, because its activation energy (corresponding to the path $\overline{4} - \overline{3}$) is increased. The activation energy for the backward reaction can be written with the help of the transfer coefficient according to

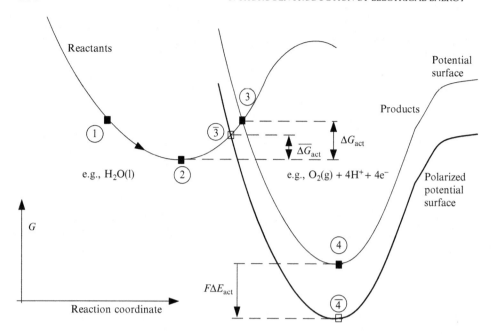

FIG. 3.3 Potential surfaces and activation energy for electrochemical oxidation reaction.

$$\overline{(\Delta G)}_{b,act} = \Delta G_{b,act} + (1 - \alpha)F\Delta E_{act} \tag{3.27}$$

The rate of reaction (either forward or backward) is proportional to the rate constant and the concentration of consumed and/or generated species (depending on the reaction order). The rate constant can be written based on the Arrhenius equation, given as

$$k = \mathcal{A}\exp(-\Delta G_{act}/RT) \tag{3.28}$$

where $\mathcal{A}$ is known as a pre-exponential factor.

The net rate of reaction is the difference between the rate of forward reaction and backward reaction, namely

$$\dot{n} = \dot{n}_f - \dot{n}_b = k_f c_f - k_b c_b \tag{3.29}$$

where c refers to the molar concentration of reactants.

Further, using Eq. (3.25), the electric current intensity becomes

$$I = zF(k_f c_f - k_b c_b) \tag{3.30}$$

The current density at the electrodes is defined by $= I/A$, where A is the effective area of the electrode. Using the Arrhenius equation and activation energy as expressed in Eqs. (3.28) and (3.30), one obtains an expression for the current density at the electrode known as the Butler-Volmer equation,

$$J = zFAk^0\left[\frac{c_R}{c_{R,0}}\exp\left(\frac{\alpha F\Delta E_{act}}{RT}\right) - \frac{c_O}{c_{O,0}}\exp\left(-\frac{(1-\alpha)F\Delta E_{act}}{RT}\right)\right] \tag{3.31}$$

where k^0 is the rate constant for a reversible operation, c_O is the concentration of oxidized species at the surface of the oxiation electrode and c_R is the concentration of the reduced species, at the surface of the electrode; the subscripts "R,0" and "O,0" refer to concentrations in bulk solution for the reduced and oxidized species, respectively.

For the simplified case, when mass transport influences can be neglected at the electrodes, then $c_R/c_{R,0} = 1$ and the Butler-Volmer equation takes the form

$$J = J_0\left[\exp\left(\frac{\alpha F\Delta E_{act}}{RT}\right) - \exp\left(-\frac{(1-\alpha)F\Delta E_{act}}{RT}\right)\right] \tag{3.32}$$

where J_0 represents the current density specific to the reversible reaction, also known as exchange current density.

The dependence of the exchange current density on temperature is related to the Arrhenius equation, as follows:

$$J_0 = zFk = zF\mathcal{A}\exp(-\Delta G_{act}/RT) \tag{3.33}$$

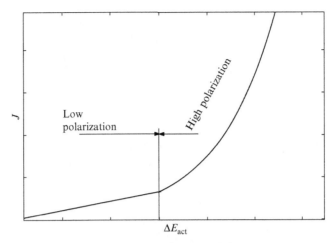

FIG. 3.4 Current density versus activation overpotential for a kinetically controlled reaction.

There are two limiting situations for Eq. (3.22), the first occurring when the electrode polarization is low (ie, ΔE_{act} is relatively small such that $\Delta E_{act} \leq 0.2RT/\alpha F$). At this limit, Eq. (3.22) can be linearized accordingly as follows:

$$\Delta E_{act} = \frac{RT}{F}\left(\frac{J}{J_0}\right) \tag{3.34}$$

The current is proportional to an overvoltage near the equilibrium. If the polarization is high, the thermodynamic system moves far from equilibrium ($\alpha \cong 1 \rightarrow \alpha\Delta E_{act}/RT \gg (1-\alpha)\Delta E_{act}/RT$). At this limit, the following approximation of Eq. (3.22) is valid:

$$\Delta E_{act} = \frac{RT}{\alpha F}\ln\left(\frac{J}{J_0}\right) \tag{3.35}$$

Fig. 3.4 shows the correlation between current density and the activation overpotential of a kinetically controlled process in an electrolytic cell. Initially, the activation overpotential grows linearly with the current for low polarization. The current density will grow exponentially with the activation overpotential at high polarizations.

For higher polarization, however, mass transport processes may become dominant (depending on the geometrical configuration of the electrolysis cell). The mass transport of ionic species within an electrolyte is influenced by the electric field (migration), hydrodynamics (convection), and concentration gradients (diffusion). The accumulation of ions in the vicinity of the electrode surface creates a potential and a concentration gradient. A driving force is generated by the concentration gradient that propels the transport of ions through the electrolyte. Table 3.4 gives a summary of the driving force of ionic species through various electrolytes.

Mass transport processes at a polarized electrode are illustrated in a simplified diagram in Fig. 3.5. The diagram depicts the case of a diffusion-controlled hydrogen evolution reaction. At the interface of any two material media, a Helmholtz layer is formed with a typical thickness on the order of 1 Å. The electrons must be transferred across the Helmholtz layer in order to reach the reaction sites at the very vicinity of the reactant molecules. Depending on the polarization and materials involved, a diffusion layer is formed at the electrode surface with a thickness directly influenced by the thickness of the Helmholtz layer. As shown in the figure, a control volume can be delimited in the vicinity of the reaction site.

TABLE 3.4 Driving Force and Coupling Coefficient for Ionic Species Mobility in Electrolytes

Transport process	Driving force (gradient)		Coupling coefficient	Equation
Conduction, J_m	Electric potential	$\dfrac{dE}{dx}$	Conductivity, σ	$J_m = \dfrac{\sigma}{\lvert z_i \rvert F}\dfrac{dE}{dx}$
Diffusion, J_d	Species concentration	$\dfrac{dc}{dx}$	Diffusivity, D	$J_d = -D\dfrac{dc}{dx}$
Convection, J_c	Pressure gradient	$\dfrac{dP}{dx}$	Viscosity, μ	$J_c = \dfrac{Ac}{\mu}\dfrac{dP}{dx}$

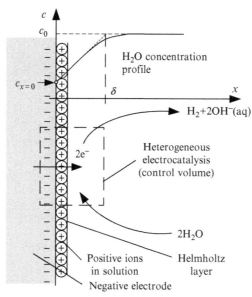

FIG. 3.5   Helmholtz layer and mass diffusion process at a polarized electrode.

The concentration of reactant molecule (e.g., water) decreases toward the electrode surface from a value specific to the bulk electrolyte c_0 to the surface value, $c_{x=0} < c_0$. Based on Fick's law of diffusion, the current density can then be written as

$$J_R = zFAD_R \left(\frac{c_0 - c_{x=0}}{\delta} \right)_R \tag{3.36}$$

for the reduction reaction and

$$J_O = zFAD_O \left(\frac{c_{x=0} - c_0}{\delta} \right)_O \tag{3.37}$$

for the oxidation reaction, where D is the diffusion coefficient, δ is the thickness of the diffusion layer and the index R refers to reduced species while the index O refers to the oxidized species.

In the preceding equation, observe that the current density displays a maximum limiting value that is obtained when the reactant concentration at electrode surface is zero. The diffusion-limited current density is therefore

$$J_{\lim,R} = zFAD_R \frac{c_{0,R}}{\delta_R} \tag{3.38}$$

for the reduction reaction and

$$J_{\lim,O} = zFAD_R \frac{c_{0,O}}{\delta_O} \tag{3.39}$$

for the oxidation reaction, where it is taken into account that both the forward and backward reactions can occur.

The limiting current density and the specific information related to actual electrolytic cell (e.g., electrode type and geometry and electrolyte type and parameters) allow for estimation of cell-specific concentration overpotential formulations. The net concentration overpotential is therefore given as

$$\Delta E_{conc} = \Delta E_{conc,R} - \Delta E_{conc,O} \tag{3.40}$$

where, $\Delta E_{conc,R}$ represents the concentration overpotential of the reduction half-reaction and $\Delta E_{conc,O}$ is the concentration overpotential of the oxidation half reaction.

In many practical cases, the kinetic process is not fast enough to reach 100% yield. In other extreme situations, when the kinetic process is relatively slow, it controls the electrochemical process. For such a case, it is assumed that because of the slow rate of the kinetics, there is no concentration gradient at the electrode surface; therefore, there is no diffusion process. Actual water electrolysis systems may operate such that neither kinetics nor transport processes (concentration gradients) are dominant. Fig. 3.6 shows the current density versus overpotential diagram for such situations.

FIG. 3.6 Current density versus overpotential in a cell with combined polarization.

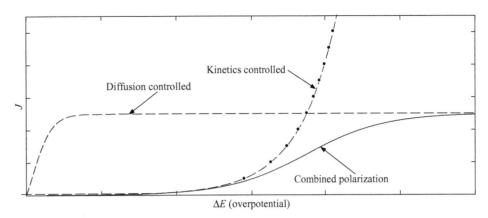

An electrical charge circuit must be closed by forcing the movement of ionic species between the anode and cathode and that of the electrons through the external circuit. Therefore, additional electric power is required to overcome these losses (overpotentials), which are ohmic in nature—namely,

$$\dot{W}_\Omega = I \sum \Delta E_\Omega \tag{3.41}$$

where the sum of all ohmic overpotentials is denoted with $\sum \Delta E_\Omega$.

The ohmic losses have different components depending on the actual type of electrolyzers (in subsequent sections, these are detailed for some specific cases). The total overpotential for an electrolysis cell becomes

$$\Delta E_{tot} = \Delta E_{pol,a} + \sum \Delta E_\Omega + \Delta E_{pol,c} \tag{3.42}$$

which can be divided by the net current I to obtain an equivalent circuit representation of the electrolysis cell as a series of electrical resistances, given as follows:

$$R_{eq} = R_a + \sum R_\Omega + R_c \tag{3.43}$$

The electric equivalent circuit is illustrated in Fig. 3.7. The circuit includes two diodes in direct polarization which suggest the reversible voltages required by the anode and cathode. Note that for an ideal diode in direct polarization, the voltage does not vary with the current; the same is the situation with reversible potentials of half-reactions. The Kirchhoff's law for the equivalent circuit is written as follows:

$$E_{tot} = E_{rev,a} + E_{rev,c} + IR_{eq} \tag{3.44}$$

Equation (3.44) can be solved for E_{tot} if the current is specified and the equivalent resistance is known. All components of electrical resistance (R_a, $\sum R_\Omega$, R_c) can be calculated provided that the current density is known. The cathodic and anodic resistances can be determined based on specific equations that give the polarization overpotential, ΔE_{pol}. As treated above, the polarization overpotential can be given according to one of the following situations:

- The cell operates in a kinetic controlled regime; then $\Delta E_{pol} = \Delta E_{act}$
- The cell operates in a concentration gradient controlled regime; then $\Delta E_{pol} = \Delta E_{conc}$
- The cell operates in a combined regime; then a combined overpotential can be derived, depending on the cell type.

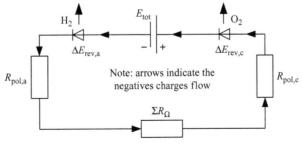

FIG. 3.7 Equivalent electrical circuit of a water electrolysis cell.

Noting that $(\Delta E_{pol})_{a,c} = J(AR_{pol})_{a,c}$, the polarization resistances at cathode and anode result as follows:

$$(AR_{pol})_{a,c} = \frac{(\Delta E_{pol})_{a,c}}{JA_{a,c}} \tag{3.45}$$

The ohmic resistances $\sum R_\Omega$ are those responsible for heat generation within the electrolyte and in the exterior electrical circuit, because of the Joule effect. Several ohmic resistances must be taken into account. One major type of ohmic loss occurs when the current passes across the electrolyte. To the same extent, the ohmic losses can be modeled, in general using the electrical conductivity (σ) and the thickness of the conductive layer (δ), according to

$$R_\Omega = \frac{\delta}{\sigma} \tag{3.46}$$

The dissipated power in the form of heat (because of the Joule's effect) can be calculated with

$$\dot{W}_{loss} = R_{eq} I \tag{3.47}$$

The design of the electrolysis system must be made such that the dissipation is minimized. If $\dot{W}_{loss} > \Delta H - \Delta G$, where ΔH and ΔG are reaction enthalpy and Gibbs free energy of water splitting at operating conditions, then the electrolyzer rejects heat to the exterior (it must be cooled) so that $\dot{Q}_{out} = \dot{W}_{loss} - (\Delta H - \Delta G)$. If $\dot{W}_{loss} < \Delta H - \Delta G$, then heat must be heated at a rate of $\dot{Q}_{in} = \Delta H - \Delta G - \dot{W}_{loss}$. If $\dot{W}_{loss} = \Delta H - \Delta G$, then the electrolyzer operates at thermoneutral conditions.

3.2.3 Efficiency Formulations for Electrolyzers

The energy efficiency of an electrolyzer can be expressed as the energy retrieved in the produced hydrogen by the energy input. The energy carried by a molar flow rate of hydrogen $\dot{n}$ is calculated based on the molar higher heating value $HHV = 283.6 \text{ kJ/mol}$, namely as the product $\dot{n}HHV$. The total electric power supplied to the electrolyzer is $E_{tot}I$. Thereafter, for a water electrolyzer, where $z = 2$, energy efficiency is defined as follows:

$$\eta_{el} = \frac{HHV}{2FE_{tot}} = \frac{1.4696}{E_{tot}} \tag{3.48}$$

If heat is supplied to the electrolyzer (such as in high-temperature electrolysis cells), then this is used as an input in the efficiency formulation; in this case, the energy efficiency is defined as follows:

$$\eta_{el} = \frac{1.4696}{E_{tot} + \dfrac{Q}{2F}} \tag{3.49}$$

The exergy efficiency of electrolysis systems is calculated using the molar chemical exergy of hydrogen ($ex^{ch} = 236.12 \text{ kJ/mol}$) as follows:

$$\psi_{el} = \frac{ex^{ch}}{zFE_{tot}} = \frac{1.224}{E_{tot}} \tag{3.50}$$

If one compares energy and exergy efficiencies from Eqs. (3.48) and (3.50), respectively, a quality factor of $CF = 0.83$ may be used, which expresses the ratio between chemical exergy and the higher heating value of hydrogen

$$CF = \frac{ex^{ch}}{HHV} = \frac{\psi_{el}}{\eta_{el}} = 0.83 \tag{3.51}$$

If the cell operation is assisted by heat input, then Eq. (3.50) can be modified to express the exergy efficiency as

$$\psi_{el} = \frac{1.224}{E_{tot} + \dfrac{ex^Q}{2F}} \tag{3.52}$$

where ex^Q is the exergy associated with the heat input.

In many practical situations, it is reasonable to determine an equivalent (or averaged) temperature (T^Q) at which the heat is supplied to the electrolyzer. Then the associated exergy is $ex^Q = (1 - T_0/T^Q)Q$, where T_0 is the reference temperature of the environment and the exergy efficiency becomes

$$\psi_{el} = \frac{1.224}{E_{tot} + \dfrac{Q}{2F}\left(1 - \dfrac{T_0}{T^Q}\right)} \tag{3.53}$$

When compared to other hydrogen production technologies, the efficiency of the water electrolysis process must account for the electrical power generation efficiency. In such a case, the efficiency equations must be written as

$$\begin{cases} \eta = \eta_{el}\eta_{pg} \\ \psi = \psi_{el}\psi_{pg} \end{cases} \tag{3.54}$$

where the subscript "pg" refers to power generation.

The efficiency of water electrolysis can be expressed by the second law efficiency, which is also known as the Faradaic efficiency. This efficiency represents the ratio between the reversible cell voltage (or reversible work) and the actual cell voltage (or actual work), given according to

$$\eta_{Faradaic} = \frac{E_{rev}}{E_{tot}} \tag{3.55}$$

where E_{rev} is the reversible cell voltage at given operating condition.

3.3 ALKALINE ELECTROLYZERS

Apparently, alkaline electrolysis is one of the major candidates for large-scale production of green hydrogen in the very near future. A review of alkaline electrolysis and its potential was done by Manabe et al. (2013), in which they note that a novel zero-gap technology was developed recently. As its name suggests, alkaline electrolyzers (AE) use an alkaline electrolyte (pH > 7). In an alkaline electrolyte, the mobile ionic species are anions (negative ions) of the hydroxyl group, OH^-. Typical electrolytes are liquid solutions of bases such as KOH or NaOH. The electrochemical half-cell reactions in an alkaline electrolyzer cell are as follows:

$$\begin{cases} 2H_2O + 2e^- \rightarrow H_2(g) + 2OH^- \text{ at cathode} \\ 2OH^- \rightarrow H_2O + 0.5O_2(g) + 2e^- \text{ at anode} \end{cases} \tag{3.56}$$

Fig. 3.8 depicts the general layout of a hydrogen production plant with alkaline electrolyzers. The plant requires a comprehensive water treatment unit. The purity of water can be a problem of much concern in water electrolysis.

The presence of metal atoms in water such as calcium or magnesium can cause reactions at the electrode surface that eventually lead to scale formation, consequently reducing an active surface area and enhancing clogging of the diaphragm. The salinity of water is important; if water contains NaCl, or other sources of chlorine ions, chlorine gas will emanate at the cathode with a highly corrosive effect.

Water is supplied at the cathode; which is made of inexpensive materials such as stainless steel coated with nickel. The anolyte solution—here water-generated according to Eqs. (3.56)—must be recirculated to reduce the concentration of overpotentials. The anode material is typically nickel. The anolyte passes through an electrolyte separator above which oxygen gas evolves. Moisture removal units (condensers) are applied at both oxygen and hydrogen evolution manifolds. The product gases also must be dried before release. The purity of hydrogen is higher than 99.7%. The electrolysis process is endothermic; therefore, the anolyte is preheated at reaction temperature to maintain good process efficiency.

Fig. 3.9 depicts the range of cell voltage for alkaline electrolyzers as a function of current density. Commercial alkaline electrolyzers have an efficiency in the range of 60–70%, while advanced systems—currently in development—can surpass 90%. Commercial alkaline electrolyzers operate at temperatures of 80–200°C with production capacities in the range of 500–30,000 Nm³/h (or equivalent 0.5–40 MW per unit with respect to the LHV of produced hydrogen). The concentration of the electrolyte must be high enough such that it assures a good mobility of ions; typically for KOH this is 30%. The energy efficiency (electricity-to-hydrogen) of commercial electrolyzers is in the range of 55–90%. When

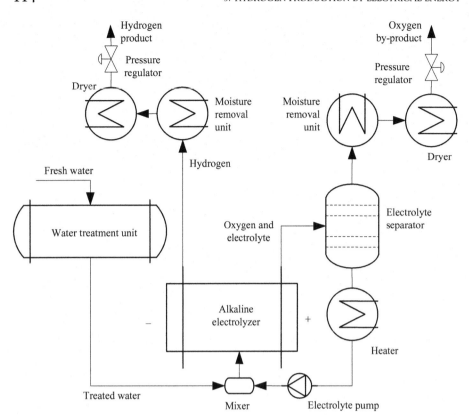

FIG. 3.8 General layout of a water alkaline electrolysis plant.

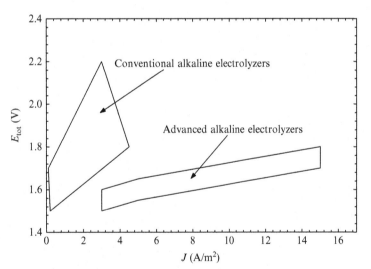

FIG. 3.9 Range of cell voltages for alkaline electrolyzers as a function of current density.

operating at higher temperatures (e.g., over 120°C), a part of the energy needed to split water can be transmitted by heat transfer from an external/sustainable heat source.

The electrodes are made on a planar geometry to enhance the contact area with the electrolyte, and most importantly, they do not require expensive catalysts (normally carbon steel-coated with nickel is used). Due to these characteristics, alkaline electrolyzers can be constructed in a cost-effective manner for higher production capacities.

Alkaline electrolyzers are constructed in two configurations: monopolar and bipolar. The monopolar configuration is sketched in Fig. 3.10a. In a monopolar configuration, several electrolysis cells are packed in series such that the voltage applied to the cell is the sum of voltages for each cell:

$$E_{AE} = \sum E_{cell} \cong (N_{cell} - 1)E_{cell} \tag{3.57}$$

whereas the current through each cell is the same.

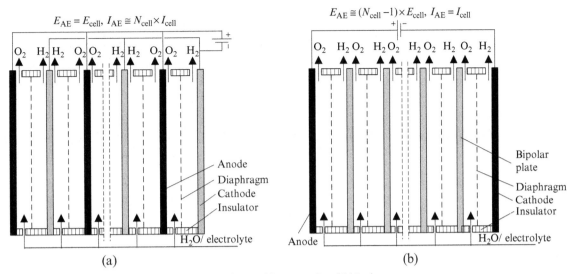

FIG. 3.10 Constructive configurations of alkaline electrolyzers: (a) monopolar; (b) bipolar.

For constructive reasons, the ohmic losses in monopolar electrolyzers are higher with respect to a bipolar configuration. In a bipolar configuration—Fig. 3.10b—the cells are connected in parallel such that the voltage is the same for all cells, while the total current is given as follows:

$$I_{AE} = \sum I_{cell} \cong N_{cell} I_{cell} \tag{3.58}$$

The gap between electrodes must be reduced as much as possible, but in a manner that avoids any surge of current between the two opposite electrodes. Cells with a zero-gap configuration were developed to minimize the ohmic losses through an electrolyte while increasing the purity of hydrogen. In a zero-gap configuration, an electrode-diaphragm assembly is devised as a sandwich-like geometry having the diaphragm permeable only to liquid and dissolved species but not to gases. The cathode and anode electrodes attached at both sides of the diaphragm have pierced holes of ~0.5 mm (diameter) to allow for the electrolyte permeation and thus permitting the ionic species transport.

Julius Tafel established experimentally in 1905 the well-known Tafel's equation, which relates the activation overpotential of a polarized electrode (a thermodynamic parameter) to the current density (a kinetic parameter). The Tafel equation is equivalent to Eq. (3.35), and it is written as follows:

$$\Delta E_{act} = a + b \ln(J) \tag{3.59}$$

The parameter a in Tafel's equation represents the value of polarization potential required for a current density equal to unity to be established. Typically, the parameter a has values in the range of 0.03–3 V. The parameter b is known as "Tafel's slope," and its range of values is $\sim 0.04 - 0.150$ V at ambient temperature. From Eqs. (3.35) and (3.59), the following can be derived for the Tafel constants:

$$\begin{cases} b = \dfrac{RT}{\alpha F} \\ a = b \ln(J_0) \end{cases} \tag{3.60}$$

In Fig. 3.11, a typical current versus activation overpotential diagram is shown for an alkaline electrolyte with a Ni electrolyte; the reaction conditions and the Tafel slope and exchange current are indicated. It is observed that at low polarization, the curve is linear, whereas at higher polarizations, the behavior becomes exponential. Note that this behavior is valid when the reaction is kinetically controlled.

The practical range of diffusion layer thickness in an alkaline electrolyzer is 50–500 μm. The limiting current density is in the range of 8000–10,000 A/m². This is about 10^6 times higher than the exchange current. If the limiting current is known, then the ratio between bulk concentrations c_0 and surface concentrations $c_{x=0}$ can be obtained by way of

$$\left(\frac{c_{x=0}}{c_0}\right)_O = 1 + \frac{J}{J_{\lim,O}} \tag{3.61}$$

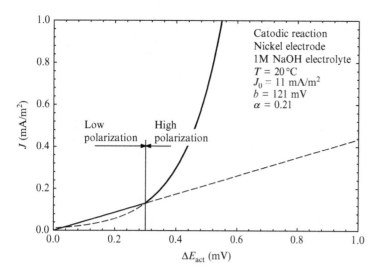

for the reduction half-reaction and

$$\left(\frac{c_{x=0}}{c_0}\right)_O = 1 + \frac{J}{J_{\text{lim},O}} \tag{3.62}$$

for the oxidation half reaction.

In the electrolyzers operating with solid electrodes immersed into a liquid electrolyte—such as the alkaline electrolyzers are—one can assume that, in the very vicinity of the electrode, the electrolyte has an ideal solution behavior. In this case, the equilibrium constant, as introduced by Eq. (3.12), must be equal to the ratio of concentrations of products $c_{x=0}$ at the surface and reactants at bulk c_0. This means that the concentration overpotential can be determined with

$$\Delta E_{\text{conc}} = \frac{RT}{zF} \ln\left(\frac{c_{x=0}}{c_0}\right) \tag{3.63}$$

Note that Eq. (3.63) is valid for both the cathode and the anode. In addition, Eq. (3.61) can be applied together with Eq. (3.63) to express the concentration overpotential at the cathode as follows:

$$\Delta E_{\text{conc},R} = \frac{RT}{zF} \ln\left(1 - \frac{J}{J_{\text{lim},R}}\right) \tag{3.64}$$

In a similar manner, Eqs. (3.62) can be applied together with Eq. (3.63) to express the concentration overpotential at the anode as follows:

$$\Delta E_{\text{conc},O} = \frac{RT}{zF} \ln\left(1 - \frac{J}{J_{\text{lim},O}}\right) \tag{3.65}$$

Since the net concentration overpotential is given by $\Delta E_{\text{conc}} = \Delta E_{\text{conc},R} - \Delta E_{\text{conc},O}$ on has that the cell overpotential due to species concentration is given as follows

$$\Delta E_{\text{conc}} = \frac{RT}{zF} \ln\left(\frac{1 + \dfrac{J}{J_{\text{lim},R}}}{1 - \dfrac{J}{J_{\text{lim},O}}}\right) \tag{3.66}$$

For a combined kinetic diffusion-controlled regime, a superposition equation for concentration and activation overpotentials can be obtained. According to Bagotski (2006), the equation that describes a combined kinetic diffusional polarization process is

$$\frac{J}{J_0} = \frac{\exp\left(\dfrac{\alpha F}{RT}\Delta E_{\text{pol}}\right) - \exp\left(-(1-\alpha)\dfrac{F}{RT}\Delta E_{\text{pol}}\right)}{1 + \exp\left(\dfrac{\alpha F}{RT}\Delta E_{\text{pol}}\right)\dfrac{J_0}{J_{\text{lim},R}} + \exp\left(-(1-\alpha)\dfrac{F}{RT}\Delta E_{\text{pol}}\right)\dfrac{J_0}{J_{\text{lim},O}}} \tag{3.67}$$

FIG. 3.12 Polarization potential at a hydrogen evolving platinum electrode.

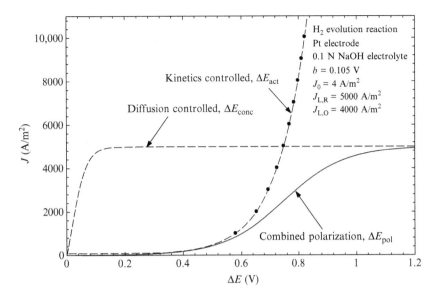

where ΔE_{pol} is the electrode polarization overpotential due to combined effects of activation energy for the reaction to occur and due to concentration gradients that need to be generated for reactants to be supplied at the reaction sites.

Fig. 3.12 shows the current versus polarization overpotential diagram for a hydrogen evolving platinum electrode and a cell configuration in which the limiting forward direct is 5000 A/m² , whereas the backward current is 4000 A/m². Eq. (3.67) is used to construct the plot.

Some typical electrode materials for a cathode in alkaline electrolyzers are Ni, Fe, Co, Zn, Pb, Pd, Pt, and Au; whereas for a cathode, Ni, Pt, Ir, Ru, Rh, TiO₂, Co, and others are used. Table 3.5 gives the main kinetic parameters of some electrodes commonly used in alkaline water electrolyzers. Typically, the operating current density of commercial alkaline electrolyzers is in the range of 1–3 kA/m², and for advanced alkaline electrolyzers, it covers 2–15 kA/m².

One of the main problems with the electrodes is the alteration in time caused by various electro-chemical reactions and deposition processes that may occur at their surface. These processes may lead to deactivation of electrocatalysts (if any are used) and a to a decrease of current density while the overpotentials will increase. With Ni-based electrodes, deactivation occurs because of the formation of a nickel hydride. The maintenance operations requiring circulation of dissolved vanadium species in the electrolyte may remove the nickel hydride and re-establish the clean-surface electrodes. The enhancement of electrode performance is typically obtained by doping its surface with components that may act as very active electrocatalysts. For example, electrodes in alkaline electrolyzers can be doped with Pt, Pd, Mo, Ru, and Ir. The structure of the electrode surface plays a major role. Nanostructured surfaces were recently engineered to obtain higher surface area and better current densities.

Regarding the electrolyte in alkaline electrolyzers, there are two specific issues: (i) mass transport must be controlled to reduce the diffusion layer and (ii) the unavoidable gas bubbles formed at an electrode surface must be removed as

TABLE 3.5 Activity Parameters for Some Electrodes Specific to Alkaline Electrolyzers

Cathode material	Electrolyte	T (°C)	J_0 (A/m²)	b (mV)	Anode material	Electrolyte	T (°C)	J_0 (A/m²)	b (mV)
Ni[a]	50% KOH	80	1.10	140	Ni	50% KOH	80	4.2×10^{-2}	95
		150	13.1	167			150	1.8	125
		208	40.0	195			208	20	110
		264	43.0	133			264	10	32
Pt	0.1 N NaOH	25	0.676	114	Pt	30% KOH	80	1.2×10^{-5}	46
Zn	9 N KOH	25	1.5×10^{-5}	124	Co	30% KOH	80	3.3×10^{-2}	126
Zn+2% Hg	9 N KOH	25	2.7×10^{-6}	116	TiO₂+0.3NiO	30% KOH	80	1.4	259
Fe+75% Ni	0.1 N NaOH	25	3.5×10^{-2}	115	Fe	30% KOH	80	0.17	191
Carbon	40% NaOH	40	0.29	148	0.75Ni+0.25Ru	30% KOH	80	8.8×10^{-3}	67

[a]*High surface area electrode* $(0.1 - 30 \, m^2/g)$.

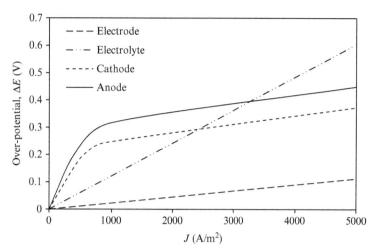

best as possible to avoid an excessive increase of ohmic losses. Both these problems can be reasonably solved with appropriate additives dissolved in the electrolyte.

The additives aim to reduce surface tension and thus to increase diffusion mechanisms at the electrode surface and to accelerate the bubble detachment. Note that the bubble detachment requires an activation energy that translates to a slight increase of overpotential. Based on liquid-vapor surface tension (γ) and the contact angle (θ), the overpotential due to bubble detachment can be approximated as follows:

$$zF\Delta E_{\text{bubble}} = \gamma(cos\theta - 1) \tag{3.68}$$

In Fig. 3.13, a typical variation of overpotentials in commercial electrolyzers is shown for industrial applications. The ohmic losses in electrodes are the smallest, and the overpotential at the anode is the highest. Furthermore, it can be observed that at the higher range of current densities, the ohmic losses in the electrolyte become significant.

The ohmic losses within the electrolyte are usually determined by the molar conductivity (Λ) in $m^2\Omega^{-1}mol^{-1}$, which is defined as electrical conductivity (σ) in $\Omega^{-1}m^{-1}$ divided to molar concentration (c) in mol/m^3. Ohmic losses are relatively insensitive to current density but may be sensitive to temperature variations. The equivalent resistances for each of the losses are presented in Fig. 3.14. The equivalent electrical resistance corresponding to polarization decreases with the current density.

Some enhanced uses of alkaline electrolyzers are known for various systems and applications. Among those uses is the application of magnetic fields, which is one of the most promising uses for large-scale hydrogen production. Magnetic fields are known to influence electrical charges through Lorentz force, which increases the convection mechanisms in the electrolyte solution due to magneto-hydrodynamic processes.

If properly applied, the magnetic field can reduce the concentration potentials substantially as well as reduce the ohmic losses of an electrolyte because of a better transport of charges. Three classes of electrodes were analyzed by

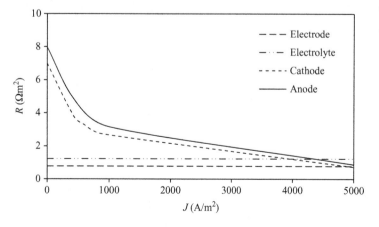

FIG. 3.14 Equivalent resistances within an alkaline electrolyzer for a range of current densities.

FIG. 3.15 Effect of magnetic field on gas bubble convection in an alkaline electrolyzer.

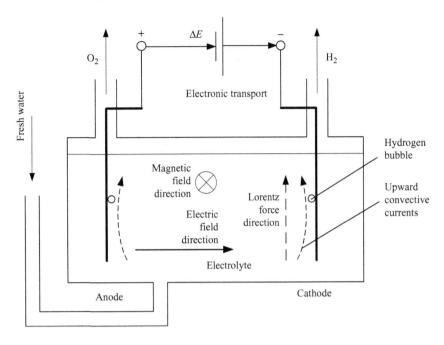

Lin et al. (2012) to investigate the influence of a magnetic field in electrolysis processes: ferromagnetic (e.g., nickel), paramagnetic (e.g., platinum), and dia-magnetic (e.g., graphite). The Lorentz force is achieved by

$$\vec{\mathbb{F}} = q\vec{\mathbb{E}} + q\vec{v} \times \vec{\mathbb{B}} \tag{3.69}$$

where q is the electrical charge of mobile ions, $\vec{\mathbb{E}}$ is the intensity vector of the electric field (in N/C), $\vec{v}$ is the velocity vector of electric charges (in m/s) and $\vec{\mathbb{B}}$ is the vector magnetic flux density (in T, Tesla).

In a magnetically assisted electrolysis cell, the magnetic field is perpendicular to the electric field. Fig. 3.15 illustrates the effect of a magnetic field on convective currents in an alkaline electrolyzer. The magnetic field is applied by using permanent magnets placed on two opposite sides of the electrolysis bath such that an upward Lorentz force is created.

The effect of the Lorentz force is a net convective current with upward direction at the electrode surface that facilitates the removal of gas bubbles. Consequently, the concentration of overpotentials decreases, and ohmic resistances caused by the presence of bubbles also decreases. The best electrodes with respect to magnetic field-assisted alkaline electrolysis appear to be those based on nickel, because of its good electrochemical activity, together with ferromagnetic behavior. The application of nickel electrodes is more beneficial than platinum electrodes when magnetic field is present, as they increase the current density two times with respect to platinum. Also, nickel is much cheaper than platinum; therefore, it appears as the preferred electrode for alkaline electrolyzers.

An interesting technique to enhance water electrolysis efficiency—applicable to alkaline electrolyzers—is to couple the heterogeneous electrocatalysis process occurring at the electrode surface with a homogenous catalytic process that takes place the electrolyte volume, near to the electrolyte-electrode interface. A net enhancement of hydrogen production is obtained from the action of homogeneous catalysts that work in parallel with the heterogeneous process.

Recently, Karunadasa et al. (2010) developed a molybdenum-based oxo-catalyst for a water reduction half reaction at a platinum electrode. The heterogeneous and homogeneous reaction mechanisms compete. The general reaction scheme can be expressed as

$$\left. \begin{cases} 2H_2O + 2e^- \rightarrow H_2(g) + 2OH^-(aq), \\ \text{at the electrode (heterogeneous catalysis)} \\ CAT + 2e^- \rightarrow CAT^{2-}, \\ \text{at electrode surface} \\ 2H_2O + CAT^{2-} \rightarrow H_2(g) \\ \quad + 2OH^-(aq), \text{ in solution} \end{cases} \right\} \text{(homogeneus catalysis cycle)} \tag{3.70}$$

The reaction mechanism from Karunadasa et al. (2010) is illustrated in Fig. 3.16. According to Eq. (3.70), the process consists of a main catalysis cycle that requires electron transfer from the electrode to the macro-molecular catalyst, which is dissolved in solution. The turnover frequency obtained with this catalyst is very high and reaches a limiting value of 1700 moles of hydrogen per mole of catalyst per hour. Moreover, this is considered as an inexpensive catalyst because it is not composed of noble metals. The catalysts can be used for seawater electrolysis.

Seawater is an alkaline solution rich in NaCl. The electrolysis of sea water is therefore an alkaline electrolysis process. It is a very challenging process because of the presence of multiple types of ions dissolved in seawater, among which are chlorine, magnesium, calcium which affect the electrode reactions in various ways. The presence of chlorine ion favors the formation of chlorine gas at anode, rather than of oxygen gas. Precipitates form at the cathode surface especially because of the formation of the hypochlorite ion ClO^-, which can combine easily with calcium and generate calcium hypochlorite—$Ca(ClO)_2$—which is a relatively stable and persistent deposit.

In a conventional seawater electrolysis cell, an anion selective membrane is used to separate the anode and cathode. As a result, the cathode alkalinity increases with the pH of the catholyte over 12 since hydrogen generates there. Furthermore, the pH at the anodic compartment decreases to ~2 because of the formation of chlorine and oxygen gas, which then protonates the solution. This process leads to an increase of the reversible cell voltage and furthermore to the chemical precipitation of magnesium, $Mg^{2+} + 2OH^- \rightarrow Mg(OH)_2$, and calcium $Ca^{2+} + 2OH^- \rightarrow Ca(OH)_2$ hydroxides at the cathode.

Seawater electrolysis received recent attention because seawater is an abundant resource from which hydrogen can be generated. Direct seawater electrolysis may be beneficial because it eliminate the need of a water desalination and purification step, which is an energetically intense process. However, the chlorine gas that evolves at anode during seawater electrolysis cannot be released freely in the atmosphere, because of its high toxicity. Manganese anodes coated with iridium dioxide show high selectivity for oxygen formation rather than chlorine formation, and they can be used to limit chlorine evolution within a seawater alkaline electrolysis process focused on hydrogen generation. The heterogeneous and homogeneous electrocatalysis processes described by Eq. (3.70) using the inexpensive molybdenum-based catalyst are promising for a direct seawater electrolysis with hydrogen-oxygen generation.

Baniasadi et al. (2013) studied a configuration of an alkaline electrolysis cell that uses cation membrane and the molybdenum oxo-catalyst previously developed by Karunadasa et al. (2010) to conduct seawater electrolysis with oxygen and chlorine evolution at a nickel electrode. In this cell, the acidic anolyte is mixed with alkaline feedstock seawater to reduce the alkalinity at the cathode and diminish the formation of electrode depositions. The following hydroxyl-consuming reactions are promoted in this way, which will reduce the cathodic pH, namely $Cl_2(g) + 2OH^- \rightarrow CLO^- + Cl^- + H_2O$ and $6ClO^- + 6OH^- \rightarrow 2ClO_3^- + 4Cl^- + 6H_2O + 6e^-$.

Power consumption reduction of alkaline electrolyzers is of major concern. Mazloomi et al. (2012) proposed a method to reduce the power consumption of alkaline water electrolyzers to generate hydrogen by targeting the

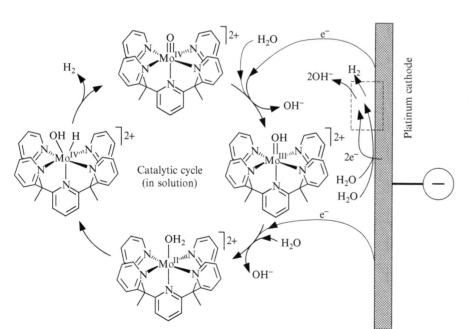

FIG. 3.16 Hybrid catalysis (heterogeneous/homogeneous) for hydrogen evolving reaction. *Modified from Karunadasa, H.I., Chang, C.J., Long, J.R., 2010. A molecular molybdenum oxo-catalyst for generating hydrogen from water. Nature 464, 1329–1333.*

resonant frequency of the cell under a pulsating DC. It is shown that, by pulsating the DC voltage, one can maintain the current density to the desired value with a correspondingly lower consumption of power. The tests indicate a 15% power reduction with this method, as compared to the conventional constant DC cell control.

3.4 PEM ELECTROLYZERS

A PEME is a device that splits water by electrolysis using an acidic electrolyte. Across acidic electrolytes, the species that are transported are positive ions (or cations). In the case of water electrolysis, the ionic transport results in a net transfer of protons from anode to cathode. The overall electrochemical reactions for acid electrolyte electrolyzers is

$$\begin{cases} H_2O \rightarrow 2H^+ + 1/2O_2(g) + 2e^- \text{ at anode} \\ 2H^+ + 2e^- \rightarrow H_2(g) \text{ at cathode} \end{cases} \tag{3.71}$$

The operating temperature range of a PEME is restricted to 25–80°C. Typical electricity consumption of a commercial PEME is 23–26 MJ electric per normal cubic meter (Nm^3) of hydrogen produced or ~540–580 MJ per kg of hydrogen produced. The energy efficiency of PEME drops quasi-exponentially with the current density and reaches 54% at ~10 kA/m^2 with 80–85% conversion. The generated hydrogen has a purity higher than 99.999%. The capacity range of an industrial PEME is ~0.2–60 Nm^3/h or from about 10 g/h to ~2.5 kg/h. The associated electrical energy consumption of a commercially available PEME can go up to a maximum of 400 kW while the water consumption reaches at most 25 L per hour.

The general layout of a PEME electrolyzer system is presented in Fig. 3.17. The water treatment process must be carefully sized to avoid impurities and therefore formation of deposits on the electrodes. The membrane is permeable to cations; therefore, some cations such as Ca$^+$ or copper ions present in water can cross over and clog the membrane.

The common stack arrangement for a PEME uses bipolar plates. Bipolar plates comprise extruded channels that facilitate a two-phase water-gas flow on both sides. Fig. 3.18 depicts the geometrical arrangement of a PEM electrolyzer stack. It comprises several bipolar plates with channels on both sides (only one is illustrated in the figure) and two end plates. Each plate is firmly attached to a planar, porous, electrically conductive layer that serves as the electrode (cathode or anode). A proton exchange membrane is placed between the electrodes.

The electrodes and bipolar plates of a PEME must have high resistance to corrosion because the membrane creates an acidic medium. Therefore, expensive platinum group metal must be used for the electrodes to coat the bipolar plates. At the cathode, the higher overpotential occurs; for its minimization, electrocatalysts such as Pt and oxides of Ir and Ru are commonly used. Thus the cost of PEME electrodes is perceived to be relatively high.

The most common acid electrolyte is a perfluorosulfonic acid polymer of which the best-known tradename is NAFION®. The sulfonate groups in a polymeric chain make the electrolyte behave as a strong acid. NAFION® presents a thin, solid membrane. The ionic species that permeate through NAFION® is the hydronium (H^{3}O$^+$) in the aqueous

FIG. 3.17   General layout of a PEM electrolyzer system.

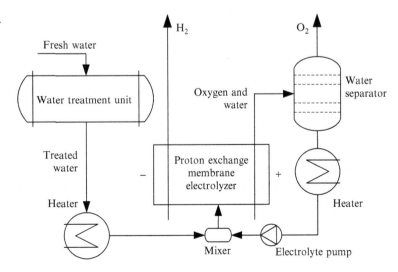

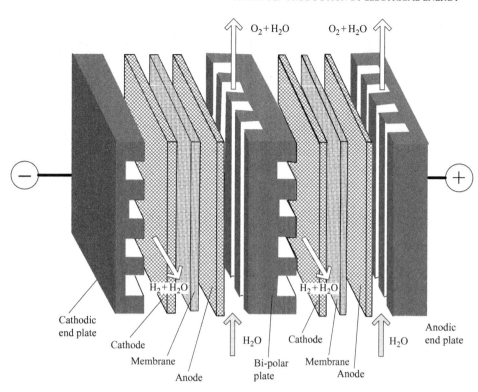

FIG. 3.18   A geometric arrangement of a PEM electrolyzer stack.

solution. The membrane must always be wet to allow for ionic transport. Moreover, one needs to compensate for a backward permeation process of molecular oxygen through the membrane by increasing the electrical current density by ~5%.

The function of an ion exchange membrane is essentially as a separator. In an electrolysis process, cation exchange membranes containing fixed negatively charged ions are used to stop the transportation of anions from cathode compartment to anode compartment. The fixed ions of the membrane are in equilibrium with the counter ions, whereas the ions that carry the same charge as the fixed ions (co-ions) are efficiently excluded from the membrane matrix. This effect is called the Donnan exclusion effect, as illustrated in the sketch shown in Fig. 3.19.

The most significant characteristics of an ion exchange membranes are the following: high permselectivity for counter-ions (excluding co-ions), high ionic conductivity, good mechanical form, and chemical stability. The permselectivity of an ion exchange membrane refers to the ability of the membrane to reject co-ions. As a result of the Donnan exclusion effect, the permselectivity is affected by the electrolyte concentration of the surrounding solution and by the ion exchange capacity of the membrane. An ideal permselective membrane should completely dismiss co-ions from the membrane matrix.

FIG. 3.19   Illustration of Donnan exclusion effect.

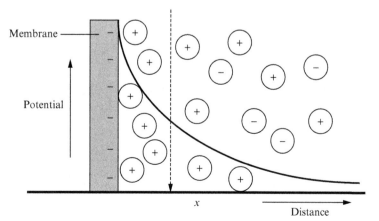

The ionic concentration of the electrolyte solution is subject to the concentration of the electrolyte solution, which affects the swelling of the membrane (water content in the membrane). Additionally, the water content of the membrane, interaction between the mobile and fixed ions, and temperature have an intense effect on the mobility of the ions through the membrane matrix. In general, water content increases the free volume inside the membrane matrix, which improves the ability of ions to move across the polymer.

The capital cost of PEME is relatively high because of the cost of the NAFION® proprietary membrane (approximately $600/m²) and the expensive metals required for the electrodes and bipolar plates. From the total electrolytic hydrogen cost, ~58% represents the operation cost and the remaining 32% reflects the capital cost. A PEM electrolyzer requires ultrapure distilled water for proper operation.

The resistance associated with the Nafion membrane also influences the efficiency of PEM electrolyzers. It was found that increasing the membrane thickness from 50 to 200 μm decreased the efficiency of a PEM electrolyzer from 60% to 56% at a current density of 9800 A/m² because of the increase in the required cell voltage.

The most significant driving force for charge transport in a membrane cell process is the electrical potential gradient that is produced because of the depletion of anions (or accumulation of cations) on the surface of anode and, conversely, the depletion of cations (or accumulation of anions) on the surface of cathode. As a result, the positively charged cations transport from the anode to the cathode compartment. The ability of a material to conduct electric current is measured by its conductivity. The following equation relates the conductivity to the resistance of a conductor:

$$R_c = \frac{l}{\sigma A} \tag{3.72}$$

Here the conductor resistance is denoted by R_c, the conductor conductivity is denoted by σ. l is the length of the conductor, and A is the cross-sectional area of the current flow. Calculations of ohmic losses through PEM make use of specific equations for electrical conductivity across the membrane, which is calculated with the following formula from Meng et al. (2008):

$$\sigma(c, T) = (0.5139c - 0.326) \exp\left(1268\left(\frac{1}{303} - \frac{1}{T}\right)\right) \tag{3.73}$$

Here c is the local molar concentration of water inside the membrane, and T is the membrane temperature.

The molar concentration varies across the membrane thickness from a high concentration Λ_a at the anode (where water is supplied) to a lower concentration Λ_c at the cathode. At any distance across the membrane, $x \in [0, \delta]$, the molar concentration is $c(x) \in [c_c, c_a]$, where $x = 0$ is the anode side and $x = \delta$ is the cathode side of the membrane. It is reasonable to assume that the molar water content varies linearly across the membrane:

$$c(x) = c_c + \frac{c_a - c_c}{\delta} x \tag{3.74}$$

The electrical conductivity is defined by $\sigma \equiv dx/dR$, where R is the electrical resistance; thus the differential equation $dR = \sigma^{-1} dx$ can be integrated from 0 to δ:

$$R_{\Omega, \text{PEM}} = \int_0^\delta \sigma^{-1}(c(x), T) dx \tag{3.75}$$

The concentration overpotentials for PEM electrolyzers must take into account the variation of the water vapor concentration across the membrane. The order of magnitude of the limiting current density is $J_{\text{lim}} = 20,000 \, \text{A/m}^2$. The concentration overpotential is given by

$$\Delta E_{\text{conc}} = J^2 \left[\beta\left(\frac{J}{J_{\text{lim}}}\right)^2\right] \tag{3.76}$$

where the factor β is defined by

$$\beta = \begin{cases} (7.16 \times 10^{-4} T - 0.622)P + (-1.45 \times 10^{-3} T + 1.68) & \text{if } P < 2 \, \text{atm} \\ (8.66 \times 10^{-5} T - 0.068)P + (-1.6 \times 10^{-4} T + 0.54) & \text{if } P \geq 2 \, \text{atm} \end{cases} \tag{3.77}$$

where $P = P_i/0.1173 + P_{\text{sat}}$ is the local pressure at the anode or cathode, P_i is the partial pressure at the anode or cathode, and P_{sat} is the saturation pressure of water at the operating temperature; the index i becomes "a" for anode and "c" for cathode.

The activation potential at anode and cathode in PEM electrolyzers accounts for the electron transfer energy at the membrane electrode assembly level. The activation potential can be determined based on the Meng et al. (2008) work,

$$\Delta E_{act,i} = \frac{RT}{F} \ln \left(0.5 \frac{J}{J_{0,i}} + \sqrt{\left(0.5 \frac{J}{J_{0,i}} \right)^2 + 1} \right) \tag{3.78}$$

where $J_{0,i}$ is the exchange current density at anode and cathode, respectively.

The exchange current density is a significant parameter in calculating the activation overpotential. It characterizes the electrode's capabilities in the electrochemical reaction. A high exchange current density implies a high reactivity of the electrode, which results in a lower overpotential. The exchange current density for electrolysis can be expressed based on the Arrhenius-type equation

$$J_{0,i} = J_{ref,i} \exp \left(-\frac{\Delta E_{act,i}}{RT} \right) \tag{3.79}$$

where $J_{ref,i}$ is a pre-exponential factor.

Fig. 3.20 shows a case study illustrating that the ohmic overpotentials in a PEM electrolyzer account for less than 5% of the total overpotentials, while the anodic activation overpotentials in a PEM electrolyzer occupy ~60–80% of the total overpotential. This is a typical case. It suggests that hydrogen transport through a NAFION® membrane is highly efficient; however, there are significant losses because of polarization.

Fig. 3.21 illustrates the magnitude of total irreversibilities of a PEM electrolyzer and their trend with operating pressure and temperature. The irreversibilities are decreasing with higher operating temperature and pressure. As a result,

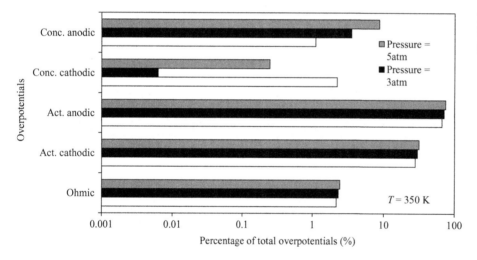

FIG. 3.20 Typical overpotential distribution in PEM electrolyzer at various pressures.

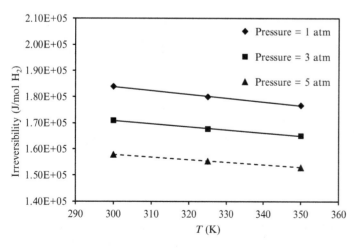

FIG. 3.21 Typical irreversibility within a PEM Electrolyzer operating at various pressures.

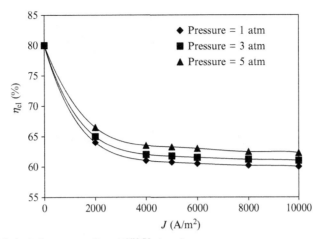

FIG. 3.22 Energy efficiency of PEM electrolyzer operating at 350 K at various pressures.

FIG. 3.23 Cathodic activation overpotentials of a PEM electrolyzer versus current densities.

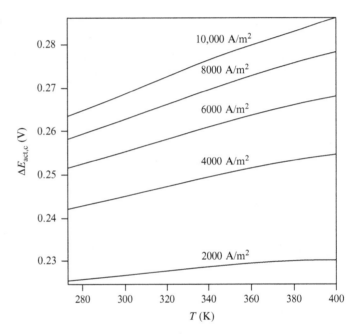

a minor increase in efficiency is occurring with higher temperature and pressure. This observation is confirmed by the efficiency profiles shown in Fig. 3.22 where the influence of current density on electrolyzer performance is also studied.

The variation of overpotentials in PEM electrolyzers with a current density is shown in Figs. 3.23–3.27. The current density varies between 2000 and 10,000 A/m^2 in increments of 2000 A/m^2. The activation overpotential ranges from 0.22 to 0.28 V at the cathode and 0.55–0.75 V at the anode. The concentration overpotentials are negligible for current densities below 10 kA/m^2; Figs. 3.25 and 3.26 shows that for 5 kA/m^2, the absolute value of concentration overpotential is below 20 mV for the cathode and below 5 mV for the anode.

3.5 SOLID OXIDE ELECTROLYZERS WITH OXYGEN ION CONDUCTION

Solid oxide electrolyzers with oxygen ion conduction (SOE-O) emerged in recent years as a promising technique to split water in the form of steam at high temperatures. They use a solid, non-porous metal oxide electrolyte, which is conductive to oxygen ions, O^{2-}. The two half reactions in a SOE-O are

$$\begin{cases} 2O^{2-} \rightarrow O_2(g) + 4e^- \text{ at anode} \\ H_2O(g) + 2e^- \rightarrow H_2(g) + O^{2-} \text{ at cathode} \end{cases} \tag{3.80}$$

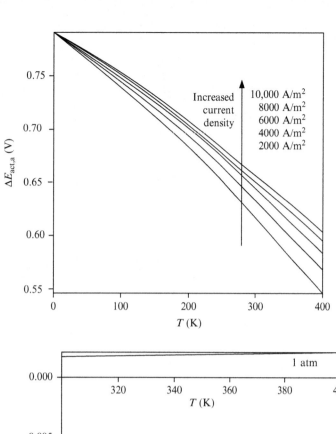

FIG. 3.24 Anodic activation overpotentials in a PEM electrolyzer versus current density.

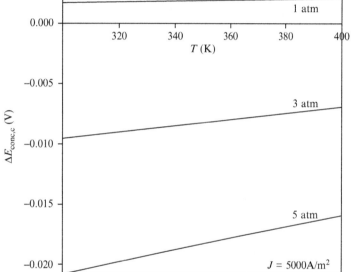

FIG. 3.25 Cathodic concentration overpotentials in a PEM electrolyzer for various pressures.

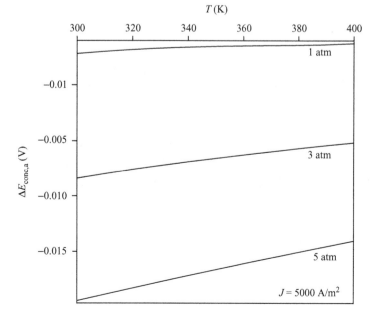

FIG. 3.26 Anodic concentration overpotentials in a PEM electrolyzer for various pressures.

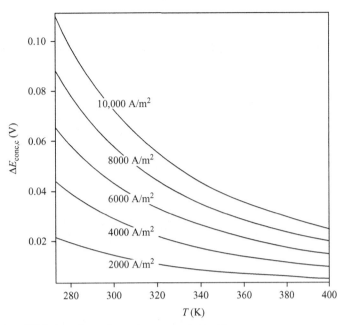

FIG. 3.27 Ohmic overpotentials in a PEM electrolyzer for various current densities.

A cross-sectional view through a SOE-O cell is illustrated in Fig. 3.28, which shows the geometrical arrangement of its components (bipolar or end plates, gas channels, anode, cathode, and electrolyte) and the flowing paths of chemical species. The electrolyte-electrode assembly consists of a thin layer of solid oxide packed in a sandwich-like structure between a porous cathode and a porous anode. The cathode is permeable to hydrogen and steam.

The cathode has a composite structure comprising a porous matrix made in yttrium-stabilized zirconium (YSZ) combined with nickel and/or cerium. The role of the electrocatalyst is served by nickel atoms, which are dispersed over the porous matrix. The hydrogen evolving reaction occurs within the porous structure of the cathode in the vicinity of active centers. Once formed, molecular hydrogen diffuses toward the collecting channel under the influence of a concentration gradient. Excess steam must be present in the cathodic channel to reduce the partial pressure of hydrogen and to facilitate its diffusion. Oxygen ions diffuse toward the positively charged anode under the influence of an electric field and concentration gradient.

Different ratio combinations of a mixture of H_2O, H_2, and CO_2 can be supplied to Ni/YSZ electrode active areas at temperatures between 750°C and 850°C. The cell operation, there is considerable change in absolute humidity of feed stream, which may be attributed to the steam diffusion at the hydrogen electrode. The area-specific resistance (ASR) of s = test SOE cells while operating at 800°C and 70% steam concentration for LSM/YSZ and LSFC is around

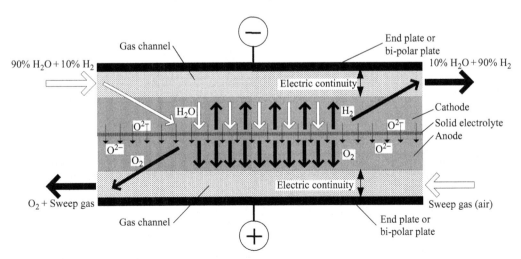

FIG. 3.28 Operating principle and geometrical configuration of a SOE-O cell.

$0.85\ \Omega\,cm^2$, respectively. A long-term degradation of 0.003–0.006 mV/h is possible while operating at –0.25 A/cm². Delamination is generally observed around the steam and air inlets in the case of LSM and LSC. Agglomeration of Ni was also reported in the hydrogen electrodes of the LSM and LSCF cells, which may lead to a reduction in the electrode active area.

The anode is normally composed of lanthanum and manganese oxides doped with strontium (LMS) with the chemical formula $La_{0.8}MnO_3Sr_{0.2}$. The oxygen ion oxidizes within the porous structure of an anode which is maintained at a positive potential by the electrical power supply. The manufacturing of the electrode-electrolyte assembly is made using the anode (~1.5 mm thickness) as a mechanical support on which is deposited first the electrolyte layer (~140 μm) and then, over the electrolyte, the cathode is deposited (~13 μm thickness). Over the cathode, a layer of pure nickel of ~10 μm thickness is applied. The bipolar plates are made in ferritic stainless steel, which is resistant to high temperatures and which presents good electric conductivity. In order to facilitate the diffusion of molecular hydrogen through the anode, a sweep gas is generally used (e.g., air), which decreases the partial pressure of oxygen.

Ceramic-based electrolytes comprising YSZ or scandium-stabilized zirconium (ScSZ) are commonly used in SOE-O construction. At temperatures of 1000–1300 K, these electrolytes allow the passage of oxygen ions O^{2-} under a driving force created by an electrical field and/or concentration gradient. Typically, zirconium dioxide (ZrO_2) is doped with 8% yttrium trioxide (Y_2O_3) or with scandium trioxide (Sc_2O_3) in order to stabilize its crystalline structure for the operating temperature range and to produce crystallographic defects that eventually allow for the passage of oxygen ions.

The electrical conductivity across the solid oxide electrolyte varies exponentially with temperature according to Meng et al. (2008):

$$\sigma = 33{,}000\ \exp\left(-10{,}300/T\right) \tag{3.81}$$

The calculation of reversible half-cell potentials and anodic and cathodic polarization overpotential, takes into account that the reactions occur in a gas phase rather than in solution. Therefore, the activity can be replaced with fugacity and molar concentrations with a partial pressure when the reversible half-cell potentials are calculated.

The activation overpotentials associated with the operation of a SOE-O are formulated as

$$\Delta E_{act,i} = \frac{RT}{zF}\sinh^{-1}\left(\frac{J}{2J_{0,i}}\right) \tag{3.82}$$

where i = a,c and the exchange current density have an order of magnitude of ~2000 A/m² for the anode and ~5000 A/m² for the anode.

The exchange current density dependence on temperature is described by the Arrhenius Eq. (3.79). For SOE-O, the order of magnitude of activation energy is around 100 kJ/mol, while the pre-exponential factors are around ~5000 s^{-1} for the anode and ~50,000 s^{-1} for the cathode.

For SOE-O, two distinct equations can be formulated for concentration overpotentials for both the anode and cathode. For the anode:

$$\Delta E_{conc,a} = \frac{RT}{4F}\ln\left(\frac{\sqrt{\left(P_{O_2}^{in}\right)^2 + \frac{JRT\mu\delta_a}{2FB_g}}}{P_{O_2}^{in}}\right) \tag{3.83}$$

Here $P_{O_2}^{in}$ is partial pressure of oxygen in the gas stream at the inlet port on the anode side, μ is the dynamic viscosity of molecular oxygen at an operating temperature, δ_a is the thickness of the anode and B_g is the flow permeability, which is determined by the Carman-Kozeny relationship as

$$B_g = \frac{2r^2\varepsilon^3}{72\xi(1-\varepsilon)^2} \tag{3.84}$$

where the notations are defined as follows: mean electrode pore radius (r), electrode porosity (ε), electrode tortuosity (ξ).

At the cathode, the concentration overpotential is a function of many parameters according to the following equations:

$$
\begin{cases}
\Delta E_{conc,c} = \dfrac{RT}{2F} \ln \left(\left[1 + \left(\dfrac{JRT\delta_c}{2FD_{H_2O}^{eff} P_{H_2}^{in}} \right) \right] \Big/ \left[1 - \left(\dfrac{JRT\delta_c}{2FD_{H_2O}^{eff} P_{H_2O}^{in}} \right) \right] \right) \\[2ex]
D_{H_2O}^{eff} = \varepsilon \Big/ \left\{ \xi \left[(D_{H_2O,H_2})^{-1} + (D_{H_2O,k})^{-1} \right] \right\} \\[2ex]
D_{H_2O,k} = \dfrac{4r}{3} \sqrt{8RT/(\pi M_{H_2O})} \\[2ex]
D_{H_2O,H_2} = 0.00133 T^{1.5} \sqrt{M_{H_2} + M_{H_2O}} \Big/ \left(P\bar{\lambda}\Omega_D \right) \\[2ex]
\Omega_D = a_0/\tau^{a_1} + a_2/\exp(a_3\tau) + a_4/\exp(a_5\tau) + a_6/a_7\tau
\end{cases}
\tag{3.85}
$$

Here the following parameters are used: effective diffusion coefficient of water ($D_{H_2O}^{eff}$), Knudsen diffusion coefficient of water ($D_{H_2O,k}$), molecular—binary diffusion coefficient of the hydrogen—water mixture (D_{H_2O,H_2}), molar mass (M) of water and hydrogen, temperature (T), pressure (P), collision-diffusion integral based on Lenard-Jones potential (Ω_D), and mean characteristic length for species H_2 and H_2O ($\bar{\lambda}$). In Eq. (3.85) one defines also

$$
\bar{\lambda} = 0.5(\lambda_{H_2} + \lambda_{H_2O})
\tag{3.86}
$$

and the collision-diffusion integral calculated based on the mean collision time with

$$
\tau = k_b T/\bar{\epsilon}
\tag{3.87}
$$

where $k_b = 1.38066 \times 10^{-23}$ J/K is the Boltzmann constant, and $\bar{\epsilon}$ is the characteristic Lenard-Jones potential for the binary system (H_2, H_2O) given by the geometrical average of the Lenard-Jones potentials of the components, $\bar{\epsilon} = (\epsilon_{H_2}\epsilon_{H_2O})^{0.5}$. Table 3.6 lists some practical values of these parameters for SOE-O modeling.

The species concentration within the porous electrodes of SOEC can be explained based on a triple-phase boundary (TPB) theory. Accordingly, each volume-averaging cell encloses the solid electrode, gas mixture and a liquid-phase electrolyte. Mass transport occurs in the liquid and gas phases. Precisely, reactants diffuse through the gaseous mixture and then transfer to the molten electrolyte so as to reach the TPB where the electrochemical reaction takes place. The TPB length affects the value of the exchange current density. A representation of the TPB process occurring in solid oxide electrolytes is shown in Fig. 3.29. The concentration overpotential accounts for losses incurred by the resistance of the porous electrode to the transport of gaseous species between the gas channel and the reaction sites at TPB. The mass transport in the electrodes is driven by the diffusion of reacting species because of the concentration gradient as well as the permeation caused by the pressure gradient.

If the partial pressure P_i^{TPB} of all species at the TPB can be determined, then the following simplified equation can be used to determine the concentration overpotential at the anode:

$$
\Delta E_{conc,a} = \frac{RT}{2F} \ln \left(\frac{P_{H_2}^{TPB} P_{H_2O}}{P_{H_2} P_{H_2O}^{TPB}} \right)
\tag{3.88}
$$

Similarly, the concentration overpotential at the cathode can be determined with

$$
\Delta E_{conc,c} = \frac{RT}{4F} \ln \left(\frac{P_{O_2}}{P_{O_2}^{TPB}} \right)
\tag{3.89}
$$

In Fig. 3.30, it can be observed that the ohmic overpotentials account for 50–70% of all overpotentials. This suggests that the hydrogen transport through the non-porous solid oxide electrolyte is not efficient. However, the evolution of hydrogen at the cathode is a relatively efficient process because of the cathodic concentration and the activation overpotentials being less than 10% of the total of overpotentials.

Figs. 3.31–3.38 exemplify the effects of operating temperature and pressure on SOE-O operation. Overall, there is a decrease in overpotentials with higher temperature and pressure. The main cause of this trend was found to be the major decrease in ohmic overpotential because of the increased ionic conductivity of the SOE-O electrolyte at higher temperatures. This suggests that H_2 production via an SOE-O plant is more efficient at higher temperatures and pressures. This makes the SOE-O an option that depends on whether or not there is an economical and reliable source of high-temperature heat available to meet the large thermal energy demand.

TABLE 3.6 Practical Parameters for SOE-O Modeling

Parameter	Symbol	Value
Activation energy (cathode and anode, about the same)	ΔG_{act}	100 kJ/mol
Pre-exponential factor for anode reaction	$\mathcal{A}_a$	5000 s^{-1}
Pre-exponential factor for cathode reaction	$\mathcal{A}_c$	50,000 s^{-1}
Electrode's average pore radius	r	0.5 μm
Electrode porosity	ε	0.4
Electrode tortuosity	ξ	5.0
Cathode thickness	δ_c	50 μm
Electrolyte thickness	δ	50 μm
Anode thickness	δ_a	500 μm
Characteristic length H$_2$	λ_{H_2}	2.827 Å
Characteristic length H$_2$O	λ_{H_2O}	2.641 Å
Lenard-Jones potential H$_2$	ϵ_{H_2}	5.144 meV
Lenard-Jones potential H$_2$O	ϵ_{H_2O}	69.72 meV
Coefficients for collision integral Ω_D	a_0	1.06036
	a_1	0.15610
	a_2	0.19300
	a_3	0.47635
	a_4	1.03587
	a_5	1.52996
	a_6	1.76474
	a_7	3.89411

In SOE-O operations, there is a large thermal energy requirement and therefore different behaviors for energy and exergy efficiencies. In general, the thermal energy input associated with higher temperature operations reduces the energy and exergy efficiencies. As an order of magnitude, a SOE-O plant operating at ~875 K shows a maximum energy efficiency at approximately 52%, 53%, and 54% for 1.0 atm, 3.0 atm and 5.0 atm, respectively, at a current density of ~2000 A/m^2. Also, the maximum exergy efficiency is ~61% for 1.0 atm with a 1–2% increase if the pressure is raised up to 5.0 atm.

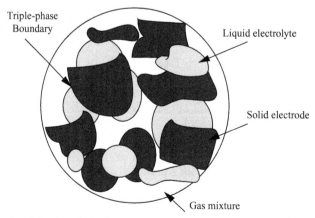

FIG. 3.29 Illustration the structure of a triple-phase boundary.

FIG. 3.30 Overpotential distribution in a SOE-O at various pressures.

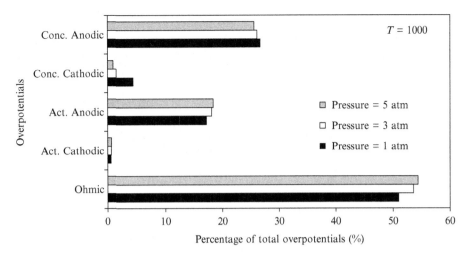

FIG. 3.31 Irreversibility variation of an SOE-O for various pressures.

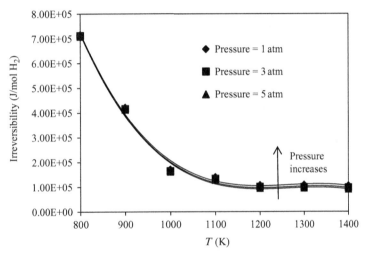

FIG. 3.32 Energy efficiency variation of SOE-O for various pressures.

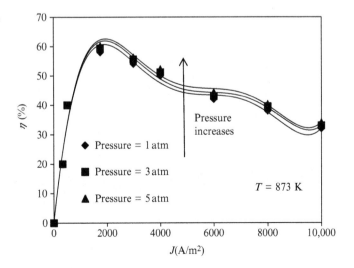

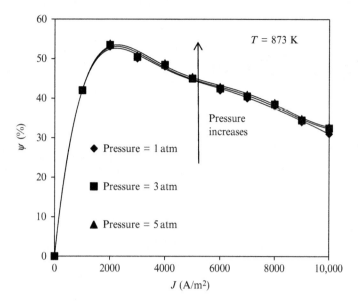

FIG. 3.33 Exergy efficiency variation of an SOE-O for various pressures.

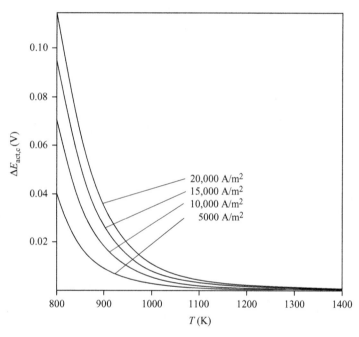

FIG. 3.34 Cathodic activation overpotentials in a SOE-O at various current densities.

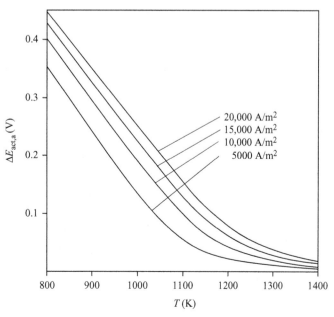

FIG. 3.35 Anodic activation overpotentials in a SOE-O at various current densities.

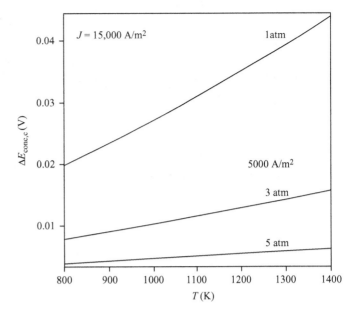

FIG. 3.36 Cathodic concentration overpotentials in a SOE-O at various pressures.

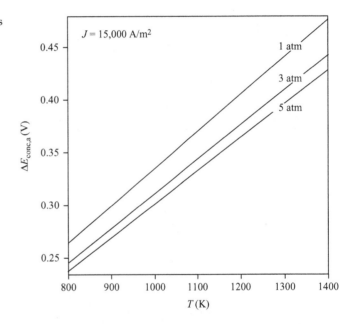

FIG. 3.37 Anodic concentration overpotentials in a SOE-O at various pressures.

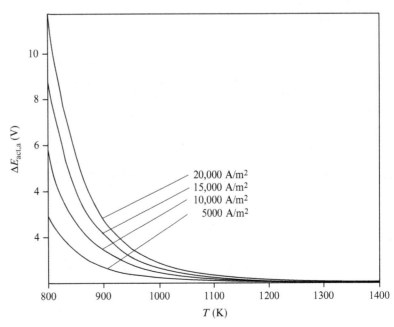

FIG. 3.38 Ohmic overpotentials in a SOE-O at various current densities.

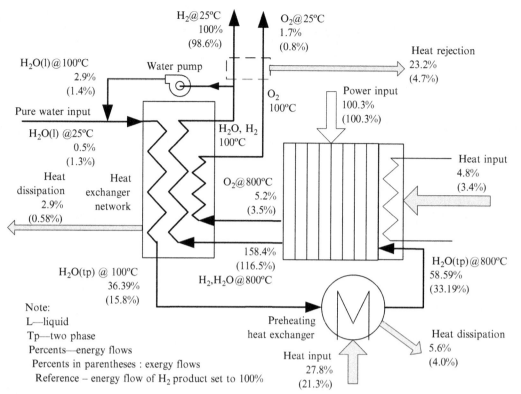

FIG. 3.39 Simplified layout of a SOE-O plant showing the energy and exergy. *Modified from Meng, N., Leung, M.K., Leung, D.Y., 2008. Energy and exergy analysis of hydrogen production by a PEM electrolyser plant. J. Energy Convers. Manage. 49, 2748–2756.*

The increase in both energy and exergy efficiency is primarily due to the increase in pressure, which caused a decrease in concentration overpotentials. This further causes a decrease in thermodynamic irreversibility, which results in an increase in both efficiencies.

A simplified layout of a water electrolysis system with SOE-O is presented in Fig. 3.39. The plant includes a water treatment station (not shown), three flow heat exchanger networks, heat exchangers for product cooling, preheating heat exchanger, water pump for water recycling, water separator and product gas dryers (not shown). The fresh water at 25°C is mixed with recycled water at 100°C (which results from condensing steam). The water boils further in a heat exchanger network.

Further water is heated from an external heat source to 800°C and supplied to the SOE-O unit. Power and high-temperature heat is also supplied to the SOE-O unit. The product gases—at 800°C temperature—are diverted to the heat exchanger network for heat recovery purposes where they are cooled to 100°C and water is fully condensed. Finally, the product hydrogen and oxygen gases are cooled to 25°C, and the heat is rejected to the ambient air.

For 100% hydrogen as equivalent energy flow, the system requires ∼100.3% electric power supply and 32.6% input energy in the form of heat and 0.5% in the form of water enthalpy; the total energy input is 133.4%. The energy efficiency of the system is 100/133.4 = 75%. If one assumes that electricity is generated with an efficiency of 30%, then the system efficiency becomes about 27%. The exergy input in the system is 126.3%. Therefore, the exergy efficiency of hydrogen production is 78%. The exergy of the output streams is 108.7%; it yields exergy destroyed within the system as 126.3 − 108.7 = 17.6%.

The data from Fig. 3.40 can be used to determine the energy and exergy portions for the SOE-O system. The energy charts from Fig. 3.40a represent percentages of output flow from the total input flow. In order to obtain these charts, first the total energy and energy inputs in the system are calculated. Then each output flow is expressed as a percentage from the total input flow of energy and exergy, respectively. In the case of the exergy diagram—Fig. 3.40—the exergy destroyed (Ex_d) for the overall system is indicated. Exergy destroyed results from the exergy balance for the overall system: $(\sum Ex)_{in} = (\sum Ex)_{out} + Ex_d$.

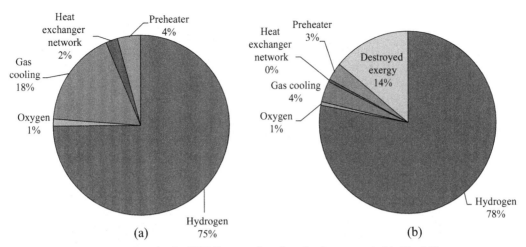

FIG. 3.40 (a) Energy and (b) exergy portions for the SOE-O system based on the data presented in Fig. 3.39.

3.6 SOLID OXIDE ELECTROLYZERS WITH PROTON CONDUCTION

Proton-conducting solid oxides operate as an electrolyte in intermediate temperature electrolysis cells for steam. One advantage of the proton-conducting electrolyte is that it allows for the absence of steam at the cathode; therefore, there is no need to separate hydrogen. The electrolyte becomes conductive to ions in a range of temperatures of $\sim 400-750°C$.

Research work on intermediate temperature electrolysis cells with proton conduction (SOE-P) is in its early stage. The electrochemical half reactions are identical to that of a PEME, but they are conducted at a much higher temperature. Also, the construction of the SOE-P is very similar to that of PEME, consisting of bipolar plates with channels for the gas flow, porous electrodes and electrolytes.

Materials based on cerium (Ce) were discovered to be good conductors of protons. One of the most-cited proton-conducting electrolytes is $BaCeO_3$; the other is $SrCeO_3$. The governing mechanism of proton conduction is based on the movement of protons between adjacent oxygen ions that are bounded to the atomic structure of the solid oxide electrolytic membrane. The protons, which are extremely small, necessitate reduced activation energy as compared to the case of oxygen ion transport through the most advanced solid electrolytes for SOE-O (e.g., electrolytes based on YSZ or doped ceria systems).

The main challenge with barium cerate results from the difficulty to sinter it in the form of a solid membrane of reduced thickness. One possible way to obtain membranes of $50\,\mu m$ or smaller is by doping barium-cerate with samarium (Sm) and a conductivity up to 0.01 S/cm. Strontium cerate-based membranes doped with ytterbium (Yb) can be packed on a layer based on strontium-zirconium-yttrium to form a SOE with proton conduction. Also anode-supported SOE with proton conduction based on a Ni-CGO|BCGO|BSCF-CGO arrangement using a dry-pressing method and applying humidified hydrogen and ammonia as fuels and operating at 650°C was a viable alternative.

The fundamental electrochemical theory of electrolysis can be applied to SOF-P modeling. Reversible potentials are calculated based on the equilibrium constant, which is estimated according to the partial pressures of the gases. The activation overvoltage is calculated with Eq. (3.82). In proton-conducting solid oxide electrolysis cells, the concentration overpotential at the cathode side modeled as

$$\Delta E_{conc,c} = \frac{RT}{2F} \ln\left(\frac{P_{H_2}(J)}{P_{H_2}^0}\right) \tag{3.90}$$

where $P_{H_2}(J)$ is the partial pressure of hydrogen at a current density J, while the exponent 0 indicates the partial pressure of hydrogen if no current flows through the circuit.

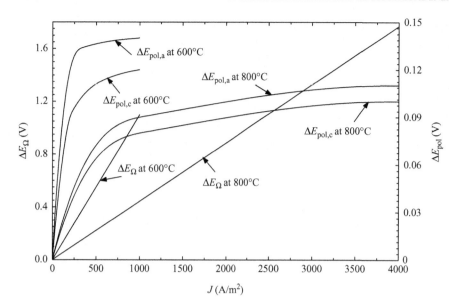

FIG. 3.41 Overpotentials in a SOE-P with SSC-55|SZCY-541|SCYb|Ni configuration [plot constructed based on the experimental data from Sakai et al. (2009)].

The anode side concentration overpotential is given by Sakai et al. (2009) as follows:

$$\Delta E_{conc,a} = -\frac{RT}{2F}\ln\left(\frac{P_{H_2O}(J)}{P_{H_2O}^0}\left(\frac{P_{O_2}^0}{P_{O_2}(J)}\right)^{0.5}\right) \tag{3.91}$$

In Fig. 3.41, overpotentials of a SOE-P cell are illustrated based on Sakai et al. (2009). The cell has an anode made of strontium-samarium and cobalt trioxide $Sr_{0.5}Sm_{0.5}CoO_3$ characterized by a conductivity of 1000 S/cm; the anode is denoted with the short notation SSC-55. The anode is in contact with a cerium-based proton-conducting strontium zirconate electrolyte SZCY-541 with a chemical formula $SrZr_{0.5}Ce_{0.4}Y_{0.1}O_{3-\alpha}$, where α denotes the oxygen vacancy in the crystalline structure. The cell has a double layer cathode SCYb|Ni formed by a strontium-cerate-ytterbium (SCYB) interlayer with a chemical formula $SrCe_{0.95}Yb_{0.05}O_{3-a}$ that is in contact with the electrolyte and a Ni substrate deposited on the SCYb.

The overpotentials were measured at 600°C and 800°C operating cell temperatures. It is observed in Fig. 3.41 that polarization overpotentials are negligible with respect to the ohmic values. Furthermore, at 800°C, overpotentials are significantly lower than at 600°C.

Based on concentration overpotentials that result from Fig. 3.41, one can calculate backward the output concentrations. The partial pressure of water, oxygen and hydrogen must be known for the condition when no current is applied to the cell. Sakai et al. (2009) used argon as sweep gas and set the idle concentrations to ~50% for water, 0.05% for oxygen, and 1% for hydrogen.

With concentrations determined for any current density, the reversible cell potential can be calculated. Further, based on the total cell potential, the efficiencies $\eta_{Faradaic}$, η_{el} and ψ_{el} can be determined. In the equations for η_{el} and ψ_{el}, it must be taken into account that the system requires heat input in order to operate. The required heat input is given by $Q_{req} = \Delta H - \Delta G$.

A part of this thermal energy is supplied by the Joule effect within the electrolyte—ie, $Q_\Omega = \Delta E_\Omega JA$, where A is the area of the electrode. The heat input provided to the electrolysis cell is therefore $Q_{inp} = Q_{req} - Q_\Omega$.

Based on these considerations, the plot from Fig. 3.42 can be constructed where it is shown that the cell efficiency depends on the current density. The same figure also presents the dependence of total and reversible potential with current density. The Faradaic efficiency drops from ~75% at low current density to 30% at 4000 A/m². The energy efficiency of the cell varies in the range of 45–95%, and the exergy efficiency varies in the range of 35–80%.

3.7 CHLORALKALI ELECTROCHEMICAL PROCESS FOR CHLORINE AND HYDROGEN PRODUCTION

The chloralkali industry is among the chief electrochemical processes producing chlorine and sodium hydroxide (caustic soda). The chloralkali process involves the electrolysis of sodium chloride solution (brine) that produces

FIG. 3.42 Efficiency and cell potentials for the SSC-55|SZCY-541|SCYb|Ni at 800°C.

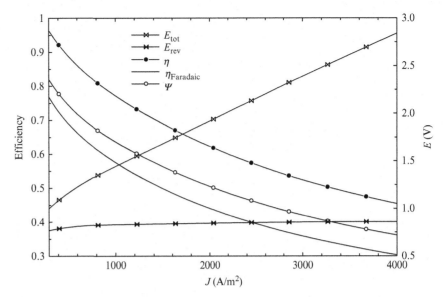

chlorine gas at the anode and sodium hydroxide (caustic soda) at the cathode. The overall chemical reaction can be written as follows:

$$2NaCl_{(aq)} + 2H_2O_{(l)} \rightarrow 2NaOH_{(aq)} + Cl_{2(g)} + H_{2(g)} \tag{3.92}$$

The chloralkali process is currently carried out by using one of three different methods:

- Mercury cell
- Diaphragm cell
- Membrane cell

In each of these three methods, chlorine is produced at the anode (positive electrode) along with hydrogen, which is produced at the cathode (negative electrode). Worldwide, 62% of chloralkali manufacturers use diaphragm cell, 10% mercury cell, 24% membrane cell, and the rest of 4% use other technologies.

The mercury cell chloralkali technology is known as the "Castner-Kellner process." The cell is made of steel coated at the interior with rubber. The feed is a saturated NaCl brine, as shown in the schematic diagram in Fig. 3.43.

Because of its light density, the saturated brine floats on top of a thin layer of mercury that, in fact, forms the cathode. The sodium ions combine with mercury to form a sodium-mercury amalgam that further reacts with water in a separate reactor called the "decomposer" where hydrogen gas and a sodium hydroxide are formed.

The chloralkali diaphragm cell is constructed as shown in Fig. 3.44. The anode is a copper plate, and the cathode is made in the form of metal screens with asbestos deposited on them. Asbestos prevents the caustic soda from reacting with the chlorine. The saturated NaCl brine is added to the anode side.

As described in Ghosh (2014), a separator is used between the anode and cathode compartments to avert the mixing of anolyte and catholyte. In the diaphragm cell process, a permeable diaphragm is used to separate the anode from the cathode compartments, which are typically made of asbestos fibers.

Fig. 3.45 shows the schematic representation of the chloralkali membrane cell process. Chlorine ions are oxidized to chlorine gas at the anode where saturated sodium chloride solution is fed. The sodium ions migrate through the membrane to the cathode compartment. In the cathode compartment, water reduction takes places and hydroxyl ions are produced. Here the migrated sodium ions combine with the hydroxyl ions. Consequently, hydrogen gas is produced in the cathode. The reactions evolved in this process can be written as follows:

At the anode:

$$2Cl^-_{(aq)} \rightarrow Cl_{2(g)} + 2e^- \tag{3.93}$$

At the cathode:

$$2H_2O_{(l)} + 2e^- \rightarrow H_{2(g)} + OH^-_{(aq)} \tag{3.94}$$

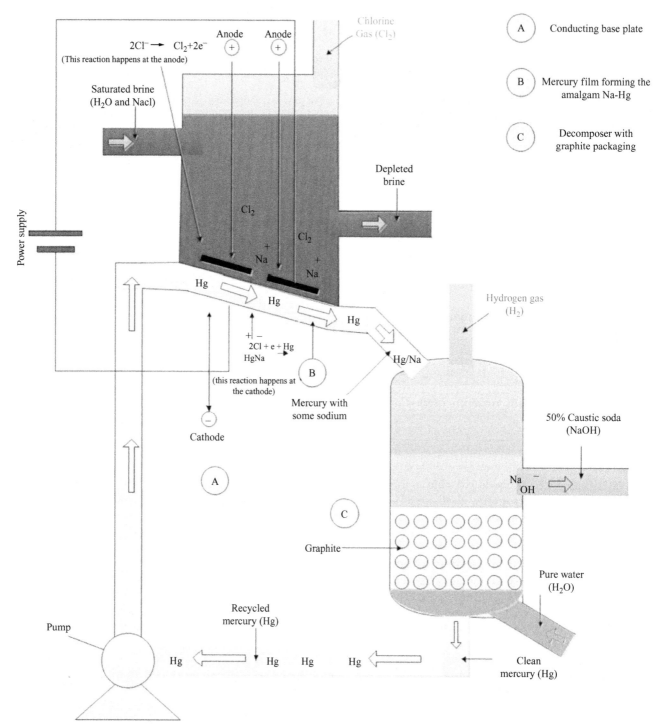

FIG. 3.43 The Castner-Kellner chloralkali process. *Modified from Rabbani, M., 2013. Design, Analysis and Optimization of Novel Photo-Electrochemical Hydrogen Production Systems. PhD Thesis, University of Ontario Institute of Technology.*

This technology yields a weak caustic stream contaminated with sodium chloride having a concentration of about 30%. Conversely, a mercury cell process is divided into two units: the electrolyzer and a secondary electrochemical reactor also commonly known as decomposer. In a mercury cell process, a high purity, strong caustic solution having a concentration of about 50% is produced. In the electrolyzer, chlorine gas is produced at the anode, and sodium amalgam is produced at the cathode. Furthermore, the sodium amalgam enters a secondary electrochemical reactor. Here water decomposes into sodium hydroxide and hydrogen gas contaminated with traces of mercury.

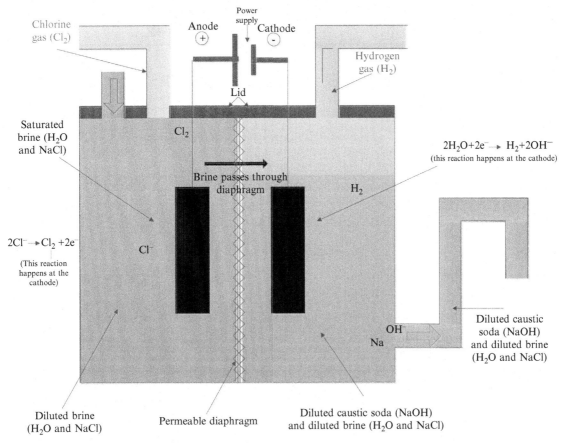

FIG. 3.44 Schematic of diaphragm. *Modified from Rabbani, M., 2013. Design, Analysis and Optimization of Novel Photo-Electrochemical Hydrogen Production Systems. PhD Thesis, University of Ontario Institute of Technology.*

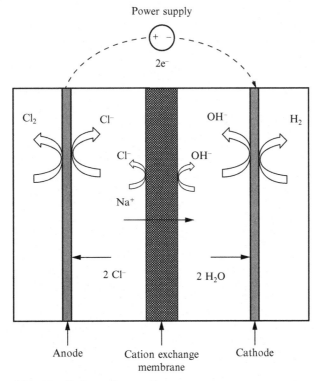

FIG. 3.45 Schematic representation of the chloralkali membrane cell process.

TABLE 3.7 Comparison of Energy Requirements of Main Three Chloralkali Technologies

	Mercury cell	Diaphragm cell	Membrane cell
Cell voltage (V)	−4.4	−3.45	−2.95
Current density (A)	1.0	0.2	0.4
Current efficiency for Cl_2 (%)	97	96	98.5
ENERGY CONSUMPTION, kWh, per ton of NaOH			
(a) Electrolysis only	3150	2550	2400
(b) Electrolysis plus evaporation to 50% NaOH	3150	3260	2520
Purity Cl_2 (%)	99.2	98	99.3
Purity H_2 (%)	99.9	99.9	99.2
O_2 in Cl_2	0.1	1.2	0.3
Cl^- in 50% NaOH (%)	0.003	1.1.20	0.005
Sodium hydroxide concentration prior to evaporation (%)	50	12	35
Mercury pollution considerations	Yes	No	No
Requirement for brine purification	Some	More stringent	Very extensive
Production rate per single cell, tons NaOH per year	5000	1000	100
Land area for plant of 10^5 tons NaOH per year (m^2)	3000	5300	2700

With the diaphragm cell process and mercury cell process having contaminated hydrogen gas, the membrane cell process appeared as an alternative to the diaphragm and mercury cells. The membrane cell process is precisely like the diaphragm cell where the permeable diaphragm is replaced by a selective ion exchange membrane. It also has several advantages over the other processes, mainly that it is highly efficient energetically with high-purity caustic soda with lesser environmental impact. This process also requires a high-quality input brine to avoid polluting the membrane.

Table 3.7 compares the three main chloralkali technologies introduced above. As shown in the table, when compared to mercury and diaphragm cells, membrane cell technology requires less energy input. Furthermore, the membrane technology provides the highest purity of hydrogen, chlorine, and sodium hydroxide. Table 3.8 gives the anode and cathode reversible potentials for typical operating conditions (90°C, 1 bar, 3.5 M of NaCl and 10 M of NaOH) in a chloralkali membrane cell process.

Therefore, the total cell voltage (except the overpotentials) is $E^0 = 2.1877$ V, which is clearly superior as compared to 1.23 V for water electrolysis. However, the chlorine and NaOH by-products are highly marketable and justify the additional expense of energy needed to operate the cell. The practical overvoltage of the membrane-based chloralkali electrochemical cell is 3.1 V at 0.4 A/cm², therefore, the cell overvoltage is of the order of 0.92 V.

The caustic soda forms in the catholyte can be written as

$$Na^+_{(aq)} + OH^-_{(aq)} \rightarrow NaOH_{(aq)} \tag{3.95}$$

Here the sodium ions migrate from the anolyte, through the cation exchange membrane (NAFION®). The sodium ions form in the anolyte by NaCl dissociation is shown by the following reaction:

$$NaCl_{(aq)} \rightarrow Na^+_{(aq)} + Cl^-_{(aq)} \tag{3.96}$$

TABLE 3.8 Reversible Half-Cell Potentials in a Typical Chloralkali Membrane Cell

	Reaction	E^0 (V)
Anode	chlorine evolving : $Cl_{2(g)} + 2e^- \rightarrow 2Cl^-_{(aq)}$	1.23
Cathode	hydrogen evolving : $2H_2O_{(l)} + 2e^- \rightarrow H_{2(g)} + 2OH^-_{(aq)}$	−0.99
Overall	$2NaCl_{(aq)} + 2H_2O_{(l)} \rightarrow 2NaOH_{(aq)} + Cl_{2(g)} + H_{2(g)}$	−2.23

The anodic reversible potential is related to chlorine ion oxidation potential as

$$E_a^o = E^\circ{}_{Cl^-/Cl_2} + \frac{RT}{2F} \ln \left(\frac{P_{Cl_2}}{(a_{Cl^-})^2} \right) \tag{3.97}$$

where $E^\circ{}_{Cl^-/Cl_2}$ results from Table 3.8, P_{Cl_2} are the partial pressure of chlorine and a_{Cl^-} is the activity of chlorine ion.

The water reduction potential results from Nernst equation according to

$$E_c^o = E^\circ{}_{H_2O/H_2+OH^-} + 2.303 \frac{RT}{2F} \ln \left(\frac{P_{H_2}}{(a_{OH^-})^2} \right) \tag{3.98}$$

where $E^\circ{}_{H_2O/H_2+OH^-}$ results from Table 3.8, P_{H_2} are the partial pressure of chlorine and a_{OH^-} is the activity of hydroxyl ion.

The overall cell reversible potential was correlated by Chandran and Chin (1986) with cell temperature, partial pressure and relevant chemical activities. The correlation is given as

$$E_{cell}^o = -2.18 + 0.0004272T + \frac{8.314T \ln \beta}{96,500} \tag{3.99}$$

where T is in K and β is dependent upon the partial pressures of hydrogen and chlorine in the gas phase and the activity coefficients of sodium chloride and sodium hydroxide, as follows:

$$\beta = \frac{a_{NaCl}\sqrt{P_{Cl_2}}\sqrt{P_{H_2}}}{a_{NaOH}} \tag{3.100}$$

The activity coefficient of sodium chloride (a_{NaCl}) is correlated to the molar concentration, M (mol/L), as given by Chandran and Chin (1986):

$$a_{NaCl} = \begin{cases} 0.63 \exp(0.028M_{NaCl}), 1.2 \leq M_{NaCl} < 2 \\ 0.575 \exp(0.07M_{NaCl}), 2 \leq M_{NaCl} < 3.5 \\ 0.5 \exp(0.112M_{NaCl}), 3.5 \leq M_{NaCl} < 6 \end{cases} \tag{3.101}$$

The activity coefficient of sodium hydroxide (a_{NaOH}) depends on its molar concentration, M (mol/L), as correlated by Chandran and Chin (1986) for $M_{NaOH} < 12$,

$$\log(a_{NaOH}) = -\frac{U\sqrt{M_{NaOH}}}{1 + \sqrt{2M_{NaOH}}} + BM_{NaOH} + CM_{NaOH}^2 + DM_{NaOH}^3 + EM_{NaOH}^4 \tag{3.102}$$

where the following correlation coefficients are given:

$$B = 0.0065 + 0.0016T - 1.8 \times 10^{-5}T^2 \tag{3.103a}$$

$$C = 0.014 - 0.0005T + 5.6 \times 10^{-6}T^2 \tag{3.103b}$$

$$D = 0.0006 + 5 \times 10^{-5}T - 6.48 \times 10^{-7}T^2 \tag{3.103c}$$

$$E = 5.96 \times 10^{-6} - 1.81 \times 10^{-6}T + 2.4 \times 10^{-8}T^2 \tag{3.103d}$$

$$U = \begin{cases} 0.00087T + 0.486, 25°C \leq T < 40°C \\ 0.00144T + 0.46, 40°C \leq T < 100°C \end{cases} \tag{3.103e}$$

When $M_{NaOH} \geq 12$, the activity coefficient is correlated by Chandran and Chin (1986) as

$$\log(a_{NaOH}) = a + bM_{NaOH} + cM_{NaOH}^2 \tag{3.104}$$

TABLE 3.9 Parameter Definitions for Eq. (3.107)

Parameter and equation	Remarks
$\dfrac{E^*}{R} = 4456.5 + ((\Omega_{NaOH} - 2)(5109.5 - 4456.5))$	$2 \leq \Omega_{NaOH} < 3$
$\dfrac{E^*}{R} = 5409.6 + ((\Omega_{NaOH} - 3)(4706.5 - 5109.8))$	$3 \leq \Omega_{NaOH} < 4$
$\dfrac{E^*}{R} = 535.2\Omega_{NaOH} + 2617.8$	$4 \leq \Omega_{NaOH} < 7$
$\dfrac{E^*}{R} = 967.5\Omega_{NaOH} + 548.2$	$7 \leq \Omega_{NaOH} < 16$
$s = -9.6 + ((\Omega_{NaOH} - 2)(11.3 - 9.9))$	$2 \leq \Omega_{NaOH} < 3$
$s = -11.3 + ((\Omega_{NaOH} - 3)(11.3 - 10))$	$3 \leq \Omega_{NaOH} < 4$
$s = -1.06\Omega_{NaOH} - 5.8$	$4 \leq \Omega_{NaOH} < 7$
$s = -2.44\Omega_{NaOH} + 5.3$	$7 \leq \Omega_{NaOH} < 16$
$\sigma_{NaOH, 100°C} = 2.6 + 40.9\Omega_{NaOH} - 5.03\Omega_{NaOH}^2 + 0.13\Omega_{NaOH}^3$	$2 \leq \Omega_{NaOH} < 7$
$\sigma_{NaOH, 100°C} = 140.9$	$7 \leq \Omega_{NaOH} < 9$
$\sigma_{NaOH, 100°C} = 156 - 1.5\Omega_{NaOH}$	$9 \leq \Omega_{NaOH} < 16$
$\Omega_{NaOH} = \dfrac{M_{NaOH}\rho_{NaOH}}{1000 + (40 M_{NaCl})}$	$M \; (mol/dm^3)$
$\rho_{NaCl} = A + BT + CT^2 + DT^3 + ET^4$	$0°C \leq T \leq 300°C$
$A = (1.001 + 0.7666x - 0.0149x^2 + 0.2663x^3 + 0.8845x^4$	x is the solution concentration (kg salt/kg solution)
$B = -0.0214 - 3.496x + 10.02x^2 - 6.56x^3 - 31.37x^4$	
$C = (-5.263 + 39.87x - 176.2x^2 + 363.5x^3 - 7.784x^4) \times 10^{-3}$	
$D = (15.42 - 167x + 980.7x^2 - 2573x^3 + 876.6x^4) \times 10^{-6}$	
$E = (-0.0276 + 0.2978x - 2.017x^2 + 6.345x^3 - 3.914x^4) \times 10^{-6}$	

where

$$\begin{cases} a = -0.327 + 0.0031T - 3.29 \times 10^{-5}T^2 \\ b = 0.0988 - 0.00059T \\ c = -2.14 \times 10^{-6} - 3.93 \times 10^{-7}T + 0.53 \times 10^{-8}T^2 \end{cases} \tag{3.105}$$

The activation overpotential can be determined as

$$\Delta E_{act,i} = 0.0277 \log\left(\frac{J}{J_0}\right) \tag{3.106}$$

where d_{Am} is the distance between the electrode and cation exchange membrane, σ_{NaCl} represents the electrical conductivity of the brine, i is the subscript representing either anode or cathode, and typically $J_0 = 0.0125 \; A/cm^2$ for anode and $J_0 = 0.0656 \; A/cm^2$ for cathode.

The electrical conductivity of NaOH is a function of both its concentration and its operating temperature and can be estimated as

$$\sigma_{NaOH} = \sigma_{NaOH, 100°C}\left(\frac{T}{373}\right)^s \exp\left(-\frac{E^*}{R}\left(\frac{1}{T} - \frac{1}{373}\right)\right) \tag{3.107}$$

where the relevant parameters and their equations are given in Table 3.9.

3.8 OTHER ELECTROCHEMICAL METHODS OF HYDROGEN PRODUCTION

Water electrolysis and chloralkali processes are the most important scalable technologies of hydrogen production. In water electrolysis, pure water is the feedstock to produce hydrogen as the main product and oxygen as the by-product. In chloralkali process, sodium chloride (NaCl) is the feedstock in the form of a saturated aqueous solution, whereas chlorine is the main product and hydrogen is the by-product. In addition to these methods, two other electrochemical processes are becoming increasingly important for green hydrogen production. These are: electrochemical extraction of hydrogen from sulfurous waters mainly containing H_2S and ammonia-containing waste water electrolysis. Other electrochemical methods to generate hydrogen and co-products are published in the open literature; some are briefly reviewed in this section.

Hydrogen sulfide is an important source of hydrogen that is available in nature, either in aqueous form (e.g., waste industrial water, deep Black Sea water) or in gas form (e.g., at natural gas wells, gas hydrate sediments). Hydrogen sulfide is a major by-product of petrochemical refineries, natural gas processing plants, heavy oil upgrade processes and metallurgical processes.

There is available technology, including electrochemical processes to decompose hydrogen sulfide and other similar sulfurous waters (e.g., SOx) to hydrogen and sulfur, two valuable commodities with a vast range of applicability. Sulfur and hydrogen are very relevant feedstocks for fertilizer production. About 80% of the sulfur production worldwide is provided from recovery processes of sulfur from sulfurous industrial wastes.

Hydrogen sulfide can be decomposed electrochemically to sulfur and hydrogen according to the following half-reactions given by Mbah et al. (2008):

$$\begin{cases} \text{anode}: H_2S \rightarrow 2H^+ + 2e^- + \frac{1}{2}S_2, \text{with } E^0 = 0.19\,V\,at\,150^\circ C \\ \text{cathode}: 2H^+ + 2e^- \rightarrow H_2 \end{cases} \tag{3.108}$$

In fact, various types of electrolytic cells do exist to split hydrogen sulfide in gas phase, liquid phase, or aqueous solution. Mbah et al. (2008) used a solid proton-conducting membrane that operates at around 150°C using hydrogen sulfide gas and sulfur in a liquid (molten) state. The Nernst equation for the reversible cell potential of Eq. (3.108) becomes

$$E_{rev} = E^0 - \frac{RT}{2F} \ln\left(\frac{P_S^{0.5} P_{H_2}^2}{P_{H_2S}}\right) \tag{3.109}$$

which assumes that the reactants and products are in gas phase.

The solid proton conductor (membrane) electrolyte used in Mbah et al. (2008) has a defected crystalline structure presenting lattice vacancies and interstitials such that ions become mobile as they are able to hop from position to position within the structure. The electrolyte is based on caesium hydrogen sulfate, namely $CsHSO_4$. At approximately 150°C, the electrolyte proton's conductivity reaches a maximum plateau, while the viscosity of molten sulfur reaches a minimum. The process requires Ruthenium and Platinum black-based catalysts, which are relatively expensive. The demonstrated process showed a rather small current density of below $20\,mA/cm^2$ at a cell voltage of approximately 1 V.

An electrochemical process to generate hydrogen and sulfur from pressurized liquid H_2S at room temperature was proposed by Gregory et al. (1980); it uses pyridine as an electrolyte and aluminum electrodes. Current density of $10\,mA/cm^2$ and overpotentials of 0.7 V were observed.

Annani et al. (1990) demonstrated hydrogen sulfide electrolysis in aqueous solution. The aqueous electrolyte was based on NaOH and NaHS in equimolar amounts; the cell operated at 80°C. A Nafion membrane was used as a separator for anode and cathode compartment. The anodic sulfur was obtained by bisulfide and sulfide oxidation. Current density of $0.3\,A/cm^2$ was obtained at relatively reduced overpotentials. Various anode materials were tested including graphite, nickel, titanium, and porous nickel-chromium. The cathode was fabricated from nickel and graphite. The electrodes were placed in the vicinity of the membrane.

Sewage and other effluents produced by anthropogenic conditions consist of ammoniated waters, either because of their content of urine, urea or other ammonium sources, and are characterized by an ammonia concentration of more than $10^{-3}\,M$ and pH 5.8. Electrolytic hydrogen production from such sources is an effective method that consumes less electrical energy than pure-water electrolysis.

Provided that the sewage waters are concentrated preliminary in urea until the limit of 0.33 M, the electrolysis is impressively efficient. The effective cell voltage of aqueous urea electrolysis in alkaline solution is as low as 1.4 V, according to Boggs et al. (2009). The reversible cell potential is −0.46 V SHE for the anodic half-cell, whereas the cathodic water reduction occurs at −0.83 V SHE. Therefore, the reversible cell potential is 0.37 V as compared with 1.23 V for water electrolysis. Inexpensive nickel catalysts can be used for the electrolysis cell.

Alkaline electrolytic cells are also devised for producing hydrogen from ammonia. For these cells, the ammonia oxidation potential is only 0.77 V over SHE whereas the cell reversible voltage is only 0.06 V.

3.9 INTEGRATED SYSTEMS FOR HYDROGEN PRODUCTION BY ELECTRICAL ENERGY

3.9.1 Hydroelectric Hydrogen

Hydropower plants can be coupled with alkaline water electrolyzers to generate green hydrogen. Depending on the head and the available volume rate, small- to large-capacity hydrogen generation plants of this kind can be installed. Hydroelectric power generation is an established technology that uses the potential energy of water to generate electricity. The main components of the hydropower plants are shown in Fig. 3.46 and comprise a dam or retaining wall, a water turbine, and an electrical generator. The dam or the retaining wall is built along the width of a river so that the water level can rise other side of the dam/retaining wall. On the other side of the dam/retaining wall, water turbines coupled with electricity generators are installed. The potential energy of water is used to run turbines, and then turbines run generators and produce electricity.

The known hydro-turbines are Pelton, Francis, Michell-Bánki, Kaplan and Deriaz; also, some water pumps can work efficiently in reverse as turbines. The Pelton machine is a turbine of free flow (action). The potential energy of the water becomes kinetic energy through injectors and control of the needles that direct and adjust the water jet on the shovels of the motive wheel. They work under approximate atmospheric pressure with a typical head in the range 400–2000 m. Compared with the Francis turbine, the Pelton head has a better efficiency curve.

The Francis turbine is used in small hydroelectric power plants with a head in the range 3–600 m and hundreds of dm^3/s flow rate of water. This turbine is very sensitive to cavitation and works well only when close to the design point. Its operation becomes unstable if the duty power becomes lower than 60% of nominal power.

The Michell-Bánki turbine works with radial thrust. The range of power goes up to 800 kW per unit, and the flow rate varies from 25 to 700 dm^3/s with a head ranging from 1 to 200 m. The number of slats installed around the rotor varies from 26 to 30, according to the wheel circumference, whose diameter is from 200 to 600 mm. This multi-cell turbine can be operated from one- to two-thirds of its capacity (in the presence of low or average flows) or at full capacity (in the presence of design flows: that is, three-thirds). The turbine can be operated even at 20% of its full power.

The Kaplan turbine is a hydraulic propeller turbine adapted to low heads from 0.8 to ~5 m. In addition, this turbine has the advantage of maintaining its electromechanical parts out of the water. This feature eases routine inspection and maintenance and adds safety in case of floods.

The Deriaz turbines were developed in the 1960s and can reach a capacity of up to 200 MW with flow rates in a broad range (from 1.5 to 250 m^3/s) and with heads of 5 to 1000 m. The runner diameter may be up to 7000 mm with

FIG. 3.46   Hydroelectric hydrogen production layout.

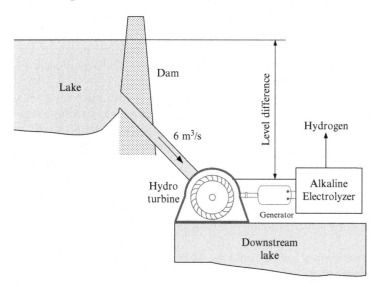

six to eight runner blades. Diagonal turbines operate very economically as either turbines or pumps. The thrusting water follows an approximately conic surface around the runner.

Thus the efficiency of hydro energy conversion is commonly 80% with respect to water head $\rho g \Delta z$. The water wheels are the least efficient devices to convert the water energy because of losses by friction, turbidity and incomplete filling of the buckets. The water pushes the shovels tangentially around the wheel. The water does not exert thrust action or shock on the shovels as is the case with turbines. The advantage of water wheels is that they can operate in dirty water or water with a suspension of solids. The power generated by a hydropower plant in an ideal reversible process is given as follows:

$$\dot{W} = \dot{V}\left(0.5\rho v^2 + \rho g \Delta z + P\right) \tag{3.110}$$

The thermodynamic limit is never reached in practice because of friction in ducts and irreversibilities in the turbine. Thus the efficiency of hydro energy conversion is commonly 80% with respect to water head $\rho g \Delta z$. The water wheels are the least efficient devices to convert the water energy due to losses by friction, turbidity, incomplete filling of the buckets. The water pushes the shovels tangentially around the wheel. The water does not exert thrust action or shock on the shovels as is the case with turbines. The advantage of water wheels is that they can operate in dirty water or water with suspension of solids.

An integrated system that couples alkaline electrolyzers with hydraulic turbines will generate hydrogen with an efficiency of ∼50%. This technology is readily available and, provided that the economics are favorable, can be applied right away.

3.9.2 Wind PEM Electrolyzer Systems

Wind turbines of horizontal-axis or vertical-axis turbines are used to convert the kinetic energy of the wind into electricity. It is one of the most cost-effective forms of renewable energy in today's technology. The electricity produced by wind energy can be supplied to the grid. The technology is beneficial for the locations where the wind velocity is high (e.g., coastal and subcoastal areas). For better functioning of a wind energy system, particularly to smooth out fluctuations, knowledge about natural geographical variations in wind speed is important.

Similar to the limitations of solar energy, wind energy generation is also affected by the intermittent nature of wind speed. Similar to hydro energy, wind energy also essentially is used to produce electricity first, and then the electricity can be used for hydrogen production. However, the problem with wind energy resides in its fluctuating availability. Fortunately, one of the remarkable features of electrolyzers is that they can adapt fast enough to fluctuations in the supplied voltage (provided that the voltage remains in an acceptable range for electrolyzer operation). Therefore, electrolysers are suitable to couple with wind turbines Fig. 3.47 shows the layout of a wind-electrolysis system for hydrogen production.

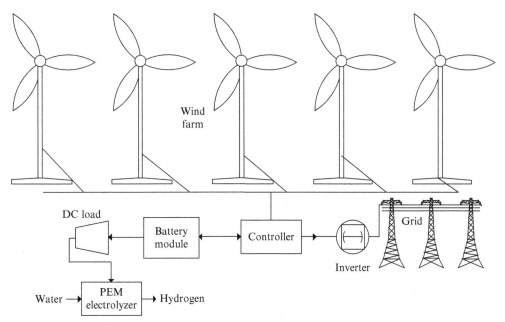

FIG. 3.47 Layout of a wind farm-PEM electrolysis system.

The annual production of hydrogen using wind energy is influenced by the capacity factor of the wind turbine and the efficiency of the PEM electrolyzer. The annually averaged hydrogen production rate by a wind turbine having the rotor area A is expressed as

$$\dot{m} = \frac{\eta_{PEM}}{HHV} \bar{\rho} A C_f (\bar{V}) V_{RMC}^3 \tag{3.111}$$

where HHV is the higher heating value of hydrogen, η_{PEM} is the energy efficiency of PEM electrolysis, $\bar{\rho}$ is the averaged air density, C_f is the capacity factor of the wind turbine, $\bar{V}$ is the annually averaged wind velocity and V_{RMC} is the root mean cube wind velocity.

The root mean cube wind velocity is defined as an average velocity that has 100% occurrence and that generates the same amount of wind energy as the probable velocity profile described by the adopted probability distribution function. The root mean cube wind velocity results from the wind speed distribution function as

$$V_{RMC}^3 = \int_{V_{ci}}^{V_{co}} h(V) \mathrm{d}(V^3) \tag{3.112}$$

where V_{ci} and V_{co} are the cut in and cut out speeds, respectively.

The so-called Weibull probability distribution can be used to model the occurrence of wind velocity based on the two parameters k and c, according to $h(V) = k/c(V/c)^{k-1} \exp\left[-(V/c)^k\right]$. For the particular case of parameter $= 2$, the Weibull probability distribution function bears the name of Rayleigh distribution and is given as follows:

$$h(V) = \frac{2V}{c^2} \exp\left[-\left(\frac{V}{c}\right)^2\right] \tag{3.113}$$

The capacity factor of the wind turbines depends on their technology and the average annual wind velocity at the site. Taking into account the low capacity factor ($\sim$25%), the expected efficiency of wind-electricity-hydrogen production is $\sim$10%.

3.9.3 Geothermally Driven Electrolysis Systems

Geothermal energy is a form of thermal energy that is available in some regions of the earth's surface at temperature levels in the range of about 35–500°C, even though most of the geothermal places provide temperature levels up to 250°C. Geothermal heat is used for process heating, or it is converted into electricity through heat engines.

Geothermal energy can be used to generate electricity through various kinds of flash cycles and organic Rankine cycles (ORCs). Depending on the level of temperature of the geothermal source, the energy efficiency of the electricity generation process may vary between 5% and 25%. Correspondingly, if a geothermal generator is coupled with an electrolyzer, the hydrogen production efficiency is expected to be in the range of 3% to 12%.

Some possible geothermal energy-driven electrolysis plants are shown in Figs. 3.48 and 3.49. When the steam pressure is considered low, or if there is no need to recycle the geothermal fluid, then after expansion down to atmospheric pressure, the seam can be released to the atmosphere (Fig. 3.48a). However, if the steam pressure and temperature are high enough, the amount of power generated allows for driving a recirculation pump and still generating satisfactory yield by the turbine. In this case (Fig. 3.48b), the steam can be expanded in vacuum, condensed, and the produced water pressurized in a pumping station and reinjected. The geothermal sources generating low pressure steam that needs to be reinjected can be coupled to ORC generators, as indicated in Fig. 3.49. These cycles are also known as binary cycles as sometimes they operate with binary mixtures (e.g., ammonia and water).

One potential integrated system of high-temperature steam electrolysis with geothermal energy is shown in Fig. 3.50. The exergy analysis of such a system was studied by Balta et al. (2009). The energy efficiency of the overall system was defined as follows:

$$\eta_{overall} = \frac{\dot{m}_{H_2} HHV_{H_2}}{\dot{W}_{e,in} + \dot{Q}_{g,in}} \tag{3.114}$$

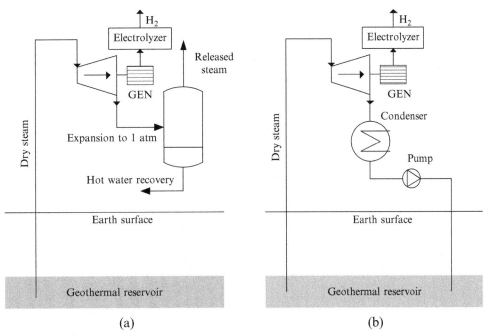

FIG. 3.48 Layout of dry geothermal hydrogen production plant (a) without reinjection and (b) with reinjection.

FIG. 3.49 Layout of a geothermal binary-cycle power plant coupled with water electrolysis.

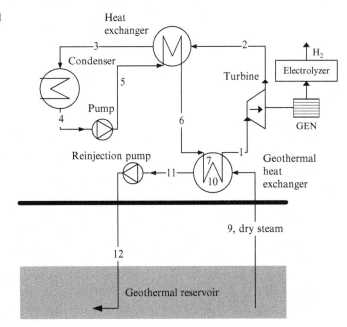

The exergy efficiency of the overall system was defined as the ratio of the exergy of the hydrogen (H_2), over the total exergy input, which can be determined as

$$\psi_{overall} = \frac{\dot{E}x_{H_2}}{\dot{W}_{e,in} + \sum \dot{E}x_{in}}$$

(3.115)

where $\dot{E}x_{H_2} = \dot{m}_{H_2} ex_{H_2}^{ch}$.

Energy efficiency values for the overall system vary between 80% and 87%, while exergy efficiency values for that range varied from 79% to 86%, respectively. The high-temperature electrolysis system consumed 3.34 kWh at 230°C and generated 573 mol/s H_2.

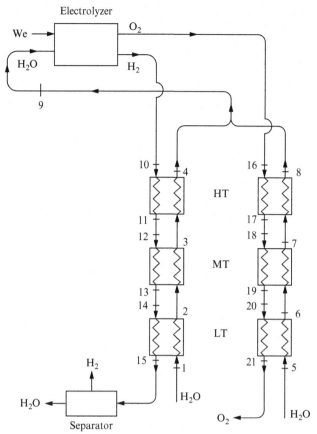

FIG. 3.50　Geothermal high-temperature steam electrolysis system. *Modified from Sigurvinsson, J., Mansilla, C., Arnason, B., Bontemps, A., Maréchal, A., Sigfusson, T.I., Werkoff, F., 2006. Heat transfer problems for the production of hydrogen from geothermal energy. Energy Convers. Manag. 47, 3543–3551.*

3.9.4 Ocean Energy Systems Integrated With Water Electrolysis

Various ocean energy technologies such as tidal, wave, and ocean thermal energy conversion (OTEC) plants that can produce electricity can be easily integrated with water electrolyzers. Tidal energy utilizes the power of tide to produce electricity, whereas wave energy systems use the waves formed in an ocean or sea. Oscillators are placed in the sea, and they oscillate when waves come in contact with them. This oscillatory motion is utilized to generate electricity. The ocean thermal technology uses the temperature difference between the upper and the deep lower-level of ocean water. This thermal difference is utilized to generate electricity.

The heat cycle of tropical solar energy affects the oceans during the earth's rotation and generates kinetic energy that could be used directly to turn submerged turbine generators. The temperature gradient of the ocean with depth results in a temperature difference just enough over reasonable depths to extract thermal energy at low efficiency. This is called ocean thermal energy conversion. Ammonia-water Rankine and Kalina cycles were proposed in some past studies for OTEC. Using deep seawater temperature as a sink at 4°C and the atmosphere at a temperature of 25–35°C as the heat source, energy efficiency of about 5% can be obtained.

An OTEC-system configuration integrated with solar thermal collectors and a PEM electrolyzer is shown in Fig. 3.51. The system is basically a Rankine cycle operating with ammonia as the working fluid (or ammonia-water). The system can be installed on a floating platform or on a ship. It uses surface water as the source. Normally, the water at the ocean's surface is at a higher temperature than water deeper in the ocean. The surface water is circulated with pumps through a heat exchanger that acts as boiler for ammonia. Water from the deep ocean at 4°C is pumped to surface and used as a heat sink in an ammonia condenser.

The numerical simulation results of Ahmadi et al. (2013) are given in Table 3.10. As shown, the net power output is ~101 kW, which leads to a hydrogen production rate of about 1.2 kg/h. In addition, the exergy efficiency of the integrated OTEC system is much higher than the energy efficiency at the ocean's surface (mainly because the work is produced using a low-grade heat, having the temperature level near to that of the reference environment).

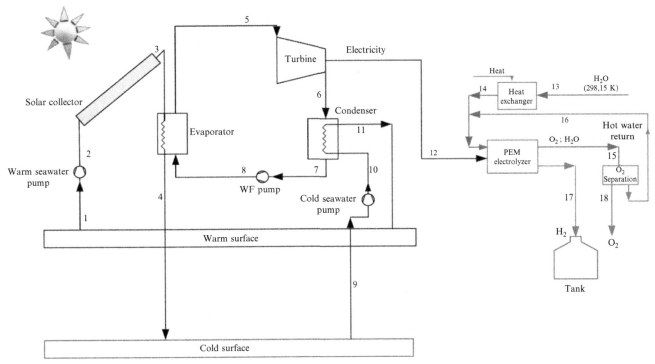

FIG. 3.51   OTEC solar-boosted H_2 production plant. *Modified from Ahmadi et al. (2013).*

Other power generation systems from ocean energy can be used in similar ways to generate hydrogen from water electrolysis. Tidal impoundment (barrage) systems can have three methods of operation, depending on the phase in which they generate power: ebb generation, flood generation, and two-way generation. When ebb generation is applied, the basin is filled during the flood tide. If it is during the night, additional water can be pumped into the basin, as a means of storing energy during the off-peak hours. When the tide ebbs low enough, water is discharged over turbine systems that generate power. In flood generation, the dam gates are closed so that the water level increases on the ocean side until it reaches the maximum level. Then water is allowed to flow through turbine systems and charge the basin while generating power. In two-way generation, electricity is generated both in flood and in ebb phases of the tide.

TABLE 3.10   Performance Parameters of an OTEC-PEM Hydrogen Production System

Parameters	Notation	Value
Net power output	$\dot{W}_{net}$	101.96 kW
Exergy efficiency	ψ	23%
Energy efficiency	η	3.6%
Total exergy destruction rate	$\dot{Ex}_d$	42.12 kW
Hydrogen production rate	$\dot{m}_{H_2}$	1.20 kg/h
PEM electrolyzer exergy efficiency	ψ_{PEM}	56%
Warm surface pump power	$\dot{W}_{WS}$	1.30 kW
Cold surface pump power	$\dot{W}CS$	3.13 kW
Working fluid pump power	$\dot{W}_{WF}$	0.88 kW

Data from Ahamdi et al. (2013).

TABLE 3.11 Some of the Major Tidal Impoundment Sites

Location	Head (m)	Mean power (MW)	Production (GWh/year)
Minas-Cobequid, North America	10.7	19,900	175,000
White Sea, Russia	5.65	14,400	126,000
Mount Saint Michel, France	8.4	9700	85,100
San Jose, Argentina	5.9	5970	51,500
Shepody, North America	9.8	520	22,100
Severn, UK	9.8	1680	15,000

Table 3.11 lists some of the main tidal power generation systems already built and their principal characteristics. The world's largest tidal power generation site is in France at "La Rance" with an installed capacity of 240 MW that operates with 24 reversible turbines and a hydrostatic head of 5 m.

Ocean water currents are generated by the action of tides, spinning of the earth and the heat cycle of tropical solar energy and superficial winds. The water currents' energy can be extracted through current turbines submerged in water. Basically, water current turbines are similar to wind turbines, with the difference that the density and viscosity of water are 100 and 1000 times higher than that of air, respectively. Therefore, there are some differences between the operating conditions of water current turbines and wind turbines. Similarly as for wind turbines, water current turbines are made in horizontal-axis or vertical-axis construction. The thrust generated by water current turbines is much higher than the thrust of wind turbines, so the construction materials for water current turbines must be more massive and resistant.

Surface winds, tides, and ocean currents contribute to the formation of ocean waves. The energy of waves can be collected with floating bodies that execute elliptical movements because of the action of gravity and wave motion. The wave energy can be measured in terms of wave power per meter of wave front. The highest wave power on the globe appears to be in southern Argentine across the Strait of Magellan where wave power density can reach 97 kW/m. Also, along south western coast of South America, the wave power density is more than 50 kW/m and often more than 70 kW/m. In southwestern Australian coasts, the wave power reaches 78 kW/m. In the Northern Hemisphere, the highest wave intensity is found along the coast of western Ireland and the United Kingdom with magnitudes around 70 kW/m. Southern coasts of Alaska record wave power density up to 65–67 kW/m. The wave power density on other coastal regions varies from about 10 kW/m to 50 kW/m. This impressive amount of energy can be converted to electricity with relatively simple mechanical systems that can be classified in two kinds: buoy and turbine types.

The principle schematic of a buoy-type wave energy converter is suggested in Fig. 3.52 and operates based on hydraulic-pneumatic systems. The buoy oscillates according to the wave movement at the ocean surface. It transmits the reciprocating movement to a double-effect hydraulic pump that is anchored rigidly on the ocean bottom. The pump generates a pressure difference between two pneumatic-hydraulic cylinders. A hydraulic motor generates shaft work by discharging the high-pressure liquid into a low pressure reservoir. The shaft work turns an electric generator that produces electricity.

The energy of waves is correlated with the energy of surface winds. Two aspects are important in determining the interrelation between the wave height and the wind characteristics, namely the wind-water fetch (ie, the length over which the superficial wind contacts the water) and the duration of the wind. For example, if the wind blows constantly for 30 h at 30 km/h and contacts with water over a length of 1200 km, the wave height can reach 20 m. The power density of the wave in deep ocean water can be approximated with

$$\dot{W} = 0.5 v \rho g h^2 \tag{3.116}$$

while in shallow waters it is about half this amount. In the equation, v is the eave speed, ρ is the water density, $g = 9.81 \, \text{m/s}^2$ is the gravity acceleration and h is the height of the wave. Thus a 10-m wave propagating with 1 m/s carries a power density of 49 kW/m.

If the waves are high, arrangement can be made so that a difference in level can be created constantly through impoundment. Thus the top of the wave carries water over the impoundment, constantly filling a small basin at a higher water level. The difference in water level is turned into shaft work by a Kaplan turbine system. The expected efficiency of hydrogen generation with wave, ocean currents and tidal systems integrated with electrolyzers should be of the same order as that obtained with hydropower-integrated systems, namely ~40–50%.

FIG. 3.52 Principle of operation of buoy-type ocean wave energy conversion system.

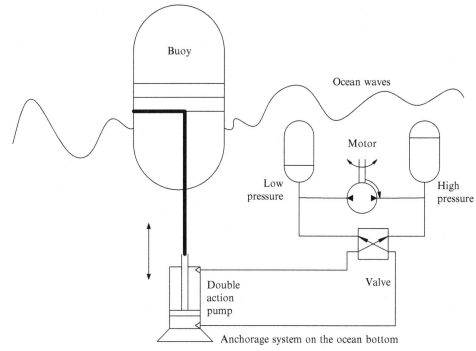

3.9.5 Solar Thermal and Biomass Power Generators Integrated With Water Electrolysis

Biomass combustion heat or solar thermal energy can be used to generate power with ORCs or other power cycles. The power plant can be integrated with water electrolysis to generate hydrogen. The efficiency of electricity generation in this case ranges from 10% to 60% depending on the feed and the technology. For example, a combined fuel cell/gas turbine cycle with biomass gasification and heat recovery can reach 60% efficiency. Therefore, the efficiency of hydrogen production with these system is in the order of 5% to 30%.

Unlike the more common steam Rankine cycles, an organic working fluid is used in the ORCs. Local and small-scale power generation as well as renewable energy systems and low-grade heat recovery systems are the best applications for ORC technology. One important feature of the organic working fluid relates to the fact that large pressure ratios over the turbine can be obtained for relatively small pressure differences. For example, with the organic fluid R123, the ORC operates at a boiling pressure of 5.87 bar and a condensation pressure of 0.85 bar. The pressure ratio is thus 6.9, and the pressure difference is 5.02 bar. This point signifies the advantage of using organic fluids as the working fluid in Rankine cycles.

An ORC of simple regenerative configuration is shown in Fig. 3.53. The regenerative schemes for ORC do not require (in general) vapor extraction (as in the case of steam cycles). Rather, an additional heat exchanger is inserted between the turbine exit and the pump exit, which allows for a transfer of heat between the hotter working fluid at the turbine exhaust and the colder liquid at the pump discharge.

A system that integrates an ORC, a proton membrane electrolyzer (PEME), hydrogen storage in metal hydrides and a fuel cell (FC) is analyzed thermodynamically for on-demand power generation from solar energy in Tarique et al. (2014). The system is destined to small-scale applications such as rooftop solar power at low concentration and total collector area on the order of 100 m². This system ensures the heating and power needs of an average residence throughout the year. Compound parabolic concentrators (CPC) are used to concentrate sunlight on line-focus solar receivers (SR) which act as vapor generators in an integrated manner with an ORC heat engine with cyclohexane working fluid. A small capacity thermo-mechanical energy storage (TMS) system is used to keep the working fluid hot during the night and to eliminate the need of a system warmup in the morning. The ORC is made to operate continuously and steadily during daylight to produce power with a solar multiplicity factor of 3. A fuel (FC) cell system is used to generate power on demand from the stored hydrogen. A part of the rejected heat is recovered and used for cogeneration of heating. In this respect a heat recovery (HR) unit is integrated within the system. The system (SYS) efficiency is determined based on thermodynamic analysis and has proven to be attractive.

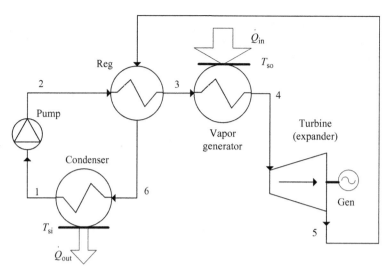

FIG. 3.53 Regenerative ORC configuration.

Solar energy resources do have some characteristics that lead to technical and economic challenges: generally diffuse, not fully accessible, sometimes fluctuating, intermittent as well as regionally variable. The solar radiation is very diffuse with sometimes less radiative power than 500 W/m².

For this reason, concentration is required. Imaging concentrators are in general easy to fabricate in the form of heliostat mirrors, solar dishes, solar through or Fresnel mirrors. They, however, lack the ability to concentrate diffuse radiation and also require expensive sun-tracking systems. Compound parabolic concentrators (CPC) are a non-imaging alternative to the common imaging concentrators. CPCs do not require sun tracking and can concentrate diffuse radiation in addition to the direct beam radiation, but their concentration ratio is limited to 3–5 because of manufacturing considerations.

A simplified diagram showing the integration of ORC with electrolyzer and fuel cell systems is given in Fig. 3.54. The ORC is the system that generates power from a fluctuating source (solar) Fig. 3.55 shows the diagram of the ORC system, which is similar to that presented previously in Case Study 1 from Chapter 1, Fig. 1.22, except that in this case the ORC is linked to water electrolysis. A thermally insulated vessel is used to keep the working fluid as a saturated liquid-vapor mixture with high temperature; this denoted as the thermo-mechanical energy storage system (TMS). Liquid (#1), which is heavier, flows downwards and reaches the solar collector. Because the liquid is hot and close to saturation, it boils (#2) in an evacuated glass tube placed at the CPC aiming spot . The boiled fraction of working fluid is in accordance to the fluctuating intensity of solar radiation. The evacuated tube is a glass tube maintained in vacuum, within which a vapor generator tube (metallic) is placed. The evacuated tube forms a solar receiver (SR). The generated vapors separate from liquid and flow further (#3) and expand in the turbine. Once expanded, the saturated vapors of cyclohexane superheat as shown in Figure 3.56. The superheated vapors (#4) of the retrograde cyclohexane working fluid cool down and start to condense (#6) reaching a saturated liquid state in (#9). Some vapor is extracted in (#5) for

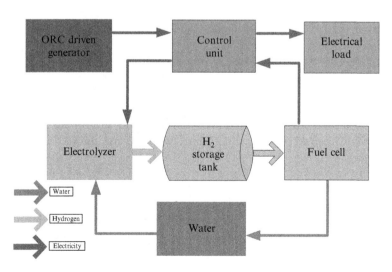

FIG. 3.54 ORC integrated with electrolyzer and hydrogen fuel cell.

FIG. 3.55 The ORC system and the thermosiphon configuration of solar collector.

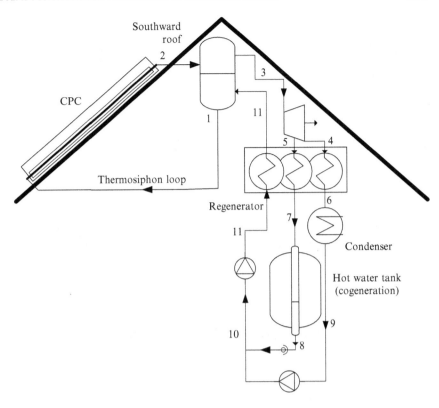

TABLE 3.12 System Parameters Assumed for a Case Study

Half acceptance angle of CPC	$\theta_{0.5} = 60°$
Average reflections number	$n_r = 1.1$
Concentration ratio CPC	$C = 5$
Total aperture area	$A = 100 \text{ m}^2$
Annual irradiation amount	$I_T = 1800 \text{ MWh/m}^2 \text{ year}$
Water heating temperature	$T_8 = 60°C$ (Fig. 3.3)
Collector temperature	$T_1 = T_2 = T_3 = 120°C$
Condensation temperature	$T_6 = T_9 = 20°C$

cogeneration where they condensate according to the heat demand in (#8). The branch 5-7-8-10 shown in Figure 3.55 forms the heat recovery (HR) system. The cycle is modeled based on balance equations written for steady operation.

A case study with the ORC/electrolyzer/fuel cell system is described by a set of reasonable values for the modeling parameters that are assumed as given in Table 3.12. The core of the system is formed by the ORC and its thermosiphon loop. For the study, the ORC is calculated first and the thermodynamic cycle is determined. This cycle operates between a ground loop temperature and the solar collector temperature. The ground loop imposes a condensation at $T_6 = T_9 = 20°C$. The solar receiver and thermo-mechanical liquid, in conjunction, impose a boiling temperature of $T_1 = T_2 = T_3 = 120°C$.

However, the thermodynamic cycle is illustrated in Fig. 3.56. The ORC cycle efficiency is $\eta_{ORC} = 28\%$ if no heat is cogenerated. The ORC efficiency depends on the amount of cogenerated heat (the heat demand). Note that the ORC self-regulates according to the heat demand. When there is no heat demand, there is no cooling in the branch 7–8 of the system. Vapor accumulates, and the extracted fraction must reduce because the pressure on the branch 5-7-8 tends to increase. When there is heat demand, there is also good condensation, and the extraction fraction tends to increase.

For estimation of the overall system efficiency, the required balance equations are solved for each system component. Table 3.13 shows the efficiency results.

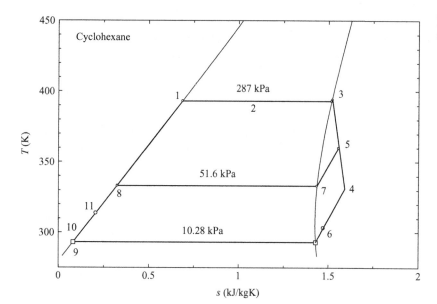

FIG. 3.56 Thermodynamic cycle of the cyclohexane-based ORC with cogeneration.

Since the half acceptance angle of CPC is 60°, one-third of the diffuse radiation is concentrated by the CPC. Because the global radiation is assumed to be 1800 MWh/m² year and, for AM 1.5, the direct beam radiation is 90% of the global radiation, it results that the CPC will accept 180 MWh/m² year diffuse radiation. The CPC orientation is toward the equator in tilt position. Since the average daylight is from 6 am to 6 pm, the CPC sees the sun from 8 am to 4 pm; therefore, it accepts two-thirds of the direct beam radiation. Therefore, 1200 MWh/m² year are received from direct beam, and 1380 MWh/m² year is the total concentrated radiation. From this point, the CPC efficiency is 1380/1800 = 76.6%.

On the basis of 1800 MWh/m² year, the generated power can be calculated if one assumes one-third of it is produced during daylight period with an efficiency of $\eta_{PG} = 14.9\%$. The efficiency ηPIC given in Table 3.13 quantifies the energy loses in the power inverter and controller (PIC) system. The hydrogen fuel cell system generates two-thirds of the daily power demand. The hydrogen electrolyser/fuel cell system generates power on-demand with an efficiency of $\eta_{ODPG} = 3.7\%$. Therefore, one calculates for 10 m² collector area $0.149 \times 0.667 \times 1800 \times 10 = 1790$ MWh power generation. For heat generation, one obtains 5652 MWh per year. Therefore, the system efficiency that uses both direct power production during daylight and fuel cell generation at night time is 41.3%, whereas the power-only efficiency of the system calculated in similar way is 9.94%.

TABLE 3.13 Energy Efficiencies of the System and Its Components

Component	Assumptions and remarks	Efficiency
CPC	AM 1.5 spectrum at 120° acceptance angle	$\eta_{CPC} = 76.7\%$
SR	Heat loss coefficient $U = 3$ W/m²K	$\eta_{SR} = 95\%$
TMS	Assume 1% heat loss	$\eta_{TMS} = 99\%$
ORC	Determined based on Fig. 3.56	$\eta_{ORC} = 25\%$
PIC	Assume electrical losses of 5% as shown in Yilanci et al. (2009)	$\eta_{PIC} = 95\%$
PEME	Assume practical value at optimal current density as in Yilanci et al. (2009)	$\eta_{PEME} = 50\%$
FC	Assume practical value at optimal current density as in Naterer et al. (2013)	$\eta_{FC} = 50$
HR	Determined based on Fig. 3.56	$\eta_{HR} = 50\%$
SYS	The following system efficiencies are calculated:	
	• Power generation efficiency $\eta_{PG} = \eta_{CPC}\eta_{SR}\eta_{TMS}\eta_{ORC}\eta_{PIC}$	$\eta_{pg} = 14.9\%$
	• Hydrogen generation efficiency $\eta_{H_2} = \eta_{pg}\eta_{PEME}$	$\eta_{H_2} = 7.5\%$
	• On-demand power generation efficiency $\eta_{ODPG} = \eta_{H_2}\eta_{FC}$	$\eta_{ODPG} = 3.7\%$
	• Heat generating efficiency $\eta_{HG} = \eta_{CPC}\eta_{SR}\eta_{TMS}\eta_{HR}$	$\eta_{HG} = 31.4\%$
	• Cogeneration efficiency $\eta_{SYS} = \eta_{ODPG} + \eta_{HG}$	$\eta_{SYS} = 35\%$

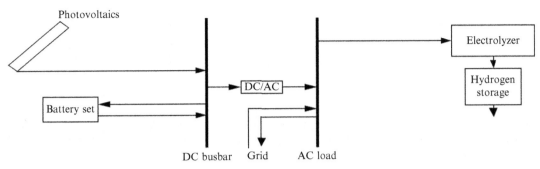

FIG. 3.57 Integrated PV-water electrolysis system for hydrogen production.

3.9.6 Solar Photovoltaic Water Electrolysis Systems

The photovoltaic-driven water electrolysis is a proven technology, which can be assembled with off-the-shelf components. A system as such is described schematically in Fig. 3.57. It comprises photovoltaic (PV) panels, DC bus bar, AC grid, accumulator battery set, electrolyzer and hydrogen storage canisters. The cost of PV-generated electricity continues to decline. For example, in 1998 the average cost was $12 per installed watt, and in 2008 the average cost was $8. This technology is currently a $10 billion business, growing 30% per annum, which shows an encouraging trend for the cost of producing PV hydrogen. The efficiency of the solar cell can range from 12% to 15%, typically for the silicon solar cell. However, it is as high as 25–30% for GaAs solar cells. The total efficiency of solar radiant energy transformed to chemical hydrogen energy is nearly 16%. The US Department of Energy projected solar-to-hydrogen conversion efficiency is in the range of 30–40% for 2025.

The exergy efficiency of the PV-electrolyzer system is calculated as the product of exergy efficiency of the PV system ψ_{PV} and the exergy efficiency of the electrolyzer ψ_{el}, which is given as

$$\psi = \psi_{PV}\psi_{el} = \frac{V_m I_m - \left(1 - \dfrac{T_0}{T_{cell}}\right)hA(T_{cell} - T_0)}{\dot{E}x_{solar}} \frac{\dot{E}x_{H_2} + \dot{E}x_{O_2}}{V_m I_m} \tag{3.117}$$

where $V_m I_m$ is the electrical power accounting for all electrical losses of the PV panel, associated electronics and electrical lines.

This power is the same as that retrieved at the input of the electrolyzer. The quantity $hA(T_{cell} - T_0)$ represents the heat losses between the PV panel and the ambient because of heat transfer; some exergy losses are associated with this heat as indicated in the numerator of above equation.

The energy and exergy efficiencies of the components of a PV-electrolysis systems are given in Table 3.14 for a representative case study with actual implementation at Pamukkale University in Turkey. The power is supplied by a solar PV array of 5 kW nameplate capacity. In this system, a line feed deionizer is selected to supply the quality of water needed for the electrolyzer. A basic particle water filter is also used before the deionizer. A PEM type electrolyzer

TABLE 3.14 Efficiencies of the PV-Electrolysis System From Pamukkale University

Components	Energy efficiency (%)	Exergy efficiency (%)
PVs	11.2–12.4	9.8–11.5
Charge regulators	85–90	85–90
Batteries	80–85	80–85
1st Inverter	85–90	85–90
Electrolyzer	56	52
Hydrogen tanksc	100	100
Fuel cells	30–44	24.5–38
2nd Inverter	85–90	85–90

Data from Yilanci, A., Dincer, I., Ozturk, H.K., 2009. A review on solar-hydrogen/fuel cell hybrid energy systems for stationary applications. Prog. Energy Combust. Sci. 35, 231–244.

is used in the system. A metal hydride (MH) storage tank is used for hydrogen. The downside is that the hydrogen produced for MH storage must be of very high purity. Six OVONIC 85G250B storage tanks are used. The system also includes a fuel cell to generate electricity from the produced hydrogen and is connected to the grid. Table 13.4 lists the efficiencies of the individual system components. Average hydrogen production of the system for a week is 4.43 kg.

Bicer et al. (2015) installed a PV electrolyzer for hydrogen production in Oshawa, Canada, and studied the impact of solar spectrum variations on the outdoor performance of the system. Various parametric studies were conducted in order to optimize the system performance.

PV modules located in the CERL Laboratory at the University of Ontario Institute of Technology (43.9448° N, 78.8917° W) were measured under a solar simulator (OAI Trisol TSS-208 Class AAA) with an irradiance of 1000 W/m^2 and under outdoor condition with the modules faced to south-west, tilted with 45° tilt angle.

The outputs from the modules were measured with a Potentiostat/Galvanostat/ZRA (Gamry Instruments Reference 3000). To analyze the spectral irradiance distribution, solar spectra with the wavelength range of 350–1000 nm were recorded by a spectrometer (Ocean Optics Red Tide USB 650). The PV surface temperature was measured with a Surface Temperature Sensor (Vernier STS-BTA), and the ambient temperature was measured with a probe temperature sensor (Vernier GO-TEMP) through a data logger unit (Vernier LabQuest). A pyranometer (Vernier PYR-BTA) was installed with same angle of PV modules to measure total global irradiance.

Electric power generated by a PV subsystem must be calculated based on the spectral response of the selected PV modules. This response can be determined with the help of cell performance parameters such as open circuit voltage (V_{oc}), short-circuit current density (I_{sc}), filling factor (FF), band-gap energy (E_g), saturation current density (I_0), internal resistance (R_s) and spectral quantum efficiency (η_{SQE}).

The energy and exergy produced by the PV module under concentrated spectral radiation is equal to the maximum power generated, as shown in the following equation:

$$\dot{Ex} = \dot{E} = \dot{W}_{max} = A_{PV} FF I_{sc} V_{oc} \tag{3.118}$$

Here I_{ph} is the current density, which is approximately equal with the photocurrent ($I_{sc} \cong I_{ph}$).

The photocurrent depends on the spectral quantum efficiency η_{SQE} and the spectral irradiance of the concentrated radiation according to the following equation:

$$I_{sc} \cong I_{ph} = \frac{eF_{field}A_h}{A_{PV}hc} \int_0^\infty \lambda \Phi_\lambda (1 - R_{\lambda,PV}) R_{\lambda,h} (1 - R_{\lambda,d}) \times (1 - R_{\lambda,g}) I_{DNI,\lambda} d\lambda \tag{3.119}$$

The open circuit voltage is related to photocurrent according and to the exchange current density (J_0) to the following equation:

$$v_{oc} = \frac{eV_{oc}}{k_B T} = \ln\left(1 + \frac{I_{ph}}{I_0}\right) \text{ with } I_0(A/m^2) = 1.5E9 \exp\left(-\frac{E_g}{k_B T_{PV}}\right) \tag{3.120}$$

The fill factor is given by the following empirical correlation:

$$FF = \frac{v_{oc} - \ln(v_{oc} + 0.72)}{V_{oc} + 1}\left(1 - \frac{R_s J_{sc}}{V_{oc}}\right) \tag{3.121}$$

Maximum power of PV is determined with fill factor definition. Hence the energy and exergy efficiencies of PV can be written as

$$\eta = \frac{FF V_{oc} I_{sc}}{I_{DNI}} \tag{3.122}$$

$$\psi = \frac{FF V_{oc} I_{sc}}{Ex_{DNI}} \tag{3.123}$$

where I_{DNI} and Ex_{DNI} represent the solar irradiance and exergy of the light in normal direction to PV surface.

A PV module can be modeled using external quantum efficiency, diode non-ideality factor, and the shunt and series resistances under any irradiance and operating temperature. The external quantum efficiency is dependent on the type of the cell and the manufacturer.

Spectral internal quantum efficiency (IQE) is an important aspect for PV modules and is defined as the ratio of electrons displaced across the semiconductor junction $\dot{N}_{e,\lambda}$ and the rate of photons absorbed by the wafer at a certain wavelength $\dot{N}_{ph,abs,\lambda}$:

$$\Phi_{IQE} = \frac{\dot{N}_{e,\lambda}}{\dot{N}_{ph,abs,\lambda}} = \left(\frac{I_\lambda}{e}\right)\frac{hc}{\lambda I_{abs,\lambda}(\lambda)} \tag{3.124}$$

Here I_λ (A/m^2) is the electric current density generated by the absorbed photons on the wafer and $I_{abs,\lambda}(\lambda)$ (W/m^2 nm) is the spectral irradiance at wavelength λ of the absorbed photons in a normal direction to PV module surface.

Since there are losses in coatings and glassing in a PV module because of transmittance and reflectance values, another quantum efficiency called external spectral quantum efficiency (EQE) is defined as

$$\Phi_{EQE} = \frac{\dot{N}_{e,\lambda}}{\dot{N}_{ph,\lambda}} = \eta_{IQE}(1 - R_\lambda - T_\lambda) = \left(\frac{I_\lambda}{e}\right)\frac{hc}{\lambda I_\lambda(\lambda)} \tag{3.125}$$

where $\dot{N}_{ph,\lambda}$ is the rate of incident photons on the surface of the PV module, R_λ is the reflectance of the module, T_λ is the transmittance of the wafer, and $I_\lambda(\lambda)$ is the spectral irradiance at the wavelength λ of the incident photons.

When EQE is known through specific measurements, the short-circuit current can be found as

$$I_{sc} \cong I_{ph} = \frac{e}{hc}\int_0^\infty \lambda\, \eta_{EQE}\, I_\lambda\, d\lambda \tag{3.126}$$

The open circuit voltage is calculated based I_{ph} and I_{Load} as follows:

$$V_{oc} \cong \frac{n_i k_B T_c}{e} \ln\left(1 + \frac{I_{ph}}{I_0}\right) \tag{3.127}$$

Note that the saturation current is proportional to the number of photons generated by blackbody radiation at the temperature of the module. Only the photons that have energy higher than the band-gap energy (E_g) of the *p-n* junction can contribute to the generation of the saturation current. Band-gap energy is defined as $E_g = k_B T_g$, where T_g is the effective temperature of the band gap. The number of photons of blackbody radiation at cell temperature having energy higher than the band gap is given by the spectral distribution of blackbody energy according to

$$\dot{N}_g = \frac{2\pi k_B T_c^3}{h^3 c}\int_{\frac{T_g}{T_c}}^\infty \frac{\chi^2}{e^\chi - 1}d\chi \tag{3.128}$$

where χ denotes dummy variable and $\dot{N}_g$ is a measured in photons/m^2 of exposed surface.

A summary of PV cell measurements under different types of filters are given in Table 3.15. The maximum power output from the PV cell is observed with the green filter after the one without the filter. Secondly, the

TABLE 3.15 Measurement Results of PV-Electrolyzer Performance With Different Filters

Parameter	No filter	UV filter	Green filter	Red filter
I_{sc} (A)	0.12	0.045	0.1	0.08
V_{oc} (V)	9.0	9.1	9.1	9
I_m (A)	0.1	0.04	0.09	0.07
V_m (V)	7.7	7	7.5	7.5
FF (%)	71.3	68.4	74.2	72.9
η (%)	10.3	3.7	9.5	7.0
ψ (%)	10.8	3.9	9.5	7.4

red filter has a maximum power output of $\sim$0.53 W. Since the UV filter absorbs high energy spectra, the amount of power generated by the PV cell is the lowest one.

Energy and exergy efficiencies are calculated as 10.3% and 10.8%, respectively, with no filter. The lowest energy and exergy efficiencies are observed when a UV filter is used. In accordance with the Faraday law, the input current required is directly proportional to the hydrogen produced as indicated in the table.

3.10 CONCLUDING REMARKS

In this chapter, the methods of hydrogen production by electrical energy are introduced and discussed. In the first section, the fundamentals of electrochemical hydrogen production are presented. The most important thermodynamic parameters and electrochemical reactions are discussed. Further, the main types of water electrolyzers, including alkaline, proton exchange membrane and solid oxide electrolyzers, are analyzed. The efficiencies of hydrogen production through electrochemical methods are formulated in terms of energy, exergy, and Faraday efficiencies. Some other electrochemical methods of hydrogen production are included, in addition to water and steam electrolysis. These methods are the chloralkali process, the hydrogen sulfide electrolysis and the ammoniated water electrolysis.

The chapter ends with a presentation of the possible integrated systems to produce green hydrogen from electrical energy. These systems include hydropower-electrolysis, wind-power electrolysis, geothermal-power electrolysis, ocean-power electrolysis, solar electrothermal electrolysis, biomass power electrolysis and solar-photovoltaic electrolysis systems.

Nomenclature

a	chemical activity ($kmol/m^3$)
a	Tafel's polarization potential (V)
A	area (m^2)
$\mathcal{A}$	pre-exponential factor
b	Tafel slope (V)
$\mathbb{B}$	magnetic flux density (T)
B_g	flow permeability
c	molar concentration ($kmol/m^3$)
c	speed of light constant (m/s)
C	concentration ratio
CF	quality factor
D	diffusivity (m^2/s)
e	elementary electric charge (C)
E	electric potential (V)
$\dot{E}$	energy rate (kW)
$\mathbb{E}$	electric field intensity (N/C)
ex	specific chemical exergy (kJ/mol)
$\dot{Ex}$	exergy rate (kW)
f	activity coefficient
F	Faraday constant (C/mol)
$\mathbb{F}$	Lorentz force (N)
FF	fill factor
G	molar Gibbs free energy (kJ/kmol)
h	Planck constant (Js)
h	probability distribution (Weibull, Rayleigh)
H	molar enthalpy (kJ/kmol)
HHV	higher heating value (kJ/mol)
I	electric current (A)
I	light intensity (W/m^2)
J	electric current density (A/m^2)
J_c	convective mass flux ($kmol/m^2 s$)
k	reaction rate constant
K	equilibrium constant
k_b	Boltzmann constant (J/K)
l	length (m)
$\dot{m}$	mass flow rate (kg/s)
M	molar mass (kg/kmol)
n	number of moles (kmol)

$\dot{n}$	molar rate (kmol/s)
n_i	diode ideality parameter
n_r	average reflections number
N	number of cells
P	pressure (kPa)
q	electric charge (C)
r	mean pore radius (m)
R	electric resistance (Ω)
R	universal gas constant (kJ/kmol K)
R_λ	spectral reflectance
S	molar entropy (kJ/kmol K)
SEC	specific energy consumption (kWh/kg H_2)
t	time (s)
T	temperature (K)
T_λ	spectral transmittance
υ	velocity (speed) (m/s)
V	air speed (m/s)
$\dot{V}$	volumetric flow rate (m^3/s)
$\dot{W}$	power (W)
x	length coordinate (m)
z	number of electric charges

Greek Letters

α	transfer coefficient
β	coefficient, Eq. (3.77), Eq. (3.100)
γ	surface tension (N/m)
δ	thickness (m)
ε	electrode porosity
ϵ	characteristic Lenard-Jones potential (meV)
η	efficiency
λ	wavelength (m)
λ	mean characteristic length (Å)
μ	chemical potential (kJ/kmol)
μ	dynamic viscosity (Pa.s)
ψ	exergy efficiency
θ	acceptance angle (rad)
θ	contact angle (rad)
ξ	electrode tortuosity
τ	mean collision time (s)
σ	electric conductivity (S)
ΩD	collision-diffusion integral

Subscripts

0	reference state
a	anode
abs	absorbed
act	activation
AE	alkaline electrolyzer
b	backward
c	cathode
c	convection
cell	electrolysis cell
ci	cut in
co	cut out
conc	concentration
CPC	compound parabolic collector
d	destroyed
d	diffusion
DNI	direct normal radiation
el	electrolysis
eq	equilibrium
eq	equivalent
EQE	external quantum efficiency
f	forward
Faradaic	Faradaic (second law) efficiency

g   gap
IQE   internal quantum efficiency
k   Knudsen diffusion
lim   limiting
loss   losses
m   advection (conduction)
oc   open circuit
oxi   oxidation
Ω   ohmic
P   products
pg   power generation
ph   photonic
pol   polarization
R   reactants
red   reduction
ref   reference
rev   reversible
RMC   root mean cube
sat   saturation
sc   short circuit
th   thermoneutral
tot   total

Superscripts

0   standard state
ch   chemical
eff   effective

References

Annani, A.A., Mao, Z., White, R.E., Srinivasan, S., Appleby, A.J., 1990. Electrochemical production of hydrogen and sulphur by low-temperature decomposition of hydrogen sulphide in an aqueous solution. J. Electrochem. Soc. 137, 2703–2709.

Ahmadi, P., Dincer, I., Rosen, M.A., 2013. Energy and exergy analyses of hydrogen production via solar-boosted ocean thermal energy conversion and PEM electrolysis. International Journal of Hydrogen Energy 38, 1795–1805.

Bagotski, V.S., 2006. Fundamentals of Electrochemistry, second ed. Wiley Interscience, Hoboken, New Jersey.

Balta, T., Dincer, I., Hepbasli, A., 2009. Thermodynamic assessment of geothermal energy use in hydrogen production. Int. J. Hydrog. Energy 34, 2925–2939.

Baniasadi, E., Dincer, I., Naterer, G.F., 2013. Electrochemical analysis of seawater electrolysis with molybdenum-oxo catalysts. Int. J. Hydrog. Energy 38, 2589–2595.

Bicer, Y., Dincer, I., Zamfirescu, C., 2015. Modeling and experimental validation of a PV-electrolyzer for hydrogen production under various solar spectra. In: Sixth International Conference on Hydrogen Production, May 3–6. University of Ontario Institute of Technology, Oshawa, ON.

Boggs, B.K., King, R.L., Botte, G.C., 2009. Urea electrolysis: direct hydrogen production from urine. Roy. Soc. Chem., Chem. Commun, 4859–4861.

Chan, S.H., Xia, Z.T., 2002. Polarization effects in electrolyte/electrode supported solid oxide fuel cells. J. Appl. Electrochem. 32, 339–347.

Chandran, R.R., Chin, D.T., 1986. Reactor analysis of a chlor-alkali membrane cell. Electrochim. Acta 31, 39–50.

Dincer, I., Zamfirescu, C., 2011. Sustainable Energy Systems and Applications. Springer, New York.

Dincer, I., Zamfirescu, C., 2012. Sustainable hydrogen production and the role of IAHE. Int. J. Hydrog. Energy 37, 16266–16286.

Dincer, I., Zamfirescu, C., 2014. Advanced Power Generation Systems. Elsevier, New York.

Ghosh, S., 2014. Experimental investigation of new light-based hydrogen production systems. MSc Thesis, University of Ontario Institute of Technology.

Gregory, T.D., Feke, D.L., Angus, J.C., Brosilow, C.B., Landau, U., 1980. Electrolysis of liquid hydrogen sulphide. J. Appl. Electrochem. 10, 405–408.

Karunadasa, H.I., Chang, C.J., Long, J.R., 2010. A molecular molybdenum oxo-catalyst for generating hydrogen from water. Nature 464, 1329–1333.

Lin, M.-Y., Hourng, L.-W., Kuo, C.-W., 2012. The effect of magnetic force on hydrogen production efficiency in water electrolysis. Int. J. Hydrog. Energy 37, 1131–1320.

Manabe, A., Kashiwase, M., Hashimoto, T., Hayashida, T., Kato, A., Hirao, K., Shimomura, I., Nagashima, I., 2013. Basic study of alkaline water electrolysis. Electrochim. Acta 100, 249–256.

Mazloomi, K., Sulaiman, N.B., Moayedi, H., 2012. An investigation into the electrical impedance of water electrolysis cells – with a view to saving energy. Int. J. Electrochem. Sci. 7, 3466–3481.

Mbah, J., Krakow, B., Stefanakos, E., Wolan, J., 2008. Electrolytic splitting of H$_2$S using CsHSO$_4$ membrane. J. Electrochem. Soc. 155, E166–E170.

Meng, N., Leung, M.K., Leung, D.Y., 2008. Energy and exergy analysis of hydrogen production by a PEM electrolyser plant. J. Energ. Convers. Manage. 49, 2748–2756.

Muradov, N., Veziroglu, T.N., 2008. "Green" path from fossil-based to hydrogen economy: an overview of carbon-neutral technologies. Int. J. Hydrog. Energy 33, 6804–6839.

Naterer, G.F., Dincer, I., Zamfirescu, C., 2013. Hydrogen Production from Nuclear Energy. Springer, New York.

Rabbani, M., 2013. Design, analysis and optimization of novel photo-electrochemical hydrogen production systems. PhD Thesis, University of Ontario Institute of Technology.

Sakai, T., Matsushita, S., Matsumoto, H., Okada, S., Hashimoto, S., Ishihara, T., 2009. Intermediate temperature steam electrolysis using strontium zirconate-based protonic conductors. Int. J. Hydrog. Energy 34, 56–63.

Sigurvinsson, J., Mansilla, C., Arnason, B., Bontemps, A., Maréchal, A., Sigfusson, T.I., Werkoff, F., 2006. Heat transfer problems for the production of hydrogen from geothermal energy. Energy Convers. Manag. 47, 3543–3551.

Tarique, M.A., Dincer, I., Zamfirescu, C., 2014. Energy and exergy analyses of solar-driven ORC integrated with fuel cells and electrolyser for hydrogen and power production. In: Dincer, I. (Ed.), Progress in Exergy, Energy, and the Environment. Springer, New York (Chapter 7).

Yilanci, A., Dincer, I., Ozturk, H.K., 2009. A review on solar-hydrogen/fuel cell hybrid energy systems for stationary applications. Prog. Energy Combust. Sci. 35, 231–244.

STUDY PROBLEMS

3.1 Explain why electrolysis process is not spontaneous.

3.2 What type of ions are conducted by solid oxide electrolytes used for electrolytic hydrogen production?

3.3 Explain why the reversible cell potential is related to Gibbs free enthalpy of the reaction and not to the reaction enthalpy.

3.4 Explain the difference between energy and exergy efficiency definitions of an electrolysis cell.

3.5 Explain the difference between Faradaic and exergy efficiencies of an electrolysis cell.

3.6 The cathodic nickel electrode half-reaction in alkaline electrolyte is characterized by a current exchange density of $J_0 = 10 \, \text{mA/m}^2$, Tafel slope of $b = 120 \, \text{mV}$ and transfer coefficient of $\alpha = 0.2$. The cell operates at the limits between low and high polarization levels. Calculate the energy and exergy efficiency of the cell when the operating temperature is 70°C.

3.7 Based on the electrode potential in Table 3.3, explain why NaCl solution is not a suitable electrolyte for water electrolysis.

3.8 Define activation overpotential in contrast to the concentration overpotential in an electrolysis cell.

3.9 What is the comparative advantage and disadvantage of alkaline and proton exchange membrane electrolyzers?

3.10 An alkaline electrolyzer operates according to the polarization curves shown in Fig. 3.12. Which is the best choice of the current density and how much is the overpotentials for the choice?

3.11 What is the influence of gas bubble occurrence in electrolyzers? Explain.

3.12 Explain the Donnan exclusion effect and its relevance in PEM electrolyzers.

3.13 Why should a sweep gas be used in high-temperature steam electrolyzers?

3.14 Explain the relevance of the triple-phase boundary in high-temperature electrolysis cells.

4

Hydrogen Production by Thermal Energy

4.1 INTRODUCTION

Our modern society depends heavily on thermal energy sources derived mostly from coal, petroleum, and natural gas combustion. The majority of thermal energy thus obtained is used to generate high pressure steam in large-scale power plants to produce electricity and to generate hydrogen as an intermediate chemical in the fertilizer industry. However, combustion of fossil fuels to generate thermal energy in the form of high temperature heat is accompanied by emissions of large amounts of GHG (greenhouse gas) and other pollutants.

Thermal energy can be used in many ways to drive thermochemical processes for hydrogen extraction from materials, such as mineral resources, water, biomass and anthropogenic wastes. Heat can be transferred to specific thermochemical processes using an indirect heat transfer configuration (ie, heat transfer through a wall). In some processes, such as pyrolysis or gasification, a direct heat transfer process is conducted, in which case, the combustion products play a direct role in the chemical reaction processes.

Two major thermochemical processes of hydrogen production are coal gasification and steam methane reforming. The overwhelming majority of hydrogen is produced worldwide based on these two processes. Both the gasification and reformation are spontaneous thermochemical processes. In both cases, the source of hydrogen is water, and the source of heat is mainly derived from the oxidation of the carbon atoms in coal and natural gas. Massive amounts of greenhouse gases are released into the atmosphere because the carbon from coal and methane is oxidized during the gasification and reformation processes, respectively and carbon dioxide is generated.

Greener methods can in principle be devised for hydrogen production through coal gasification and steam methane reforming. This is possible when the resulting carbon dioxide is captured and sequestrated, rather than being released into the atmosphere. Of course, an energy penalty must be considered, which is commensurable with the effort of safely-sequestrating the polluting gases rather than allowing them to escape in the atmosphere.

Other green thermochemical processes are known to produce hydrogen. Biomass gasification is similar to coal gasification, but since biomass is CO_2 neutral, this is a thermally driven green method. Similarly, biogas and biofuels can be reformed to generate hydrogen, using specific thermochemical methods.

Many types of anthropogenic wastes can be recovered as valuable materials to generate hydrogen via open-loop thermochemical processes. One example is aluminum, which is used heavily in the canning industry and other manufacturing. Recovery of aluminum waste represents a feedstock for an aluminum-water, thermo-catalytic process that generates pure hydrogen and alumina. Combustible waste materials can, in principle, be reacted in a modified incinerator to derive hydrogen. Ammonia represents a source of hydrogen or a chemical storage medium for hydrogen in a very compact form. Thermocatalytic processes are required to release hydrogen from ammonia.

When thermal energy is made available through an indirect heat transfer process, thermochemical cycles can be applied to generate hydrogen from water. Thermochemical cycles represent closed loop processes devised to generate hydrogen and oxygen from water in separate manifolds. In this cycle, a set of chemical reactions—mainly thermochemical as the name implies—are conducted in a closed loop with the net result of water splitting and the rest of the chemicals being recycled. The source of thermal energy must be green (clean) such that a green hydrogen can be produced with thermochemical cycles. These heat sources are mainly derived from renewable energy, nuclear energy and waste recovery. Minimal GHG emissions can be associated with these thermal energy sources.

The thermal energy sources derived from renewable energy are as follows: solar thermal, geothermal, ocean thermal, biomass, biofuels, biogas, and landfill gas. In addition, anthropogenic recovered wastes can be used to sustainably generate thermal energy as follows: municipal waste, industrial combustible waste, recoverable heat releases from industrial processes, and certain hospital wastes. A special, sustainable thermal energy source is nuclear energy. With current nuclear power plant technology, nuclear energy is made available in the form of high pressure, superheated steam at 300–350°C. The next generation of nuclear reactors will make thermal energy available at temperatures in a controlled manner, in the range of approximately 600–1000°C.

In this chapter, the hydrogen production methods by thermal energy are presented. To this purpose, the thermodynamic and kinetic fundamentals of thermochemical reactions are reviewed first. The assessment criteria through energy and exergy efficiency for thermally driven processes, including hybrid thermochemical processes for hydrogen production, are introduced. Certain once-through (or open loop) hydrogen production processes, such as biomass gasification, biofuel reformation, clean coal hydrogasification with carbon dioxide capture, are presented. Other methods include hydrogen extraction from ammonia and aluminum-water reaction. Thermochemical cycles for hydrogen production are of special interest in the chapter and are discussed with a focus on the major ones: sulfur-iodine, hybrid sulfur, copper-chlorine and magnesium-chlorine cycles.

4.2 FUNDAMENTALS OF THERMOCHEMICAL HYDROGEN PRODUCTION

With a focus on hydrogen production processes using thermochemical cycles or relevant open-cycle methods (e.g., gasification), the specific fundamentals are here reviewed, and the criteria for assessing the performance based on energy and exergy efficiencies are introduced. The review mainly includes aspects of chemical thermodynamics and kinetics. In this section, specific notions of chemical equilibrium, reaction extend, product yield, rate of reaction and so on are introduced; therefore, this section is complementary to the general "Fundamental Aspects" as presented in Chapter 1.

4.2.1 Thermodynamic Analysis of Thermochemical Reactions

In the context of this chapter, it is very important to apply thermodynamics analysis to chemical reactions to determine their feasibility in terms of spontaneity, reaction completion, and species concentrations when a state of chemical equilibrium is reached. The importance of standard reaction enthalpy, entropy, Gibbs free energy, equilibrium constant, and reaction quotient is illustrated.

To begin, let us consider a simple chemical reaction as shown in Fig. 4.1. Assume that, in a thermodynamic system, two chemically reacting species are present at the initial moment ($t = 0$) in the molar amounts a_0 and b_0, respectively. After a sufficient time, the species react up to a certain extent and generate the products $cC + dD$. Of course, the initial species consume such that the remaining amounts at the latter moment ($t > 0$) are $aA + bB$, where $a < a_0$ and $b < b_0$. One may assume that the process occurs at a constant temperature and pressure. The volume of the thermodynamic system is free to vary (enlarge or shrink). In addition, the system boundary is not insulated; therefore, it can release or receive heat Q to/from the surroundings. The chemical reaction that occurs when the process goes to completion is as follows:

$$a_0 A + b_0 B \rightarrow c_0 C + d_0 D \tag{4.1}$$

From a thermodynamic point of view, several aspects need to be analyzed, and several questions need to be answered: namely, is the reaction spontaneous? For example, is it endothermic or exothermic? Does the chemical equilibrium tend toward generating more products? Or how carefully does one set needs to initially the amounts of reactants in the mixture such that more products are formed at equilibrium conditions?

In order to tackle with such issues, some elementary notions must be introduced. The first is chemical equilibrium, also discussed in Chapter 1. For the purpose of this chapter, we will define chemical equilibrium as a state when the concentration of reactants and products remains constant over time. The spontaneity feature of a chemical reaction is defined as follows: if a chemical reaction occurs without any external influence, then it is denoted as spontaneous. A nonspontaneous process is defined as the reverse of a spontaneous process. From a thermodynamic viewpoint, a spontaneous reaction is one that occurs in accordance to the second law of thermodynamics, which states that, for a process to be possible, it must evolve such that the generated entropy is positive.

The spontaneity of a reaction does not have anything to do with the reaction rate. Iron oxide (or rust) is a spontaneous reaction that occurs at a very slow rate in usual atmospheric conditions. Combustion of a fuel, once ignited, occurs without any other external influence; therefore, it is a spontaneous process, and it occurs at a very high rate. In a spontaneous chemical reaction, the mixture of species displaces toward the equilibrium composition. Conversely, in a nonspontaneous chemical reaction, the composition of species moves away from the equilibrium.

If during a reaction process, heat is released into the ambient, then the reaction is referred to as exothermic; otherwise, if the reaction receives heat, it is denoted as endothermic. The spontaneity of a reaction does not have anything to do with its endothermic or exothermic character. For example, the table salt dissociation in water is spontaneous and endothermic. However, the dissociation of $CaSO_4$ in water is spontaneous and exothermic. In essence, the reaction spontaneity is a matter of probability: namely, the system displaces spontaneously toward a state of maximum randomness (or less order) because that the type of state is the most probable.

The entropy change of a reaction process represents the entropy at the final (or latter) state minus the entropy at the initial state. Referring to the system considered in Fig. 4.1, the entropy change is expressed as follows:

$$\Delta S = S_{\text{final}} - S_{\text{initial}} = (aS_A + bS_b + cS_C + dS_d) - (a_0 A + b_0 B) \tag{4.2}$$

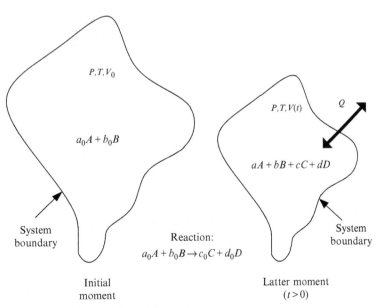

FIG. 4.1 Thermodynamic representation of a chemical reaction process.

If the reaction goes to completion, then the reaction entropy becomes

$$\Delta S = (c_0 S_C + d_0 S_d) - (a_0 A + b_0 S_b) \tag{4.3}$$

When determining the entropy of any reaction, the standard molar entropy of involved substances becomes an important quantity since standard entropies are known for numerous substances. Standard molar entropy—denoted with S^0—represents the entropy of one mol of pure substance at 25°C. Furthermore, at 0 K, the standard molar entropy is nil because at this state the substance is represented by a perfect crystalline structure of zero randomness state. Moreover, the standard state condition is specified in function of the state of aggregation as follows:

- For pure solids, liquids, and gases, the pressure at standard state is stated at 1 atm.
- For solutions, the concentration is stated at 1 M.

The standard entropy of the reaction from Eq. (4.3) is written as follows:

$$\Delta S^0 = \left(c_0 S_C^0 + d_0 S_d^0\right) - \left(a_0 A^0 + b_0 S_b^0\right) \tag{4.4}$$

Then the reaction entropy at a given state is determined based on the standard reaction entropy ΔS^0 and the reaction quotient K_Q as follows:

$$\Delta S = \Delta S^0 + R \ln K_Q \tag{4.5}$$

Here, the reaction quotient for the reaction given by Eq. (4.1) is defined as

$$K_Q = \frac{\{C\}^{c_0} \{D\}^{d_0}}{\{A\}^{a_0} \{B\}^{b_0}} \tag{4.6}$$

where {} denotes chemical activity of species in the actual reaction conditions, but only species that are in gas phase or dissolved molecules or ions in solution.

In cases when a species is phase, the activity is preplaced with fugacity. In many situations, the chemical activity depends on an activity coefficient and molar concentrations. Provided that the activity coefficient is 1, activities can be replaced by concentrations. As an example, the thermochemical decomposition of ammonia to hydrogen evolves according to the following equilibrium reaction in gas phase:

$$NH_3(g) \rightleftharpoons 3H_2(g) + N_2(g)$$

This has a reaction quotient defined based on their partial pressures of species as follows:

$$K_Q = \frac{P_{H_2}^3 P_{N_2}}{P_{NH_3}} \tag{4.7}$$

The entropy of the surroundings depends on the molar reaction enthalpy (ΔH) and reaction temperature T as

$$S_{surr} = -\frac{\Delta H}{T} \tag{4.8}$$

where

$$\Delta H = H_{final} - H_{initial} \tag{4.9}$$

where $H = \sum c_i H_i$ is the sum of the molar enthalpies of species weighted by their molar concentrations c_i, with $\sum c_i = 1$. The total entropy of the system and surroundings becomes

$$\Delta S_{tot} = \Delta S - \frac{\Delta H}{T} \tag{4.10}$$

Equation (4.10) can be rearranged as

$$\Delta G = \Delta H - T\Delta S = -T\Delta S_{tot} \tag{4.11}$$

where ΔG represents the Gibbs free energy of the reaction.

The following possibilities do exist regarding the spontaneity and equilibrium state of a chemically reacting system, as given in Table 4.1. In general both the enthalpy and entropy of reactants and products increase with the

TABLE 4.1 Criteria for Spontaneity and Equilibrium of Chemical Reactions

ΔG	ΔS_{tot}	Spontaneity	ΔH	ΔS	T	ΔG	Spontaneity	ΔH	ΔS	T	ΔG	Spontaneity
<0	>0	Spontaneous (S)	<0	>0	any	<0	S	>0	<0	any	>0	NS
>0	<0	Nonspontaneous (NS)	<0	<0	low	<0	S	>0	<0	low	>0	NS
0	0	Equilibrium (E)	<0	<0	high	>0	NS	>0	>0	high	<0	S

temperature. Consequently, the temperature variation of reaction enthalpy and entropy are very slow. In most cases, temperature is a key parameter that influences the reaction equilibrium. As given in Table 4.1, if a system is at equilibrium, then its Gibbs free energy is zero. From Eq. (4.11), the following approximate estimation of the temperature at equilibrium can be derived, by imposing $\Delta G = 0$, namely:

$$T_{eq} \cong \frac{\Delta G}{\Delta S} \qquad (4.12)$$

The Gibbs free energy at any conditions can be expressed based on the reaction quotient and the Gibbs free energy at standard conditions ΔG^0, as follows:

$$\Delta G = \Delta G^0 + RT \ln K_Q \qquad (4.13)$$

Since ΔG^0 is fixed, the value of ΔG can be influenced only by the temperature and reaction quotient. Depending on the reaction, equilibrium can be attained at a temperature T, in which case, the reaction quotient K_Q becomes the same as the equilibrium constant of the reaction, K_{eq}. One has the following: $0 = \Delta G^0 + RT \ln K_{eq}$. Therefore, the equilibrium constant becomes

$$K_{eq} = \exp\left(-\frac{\Delta G^0}{RT}\right) \qquad (4.14)$$

Furthermore, the equilibrium constant can help determing the reaction mixture composition at the equilibrium, since at equilibrium, the reaction quotient equals the equilibrium constant. Therefore, the equilibrium conditions can be derived from Eq. (4.6) where partial pressures can replace activities for the species in gas phase and concentrations can replace activities for the species in solution. The following situations may occur:

- If equilibrium constant is $K_{eq} > 10^3$, then the equilibrium mixture comprises mainly reaction products.
- If equilibrium constant is $K_{eq} < 10^{-3}$, then the equilibrium mixture comprises mainly reactants.
- If equilibrium constant is $10^3 > K_{eq} > 10^{-3}$, then the equilibrium mixture comprises both reactants and products.

According to Le Châtelier's principle, if a reaction mixture at equilibrium is placed upon a stress, the system reacts in the direction that relieves the stress. For a reaction that reduces the moles of gas, this means that

- An increase of pressure produced by shrinking the volume will move the reaction toward reducing the moles of gas.
- An expansion of volume that produces a decrease of pressure will move the reaction toward an increase of the mole of gas.

Similarly to the reaction quotient, the notion of extent of reaction—denoted as ξ—is very useful for calculating the equilibrium conditions. The extent of reaction is a parameter between 0 and some positive value that depends on the reaction stoichiometry. When the reaction quotient is smaller than one, the extent of reaction is near zero. Total free energy of reaction mixture decreases when the reaction proceeds spontaneously in the forward direction. The total Gibbs energy of the reaction mixture can be defined as the sum of moles of chemical species "i" multiplied by the respective molar specific Gibbs free energy at constant temperature and pressure. Therefore, the Gibbs free energy of a reaction mixture is written as

$$G = \sum n_i G_i \qquad (4.15)$$

where G_i is the molar-specific free energy of species i (reactant of product) and n_i is the number of moles of species "i."

When the reaction mixture consists mostly in products, then the reaction quotient is larger than 1, and the extent of reaction approaches unity. In this case, the total free energy of the reaction decreases as the reaction proceeds spontaneously in the reverse direction. Fig. 4.2 shows a plot of the total Gibbs free energy of the reaction mixture versus the extent of reaction. At constant pressure and temperature, Gibbs free energy is a complete differential; therefore, one has

$$dG = \sum \left(\frac{\partial G}{\partial n_i}\right)_{T,P} dn_i \qquad (4.16)$$

At equilibrium, one must have $dG = 0$ with a minimum as suggested in Fig. 4.2. Therefore, Gibbs free energy minimization will determine the reaction mixture composition at the equilibrium, in terms of the number of moles n_i. Gibbs free energy of a general reaction mixture comprising gas, condensed phases, and solutions can be written as

$$G = \left(\sum n_i \left(G_i^0 + RT \ln (y_i P)\right)\right)_{ig} + \left(\sum n_i \left(G_i^0 + RT \ln (a_i)\right)\right)_{\substack{rg\,or\\sol}} + \left(\sum n_i G_i^0\right)_{cond} \qquad (4.17)$$

where subscript ig stands for ideal gas, rg represents real gas, sol stands for solution, and cond represents condensed phase.

The minimization of total Gibbs free energy of the reaction mixture will lead to determination of the mixture composition in terms of number of moles of each species, n_i. The minimization of Gibbs free energy and the search for equilibrium state is well facilitated by the use of the reaction extent. For the reaction given in Eq. (4.1), the reaction extent can be defined as follows:

$$\frac{d\xi}{dn_A} = -\frac{1}{a_0} \qquad (4.18)$$

This equation can be integrated easily from $\xi = 0$ when the number of moles of reactant A in the reaction mixture is defined as $n_{A,0}$ and a later moment when $\xi > 0$, for which the number of moles of reactant A is $n_A < n_{A,0}$. It results in

$$\xi = -\frac{1}{a_0}(n_A - n_{A,0}) \qquad (4.19)$$

From Eq. (4.19), the maximum value of the extent of reaction is obtained when all reactant A is consumed, that is, when $n_A = 0$. Consequently, $a_0 \xi_{max} = n_{A,0}$. Furthermore, because the rate of the reactant's consumption has the opposing sign of the rate of the product's generation for reaction in Eq. (4.1), one must have $-a_0 dn_A = -b_0 dn_B = c_0 dn_C = d_0 dn_D$. Therefore, one has

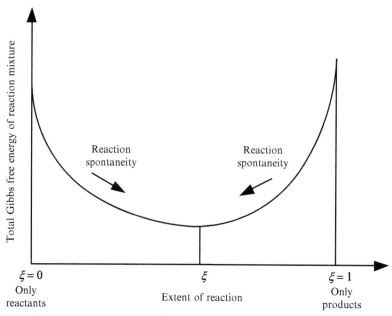

FIG. 4.2   Gibbs free energy variation versus the extent of reaction.

$$d\xi = -\frac{dn_A}{a_0} = -\frac{dn_B}{b_0} = \frac{dn_C}{c_0} = \frac{dn_D}{d_0} \tag{4.20}$$

Using the chemical potentials, the total infinitesimal change of Gibbs free energy of the reaction mixture becomes

$$dG = VdP - SdT + \mu_A dn_A + \mu_B dn_B + \mu_C dn_C + \mu_D dn_D \tag{4.21}$$

At fixed pressure and temperature and for a given extent of reaction, the infinitesimal change of Gibbs free energy of the reaction mixture becomes

$$dG = (-a_0\mu_A - b_0\mu_B + c_0\mu_C + d_0\mu_D)d\xi \tag{4.22}$$

From Eq. (4.22), accounting from the fact that $G = G(T, P, \xi)$, and that the Gibbs free energy of the reaction is identified as $\Delta G = c_0\mu_C + d_0\mu_D - a_0\mu_A - b_0\mu_B$, the following equation, which relates the extent of reaction, the total Gibbs free energy of the reaction mixture and the Gibbs free energy of the reaction, will result:

$$\Delta G = \left(\frac{\partial G}{\partial \xi}\right)_{T,P} \tag{4.23}$$

A reaction achieves full conversion when all the reactants are consumed and products are generated in a stoichiometric amount. This does not happen with actual reactions for several reasons. When a reaction has an equilibrium constant around one, there will not be a full conversion of the reactants into products because both reactants and products will be present in the reaction mixture. In addition, two or more reactions can occur simultaneously thereby creating undesired products. Also, several types of losses occur with purification and separation of the products and the presence of impurities. The theoretical yield is the amount of product that can be generated by a reaction according to the stoichiometry. The actual yield will be smaller than the theoretical yield. The percent yield is defined as the ratio of actual to theoretical yield.

4.2.2 Chemical Kinetics Aspects

The rate of chemical reactions and their enhancement through catalysis as well as the identification of a detailed reaction mechanism are considered key matters of chemical kinetics. In the context of thermochemical hydrogen production, chemical kinetics is of crucial importance, influencing the reaction selectivity, the reactor size, residence time and other relevant design and process parameters.

Here a brief summary of chemical kinetics is given. To begin, let us define the reaction rate in terms of speed of the concentration change of a reactant or a product as

$$k = \frac{dc}{dt} \tag{4.24}$$

where c represents the concentration.

Of course, the changing rate of reactants concentration (which are consumed) has the opposite sign to the changing rate of products concentration (which are generated). In general, the reaction rate must be influenced by the driving force of the reaction, which is represented by the chemical potential. According to the Le Chatelier principle, the reaction will be driven in the direction that relieves the stress. Therefore, the reaction rate will depend on the concentration of species. Based on these considerations, a law of reaction rate can be written for the generic reaction given in Eq. (4.1) as follows

$$k = \hbar c_A^m c_B^n \tag{4.25}$$

where $\hbar$ bears the name of rate constant.

In Eq. (4.25) the exponents m and n characterize the reaction order. More specifically, the reaction order is given by the sum $m + n$. The reaction order with respect to reactant A will be given by m only; whereas n gives the reaction order with respect to the reactant B. The reaction rate exponents are empirical constants that depend on the reaction itself, the reactor type and the reaction conditions. Table 4.2 shows particular forms of the reaction rate equation in function of the reaction order. As shown, the unit of measure for the rate constant differs in function of the overall reaction order.

TABLE 4.2   Definitions for Reaction Order and the Rate Constant

Reaction order	Rate constant	Rate constant units
Zeroth order	$\mathscr{k} = k$	$\dfrac{\text{mol}}{\text{s}}$
First order	$\mathscr{k} = \dfrac{k}{c_A}$	$\dfrac{1}{\text{s}}$
Second order	$\mathscr{k} = \dfrac{k}{c_A c_B}$	$\dfrac{1}{\text{mol} \cdot \text{s}}$
Third order	$\mathscr{k} = \dfrac{k}{c_A^2 c_B}$	$\dfrac{1}{\text{mol}^2 \text{s}}$

Note that chemical reactions can proceed in both directions, forward and backward, but at different rates such that typically the net reaction rate is forward. Denote k_f the forward reaction rate and k_b the backward reaction rate. Then the net reaction rate is given as follows:

$$k = k_f - k_b \tag{4.26}$$

Let us assume that reaction of Eq. (4.1) is of second order for which $a_0 = b_0 = c_0 = d_0$. Therefore, the rate of forward and backward reactions and the net rate of reaction can be written as follows:

$$k_f = \mathscr{k}_f c_A c_B \tag{4.27a}$$

$$k_b = \mathscr{k}_b c_C c_D \tag{4.27b}$$

$$k = \mathscr{k}_f c_A c_B - \mathscr{k}_b c_C c_D \tag{4.27c}$$

At equilibrium, there is no net rate of reaction; therefore, $k = 0$. It follows from Eqs. (4.6) and (4.27c) that at equilibrium one has

$$K_{eq} = \frac{\mathscr{k}_f}{\mathscr{k}_b} = \frac{c_C c_D}{c_A c_B} \tag{4.28}$$

If several reactions occur concurrently, then the equilibrium constant for the net reaction becomes

$$K_{eq} = \Pi \left(\frac{\mathscr{k}_f}{\mathscr{k}_b} \right)_i \tag{4.29}$$

where index "i" refers to each individual reaction.

From Eqs. (4.27c) and (4.28), the net reaction rate becomes

$$k = \mathscr{k}_f \left(c_A c_B - \frac{c_C c_D}{K_{eq}} \right) \tag{4.30}$$

Because according to Eq. (4.14) the equilibrium constant is fixed once the temperature is fixed, it results from Eq. (4.30) that at a fixed temperature, the reaction rate will be maximum provided that no products are available in the reaction mixture (that is $c_C = 0$ and/or $c_D = 0$). In such a case, the reaction quotient tends to unity; therefore, the Gibbs free energy of the reaction equals $\Delta G^0(T)$, which for a forward-proceeding reaction must have a negative value. Furthermore, since there is no product in the reaction mixture, there is no net product generation. Moreover, at this state, the chemical potential of the forward reaction can be set by convention to a zero value.

Once the reaction starts generating products, the chemical potential will increase until it reaches a maximum value at equilibrium state. Concurrently, Gibbs free energy of the reaction increases from a more negative value toward a zero value, at equilibrium. Again, at equilibrium, there will not be a net generation of product, because the backward reaction proceeds at the same rate as the forward reaction.

FIG. 4.3 Product generation rate and net reaction rate variations with chemical potential of a reaction system.

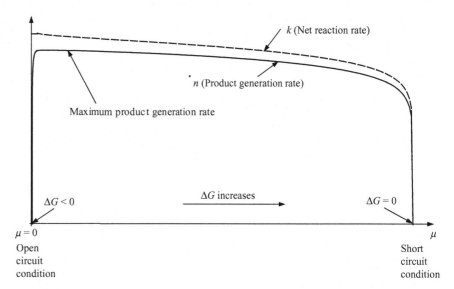

The described processes are illustrated graphically in the plot shown in Fig. 4.3. When Gibbs energy of the reaction is negative such that the reaction quotient is $K_Q = 1$, products are not yet generated, although the reaction rate is maximum. This situation can be denoted as an open circuit condition. Once product starts to appear in the reaction mix, the reaction rate diminishes, but the product generation increases, until it reaches a maximum. Further presence of products at the reaction site induces more decrease of the reaction rate. Therefore, the product generation starts to decrease because the backward reaction increases its rate. Eventually, at equilibrium, there will be no net product generation; this situation can be referred to as short circuit condition.

The rate constant involved in a chemical reaction rate is usually determined based on the Arrhenius equation. Based on the Arrhenius model, the rate constant must be proportional with molecular collision frequency and will depend also on the activation energy and temperature. The molecular collision frequency is accounted for with the help of a proportionality factor denoted as "pre-exponential" factor, $\mathcal{A}$. The Arrhenius equation is as follows:

$$k = \mathcal{A} \exp\left(-\frac{E_a}{RT}\right) \tag{4.31}$$

The rate constant and therefore the reaction rate can often be increased with the use of a catalyst. By definition, a catalyst will not be consumed by the process (although it may degrade over time). A catalysit is denoted as homogeneous when it is of the same phase as the reactants, otherwise it is heterogeneous. The catalyst lowers the activation energy of the forward and reverse directions in the same amount; therefore, the rates of forward and backward reaction will increase with the same factor. This means that the presence of a catalyst does not change the equilibrium constant. Fig. 4.4 shows qualitatively the variation of the potential energy with the reaction coordinate as one progresses from reactants to products. The activation energy will be reduced by the presence of a catalyst with the amount ΔE_a.

FIG. 4.4 Illustrating the effect of catalyst in reducing the activation energy.

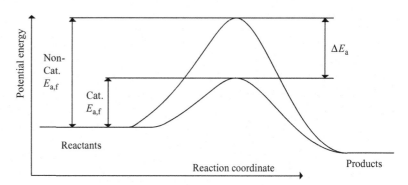

4.2.3 Efficiency Formulations

The general efficiency formulations through energy and exergy take various particular forms depending on the specific thermally driven hydrogen production system under the analysis. Here some relevant efficiency equations are introduced based on black-box representations.

First, pure thermochemical cycles are considered. Thermal energy at high temperature is the only essential energy supply to these processes, as shown in Fig. 4.5. Pure water is the usual feedstock. The cycle recycles all other chemicals except the water feedstock, which is split into hydrogen and oxygen. The energy efficiency is defined commonly as the energy retrieved in hydrogen product divided by the consumed thermal energy input. Let us denote $\dot{n}_{H2}$ as the molar rate of hydrogen generated by the system. Then the energy efficiency is defined as

$$\eta = \frac{\dot{n}_{H_2} HHV}{\dot{Q}_{in}} \tag{4.32}$$

where HHV represents the higher heating value of hydrogen and $\dot{Q}_{in}$ is the heat transfer rate provided to the system through a heat transfer operation.

Assume that the heat transfer boundary is at a temperature T_H, as shown in Figure 4.5. In this case, the exergy input rate is represented by heat transfer rate multiplied by Carnot factor relative to T_H and T_0, where T_0 is the reference temperature. Therefore, the exergy efficiency of the process is defined by

$$\psi = \frac{\dot{n}_{H_2} ex_{H_2}^{ch}}{\dot{Q}_{in}\left(1 - \dfrac{T_0}{T_H}\right)} \tag{4.33}$$

where $ex_{H_2}^{ch}$ represents the chemical exergy of hydrogen.

The efficiency formulations according to Eqs. (4.32) and (4.33) assume that hydrogen is valorized as a fuel. This may not always be the case because hydrogen may be used as a chemical. Some alternative efficiency definitions are possible and relevant when hydrogen is viewed as a chemical. For example, the efficiency according to the first law may be defined by the ratio of the negative of the standard formation enthalpy of water to the consumed thermal energy input for generation of a mole of hydrogen. The formation enthalpy of liquid water is negative, namely $\Delta H^0 = -285,813 \text{ kJ/kmol}$. Assume that water in liquid phase is provided as process input at standard conditions. The minimum energy required to split a water molecule into molecular hydrogen and oxygen becomes $-\Delta H^0$. Therefore, in a rate form, the hydrogen production efficiency according to the first law becomes

$$\eta = \frac{-\dot{n}_{H_2} \Delta H^0}{\dot{Q}_{in}} \tag{4.34}$$

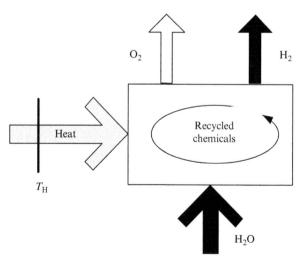

FIG. 4.5   Black-box representation of a pure thermochemical water splitting cycle.

Similarly, the exergy efficiency can be defined in terms of the reversible work required to split water at standard state divided by the actual work required by the process, namely the work potential associated with $\dot{Q}_{in}$, or the exergy input. The reversible work for water splitting is thence equal to the opposite of standard Gibbs free energy of water formation: $-\Delta G^0 = 237,173$ kJ/kmol. Therefore, the exergy efficiency of the hydrogen production process becomes

$$\psi = \frac{-\dot{n}_{H_2}\Delta G^0}{\dot{Q}_{in}\left(1-\frac{T_0}{T_H}\right)} \tag{4.35}$$

Let us consider now a hybrid thermochemical cycle represented as shown in Fig. 4.6. Assume again that thermal energy is provided to the system via heat transfer through a boundary at temperature T_H. Some relatively smaller amount of electrical energy is also provided to the system from a power generator. For this system, the energy efficiency is usually defined as

$$\eta = \frac{\dot{n}_{H_2}HHV}{\dot{Q}_{in} + \dfrac{\dot{W}_{el}}{\eta_{el}}} \tag{4.36}$$

where η_{el} represents the energy efficiency of the power generator (e.g., efficiency of the regional grid), and $\dot{W}_{el}$ is the electrical energy rate consumed by the system.

The electrical power input in Eq. (4.36) is required for sustaining electrochemical processes ($\dot{W}_{elch}$), for supplying the required power for separation operations ($\dot{W}_{sep}$) and for any power requirement for the auxiliary systems ($\dot{W}_{aux}$) for mechanical/electrical processes such as for material conveying, pumping, compressing and mixing of materials. Therefore, the electric power becomes

$$\dot{W}_{el} = \dot{W}_{elch} + \dot{W}_{sep} + \dot{W}_{aux} \tag{4.37}$$

If one assumes that the power is generated from thermal energy at a very high temperature, then in Eq. (4.36), one can approximate $\eta_{el} \cong 1$. This may be an acceptable assumption for thermal energy sources such as solar and nuclear energy. In such a case, the energy efficiency equation becomes

$$\eta = \frac{\dot{n}_{H_2}HHV}{\dot{Q}_{in} + \dot{W}_{el}} \tag{4.38}$$

The exergy efficiency for hybrid thermochemical water splitting cycles can be defined based on exergy delivered in form of hydrogen to the exergy input in all forms. Therefore, the following equation expresses the exergy efficiency:

$$\psi = \frac{\dot{n}_{H_2}ex_{H_2}^{ch}}{\dot{Q}_{in}\left(1-\frac{T_0}{T_H}\right) + \dot{W}_{el}} \tag{4.39}$$

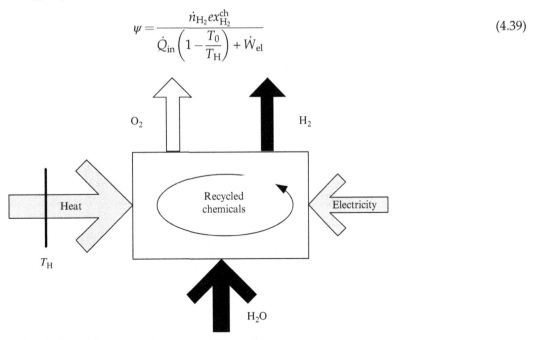

FIG. 4.6 Black-box representation of a hybrid thermochemical water splitting cycle.

The efficiency formulations according to the first law of thermodynamics for hybrid thermochemical cycles—given in analogy to Eq. (4.34)—are written as

$$\eta = \frac{-\dot{n}_{H_2}\Delta H^0}{\dot{Q}_{in} + \dfrac{\dot{W}_{el}}{\eta_{el}}} \tag{4.40a}$$

$$\eta = \frac{-\dot{n}_{H_2}\Delta H^0}{\dot{Q}_{in} + \dot{W}_{el}} \tag{4.40b}$$

whereas the efficiency according to the second law of thermodynamics, given in analogy to Eq. (4.35), becomes

$$\psi = \frac{-\dot{n}_{H_2}\Delta G^0}{\dot{Q}_{in}\left(1 - \dfrac{T_0}{T_H}\right) + \dot{W}_{el}} \tag{4.41}$$

Consider now once-through processes (or open cycles) such as gasification, liquefaction and reformation in which the thermal energy is provided by a fuel: biomass, coal, natural gas. Figure 4.7 shows a representation of an open-cycle hydrogen production process in which fuel is usually a feedstock for the chemical processes involved. The energy efficiency will be

$$\eta = \frac{\dot{n}_{H_2}HHV}{\dot{n}_f LHV} \tag{4.42}$$

where $\dot{n}_f$ is the rate of feedstock input (the fuel) and LHV represents the lower heating value of the fuel.

The exergy efficiency of once-through processes is expressed as

$$\psi = \frac{\dot{n}_{H_2}ex_{H_2}^{ch}}{\dot{n}_f ex_f^{ch}} \tag{4.43}$$

where ex_f^{ch} represents the chemical exergy of the fuel/feedstock.

The alternative expressions for efficiencies according to first and second laws for the once-through hydrogen producing processes are given as follows:

$$\eta = \frac{\dot{n}_{H_2}\left|\Delta H_P^0\right|}{\dot{n}_f LHV} \tag{4.44a}$$

$$\psi = \frac{\dot{n}_{H_2}\left|\Delta G_P^0\right|}{\dot{n}_f ex_f^{ch}} \tag{4.44b}$$

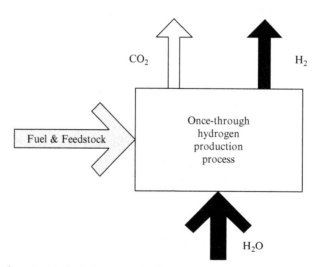

FIG. 4.7 Black-box representation of an open cycle hydrogen production process.

Here the standard enthalpy and Gibbs free energy for the overall process are used at the numerator in absolute value. Let us consider an example of hydrogen production from a steam methane reforming process. This process occurs in two chemical stages, with the net reaction described as follows: $CH_4 + 2H_2O \rightarrow CO_2 + 4H_2$. Consider that the reactants are gaseous methane and liquid water. Therefore, the standard reaction enthalpy and entropy can be easily calculated using thermodynamic tables, namely $\Delta H_p^0 = 252,735$ kJ/kmol and $\Delta G_p^0 = 130,555$ kJ/kmol.

4.3 WATER THERMOLYSIS

Chemical decomposition of a compound caused by the addition of thermal energy is denoted as thermolysis (note that in analogy, electrolysis represents decomposition of a compound caused by electric energy). Water thermolysis is technically possible at temperatures above 2500 K, although it is a rather difficult process to conduct. One of the major problems involves separation of the desired product, namely hydrogen, which easily can react with oxygen present in the reaction mixture and form water. To avoid this situation, special process conditions, catalysts and separation devices must be included in an actual process. Water thermolysis occurs according to the following reaction:

$$H_2O(l) \xrightarrow{\text{Heat, } T > 2500\,K} H_2(g) + 0.5O_2(g)$$

Water thermolysis is a once-through process (open cycle) as shown in the black-box representation from Fig. 4.8. The thermolysis reactor itself will generate a mixture consisting mainly of hydrogen, oxygen and steam at high temperature. Other intermediate species are also involved at that very high temperature, such as OH, atomic H and O, and ions. Let us consider first a simplified thermodynamic treatment, assuming an output mixture of only H_2O, H_2, and O_2.

Denote $y_{H_2}, y_{O_2}, y_{H_2O}$ the molar fraction of the gaseous constituents of the product mixture generated by the thermolysis reaction. The following constraint is then imposed:

$$y_{H_2} + y_{O_2} + y_{H_2O} = 1 \tag{4.45}$$

Here we denote with ξ the number of moles in the product mixture per mole of water reactant. The parameter ξ takes values between $\xi = 1$ when no reaction occurs and $\xi = 1.5$ when the reaction goes to completion ($H_2O \rightarrow H_2 + 0.5O_2$); note that ξ quantifies the extent of reaction. An additional constraint requires that the number of atoms of chemical elements in the process is preserved. Therefore, one has

$$\begin{cases} 2\xi(y_{H_2} + y_{H_2O}) = 2 \\ \xi(y_{O_2} + y_{H_2O}) = 1 \end{cases} \tag{4.46}$$

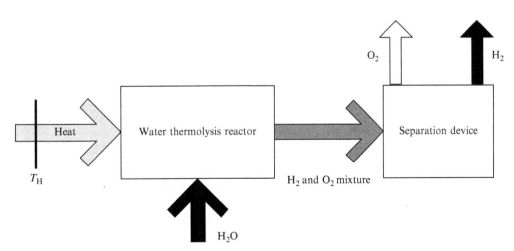

FIG. 4.8 Black-box representation of a water thermolysis process.

Using the parameter ξ, the Gibbs free energy of the reaction can be calculated as follows:

$$\Delta G = \Delta G^0 + RT \ln \left(y_{H_2}^{(\xi y_{H_2})} y_{O_2}^{(\xi y_{O_2})} y_{H_2O}^{(\xi y_{H_2O}-1)} (P/P_0)^{\xi-1} \right) \tag{4.47}$$

In order to determine the equilibrium concentration, ΔG is imposed to zero in Eq. (4.47) and the Eqs. (4.45)–(4.47) are solved simultaneously for a given pressure and given temperature. Fig. 4.9 shows the chemical equilibrium results obtained using the Engineering Equation Solver (Klein, 2015) for $P = 1$ atm. As seen in the figure, at a temperature approximately 3820 K, the reaction completion is nearly achieved with ~0% unreacted steam and 66% hydrogen in the output stream. However, at 2500 K, the equilibrium concentrations are 10% H_2, 6% O_2, and 84% H_2O.

The reaction yield—denoted here with ζ—is defined as the number of moles of hydrogen in the product stream per number of moles of reacted water. Ideally, for one mole of water, there is one mole of hydrogen; therefore, the ideal yield is one. The yield for the thermolysis reaction just analyzed can be calculated as follows:

$$\zeta = \xi \frac{y_{H_2}}{\xi_{max}} \tag{4.48}$$

with $\xi_{max} = 1.5$.

A high temperature is necessary to generate the maximum hydrogen yield. The yield reaches the maximum of 100% at 3830 K when the hydrogen concentration in the product stream is 0.66. Below 1500 K, the yield becomes zero. Fig. 4.10 shows the variation of thermolysis reaction yield with temperature for a pressure of 1 atm.

At equilibrium conditions, there is no net production of any chemical because the reaction rate of the forward reaction ($\dot{n}_f$) is equal to the reaction rate of the backward reaction ($\dot{n}_b$). In other words, the produced chemical species are immediately consumed to generate the reactants. At equilibrium, no net change is possible. If $\Delta G = 0$ at standard pressure P_0, then the equilibrium constant becomes

$$K_{eq} = \frac{y_{H_2}^{(\xi y_{H_2})} y_{O_2}^{(\xi y_{O_2})} y_{H_2O}^{(\xi y_{H_2O})}}{y_{H_2O}} = \exp\left(-\frac{\Delta G^0}{RT}\right) \tag{4.49}$$

Fig. 4.11 shows the variation of equilibrium constant with temperature as well as the variation of molar Gibbs energy of the reaction for three pressures (low, standard and high). The reactant and products are assumed in the gas phase, and the products exist at the stoichiometric molar fraction $y_{H_2} = 2/3$ and $y_{O_2} = 1/3$.

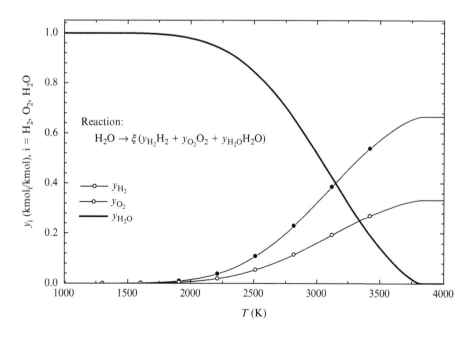

FIG. 4.9 Thermodynamic equilibrium of H_2, O_2, and H_2O species in water thermolysis reaction.

FIG. 4.10 Variation of hydrogen yield during water thermolysis at equilibrium and $P = 1$ atm.

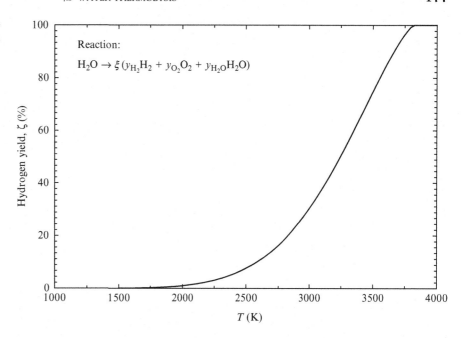

FIG. 4.11 Equilibrium constant and molar Gibbs free energy for water thermolysis reaction.

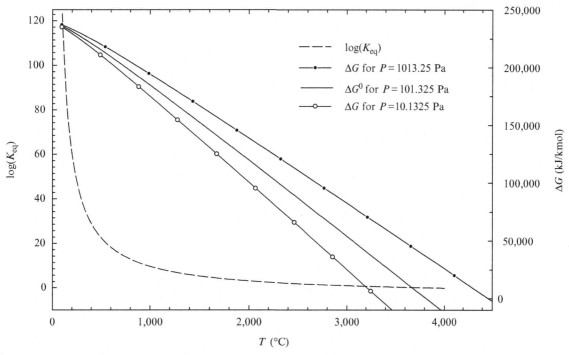

The forward reaction rate is increased if the concentration of reactant (y_{H_2O}) is higher. The backward reaction rate increases if the concentration of products is higher. The following proportionality must be established between the reaction rates and concentrations in the vicinity of equilibrium:

$$\begin{cases} \dot{n}_f = k_f y_{H_2O} \\ \dot{n}_b = k_b y_{H_2}^{(\xi y_{H_2})} y_{O_2}^{(\xi y_{O_2})} y_{H_2O}^{(\xi y_{H_2O})} \end{cases} \quad (4.50)$$

At equilibrium, in Eq. (4.50), the forward and backward rates must be equal, so

$$k_f y_{H_2O} = k_b y_{H_2}^{(\xi y_{H_2})} y_{O_2}^{(\xi y_{O_2})} y_{H_2O}^{(\xi y_{H_2O})} \tag{4.51}$$

and the net rate of reaction $\dot{n} = \dot{n}_f - \dot{n}_b$ becomes

$$\dot{n} = k_f \left[y_{H_2O} - \frac{y_{H_2}^{(\xi y_{H_2})} y_{O_2}^{(\xi y_{O_2})} y_{H_2O}^{(\xi y_{H_2O})}}{K_{eq}} \right] \tag{4.52}$$

A rate-equilibrium diagram for water thermolysis is shown in Fig. 4.12 for 3400 K and based on Eq. (4.52). For this studied case at a short circuit condition, one has $\Delta G = 13{,}370$ J/mol. Furthermore, based on $\Delta G^0 = -5828$ J/mol (for 3400 K), one can set the reference chemical potential at $\mu^0 = 7542$ J/mol. Therefore, at equilibrium condition, where $\Delta G = 0$, $\mu = 13{,}370$ J/mol. With known concentrations and reaction extents, the dimensionless quantity $\dot{n}/k_f$ can be calculated. This quantity expresses the reaction rate. Furthermore, the dimensionless quantity $\dot{n}_{H_2}/k_f = y_{H_2} \dot{n}/k_f$ is calculated. This represents the hydrogen production rate. In a steady state operation, the actual molar flow rate of hydrogen that leaves the unit of the control volume will be $\dot{n}_{H_2} = y_{H_2} \dot{n}$.

Many technical aspects must be considered for a system's practical design: selection of materials, heat recovery systems, heat sources, possibility of using catalysts, product separation and purification and so on. One primary issue is the influence of side reactions. Lede et al. (1982) found that when water is heated at temperatures over 2000 K, it dissociates thermally into many species, with the significant molecular compounds, such as: H_2O, H_2, O_2, O, H, and OH. The most representative elementary reactions for water dissociation and their thermodynamic and kinetic parameters at temperatures between 2000 and 4000 K are given in Table 4.3. For each case, the activation energy and pre-exponential factor of the Arrhenius Eq. (4.31) are specified.

Fig. 4.13 shows the equilibrium concentration and hydrogen yield versus temperature for the water thermolysis reaction system based on chemical reactions described as given in Table 4.3. It can be observed that the hydrogen concentration reaches a maximum at ~3400 K, while the maximum hydrogen yield occurs at approximately 3600 K. The yield is slightly over 26%, which means that maximum 0.26 moles of hydrogen can be obtained ideally from 1 mole of water in a thermolysis reactor.

The upper-bound limit of efficiencies can be calculated assuming that hydrogen is extracted at an equilibrium yield and the heat input required by the process is equal to the enthalpy of reaction (there are no heat losses). Based on the reaction data given in Table 4.3, the calculated efficiency results of water thermolysis are shown in Fig. 4.14.

It appears that the ideal maximum efficiencies are approximately 60% at 2500 K. The exergy efficiency is slightly higher than the energy efficiency. Note that at 2500 K, the Carnot factor is 88%. The only irreversibilities considered in

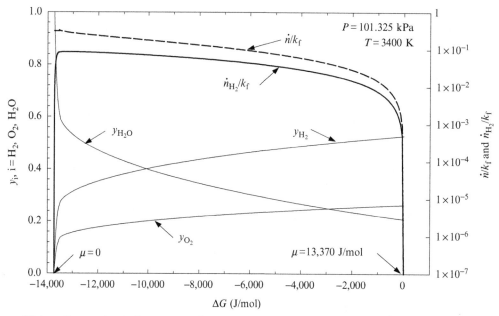

FIG. 4.12   Rate-equilibrium diagram for single step water dissociation.

TABLE 4.3 Representative Elementary Reactions for Water Thermolysis

Reaction[a]	ΔH^0 (kJ/mol)	ΔS^0 (J/mol K)	ΔG^0 (kJ/mol)	Direction	$E_{act}{}^a$ (kJ/mol)	$\mathcal{A}^a$ (m³/mol.s)
$2H_2O \rightleftarrows H + OH + H_2O$	231.8	120.2	196.0	F	439.8	2.2×10^{10}
				B	$2RT \ln(T)$	$1.4 \times 10^{11}{}^b$
$H + H_2O \rightleftarrows H_2 + OH$	−200.6	−51.5	−185.2	F	85.2	9.0×10^7
				B	21.5	2.2×10^7
$H + OH \rightleftarrows H_2 + O$	−202.9	−52.62	−187.2	F	29.1	8.3×10^3
				B	37.2	1.8×10^4
$O + OH \rightleftarrows O_2 + H$	198.5	104.5	167.3	F	0.0	2.0×10^7
				B	70.0	2.2×10^8

[a]Source: Lede et al. (1982); data valid for temperature range of 2000–4000 K.
[b]Unit is m⁶/mol s; F = forward; B = backward.

this process model are due to side reactions, because the reaction does not go to completion. Nevertheless, there are other irreversibilities in this process due to heat transfer, departure from equilibrium, hydrogen separation from the output stream and efficiency losses due to gas separation.

Gas separation technology at high temperatures necessitates high-pressure gradients to generate a sufficient driving force across the separation membranes. Another issue is quenching the product gases, which is required as a measure to avoid a recombination of hydrogen and oxygen. According to Nakamura (1977), it is estimated that the energy efficiency of practical water splitting systems based on direct water thermolysis is in the range of 25–55%.

A water thermolysis system is suggested in Fig. 4.15. A thermal energy at two temperature levels is required to drive the process. The reactor is supplied with thermal energy $\dot{Q}_{in,1}$ at highest temperature. Water is preheated, steam generated and superheated using heat recovered from product stream, $Q_{rec,1}$; from oxygen stream, $Q_{rec,2}$; and from hydrogen stream. Additional heat is transferred from external sources as shown with $\dot{Q}_{in,2}$ in the figure. Selective membranes are required to extract pure products from the reaction mixture. The arrangement is such that the reaction temperature is superior to the atmospheric pressure. In addition, the pure products are extracted under vacuum such that sufficient permeation conditions are created across the membrane. In theory, excess water, if any, can be recovered from the reaction mixture after cooling and recycling.

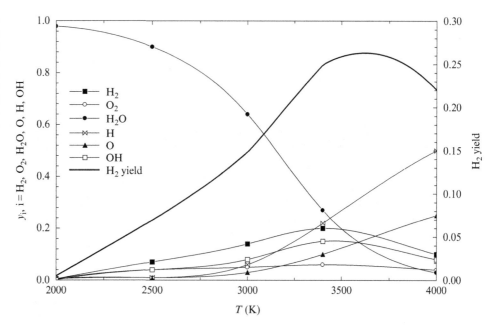

FIG. 4.13 Equilibrium concentrations versus temperature for the water thermolysis reaction system defined in Table 4.3. *Data from Lede, J., Lapicque, F., Villermaux, J., Gales, B., Ounalli, A., Baumard, J.F., Anthony, A.M., 1982. Production of hydrogen by direct thermal decomposition of water: preliminary investigations. Int. J. Hydrog. Energy 7, 939–950.*

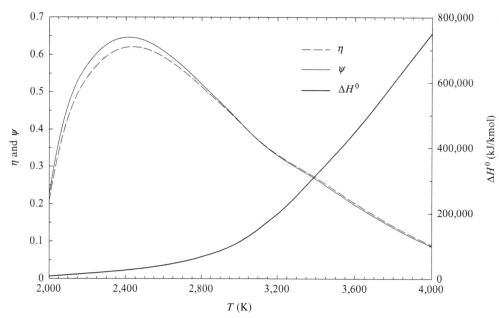

FIG. 4.14 Upper-bound efficiency efficiencies and reaction enthalpy versus temperature for water thermolysis in the reaction system described in Table 4.3.

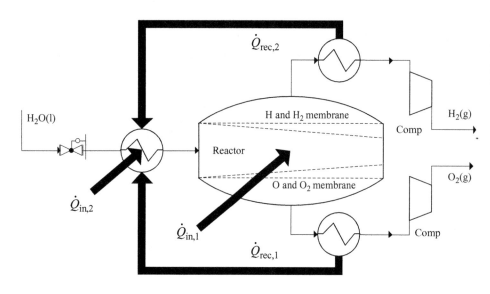

FIG. 4.15 Simplified schematic diagram of a water thermolysis system. *Modified from Nakamura, T., 1977. Hydrogen production from water utilising solar heat at high temperatures. Sol. Energy 19, 467–475.*

4.4 PURE THERMOCHEMICAL WATER SPLITTING CYCLES

Thermochemical water splitting cycles represent technological processes that decompose the water molecule while separate hydrogen and oxygen gases are released via a closed sequence of chemical reactions. Besides the chemical elements constituting the water molecule (H_2 and O_2), the chemical compounds in multistep thermochemical water splitting cycles comprise other elements

Pure thermochemical cycles, which are promising hydrogen production methods driven by thermal energy, are introduced in this section. Thermochemical cycles of so-called pure kind are supplied only with heat and water to operate. Their overall process is

$$H_2O + heat \rightarrow \left[\text{Cyclic chemical reactions} \right] \rightarrow H_2 + O_2$$

A historical perspective is now provided that presents the origins of concepts and further developments. The most relevant prototypical reactions are thereafter discussed. Many cycles are compiled herein and grouped as two-step, three-step, four-step, five-step, and six-step cycles (and more) based on the previous group work by Naterer et al. (2013), which is here expanded. A large number of reactions and thermochemical cycles are compiled, categorized and discussed. Analysis of the practicality of chemical reactions is established based on thermodynamic, kinetic analyses and other considerations. The sulfur-iodine cycle is discussed in more detail as being one of the most-recognized pure thermochemical cycles.

4.4.1 Historical Perspective

The U.S. Department of Defense started developing thermochemical cycles by 1963 in a research program called Energy Depot and was concluded in a first phase in a report by Funk and Reinstorm (1964). This project focused on the development of a nuclear military combat reactor (MCR) facility to generate high temperature heat in a combat zone, whereby the heat is used with appropriate technologies to fabricate synthetic fuels from air and water. The envisioned fuels were hydrogen, ammonia and hydrazine as they were believed to be applicable to the existing military equipment propelled by internal combustion engines, without major modifications.

The researchers of the Energy Depot project proposed the idea of a pure thermochemical water splitting cycle which potentially avoids the electricity generation step and can therefore directly produce hydrogen out of thermal energy, with a net gain of efficiency. These conclusions were published widely in Grimes (1966) and Funk and Reinstorm (1966). Two-step thermochemical water splitting cycles were first proposed in the seminal paper by Funk and Reinstorm (1966). Two types of two-step cycles were originally specified, namely the oxide type and the hydride type.

A first international roundtable on hydrogen production from thermal energy by thermochemical cycles was held at Ispra, Italy, in 1969, where a process, called Mark 1, was proposed as a promising thermochemical cycle. Mark 1 operates with chemical elements based on Ca, Br, and Hg and requires high temperature heat at ~1000 K.

At The Hydrogen Economy Miami Energy (THEME) conference held in Miami in March 1974, approximately 30 established thermochemical water splitting cycles were described, which were proposed and developed at seven leading laboratories: ISPRA (Euratom); Jülich Center (University of Aachen); Argonne National Laboratory; Los Alamos Scientific Laboratory; General Atomics; General Electric; and Allison Laboratory of General Motors. The European Commission initiated a Joint Research Center at Ispra in Italy in conjunction with the University of Aachen in Germany, with an annual budget for the period 1973–1983 dedicated to Mark cycles development. After the first phase of research—with mercury-based compounds—better subsequent cycles were developed within the Ispra-Mark program, ranging from iron-chlorine cycles, to hybrid sulfuric acid cycles, to sulfur-iodine cycles. In total, 24 versions of Mark cycles were reported by the end of the program.

The Ispra Mark 13 cycle—a hybrid type with Br and S compounds—was also studied at the Los Alamos National Laboratory. Sulfur and iodine compounds were used in the Mark 11 cycle. Mark 11 used sulfuric acid decomposition to evolve oxygen at 1120 K, according to the reaction: $H_2SO_4 \rightarrow H_2O + SO_2 + 0.5O_2$. In Ispra's Mark 11 version, there was an electrochemical H_2 evolving reaction, as follows: $H_2O + SO_2 \rightarrow H_2SO_4 + H_2$. Exactly the same cycle as Ispra's Mark 11 was pursued by Westinghouse for further development (Funk, 2001). The Ispra cycle Mark 16 was a variant of Mark 11 that used another hydrogen evolving scheme—which is purely thermochemical—consisting of two reaction steps, namely $2H_2O + I_2 + SO_2 \rightarrow H_2SO_4 + 2HI$ and $2HI \rightarrow I_2 + H_2$. This cycle (Mark 16) was adopted and pursued for development by General Atomics in the United States, which contributed significantly to process engineering and scaling up.

By the end of 1976, the University of Tokyo proposed a cycle based on Ca, Br and Fe compounds called UT-3. This cycle has been extensively developed, as it shows a promising energy efficiency of 49%. Another cycle is the copper-chlorine (Cu-Cl) cycle that has recently been proven at a laboratory scale, and several commercially appealing variants are being developed for heat sources of approximately 350–550°C. For this temperature range, the Cu-Cl cycle appears to be one of the most promising cycles among other options. This cycle has an estimated efficiency of 40–50%.

The development of thermochemical cycles implies an identification of certain intermediate chemical compounds and chemical or electro-chemical reactions that facilitate the overall water splitting and product separation processes. This involves a trial-and-error search of the most feasible cycle that should be conducted and searched in a systematic way. Many multistep processes involving recycled chemicals were developed such as two-step, three-step, four-step, five-step, six-step, and so on, where "steps" refers to the number of chemical reactions (or sometimes to important physical processes such as drying, extraction, crystallization).

In Naterer et al. (2013), a comprehensive review of thermochemical cycles is provided that shows that there is intensive research worldwide in this direction. Three types of heat sources for hydrogen production with thermochemical

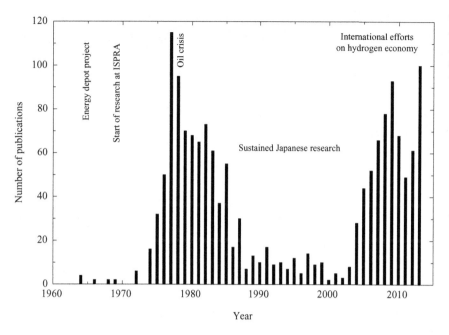

FIG. 4.16 Publication on thermochemical water splitting research. *Actualized data from Naterer, G.F., Dincer, I., Zamfirescu, C., 2013. Hydrogen Production from Nuclear Energy. Springer, New York.*

cycles are envisioned: concentrated solar heat, nuclear heat and geothermal heat. In general, solar radiation can operate at lower production scales and higher temperatures, whereas nuclear systems are more suitable for intermediate temperatures and higher production scales.

Fig. 4.16 shows the number of publications on thermochemical water splitting for each year from 1964 to 2015. After a comprehensive screening analysis, only a few thermochemical cycles have been pursued for further development at a flow-sheet plant simulation level for nuclear hydrogen plants. Very few cycles were developed up to lab-scale demonstrations; even fewer have bench-scale or pilot plant-proof-of-concept facilities. To date, no full-scale thermochemical plant have been constructed, although the technology shows much future promise.

The pioneering research at Ispra (Mark cycles) and other research centers in the 1970s led to the establishment of a number of reaction schemes for thermochemical water splitting. Table 4.4 lists the main research centers active in the 1970s in the field of thermochemical water splitting and the most relevant processes proposed at each center.

The ISPRA Mark 9 thermochemical cycle requires 650°C as a maximum temperature to drive a hydrolysis reaction of iron dichloride according to $6FeCl_2 + 8H_2O \rightarrow 2Fe_3O_4 + 12HCl + 2H_2$. It follows a chlorination step of magnetite occurring at 150°C, and a decomposition step of iron tri-chloride $6FeCl_3 \rightarrow 3Cl_2 + 6FeCl_2$ conducted at 430°C.

The lithium-nitrite cycle proposed by the Argonne National Laboratory (ANL) comprises three steps, namely (i) hydrogen iodide production, $LiNO_2 + I_2 + H_2O \rightarrow LiNO_3 + 2HI$ by oxidation of lithium nitrite at about 25°C; (ii) hydrogen production through thermal decomposition of hydrogen iodide, $2HI \rightarrow H_2 + I_2$ at 425°C; and (iii) lithium nitrate decomposition, $LiNO_3 \rightarrow LiNO_2 + 0.5O_2$ at 475°C.

The sulfur-iodine (S-I) cycle is a leading example that has been scaled up from proof-of-principle tests to a larger engineering scale by the Japan Atomic Energy Agency. General Atomics was also investigating the S-I cycle, and a capacity of 2 kg/day was developed. Commissariat à l'Energie Atomique (CEA, France) and the Sandia National Laboratory (SNL) have also actively developed the S-I cycle.

4.4.2 Prototypical Reactions for Thermochemical Cycles

The reactions for thermochemical water splitting cycles can be categorized as proposed by Sato (1979) in four categories as given in Table 4.5. In a hydrolysis reaction, water combines with a reactant and generates some intermediary product(s). In a hydrogen evolving reaction, reactant(s) are combined with a hydrogen containing compound—denoted with HX in the table—and hydrogen is generated and some recyclable products are formed. In an oxygen evolving reaction there will be an oxygen containing compound—denoted with XO in the table—that is combined with certain reactants to generate oxygen gas and recyclable products. A regeneration reaction converts the intermediate products back to reactants.

TABLE 4.4 Main Institutions That Developed Thermochemical Cycles in the 1970s

Country	Institution	Developed process	Elements
France	Gas de France	K-process	K
		"Souriau" process	Sn
Germany	Rheinisch-Westfaelische Technische Hochschule Aachen	"Jülich Fe-S"	Fe, S
Italy	Joint Research Center, Ispra	Mark series	I, S
Japan	Japan Atomic Energy Research Institute (JAERI)	NiIS-Process	I, Ni, S
	Mitsubishi Heavy Industry	Fe-Cu-Cl	Cl, Cu, Fe
	University of Osaka	Osaka-75	Ba, C, I, N
	University of Tokyo	UT-3	Br, Ca, Fe
USA	Argonne National Laboratory	Ag-process	Ag, Br, Na
	General Atomic Company	GA-process	I, S
	General Electric	Agnes process	Cl, Fe, Mg
		Catherine process	I, K, Li, Ni
	Institute of Gas Technology (IGT)	A-2	Cl, Fe
	Lawrence Livermore Laboratory, University of California (LLLC)	MgSeO$_4$-Process	Mg, Se
	Los Alamos Scientific Laboratory (LASL)	Li-Mn-process	Li, Mn
	Oak Ridge National Laboratory (ORNL)	Fe-process	Ba, Cr
	University of Kentucky (UoK)	UoK-8	Ba, Fe, S
	Allison Laboratory of General Motors	Funk, Ta-process	Cl, Ta
		Funk, V-process	Cl, V

Table 4.6 shows some prototypical hydrolysis reactions. Water is combined with a reactant, and it forms intermediate reactants and/or hydrogen or oxygen depending on the actual reaction. The reverse Deacon reaction, for example, generates oxygen while hydrolyzing chlorine; the iron-steam reaction produces hydrogen; and the hydrolysis reaction of a halogen such as iodine in the presence of an oxygen acceptor such as SO$_2$ generates neither hydrogen nor oxygen but rather compounds that must be recycled.

The reverse Deacon reaction is included in many multistep thermochemical cycles such as ISPRA Mark 15. The reverse Deacon reaction is conducted at a temperature of approximately 873 K. This reaction, also included in Table 4.6, is written as follows:

$$\text{Reverse Deacon reaction}: H_2O(g) + Cl_2(g) \rightarrow 0.5O_2(g) + 2HCl(g) \tag{4.53}$$

The reverse Deacon reaction can be carried out in quartz reactors without catalysts at temperatures over 1273 K with a feed ratio of steam/chlorine of ~5, residence time of 8–24 s, and a departure from equilibrium of 13–25%. In order to improve the yield at lower temperatures, a catalyst (e.g., CuCl$_2$) must be used.

TABLE 4.5 Main Reaction Types in Thermochemical Water Splitting Cycles

Reaction type	Generic equation
Hydrolysis	$H_2O + \sum R \rightarrow \sum P$
Hydrogen evolving	$HX + \sum R \rightarrow H_2 + \sum P$
Oxygen evolving	$XO + \sum R \rightarrow 0.5O_2 + \sum P$
Regeneration	$\sum P \rightarrow \sum R$

Note: R, reactant; P, product.

TABLE 4.6   Prototypical Water Hydrolysis Reactions

Reaction	Examples
$nH_2O + MO_m \rightarrow nH_2 + MO_{m+n}$, reaction of a metallic oxide MO_m with water to form a higher oxide MO_{m+n}	$4H_2O + 3Fe \rightarrow 4H_2 + Fe_3O_4$ at 773 K
	$H_2O + 3FeO \rightarrow H_2 + Fe_3O_4$ at 450 K
	$2H_2O + 2Cs \rightarrow H_2 + 2CsOH$ at 450 K
$nH_2O + MX_m \rightarrow (n - m/2)H_2 + mHX + MO_n$, $2n \geq m$, reaction of a metallic halide with water to form hydrogen, hydrogen halide and metallic oxide	$4H_2O + 3FeCl_2 \rightarrow H_2 + 6HCl + Fe_3O_4$ at 923 K
	$4H_2O + 3MnCl_2 \rightarrow H_2 + 6HCl + Mn_3O_4$ at 923 K
	$4H_2O + SrBr_2 \rightarrow 2HBr + SrO$ at 1073 K
$H_2O + X_2 \rightarrow 0.5O_2 + 2HX$, hydrolysis with a halogen	$H_2O + Cl_2 \rightarrow 0.5O_2 + 2HCl$ at 873 K
	$H_2O + Br_2 \rightarrow 0.5O_2 + 2HBr$ at 873 K
$H_2O + X_2 + A \rightarrow 2HX + AO$, reaction of steam with a halogen in the presence of an oxygen acceptor	$H_2O + 0.5Br_2 + 0.5SO_2 \rightarrow HBr + 0.5H_2SO_4$ at 373 K
	$H_2O + 0.5I_2 + 0.5SO_2 \rightarrow HI + 0.5H_2SO_4$ at 373 K

The reverse Deacon process temperature can be lowered to roughly 1000 K when an acceptable yield of $\sim$60% can be obtained. At lower temperatures, the reaction rate decreases as the backward reaction is more favored. At 900 K, the yield becomes $\sim$50%; in the Mark 15 cycle, this reaction is performed at 873 K. An alternative way to conduct the reverse Deacon reaction is via a three-step reaction process with magnesium based compounds as follows:

$$\begin{cases} Cl_2 + Mg(OH)_2 \rightarrow 0.5O_2 + MgCl_2 + H_2O \text{ at } 623\,K \\ MgCl_2 + H_2O \rightarrow Mg(OH)Cl + HCl \text{ at } 570\,K \\ Mg(OH)Cl + H_2O \rightarrow MgCl_2 + HCl \text{ at } 623\,K \end{cases} \tag{4.54}$$

Table 4.7 presents some prototypical hydrogen evolving reactions. Thermolysis (thermal decomposition) of a hydrogen halide is a typical example of a gas phase reaction of hydrogen evolution. Reaction of a hydrogen halide (e.g., HCl) with salt can evolve hydrogen. Some hydrolysis reactions of metallic oxides generate hydrogen and a higher metal oxide. Table 4.8 presents some oxygen evolution reactions such as a metallic oxide or oxyacid thermolysis, or a reaction of a metallic oxide with a halogen or the reverse Deacon reaction.

TABLE 4.7   Some Types of Thermochemical Reactions for Hydrogen Generation

Reaction	Examples
$MX_m + HX_n \rightarrow H_2 + MX_{m+n}$, reaction of hydrogen halide with a metallic salt or a metal to form metallic halide	$2Cu + 2HCl \rightarrow H_2 + 2CuCl$ at 703 K
	$2VCl_2 + 2HCl \rightarrow H_2 + 2VCl_3$ at 298 K
	$Hg_2Br_2 + 2HBr \rightarrow H_2 + 2HgBr_2$ at 393 K
$2HX \rightarrow H_2 + X_2$, thermolysis of a hydrogen halide to form hydrogen and halogen	$2HI \rightarrow H_2 + I_2$ at 773 K
	$2HBr \rightarrow H_2 + Br_2$ at 1600 K
$nH_2O + MO_m \rightarrow nH_2 + MO_{m+n}$, hydrolysis of a metallic oxide to generate hydrogen and a higher oxide MO_{m+n} (see Table 5.6)	$H_2O + 3MnO \rightarrow H_2 + Mn_3O_4$ at 1123 K
	$H_2O + 3FeO \rightarrow H_2 + Fe_3O_4$ at 450 K
	$H_2O + 3CoO \rightarrow H_2 + Co_3O_4$ at 1273 K
$nH_2O + MX_m \rightarrow (n - m/2)H_2 + mHX + MO_n$, $2n > m$, hydrolysis of a metallic halide to form H_2, hydrogen halide and metallic oxide (see Table 5.6)	$4H_2O + 3FeCl_2 \rightarrow H_2 + 6HCl + Fe_3O_4$ at 923 K
	$4H_2O + 3MnCl_2 \rightarrow H_2 + 6HCl + Mn_3O_4$ at 923 K

TABLE 4.8 Some Types of Thermochemical Reactions for Oxygen Generation

Reaction	Examples
$MO_{m+1} \rightarrow 0.5O_2 + MO_m$, completed decomposition of an oxide or partial decomposition of a higher oxide	$HgO \rightarrow 0.5O_2 + Hg$ at 873 K
	$2CuO \rightarrow 0.5O_2 + Cu_2O$ at 1173 K
	$SO_3 \rightarrow 0.5O_2 + SO_2$ at 1073 K
$AO \rightarrow 0.5O_2 + A$, thermolysis of oxyacid (or its salt, e.g., chlorate, iodate, sulfate, and nitrate) to form a halide (e.g., chloride, iodide, bromide) or oxide	$MgSO_4 \rightarrow 0.5O_2 + MgO + SO_2$ at 1400 K
	$H_2SO_4 \rightarrow 0.5O_2 + H_2O + SO_2$ at 1123 K
	$Fe_3O_4 \rightarrow 0.5O_2 + 3FeO$ at 2473 K
$MO + X_2 \rightarrow 0.5O_2 + MX_2$, reaction of a metallic oxide with a halogen to form oxygen and halide	$CuO + 0.5I_2 \rightarrow 0.5O_2 + CuI$, at 973 K
	$0.25Fe_3O_4 + 2.25Cl_2 \rightarrow 0.5O_2 + 0.75FeCl_3$, at 1273 K
$H_2O + Cl_2 \rightarrow 0.5O_2 + 2HCl$, reverse Deacon reaction of chlorine hydrolysis	Hydrogen evolution in: Mark 3, 4, 6, and 6C at 1073 K
	Mark 14 at 1003 K, Mark 15 cycle at 873 K

Table 4.9 exemplifies some reactions for regeneration of intermediate reactants in thermochemical cycles. Chlorination reactions are important examples of regeneration reactions for intermediated reactants in iron or copper-based thermochemical processes. Chlorination of ferrites or magnetite is used as an oxygen evolving process or intermediate ferrite-based reagent recycling. The chlorination can be performed directly with chlorine or with hydrochloric acid.

Fig. 4.17 shows the product yield (y) and residence time (τ) at the chlorination process of Fe_3O_4 with HCl for a range of temperatures. The conversion of Fe_3O_4 to $FeCl_2$ increases with temperature, reaching about 60% at 923 K. Depending on the temperature and on how the reaction is conducted, this chlorination can evolve according to one of the following reactions:

$$Fe_3O_4(s) + 8HCl \rightarrow 3FeCl_2(s) + Cl_2(g) + 4H_2O(g) \tag{4.55a}$$

$$Fe_3O_4(s) + 8HCl(g) \rightarrow FeCl_2(s) + Fe_2Cl_6(g) + 4H_2O(g) \tag{4.55b}$$

$$Fe_3O_4(s) + 6HCl(g) \rightarrow Fe_2Cl_6(g) + 3H_2O(g) \tag{4.55c}$$

Note that chlorination of ferrous ferric oxide can also be made with gaseous chlorine, a case in which higher temperatures are required (700–1300 K). Fig. 4.18 presents the equilibrium concentrations for the decomposition of ferric chloride at standard pressure. It can be observed that this reaction shows a maximum yield at approximately 550 K. At higher temperatures than 600 K, vapors of ferric chloride start to evolve, and the concentration of Fe_2Cl_6 in the gas phase increases sharply, thereby leading to a decrease of $FeCl_2$ yield.

TABLE 4.9 Some Types of Reactions to Recycle Intermediary Reagents in Thermochemical Cycles

Reaction	Examples
$M_3O_4 + 8HX \rightarrow 4H_2O + 3MX_2 + X_2$, (or $\rightarrow MX_2 + 2MX_3$) reaction of a metal oxide with a hydrogen halide	$Fe_3O_4 + 8HCl \rightarrow 4H_2O + FeCl_2 + 2FeCl_3$ at 573 K
	$Fe_3O_4 + 8HBr \rightarrow 4H_2O + 3FeBr_2 + Br_2$ at 573 K
$MX_{n+1} \rightarrow MX_n + 0.5X_2$, thermolysis of a metallic halide to form a lower halide and a halogen	$FeCl_3 \rightarrow FeCl_2 + 0.5Cl_2$ at 693 K
	$CuCl_2 \rightarrow CuCl + 0.5Cl_2$ at 773 K
$MX_{n+1} + NX_n \rightarrow MX_n + NX_{n+1}$, reaction of two metallic halides that exchange one halogen atom	$CrCl_3(s) + FeCl_2(s) \rightarrow CrCl_2(s) + FeCl_3(g)$ at 973 K
	$FeCl_3(s) + CuCl(s) \rightarrow FeCl_2(s) + CuCl_2(s)$ at 423 K
$M_{n+1}O_{n+2} + 1/2nO_2 \rightarrow (1+1/n)M_nO_{n+1}$, oxidation of a metal oxide to form an oxide or oxidation of a metal sulfide to form a sulfate $M_2S + SO_2 + 3O_2 \rightarrow 2MSO_4$	$Fe_3O_4(s) + 0.25O_2(g) \rightarrow 1.5Fe_2O_3(s)$ at 623 K
	$2Fe_3O_4 + CO \rightarrow 3Fe_2O_3 + C$ at 523 K
	$Cu_2S + SO_2 + 3O_2 \rightarrow 2CuSO_4$ at 573 K

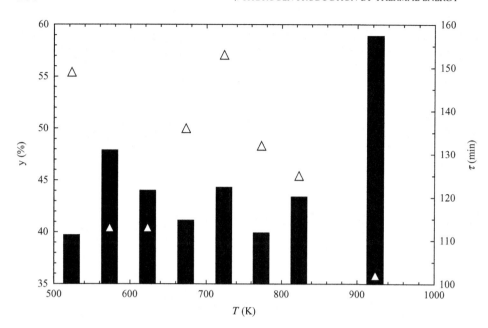

FIG. 4.17 Chlorination of Fe_3O_4 with hydrochloric acid. *Data from Knoche, K. F., Cremer, H., Breywisch, D., Hegels, S., Steinborn, G., Wüster, G., 1978. Experimental and theoretical investigation of thermochemical hydrogen production. Int. J. Hydrog. Energy 3, 209–216.*

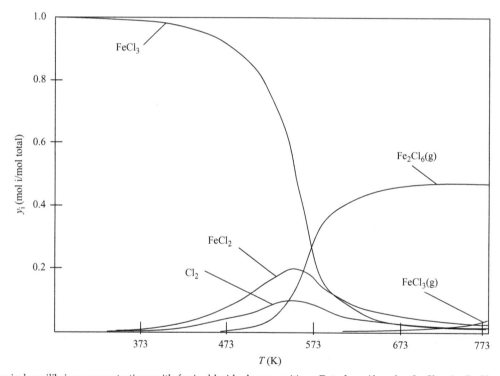

FIG. 4.18 Chemical equilibrium concentrations with ferric chloride decomposition. *Data from Abanades, S., Charvin, P., Flamant, G., Neveu, P., 2006. Screening of water-splitting thermo-chemical cycles potentially attractive for hydrogen production by concentrated solar energy. Energy 31, 2469–2486.*

Apart from chlorination, other halogenation reactions (with bromine or iodine instead of chlorine) are important in some thermochemical cycles, when after hydrolysis, a halide must be recycled. The reaction can be conducted either in the gas phase or in an aqueous solution. Some reactions are shown as follows:

$$3Fe_3O_4 + I_2 \rightarrow FeI_2 + 4Fe_2O_3 \tag{4.56a}$$

$$3Fe_3O_4 + Br_2 \rightarrow FeBr_2 + 4Fe_2O_3 \tag{4.56b}$$

$$Fe_3O_4 + 8HI \rightarrow 3FeI_2 + 4H_2O + I_2 \tag{4.56c}$$

Other reactions of interest of intermediates regeneration are those that lead to mixed acid formation. For example, hydrolysis of iodine in the presence of SO_2 can lead to a mixture of two acids, namely sulfuric acid and hydroiodic acid, according to the reaction

$$2H_2O + I_2 + SO_2 \rightarrow H_2SO_4 + 2HI$$

This reaction poses problems of acid separation in the liquid phase. The same type of reaction is possible with bromine, namely

$$2H_2O + Br_2 + SO_2 \rightarrow H_2SO_4 + 2HBr$$

One of the most stringent conditions of a thermochemical cycle is that the yield (or fractional conversion) of each reaction step must be sufficiently high. For example, if the cycle has three steps and each reaction in the sequence has 90% yield, then the overall yield is 73%; if the cycle has four steps, then the overall yield is 66%; if the yield of each individual reaction in the sequence is 80% for a four-step process, the overall yield is 41%. A low yield poses problems with chemical separation and recycling. For smaller yields, the process may become either infeasible or not competitive. Therefore, it is important to reduce the number of steps as much as possible or to obtain high yield for each reaction step. As a general indication, a thermochemical reaction can be judged as practical or impractical according to the following considerations:

- If the reaction yield is smaller than ~ 0.001, the reaction is unfeasible.
- The reaction has a marginal yield if the reactant conversion is between ~ 0.001 and ~ 0.1.
- If the conversion is higher than ~ 0.1, the reaction has a good yield, and it can be accomplished with ease.

Consider—as an example—a thermochemical reaction $A(s) \rightarrow B(s) + 0.5C(g)$, which is the thermal decomposition of a single solid reactant $A(s)$ to form a solid product $B(s)$ and evolve a gas $C(g)$. An actual example is the reaction $CuCl(s) \rightarrow Cu(s) + 0.5Cl_2(g)$ of cuprous chlorine decomposition given in Table 4.9. The Gibbs free energy for the reactions of this type is given by

$$\Delta G = \Delta G^0 + 0.5RT \ln(y) \tag{4.57}$$

where $y = P/P_0$ quantifies the product yield, P is the pressure of the gaseous product evolved from the reaction, and $P_0 = 1$ bar.

If the reaction is at equilibrium, then imposing $\Delta G = 0$ in Eq. (4.55) one obtains

$$\Delta G^0 = -0.5RT \ln(y) \tag{4.58}$$

Now, let's consider three range values for gaseous product molar fraction, namely $y \leq 0.001$ (impractical yield), $0.001 < y < 0.1$ (marginal yield), and $y \geq 0.1$ (very good yield). It turns out from Eq. (4.58) that for the first range, one must have $\Delta G^0_{0.001} \geq 28.7T$ when $y \leq 0.001$. If the yield is very good, then one must have $\Delta G^0_{0.1} \leq 9.6T$. This scenario allows us to construct the chart shown in Fig. 4.19 that helps analyzing the practicality of the prototypical reactions of type $A(s) \rightarrow B(s) + 0.5C(g)$, which is a thermolysis reaction.

As presented in the figure, Gibbs free energy variation with temperature at standard pressure for some relevant thermolysis reactions is plotted against the reaction temperature. The obtained diagram is divided into three regions by the lines $\Delta G^0_{0.001}$ and $\Delta G^0_{0.1}$: impractical yield, marginal yield and good yield. Among reaction types analyzed in Fig. 4.19, only reactions with $\Delta G^0 < 10{,}000$ J/reaction give a good yield at temperatures lower than 1200 K. Based on this criterion, reactions listed as #6, #8 and #1 may be selected as the most promising reactions. Reaction #2 is impractical for temperatures below 1100 K, while the other reactions show a marginal yield.

Let us consider now some prototypical gas phase reactions, mostly encountered in thermochemical water splitting cycles, and apply the same procedure to determine the reaction practicality in terms of yield and conversion. Here nine reaction were selected as shown in Fig. 4.20. Among those, reactions listed as #1 and #2 evolve oxygen by thermolysis of a single gaseous reactant. For both reactions, from 1 mole of reactant, it yields 1.5 moles total of gaseous products.

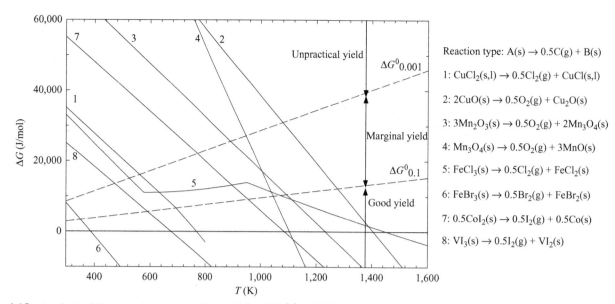

FIG. 4.19 Analysis of thermolysis reactions of type $A(s) \rightarrow 0.5C(g) + B(s)$ based on the Gibbs free energy magnitude at various temperatures.

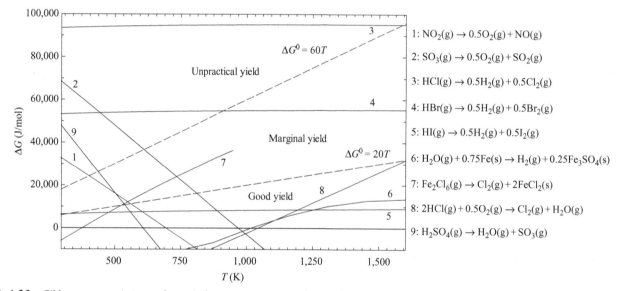

FIG. 4.20 Gibbs energy variation and practical temperature range of some chemical reactions used in multistep thermochemical cycles.

If one denotes with y the yield (moles of converted reactant per mole of reactant fed) then at equilibrium, one has

$$\Delta G^0 = -RT \ln \left(\frac{y^{1.5}}{1-y} \right) \tag{4.59}$$

where $1 - y$ represents the unconverted fraction of gaseous reactant.

The reactions #3 to #4 are hydrogen evolving reactions, among which reaction #6 is a hydrolysis; reaction #7 is an example of reagents recycling. The reactions #3 to #7 in Fig. 4.20 generate 1 mole of gas in the total products from 1 mole of feed. Therefore, for those reactions at equilibrium, one has

$$\Delta G^0 = -RT \ln \left(\frac{y}{1-y} \right) \tag{4.60}$$

Reaction #8 represents the Deacon process that has been used industrially to produce chlorine, namely $2HCl(g) + 0.5O_2(g) \rightarrow Cl_2(g) + H_2O(g)$. This reaction requires a $CuCl_2$ catalyst and operates at a typical temperature of 600–723 K. This reaction is used in Mark 7 and 7B cycles to recycle chlorine. For Deacon's reaction, the equilibrium condition imposes that

$$\Delta G^0 = -RT \ln \left(\frac{y^2}{(1-y)^{2.5}} \right) \qquad (4.61)$$

For reaction #9, namely $H_2SO_4(g) \rightarrow H_2O(g) + SO_3(g)$ from the equilibrium condition, one obtains that

$$\Delta G^0 = -RT \ln \left(\frac{y^2}{1-y} \right) \qquad (4.62)$$

In Fig. 4.20, above the line (dashed) $\Delta G^0 = 60T$, the yield for reactions #1 and #2 is therefore 0.008, while for reaction group #3 to #7, the yield is 0.0007, and for #8 and #9, it is 0.026 and 0.027, respectively. Therefore, above this line, the reactions #1–#7 are infeasible, while reactions #8 and #9 are difficult (marginal yield). Below line $\Delta G^0 = 20T$ (dashed), the reaction yields can be characterized as good because the yield is higher than 0.18 for reactions #1 and #2, higher than 0.083 for reactions #3–#7, higher than 0.22 for reaction #8 and higher than 0.23 for reaction #9. From this plot, one learns that the decomposition of HCl and HBr in the gaseous phase are impractical processes, but decomposition of HI (reaction #5) gives a good yield, thus it is practical, even for temperatures well below 1000 K. The reaction #5 is used in the sulfur-iodine cycle and the Mark 16 cycle where it is conducted at 773 K.

4.4.3 Two-Step Thermochemical Cycles

Thermochemical cycles of the two-step kind comprise at least two distinct chemical reactors. The product gases (hydrogen and oxygen) are chemically separated as they are produced in distinct chemical reactors; therefore, the risk of hydrogen oxidation is greatly reduced. This is a major advantage as compared to the direct water thermolysis process. Another advantage is derived from the fact that the temperature level for the reactions is significantly lower than that of water thermolysis (below ~2500 K); this is better for material selection. At the same time, a larger range of sustainable heat sources may be identified at lower temperatures.

Fig. 4.21 shows a classification of two-step thermochemical water splitting cycles, with respect to the categories of chemical reactions. In the hydride type of cycle, water is hydrolyzed with a metal or metallic compound to form a metallic hydride and evolve oxygen. Subsequently, the metal hydride is decomposed thermally to regenerate the metal and produce hydrogen gas.

In an oxide-type cycle, water is hydrolyzed such that it releases hydrogen and produces a metal oxide. Then a thermolysis of metal oxide will generate oxygen gas and regenerate the metal. The hydroxide scheme was proposed by Nakamura (1977) for two-step thermochemical cycles. As shown in the figure, in a hydroxide type of cycle, the hydrolysis reaction leads to the formation of a metal hydride and hydrogen generation as a first step. Again, a thermolysis reaction is required to decompose the metal hydroxide and generate oxygen.

Bilgen and Bilgen (1982) suggested an oxide-sulfate scheme for two-step thermochemical water splitting cycles. In this case, sulfate and a metal oxide are required to generate a metallic sulfate and generate hydrogen from water at a

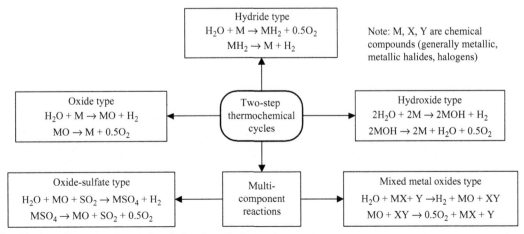

FIG. 4.21 Classification of two-step thermochemical water splitting cycles.

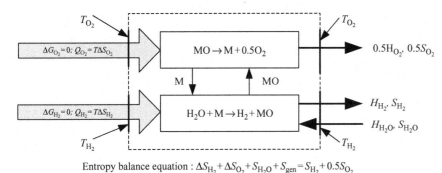

FIG. 4.22 Energy flows and entropy balance for a generic two-step water splitting process assuming both reactions at equilibrium.

first step. A thermolysis reaction is conducted thereafter to generate oxygen and regenerate the sulfate and metal oxide. The mixed metal oxide type was introduced in McQuillan et al. (2010) and involves metallic halides and halogens.

Consider a basic thermodynamic representation of a two-step thermochemical cycle as shown in Fig. 4.22 in which both reaction steps are assumed at equilibrium. The entropy balance equation for the overall system is suggested on the same figure. Assume that the temperature of each reaction can be set such that the forward reactions are spontaneous. This means that for hydrogen evolving reaction $\Delta G_{H_2} \lesssim 0$ at T_{H_2} and for oxygen evolving process $\Delta G_{O_2} \lesssim 0$ at T_{O_2}.

Since ΔG is zero for both reactions, the heat input in the process is given by $Q_{H_2} = T_{H_2}\Delta S_{H_2}$ at T_{H_2} and $Q_{O_2} = T_{O_2}\Delta S_{O_2}$ at T_{O_2}. The first law of thermodynamics (energy balance) for this system yields

$$Q_{H_2} + Q_{O_2} + H_{H_2O} = H_{H_2} + 0.5H_{O_2} \tag{4.63}$$

where H is the molar enthalpy and Q_{reheat} is the net heat input necessary to reheat the chemicals circulated between the lower temperature reaction to the high temperature reaction.

The second law of thermodynamics is written as an entropy balance equation using molar entropies (S) as follows,

$$\Delta S_{H_2} + \Delta S_{O_2} + S_{H_2O} + S_{gen} = S_{H_2} + 0.5S_{O_2} \tag{4.64}$$

where ΔS_{H_2}, ΔS_{O_2} are the reaction entropies for the hydrogen and oxygen evolving reaction, respectively.

Now, recall that the practical interest in two-steps thermochemical cycles is mainly due to a lower temperature level requirement as compared with direct water thermolysis. The temperature at which the thermolysis reaction $H_2O \rightarrow H_2 + 0.5O_2$ has a nil ΔG is easily determined to 3815 K for standard conditions. Thence, one wants the temperature requirement level of thermal energy input for the two-step cycle to be inferior to 3815 K. Therefore, one must search for reaction conditions in which $T_{H_2} < 3815\,K$ and $T_{O_2} < 3815\,K$ are feasible conditions. In addition to this, the condition that $S_{gen} \geq 0$ and Eq. (4.64) led to the following thermodynamic requirements for the two-step thermochemical cycles:

$$\begin{cases} \Delta S_{H_2} + \Delta S_{O_2} \leq S_{H_2} + 0.5S_{O_2} - S_{H_2O} \\ T_{H_2}\Delta S_{H_2} + T_L\Delta S_L = H_{H_2} + 0.5H_{O_2} - H_{H_2O} \\ \Delta G_{H_2} \cong 0 \\ \Delta G_{O_2} \cong 0 \\ T_{H_2} < 3815K \\ T_{O_2} < 3815K \end{cases} \tag{4.65}$$

Funk and Reinstorm (1966) identified 22 oxide-type and 16 hydride-type reactions that satisfy Eq. (4.60), and they are therefore possible candidates for two-step thermochemical water splitting cycle development. However, very few of these cycles have been attractive for practical implementation. An example of a hydride-type cycle that was proposed early on and that may be practically attractive is based on boron hydride (HB) and borane (H_3B), both reacting in a gas phase at high temperature. Boron hydride is reacted with steam at 1335 K to generate oxygen and borane via a chemical reaction at equilibrium. Direct borane thermolysis is performed at 2503 K with production of hydrogen and boron hydride. The cycle is described by these equations:

$$\begin{cases} H_2O + HB \rightarrow 0.5O_2 + H_3B, \text{ at } 1335\,K \\ H_3B \rightarrow H_2 + HB, \text{ at } 2503\,K \end{cases} \tag{4.66}$$

FIG. 4.23 Free Gibbs energy versus temperature for a hydride-type cycle with boronhydride and for water thermolysis reaction.

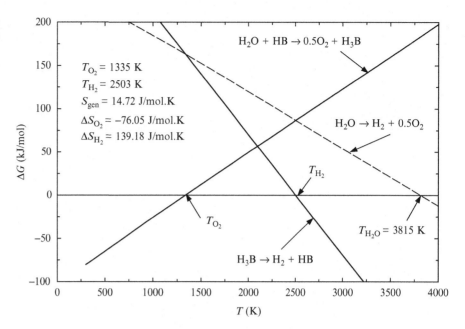

Thermodynamic calculations for entropy, enthalpy and free energy of the reactions from Eq. (4.66) are easy to perform. Fig. 4.23 shows the variation of the standard Gibbs free energy with the temperature for Eq. (4.66) and the water thermolysis reaction. As shown, the slope of the Gibbs free energy curve for oxygen evolving reaction is positive meaning that the reaction is exothermic and spontaneous at temperatures below equilibrium. The hydrogen evolving reaction is not spontaneous at lower temperatures. However, because the Gibbs free energy decreases with temperature, the equilibrium is eventually reached at 2503 K in standard conditions. The reaction entropy is positive and high, which indicates that the reaction is endothermic. The generated entropy by the two-step water splitting cycle is positive (14.72 J/mol K), implying that the process is feasible according to the second law of thermodynamics. In the figure, for comparison purposes, the Gibbs free energy curve is superimposed, characterizing the water decomposition reaction.

Nakamura (1977) discusses an oxide-type two-step cycle for thermochemical water splitting that operates with ferrous oxide. Steam reacts with ferrous oxide and generates hydrogen at a lower temperature, 700 K, while ferrous ferric oxide (Fe_3O_4) is formed. At a higher temperature of 2475 K, the direct thermal decomposition of Fe_3O_4 occurs with a release of oxygen. The reactions for this cycle are written as

$$\begin{cases} H_2O + 3FeO \rightarrow H_2 + Fe_3O_4, \text{ at } 700\,K \\ Fe_3O_4 \rightarrow 0.5O_2 + 3FeO, \text{ at } 2475\,K \end{cases} \tag{4.67}$$

Fig. 4.24 shows the variation of the standard Gibbs free energy for reactions described by Eq. (4.67). The hydrogen evolving reaction is exothermic at equilibrium (with a negative reaction entropy) and spontaneous at low temperature. The oxygen evolving reaction is endothermic and not spontaneous at low temperature with a negative Gibbs free energy. The generated entropy of the overall two-step process is positive and high (209 kJ/mol K), which is an indication that the process is feasible according to the second law of thermodynamics, but it suggests significant irreversibilities.

One promising two-step cycle that was investigated recently by the Department of Energy is the ferrite process described in Kromer et al. (2011). In this cycle, the high temperature step is the reduction of a ferrite ($NiFe_2O_4$) to form a metal oxide at 1178 K and evolve oxygen. The second step is the oxidation of the metal oxide within a hydrolysis reaction with hydrogen evolution and regeneration of the ferrite at 727 K.

A nickel-manganese-ferrite cycle appears to have an even lower level of high temperature requirement with 1093 K, as shown in McQuillan et al. (2010). The nickel-manganese-ferrite cycle plant is described by the simplified diagram from Fig. 4.25. Liquid water enters in #1 at T_0, and it is heated and partially boiled (#1–#2) using heat recovered from product gases, hydrogen and oxygen (#4–#5 and #12–#13). Water is then boiled and superheated using heat recovered from a hydrogen evolving reactor (#2–#3). Nickel-manganese-wustite ($NiMnFe_4O_8$) is heated (#7–#8) using heat recovered from the oxygen stream (#11–#12) and nickel-manganese-ferrite (#9–#10) to reach the reaction temperature

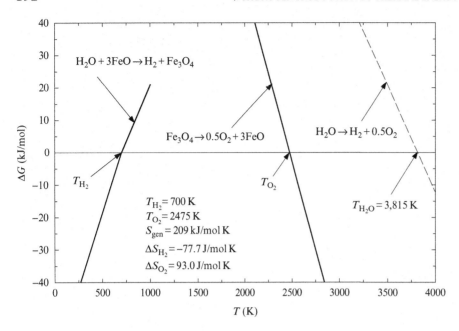

FIG. 4.24 Free Gibbs energy versus temperature for an oxide-type cycle with ferrous oxide and for water thermolysis reaction.

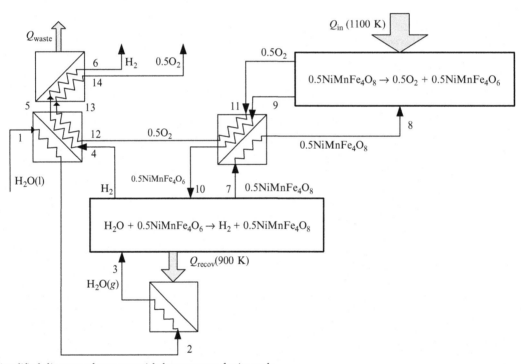

FIG. 4.25 Simplified diagram of two-step nickel-manganese-ferrite cycle.

($\sim$1100 K) of the oxygen evolving reactor. The product stream (hydrogen and oxygen), still hot, is cooled (without heat recovery) to reach a temperature close to T_0 prior to further compression (#5–#6 and #13–#14). This cycle is attractive also because it involves simpler solid-gas separation of reaction products with the possibility of improved heat recovery. Fluidized bed reactors can be used for both reactions with solid-gas separation; water is separated by condensation.

The steps of the nickel-manganese-ferrite cycle are given by the following reactions:

$$\begin{cases} H_2O + 0.5NiMnFe_4O_6 \rightarrow H_2 + 0.5NiMnFe_4O_8, \text{ at } 873\,K \\ 0.5NiMnFe_4O_8 \rightarrow 0.5O_2 + 0.5NiMnFe_4O_6, \text{ at } 1093\,K \end{cases} \qquad (4.68)$$

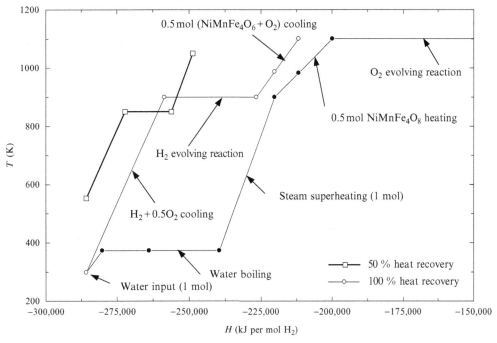

FIG. 4.26   Temperature versus total enthalpy of material streams within the two-step nickel-manganese-ferrite cycle.

Fig. 4.26 shows a temperature versus total enthalpy diagram that represents heating and cooling processes within the thermochemical cycle described by Eq. (4.68). The diagram is useful for determining the efficiency and the heat recovery potential within the cycle. Assume that water is supplied as a liquid at standard pressure and temperature and then heated to the normal boiling point, boiled and then superheated to generate steam at 900 K. An amount of 0.5 moles of $NiMnFe_4O_8$ per mole of generated hydrogen must be heated from 900 to 1100 K using heat recovered internally.

If 100% heat recovery and reuse within the cycle can be obtained, then the cycle's energy efficiency is 100% because 1 mole of liquid water generates 1 mole of hydrogen and a half mole of oxygen at standard conditions; therefore, the heat required for the process is equal to higher heating value of hydrogen (no heat is lost and other irreversibilities are neglected). Fig. 4.26 shows the temperature-enthalpy curves of processes that reject heat.

These heat rejecting processes are as follows: cooling of a half mole of O_2 and a half mole of $NiMnFe_4O_6$ from 1100 to 900 K, heat rejected by the hydrogen evolving reaction at 900 K, and cooling of hydrogen and oxygen streams from 900 down to 298.15 K.

Assume that only 50% of the heat rejected can be recovered and reused internally within the cycle. In this assumption, the heat recovered and reused is represented with a bold line in Fig. 4.26. It can be observed that the recovered heat cannot completely boil the water. The estimated heat input requirement in this case is 323 kJ/mol. According to McQuillan et al. (2010), the estimated efficiency of a hydrogen production plant operating under this nickel-manganese-ferrite cycle is 48%.

Table 4.10 gives a compilation of 24 thermochemical cycles of two steps. These cycles are grouped as oxide, mixed metal oxides, hydride and oxide-sulfate types. The main chemical elements involved are also given. The cycle with ID #10, namely the just-discussed nickel-manganese-ferrite shows the lowest temperature level requirement.

4.4.4 Thermochemical Cycles With More Than Two Steps

Although two-step cycles are an attractive option for water splitting due to their unique advantage of simplicity of having only two chemical reactors and intrinsic hydrogen-oxygen separation features, the required working temperature is too high. If the cycle comprises three or more steps, then the temperature level for the heat sources can be reduced below 1500 K. A three-step cycle can comprise, for example, a hydrolysis reaction with oxygen evolution,

TABLE 4.10   Two-Step Thermochemical Water Splitting Cycles

Type	TID: elements	Reaction steps	T (K)
Oxide-type	1: Si	$H_2O + SiO \rightarrow H_2 + SiO_2$	2929
		$SiO_2 \rightarrow 0.5O_2 + SiO$	3250
	2: Ce	$H_2O + Ce_2O_3 \rightarrow H_2 + 2CeO_2$	700
		$2CeO_2 \rightarrow 0.5O_2 + Ce_2O_3$	2573
	3: Fe	$H_2O + 3FeO \rightarrow H_2 + Fe_3O_4$	450
		$Fe_3O_4 \rightarrow 0.5O_2 + 3FeO$	2475
	4: C	$H_2O + CO \rightarrow H_2 + CO_2$	423
		$CO_2 \rightarrow 0.5O_2 + CO$	2473
	5: Zn	$H_2O + Zn \rightarrow H_2 + ZnO$	1489
		$ZnO \rightarrow Zn + 0.5O_2$	2337
	6: Mo	$H_2O + 0.5Mo \rightarrow H_2 + 0.5MoO_2$	1816
		$0.5MoO_2 \rightarrow 0.5O_2 + 0.5Mo$	2000
	7: W	$H_2O + 1/3W \rightarrow H_2 + 1/3WO_3$	1157
		$1/3WO_3 \rightarrow 0.5O_2 + 1/3W$	2000
	8: In	$H_2O + 2/3In \rightarrow H_2 + 1/3In_2O_3$	1116
		$1/3In_2O_3 \rightarrow 0.5O_2 + 2/3In$	2000
	9: Sn	$H_2O + 0.5Sn \rightarrow H_2 + 0.5SnO_2$	873
		$0.5SnO_2 \rightarrow 0.5O_2 + 0.5Sn$	2000
	10: Fe, Mn, Ni	$H_2O + 0.5NiMnFe_4O_6 \rightarrow H_2 + 0.5NiMnFe_4O_8$	873
		$0.5NiMnFe_4O_8 \rightarrow 0.5O_2 + 0.5NiMnFe_4O_6$	1093
Oxide/sulfate-type	11: Sr, S	$H_2O + SrO + SO_2 \rightarrow H_2 + SrSO_4$	1360
		$SrSO_4 \rightarrow 0.5O_2 + SrO + SO_2$	2038
	12: Ca, S	$H_2O + CaO + SO_2 \rightarrow H_2 + CaSO_4$	1150
		$CaSO_4 \rightarrow 0.5O_2 + CaO + SO_2$	1880
	13: Fe, S	$H_2O + FeO + SO_2 \rightarrow H_2 + FeSO_4$	522
		$FeSO_4 \rightarrow 0.5O_2 + FeO + SO_2$	1373
	14: Co, S	$H_2O + CoO + SO_2 \rightarrow H_2 + CSO_4$	473
		$CoSO_4 \rightarrow 0.5O_2 + CoO + SO_2$	1373
	15: Mn, S	$H_2O + MnO + SO_2 \rightarrow H_2 + MnSO_4$	563
		$MnSO_4 \rightarrow 0.5O_2 + MnO + SO_2$	1373
	16: Cd, S	$H_2O + CdO + SO_2 \rightarrow H_2 + CdSO_4$	473
		$CdSO_4 \rightarrow 0.5O_2 + CdO + SO_2$	1273
Mixed metal oxides-type	17: Fe, Zn	$H_2O + 2FeO + ZnO \rightarrow H_2 + Fe_2ZnO_4$	694
		$Fe_2ZnO_4 \rightarrow 0.5O_2 + 2FeO + ZnO$	2358
	18: C, Na, Fe	$H_2O + 2Na_2FeO_2 + CO_2 \rightarrow H_2 + Na_2CO_3 + 2NaFeO_2$	1073
		$Na_2CO_3 + 2NaFeO_2 \rightarrow 0.5O_2 + 2Na_2FeO_2 + CO_2$	2273
	19: Cl, Fe	$4H_2O + 3FeCl_2 \rightarrow H_2 + Fe_3O_4 + 6HCl$	973
		$Fe_3O_4 + 6HCl \rightarrow 0.5O_2 + 3FeCl_2 + 3H_2O$	1773

TABLE 4.10 Two-Step Thermochemical Water Splitting Cycles—cont'd

Type	TID: elements	Reaction steps	T (K)
	20: Fe,Zn	$H_2O + 1/2Fe_3O_4 + 3/4Zn \rightarrow H_2 + 3/4Fe_2ZnO_4$	873
		$3/4Fe_2ZnO_4 \rightarrow 0.5O_2 + 1/2Fe_3O_4 + 3/4Zn$	1573
	21: Fe, Mn, Zn	$H_2O + MnFe_2O_4 + 3ZnO \rightarrow H_2 + Zn_3MnFe_2O_8$	1273
		$Zn_3MnFe_2O_8 \rightarrow 0.5O_2 + MnFe_2O_4 + 3ZnO$	1473
Hydride type	22: B	$H_2O + HB \rightarrow 0.5O_2 + H_3B$	1335
		$H_3B \rightarrow H_2 + HB$	2500
	24: Cl	$H_2O + Cl_2 \rightarrow 0.5O_2 + 2HCl$	873
		$2HCl \rightarrow H_2 + Cl_2$	1073

[a]Note: data calculated with Engineering equation Solver (Klein, 2015).
Source: Naterer, G.F., Suppiah, S., Stolberg, L., Lewis, M., Ferrandon, M., Wang, Z., Dincer, I.,Gabriel, K., Rosen, M.A., Secnik, E., Easton, E.B., Trevani, L., Pioro, I., Tremaine, P., Lvov, S., Jiang, J., Rizvi, G., Ikeda, B.M., Luf, L., Kaye, M., Smith, W.R., Mostaghimi, J., Spekkens, P., Fowler, M., Avsec, J., 2011a. Clean hydrogen production with the Cu-Cl cycle – progress of international consortium, I: Experimental unit operations. Int. J. Hydrog. Energy 36, 15472–15485.

a hydrogen evolution reaction at lower temperatures and a reagent regeneration reaction that is typically a thermolysis reaction conducted at highest temperature. This type of process is generically described as follows:

$$\begin{cases} H_2O + A \rightarrow 0.5O_2 + AH_2 \\ AH_2 + B \rightarrow H_2 + AB \\ AB \rightarrow A + B \end{cases} \tag{4.69}$$

An alternative to the generic three-step process from Eq. (4.64) will include a hydrogen evolving hydrolysis with formation of an oxide, followed by an oxygen evolving reaction, and then a thermolysis reaction to regenerate the intermediate reagents as follows:

$$\begin{cases} H_2O + A \rightarrow 0.5H_2 + AO \\ AO + B \rightarrow 0.5O_2 + AB \\ AB \rightarrow A + B \end{cases} \tag{4.70}$$

Three-step processes can be constructed starting from two-step processes in which the highest temperature reaction is replaced by a two-step reaction process. The effect of this change will be a reduction of the maximum temperature requirement for the cycle. Let us give an example. Consider the two-step iron oxide cycle (#6 in Table 4.10) in which the hydrogen evolution reaction is at lower temperature, namely

$$H_2O + 3FeO \rightarrow H_2 + Fe_3O_4 \text{ at 450 K} \tag{4.71}$$

and the oxygen evolving reaction is at a very high temperature of 2475 K as follows:

$$Fe_3O_4 \rightarrow 0.5O_2 + 3FeO \tag{4.72}$$

If the oxygen evolution reaction, which is the thermolysis of Fe_3O_4 in Eq. (4.72), is replaced with a two-step chemical process having the same net effect, then the maximum temperature level requirement for the cycle can be reduced from 2475 to 1073 K. This is very significant and opens the door of many applications in which a green thermal energy source may be available. The cycle developed by Jülich Center EOS in Germany performs according to the three-step process in which Eq. (4.72) is performed in two steps introducing an additional chemical compound, namely $FeSO_4$. More exactly, the Fe_3O_4 produced by the hydrogen evolving reaction Eq. (4.66) is combined with $FeSO_4$ at 1073 K, which additionally forms Fe_2O_3 and SO_2

$$Fe_3O_4 + 3FeSO_4 \rightarrow 0.5O_2 + 3Fe_2O_3 + 3SO_2 \text{ at 1073 K} \tag{4.73}$$

The new intermediate reagents can be recycled at a temperature as low as 473 K as follows:

$$3Fe_2O_3 + 3SO_2 \rightarrow 3FeSO_4 + 3FeO \text{ at } 473 \text{ K} \tag{4.74}$$

Let us consider another three-step process derived from a two-step one. The iron chloride cycle #19 from Table 4.10 has the hydrogen evolving hydrolysis at 923 K as follows:

$$4H_2O + 3FeCl_2 \rightarrow H_2 + Fe_3O_4 + 6HCl \tag{4.75}$$

The closing reaction of the two-step process evolves oxygen at a very high temperature of 1773 K as follows:

$$Fe_3O_4 + 6HCl \rightarrow 0.5O_2 + 3FeCl_2 + 3H_2O \tag{4.76}$$

The cycle Mark 9 developed at ISPRA (Italy) replaces the closing reaction from Eq. (4.76) with a two-stage process having the maximum temperature requirement of 923 K. The additional chemical used is chlorine, and the oxygen evolving reaction becomes

$$Fe_3O_4 + 6HCl + 1.5Cl_2 \rightarrow 0.5O_2 + 3FeCl_3 + 3H_2O \text{ at } 473 \text{ K} \tag{4.77}$$

The intermediate chemicals to be regenerated are $FeCl_2$ and Cl_2. The closing reaction becomes

$$3FeCl_3 \rightarrow 3FeCl_2 + 1.5Cl_2 \text{ at } 923 \text{ K} \tag{4.78}$$

Table 4.11 gives a compilation of representative three-step thermochemical cycles sorted according to the maximum temperature requirement in ascending order. The first thermochemical cycle of three steps having an oxygen evolving hydrolysis process was proposed by Funk and Reinstorm (1964). This is the well-known Tantalum cycle as given in Table 4.11 as the GM-Funk process with a maximum temperature requirement of 1366 K. The cycle with minimum temperature requirement appears to be the Shell process with 773 K.

TABLE 4.11   Representative Three-Step Thermochemical Cycles

Cycle	T_{max} (K)	Reaction steps
Shell	773	$H_2O + 2Cu \rightarrow H_2 + Cu_2O$, 773 K
		$1/3Cu_2O + 4/3CuSO_4 \rightarrow 0.5O_2 + 2Cu + 4/3SO_2 + O_2$, 773 K
		$2/3Cu_2O + 4/3SO_2 + O_2 \rightarrow 4/3CuSO_4$, 573 K
KIER-B Cu-Cl	873	$H_2O + 2CuCl_2 \rightarrow 0.5O_2 + 2CuCl + 2HCl$, 873 K
		$2HCl + 6CuCl + 2Cl_2 \rightarrow H_2 + 6CuCl_2$, 500 K
		$4CuCl_2 \rightarrow 4CuCl + 2Cl_2$, 773 K
Mark 9	923	$4H_2O(g) + 3FeCl_2(s) \rightarrow H_2(g) + Fe_3O_4(s) + 6HCl(g)$ at 923 K
		$Fe_3O_4(s) + 1.5Cl_2(g) + 6HCl(g) \rightarrow 0.5O_2(g) + 3FeCl_3(g) + 3H_2O(g)$ at 473 K
		$3FeCl_3(g) \rightarrow 3FeCl_2(s) + 1.5Cl_2(g)$ at 693 K
GIRIO Cu-I-N	973	$H_2O + 0.5O_2 + 2CuI + 2NH3 \rightarrow 2CuO + 2NH_4I$, 298 K
		$2NH_4I \rightarrow H_2 + I_2 + 2NH_3$, 773 K
		$2CuO + I_2 \rightarrow 0.5O_2 + 2CuI + 0.5O_2$, 973 K
Yokohama Mark 3	973	$2H_2O + I_2 + 2FeSO_4 \rightarrow 2HI + 2Fe(OH)SO_4$, 973 K
		$2HI \rightarrow H_2 + I_2$, 573 K
		$2Fe(OH)SO_4 \rightarrow 0.5O_2 + 2FeSO_4 + H_2O$, 523 K
Hitachi	973	$H_2O + 2NaI + 2NH_3 + CO_2 \rightarrow Na_2CO_3 + 2NH_4I$, 298 K
		$2NH_4I \rightarrow H_2 + I_2 + 2NH_3$, 773 K
		$Na_2CO_3 + I_2 \rightarrow 0.5O_2 + CO + 2NaI$, 973 K
Euratom Fe-Cl	973	$H_2O + Cl_2 \rightarrow 0.5O_2 + 2HCl$, 973 K
		$2HCl + 2FeCl_2 \rightarrow H_2 + 2FeCl_3$, 873 K
		$2FeCl_3 \rightarrow 2FeCl_2 + Cl_2$, 723 K

TABLE 4.11 Representative Three-Step Thermochemical Cycles—cont'd

Cycle	T_{max} (K)	Reaction steps
UNLV-159	1023	$2H_2O + SO_2 + C_2H_4 \rightarrow C_2H_6 + H_2SO_4$, 623 K
		$C_2H_6 \rightarrow H_2 + C_2H_4$, 1073 K
		$H_2SO_4 \rightarrow 0.5O_2 + SO_2 + H_2O$, 1123 K
Mark 2	1073	$Mn_2O_3(g) + 4NaOH(l) \rightarrow H_2(g) + H_2O(g) + 2Na_2O \cdot MNO_2(s)$ at 1073 K
		$2MnO_2(s) \rightarrow 0.5O_2(g) + Mn_2O_3(s)$ at 873 K
		$2H_2O(g) + 2Na_2O \cdot MnO_2(s) \rightarrow 4NaOH(aq) + 2MnO_2(s)$ at 373 K
Mark 1S	1073	$2HBr(g) + Hg(l) \rightarrow H_2(g) + HgBr_2(s)$ at 473 K
		$SrO(s) + HgBr_2(s) \rightarrow 0.5O_2 + Hg(g) + SrBr_2(s)$ at 773 K
		$H_2O(g) + SrBr_2(s) \rightarrow SrO(s) + 2HBr(g)$ at 1073 K
Theme-S3	1073	$H_2O + I_2 + SO_2 \rightarrow 2HI + SO_3$, 373 K
		$2HI \rightarrow H_2 + I_2$, 573 K
		$SO_3 \rightarrow 0.5O_2 + SO_2$, 1073 K
GDF K-process	1100	$H_2O + K_2O_2 \rightarrow 0.5O_2 + 2KOH$, 400 K
		$2KOH + 2K \rightarrow H_2 + 2K_2O$, 1000 K
		$2K_2O \rightarrow K_2O_2 + 2K$, 1100 K
KIER-A Cu-Cl	1100	$H_2O + 2CuCl_2 \rightarrow 0.5O_2 + 2CuCl + 2HCl$, 900 K
		$4CuCl_2 + 2HCl \rightarrow H_2 + 4CuCl + 3Cl_2$, 1100 K
		$6CuCl + 3Cl_2 \rightarrow 6CuCl_2$, 873 K
KIER-1 Cu-Cl	1100	$2H_2O + 2Cl_2 \rightarrow 0.5O_2 + 4HCl + 0.5O_2$, 873 K
		$Cu_2O + 4HCl \rightarrow H_2 + 2CuCl_2 + H_2O$, 500 K
		$2CuCl_2 + 0.5O_2 \rightarrow 2Cl_2 + Cu_2O$, 1100 K
KIER-2 Cu-Cl	1100	$H_2O + CuCl_2 \rightarrow 2HCl + CuO$, 1100 K
		$2HCl + 4CuCl_2 \rightarrow H_2 + 4CuCl + 3Cl_2$, 1100 K
		$4CuCl + 3Cl_2 + CuO \rightarrow 0.5O_2 + 5CuCl_2$, 873 K
KIER-3	1100	$H_2O + CuSO_4 \rightarrow CuO + H_2SO_4$, 1100 K
		$CuO + SO_2 + H_2O \rightarrow H_2 + CuSO_4$, 473 K
		$H_2SO_4 \rightarrow 0.5O_2 + SO_2 + H_2O$, 1100 K
Mark 16 S-I cycle	1123	$2HI(g) \rightarrow H_2(g) + I_2(g)$ at 773 K
		$H_2SO_4(g) \rightarrow 0.5O_2(g) + SO_2(g) + H_2O(g)$ at 1123 K
		$2H_2O(l) + SO_2(g) + I_2(s) \rightarrow H_2SO_4(aq) + 2HI(aq)$ at 373 K
Williams, US-chlorine	1123	$H_2O + Cl_2 \rightarrow 0.5O2 + 2HCl$, 1123 K
		$2HCl + 2CuCl \rightarrow H_2 + 2CuCl_2$, 473 K
		$2CuCl_2 \rightarrow 2CuCl + Cl_2$, 773 K
GIRIO Fe-Br	1123	$H_2O + Br_2 \rightarrow 0.5O_2 + 2HBr$, 923 K
		$4H_2O + 3FeBr_2 \rightarrow H_2 + 6HBr + Fe_3O_4$, 1123 K
		$8HBr + Fe_3O_4 \rightarrow 3FeBr_2 + 4H_2O + Br_2$, 523 K
Mark 8	1173	$4H_2O(g) + 3MnCl_2(s) \rightarrow H_2(g) + Mn_3O_4(s) + 6HCl(g)$ at 973 K
		$1.5MnO_2(s) \rightarrow 0.5O_2(g) + 0.5Mn_3O_4(s)$ at 1173 K
		$1.5Mn_3O_4(s) + 6HCl(g) \rightarrow 3MnCl_2(aq) + 1.5MnO_2(s) + 3H_2O(l)$ at 373 K

Continued

TABLE 4.11 Representative Three-Step Thermochemical Cycles—cont'd

Cycle	T_{max} (K)	Reaction steps
IGT	1200	$H_2O + Cl_2 \rightarrow 0.5O_2 + HCl$, 1123 K
		$4H_2O + 2CeCl_3 \rightarrow H_2 + 6HCl + 2CeO_2$, 1200 K
		$2CeO_2 + 8HCl \rightarrow 2CeCl_3 + Cl_2$, 383 K
Jülich C-S	1223	$H_2O + CO \rightarrow H_2 + CO_2$, 773 K
		$H_2SO_4 \rightarrow 0.5O_2 + SO_2 + H_2O$, 1223 K
		$CO_2 + SO_2 + H_2O \rightarrow CO + H_2SO_4$, 623 K
GA Cl-Cs	1273	$H_2O + Cl_2 \rightarrow 0.5O_2 + 2HCl$, 1273 K
		$2CeCl_3 + 4H_2O \rightarrow H_2 + 2CeO_2 + 6HCl$, 1073 K
		$2CeO_2 + 8HCl \rightarrow 2CeCl_3 + 4H_2O + Cl_2$, 523 K
IGT-C7	1273	$H_2O + Fe_2O_3 + 2SO_2 \rightarrow H_2 + 2FeSO_4$, 400 K
		$SO_3 \rightarrow 0.5O_2 + SO_2$, 1273 K
		$FeSO_4 \rightarrow Fe_2SO_3 + SO_2 + SO_3$, 973 K
Miura	1273	$H_2O + I_2 + 0.5Sb_2O_3 \rightarrow 0.5Sb_2O_5 + 2HI$, 278 K
		$2HI \rightarrow H_2 + I_2$, 823 K
		$0.5Sb_2O_5 \rightarrow 0.5O_2 + 0.5Sb_2O_3$, 1273 K
LASL	1273	$3H_2O + 3Li_2O \cdot Mn_2O_3 \rightarrow 6LiOH + 3Mn_2O_3$, 355 K
		$6LiOH + 2Mn_3O_4 \rightarrow H_2 + 3Li_2O \cdot Mn_2O_3 + 2H_2O$, 973 K
		$3Mn_2O_3 \rightarrow 0.5O_2 + 2Mn_3O_4$, 1273 K
GM-Funk Ta-Cl	1366	$H_2O + Cl_2 \rightarrow 0.5O_2 + 2HCl$, 1000 K
		$2HCl + 2TaCl_2 \rightarrow H_2 + 2TaCl_3$, 298 K
		$2TaCl_3 \rightarrow 2TaCl_2 + Cl_2$, 1366 K
CNRS, UNLV-133	1373	$H_2O + 3FeO \rightarrow H_2 + Fe_3O_4$, 823 K
		$Fe_3O_4 + 3SO_3 \rightarrow 0.5O_2 + 3FeSO_4$, 1073 K
		$3FeSO_4 \rightarrow 3FeO + 3SO_3$, 1373 K
ORNL	1573	$3H_2O + 3MFeO_2 \rightarrow 6MOH + 3Fe_2O_3$, 373 K
		$6MOH + 2Fe_3O_4 \rightarrow H_2 + 3MFeO_2 + 2H_2O$, 773 K
		$3Fe_2O_3 \rightarrow 0.5O_2 + 2Fe_3O_4$, 1573 K
Euratom C-Fe	1573	$H_2O + C \rightarrow H_2 + CO$, 973 K
		$3Fe_2O_3 \rightarrow 0.5O_2 + 2Fe_3O_4$, 1573 K
		$CO + 2Fe_3O_4 \rightarrow C + 3Fe_2O_3$, 523 K

Many thermochemical cycles with three steps pose practical problems related to separation of species and chemical recycling because often chemical reactions do not go to completion and side reactions occur. The work required to separate the components in a gaseous mixture can be determined based on Gibbs free energy of the separation process given by

$$\Delta G_{sep} = -RT \ln \left(\Pi_i y_i^{n y_i} \right) \tag{4.79}$$

where y_i is the molar fraction of component i and n is the number of moles of the gas mixture.

The clear result from Eq. (4.79) is that the separation work will be greater when more compounds are involved. However, in some instances, the addition of one more step can be beneficial for thermochemical cycles. Although there is no general rule, four-step thermochemical cycles with carefully selected reactions may be overall better with higher efficiency and lower temperature level requirements, more complete reaction and less separation and work.

Thermochemical cycles with four steps comprise in general a hydrolysis reaction, a hydrogen evolving reaction, an oxygen evolving reaction, and a reagent recycling reaction. The general scheme of a four-step thermochemical cycle is described by the following equation:

$$\begin{cases} H_2O + AB \rightarrow AH_2 + BO \\ AH_2 \rightarrow H_2 + A \\ BO \rightarrow 0.5O_2 + B \\ A + B \rightarrow AB \end{cases} \tag{4.80}$$

Table 4.12 gives the representative four-step thermochemical water splitting cycles. Funk and Reinstorm (1964) developed the first four-step thermochemical cycle based on this type of reaction sequence, known as a vanadium-chlorine cycle. In a similar manner, there are cycles that consist of hydrolysis and hydrogen evolution plus a separate oxygen generation reaction followed by two che`mical steps for reagent recycling as given in the table.

There are many compilations of thermochemical cycles in the open literature. Here the representative cycles were selected from the group work of the authors published in Naterer et al. (2013). Table 4.13 gives the representative five-step cycles and Table 4.14 the six-step cycles.

A major landmark of thermochemical cycle development is represented by the sequence of Mark cycles developed at ISPRA in Italy during the 1970s. According to Savage et al. (1973), the first cycle established at ISPRA was defined by De Beni in 1970 and was called Mark 1. Some active US research centers in the area were the Institute of Gas Technology, the University of Nevada at Las Vegas, U.S. Department of Energy, Sandia National Laboratory, Argonne National Laboratory, Savannah River National Laboratory, National Renewable Energy Laboratory, University of Colorado, General Electric and General Motors.

The cycle ISPRA Mark 1 successfully demonstrated the feasibility of the thermochemical water splitting process at lab-pilot scale. Mark 1 is a four-step cycle with a maximum temperature requirement of ~1000 K. The additional chemical elements involved in the cycle (except H and O) are Br, Ca and Hg. Fig. 4.27 shows a steam temperature versus total stream enthalpy diagram for the Mark 1 cycle.

In Fig. 4.27, state #1 in the diagram represents liquid water at $T_0 = 298.15$ K, while state #16 represents the final product state at T_0 and P_0. In order to drive water to evolve from state #1 to state #16, an enthalpy change of $\Delta H^0 = 285.813$ kJ must occur for 1 mole of liquid water. The dashed line splits the zone between the enthalpy line of reactant and that of the products into two parts, indicating the amount that must be given in the form of the Gibbs free energy (work) and the amount representing $T\Delta S$ (heat). If the process is conducted in an ideal electrolyzer operating at T_0, the amount of necessary Gibbs energy is $\Delta G^0 = 237,172$ kJ/mol, representing ~83% of the total enthalpy required for the process.

The line from 3 to 4 represents the enthalpy line for reactants of the highest temperature reaction in the Mark 1 cycle, which is $2H_2O(g) + CaBr_2(s) \rightarrow Ca(OH)_2(s) + 2HBr(g)$ at 1003 K. The enthalpy line of the products (line passing through state point #5) and the line representing the Gibbs energy requirement of this reaction (dashed line passing through state #4) are also indicated on the diagram. As observed, the highest temperature reaction is spontaneous at 1003 K because its associated Gibbs energy is slightly negative. The processes and state points in the diagram from Fig. 4.27 are described as follows:

- State #1: one mole of liquid water at T_0 and P_0.
- Process 1–2: water heating, boiling and steam superheating up to 473 K
- State #3: mixing of 1 mol H_2O with 1 mole of recycled H_2O and 1 mol $CaBr_2$
- Process 3–4: heating $2H_2O + CaBr_2$ from 473 up to 1003 K
- Process 4–5: reaction $2H_2O(g) + CaBr_2(s) \rightarrow Ca(OH)_2(s) + 2HBr(g)$ at 1003 K
- Process 5–6: cooling reaction products $Ca(OH)_2(s) + 2HBr(g)$ to 523 K
- State #6: separation of 2HBr from $Ca(OH)_2$ at 523 K
- Process 6–7: reacting 2HBr from state #6 with Hg from state #14, at 523 K, according to $2HBr(aq) + Hg(l) \rightarrow H_2(g) + HgBr_2(aq)$.
- State #7: separation of H_2 from $HgBr_2$ from aqueous solution
- State #8: mixing $HgBr_2(s)$ from state #7 with $Ca(OH)_2$ from state #6, at 523 K

TABLE 4.12 Representative Four-Step Thermochemical Cycles

Cycle	T_{max} (K)	Reaction steps
ANL Br-Eu-U	573	$2H_2O + Br_2 + UO_3 \cdot H_2O \rightarrow 0.5O_2 + UO_2 \cdot 3H_2O$, 298 K
		$2EuBr_2 + 2HBr \rightarrow H_2 + 2EuBr_3$, 298 K
		$2EuBr_3 \rightarrow 2EuBr_2 + Br_2$, 573 K
		$UO_2 \cdot 3H_2O \rightarrow 2UO_3 \cdot H_2O + 2HBr + H_2O$, 573 K
UNLV-82	710	$2H_2O + MgI_2 \rightarrow 2HI + Mg(OH)_2$, 500 K
		$2HI \rightarrow H_2 + I_2$, 573 K
		$2MnO_2 \rightarrow 0.5O_2 + Mn_2O_3$, 710 K
		$Mg(OH)_2 + I_2 + Mn_2O_3 \rightarrow MgI_2 + 2MnO_2 + H_2O$, 300 K
UNLV-162	873	$H_2O + CO \rightarrow H_2 + CO_2$, 573 K
		$3UO_3 \rightarrow 0.5O_2 + U_3O_8$, 873 K
		$4CO_2 + U_3O_8 \rightarrow CO + 3UO_2CO_3$, 423 K
		$3UO_2CO_3 \rightarrow 3CO_2 + 3UO_3$, 873 K
LLL-UoC	873	$2H_2O + Cs_2Hg \rightarrow H_2 + Hg + 2CsOH$, 873 K
		$HgO \rightarrow 0.5O_2 + Hg$, 773 K
		$2CsOH \rightarrow Cs_2O + H_2O$, 683 K
		$Cs_2O + 2Hg \rightarrow Cs_2Hg + HgO$, 573 K
LASL Ce-Cl	883	$2H_2O + 2CeClO \rightarrow H_2 + 2CeO_2 + 2HCl$, 723 K
		$H_2O + Cl_2 \rightarrow 0.5O_2 + 2HCl$, 883 K
		$2CeO_2 + 8HCl \rightarrow 2CeCl_3 + Cl_2 + 4H_2O$, 298 K
		$2CeCl_3 + 2H_2O \rightarrow 2CeClO + 4HCl$, 700 K
UNLV-43	900	$2H_2O + 2CO_2 + 2NH_3 + 2NaBr \rightarrow 2NaHCO_3 + 2NH_4Br$, 298 K
		$2Ag + 2NH_4Br \rightarrow H_2 + 2AgBr + 2NH_3$, 700 K
		$2AgBr + Na_2CO_3 \rightarrow 0.5O_2 + 2Ag + CO_2 + 2NaBr$, 900 K
		$2NaHCO_3 \rightarrow CO_2 + Na_2CO_3 + H_2O$, 400 K
Mark 15	923	$4H_2O(g) + 3FeCl_2(s) \rightarrow H_2(g) + Fe_3O_4(s) + 6HCl(g)$ at 923 K
		$H_2O(g) + Cl_2(g) \rightarrow 0.5O_2(g) + 2HCl(g)$ at 873 K
		$Fe_3O_4(s) + 8HCl(g) \rightarrow FeCl_2(s) + 2FeCl_3(s) + 4H_2O(g)$ at 573 K
		$2FeCl_3(s) \rightarrow 2FeCl_2(s) + Cl_2(g)$ at 673 K
GM-Funk Cl-V	973	$H_2O + Cl_2 \rightarrow 0.5O_2 + 2HCl$, 883 K
		$2HCl + 2VCl_2 \rightarrow H_2 + 2VCl_3$, 298 K
		$2VCl_3 \rightarrow VCl_4 + VCl_2$, 973 K
		$2VCl_4 \rightarrow Cl_2 + 2VCl_3$, 298 K
NCLI	973	$H_2O + MgI_2 \rightarrow 2HI + MgO$, 673 K
		$2HI \rightarrow H_2 + I_2$, 573 K
		$0.2Mg(IO_3)_2 \rightarrow 0.5O_2 + 0.2I_2 + 0.2MgO$, 973 K
		$1.2I_2 + 1.2MgO \rightarrow MgI_2 + 0.2Mg(IO_3)_2$, 423 K
ANL	1000	$2(H_2O + NH_3 + CO_2 + NaBr) \rightarrow 2(NaHCO_3 + NH_4Br)$, 298 K
		$2NH_4Br + 2Ag \rightarrow H_2 + 2NH_3 + 2AgBr$, 750 K
		$2AgBr + Na_2CO_3 \rightarrow 0.5O_2 + 2Ag + 2NaBr + CO_2$, 1000 K
		$2NaHCO_3 \rightarrow Na_2CO_3 + H_2O$, 400 K

TABLE 4.12   Representative Four-Step Thermochemical Cycles—cont'd

Cycle	T_{max} (K)	Reaction steps
UNLV-164	1000	$H_2O + CH_4 \rightarrow 3H_2 + CO$, 973 K
		$ZnSO_4 \rightarrow 0.5O_2 + SO_2 + ZnO$, 1000 K
		$CO + 2H_2 \rightarrow CH_3OH$, 500 K
		$CH_3OH + SO_2 + ZnO \rightarrow CH_4 + ZnSO_4$, 600 K
Mark 1	1003	$2HBr(aq) + Hg(l) \rightarrow H_2(g) + HgBr_2(aq)$ at 523 K
		$HgO(s) \rightarrow 0.5O_2(g) + Hg(g)$ at 873 K
		$2H_2O(g) + CaBr_2(s) \rightarrow Ca(OH)_2(s) + 2HBr(g)$ at 1003 K
		$HgBr_2(s) + Ca(OH)_2(s) \rightarrow CaBr_2(s) + HgO(s) + H_2O(g)$ at 473 K
UT-3	1023	$H_2O + CaBr_2 \rightarrow CaO + 2HBr$, 1023 K
		$3FeBr_2 + 4H_2O \rightarrow H_2 + Fe_3O_4 + 6HBr$, 873 K
		$CaO + Br_2 \rightarrow 0.5O_2 + CaBr_2$, 873 K
		$Fe_3O_4 + 8HBr \rightarrow 3FeBr_2 + Br_2 + 4H_2O$, 573 K
UNLV-3	1023	$H_2O + CaBr_2 \rightarrow CaO + 2HBr$, 1023 K
		$2HBr + Hg \rightarrow H_2 + HgBr_2$, 473 K
		$2HgO \rightarrow 0.5O_2 + 2Hg$, 773 K
		$CaO + HgBr_2 \rightarrow CaBr_2 + HgO$, 298 K
Mark 3	1073	$2HCl(g) + 2VOCl(s) \rightarrow H_2(g) + 2VOCl_2(s)$ at 443 K
		$H_2O(g) + Cl_2(g) \rightarrow 0.5O_2(g) + 2HCl(g)$ at 1073 K
		$4VOCl_2(s) \rightarrow 2VOCl(s) + 2VOCl_3(g)$ at 873 K
		$2VOCl_3(g) \rightarrow 2VOCl_2(s) + Cl_2(g)$ at 473 K
Mark 4	1073	$H_2S(g) \rightarrow H_2(g) + 0.5S_2(g)$ at 1073 K
		$H_2O(g) + Cl_2(g) \rightarrow 0.5O_2(g) + 2HCl(g)$ at 1073 K
		$2HCl(g) + S(s) + 2FeCl_2(s) \rightarrow H_2S(g) + 2FeCl_3(s)$ at 373 K
		$2FeCl_3(g) \rightarrow 2FeCl_2(s) + Cl_2(g)$ at 693 K
Mark 6	1073	$2HCl(g) + 2CrCl_2(s) \rightarrow H_2(g) + 2CrCl_3(s)$ at 433 K
		$H_2O(g) + Cl_2(g) \rightarrow 0.5O_2(g) + 2HCl(g)$ at 1073 K
		$2CrCl_3(s) + 2FeCl_2(s) \rightarrow 2CrCl_2(s) + 2FeCl_3(g)$ at 973 K
		$2FeCl_3(g) \rightarrow 2FeCl_2(s) + Cl_2(g)$ at 623 K
UNLV-85	1100	$2H_2O + MgI_2 \rightarrow 2HI + Mg(OH)_2$, 500 K
		$2HI \rightarrow H_2 + I_2$, 573 K
		$0.5As_2O_5 \rightarrow 0.5O_2 + 0.25As_4O_6$, 1100 K
		$0.25As_4O_6 + I_2 + Mg(OH)_2 \rightarrow 0.5As_2O_5 + MgI_2 + H_2O$, 300 K
UNLV-95	1100	$H_2O + WO_2 \rightarrow H_2 + WO_3$, 400 K
		$1.5Ce(SO_4)_2 \rightarrow 0.5O_2 + 1.5CeO_2 + 3SO_2 + O_2$, 1100 K
		$2CeO_2 + 3SO_2 + O_2 \rightarrow Ce_2(SO_4)_3$, 825 K
		$Ce_2(SO_4)_3 + WO_3 \rightarrow 0.5CeO_2 + 1.5Ce(SO_4)_2 + WO_2$, 680 K
Mark 2C	1123	$H_2O(g) + CO(g) \rightarrow H_2(g) + CO_2(g)$ at 773 K
		$2MnO_2(s) \rightarrow 0.5O_2(g) + Mn_2O_3(s)$ at 873 K
		$Mn_2O_3(s) + 2Na_2CO_3(s) \rightarrow 2NaO \cdot MnO_2(s) + CO_2(g) + CO(g)$ at 1123 K
		$nH_2O(l) + 2Na_2O \cdot MnO_2(s) + 2CO_2(g) \rightarrow 2Na_2CO_2 \cdot nH_2O(aq) + 2MnO_2(s)$ at 373K

Continued

TABLE 4.12 Representative Four-Step Thermochemical Cycles—cont'd

Cycle	T_{max} (K)	Reaction steps
Mark 12	1123	$2NH_4I(s) \rightarrow H_{2(g)} + 2NH_3(g) + I_2(g)$ at 973 K
		$ZnSO_4(s) \rightarrow 0.5O_2(g) + ZnO(s) + SO_2(g)$ at 1123 K
		$2H_2O(l) + I_2(s) + SO_2(g) + 4NH_3(g) \rightarrow 2NH_4I(s) + (NH_4)_2SO_4(s)$ at 323 K
		$(NH_4)_2SO_4(s) + ZnO(s) \rightarrow ZnSO_4(s) + 2NH_3(g) + H_2O(g)$ at 773 K
GA-23	1123	$H_2O + 1.5SO_2 \rightarrow H_2SO_4 + 0.5S$, 298 K
		$H_2S \rightarrow H_2 + S$, 1073 K
		$H_2SO_4 \rightarrow 0.5O_2 + SO_2 + H_2O$, 1123 K
		$1.5S + H_2O \rightarrow H_2S + 0.5SO_2$, 973 K
UNLV-127	1123	$H_2O + Cl_2 \rightarrow 0.5O_2 + 2HCl$, 1123 K
		$2HCl + 2VOCl \rightarrow H_2 + 2VOCl_2$, 443 K
		$4VOCl_2 \rightarrow 2VOCl + 2VOCl_3$, 873 K
		$2VOCl_3 \rightarrow Cl_2 + 2VOCl_2$, 473 K
UNLV-142	1123	$H_2O + CH_4 \rightarrow 3H_2 + CO$, 1100 K
		$H_2SO_4 \rightarrow 0.5O_2 + SO_2 + H_2O$, 1123 K
		$CO + 2H_2 \rightarrow CH_2OH$, 500 K
		$CH_3OH + SO_2 + H_2O \rightarrow CH_4 + H_2SO_4$, 473 K
UNLV-88	1160	$H_2O + 2/3ScI_3 \rightarrow 2HI + 1/3Sc_2O_3$, 500 K
		$2HI \rightarrow H_2 + I_2$, 573 K
		$Co_3O_4 \rightarrow 0.5O_2 + 3CoO$, 1160 K
		$3CoO + I_2 + 1/3Sc_2O_3 \rightarrow Co_3O_4 + 2/3ScI_3$, 300 K
UNLV-30	1173	$H_2O + 0.5Fe_3O_4 + 1.5SO_2 \rightarrow H_2 + 1.5FeSO_4$, 298 K
		$SO_3 \rightarrow 0.5O_2 + SO_2$, 1073 K
		$0.75Fe_2O_3 + 0.25SO_2 \rightarrow 0.5Fe_3O_4 + 0.25SO_3$, 298 K
		$1.5FeSO_4 \rightarrow 0.75Fe_2O_3 + 0.75SO_2 + 0.75SO_3$, 1173 K
UNLV-81	1173	$2H_2O + MgI_2 \rightarrow 2HI + Mg(OH)_2$, 500 K
		$2HI \rightarrow H_2 + I_2$, 573 K
		$2CuO \rightarrow 0.5O_2 + Cu_2O$, 1173 K
		$Cu_2O + I_2 + Mg(OH)_2 \rightarrow 2CuO + MgI_2 + H_2O$, 298 K
Mark 1C	1173	$4HBr(aq) + Cu_2O(s) \rightarrow H_2(g) + 2CuBr_2(aq) + H_2O(l)$ at 373 K
		$CuO(s) \rightarrow 0.5O_2(g) + Cu_2O(s)$ at 1173 K
		$4H_2O(g) + 2CaBr_2(s) \rightarrow 2Ca(OH)_2(s) + 4HBr(g)$ at 730 K
		$2CuBr_2(aq) + 2Ca(OH)_2(aq) \rightarrow 2CuO(s) + 2CaBr_2(aq) + 2H_2O(l)$ at 373 K
UNLV-150	1273	$H_2O + Ge \rightarrow H_2 + GeO$, 1173 K
		$Co_3O_4 \rightarrow 0.5O_2 + 3CoO$, 1273 K
		$GeO + SO_2 \rightarrow Ge + SO_3$, 1073 K
		$3CoO + SO_3 \rightarrow Co_3O_4 + SO_2$, 673 K
NCLI	1273	$H_2O + CaI_2 \rightarrow 2HI + CaO$, 1273 K
		$2HI \rightarrow H_2 + I_2$, 573 K
		$0.2Ca(IO_3)_2 \rightarrow 0.5O_2 + 0.2I_2 + 0.2CaO$, 973 K
		$1.2CaO + 1.2I_2 \rightarrow CaI_2 + 0.2Ca(IO_3)_2$, 300 K

TABLE 4.12 Representative Four-Step Thermochemical Cycles—cont'd

Cycle	T_{max} (K)	Reaction steps
UNLV-144	1293	$H_2O + 0.25BaS \rightarrow H_2 + 0.25BaSO_4$, 1143 K
		$FeSO_4 \rightarrow 0.5O_2 + 0.5Fe_2O_3 + 2SO_2$, 1200 K
		$Fe_2O_3 + 2.5SO_2 \rightarrow 2Fe_{SO_4} + 0.5S$, 700 K
		$0.25BaSO_4 + 0.5S \rightarrow 0.25BaS + 0.5SO_2$, 1293 K
UNLV-99	1300	$2/3(H_2O + Ag_2O + K_2CrO_4) \rightarrow 2/3(Ag_2CrO_4 + 2KOH)$, 298 K
		$1/3(H_2O + Cr_2O_3 + 4KOH) \rightarrow H_2 + 2/3K_2CrO_4$, 1300 K
		$8/3Ag_2CrO_4 \rightarrow 0.5O_2 + 16/3Ag + 1/3Cr_2O_3 + 1/3O_2$, 980K
		$4/3Ag + 1/3O_2 \rightarrow 2Ag_2O$, 410 K
UNLV-140	1300	$H_2O + 3FeO \rightarrow H_2 + Fe_3O_4$, 673 K
		$H_2O + Cl_2 \rightarrow 0.5O_2 + 2HCl$, 1300 K
		$Fe_3O_4 + 8HCl \rightarrow 3FeCl_2 + Cl_2 + 4H_2O$, 653 K
		$3FeCl_2 + 3H_2O \rightarrow 3FeO + 6HCl$, 963 K
UNLV-92	1400	$H_2O + WO_2 \rightarrow H_2 + WO_3$, 400 K
		$MgSO_4 \rightarrow 0.5O_2 + MgO + SO_2$, 1400 K
		$MgO + SO_2 \rightarrow MgSO_3$, 610 K
		$WO_3 + MgSO_3 \rightarrow WO_2 + MgSO_4$, 800 K
UNLV-99	1400	$H_2O + Na_2SO_3 \rightarrow H_2 + Na_2SO_4$, 500 K
		$MgSO_4 \rightarrow 0.5O_2 + MgO + SO_2$, 1400 K
		$MgO + SO_2 \rightarrow MgSO_3$, 610 K
		$MgSO_3 + Na_2SO_4 \rightarrow MgSO_4 + Na_2SO_3$, 298 K
UNLV-115	1400	$H_2O + MgI_2 \rightarrow 2HI + MgO$, 673 K
		$2HI \rightarrow H_2 + I_2$, 573 K
		$MgSO_4 \rightarrow 0.5O_2 + MgO + SO_2$, 1400 K
		$2MgO + SO_2 + I_2 \rightarrow MgSO_4 + MgI_2$, 373 K
Aerojet General	1473	$2H_2O + 2Cs \rightarrow H_2 + 2CsOH$, 373 K
		$Cs_2O \rightarrow 0.5O_2 + 2Cs$, 1473 K
		$2CsOH + 1.5O_2 \rightarrow 2CsO_2 + H_2O$, 773 K
		$2CsO_2 \rightarrow Cs_2O + 1.5O_2$, 973 K

- Process 8–9: cooling $HgBr_2(s)$ and $Ca(OH)_2$ from 523 to 473 K
- Process 9–10: reaction $HgBr_2(s) + Ca(OH)_2(s) \rightarrow CaBr_2(s) + HgO(s) + H_2O(g)$ at 473 K
- State #11: extraction of $HgO(s)$ from product stream at 473 K
- Process 11–12: heating HgO from 473 to 873 K
- Process 12–13: reaction $HgO(s) \rightarrow 0.5O_2(g) + Hg(g)$ at 873 K
- Process 13–14: cooling reaction products $0.5O_2$ and Hg from 873 to 523 K
- State #14: separation of $0.5O_2$ from Hg at 523 K
- Process #15–16: cooling 1 mole H_2 and 0.5 moles O_2 from 523 K to T_0.

Beside its basic version, the Mark 1 cycle has three additional variants: 1B, 1C, and 1S. In version 1B, the hydrogen evolution reaction occurs at 393 K, and one additional chemical reaction is required to recycle the chemicals. In the Mark 1S cycle, the number of steps is reduced to three. The need for calcium compounds is eliminated; however, the maximum temperature required is slightly increased to 1073 K. Although shown to be technically feasible, the Mark 1 cycle was abandoned because several side reactions occurred, and there was a very large quantity of mercury that had to be handled for an industrial-scale plant to safely operate.

TABLE 4.13 Representative Five-Step Thermochemical Cycles

Cycle	T_{max} (K)	Reaction steps
GE Cu-Cl-Mg	773	$2H_2O + MgCl_2 \rightarrow 2HCl + Mg(OH)_2$, 723 K
		$2HCl + 2Cu \rightarrow H_2 + 2CuCl$, 373 K
		$Mg(OH)_2 + Cl_2 \rightarrow 0.5O_2 + MgCl_2 + H_2O$, 298 K
		$4CuCl \rightarrow 2Cu + 2CuCl_2$, 373 K
		$2CuCl_2 \rightarrow 2CuCl + Cl_2$, 773 K
LLL–UoC As-C	973	$H_2O + CH_4 \rightarrow 3H_2 + CO$, 973 K
		$0.5As_2O_5 \rightarrow 0.5O_2 + 0.5As_2O_3$, 973 K
		$CO + 2H_2 \rightarrow CH_3OH$, 500 K
		$CH_3OH + As_2O_4 \rightarrow CH_4 + AS_2O_5$, 500 K
		$0.5As_2O_3 + 0.5As_2O_5 \rightarrow As_2O_4$, 298 K
LLL–UoC: K-Se-V	973	$2H_2O + K_2Se \rightarrow 2KOH + H_2Se$, 373 K
		$H_2Se \rightarrow H_2 + Se$, 473 K
		$V_2O_5 \rightarrow 0.5O_2 + V_2O_4$, 773 K
		$V_2O_4 + 0.5SeO_2 \rightarrow Ve_2O_5 + 0.5Se$, 500 K
		$1.5Se + KOH \rightarrow K_2Se + 0.5SeO_2 + H_2O$, 973 K
ANL I-K-N	973	$H_2O + NO_2 + 0.5O_2 \rightarrow 2HNO_3$, 298 K
		$2NH_4I \rightarrow H_2 + 2NH_3 + I_2$, 773 K
		$I_2 + 2KNO_3 \rightarrow 0.5O_2 + 2KI + 2NO_2 + 0.5O_2$, 973 K
		$2KI + 2NH_4NO_3 \rightarrow 2KNO_3 + 2NH_4I$, 298 K
		$2HNO_3 + 2NH_3 \rightarrow 2NH_4NO_3$, 298 K
Mark 1B	1003	$2HBr(aq) + Hg_2Br_2(s) \rightarrow H_2(g) + 2HgBr_2(s)$ at 393 K
		$HgO(s) \rightarrow 0.5O_2(g) + Hg(g)$ at 873 K
		$2H_2O(g) + CaBr_2(s) \rightarrow Ca(OH)_2(s) + 2HBr(g)$ at 1003 K
		$HgBr_2(s) + Hg(l) \rightarrow Hg_2Br_2(s)$ at 393 K
		$HgBr_2(s) + Ca(OH)_2(s) \rightarrow CaBr_2(s) + HgO(s) + H_2O(g)$ at 473 K
Mark 14	1003	$4H_2O(g) + 3FeCl_2(s) \rightarrow H_2(g) + Fe_3O_4(s) + 6HCl(g)$ at 923 K
		$H_2O(g) + Cl_2(g) \rightarrow 0.5O_2(g) + 2HCl(g)$ at 1003 K
		$Fe_3O_4(s) + 0.5Cl_2(g) \rightarrow 1/3FeCl_2(g) + 4/3Fe_2O_3(s)$ at 573
		$4/3Fe_2O_3(s) + 8HCl(g) \rightarrow 8/3FeCl_2(s) + 4H_2O(g)$ at 443 K
		$3FeCl_2(s) \rightarrow 3FeCl_2(s) + 1.5Cl_2(g)$ at 553 K
GE-Agnes	1023	$4H_2O + 3FeCl_2 \rightarrow H_2 + Fe_3O_4 + 6HCl$, 1023 K
		$Mg(OH)_2 + Cl_2 \rightarrow 0.5O_2 + MgCl_2 + H_2O$, 363 K
		$MgCl_2 + 2H_2O \rightarrow Mg(OH)_2 + 2HCl$, 623 K
		$Fe_3O_4 + 8HCl \rightarrow FeCl_2 + 2FeCl_3 + 4H_2O$, 383 K
		$2FeCl_3 \rightarrow 2FeCl_2 + Cl_2$, 573 K
Mark 6C	1073	$2HCl(g) + 2CrCl_2(s) \rightarrow H_2(g) + 2CrCl_3(s)$ at 443 K
		$H_2O(g) + Cl_2(g) \rightarrow 0.5O_2(g) + 2HCl(g)$ at 1073 K
		$2CrCl_3(s) + 2FeCl_2(s) \rightarrow 2CrCl_2(s) + 2FeCl_3(g)$ at 973 K
		$2FeCl_3(s) + 2CuCl(s) \rightarrow 2FeCl_2(s) + 2CuCl_2(s)$ at 423 K
		$2CuCl_2(s) \rightarrow 2CuCl(l) + Cl_2(g)$ at 773 K

TABLE 4.13 Representative Five-Step Thermochemical Cycles—cont'd

Cycle	T_{max} (K)	Reaction steps
Mark 7	1073	$4H_2O(g) + 3FeCl_2(s) \rightarrow H_2(g) + Fe_3O_4(s) + 6HCl(g)$ at 923 K
		$1.5H_2O(g) + 1.5Cl_2(g) \rightarrow 0.5O_2(g) + 3HCl(g) + 0.25O_2(g)$ at 1073 K
		$Fe_3O_4(s) + 0.25O_2(g) \rightarrow 1.5Fe_2O_3(s)$ at 623 K
		$1.5Fe_2O_3(s) + 9HCl(g) \rightarrow 3FeCl_3(s) + 4.5H_2O(g)$ at 423 K
		$3FeCl_2(g) \rightarrow 3FeCl_2(s) + 1.5Cl_2(g)$ at 693 K
Hitachi	1073	$2H_2O + 2NaI + 2NH_3 + 2CO_2 \rightarrow 2NaHCO_3 + 2NH_4I$, 733 K
		$2NH_4I + Ni \rightarrow H_2 + NiI_2 + 2NH_3$, 773 K
		$Na_2CO_3 + I_2 \rightarrow 0.5O_2 + 2NaI + CO_2$, 973 K
		$2NaHCO_3 \rightarrow Na_2CO_3 + CO_2 + H_2O$, 573 K
		$NiI_2 \rightarrow Ni + I_2$, 1073 K
MHI	1073	$4H_2O + 3FeCl_2 \rightarrow H_2 + Fe_3O_4 + 6HCl$, 873 K
		$H_2O + Cl_2 \rightarrow 0.5O_2 + HCl$, 1073 K
		$Fe_3O_4 + 8HCl \rightarrow FeCl_2 + 2FeCl_3 + 4H_2O$, 373 K
		$2FeCl_3 + 2CuCl \rightarrow 2FeCl_2 + 2CuCl_2$, 373 K
		$2CuCl_2 \rightarrow 2CuCl + Cl_2$, 773 K
JAERI	1123	$2H_2O + I_2 + SO_2 \rightarrow 2HI + H_2SO_4$, 298 K
		$2Ni + 2HI + H_2SO_4 \rightarrow 2H_2 + NiI_2 + NiSO_4$, 600 K
		$NiSO_4 \rightarrow 0.5O_2 + NiO + SO_2$, 1123 K
		$NiO + H_2 \rightarrow Ni + H_2O$, 673 K
		$NiI_2 \rightarrow Ni + I_2$, 873 K
Mark 5	1173	$2HBr(g) + Hg(l) \rightarrow H_2(g) + HgBr_2(s)$ at 873 K
		$HgO(s) \rightarrow 0.5O_2(g) + Hg(g)$ at 873 K
		$H_2O(g) + CaBr_2(s) + CO_2(g) \rightarrow CaCO_3(s) + 2HBr(g)$ at 873 K
		$CaCO_3(s) \rightarrow CaO(s) + CO_2(g)$ at 1173 K
		$HgBr_2(s) + CaO(s) + nH_2O(l) \rightarrow CaBr_2 \cdot nH_2O(aq) + HgO(s)$ at 473 K
LLL-UoC Cl-S-Se-Zn	1200	$H_2O + ZnCl_2 \rightarrow 2HCl + ZnO$, 883 K
		$H_2Se \rightarrow H_2 + Se$, 473 K
		$ZnSO_4 \rightarrow 0.5O_2 + SO_2 + ZnO$, 1200 K
		$2HCl + ZnSe \rightarrow H_2Se + ZnCl_2$, 350 K
		$Se + SO_2 + 2ZnO \rightarrow ZnSe + ZnSO_4$, 883 K
Mark 7A	1273	$4H_2O(g) + 3FeCl_2(s) \rightarrow H_2(g) + Fe_3O_4(s) + 6HCl(g)$ at 923 K
		$0.5Fe_2O_3(s) + 1.5Cl_2(g) \rightarrow 0.5O_2(g) + FeCl_3(g) + 0.25O_2(g)$ at 1273 K
		$Fe_3O_4(s) + 0.25O_2(g) \rightarrow 1.5Fe_2O_3(s)$ at 623 K
		$Fe_2O_3(s) + 6HCl(g) \rightarrow 2FeCl_3(s) + 3H_2O(g)$ at 423 K
		$3FeCl_2(g) \rightarrow 3FeCl_2(s) + 1.5Cl_2(g)$ at 693 K
Mark 7B	1273	$4H_2O(g) + 3FeCl_2(s) \rightarrow H_2(g) + Fe_3O_4(s) + 6HCl(g)$ at 923 K
		$1.5Fe_2O_3(s) + 4.5Cl_2(g) \rightarrow 0.5O_2(g) + 3FeCl_2(s) + 1.75O_2(g)$ at 1273 K
		$Fe_3O_4(s) + 0.25O_2(g) \rightarrow 1.5Fe_2O_3(s)$ at 623 K
		$6HCl(g) + 1.5O_2(g) \rightarrow 3Cl_2(g) + 3H_2O(g)$ at 673 K
		$3FeCl_2(g) \rightarrow 3FeCl_2(s) + 1.5Cl_2(g)$ at 693 K

Continued

TABLE 4.13 Representative Five-Step Thermochemical Cycles—cont'd

Cycle	T_{max} (K)	Reaction steps
IGT A-2	1273	$4H_2O + 3Fe \rightarrow H_2 + Fe_3O_4 + 3H_2$, 773 K
		$Fe_3O_4 + 4.5Cl_2 \rightarrow 0.5O_2 + 3FeCl_3 + 1.5O_2$, 1273 K
		$3FeCl_3 \rightarrow 3FeCl_2 + 1.5Cl_2$, 623 K
		$3FeCl_2 + 3H_2 \rightarrow 3Fe + 6HCl$, 1273 K
		$6HCl + 1.5O_2 \rightarrow 3Cl_2 + 3H_2O$, 773 K
LLL-UoC	1373	$0.5H_2O + 0.25MgSe \rightarrow 0.25H_2Se + 0.25Mg(OH)_2$, 373K
		$0.5H_2O + 1/8MgSe \rightarrow 0.5H_2 + 1/8MgSeO_4$, 673 K
Mg-Se		$0.5H_2Se \rightarrow 0.5H_2 + 0.5Se$, 473 K
		$0.25MgSeO_4 \rightarrow 0.5O_2 + 0.25MgSe$, 1373 K
		$0.25Mg(OH)_2 + 0.5Se \rightarrow 0.25H_2Se + 1/8MgSe + 1/8MgSeO_4$, 773 K
NCLI	1400	$H_2O + MgI_2 \rightarrow 2HI + MgO$, 673 K
		$2HI \rightarrow H_2 + I_2$, 573 K
		$MgSO_4 \rightarrow 0.5O_2 + MgO + SO_2$, 1400 K
		$I_2 + SO_2 + 2H_2O \rightarrow 2HI + H_2SO_4$, 373 K
		$2HI + H_2SO_4 + 2MgO \rightarrow MgI_2 + MgSO_4 + 2H_2O$, 350 K

TABLE 4.14 Representative Six-Step Thermochemical Cycles

Cycle	T_{max} (K)	Reaction steps
GE, Catherine process	973	$2H_2O + 2LiI \rightarrow 2HI + 2LiOH$, 873 K
		$2HI + Ni \rightarrow H_2 + NiI_2$, 423 K
		$1/3KIO_3 \rightarrow 0.5O_2 + 1/3KI$, 923 K
		$I_2 + 2LiOH \rightarrow 5/3LiI + 1/3LiIO_3 + H_2O$, 463 K
		$1/3LiO_3 + 1/3KI \rightarrow 1/3KIO_3 + 1/3LiI$, 298 K
		$NiI_2 \rightarrow Ni + I_2$, 973 K
Mark 10	1123	$2NH_4I(s) \rightarrow H_2(g) + 2NH_3(g) + I_2(g)$ at 900 K
		$SO_3(g) \rightarrow 0.5O_2(g) + SO_2(g)$ at 1123 K
		$2H_2O(l) + I_2(s) + SO_2(g) + 4NH_3 \rightarrow 2NH_4I(s) + (NH_4)_2SO_4(s)$ at 323 K
		$(NH_4)_2SO_4(s) + Na_2SO_4(s) \rightarrow 2NaHSO_4(s) + 2NH_3(g)$ at 673 K
		$2NaHSO_4(s) \rightarrow Na_2S_2O_7(s) + H_2O(g)$ at 673 K
		$Na_2S_2O_7(s) \rightarrow Na_2SO_4(s) + SO_3(g)$ at 825 K
UNLV-116	1123	$H_2O + CH_4 \rightarrow H_2 + CO + 2H_2$, 973 K
		$H_2SO_4 \rightarrow 0.5O_2 + SO_2 + H_2O$, 1123 K
		$CH_3I + HI \rightarrow CH_4 + I_2$, 298 K
		$CH_3OH + HI \rightarrow CH_3I + H_2O$, 298 K
		$CO + 2H_2 \rightarrow CH_3OH$, 500 K
		$I_2 + SO_2 + 2H_2O \rightarrow 2HI + H_2SO_4$, 373 K
Hitachi	1223	$H_2O + 2NH_3 + 2CuCl \rightarrow Cu_2O + 2NH_4Cl$, 353 K
		$2NH_4Cl + 2Cu \rightarrow H_2 + 2CuCl + 2NH_3$, 873 K
		$2CuO \rightarrow 0.5O_2 + Cu_2O$, 1223 K
		$2Cu_2O + 2H_2SO_4 \rightarrow 2Cu + 2CuSO_4 + 2H_2O$, 373 K
		$2CuSO_4 \rightarrow 2CuO + 2SO_3$, 1123 K
		$2SO_3 + 2H_2O \rightarrow 2H_2SO_4$, 573 K

FIG. 4.27 Representation of stream enthalpy versus temperature in Mark 1 thermochemical cycle.

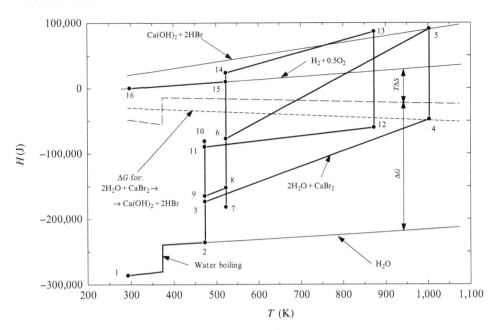

The cycle Mark 1C eliminates the use of mercury and reduces significantly the ecological risks in case of accidents in a large-scale chemical plant. This cycle uses copper compounds instead of mercury compounds. The main drawback of the Mark 1C variant with respect to the basic Mark 1 cycle is the higher temperature level requirement, which is increased to 1173 K. The Mark 2 cycles are based on manganese and sodium elements.

Mark 3 uses chlorine and vanadium and requires a maximum temperature of 1073 K. This cycle uses a prototypical oxygen evolving reaction, namely

$$H_2O(g) + Cl_2(g) \rightarrow 0.5O_2(g) + 2HCl(g)$$

that is also employed in many other cycles. This reaction is the reverse Deacon reaction, also mentioned in the previous section. Sulfur, iron and chlorine are the elements used in the Mark 4 cycle, whereas the same oxygen evolving reaction as with Mark 3 is employed. Mark 5 was a five-step trial at Ispra, but was not pursued because it used mercury compounds. Mark 6 cycles with two versions are based on chromium, chlorine and iron.

According to Beghi (1986), three iron-chlorine cycles were also developed as five-step reaction processes: Mark 7, 7A, and 7B. Iron-chlorine cycles refer to a second phase of research and development on thermochemical water splitting at ISPRA. In this phase, a manganese-chlorine cycle, denoted Mark 8, was studied with three reaction steps. This research phase culminated with the development of a Mark 9 cycle with three reaction steps. This cycle, together with Mark 15 (also based on iron-chloride), has the lowest temperature requirement of 920 K. A hydrogen yield of more than 80% was obtained in a continuous bench-scale unit. For the Mark 15 cycle, a semi-continuous reactor for the hydrogen evolving reaction was developed and tests were conducted for a production rate of 25 standard liters of hydrogen per hour with 90% yield. The reverse Deacon reaction was successfully demonstrated at 873 K. The Mark 14 cycle is one of the same iron-chlorine series with a maximum temperature of 1003 K. However, the researchers at ISPRA abandoned Mark 14 and Mark 15 cycles because they found that the thermal efficiency was not satisfactory (~20% for Mark 14 and ~35% for Mark 15).

The last phase of research at ISPRA, starting with Mark 10, focused on sulfur family cycles, which are a six reaction step process, followed by the development of Mark 12, which is similar to Mark 10, but it includes an additional chemical element (Zn) in a four-step process operating at a similar temperature. The last cycle in the Mark series is the sixteenth cycle, which is the well-known, three reaction step, sulfur-iodine (S-I) cycle. This was further developed by General Atomics. The efficiency of the Mark 16 (sulfur-iodine) cycle is in the order of ~44%. This is perhaps the most developed thermochemical cycle worldwide and for this reasons, a detailed section on the S-I cycle is provided later in this chapter.

Another promising cycle for nuclear hydrogen production is the UT-3 cycle developed at the University of Tokyo that includes four reaction steps, requires a heat source of 1023 K and has a theoretical efficiency of 45%. The prominent examples among the "pure" thermochemical cycles with the lowest temperature of heat source are the five-step cycle of General Electric with Cl, Cu, and Ng compounds (773 K), the three-step cycle of Shell operating with Cu and S compounds (773 K), the UNLV-82 cycle of four steps that requires 710 K, and the heavy element halide cycle of the Argonne National Laboratory with Br, Eu, and U that requires 573 K.

4.4.5 Sulfur-Iodine Cycle

The sulfur-iodine cycle is a pure thermochemical water splitting process consisting of three steps as defined in Table 4.11 under Mark 16 cycle. This cycle was initiated at ISPRA and is currently studied in many research centers worldwide. General Atomics developed the cycle on a small-plant scale in 1970s and early 1980s. There is no solid chemical involved in the cycle, the involved phases being either liquid or solid. The sulfur-iodine cycle comprises a high temperature endothermic reaction of sulfuric acid decomposition, which is conducted in bayonet-type heat exchanger/reactors at approximately 1123 K. This thermolysis reaction is as follows:

$$H_2SO_4(g) \rightarrow 0.5O_2(g) + SO_2(g) + H_2O(g) \text{ at } 1123 \text{ K} \tag{4.81}$$

The oxygen evolves as gas and separates from the reaction mixture of Eq. (4.81), which is cooled at temperature below 393 K where water condenses and SO_2 dissolves in water. The condensed products are combined with iodine in solution, and sulfuric acid and hydriodic acid are spontaneously formed according to the following hydrolysis reaction

$$2H_2O(l) + SO_2(aq) + I_2(aq) \rightarrow H_2SO_4(aq) + 2HI(aq) \text{ at less than } 393 \text{ K} \tag{4.82}$$

Equation (4.77) is known as the Bunsen reaction. The hydriodic acid is separated from the reaction mixture and decomposed to evolve hydrogen in a third step at 573 K as follows:

$$2HI(g) \rightarrow H_2(g) + I_2(g) \text{ at } 573 \text{ K} \tag{4.83}$$

The cycle is represented in a simplified manner, as shown in Fig. 4.28. Each of the three reaction steps poses special problems regarding the chemical separation, avoiding side products, and in addition there are important issues regarding heat recycling within the cycle. The system requires some separation processes. Steam and SO_2 must be separated as well as possible from the reaction mixture of the thermolysis process of sulfuric acid. Hydriodic acid and sulfuric acid must be separated from each other at the Bunsen process reactor. Iodine must be separated out of the hydrogen evolving reactor where hydrogen gas and hydriodic acid coexist.

A sulfur-iodine (S-I) plant must have at least nine functional units as shown in Fig. 4.29 where a simplified flowsheet is presented without emphasizing the internal heat regeneration within the plant. In this plant, water is fed to the Bunsen reactor where it spontaneously reacts with iodine and sulfur dioxide and decomposes to release oxygen and generate the mixed acids. The acids—which are in liquid phase—are collected at the bottom of the Bunsen reactor and actively mixed mechanically.

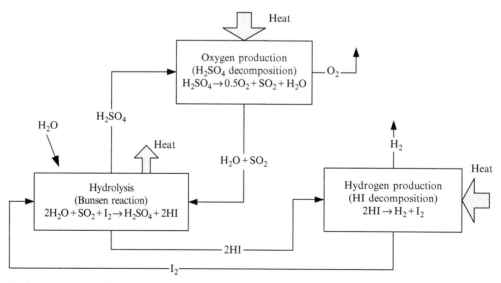

FIG. 4.28   Simplified representation of the thermochemical sulfur-iodine cycle.

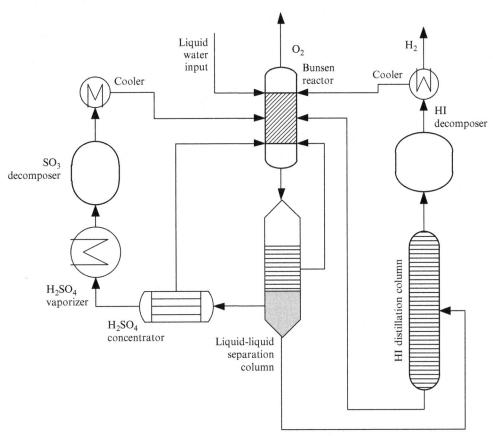

FIG. 4.29 Simplified flowsheet of S-I plant. *Modified from Kasahara S., Kubo S., Hino R., Onuki K., Nomura M., Nakao S. 2007. Flow-sheet study of the thermochemical water splitting iodine-sulfur process for effective hydrogen production.* International Journal of Hydrogen Energy 32:489-496.

In the liquid-liquid separation column, the HI acid collects gravitationally at the bottom. The hydrioidic acid in water does exist in lighter and heavier forms generally represented by the formula HI_x. In fact, only HI is required for further reactions at the HI decomposer. The HI_x must be eliminated. A fractional distillation process is conducted to extract the lighter HI phase such that hydrogen concentration is maximized. The distillation process requires the addition of heat, which is provided through heat transfer to the distillation column. The heavier phase of HI_x is returned back to the Bunsen reactor after a cooling down process to the reactor's temperature level.

The sulfuric acid extracted from the upper part of the liquid-liquid separation column is first concentrated by removing the water that is returned to the Bunsen reactor. Further, the sulfuric acid is vaporized under the heat addition process during which the acid decomposes thermally releasing water according to the following reaction:

$$H_2SO_4 \rightarrow SO_3 + H_2O \tag{4.84}$$

Further heat is added to SO_3, which decomposes and generates oxygen gas according to Eq. (4.81). The separation of oxygen from SO_2 and steam is done easily by a first phase of cooling, when the process stream will reach the temperature of the Bunsen reactor. Water condenses and absorbs SO_2. The process stream is injected into the Bunsen reactor where oxygen gas separates and is collected at the top.

4.4.5.1 Bunsen Reaction Unit

The reaction of sulfur dioxide and iodine in aqueous solution bears the name of the German chemist R. Bunsen who discovered it. Although this reaction has an intrinsic separation in liquid phase, with sulfuric acid being lighter than hydriodic acid, the separation is not perfect and some unavoidable miscibility effects pose difficult technical problems, including sulfur precipitation. One of the major difficulties with the Bunsen reaction within a sulfur-iodine cycle is due to the side reaction that may occur especially when the iodine and sulfate dioxide concentrations are not accurately set. The following are the main side reactions in

$$6HI + H_2SO_4 \rightarrow S + 3I_2 + 4H_2O \tag{4.85a}$$

$$8HI + H_2SO_4 \rightarrow H_2S + 4I_2 + 4H_2O \tag{4.85b}$$

$$6HI + SO_2 \rightarrow H_2S + 3I_2 + 2H_2O \tag{4.85c}$$

$$4HI + SO_2 \rightarrow S + 2I_2 + 2H_2O \tag{4.85d}$$

Iodine must be added well in excess to avoid formation of precipitated sulfur and hydrogen sulfide products that are very undesirable in a thermochemical cycle. If this reactions happens, the cycle cannot be closed. According to Brown et al. (2002), the molar fraction of iodine must be larger than 10%, and the sulfur dioxide molar fraction must be higher than 5–7%, whereas the molar fraction of water is ~90%.

The Bunsen reaction with liquid-liquid phase separation (as proposed by General Atomics) can be represented, according to Ewan and Allen (2005), by the following two-reaction sets evolving at 398 K:

$$\begin{cases} SO_2(g) + 9I_2(l) + 16H_2O(l) \rightarrow H_2SO_4 \cdot 4H_2O(l) + 2HI \cdot 10H_2O \cdot 8I_2(l) \\ H_2SO_4 \cdot 4H_2O(l) + 2HI \cdot 10H_2O \cdot 8I_2(l) \rightarrow H_2SO_4 \cdot 4H_2O(l) + 8I_2 \cdot 10H_2O + 2HI(g) \end{cases} \tag{4.86}$$

A short-circuit electrochemical cell was proposed Dokyia et al. (1979) to conduct Bunsen reaction in a potentially better way. This cell does not require electricity input because the reaction is spontaneous. Fig. 4.30 shows a sort-circuit electrochemical cell for Bunsen reaction. Polymer proton exchange membranes (PEMs) were used in construction of the short-circuit electrochemical cell. However, these membranes allowed for sulfur dioxide crossover that led to formation of elemental sulfur. This is reported as a potential drawback of the device.

The anode is a composite made as a platinized electrode and activated charcoal with very low overpotentials. The cathode is an electro-catalytic electrode with a platinum substrate doped with metal oxides (e.g., TiO_2, Fe_3O_4). Diluted hydroiodic acid must be supplied at the anode, while diluted sulfuric acid must be supplied at the cathode for optimized cell operation. The half-reaction at the cathode is

$$I_2 + 2H^+ + 2e^- \rightarrow 2HI \tag{4.87}$$

while the half-reaction at the anode is

$$2H_2O + SO_2 \rightarrow H_2SO_4 + 2H^+ + 2e^- \tag{4.88}$$

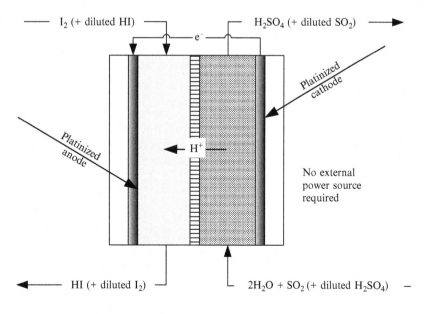

FIG. 4.30 Short-circuit electrochemical cell for Bunsen reaction. *Modified from Dokyia, M., Kameyama, T., Fukuda, K., 1979. Thermochemical hydrogen preparation – Part V. A feasibility study of the sulphur iodine cycle. Int. J. Hydrog. Energy 4, 267–277.*

The overall process of the short electrochemical cell operating at 298 K as devised in Dokyia et al. (1979) is as follows:

$$SO_2 + 0.25(H_2SO_4 \cdot 26H_2O) + 19.2(HI \cdot 6H_2O) + 10.6I_2$$
$$\rightarrow 1.25(H_2SO_4 \cdot 3.6H_2O) + 21.2(2HI \cdot 5.43H_2O) + 9.6I_2$$

(4.89)

The process of Eq. (4.89) is obtained when the cell is supplied with 20% sulfuric acid by weight, whereas the $H_2O/HI/I_2$ mixture feed at the anode has a molar composition of 115/19.2/10.6 that will concentrate to 115/21.2/9.6. Note that the Gibbs free energy of the Bunsen reaction is -41 kJ/mol at 393 K and releases heat equivalent to the reaction enthalpy of -216 kJ/mol.

4.4.5.2 Hydriodic Acid Decomposer

General Atomics devised a process to decompose hydriodic acid in a liquid phase within a modified distillation column, referred to as reactive distillation column. According to Brown et al. (2002), the decomposition can be conducted at 423 K using noble metal catalysts. The liquid phase in the distillation column will act as an iodine absorber for the mixture system $HI/H_2/I_2$. The reactive distillation column performs three functions:

- Extraction of HI from HI_x
- Decomposition of HI
- Separation of H_2 from I_2

There are two major challenges to the reactive distillation column process, namely the concentration of HI prior to its decomposition is difficult and the distillation columns normally have low efficiency. An alternative HI decomposition technology development was recently reported by Kasahara et al. (2007), namely the concentration of HI from HI_x by electro-dialysis technology. For an electro-dialysis system, a detailed flowsheet of the S-I process predicts 40% energy efficiency.

The membrane reactors for HI decomposition were investigated by Dokyia et al. (1979). Recent research by Kasahara et al. (2007) reported that silica membranes can be used for selective hydrogen extraction and enhanced HI decomposition conversion up to a demonstrated conversion of 61.3% at 723 K. With this technology, HI is decomposed into a gas phase. Fig. 4.31 shows the concept of a membrane catalytic reactor. Hydrogen iodide will decompose thermos-catalytically in the annular space of the reactor. Hydrogen will pass through the hydrogen selective membrane. The removal of products allows the forward reaction to go further toward completion by Le Châtelier's Principle. With membrane reactors developed in Japan at NCLI (National Chemical Laboratory for Industry), it was found that the yield may reach >65% molar for one pass conversion.

It is also generally known that a catalytic membrane reactor equipped with a hydrogen selective palladium composite membrane performs better than an equivalent reactor equipped with a porous ceramic membrane due to increases in hydrogen selectivity and separation factor. Membrane reactors have been studied since the late 1960s to further understand the equilibrium shift associated with the removal of products from the product stream.

There is a large body of literature on the study of hydrogen diffusion through palladium. A great deal of study in the area of surface science has been done with particular focus on the palladium-hydrogen system due to the potential of palladium as a hydrogen storage and transport medium. Gas pressurization is beneficial for the reaction and membrane transport. Dokyia et al. (1979) conducted experiments of catalytic decomposition of hydrogen iodide at 523 K under increased pressure. As shown in Fig. 4.32 the yield of the reaction may be increased from 19% up to 45% when the pressure is varied from 3 to 10 MPa. Simultaneously, the production rate of hydrogen corresponds to the reaction conducted at 3 MPa.

FIG. 4.31 Membrane reactor for hydrogen iodide decomposition in the gas phase.

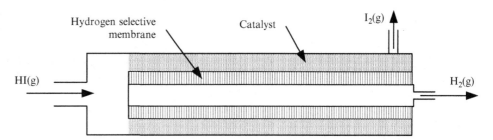

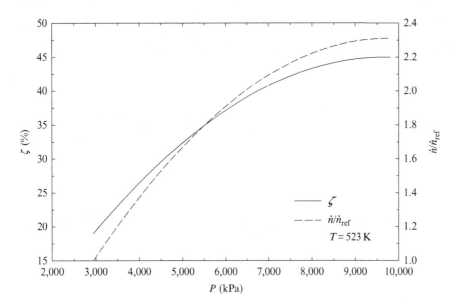

FIG. 4.32 Improvement of reaction yield (ζ) and hydrogen production rate of catalytic hydrogen iodide decomposition at elevated pressures. *Data from Dokyia, M., Kameyama, T., Fukuda, K., 1979. Thermochemical hydrogen preparation – Part V. A feasibility study of the sulphur iodine cycle. Int. J. Hydrog. Energy 4, 267–277.*

4.4.5.3 Sulfuric Acid Decomposer

The decomposition process of sulfuric acid at high temperature is thermo-catalytic and evolves at 1123 K. Iron-based catalysts may be used, although the more expensive platinum catalysts on titanium dioxide supports show much better activity. The process occurs in several stages. First, a dehydration of aqueous sulfuric acid is necessary, which evolves according to the following reaction:

$$1.25(H_2SO_4 \cdot 3.6H_2O) \rightarrow 1.25H_2SO_4 + 4.5H_2O, \text{ at } 298 \text{ K} \tag{4.90}$$

Reaction (4.90) is followed by dehydration of sulfur trioxide, namely

$$1.25H_2SO_4 + 4.5H_2O \rightarrow 5.25SO_3 + 5.75H_2O, \text{ at } 298 \text{ K} \tag{4.91}$$

which is conducted at 700 kPa.

Aqueous sulfuric acid is first concentrated up to 90% and heated up to 675 K to release water and SO_3 according to the following spontaneous reaction:

$$H_2SO_4(g) \rightarrow SO_3(g) + H_2O(g), \text{ at } 675 \text{ K} \tag{4.92}$$

The water-sulfur trioxide mixture is heated up to 1123 K where the following reaction occurs:

$$5.25SO_3 + 5.75H_2O \rightarrow 0.5O_2 + SO_2 + 0.25SO_3 + 5.75H_2O, \text{ at } 1123 \text{ K} \tag{4.93}$$

The overall final decomposition phase occurs in gas phase can be expressed as follows:

$$SO_3(g) \rightarrow 0.5O_2(g) + SO_2(g), \text{ at } 1123 \text{ K} \tag{4.94}$$

Significant advances were reported by the Japan Atomic Energy Agency (JAEA) in the development of the sulfuric acid decomposition unit for the S-I hydrogen production cycle. JAEA reported an emerging decomposer design to be coupled with a very high temperature reactor (VHTR), with a demonstration prototype presented in Ogawa et al. (2009).

The decomposition reactor recovers heat from the secondary helium loop and provides the required thermal energy input for the chemical reaction. The main challenge has been to cope with the highly corrosive behavior of the reactant, while facilitating a high heat transfer rate through the wall material. In a JAEA design, the reactor is made of two solid blocks of silicon carbide (SiC). The two blocks are piled vertically using a pure gold plate with holes as the sealant. Both the upper and lower blocks have two types of holes, namely through-holes for the sulfuric acid decomposition side and shorter bores for the helium heat transfer fluid side. The rows of through-holes and bore-holes alternate such that heat is transferred from helium to the reacting stream.

In this reactor configuration, hot helium enters from above in a ring-type distributor that is detailed in a cross-section at the upper left of Fig. 4.33. After its distribution, the hot helium flows downward through the vertical tubes

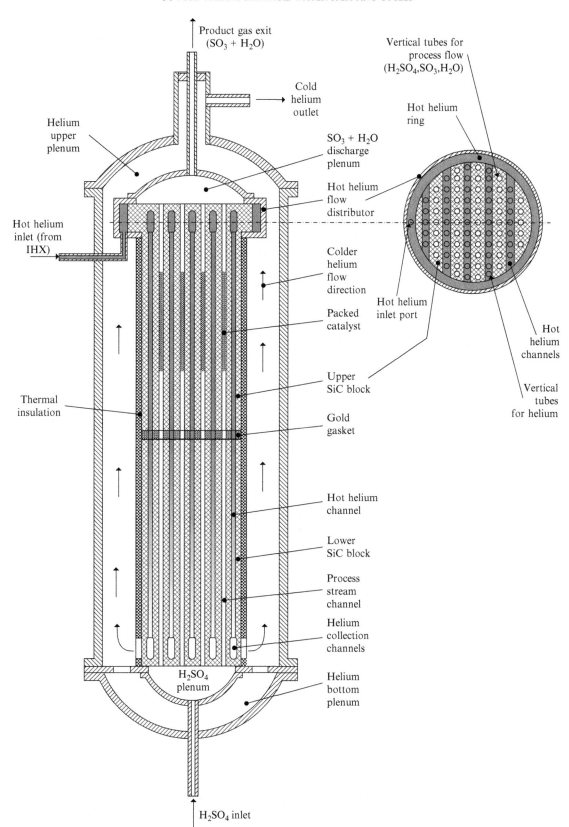

FIG. 4.33 Conceptual design of a sulfuric acid decomposition reactor linked to the secondary helium loop of a VHTR. *Modified from Ogawa, M., Hino, R., Inagaki, Y., Kunitomi, K., Onuki, K., Takegami, H., 2009. Present status of HTGR and hydrogen production development at JAFA.*

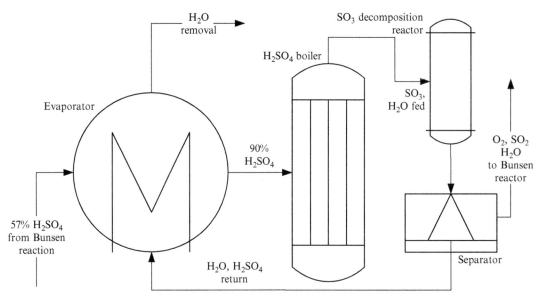

FIG. 4.34 Sulfuric acid decomposition system. *Adapted from Brown, L.C., Besenbruch, G.E., Schultz, K.R., Showalter, S.K., Marshall, A.C., Pickard, P.S., Funk, J.F., 2002. High Efficiency Generation of Hydrogen Fuels Using Thermochemical Cycles and Nuclear Power. General Atomics Report GA-A24326.*

(see figure). Heat is transferred to the SiC block and eventually to the process stream. At the bottom of the lower SiC block, there is a collector for the colder helium stream that diverts the flow toward the annular space between the reactor block and outer mantel. Furthermore, helium flows upward in the annular space.

The core reactor block is thermally insulated to impede heat transfer between the uprising helium and the process flow inside the SiC block. Cold helium is extracted from the upper side of the reactor's outer shell. The reactant—sulfuric acid—is fed at the bottom and distributed to each of the vertical tubes via a plenum-type distributor. In principle, the catalyst can be packed in the vertical tubes to enhance the reaction rate. The product stream is collected in an upper plenum (see the figure) and delivered at an exit port.

A sulfuric acid decomposition system—with a minimal number of functional units—is shown in the simplified schematics from Fig. 4.34, whereas the required pumps, auxiliary heat exchangers and auxiliary units are not represented. After extraction from the Bunsen reaction system, sulfuric acid is directed to an evaporator that concentrates the solution by the addition of heat.

A vapor phase, mostly containing steam, is removed and directed back to the Bunsen reactor for recycling. A concentrated solution of 90% H_2SO_4 is obtained and fed into an acid boiler that heats, boils, and partially decomposes acid by releasing sulfur trioxide. Special heat exchanger designs were adopted for this boiler at General Atomics, comprising printed channels coated with catalysts for reaction rate enhancement.

After partial decomposition at 675 K, the product gases comprising sulfur trioxide and steam are fed in a catalytic reactor for complete decomposition, where sulfur trioxide is converted into sulfur dioxide and steam. A separation step is necessary to recycle the unconverted sulfur trioxide. The final products of the sulfuric acid decomposition step are extracted in the form of hot oxygen, sulfur dioxide and steam in the gas phase; this mixture, after cooling, is fed back into the Bunsen reactor.

4.4.5.4 Sulfur-Iodine Plant Development

Detailed flowsheet diagrams of a sulfur-iodine thermochemical water splitting plant were enhanced by many research centers worldwide. The main functional units are approximately 7 chemical reactors for a full-scale plant. In addition, at least four separation units and many heat exchangers are required. General Atomics has developed for the very high temperature nuclear reactor (VHTR). Some relevant S-I plant flowsheets and analyses are given by Dokyia et al. (1979), Brown et al. (2002), Ewan and Allen (2005), and Kasahara et al. (2007).

In the plant diagram as shown in Fig. 4.35, the Bunsen reactor is a short-circuit electrochemical unit, with no requirement of power to run. As shown, liquid water is fed into a hydrating reactor R_1 where it hydrates the unreacted sulfur trioxide and the unreacted sulfuric acid in the presence of dissolved sulfur dioxide. The reactor R_1 is a hydration reactor where the material stream resulting from the sulfuric acid thermolysis—after oxygen separation—is mixed with water. The short circuit electrochemical cell is denoted with R_2. The catholyte flow is recycled through a SO_2 gas-liquid separator S_1, which is necessary because sulfur dioxide dissociates as gas in the catholyte stream.

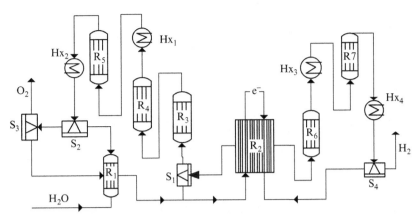

FIG. 4.35 Sulfur-iodine plant according to the process by Dokyia et al. (1979).

The separator S_1 returns SO_2 to the Bunsen reactor. The separated liquid phase out of S_1, concentrated in sulfuric acid, goes toward the sulfuric acid decomposition process. The anolyte comprising an iodine-rich solution is fed as shown by the stream coming from hydrogen separator S_4 toward the electrochemical reactor R_2.

The concentrated sulfuric acid is passed to a dehydration reactor (R_3) where heat is added. By further heat addition in an isothermal process, sulfur trioxide is dehydrated next in R_4. In heat exchanger H_1, the sulfur trioxide and water present in molar proportions $1.25SO_3$ and $5.75H_2O$ are heated up to a decomposition temperature of 1100 K. The sulfur trioxide decomposes in the reactor R_5. In heat exchanger H_2 the reaction products are cooled to a low temperature (assumed 298 K for the idealized analysis), while the heat release is recovered. The material stream is then directed to the separator S_2 where the gas phase comprising oxygen and sulfur trioxide is separated from the liquid phase of aqueous sulfuric acid.

Some sulfur trioxide remains residual in the liquid phase, but it is further recycled in a separator by hydration in reactor R_1. In the separator S_3 sulfur dioxide condenses and separates. At 25°C, the saturation pressure of sulfur dioxide is 392.2 kPa.

At the electrochemical cell anode, hydroiodic acid is produced in aqueous solution. There is no need for fractional distillation, as the case of the GA process, because there is little or no HI_x, $x > 1$ produced by the electrochemical cell. The products are first dehydrated by isothermal heat addition in reactor R_6 according to

$$3HI \cdot H_2O \rightarrow 3HI + 16.2H_2O \tag{4.95}$$

4.4.5.5 The products from R_6 are heated up to the HI decomposition temperature, which is about 900 K; heating occurs in the heat exchanger denoted by H_3 on the figure. After heating, hydrogen iodide is decomposed in R_7 according to

$$3HI + 1.5I_2 + 16.2H_2O \rightarrow H_2 + HI + 2.5I_2 + 16.2H_2O \tag{4.96}$$

The product stream resulting from R_7 is cooled to an ambient temperature in the heat exchanger H_4 with heat recovery. A liquid-gas separator follows, where the gas phase separates as (assumed) pure hydrogen. In the liquid phase, it remains hydroiodic acid, iodine, and water. This separator is illustrated on the figure as S_4. The liquid is fed thereafter at the anodic side of the electrochemical cell. Note also that in R_1, the unreacted sulfur trioxide is rehydrated by isothermal heat absorption according to the following chemical equation:

$$0.25SO_3 + 0.25H_2O \rightarrow 0.25H_2SO_4 \tag{4.97}$$

In the same reactor (R_1), the unreacted sulfuric acid is rehydrated as described by

$$0.25H_2SO_4 + 26H_2O \rightarrow 0.25(H_2SO_4 \cdot 26H_2O) \tag{4.98}$$

In addition, once the recycled hydrogen iodide is fed into the electrochemical cell (R_2), a spontaneous hydration process occurs isothermally with heat rejection, according to

$$HI + 16.2H_2O \rightarrow HI \cdot 16.2H_2O \tag{4.99}$$

Table 4.15 gives the process enthalpies of material streams within the S-I plant described in Fig. 4.35. The process 1–2 takes place in reactor R_3, where 1 represents the input of a sulfuric acid solution and 2 is the output of concentrated

TABLE 4.15 Enthalpies of Specific Processes Within the S-I Plant Described in Fig. 4.35

Process	Device	Process description	ΔH (kJ/molH$_2$)
$1 \rightarrow 2$	R$_3$	Dehydration of H$_2$SO$_4$ at 298 K, $1.25(\text{H}_2\text{SO}_4 \cdot 3.6\text{H}_2\text{O}) \rightarrow 1.25\text{H}_2\text{SO}_4 + 4.5\text{H}_2\text{O}$	65.3
$2 \rightarrow 3$	R$_4$	Dehydration of SO$_3$ at 298 K, $1.25\text{H}_2\text{SO}_4 + 4.5\text{H}_2\text{O} \rightarrow 5.25\text{SO}_3 + 5.75\text{H}_2\text{O}$	165.8
$3 \rightarrow 4$	R$_6$	Dehydration of HI at 298 K, $3\text{HI} \cdot \text{H}_2\text{O} \rightarrow 2\text{HI} + 16.2\text{H}_2\text{O}$	1563
$4 \rightarrow 5$	Hx$_1$ Hx$_3$	Heating $[1.25\text{SO}_3 + 5.75\text{H}_2\text{O} \text{ and } 21.2\text{HI} + 115\text{H}_2\text{O} + 9.6\text{I}_2]$ from 298 to 900 K	9099
$5 \rightarrow 6$	R$_7$	Hydrogen iodide decomposition at 900 K, $3\text{HI} + 1.5\text{I}_2 + 16.2\text{H}_2\text{O} \rightarrow \text{H}_2 + \text{HI} + 2.5\text{I}_2 + 16.2\text{H}_2\text{O}$	13.8
$6 \rightarrow 7$	R$_5$	Heating $1.25\text{SO}_3 + 5.75\text{H}_2\text{O}$ from 900 to 1100 K	122.6
$7 \rightarrow 8$	R$_5$	Sulfur trioxide decomposition at 1100 K, $5.25\text{SO}_3 + 5.75\text{H}_2\text{O} \rightarrow 0.5\text{O}_2 + \text{SO}_2 + 0.25\text{SO}_3 + 5.75\text{H}_2\text{O}$	93.4
$9 \rightarrow 10$	Hx$_2$	Heat rejection $0.25\text{SO}_3 + \text{SO}_2 + 0.5\text{O}_2 + 5.75\text{H}_2\text{O}$ from 1100 to 900 K	-121.9
$10 \rightarrow 11$	Hx$_2$ Hx$_3$	Heat rejection $[0.25\text{SO}_3 + \text{SO}_2 + 0.5\text{O}_2 + 5.75\text{H}_2\text{O} \text{ and } 16.2\text{HI} + 115\text{H}_2\text{O} + 9.6\text{I}_2]$ from 900 to 298 K	-9163
$11 \rightarrow 12$	R$_2$	Heat rejection from Bunsen reaction at 298 K, $\text{SO}_2 + 0.25(\text{H}_2\text{SO}_4 \cdot 26\text{H}_2\text{O}) + 19.2(\text{HI} \cdot 6\text{H}_2\text{O}) + 10.6\text{I}_2 \rightarrow 1.25(\text{H}_2\text{SO}_4 \cdot 3.6\text{H}_2\text{O}) + 21.2(2\text{HI} \cdot 5.43\text{H}_2\text{O}) + 9.6\text{I}_2$	-64.5
$12 \rightarrow 13$	R$_1$	Heat rejection at hydration of SO$_3$, $0.25\text{SO}_3 + 0.25\text{H}_2\text{O} \rightarrow 0.25\text{H}_2\text{SO}_4$	-33.1
$13 \rightarrow 14$	R$_1$	Heat rejection at hydration of H$_2$SO$_4$, $0.25\text{H}_2\text{SO}_4 + 26\text{H}_2\text{O} \rightarrow 0.25(\text{H}_2\text{SO}_4 \cdot 26\text{H}_2\text{O})$	-18
$14 \rightarrow 15$	R$_2$	Heat rejection at hydration of HI, $\text{HI} + 16.2\text{H}_2\text{O} \rightarrow \text{HI} \cdot 16.2\text{H}_2\text{O}$	-1439

Note: (+) sign means heat addition, (−) is heat rejection.
Data from Dokyia, M., Kameyama, T., Fukuda, K., 1979. Thermochemical hydrogen preparation – Part V. A feasibility study of the sulphur iodine cycle. Int. J. Hydrog. Energy 4, 267–277.

sulfuric acid solution. This dehydration releases 4.5 moles of water for every 1.25 moles of H$_2$SO$_4$. Two low-grade heat addition processes are the process $2 \rightarrow 3$ occurring in reactor R$_4$ and $3 \rightarrow 4$ dehydration of HI occurring at reactor R$_6$. The heat addition is required for the sulfur trioxide and hydriodic acid streams that must be heated from about room temperature (Bunsen reactor) up to 900 K in Hx$_1$ and Hx$_3$, respectively, forming the heat addition process $4 \rightarrow 5$.

Further, the sulfur trioxide is heated to the decomposition temperature of 1100 K in decomposition reactor R$_5$ and decomposed, according to the processes $6 \rightarrow 7$ and $7 \rightarrow 8$, respectively. Heat is added to the HI decomposition reactor at a temperature of 900 K as process $5 \rightarrow 6$.

The heat rejection from the cycle units is a combination of processes $9 \rightarrow 10 \rightarrow 11 \rightarrow 12 \rightarrow 13 \rightarrow 14 \rightarrow 15$ that release heat from 1100 K (the SO$_3$ decomposition temperature) down to the ambient grade. The heat exchangers Hx$_2$ and Hx$_3$ release heat at the highest temperature end.

The pinch analysis of the S-I plant is illustrated on the temperature versus stream enthalpy diagram shown in Fig. 4.36. This diagram shows that the heat rejecting processes $9 \rightarrow 13$ cover a good portion of the heat demand processes, namely $1 \rightarrow 2 \rightarrow 3 \rightarrow 4 \rightarrow 5 \rightarrow 6 \rightarrow 7 \rightarrow 8$, of which only the $4 \rightarrow 8$ overall process is the essential heating portion. The remaining heat input must be provided by an external thermal energy supply, as illustrated by the process $16 \rightarrow 17$ in the figure.

Based on an assumed 20% loss, the predicted energy efficiency of the plant is 55%, and exergy efficiency is 65% (see Naterer et al., 2013). These figures are quite optimistic because they do not deal with the large amount of energy requirement for the separation processes, pumping, gas circulation, streams conditioning and the large amount of excess water that is recirculated through the system. Hydration of one mole H$_2$SO$_4$ is made with more than 4 moles of water, whereas one mole of HI is hydrated with 16.2 moles of water.

Investigations by Nomura et al. (2004) found that the S-I process with a short-circuit electrochemical cell can reach 42% efficiency with an optimum HI concentration on the catholyte of $\sim$14.5 mol/kg H$_2$O. The current density of the electrochemical cell is close to 1 A/cm^2 at the highest efficiency. If the cell is set to a higher current density of 500 A/cm^2, then the cycle efficiency will decrease to 25%. The sulfuric acid concentration in the cell anolyte must be set in the range of 10–13 mol/kgH$_2$O, better efficiencies being obtained at lower concentrations.

FIG. 4.36 Temperature versus stream enthalpy diagram for pinch analysis of the S-I plant shown in Fig. 4.35. *Data from Dokyia, M., Kameyama, T., Fukuda, K., 1979. Thermochemical hydrogen preparation – Part V. A feasibility study of the sulphur iodine cycle. Int. J. Hydrog. Energy 4, 267–277.*

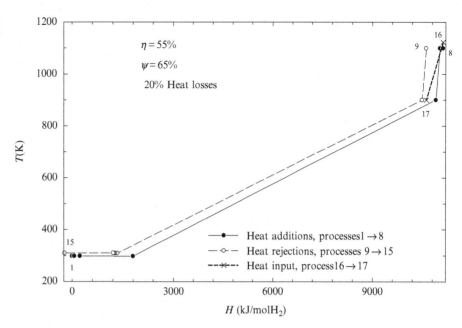

4.5 HYBRID THERMOCHEMICAL CYCLES

4.5.1 Electrochemical Closure Processes for Thermochemical Cycles

One of the big technological challenges with the pure thermochemical cycles relates to the high temperature requirement, which is normally 1000 K or more. Therefore, issues about material selection become very important. In addition, green hydrogen production will be limited to sources such as concentrated solar radiation and next-generation nuclear reactors for high temperature process heat. Those technologies are expensive and not yet available commercially.

Electrochemical reactions have three main advantages that make them preferable (in some conditions) from a technical and economical point of view: (i) the temperature of the reaction may be lowered with respect to a pure thermochemical process; (ii) reactions that have infeasible thermochemical conversion can be made possible when performed electrochemically; and (iii) the reaction products are intrinsically separated.

Fig. 4.37 presents the reversible potential of some significant electrochemical processes as a function of temperature, assuming reactions occur in the gas phase. It can be observed that some reactions, such as 1–3 and 7, are favored at low temperature where their reversible potential is reduced. Some other reactions (4–6) are favored at higher temperatures as they have thermo-neutral points.

4.5.1.1 Bayern-Hoechst-Uhde Process

The reaction #1 in Fig. 4.37 is the hydrochloric acid decomposition step that is a known industrial process, referenced as Bayern-Hoechst-Uhde (patented in 1964). It uses a feed of 22% aqueous HCl by weight and generates wet chlorine gas with 1–2% moisture. The moisture does not need to be removed from chlorine. The process uses a reverse Deacon reaction in conjunction with electrolytic decomposition of hydrochloric acid, given as follows:

$$2HCl(g) \rightarrow H_2(g) + Cl_2(g) \tag{4.100}$$

The electrochemical process requires inexpensive electrodes made of graphite. Current technological developments have shown improvement of the process with a 25% reduction of reversible potential as compared to the original process, 1.5 versus 2 V, respectively. The electrolyzer can generate hydrogen at 5 bars with an efficiency of approximately 88% and a requirement of electricity input of 289.5 MJ per kg of hydrogen.

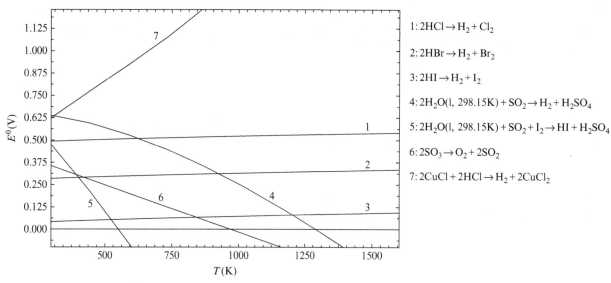

FIG. 4.37 Reversible cell potential of some significant electrochemical processes as a function of temperature, assuming reactions in gas phase.

4.5.1.2 *Electrolysis of Hydrobromic Acid*

The reaction #2 is the electrochemical decomposition of hydrobromic acid. The chemical activity of HBr is typically assumed as unity in an electrolyte solution, while bromine evolves under a vacuum at 0.28 bar absolute, according to the reaction

$$2HBr \rightarrow H_2(g) + Br_2(g) \tag{4.101}$$

At such operating conditions, the reversible cell potential becomes 1.07 V. A remarkable property of bromine is that it is highly soluble, as it forms the group Br_3^- in the electrolyte according to the process

$$Br_2 + Br^- \rightarrow Br_3^- \tag{4.102}$$

Due to the process from Eq. (4.97), gaseous bromine hardly evolves. In a typical process, the concentration of HBr in solution is maintained at ~50%, while the current density with graphite electrodes approaches 1 kA/m². When platinum is used, the current density can surpass 12 kA/m².

In order to appreciate the opportunity of a hydrogen halide electrolysis, we can compare the process to a competing one, namely the hydrogen halide thermolysis. For the case of hydrogen bromide, Fig. 4.38 shows that very high temperatures are required at over 1500 K to get only a marginal yield. Therefore, thermolysis is not favorable for HBr.

The same situation is valid for HCl, which should be better electrolyzed rather than thermally decomposed. However, the figure shows that hydriodic acid thermolysis is certainly feasible at over 500 K and will compete with an electrolysis process. Accurate estimations show the following yield at HI thermolysis: 19% at 300 K, 22% at 723 K and 33% at 1273 K.

4.5.1.3 *Hydrogen Iodide Electrolysis*

Hydrogen iodide electrolysis is similar to that of HBr. When HI is in aqueous solution, the reversible cell potential is 0.535 V. This electrochemical reaction must be conducted over the melting point of iodide (321 K) to avoid deposition of solid iodine on the electrode due to the following reaction:

$$I_2(aq) + I^- \rightarrow I_3^-(s) \tag{4.103}$$

FIG. 4.38 Comparison of equilibrium conversion (ζ) at thermolysis of hydrogen halides.

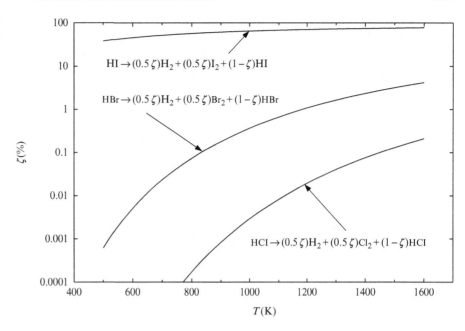

4.5.1.4 Sulfurous Acid Electrolysis

Electrolysis of sulfurous acid (H_2SO_3) is a crucial process in many hybrid cycles, e.g., Mark 11/ Westinghouse cycle and the NCLI I-S cycle (NCLI, National Chemical Laboratory for Industry at Tokyo). When sulfur dioxide dissolves in water, sulfurous acid is formed according to

$$SO_2 + H_2O \rightarrow H_2SO_3(aq) \tag{4.104}$$

The sulfurous acid exists in solution only; it cannot be insulated as it evolves sulfur dioxide. Therefore, an aqueous solution of sulfur dioxide can be electrolyzed in an acid electrolyte according to the following half-reactions:

$$\begin{cases} H_2O + H_2SO_3(aq) \rightarrow H_2SO_4(aq) + 2H^+ + 2e^-, \text{ at anode, } E^0 = -0.17\,\text{V} \\ 2H^+ + 2e^- \rightarrow H_2(g), \text{ at cathode, } E^0 = 0.0\,\text{V} \end{cases} \tag{4.105}$$

The overall process described by Eqs. (4.99) and (4.105) therefore becomes

$$2H_2O + SO_2(aq) \rightarrow H_2 + H_2SO_4(aq) \tag{4.106}$$

The reaction from Eq. (4.106) facilitates a hydrogen production process that reduces substantially the electricity requirement. The water splitting energy input becomes one third with respect to direct water electrolysis. Precautions are made to impede the migration of sulfurous acid from the anode (where it is generated) toward the cathode. The possible presence of sulfurous acid at the cathode favors side reactions that eventually lead to sulfur deposition on the electrode or emanation of hydrogen sulfide. The main side reactions at the cathode are as follows

$$\begin{cases} H_2SO_3 + 4H^+ + 4e^- \rightarrow 3H_2O + S, E^0 = +0.45\,\text{V} \\ S + 2H^+ + 2e^- \rightarrow H_2S, E^0 = +0.141\,\text{V} \end{cases} \tag{4.107}$$

The practical experience with the electrolytic process described by Eq. (4.107) shows that the current density of $2\,\text{kA/m}^2$ is readily obtained with a cell voltage less than 0.8 V. It uses diaphragms around the cathode where the electrolyte is slightly over-pressurized to reduce the probability of H_2SO_3 migration and the occurrence of side reactions.

A variant of the sulfurous acid electrolysis process has been proposed by the National Chemical Laboratory of Industry (NCLI) in Tokyo, according to Dokyia et al. (1977). Aqueous sulfurous acid is electrolyzed in the presence of iodine according to the overall reaction

$$2H_2O + SO_2 + I_2 \rightarrow H_2SO_4 + 2H \text{ at } 353\,\text{KI} \tag{4.108}$$

which is performed based on the following half-reactions:

$$\begin{cases} H_2O + H_2SO_3(aq) \rightarrow H_2SO_4 + 2H^+ + 2e^- \text{ at anode} \\ I_2 + 2H^+ + 2e^- \rightarrow 2HI \text{ at cathode} \end{cases} \quad (4.109)$$

This electrolysis process uses activated carbon as the anode and a platinum electrode as the cathode. It does not require electrical energy input; actually, this cell behaves like a separator of sulfuric and hydriodic acid. In order to accelerate the reaction to obtain a better production rate, some potential difference must be applied to the electrodes.

4.5.1.5 *Cuprous Chloride Electrolysis*

Cuprous chloride electrolysis is also an important process. It occurs in the Cu-Cl hybrid cycle family. One variant of this reaction is as follows:

$$2CuCl(aq) + 2HCl(aq) \rightarrow H_2(g) + 2CuCl_2(aq) \quad (4.110)$$

This process was actively researched recently at AECL (Atomic Energy of Canada Limited) and ANL (Argonne National Laboratory). The copper (I) oxidation reaction is presently conducted in a 0.5 mol/L CuCl/11 mol/L HCl electrolyte-anode reaction. In 11 mol/L HCl, copper metal deposition at the cathode appears to be absent at the 0.1 A/cm^2 current density used in the experiments. In 6 mol/L HCl, the CuCl$^+$ that is formed can be transported across the membrane by diffusion and migration. In 11 mol/L HCl, CuCl$_2$ is transported across the membrane by diffusion only. Thus the flux of copper species through the membrane is dependent upon the HCl concentration. Experimental data show that for similar experimental conditions, the catholyte copper species concentration decreases with increasing HCl concentration. This is consistent with the dependence of the flux on HCl concentration.

Another version of CuCl(aq) electrolysis occurs based on the following reaction:

$$2CuCl(aq) \rightarrow 2CuCl_2(aq) + Cu(s) \quad (4.111)$$

Reaction (4.111) can be conducted with 1mM solution of CuCl in 1.5 M HCl and platinum and/or glassy carbon electrodes. Although the glassy carbon electrode shows lower current density, it has a very high surface area that compensates for the lower activity such that the reaction rate is higher than with platinum. Therefore, glassy carbon appears to be the recommended choice.

4.5.2 Representative Hybrid Thermochemical Cycles

One of the most important hybrid cycles is HyS or hybrid sulfur cycle developed initially at ISPRA as Mark 11. This cycle focuses on commercial-scale development by Westinghouse. The HyS will be discussed in an upcoming section. Two other major hybrid processes are the Cu-Cl and the Mg-Cl; both are discussed in separate sections.

Tables 4.16–4.18 give a compilation of representative hybrid cycles with two steps, three steps, and four steps. One representative two-step cycle is copper-chlorine as proposed in Dokyia and Kotera (1976) at NCLI. The electrochemical reaction of this cycle is coupled with a thermochemical reaction at higher temperature, which is the hydrolysis of CuCl$_2$ at 823 K.

The two-step cycle with Bi and K developed at ANL shows a theoretical efficiency of approximately 46%. At ANL, other hybrid cycles are under development, such as the three-step Br-Ca cycle at 1043 K maximum temperature, the three-step Cl-Mg cycle (studied in conjunction with the Idaho National Laboratory), and the three-step LiNO$_3$ cycle with a 750 K temperature requirement (see UNVL-24 in the table). The Institute of Gas Technology—in the 1970s— proposed a cadmium-based cycle with 53% energy efficiency, which was considered as promising in recent studies by McQuillan et al. (2010). Other cycles inventoried and assessed at the University of Nevada at Las Vegas are the UNLV-184 with barium and antimony; the UNLV-4, which is a hybrid version of the Mark 9 cycle; and the UNLV-185 of four steps working with bromine and cobalt compounds.

In the mid-1970s, the cycles with iodine and antimony were studied in Japan at Kyushu University (KU) and iodine, potassium and nitrogen at Iwate University of Ueda (IUU). The Los Alamos Scientific Laboratory (LASL) proposed a four-step bismuth-sulfur cycle. ISPRA proposed a four-step hybrid version of the sulfur-iodine cycle that operates at 1073 K. A recent development was reported at the Japan Atomic Energy Agency, which developed a lower temperature version of the Westinghouse process that includes two electrochemical steps. In this cycle, H$_2$SO$_4$ is decomposed thermally at 673 K in steam and SO$_3$; thereafter, SO$_3$ is decomposed electrochemically in a solid oxide membrane electrolysis cell operating at 773 K. The cell generates oxygen gas and SO$_2$. The second electrochemical process is the same as the Westinghouse cycle, namely it performs the electrolysis of sulfurous acid. The process efficiency is estimated at 44%.

TABLE 4.16 Representative Hybrid Thermochemical Cycles With Two Steps

Cycle	T_{max} (K)	Reaction steps
NCLI	823	E: $2CuCl + 2HCl \rightarrow H_2 + 2CuCl_2$, 473 K
Cu-Cl-C		T: $KH_2O + 2CuCl_2 \rightarrow 0.5O_2 + 2HCl + 2CuCl$, 823 K
ANL	850	E: $KOH + 1/3Bi \rightarrow 0.5H_2 + 0.5O_2 + 1/3K_3Bi$, 850 K
Bi, K		T: $H_2O + 1/3K_3Bi \rightarrow 0.5H_2 + KOH + 1/3Bi$, 850 K
UNLV-53	873	E: $2HCl \rightarrow H_2 + Cl_2$, 363 K
		T: $H_2O + Cl_2 \rightarrow 0.5O_2 + 2HCl$, 873 K
Mark 11, Hy-S	1173	E: $2H_2O + SO_2 \rightarrow H_2 + H_2SO_4$, 310 K, 22 bar, 0.48 V
Westinghouse		T: $H_2SO_4 \rightarrow 0.5O_2 + SO_2 + H_2O$, 1173 K

TABLE 4.17 Representative Three-Step Hybrid Thermochemical

Cycle	T_{max} (K)	Reaction steps
UNLV-24	750	E: $H_2O + I_2 + LiNO_2 \rightarrow LiNO_3 + 2HI$, 300 K, 1.2 V
		T: $2HI \rightarrow H_2 + I_2$, 573 K
		T: $LiNO_3 \rightarrow 0.5O_2 + LiNO_2$, 700 K
ANL-INL	773	E: $2HCl \rightarrow H_2 + Cl_2$, 363 K
		T: $H_2O + MgCl_2 \rightarrow 2HCl + MgO$, 723 K
		T: $MgO + Cl_2 \rightarrow 0.5O_2 + MgCl_2$, 773 K
JAEA-HHLT	773	E: $2H_2O + SO_2 \rightarrow H_2 + H_2SO_4$, 350 K
		E: $SO_3 \rightarrow 0.5O_2 + SO_2$, 773 K
		T: $H_2SO_4 \rightarrow SO_3 + H_2O$, 673 K
ANL, CuCl-B	800	E: $2CuCl + 2HCl \rightarrow H_2 + 2CuCl_2$, 373 K, 24 bar
		T: $H_2O + 2CuCl_2 \rightarrow Cu_2OCl_2 + 2HCl$, 700 K
		T: $Cu_2OCl_2 \rightarrow 0.5O_2 + 2CuCl$, 800 K
UNLV-191	823	E: $4CuCl \rightarrow 2Cu + 2CuCl_2$, 353 K
		T: $H_2O + 2CuCl_2 \rightarrow 0.5O_2 + 2CuCl + 2HCl$, 823 K
		T: $2Cu + 2HCl \rightarrow H_2 + 2CuCl$, 700 K
UNLV-4	923	E: $1.5Cl_2 + Fe_3O_4 + 6HCl \rightarrow 0.5O_2 + 3FeCl_3 + 3H_2O$, 423 K
		T: $3FeCl_2 + 4H_2O \rightarrow H_2 + Fe_3O_4 + 6HCl$, 823 K
		T: $3FeCl_3 \rightarrow 1.5Cl_2 + 3FeCl_2$, 693 K
ANL Br-Ca	1043	E: $2HBr \rightarrow H_2 + Br_2$, 323 K, 0.58 V
		T: $H_2O + CaBr_2 \rightarrow CaO + 2HBr$, 1043 K
		T: $CaO + Br_2 \rightarrow 0.5O_2 + CaBr_2$, 853 K
UNLV-56	1073	E: $2CuCl + 2HCl \rightarrow H_2 + 2CuCl_2$, 473 K
		T: $H_2O + Cl_2 \rightarrow 0.5O_2 + 2HCl$, 1073 K
		T: $2CuCl_2 \rightarrow 2CuCl + Cl_2$, 773 K
Mark 13, LASL	1123	E: $2HBr \rightarrow H_2 + Br_2$, 298 K, 0.58 V
		T: $2H_2O + Br_2 + SO_2 \rightarrow H_2SO_4 + 2HBr$, 373 K
		T: $H_2SO_4 \rightarrow 0.5O_2 + SO_2 + H_2O$, 1123 K
NCLI	1173	E: $2H_2O + I_2 + SO_2 \rightarrow 2HI + H_2SO_4$, 353 K
		T: $2HI \rightarrow H_2 + I_2$, 773 K
		T: $H_2SO_4 \rightarrow 0.5O_2 + SO_2 + H_2O$, 1173 K
KU	1273	E: $2HI \rightarrow H_2 + I_2$, 373 K, 0.5 V
		T: $H_2O + 0.5Sb_2O_3 + I_2 \rightarrow 2HI + 0.5Sb_2O_5$, 298 K
		T: $0.5Sb_2O_5 \rightarrow 0.5O_2 + 0.5Sb_2O_3$, 1273 K
UNLV-184	1273	E: $2HBr \rightarrow H_2 + Br_2$, 373 K, 0.58 V
		T: $H_2O + 0.5Sb_2O_3 + Br_2 \rightarrow 2HBr + 0.5Sb_2O_5$, 353 K
		T: $0.5Sb_2O_5 \rightarrow 0.5O_2 + 0.5Sb_2O_3$, 1273 K
UNLV-5	1473	E: $2H_2O + Cd \rightarrow H_2 + Cd(OH)_2$, 298 K, 0.02 V
		T: $CdO \rightarrow 0.5O_2 + Cd$, 1473 K
		T: $Cd(OH)_2 \rightarrow CdO + H_2O$, 650 K

TABLE 4.18 Representative Four-Step Hybrid Thermochemical

Cycle	$T_{\max}$ (K)	Reaction steps
UOIT Cu-Cl	800	E : $2CuCl + 2HCl \rightarrow H_2 + 2CuCl_2(aq)$, 353 K, 1 bar
		T : $H_2O + 2CuCl_2(s) \rightarrow Cu_2OCl_2 + 2HCl$, 700 K
		T : $Cu_2OCl_2 \rightarrow 0.5O_2 + 2CuCl$, 800 K
		T : $CuCl_2(aq) \rightarrow CuCl_2(s)$, 400 K
IGT, CuCl-A	800	E : $4CuCl \rightarrow 2Cu + 2CuCl_2$, 353 K
		T : $H_2O + 2CuCl_2 \rightarrow Cu_2OCl_2 + 2HCl$, 700 K
		T : $2Cu + 2HCl \rightarrow H_2 + 2CuCl$, 700 K
		T : $Cu_2OCl_2 \rightarrow 0.5O_2 + 2CuCl$, 800 K
UNLV-114	973	E : $2HNO_3 + 2KI \rightarrow H_2 + I_2 + 2KNO_3$, 298 K
		T : $H_2O + 3NO_2 \rightarrow 2HNO_3 + NO$, 373 K
		T : $I_2 + 2KNO_3 \rightarrow 0.5O_2 + 2KI + 2NO + 1.5O_2$, 973 K
		T : $3NO + 1.5O_2 \rightarrow 3NO_2$, 373 K
ISPRA-S	1073	E : $H_2O + H_2SO_3 \rightarrow H_2 + H_2SO_4$, 300 K
		T : $SO_3 \rightarrow 0.5O_2 + SO_2$, 1073 K
		T : $H_2SO_4 \rightarrow SO_3 + H_2O$, 1073 K
		T : $SO_2 + H_2O \rightarrow H_2SO_3$, 298 K
IGT Cu-S	1100	E : $2H_2O + SO_2 \rightarrow H_2 + H_2SO_4$, 298 K, 0.17 V
		T : $CuSO_4 \rightarrow 0.5O_2 + CuO + SO_2$, 1100 K
		T : $H_2SO_4 + CuO + 3H_2O \rightarrow CuSO_4 \cdot 4H_2O$, 298 K
		T : $CuSO_4 \cdot 4H_2O \rightarrow CuSO_4 + 4H_2O$, 800 K
UNLV-185	1123	E : $2HBr \rightarrow H_2 + Br_2$, 298 K, 0.58 V
		T : $H_2O + CoBr_2 \rightarrow CoO + 2HBr$, 1023 K
		T : $Co_3O_4 \rightarrow 0.5O_2 + 3CoO$, 1023 K
		T : $Br_2 + 4CoO \rightarrow CoBr_2 + Co_3O_4$, 773 K

A five-step electrochemical cycle with a maximum temperature of 773 was studied at ANL-UOIT-AECL in 2007–2010. The five steps are as follows:

$$E : 4CuCl(aq) \rightarrow 2Cu(s) + 2CuCl_2(aq), 353 \text{ K}$$
$$T : H_2O + 2CuCl_2(s) \rightarrow CuO \cdot CuCl_2 + 2HCl, 648 \text{ K}$$
$$T : 2Cu + 2HCl \rightarrow H_2 + 2CuCl, 723 \text{ K}$$
$$T : CuO \cdot CuCl_2 \rightarrow 0.5O_2 + 2CuCl(l), 773 \text{ K}$$
$$T : 2CuCl_2(aq) \rightarrow 2CuCl_2(s), 373 \text{ K}$$

The research on this five-step cycle was abandoned in favor of the more promising three-step version of the same cycle that includes a hydrolysis reaction of $CuCl_2$, thermolysis of copper oxychloride (Cu_2OCl_2) and a drying step of aqueous cupric chloride.

4.5.3 Hybrid Sulfur Cycle

The hybrid sulfur cycle is one of the major water splitting cycles; it was developed by Westinghouse for large-scale hydrogen production. The processes in the hybrid sulfur cycle (HyS) are described as shown in the simplified diagram from Fig. 4.39. In this cycle, power is applied to an electrochemical cell that electrolyzes water in the presence of

FIG. 4.39 Simplified schematic of hybrid sulfur cycle.

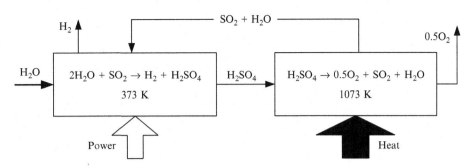

dissolved SO_2. Aqueous sulfuric acid is generated in this way. The sulfuric acid is then heated at high temperature and supplied to a thermolysis reactor. Oxygen is evolved out of the thermolysis reaction which is conducted at 1073 K. Oxygen is easily separable from water and SO_2. Therefore, the recycled water and SO_2 are returned to the electrochemical cell.

The electrochemical process in HyS cycle evolves as follows:

$$2H_2O + SO_2(aq) \xrightarrow{\text{Electrolytic}} H_2(g) + H_2SO_4(aq), E^0 = 0.158\,V \tag{4.112}$$

The thermolysis of sulfuric acid evolves as follows:

$$H_2SO_4(g) \rightarrow 0.5O_2(g) + SO_2(g) + H_2O(g) \tag{4.113}$$

The main process units of these cycles are the electrolytic cell and the bayonet reactor for sulfuric acid thermolysis. The HyS plant will contain many auxiliary units, including the acid vapor generators (primary and secondary), acid separator, pumps, compressors, expanders, steam condenser, sulfur dioxide condenser, and oxygen separator. In the subsequent part of this section, the main plant units and the HYS plant system are introduced.

4.5.3.1 Electrochemical Cell

The presence of sulfur dioxide in the cell electrolyte depolarizes the electrochemical process, which means that the reversible electrode potential at anode is well reduced as compared to pure water electrolysis. Westinghouse invented a pressurized parallel plate electrochemical unit which uses diaphragms for gas product separation. Fig. 4.40 shows the configuration of the Westinghouse electrochemical process.

Both the anolyte and catolyte are recirculated (separately) and formed from a sulfuric acid aqueous solution (about 70% by weight concentration). The process operates at ~25 bar with a slightly pressurized catolyte in order to avoid sulfur dioxide penetration in the catholyte. Recycled sulfur dioxide and water are mixed with supply water to form the anolyte. Under the influence of an electric field and electro-osmotic force, protons transfer to the anolyte and form

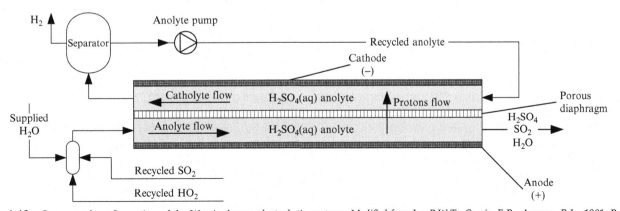

FIG. 4.40 Conceptual configuration of the Westinghouse electrolytic system. *Modified from Lu, P.W.T., Garcia, E.R., Ammon, R.L., 1981. Recent developments in the technology of sulfur dioxide depolarized electrolysis. J. Appl. Electrochem. 11, 347–355.*

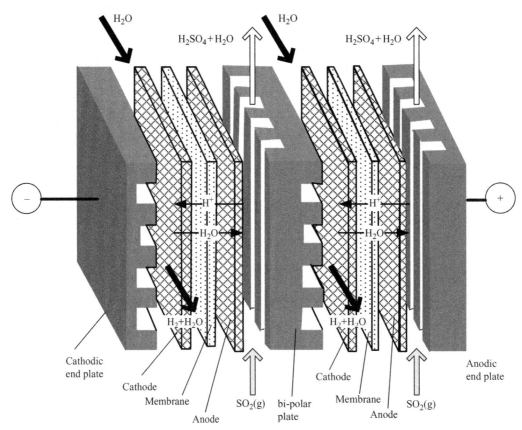

FIG. 4.41 Electrolysis cell for hybrid sulfur cycle with proton exchange membrane and water fed at the cathode. *Modified from Sivasubramanian, P., Ramasamy, R.P., Freire, F.J., Holland, C.E., Weidner, J.W., 2007. Electrochemical hydrogen production from thermochemical cycles using a proton exchange membrane electrolyzer. Int. J. Hydrog. Energy 32, 463–468.*

dissolved molecular hydrogen. At the exit of the anolyte, a gas-liquid separator is placed where hydrogen is separated and released while the remaining anolyte is recirculated by a pump.

The Westinghouse electrochemical cell shows effective measured potentials of 0.61.05 V for current densities of 2–4 kA/m^2. The purity of hydrogen produced from the cell is 98.7%. The optimum concentration of sulfur dioxide in the feed stream is 30%. Platinum-based electrodes were used with various substrates.

The more recent design is based on a PEM configuration as shown in Fig. 4.41. Gaseous sulfur dioxide is the only feed stream at the bottom of the anodic side of the bipolar plate or the anodic end plate. Liquid water migrates through the PEM and at the anode encounters activated electro-catalysts under adequate electrode polarization that produces water decomposition according to the half-reaction:

$$2H_2O(l) + SO_2(g) \rightarrow H_2SO_4(aq) + 2H^+ + 2e^- \tag{4.114}$$

The process from Eq. (4.114) is the depolarized electrochemical reaction. The protons migrate toward the cathode across the PEM and combine with electron reduction in a half-reaction described as flows:

$$2H^+ + 2e^- \rightarrow H_2(g) \tag{4.115}$$

The electrolytic process with a PEM membrane cell is depicted as shown in Fig. 4.42. The supplied water and the recycled water are fed at the cathode. The anolyte consists of only demineralized water in which the hydrogen product is dissolved. Hydrogen is separated with the help of a gas-liquid separator. The recycled sulfur dioxide is fed at the anode. While flowing through the anode's channels, the dissolved sulfur dioxide enters in a reaction with protons and forms sulfuric acid. The anode's product consists of unreacted sulfur dioxide, water and sulfuric acid.

Fig. 4.43 reports the potential versus current density curve of the Westinghouse cell, compared with more recent results by Sivasubramanian et al. (2007); compare this to the potential of the older diaphragm-based cell developed in the 1980s. It can be observed from the potential versus current density diagram that a novel design has better efficiency because voltages are lower for the same current density. For 400 A/m^2, the cell voltage was decreased from 1.05 V in 1980 to 0.87 V in 2007 (180 mV reduction).

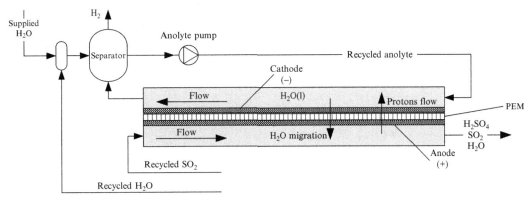

FIG. 4.42 Electrolytic process for hybrid sulfur cycle with a PEM electrolysis cell.

FIG. 4.43 Cell potential versus current density for early Westinghouse design compared with newer PEM cell design. *Data from Sivasubramanian, P., Ramasamy, R.P., Freire, F.J., Holland, C.E., Weidner, J.W., 2007. Electrochemical hydrogen production from thermochemical cycles using a proton exchange membrane electrolyzer. Int. J. Hydrog. Energy 32, 463–468 and Lu, P.W.T., Garcia, E.R., Ammon, R.L., 1981. Recent developments in the technology of sulfur dioxide depolarized electrolysis. J. Appl. Electrochem. 11, 347–355.*

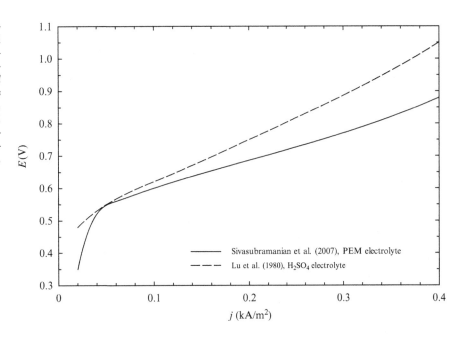

A third electrochemical cell design was recently proposed by Gorensek and Summers (2009). This process is described as shown in Fig. 4.44. The cell is also a PEM electrolysis cell. However, in this case, water is fed at the anode. Consequently, there is no net migration of water toward the cathode, as required by electrochemical reactions wherein only protons need to transfer to cathodic side. However, due to the functioning issues of the PEM, protons migrate as H_3O^+ and water transferred to the cathodic side must be transferred back to the anode.

4.5.3.2 Sulfuric Acid Thermolysis Reactor

Sandia National Laboratory (SNL) proposed a silicon carbide-made bayonet decomposition reactor for sulfuric acid within the HyS cycle. According to Gorensek and Summers (2009), this technology aims to decompose sulfuric acid at 9 MPa and 1073 K. Silicon carbide was found as the best material choice of ceramic type that confers good heat transfer properties, satisfactory mechanical stability and corrosion resistance.

A bayonet reactor has a tube-in-tube configuration, and it is immersed in a hot fluid. The temperature in the annulus increases from approximately 400 K at the fed inlet up to 950 K at the location where the catalytic decomposition process starts. The stream reaches 1143 K after passing through the catalytic bed. Once reaching the bayonet tip, the stream returns back through the core. The product stream exits the reactor with a temperature of approximately 528 K, which corresponds to the pinch point. The schematic of the bayonet reactor-based system is illustrated in Fig. 4.45.

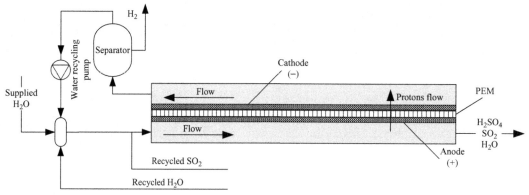

FIG. 4.44 PEM-based electrolytic cell and its system when water is fed at the anode. *Modified from Gorensek, M.G., Summers, W.A., 2009. Hybrid sulfur flowsheets using PEM electrolysis and a bayonet decomposition reactor. Int. J. Hydrog. Energy 34, 4097–4114.*

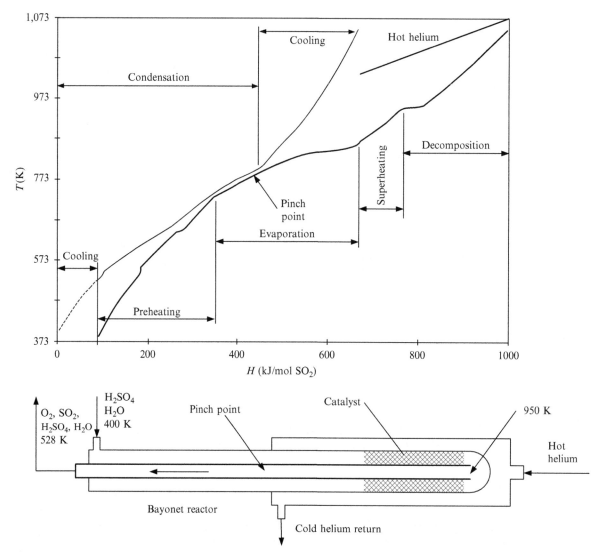

FIG. 4.45 Bayonet reactor for sulfuric acid decomposition and its temperature profiles.

The pinch point diagram of the system is shown. The bayonet reactor facilitates the heat transfer between the feed gas and the return gas at lower temperatures, as well as between the heat transfer fluid of the nuclear reactor (e.g., helium) and the feed stream at higher temperature. According to data from the figure, more than 80% of heat rejected is recovered and used for preheating of a cold stream delivered by the electrolytic subsystem.

4.5.3.3 Hybrid Sulfur Plant Development

The hybrid sulfur process is a relatively simplified plat with respect to sulfur-iodine plant, many units being similar for both plants. More specifically, the decomposition of sulfuric acid is practically the same as in the S-I cycle. In brief, this process requires heating the sulfuric acid which first dehydrates and eventually forms sulfur trioxide. Subsequently sulfur trioxide decomposes thermally.

The process proposed by Westinghouse comprises many functional units, as follows: surge tank (ST), acid separator (AS), stream condenser (SC), compressor (C), primary acid vaporizer (PAV), secondary acid vaporizer (SAV), sulfur trioxide reduction reactor (RR), primary and secondary SO_2 condenser (PCND, SCND), oxygen separator (OS), turbo-expander (TE), and electrochemical cell (E).

Fig. 4.46 shows a flowsheet diagram of the Westinghouse-based HyS process described in Gorensek and Summers (2009). The electrolytic cell generates hydrogen and recovers water from the catholyte. The generated sulfuric acid stream is in aqueous form and contains SO_2. Therefore, the stream passes through a surge tank (ST) where it is combined with recycled acid from the acid separator AS. The mixed stream is then fed into the primary acid vaporizer (PAV) to boil off the sulfuric acid as a first stage. In the secondary acid vaporizer (SAV), the boiling process is completed, and the material stream flows into the bayonet reactor (RR) where stream receives the required high temperature heat for decomposition.

Heat is recovered from the product stream and transferred to PAV, and the water is separated by AC and fed into the steam condenser (SC) which recycles water. In the condenser, SO_2 and O_2 are separated from the moisture. Further, SO_2 and O_2 in gaseous form are compressed and fed into a primary SO_2 condenser (PCND). Sulfur dioxide condenses further in SCND and separates completely from oxygen. Then SO_2 is returned at the electrochemical cell while O_2 is depressurized through a turbo-expander.

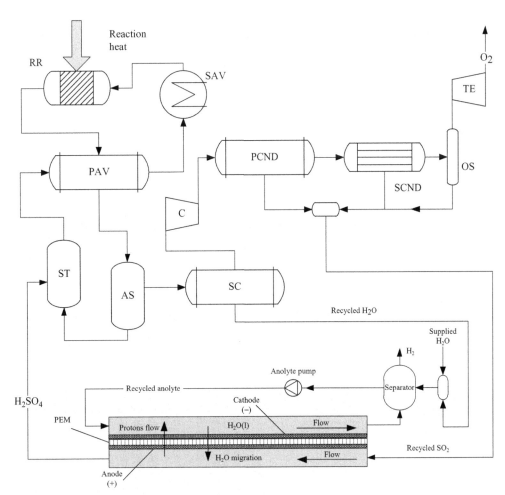

FIG. 4.46 Process flow diagram of a hybrid sulfur hydrogen production plant.

Based on a detailed flowsheet analysis, under the assumption that electricity is generated with 45% efficiency, Gorensek and Summers (2009) determined that the net energy efficiency of the hybrid sulfur cycle is ~42%, being thus superior to water electrolysis.

4.5.4 Magnesium-Chlorine Cycle

The hybrid magnesium-chlorine cycle comprises three or four reaction steps among which one is electrochemical depending on the configuration. Older work reviewed in Hesson (1979) reports two configuration of the three-step cycle, one referred to as $MgCl_2$-MgO and the other as $MgCl_2$-MgOHCl. A very recent development at the University of Ontario Institute of Technology led to a four-step configuration of the cycle, as specified by Ozcan (2015). In this cycle, the electrochemical step is the electrolysis of hydrochloric acid that has a reversible cell potential of 0.99 V.

The $MgCl_2$-MgO configuration of the magnesium chlorine cycle is represented in a simplified manner in Fig. 4.47. The cycle comprises three processes. The first process is hydrolysis of $MgCl_2$ at high temperature. This process requires heat addition at 773 K and evolves as follows:

$$H_2O + MgCl_2 \rightarrow MgO + 2HCl \text{ at } 773 \text{ K} \tag{4.116}$$

Hydrochloric acid is separated from MgO and fed into the electrochemical process where hydrogen and chlorine are generated in separate manifolds, as follows:

$$2HCl \xrightarrow{\text{Electrochemical}} H_2(g) + Cl_2(g) \text{ at } 363 \text{ K} \tag{4.117}$$

In the third step, the magnesium oxide (MgO) from hydrolysis and chlorine gas (Cl_2) from electrolysis are combined into a chlorination reactor that regenerates the magnesium chloride ($MgCl_2$). This reaction is slightly exothermic and evolves oxygen as follows:

$$MgO + Cl_2(g) \rightarrow MgCl_2 + 0.5O_2(g) \text{ at } 723 \text{ K} \tag{4.118}$$

The $MgCl_2$-MgOHCl cycle will operate hydrolysis at an essentially lower temperature and form the compound MgOHCl. This cycle is represented as shown in Fig. 4.48. The hydrolysis is, in this case, an exothermic reaction occurring at 573 K with release of HCl and MgOHCl as follows:

$$2H_2O + 2MgCl_2 \rightarrow 2MgOHCl + 2HCl \text{ at } 573 \text{ K} \tag{4.119}$$

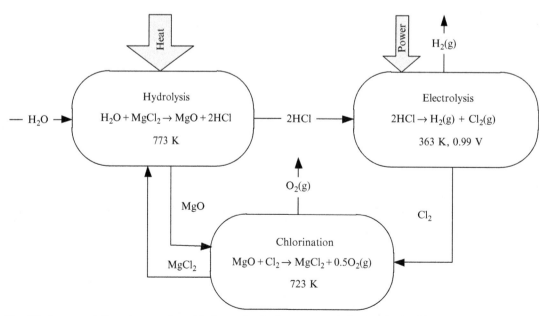

FIG. 4.47 Simplified representation of magnesium-chlorine cycle of a three-step cycle having the $MgCl_2$-MgO configuration.

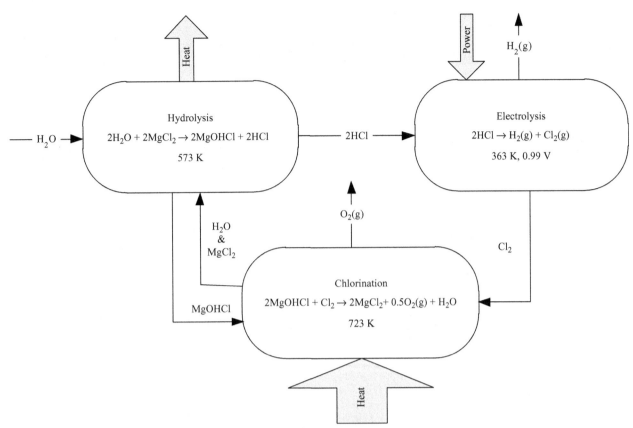

FIG. 4.48 Simplified representation of magnesium-chlorine cycle of a three-step having the MgCl$_2$-MgOHCl configuration.

The electrolysis reaction is the same as given in Eq. (4.117). The chlorination reaction is endothermic evolving at 723 K, as follows:

$$2MgOHCl + Cl_2 \rightarrow 2MgCl_2 + 0.5O_2 + H_2O \text{ at } 723 \text{ K} \tag{4.120}$$

A new version of the magnesium-chlorine cycle was proposed by Ozcan (2015) with a four-step and lower temperature requirement. This cycle is denoted as MgCl-MgO-MgOHCl and requires a maximum temperature of 723 K. Fig. 4.49 shows a simplified diagram of the four-step magnesium-chlorine cycle. There is only one endothermic reaction, namely the thermal decomposition of MgOHCl into MgO particles and HCl that has a very fast and favorable kinetics. The hydrolysis process is described by Eq. (4.112), which, as already discussed, is slightly exothermic. However, the stoichiometry of this reaction is reduced, requiring less steam, namely

$$H_2O + MgCl_2 \rightarrow MgOHCl + HCl \text{ at } 573 \text{ K} \tag{4.121}$$

The electrolysis process is the same as in Eq. (4.117) with the opportunity to be conducted with only anhydrous HCl stream, which will lead to less power consumption. Chlorination is slightly exothermic as mentioned before in Eq. (4.118). The only endothermic reaction of the cycle is the new fourth step that decomposes MgOHCl thermally, as follows:

$$MgOHCl \rightarrow MgO + HCl(g) \text{ at } 723 \text{ K} \tag{4.122}$$

This reaction will generate a dry HCl gas. The four-step cycle appears to require heat at the lowest temperature level and less electrical energy input for its electrochemical step.

4.5.4.1 Electrochemical Process

The electrochemical process evolves typically according to the well-established Bayern-Hoechst-UHDE process in aqueous HCl electrolyte, described also in Eq. (4.68). The aqueous form of HCl limits the current density and introduces technical problems related to chlorine gas separation from moisture. In this process, when HCl concentration is higher than 22%, the overpotential losses increase greatly. When HCl concentration is lower than 17%, there are

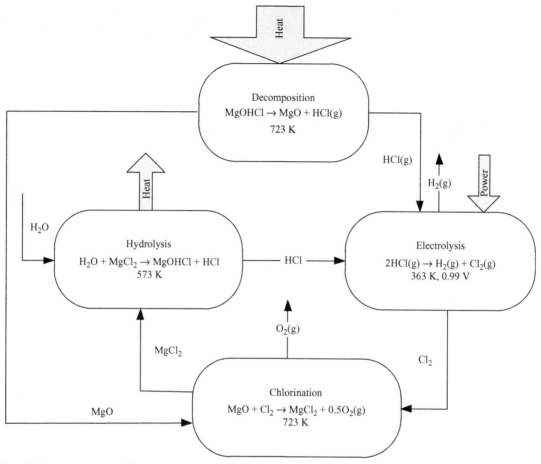

FIG. 4.49   Simplified representation of four-step magnesium-chlorine cycle.

various unwanted side reactions. The reversible cell potential of a Bayern-Hoechst-UHDE cell is 0.99 V; however, the actual potential is 1.8 V at 0.5 A/cm².

An alternative to the Bayern-Hoechst-UHDE process is the anhydrous HCl electrolysis as shown in the US patent by Bulan et al. (2006). This is an emerging technology that is not yet established and at present focuses mostly on chlorine recovery from HCl, rather than hydrogen production from the process—because of the technical difficulty in keeping hydrogen isolated from air in reaction conditions in a gas phase, where it can combine with oxygen and form steam. One method to electrolyze gaseous HCl at low temperature is in a diaphragm-based electrolysis cell operating at 343 K in which steam is provided in a volumetric fraction of 21%. Therefore, HCl, Cl_2, and H_2 will not be dry but humid. Nevertheless, Messner (1974) showed a patented device of this kind in which the product gases can be easily purified up to a purity of 99.7% by volume.

A polymer electrolyte membrane-based cell was proposed by Eames and Newman (1995), who obtained a current density of 0.9 A/cm² at below 2 V polarization. A cell that utilizes essentially a dry HCl gas was reported by Zimmerman et al. (2001). This device comprises a membrane electrode assembly (MEA) that has two porous gas diffusion electrode layers at anode and cathode and a cation exchange membrane. The MEA can run at 0.4 A/cm² for a polarization of 1.4 V. The selected cation exchange membrane will have sulfonate, phosphonate, imide, sulfonimide, and sulfonamide groups able to transport protons including a mixture of hydrophobic-hydrophilic polymers. Recently, Huskinson et al. (2012) used a modified commercial PEM fuel cell in reverse to conduct electrolysis of HCl. The chlorine electrode was doped with a low precious metal alloy. The operation pressure was increased to higher values. An energy efficiency of 60% and a power density of 1 W/cm² were obtained and are encouraging results.

The reversible cell potential for the anhydrous HCl electrolysis cell can be determined using Nernst equation and partial pressures of hydrogen, chlorine, and hydrochloric acid, as follows:

$$E = E^0 - \frac{RT}{2F} \ln \left(\frac{P_{H_2} P_{Cl_2}}{P_{HCl}^2} \right)$$

(4.123)

The Butler-Volmer equation for the anhydrous HCl electrolysis cell can be applied as

$$J = J_0 \left[\exp\left(\frac{0.5FE_{pol}}{RT} \right) - \frac{k_c P_{H2}}{k_a P_{HCL}} \exp\left(-\frac{0.5FE_{pol}}{RT} \right) \right] \tag{4.124}$$

where k is the rate constants at cathode (c) and anode (a) and P is the partial pressure.

Ozcan (2015) calculated the reversible cell voltage variation with HCl utilization ratio (or fractional conversion) for three temperatures obtaining the results shown in Fig. 4.50. Furthermore, based on the model developed in Eames and Newman (1995), the current density variation with cell voltage (the J-V curve) for three HCl utilization ratios is shown in Fig. 4.51.

With lower partial pressure, a higher HCL utilization ratio occurs. Therefore, higher utilization ratio of HCl results in low current densities regardless of the applied voltage. However, the current density shows a similar trend up to

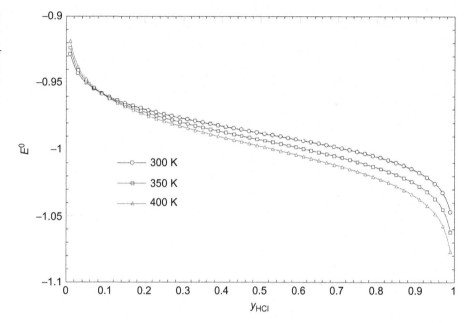

FIG. 4.50 Reversible cell potential of the anhydrous HCl electrolysis cell versus HCl utilization ratio. *Data from Ozcan, H., 2015. Experimental and Theoretical Investigations of Magnesium-Chlorine Cycle and Its Integrated Systems. PhD Thesis, University of Ontario Institute of Technology.*

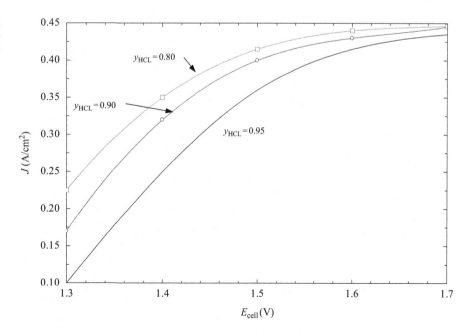

FIG. 4.51 The current density versus cell potential curve predicted for the HCl anhydrous electrolysis process for three HCl utilization ratio.

80% HCl utilization. The change in current density slows down after 1.6 V of applied voltage, where a limiting current density approximately 0.42–0.45 A/cm^2 can be observed from the results shown in the figure.

4.5.4.2 Hydrolysis Process and Reactor

Depending of the cycle configuration, two hydrolysis processes are possible, namely the low temperature one according to Eq. (4.112) and the high temperature one according to Eq. (4.109). Regardless of the actual process, during hydrolysis of magnesium chloride, the metal tends to bond both to the chlorine and to the hydroxyl anions. This is why MgOHCl forms at an intermediate temperature range, whereas at high temperature, chlorine is completely released to generate hydrochloric acid and magnesium oxide.

A review on the chemistry of MgOHCl formation is provided in Ozcan (2015). Equilibrium thermodynamics calculations for a low temperature hydrolysis process forming MgOHCl are reported in Kelley (1945). The hydrolysis of MgCl$_2$ starts at approximately 473 K with the generation of MgOHCl at a relatively low rate. A conversion of 90% can be obtained at 566 K and 95% at 649 K using a steam/MgCl$_2$ molar ratio of 11.

The reaction rate increases with temperature. At approximately 650 K, the MgO starts to form at a slow rate, concurrently with MgOHCl until a temperature threshold of 773 K is reached. At 773 K, an over pure MgO start forming very fast, instead of MgOHCl. With a steam to MgCl$_2$ molar ratio of 6, a 100% conversion can be obtained. The generation of MgO can be accelerated with silicate supported catalysts.

Hydrolysis has been conducted in fixed bed reactors. In a fixed bed reactor, the gaseous reactant flows over the porous bed of particle reactant. A reaction occurs when appropriate conditions are set and the residence time is sufficiently high. In the fixed bed configuration, the gas phase velocity must be set well below the fluidization threshold. Fig. 4.52 shows a suggested configuration of a fixed bed reaction setup of MgCl$_2$ powder. In the figure, u_{mf} represents the fluidization velocity, and ΔP is the pressure drop of the gas phase across the bed. In order to avoid side reactions, both the solid and gas reactants should be heated separately up to the desired reaction temperature, before feeding them into the reactor. With a fixed bed experiment the reactor can be heated up first, and the gas reactant feed at the specified temperature.

The temperature must be maintained below the melting point of MgCl$_2$, namely 973 K. At a standard pressure of 86%, HCl yield was reported by Ozcan (2015) at 773 K. In addition, a higher steam/MgCl$_2$ ratio helps increase the production of HCl at lower temperatures. The equilibrium thermodynamic analysis of the reaction illustrates the importance of steam excess. Fig. 4.53 shows the equilibrium molar fraction of steam and HCl for a range of temperature and pressure.

At a lower temperature, the thermodynamic equilibrium favors the formation of MgOHCl during hydrolysis. The diagram in Fig. 4.54 shows the equilibrium molar fraction of MgOHCl and MgO during hydrolysis of MgCl$_2$ in the range of 473–773 K. Two extreme molar fractions of steam versus MgCl$_2$ are shown as 1.0 and 2.0 and intermediate values of 1.25, 1.5, and 1.75.

FIG. 4.52 Fixed bed reactor configuration for MgCl$_2$ hydrolysis.

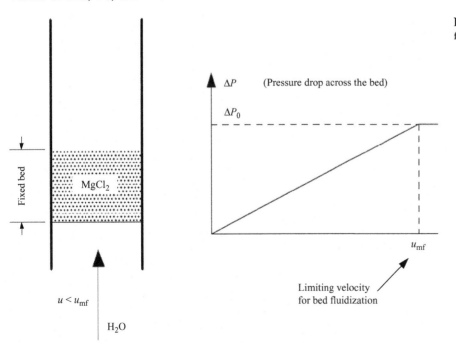

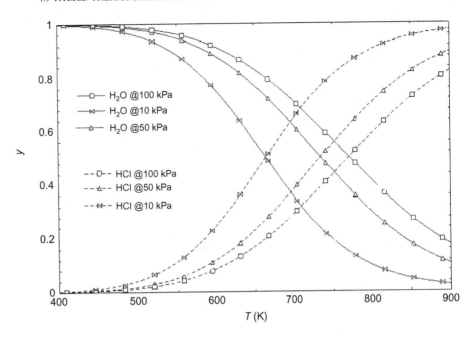

FIG. 4.53 Equilibrium molar fraction of steam and hydrochloric acid at high temperature hydrolysis of MgCl₂.

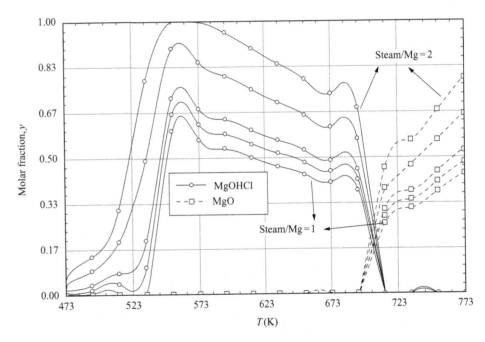

FIG. 4.54 Equilibrium molar fractions in solids MgOHCl and MgO at MgCl₂ hydrolysis.

The specific heat of anhydrous magnesium chloride (MgCl₂) in solid form depends on temperature. In Ozcan (2015), an equation of the specific heat is given, as follows:

$$\text{MgCl}_2(s): C_p(T[^\circ C > 0]) \left[\frac{\text{J}}{\text{molK}}\right] = 79.1 + 0.0059T - 860,000T^2 \tag{4.125}$$

The specific heat of MgOHCl is also given in Ozcan (2015) as follows:

$$\text{MgOHCl}(s): C_p(T[^\circ C > 0]) \left[\frac{\text{J}}{\text{molK}}\right] = 56.8 + 0.0605T \tag{4.126}$$

The specific heat of the solid magnesium oxide is determined in Ozcan (2015) as follows:

$$MgO(s) : C_p(T[^\circ C > 0]) \left[\frac{J}{mol K}\right] = 45.45 + 0.005T - 873,000T^2 \tag{4.127}$$

Ozcan (2015) reported that low temperature hydrolysis is advantageous for lower steam consumption and, therefore, lower required energy input. The full hydrolysis can be conducted in two steps, namely after MgOHCl is obtained, a heat addition process will release HCL without the addition of steam. It is necessary to first remove any contamination with HCl of the MgOHCl product, before any heat addition; otherwise, an undesired MGCl$_2$ may be formed again. However, in the MgCl$_2$-MgOHCl cycle, there is no need for MgO generation because the MgOHCl is to be chlorinated, according to Eq. (4.120).

4.5.4.3 Chlorination Reactor

Two types of chlorination processes are possible, depending on the cycle configuration. Chlorination of MgO is slightly exothermic and evolves according to Eq. (4.118). Chlorination reaction of MgOHCl by Eq. (4.120) is slightly exothermic with a standard heat reaction of 51 kJ/mol at 825 K. This reaction might be feasible in terms of chlorination; however, the additional heat requirement drastically increases and results in lower efficiency of the cycle.

Regarding the MgO chlorination, Ino and Ochiai (1961) concluded that the reaction must be catalyzed with charcoal in order to obtain complete conversion at 673–773 K. The main challenge with MgO chlorination is due to the formation of an ash layer of MgCl$_2$ on the surface of MgO conversion that slows down the reaction. Preparation of the MgO particles also has a very influential effect on the chlorination process. MgO produced from high temperature MgCl$_2$ hydrolysis has been less reactive and has had less surface area than the MgO produced by Mg(OH)$_2$ decomposition. The MgO produced from decomposition of MgOHCl has shown better surface and activity characteristics in a chlorination environment compared to direct hydrolysis of MgCl$_2$. Diffusion of HCl into an ash layer of MgCl$_2$ particles has been much better than Cl$_2$ gas alone, where the conversion rate has been increased for MgO.

The gas-solid reactions—such as that of MgO or MgOHCl chlorination—can be analyzed numerically based on two well-known models: the uniform conversion model and the shrinking core model. These mechanistic models are briefly explained here and are shown in Fig. 4.55. The uniform-conversion model has a reaction process that proceeds throughout the solid particle, as shown in Fig. 4.55a. The uniform-reaction model is applicable when the diffusion of a gaseous reactant into a particle is much faster than the chemical reaction. The solid reactant is consumed nearly uniformly throughout the particle.

On the other hand, the shrinking core model is applicable if the diffusion of gaseous reactant is much slower and restricts the reaction zone to a thin layer that advances from the outer surface into the particle. The solid layer configuration of a reacting particle for a shrinking core model and the overall process is shown in Fig. 4.55b. Diffusion of a gaseous reactant occurs through a film surrounding the particle, after which the reactant penetrates and diffuses through a layer of ash to the surface of the unreacted core. The shrinking-core model for spherical particles predicts the reaction of the gaseous reactant with the solid at the particle surface, and diffusion of gaseous products through the ash, back to the exterior surface of the solid. The reaction conditions are assumed to be isothermal, with a constant pellet size, and pseudo-steady state conditions that are valid for gas-solid reactions.

Regarding the MgOHCl chlorination process, Eom et al. (2010) conducted an experiments that demonstrated that, at a temperature of 473 K, anhydrous MgCl$_2$ was produced with chemical residues in a proportion of 35%.

Packed bed and fluidized bed reactors were considered for the chlorination reaction. A chlorination simulation was conducted by Ozcan (2015) with the help of ASPEN Plus software. The results, as shown in Fig. 4.56, demonstrate that

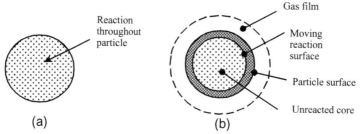

FIG. 4.55 Gas-solid reaction models: (a) uniform conversion model (UCM) and (b) shrinking core model (SCM).

FIG. 4.56 Aspen Plus simulations of MgO chlorination showing the reactant and product molar rates for a range of temperatures. *Modified from Ozcan, H., 2015. Experimental and Theoretical Investigations of Magnesium-Chlorine Cycle and Its Integrated Systems. PhD Thesis, University of Ontario Institute of Technology.*

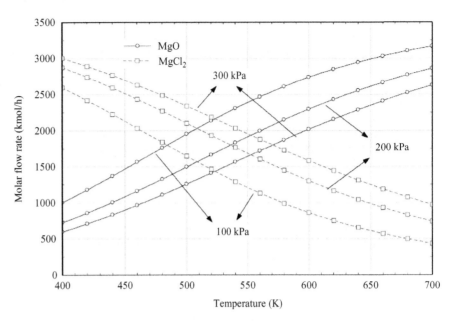

a decreased pressure and increased temperature influence favorably the formation of magnesium chloride. At 773 K, the MgO chlorination reaction heat is −31.51 MJ per kmol of H_2. This heat can be compared to the reaction heat of MgOHCl chlorination that is 116.3 MJ per kmol H_2 at 810 K.

4.5.4.4 Dry hydrochloric Acid Capture Process and System

In the magnesium-chlorine cycle, it is important to capture hydrochloric acid from steam (water) and thus to obtain a pure HCl gas for the electrochemical process. A pure HCl electrolysis is very advantageous. The magnesium chloride itself is a desiccant able to capture up to 12 moles of water for one mole of $MgCl_2$. The interaction of magnesium chloride with water can occur in four steps. The first step is hydration of the pure magnesium chloride, as follows:

$$MgCl_2(s) + H_2O(g) \leftrightarrow MgCl_2 \cdot H_2O \tag{4.128}$$

The water absorption reaction form Eq. (4.128) is exothermic and spontaneous and evolves at equilibrium. More water can be bonded gradually to the $MgCl_2$-H_2O system as

$$MgCl_2 \cdot H_2O(s) + H_2O(g) \leftrightarrow MgCl_2 \cdot 2H_2O(s) \tag{4.129}$$

then

$$MgCl_2 \cdot 2H_2O(s) + 2H_2O(g) \leftrightarrow MgCl_2 \cdot 4H_2O(s) \tag{4.130}$$

and then

$$MgCl_2 \cdot 4H_2O(s) + 2H_2O(g) \leftrightarrow MgCl_2 \cdot 6H_2O(s) \tag{4.131}$$

Lowering the temperature helps the water absorption process or the $MgCl_2$ hydration. However, the tendency of $MgCl_2$ to absorb either more H_2O or more HCl is not a well-known process, even if solubility of these binary and ternary systems were reported previously. Some literature sources suggest that, in the presence of HCl, the $MgCl_2$-H_2O system forms a ternary mixture that can bond HCl. Based on these observations, Ozcan (2015) conducted an experimental study focused on capturing HCl through the ternary HCl-$MgCl_2$-H_2O system to release a dry hydrochloric acid within the magnesium-chlorine cycle. The expected form of the $MgCl_2$ hydrate is as follows:

$$MgCl_2(s) + HCl(aq, in\ mH_2O) \leftrightarrow MgCl_2 \cdot nH_2O \cdot mH_2O(s) + (1-n)HCl(g) \tag{4.132}$$

This ternary mixture was generally studied by Li et al. (2006) based on the Pitzer ion interaction model. The solubility of HCl was lower at a higher rate of $MgCl_2$ in the ternary system. The ternary system led to production of $HCl \cdot MgCl_2 \cdot 7H_2O$. The resulting data suggested that it might be possible for $MgCl_2$ to absorb more H_2O than HCl with a possible concentrated mixture. Higher HCl concentration leads to $HCl \cdot MgCl_2 \cdot 7H_2O$ formation. It has been suggested that the HCl concentration should be controlled to prevent double salt formation. Existing studies have not validated any results for the higher temperature absorption of HCl gas, or occurrence of double salts.

Ozcan (2015) conducted a preliminary experiment of $MgCl_2$ hydration with an aqueous HCl solution aimed to measure the bonding behavior of HCl and therefore to assess the applicability of $MgCl_2$ to HCl capture and release as dry gas. In this experiment, a fixed bed reactor is set up with a HCl gas capturing flask as shown in Fig. 4.57. In the reactor, an amount of 0.46 mol of anhydrous $MgCl_2$ powder is placed under dry glovebox conditions. A hydrochloric acid solution of 5.5 M was added with a syringe. As soon as the solution passed through the distributor, a very fast reaction at the top of solid particles occurred resulting in a solidification of the powder. The formed and saturated product at the top, being mostly $MgCl_2 \cdot nH_2O \cdot mH_2O(s)$, did not let the remaining solution pass through the powder and stopped the reaction. The clog is shown in the detail in the photograph in Fig. 4.58.

Furthermore, a mixer was used to locate holes on the solidified product. Another fast reaction occurred with the collected solution resulting in a very fast temperature increase up to 36°C after mixing. This resulted in the release of some solution in gaseous form in the glovebox ambient and failed to pass through the lower side distributor.

As soon as the reactor is kept under ambient air, the temperature of the reactor does not decrease to the desired value; this results in the release of more HCl gas from the solution. Thus, the reactor is kept under 20°C to decrease the gas release.

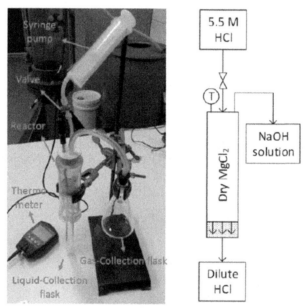

FIG. 4.57 Dry HCl capture experiment in $MgCl_2$ salt at UOIT. *Courtesy photo from Ozcan, H., 2015. Experimental and Theoretical Investigations of Magnesium-Chlorine Cycle and Its Integrated Systems. PhD Thesis, University of Ontario Institute of Technology.*

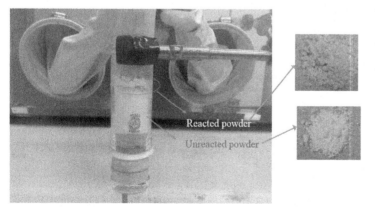

FIG. 4.58 Detail photo of $MgCl_2$ hydration and clog formation of $MgCl_2 \cdot nH_2O \cdot mH_2O(s)$. *Courtesy photo from Ozcan, H., 2015. Experimental and Theoretical Investigations of Magnesium-Chlorine Cycle and Its Integrated Systems. PhD Thesis, University of Ontario Institute of Technology.*

Another HCl capturing process relevant to the magnesium-chlorine cycle is the hydrochlorination of magnesium oxide. Magnesium oxide has a very low solubility to water and preferably bonds HCl chemically, forming MgOHCl as follows:

$$MgO(s) + HCl(Humid) \leftrightarrow MgOHCl(s) + Moisture \tag{4.133}$$

The separated MgOHCl(s) can now be decomposed through a thermolysis reaction, namely:

$$MgOHCl(s) \leftrightarrow MgO(s) + HCl(g) \tag{4.134}$$

There is a competing hydrochlorination process of magnesium oxide that is devoured when steam is given in excess to the reaction mixture as follows:

$$MgO(s) + 2HCl \leftrightarrow MgCl_2(s) + H_2O(g) \tag{4.135}$$

Another concurrent reaction may be written as follows:

$$MgO(s) + H_2O \leftrightarrow Mg(OH)_2 \tag{4.136}$$

The reaction from Eq. (4.136) was studied by Lamy et al. (2004) in a molten MgOHCl reactor comprising MgO solid particles. The hydrochloric acid gas was bubbled through the bed and interacted with an MgO particles with which HCl reactes at the surface. Another separation process is then required to recover the unreacted MgO particles and recycle them. The simulations in ASPEN also reported by Ozcan (2015) show that the main products of hydrochlorination reaction by Eq. (4.130) are MgOHCl, $Mg(OH)_2$, and $MgCl_2$ and its hydrates $MgCl_2 \cdot H_2O$, $MgCl_2 \cdot 4H_2O$ and $MgCl_2 \cdot 6H_2O$. Formation of MgOHCl initiates slightly at 398 K and starts decomposing into HCl and MgO after 523 K. The formed $Mg(OH)_2$ can be decomposed into H_2O and MgO at 605 K. A controlled dehydration process can be made at this temperature range without jeopardizing a possible HCl/steam mixture in the decomposition reaction of MgOHCl.

Ozcan (2015) experimented with the hydrochlorination process in a fixed bed reactor made of Pyrex cylindrical glass, as shown in Fig. 4.59. The setup comprises an aqueous HCl supply vessel, a nitrogen supply, a preheater, a reactor, an electric heater, a Digi-Sense temperature controller, an air-cooled pre-cooler, a water-cooled condenser and a HCl capturing vessel. When the reactor temperature is cooler than the desired temperature, the temperature controller assures that the reactor temperature is kept at the desired level.

The aqueous HCl solution is enclosed in a glass vessel placed on a hot plate. Nitrogen gas at a known flow rate is bubbled through the liquid HCl solution such that a gaseous mixture of $HCl/H_2O/N_2$ can be extracted above the liquid at a desired temperature. The aim of the experiment is to extract dry HCl out of the humid mixture. Therefore, the gas mixture is preheated and then heated into the fixed bed reactor. An electric tape heater is used, and the reactor is thermally insulated. The temperature level is adjusted and maintained by a temperature controller.

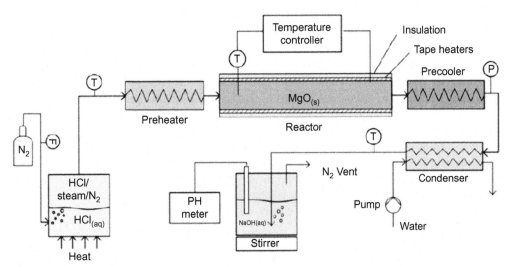

FIG. 4.59 Fixed bed reactor and its setup for MgO hydrochlorination with HCl gas capture developed at UOIT. *Modified from Ozcan, H., 2015. Experimental and Theoretical Investigations of Magnesium-Chlorine Cycle and Its Integrated Systems. PhD Thesis, University of Ontario Institute of Technology.*

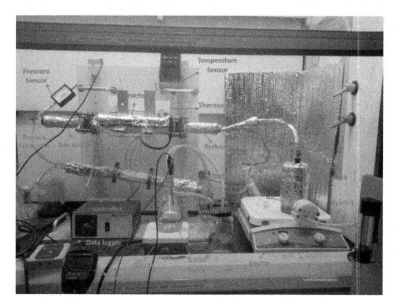

FIG. 4.60 Hydrochlorination reactor setup at UOIT. *Courtesy photo from Ozcan, H., 2015. Experimental and Theoretical Investigations of Magnesium-Chlorine Cycle and Its Integrated Systems. PhD Thesis, University of Ontario Institute of Technology.*

Temperature probes are placed in three points as shown in the figure. In the reactor, a fixed bed of MgO(s) powder is placed at a known amount. After passing through the bed at the desired temperature, the gaseous mixture reacts and retains mostly HCl. Afterward, the stream is cooled, and the moisture condensed and discharged into a titration vessel comprising sodium hydroxide solution. The nitrogen separates and releases as gas; water remains in a condensed form, and the HCl still present in the stream will dissolve in the solution. The flow rate of hydrogen and pH of the titration solution are measured with an Oakton Ion-2700 pH meter.

An overall photograph of the setup, which was placed in a fume hood, is shown in Fig. 4.60. The reactor was filled with approximately 5 g of MgO powder that were distributed evenly. The titration medium was 200 ml of 0.5 M solution of NaOH for which the measured pH was 12.7. The gaseous reactants were preheated to 418 K to prevent water condensation. The reactor temperature was measured with a J-type thermocouple. A vane was located to relieve possible overpressures during the experiment. The pressure of the reactor was also measured to prevent the reactor from instant pressure changes.

The precipitation titration method was selected to measure the amount of Cl^- ions in the solution. For this purpose, 1 M $AgNO_3$ was selected as the precipitating agent. The reaction between the silver cation and the chlorine halide was

$$Ag^+(aq) + Cl^-(aq) \rightarrow AgCl_2(s) \tag{4.137}$$

where the reaction of NaCl and $AgNO_3$ evolved as follows:

$$AgNO_3 + NaCl \rightarrow AgCl(s) + NaNo_3(aq) \tag{4.138}$$

The precipitation titration has been used with a burette and pH meter on a 5 ml sample of analyzed solution. Determination of the total HCl capture by the MgO particles was made by the difference of HCl amount in the 5.5 M HCl and the NaOH solution.

Further, thermogravimetric analysis was conducted to determine amounts of captured species on the surface of MgO particles. These results are summarized as shown in Figure 4.61. A Hitachi-STA 7300 TGA/DTA device with uncertainty values of 0.2 μg for TGA and a 0.06 μV for DTA was used for testing the sample under nitrogen environment. A random 16.5 mg wet sample was first heated up to 120°C and the temperature kept at this rate approximately 20 min for a fully dried sample. In 30 min, the sample lost 43.7% of its weight that corresponded to the unreacted water and HCl content. The dried sample was further heated to 623 K with a heating rate of 10 K/min. The $MgCl_2 \cdot mH_2O$ hydrates were dehydrated and approximately 14.4% weight loss was observed. After 533 K, there was no change of weight, which meant that all hydrates of $MgCl_2$ were liberated. Reaction of the MgO particles with the HCl solution generally leads to high amounts of $MgCl_2$ hydrates production, which are regarded as unwanted side reactions. Thus the reactor temperature should be kept above 423 K, at least to prevent hydrate formations.

The characteristics of the reacted solid particles were also evaluated using the Rigaku Ultra IV XRD analyzer. Continuous scanning was conducted at a scan range of 10–90° at 4000 deg/min. Scanned substances were limited to Mg-based compounds and all possible hydrates of $MgCl_2$. Five samples taken from an experiment, conducted at 548 K

FIG. 4.61 TGA analysis results of dry hydrogen capture process conducted via hydrochlorination followed by a heat addition process.

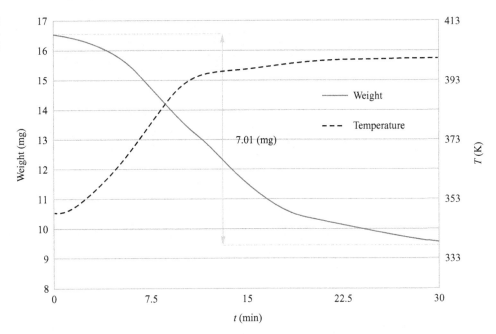

were further heated up to 673 K and kept at this temperature for 30 min in order to obtain a preliminary estimation for the dehydration of Mg(OH)$_2$. XRD analyses results shown in Figure 4.62 suggest that keeping the temperature of the reaction under 498 K is not a feasible option due to side reactions and the unwanted formation of HCl adsorbents.

The temperature range of this reaction should be kept at a specific temperature range due to the tendency of MgO to react with both steam and HCl resulting in various formations of Mg compounds. The only species at 518 K are found to be MgO, MgCl$_2$, and the remaining Mg(OH)Cl. The intensity of Mg(OH)Cl is still at a relatively high level, which shows that decomposition of this substance is not completed yet and that HCl adsorption is succeeded at a reasonable rate.

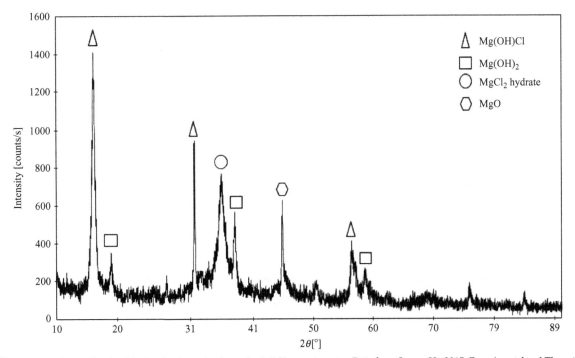

FIG. 4.62 XRD analysis of hydrochlorination samples from the 548 K experiments. *Data from Ozcan, H., 2015. Experimental and Theoretical Investigations of Magnesium-Chlorine Cycle and Its Integrated Systems. PhD Thesis, University of Ontario Institute of Technology.*

FIG. 4.63 SEM analysis of hydrochlorination samples from the 548 K experiments. *Courtesy photo from Ozcan, H., 2015. Experimental and Theoretical Investigations of Magnesium-Chlorine Cycle and Its Integrated Systems. PhD Thesis, University of Ontario Institute of Technology.*

The SEM image of the 549 K sample is shown Fig. 4.63. Carl Zeiss Gemini Fesem–type imaging equipment is used with the sample coated by Au-Pd. Imaging is made at 10K, 20K, and 30K zoom at 5 kV. At 30K zoom, it is possible to differentiate the porous surface of the MgO particles holding HCl at its surface with denser and more homogenous structure in the background.

4.5.4.5 Flowsheet Development of Magnesium-Chlorine Cycle

Two flowsheet diagrams for the magnesium-chlorine cycle of three steps were developed by Ozcan (2015) among which one is for the MgCl$_2$-MgO cycle and the other for the MgCl$_2$-MgOHCl cycle.

The ASPEN Plus software has been used to simulate the flowsheets. The results showed that the energy end exergy efficiencies of those cycles can be improved if the forth step is added. Therefore, in Ozcan (2015), a four-step cycle flowsheet was developed and assessed with ASPEN Plus accounting also for the experimental results of a dry-capture step of HCl, which were used for simulation tuning and validation.

The cycle flowsheet is shown in the diagram as illustrated in Fig. 4.64. The meaning of each component and the associated processes are given in Table 4.19. The reaction enthalpies are also in the table. For the electrochemical reaction, the Gibbs free energy of the reaction is given instead of the reaction enthalpy. Since the required steam is lower than direct hydrolysis at elevated temperatures, energy required for water superheating is significantly lower than the previous options, and the recovery potential is also higher due to an internal heat exchanging process. Using the same method for the previous options, a total external heat requirement for this cycle is found to be 244.98 MJ/kmol H$_2$.

Table 4.20 gives the content of each material stream, its temperature, specific enthalpy, specific entropy and specific total exergy. According to Ozcan (2015), the four-step cycle carries the potential to be more than 45% efficient with proper HCl electrolysis at the desired voltage values and heat exchangers having an effectiveness better than 85%. Since one of the major energy-consuming devices is the aqueous electrolysis step, a proper dry HCl capturing process can be a promising option for reduced power consumption of the cycle. The HCl capturing process can be another energy-intensive process that is expected to contribute to the total thermal energy of the cycle. Based on the flowsheet simulations of Ozcan (2015), the exergy efficiency of the cycle for an assumed 85% effectiveness of the heat exchanger is better than 53%.

4.5.5 Copper-Chlorine Cycle

The copper-chlorine thermochemical cycle—with several variants—is under development in several institutions worldwide, including the following: University Ontario Institute of Technology, Canadian National Laboratories (former AECL) and "Commisariat à Energy Atomique et aux Energies Alternatives" (CEA) in France. The U.S. Department of Energy envisions the use of this cycle in conjunction with concentrated solar power systems to generate hydrogen. The Canadian Nuclear Laboratories is interested particularly in applying the Cu-Cl cycle for hydrogen production with the next-generation supercritical water cooled reactor (SCW-CANDU) that generates process heat at 800–900 K.

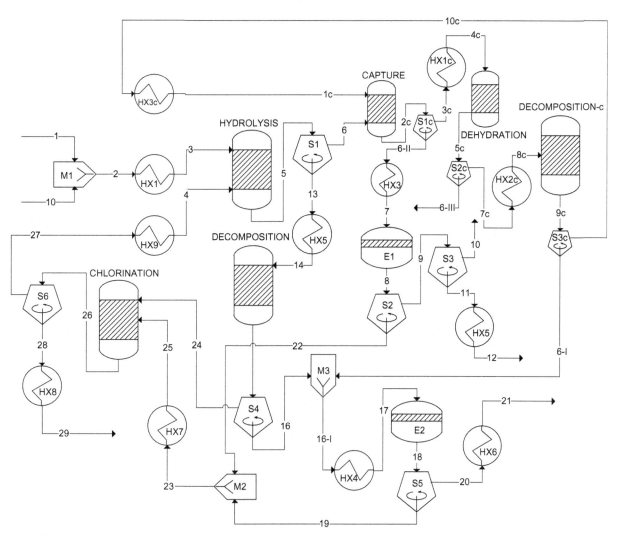

FIG. 4.64 Flowsheet of a four-step MgCl₂-MgOHCl thermochemical cycle with wet and dry HCl electrolysis. *Modified from Ozcan, H., 2015. Experimental and Theoretical Investigations of Magnesium-Chlorine Cycle and Its Integrated Systems. PhD Thesis, University of Ontario Institute of Technology.*

The copper-chlorine hybrid cycle developed from a precursor pure thermochemical cycle proposed at General Electric as discussed by Wentorf and Hanneman (1974). This cycle used compounds of copper, chlorine and magnesium. In this cycle, a cuprous chlorine complexation is a first reaction step which evolves as

$$4CuCl + \chi \rightarrow 2Cu(s) + 2CuCl_2 \cdot \chi, \text{ at around 338 K} \tag{4.139}$$

where χ is a chelating agent.

The chelating agent can be hydrochloric acid with a moderate concentration so that hydrolysis is prevented and the solubility of CuCl is reduced, or it can be the propylenediamine. In an aqueous solution comprising copper ions and the chelation agent, coordinate bonds are formed between copper and ligands of the organic compound at equilibrium, which will selectively link and dissolve the copper II ions, as follows:

$$Cu^{2+}(aq) + \chi \leftrightarrows [Cu(\chi)]^{2+}(aq) \tag{4.140}$$

The reaction step described by Eq. (4.132) is followed by a thermolysis of the formed cupric chloride complex as follows:

$$2CuCl_2 \cdot \chi \rightarrow 2CuCl_2(aq) + \chi, \text{ at around 338 K} \tag{4.141}$$

The chelation agent, the solid copper and the aqueous CuCl₂ are separated from processes media described by Eqs. (4.132) and (4.134). The overall enthalpy of reaction of this process is given as approximately 125 kJ per

TABLE 4.19 Component and Process and Parameters of Four-Step Mg-Cl Cycle

Component	Process	T (°C)	ΔH (kJ/mol H_2)
Hydrolysis	$MgCl_2$ hydrolysis	280	−17.82
Chlorination	MgO chlorination	500	−31.51
Decomposition	MgOHCl decomposition	450	118.1
E1	HCl(aq) electrolysis	70	ΔG:173.67
E2	HCl(dry) electrolysis	70	ΔG: 135.08
HX1	Water superheating	280	546.8
HX2	MgOHCl heating	450	7.8
HX3	HCl(aq) cooling	70	−483.8
HX4	HCl(dry) cooling	70	−11.2
HX5	H_2 cooling	25	−0.65
HX6	H cooling	25	−0.65
HX7	Cl_2 heating	500	15.61
HX8	O_2 cooling	25	−7.46
HX9	$MgCl_2$ cooling	280	−17.8

TABLE 4.20 Stream Parameters in the Four-Step $MgCl_2$-MgOHCl Cycle

State	Material	T (°C)	h (kJ/kmol)	s (kJ/kmol.K)	ex (kJ/kmol)
1	H_2O	25	−285,820	−163.142	3086.227
2	H_2O	65.9	−282,750	−153.471	3274.298
3	H_2O	280	−233,040	−22.9991	14143.74
4	$MgCl_2$	280	−621,940	−119.425	157298.5
5	Mixture	280	−266,930	−31.1178	n/a
6	H_2O/HCl	280	−219,570	−15.8138	13762.5
7	H_2O/HCl	70	−263,540	−132.758	13,192
8	Mixture	70	−255,400	−133.775	n/a
9	H_2/H_2O	70	−268,010	−142.293	n/a
10	H_2O	70	−282,440	−152.569	3315.502
11	H_2	70	1301.857	4.171019	238547.4
12	H_2	25	0.987862	0.107765	238457.4
13	MgOHCl	280	−787,930	−199.462	148329.7
14	MgOHCl	450	−780,160	−187.201	152,446
15	Mixture	450	−331,050	−16.4064	n/a
16	HCl	450	−79812.6	36.14458	90679.49
17	HCl	70	−91027.6	14.17917	86010.2
18	Mixture	70	1406.357	10.28291	n/a
19	Cl_2	70	1484.522	4.822405	117595.9
20	H_2	70	1301.857	4.171019	238547.4

TABLE 4.20   Stream Parameters in the Four-Step MgCl$_2$-MgOHCl Cycle—cont'd

State	Material	T (°C)	h (kJ/kmol)	s (kJ/kmol.K)	ex (kJ/kmol)
21	H$_2$	25	0.987862	0.107765	238458.4
22	Cl$_2$	70	1484.522	4.822405	117595.9
23	Cl$_2$	70	1484.522	4.822405	117595.9
24	MgO	450	−582,290	−68.9573	66629.28
25	Cl$_2$	500	17097.69	34.19978	124454.6
26	Mix	500	−397,800	−51.694	n/a
27	MgCl$_2$	500	−604,170	−92.4003	167015.3
28	O$_2$	500	14926.99	29.71871	10042.7
29	O$_2$	25	−7.95734	0.088981	3937.408

reaction. However, the overall Gibbs free energy is very positive, namely approximately 160 kJ per reaction, indicating that relatively high mechanical work is required to conduct the involved separation processes. After these processes, the aqueous cupric chloride is dried to remove water at 373 K

$$2CuCl_2(aq) \rightarrow 2CuCl_2(s), \text{ at } 373 \text{ K} \tag{4.142}$$

and then cupric chloride is thermolyzed to release chlorine at 773 K, as follows:

$$2CuCl_2(s) \rightarrow 2CuCl(s) + Cl_2(g) \text{ at } 773 \text{ K} \tag{4.143}$$

The reaction from Eq. (4.143) evolves at equilibrium with a heat of reaction of 125 kJ. The resulted chlorine gas is further reacted in a low temperature, spontaneous and exothermic process involving magnesium hydroxide in aqueous solution, as follows:

$$Cl_2(g) + Mg(OH)_2 \rightarrow MgCl_2(aq) + H_2O(l) + 0.5O_2(g) \text{ at } 353 \text{ K} \tag{4.144}$$

The resulting magnesium chloride is separated (e.g., by drying) and heated up to 623 K where a reaction step—the hydrolysis—is conducted, as follows:

$$MgCl_2(s) + 2H_2O(g) \rightarrow Mg(OH)_2(s) + 2HCl(g) \text{ at } 623 \text{ K} \tag{4.145}$$

The reaction enthalpy of Eq. (4.146) is 16.7 kJ per reaction, and Gibbs free energy is positive of 33.4 kJ, meaning that the yield is marginal; therefore, intense recycling may be needed. The hydrochloric acid that evolves as gas is easily separated from the preceding reaction and is used as a reactant in a final reaction step, described as follows:

$$2Cu(s) + 2HCl(aq) \rightarrow 2CuCl(s) + H_2(g) \text{ at } 373 \text{ K} \tag{4.146}$$

The preceding hydrochlorination reaction evolves at rather low temperature and generates pure hydrochloric acid, exothermically. The heat of reaction is −33.5 kJ per reaction with a slightly positive Gibbs free energy of 8.3 kJ. The reaction requires fine copper particles immersed in water and HCl gas that passed through water at vicinity of its normal boiling point. The resulted solid product will then dissolve in hydrochlorinated water forming a CuCl(aq) that is separable from the solid copper. About nine moles of water and two moles of HCl will dissolve a mole of CuCl at a temperature in the range of 353–373 K. The aqueous solution of CuCl can then be cooled and the solubility essentially reduced, therefore, CuCl will crystallize and will be easily separated.

The process described by Eqs. (4.132)–(4.139) is a closed loop process with mingled reactions between what we today categorize as Mg-Cl and Cu-Cl cycles. The presence of magnesium may not be beneficial when side reactions are considered because products such as MgO, MgOHCl can be formed.

A later, similar cycle that, instead of Mg, has intermediary Fe compounds was developed at the US Institute of Gas Technology in 1980 as reported in Carty et al. (1981), namely the cycle denoted as B-16. Nevertheless, simplifying those Cu-Cl-Fe or Cu-Cl-Mg cycles by eliminating the Fe, Mg from the chemical loop will simplify the process.

Carty et al. (1981) investigated the hydrolysis of cupric chloride aiming at eliminating iron from the B-16 cycle. Ferrous chloride was hydrolyzed in the presence of solid cupric chloride and formed a novel mixed copper oxide-cupric chloride, which is a coordination complex, namely $CuO \cdot CuCl_2$, which today is known as copper oxychloride. They also found that the presence of iron is not necessary in the hydrolysis of cupric chloride, which goes to completion at 675 K. Thereafter, IGT proposed an oxygen evolution step by thermolysis of copper oxychloride at 800 K without pursuing the validation tests. The resulting Cu-Cl cycle incorporating this hydrolysis was a five-step process denoted by IGT as H-6. The thermolysis of copper oxychloride was confirmed much later by experimental work performed at Argonne National Laboratory (ANL), which was reported in Lewis et al. (2003).

Several versions of Cu-Cl cycles were evolved and particular processes studied in detail. An integrated engineering scale demonstration unit of $50 \, Nm^3$ hydrogen per day is presently under development at UOIT. In past literature, a total of seven versions of the copper-chlorine cycle were reported: one cycle with two steps, two versions with three steps, two cycles with four steps and one five-step cycle. The cycle and its variants comprise thermochemical, or electrochemical, or physicochemical, or thermophysical processes. These processes were categorized in Naterer et al. (2013) in 12 types that are discussed next.

The first process is cuprous chlorine electrochemical disproportionation evolving at 350 K in an electrochemical cell with acidic electrolyzer consisting of a HCl solution. In this process, a complexation of cupric chloride occurs, namely hydrated cupric chloride is produced in the form of the complex $CuCl_2 \cdot nH_2O$ that is present in the aqueous phase (dissolved). Copper precipitates as solid particles of micron size. This process—denoted as Process 1—is described as follows:

$$P1 : 4CuCl(s) + nH_2O(l) \rightarrow 2Cu(s) + 2CuCl_2 \cdot nH_2O(aq) \text{ at } 350 \text{ K} \tag{4.147}$$

The second process (P2) is the low temperature hydrolysis of cupric chloride. This process involves gas and solid phases and evolves at 650 K, as follows:

$$P2 : H_2O(g) + 2CuCl_2 \cdot nH_2O(s) \rightarrow CuO \cdot CuCl_2(s) + 2HCl(g) + nH_2O(g) \text{ at } 650 \text{ K} \tag{4.148}$$

In Eq. (4.148), $2CuCl_2 \cdot nH_2O(s)$ is a hydrated cupric chloride in solid form. When hydrolyzed with steam, the hydrated cupric chloride solid copper oxychloride is formed together with a hydrochloric acid. The third process is dry hydrolysis, which requires preliminary drying of the hydrated cupric chloride. No steam is released in this case. The reaction occurs at 673 K, as follows:

$$P3 : H_2O(g) + 2CuCl_2(s) \rightarrow CuO \cdot CuCl_2(s) + 2HCl(g) \text{ at } 673 \text{ K} \tag{4.149}$$

A high temperature hydrolysis process is inventoried as Process 4 (P4) and occurs at 823 K using hydrated cupric chloride as reactant and generates oxygen, hydrochloric acid and molten cuprous chloride. This thermochemical reaction is described, as follows:

$$P4 : H_2O(g) + 2CuCl_2 \cdot nH_2O(s) \rightarrow \frac{1}{2}O_2(g) + 2HCl(g) + 2CuCl(l) + nH_2O(g) \text{ at } 823 \text{ K} \tag{4.150}$$

There is also a high temperature hydrolysis process of chlorine—encountered in the three-step cycle of Dokyia and Kotera (1976)—that is a reverse Deacon reaction at 1073 K, noted as Process 5 and described by the following equation evolving at 1073 K:

$$P5 : H_2O(g) + Cl_2(g) \rightarrow 0.5O_2(g) + 2HCl(g) \text{ at } 1073 \text{ K} \tag{4.151}$$

Process six (P6) is inventoried as a hydrochlorination of copper at 723 K described by the following thermochemical reaction:

$$P6 : 2Cu(s) + 2HCl(g) \rightarrow H_2(g) + 2CuCl(l) \text{ at } 723 \text{ K} \tag{4.152}$$

The electrochemical process of hydrochlorination of cuprous chloride evolves at approximately 350 K and generates hydrogen gas and hydrated cupric chloride as the process seven (P7) described as follows:

$$P7 : nH_2O(l) + 2CuCl(s) + 2HCl(aq) \rightarrow H_2(g) + 2CuCl_2 \cdot nH_2O(aq) \text{ at } 350 \text{ K} \tag{4.153}$$

A complexation process occurs in the electrolyte of Eq. (4.153) and forms a hydrated cupric chloride complex in the aqueous phase at the anode, namely $CuCl_2 \cdot nH_2O(aq)$, $n > 0$. Electrochemical cells for this reaction are in development at UOIT, ANL, AECL, and GTI (Gas Technology Institute, USA).

The thermolysis of hydrated cupric chloride is denoted as oxychlorination and inventoried as process eight (P8). The sole reactant is the complex $CuCl_2 \cdot nH_2O(s)$. In this reaction, copper oxychloride is formed together with gaseous hydrochloric acid. This reaction is conducted at 800 K, as follows:

$$P8 : 2CuCl_2 \cdot nH_2O(s) \rightarrow CuO \cdot CuCl_2(s) + 2HCl(g) + (n-1)H_2O(g) \text{ at } 800 \text{ K} \tag{4.154}$$

Thermolysis of copper oxychloride is referred as process nine (P9). This is a thermal decomposition that releases oxygen and forms molten cuprous chloride. The reaction is conducted at 800 K as given by the following equation, namely:

$$P9 : CuO \cdot CuCl_2(s) \rightarrow 0.5O_2(g) + 2CuCl(l) \text{ at } 800 \text{ K} \tag{4.155}$$

The thermolysis of the hydrated cupric chloride is denoted as process ten (P10). This thermochemical process occurs at 773 K and evolves chlorine gas, steam and molten cuprous chloride from a hydrated cupric chloride. The process is described as follows:

$$P10 : 2CuCl_2 \cdot nH_2O(s) \rightarrow 2CuCl(l) + Cl_2(g) + nH_2O(g) \text{ at } 773 \text{ K} \tag{4.156}$$

Dehydration of wet cupric chloride can be achieved by spray drying, in which the moisture content is reduced partially or totally as described next for the process denoted P11—spray drying:

$$P11 : 2CuCl_2 \cdot nH_2O(s) \rightarrow 2CuCl_2 \cdot mH_2O(s) + (n-m)H_2O(g), n > m \text{ at } 530 \text{ K} \tag{4.157}$$

Finally, a crystallization drying process (P12) does exist in which anhydrous cupric chloride is obtained at less than 373 K, as follows:

$$P12 : 2CuCl_2 \cdot nH_2O(aq) \rightarrow 2CuCl_2(s) + nH_2O(l) \text{ at } < 373 \text{ K} \tag{4.158}$$

The five-step cycle version of the copper-chlorine cycle was the first proposed and studied. The Institute of Gas Technology inventoried this cycle as H-5 in Carty et al. (1981). In Lewis and Masin (2009), this cycle was denoted as Cu-Cl-A. This cycle has a maximum temperature requirement of 800 K. This cycle is denoted here as Cu-Cl-5 and is described in Table 4.21 and in the simplified diagram shown in Fig. 4.65.

Fig. 4.66 shows the simplified flowchart of the hybrid Cu-Cl-5 hydrogen production plant. This cycle comprises a hydrolysis reactor R1, a thermolysis reactor R2, a chlorination reactor R3, a dehydration reactor D, a separator S and nine heat exchangers. Water is supplied as liquid in #1, then heated, boiled and superheated in Hx1 and provided as gas—stream #2—to hydrolysis reactor R1. In the hydrolysis reactor, solid copper oxychloride is formed and collected

TABLE 4.21 The Five-Step Hybrid Copper-Chlorine Cycle, Cu-Cl-5 Configuration

Cycle step	Chemical equation	T (K)
Electrochemical disproportionation	$4CuCl(s) + nH_2O(l) \rightarrow 2Cu(s) + 2CuCl_2 \cdot nH_2O(aq)$	350
Hydrolysis	$H_2O(g) + 2CuCl_2 \cdot mH_2O(s) \rightarrow CuO \cdot CuCl_2(s) + 2HCl(g) + mH_2O(g)$	650
Hydrochlorination	$2Cu(s) + 2HCl(g) \rightarrow H_2(g) + 2CuCl(l)$	723
Thermolysis	$CuO \cdot CuCl_2(s) \rightarrow 0.5O_2(g) + 2CuCl(l)$	800
Spray drying	$2CuCl_2 \cdot nH_2O(aq) \rightarrow 2CuCl_2 \cdot mH_2O(s) + (n-m)H_2O(g), n > m$	<530

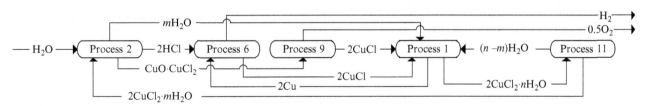

FIG. 4.65 Simplified representation of Cu-Cl-5 cycle process.

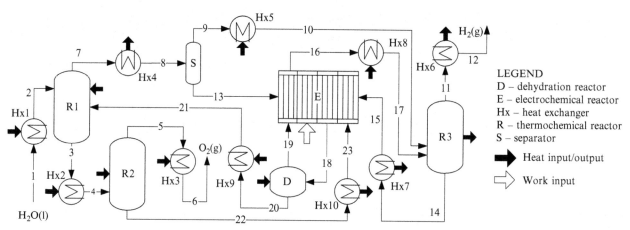

FIG. 4.66 Simplified flowchart of Cu-Cl-5 hydrogen production plant.

at the reactor bottom as stream #3. Additional heating is supplied to the copper oxychloride stream in Hx2 prior to the supply the thermolysis reactor R3 as stream #4.

Oxygen evolves from the thermolysis process—stream #5—which is cooled in heat exchanger system Hx3 and delivered in #6. The hydrolysis reaction also produces a gaseous mixture of hydrochloric acid and steam in #7, which is cooled in Hx4 in view of separation. It is assumed that stream #8 comprises liquid water and aqueous hydrochloric acid that are separated by a selected method (e.g., cascaded distillation) in separator S. Hydrochloric acid results as stream #9, further heated in Hx5 and supplied to the chlorination reactor R3 in a gas phase as stream #10.

The chlorination process releases pure hydrogen that is extracted at #11, cooled in Hx6 and delivered as product in #12. Water from separator S—stream #13—is supplied to the electrochemical cell E. In the chlorination process, molten cuprous chloride is generated and collected at the bottom of the reactor in #14. It is further cooled and solidified in Hx7 and supplied as solid particles to the electrochemical cell in #15. The electrochemical cell disproportionates CuCl and generates particulate copper—stream #16—which is heated to the chlorination reaction temperature in Hx and supplied as stream #17 to R3.

The electrochemical process also produces hydrated cupric chloride in #18, which is dehydrated in the spray drier, D, from which water is returned as stream #19 to the electrochemical cell, E. The dehydrated cupric chloride—stream #20—is heated in Hx9 to the temperature level of the hydrolysis reaction and supplied as hot stream #21 of metallic particles to R1. The thermolysis reactor R2 produces molten cuprous chloride in #22 that is cooled and solidified in Hx10. A stream of solid CuCl particles is provided to the electrochemical cell.

Two-cycle versions do exist for the four-step Cu-Cl cycle. First, the so-called Cu-Cl-4A cycle is introduced. This cycle was proposed in Wang et al. (2009), having a maximum temperature requirement of 800 K. The cycle steps are described in Table 4.22.

The four-step Cu-Cl-4A process is shown in a simplified manner in Fig. 4.67. The simplified flowchart of a hydrogen generation plant operating with Cu-Cl-4A cycle is shown in Fig. 4.68. In this plant, supply water in the liquid phase is fed to the electrochemical cell E as stream #1. The electrochemical cell produces hydrated cupric chloride in the aqueous phase represented by stream #2 in the flowchart. This is heated in the heat exchanger system Hx1 and fed into oxychlorination reactor R1 (process P8). In this reactor, a stream of gaseous products consisting of hydrochloric acid and steam is released as stream #4. This is cooled to a lower temperature in heat exchanger system Hx2 and is provided as stream #5 to a separator where hydrochloric acid is separated from water. This separator presents significant engineering challenges. Dehydrated hydrochloric acid results in #6, which is heated in Hx3 to a temperature level for the chlorination reactor R3 where it enters as stream #7.

TABLE 4.22 The Four-Step Hybrid Copper Chlorine Cycle, Cu-Cl-4A Configuration

Cycle step	Chemical equation	T (K)
Electrochemical disproportionation	$4CuCl(s) + nH_2O(l) \rightarrow 2Cu(s) + 2CuCl_2 \cdot nH_2O(aq)$	350
Hydrochlorination	$2Cu(s) + 2HCl(g) \rightarrow H_2(g) + 2CuCl(l)$	723
Thermolysis	$CuO \cdot CuCl_2(s) \rightarrow 0.5O_2(g) + 2CuCl(l)$	800
Oxychlorination	$2CuCl_2 \cdot nH_2O(s) \rightarrow CuO \cdot CuCl_2(s) + 2HCl(g) + (n-1)H_2O(g)$	650

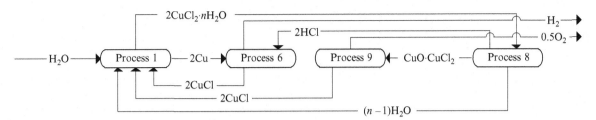

FIG. 4.67 Simplified representation of Cu-Cl-4A cycle process.

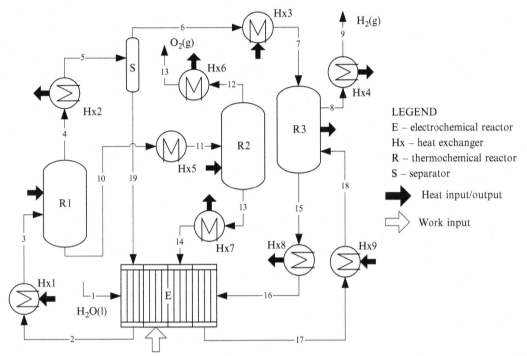

FIG. 4.68 Simplified flowchart of Cu-Cl-4A hydrogen production plant.

In this reactor, process P6 occurs and gaseous hydrogen is formed, which is extracted as stream #8. This stream is cooled in HX4 and delivered as product in #9. The other product of oxychlorination is copper oxychloride in a solid phase that separates at the bottom of the reactor as stream #10. This is heated in Hx5 and provided to the thermolysis reactor R2 at port #11. In the thermolysis reactor, the process P9 is conducted which generates hot oxygen in #12.

Furthermore, the oxygen stream is cooled by Hx6 and delivered at #13. The other product of thermolysis is molten cuprous chloride that is collected at the bottom of the reactor as stream #13. This stream is cooled by Hx7 and supplied as solid to the electrochemical cell in #14. In the chlorination reactor R3, molten CuCl is also formed at the bottom and collected as stream #15, which is cooled and solidified in Hx8 before being provided to the electrochemical cell as stream #16. The electrochemical process generates particulate copper in #17, which is heated in Hx9 to the temperature level corresponding to the chlorination reaction and further supplied to reactor R3. Water separated from hydrochloric acid—stream #19—is returned to the electrochemical cell.

The second version of a four-step Cu-Cl cycle was proposed in Naterer et al. (2010) and denoted Cu-Cl-4B. The cycle process is described as shown in Table 4.23 and in the simplified representation from Fig. 4.69. A simplified flowchart of a Cu-Cl-4B thermochemical water splitting plant for hydrogen production is shown in Fig. 4.70. In this plant, water is supplied in #1 as liquid and heated in Hx1 to a temperature level corresponding to the hydrolysis reaction that occurs in reactor R1 according to process P3.

Solid copper oxychloride is generated—stream #3—and further heated in Hx2 and then supplied as stream #4 to the thermolysis reactor R1 where it decomposes thermally according to the chemical equation described by Eq. (4.147) as process P9. During decomposition, oxygen gas is released in #5, which is cooled in Hx3 and delivered at #6. The other decomposition process is molten cuprous chloride that is collected at the bottom of R1—stream #7. Furthermore, the cuprous chloride is cooled in Hx4. It solidifies—stream #8—and is fed to the electrochemical cell E. The electrochemical

TABLE 4.23   The Four-Step Hybrid Copper-Chlorine Cycle, Cu-Cl-4B Configuration

Cycle step	Chemical equation	T (K)
Electrochemical chlorination	$2CuCl(s) + 2HCl(aq) + nH_2O(l) \rightarrow H_2(g) + 2CuCl_2 \cdot nH_2O(aq)$	350
Dry hydrolysis	$H_2O(g) + 2CuCl_2(s) \rightarrow CuO \cdot CuCl_2(s) + 2HCl(g)$	673
Thermolysis	$CuO \cdot CuCl_2(s) \rightarrow 0.5O_2(g) + 2CuCl(l)$	800
Crystallization	$2CuCl_2 \cdot nH_2O(aq) \rightarrow 2CuCl_2(s) + nH_2O(l)$	<373

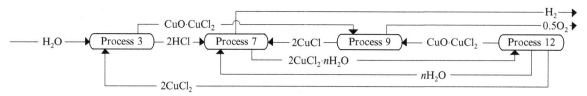

FIG. 4.69   Simplified representation of Cu-Cl-4B cycle process.

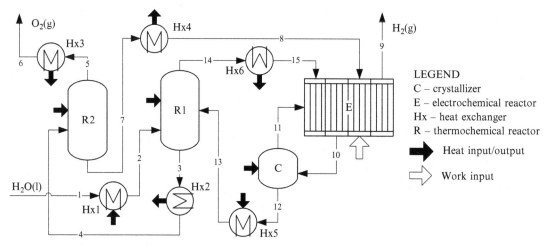

FIG. 4.70   Simplified flowchart of Cu-Cl-4B hydrogen production plant.

process P7 generates a hydrogen product stream #9. In addition, the electrochemical process produces a hydrated cupric chloride output stream, namely #10, which is directed toward a crystallization system C. The moisture is removed and dehydrated cupric chloride is obtained according to process P12. Water is returned to the electrochemical cell as stream #11. The dehydrated cupric chloride in stream #12 is heated in Hx5 and supplied as stream #13 to the hydrolysis reactor R1.

Two configurations of a three-step copper-chlorine cycle were proposed in 2005, and one older three-step cycle configuration was proposed in 1976. The cycle inventoried in Naterer et al. (2013) as Cu-Cl-3A appeared for the first time in Perret et al. (2005) and was proposed at Argonne National Laboratory. This cycle is also known as UNLV-191 and ANL-287. The electrochemical process in Cu-Cl-3A evolves according to process P1, given by Eq. (4.140).

Table 4.24 gives the cycle processes, and Fig. 4.71 shows a simplified representation of the three-step Cu-Cl-3A cycle configuration. The simplified flowchart of a Cu-Cl-3A cycle plant is shown in Fig. 4.72. This cycle comprises three reactors: R1 (hydrolysis reactor, process P4), R2 (chlorination reactor, P6), and E (electrolytic disproportionation cell, P1).

Liquid water is fed at #1 of liquid-gas separator S1, where it helps scrubbing of oxygen gas, which is delivered at #2 and mixes with aqueous hydrochloric acid and recycled water, which are then supplied to #2 of the separator. A liquid solution of water and aqueous HCl leaves the separator from the bottom location #4 and mixes in mixer M with a slurry formed of solid copper particles in liquid water, shown on the flowchart as stream #5. The resulting slurry in #5 comprises copper particles, water and hydrochloric acid. This mixture is heated in heat exchanger Hx1 to a suitable temperature for chlorination (process P6).

TABLE 4.24 The Three-Step Hybrid Copper-Chlorine Cycle, Cu-Cl-3A Configuration

Cycle step	Chemical equation	T (K)
Electrochem. disproportionation	$nH_2O(l) + 4CuCl(s) \rightarrow 2Cu(s) + 2CuCl_2 \cdot nH_2O(aq)$	350
Hydrochlorination	$2Cu(s) + 2HCl(g) \rightarrow H_2(g) + 2CuCl(l)$	723
Hydrolysis	$H_2O(g) + 2CuCl_2 \cdot nH_2O(s) \rightarrow \frac{1}{2}O_2(g) + 2HCl(g) + 2CuCl(l) + nH_2O(g)$	823

FIG. 4.71 Simplified representation of Cu-Cl-3A cycle process.

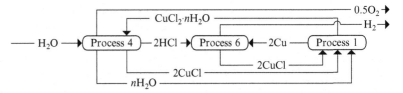

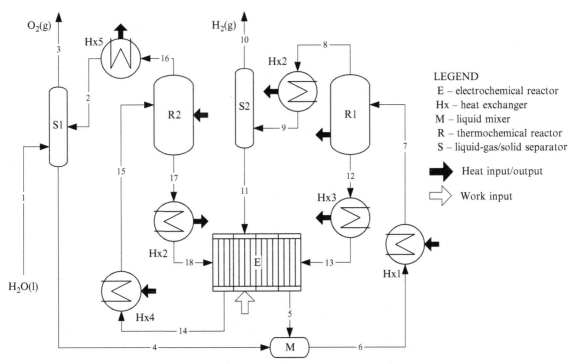

FIG. 4.72 Simplified flowchart of Cu-Cl-3A hydrogen production plant.

The stream #7 consists of steam, gaseous hydrochloric acid and copper particles. This is fed into chlorination reactor R1. At the top of the reactor, a gaseous product is generated, consisting of hydrogen and steam. This is extracted from topmost #8. The gaseous mixture is cooled in the heat exchanger system Hx2 where water condenses fully while hydrogen remains gaseous. This two-phase mixture is fed at #9 into the gas-liquid separator S2 where hydrogen separates at the top and is extracted at port #10. Liquid water is extracted from the bottom #11 and fed to the electrolytic cell E. At the bottom of chlorination reactor R1, molten cuprous chloride is formed; then collected and extracted at #12.

The molten CuCl is cooled in the heat exchanger system Hx3 such that cuprous chloride solidifies and moves as solid particles—stream #13—to the electrochemical cell E. The electrochemical cell generates solid copper in an aqueous slurry in #5. In addition, an aqueous mixture of wet cupric chloride, $CuCl_2 \cdot nH_2O(aq)$, is formed, which is extracted at #14. This aqueous solution is heated in a heat exchanger system Hx4 such that all water boils and forms a mixture of solid $CuCl_2 \cdot nH_2O(s)$ and steam, represented on the flowchart as stream #15.

This stream is fed to the hydrolysis reactor R2 where process P4 occurs. In this process, oxygen is released together with other gaseous products that are steam and hydrochloric acid. This gas mixture is indicated on the diagram as stream #16, which after cooling in heat exchanger system Hx5, becomes a two-phase mixture in #2 that comprises liquid water, aqueous hydrochloric acid and gaseous oxygen.

The Cu-Cl-3B configuration was proposed in Suppiah et al. (2005); the cycle was also denoted Cu-Cl-N in Lewis and Masin (2009). The maximum temperature requirement for this cycle configuration is 800 K. Table 4.25 gives the description of cycle steps, and Fig. 4.73 shows a simplified representation of the processes within the Cu-Cl-3B cycle. A simplified flowchart of the Cu-Cl-3B hydrogen production plant is shown in Fig. 4.74. In the Cu-Cl-3B plant, liquid water is supplied as stream #1 directly to the electrochemical cell E where process 7 occurs. During this process, hydrogen is generated and delivered as product, stream #2. In addition, wet cupric chloride is generated in aqueous solution $CuCl_2 \cdot nH_2O(aq)$, which is extracted as stream #3.

The wet cupric chloride is heated in heat exchanger system Hx1 such that all free water boils and superheats while cupric chloride remains in hydrated form as solid, $CuCl_2 \cdot nH_2O(s)$. This stream, namely stream #4, is directed toward the oxychlorination reactor R1. In this reactor, process P8 occurs, which forms solid copper oxychloride, gaseous hydrochloric acid and steam.

The gaseous phase is extracted from the top of the reactor as stream #5 and further directed toward the heat exchanger system Hx2 where water condenses and hydrochloric acid completely dissolves in the aqueous solution. The resulting liquid solution is fed to the electrochemical cell E as stream #6. The solid particles of copper oxychloride, resulting from the oxychlorination reactor, are extracted at the bottom stream #7 and heated in heat exchanger system Hx3 to the temperature level compatible with the thermolysis reaction in reactor R2.

TABLE 4.25 The Three-Step Hybrid Copper-Chlorine Cycle, Cu-Cl-3B Configuration

Cycle step	Chemical equation	T (K)
Electrochemical chlorination	$nH_2O(l) + 2CuCl(s) + 2HCl(aq) \rightarrow H_2(g) + 2CuCl_2 \cdot nH_2O(aq)$	350
Thermolysis	$CuO \cdot CuCl_2(s) \rightarrow 0.5O_2(g) + 2CuCl(l)$	800
Oxychlorination	$2CuCl_2 \cdot nH_2O(s) \rightarrow CuO \cdot CuCl_2(s) + 2HCl(g) + (n-1)H_2O(g)$	673

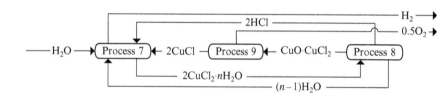

FIG. 4.73 Simplified representation of Cu-Cl-3B cycle process.

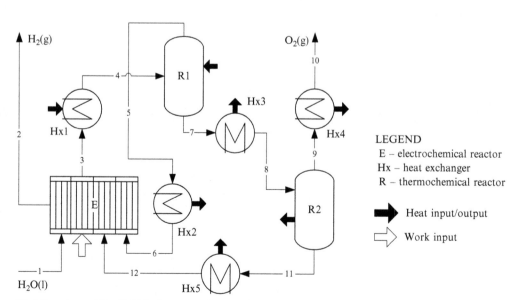

FIG. 4.74 Simplified flowchart of Cu-Cl-3B hydrogen production plant.

The copper oxychloride is supplied to the thermolysis reactor R2 as stream #8. In the thermolysis reactor, process #9 occurs and releases gaseous oxygen and forms liquid cuprous chloride. The two phases—gas and liquid—separate inside the reactor. Oxygen is extracted as gas at the top of the reactor—stream #9—and further cooled to the ambient temperature in heat exchanger system Hx4 that delivers the oxygen product as stream #10. The molten cuprous chloride, collected from the bottom of the reactor as stream #11, is cooled and solidified in the heat exchanger system Hx5 and then supplied to the electrolytic cell E as stream #12.

The three-step cycle inventoried in Naterer et al. (2013) as Cu-Cl-3C was proposed in Dokyia and Kotera (1976) and has a maximum temperature requirement of 1073 K. This cycle is denoted as UNLV-56 in McQuillan et al. (2010) and as Cu-Cl-D in Lewis and Masin (2009). The involved processes are described in Table 4.26. The cycle is represented in a simplified manner, as shown in Fig. 4.75. A simplified Cu-Cl3C flowchart is shown in Fig. 4.76.

In this cycle, water fed as liquid in #1 is heated and boiled in HX1 to approximately 1000 K to state #2 prior to entering into the hydrolysis reactor R1 that performs the reverse Deacon reaction, process P5. From this reactor, a stream #3 of gaseous products comprising oxygen and hydrochloric acid is generated.

TABLE 4.26 The Three-Step Hybrid Copper-Chlorine Cycle, Cu-Cl-3C Configuration

Cycle step	Chemical equation	T (K)
Electrochemical hydrochlorination	$2CuCl(s) + 2HCl(aq) + nH_2O(l) \rightarrow H_2(g) + 2CuCl_2 \cdot nH_2O(aq)$	350
Hydrolysis	$H_2O(g) + Cl_2(g) \rightarrow 0.5O_2(g) + 2HCl(g)$	1073
Thermolysis	$2CuCl_2 \cdot nH_2O(s) \rightarrow 2CuCl(l) + Cl_2(g) + nH_2O(g)$	773

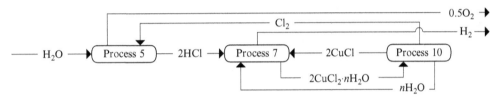

FIG. 4.75 Simplified representation of Cu-Cl-3C cycle process.

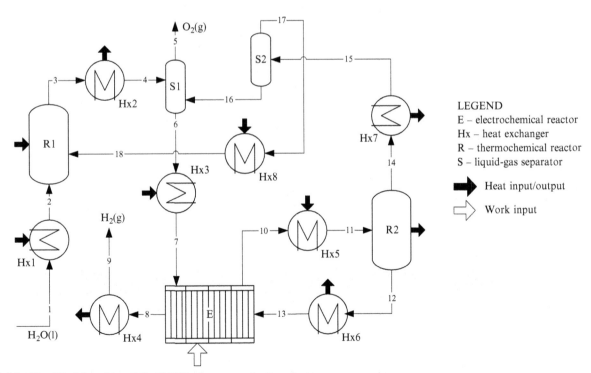

FIG. 4.76 Simplified flowchart of Cu-Cl-3C hydrogen production plant.

The gas mixture is cooled down in Hx2 to the ambient temperature and fed into a liquid-gas separator S1 where liquid water is also supplied. The hydrochloric acid separates as an aqueous solution, while gaseous oxygen is collected from the top of the separator as product stream #5. The aqueous hydrochloric acid solution—stream #6—is heated to approximately 473 K in Hx3 prior to feeding the electrochemical cell E at #7. The electrochemical cell generates pure hydrogen gas in #8, which is cooled to ambient temperature in Hx4 and delivered at #9.

The other product of the electrochemical process is the hydrated cupric chlorine—stream #10—which is heated in Hx5 prior to entering the thermolysis reactor R2 at #11. The reaction in the thermolysis reactor occurs according to process P10. It generates molten cuprous chloride at #12, which is cooled and solidified in heat exchanger system Hx6 prior to flowing into the electrochemical cell as stream #13.

The gaseous products of thermolysis consist of steam and chlorine collected from the top of the reactor as stream #14. This stream is cooled in Hx7 such that water condenses completely while chlorine remains as gas. The separation of chlorine from liquid water can be performed by distillation-based methods performed as successive boiling/condensation and gas-liquid separations, or other methods. In the flowchart, the separation process is shown in a simplified manner, such as stream #15 of water + gaseous chlorine is fed to the liquid-gas separator S2 where water separates at the bottom (stream #16) and chlorine at the top (stream #17). Water is fed to separator S1 where it dissolves the hydrochloric acid, while chlorine is heated to approximately 1000 K and supplied to the reverse Deacon reaction reactor at #18.

There is a two-step copper-chlorine cycle proposed in Dokyia and Kotera (1976) and inventoried in Lewis and Masin (2009) as Cu-Cl-C cycle. In Naterer et al. (2013), this cycle was denoted as Cu-Cl-2. The maximum temperature requirement is 823 K. The cycle steps are described in Table 4.27, and the cycle process is represented in a simplified manner in the diagram in Fig. 4.77. The simplified plant flowchart is shown in Fig. 4.78.

In this system, liquid water is supplied in #1, which is converted to superheated steam in #2 at a temperature that corresponds to the hydrolysis reaction. In the reactor R, the hydrolysis of $CuCl_2$ occurs at a high temperature according to process P4. Hot gases—namely, oxygen, hydrochloric acid and steam—are extracted at location #3. This mixture is cooled to a temperature close to the ambient in HX2 and delivered as stream #4 to the liquid gas separator S.

TABLE 4.27 The Two-Step Hybrid Copper-Chlorine Cycle, Cu-Cl-2 Configuration

Cycle step	Chemical equation	T (K)
Electrochem. hydrochlorination	$2CuCl(s) + 2HCl(aq) + nH_2O(l) \rightarrow H_2(g) + 2CuCl_2 \cdot nH_2O(aq)$	350
Hydrolysis	$H_2O(g) + 2CuCl_2 \cdot nH_2O(s) \rightarrow 0.5O_2(g) + 2HCl(g) + 2CuCl(l) + nH_2O(g)$	823

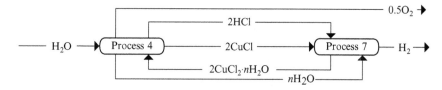

FIG. 4.77 Simplified representation of Cu-Cl-2 cycle process.

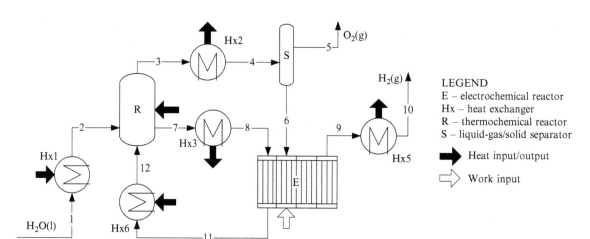

FIG. 4.78 Simplified flowchart of Cu-Cl-2 hydrogen production plant.

Thereafter, water is present as liquid and hydrochloric acid in the aqueous phase, while oxygen separates as a gas and is extracted at #5. The liquid of the aqueous hydrochloric solution and water are transferred to the electrochemical cell, E. In the hydrolysis reactor R, molten cuprous chloride is generated at the outlet #7, which is then cooled and solidified in the heat exchanger system Hx3 and then delivered as solid particles to the electrolytic cell E as stream #8.

In the electrolytic cell, the solid CuCl particles dissolve and form an aqueous phase. The electrolytic cell E generates gaseous hydrogen in #9 at approximately 473 K, which is cooled in Hx5 and delivered as a product at a temperature close to the ambient in #10. The other product of the electrolytic step is aqueous cupric chloride, which results at #11; this is heated in Hx6 to the reaction temperature of the hydrolysis process and supplied at #12 of the hydrolysis reactor R.

4.5.5.1 Electrochemical Process and Cell Development

The electrochemical process within the Cu-Cl cycle has several variants depending on the cycle version and the type of membrane used within the cell. The cell can produce at the cathode either metallic copper or molecular hydrogen; also, one can use a cation exchange membrane that conducts protons or an anion exchange membrane conductive to chlorine ions.

Regardless of the cell variant, the anodic half-cell reaction is the same for all cases. Hydrochloric acid must be present in the anolyte and catholyte where it has three roles: (1) it impedes precipitation of CuCl that otherwise is sparingly soluble in water; (2) it facilitates complexation of copper ions with water molecules with an eventual formation of a hydrated cupric chloride within the catholyte; (3) it contributes to an ionic balance in electrolyte, being either a source of protons or Cl^-.

The most developmental efforts were focused toward the development of a cation exchange membrane cell with limited, to no, crossover of copper ions. Copper crosses through membranes whether as ions (if a cation exchange membrane is used) or as a complex in the aqueous solution. Due to the copper complexation process at the anode side, copper complexes in the aqueous phase exist. Such complexes can cross the anion exchanger membrane. The occurrence of ionic copper or copper complexes to the cathode side lead to the unavoidable formation of metallic copper, which greatly reduces the activity of catalysts and leads to a significant decrease of cell performance. It is, therefore, a topic of intense research aimed at finding a suitable membrane that impedes copper crossover in the hydrogen producing electrochemical cell.

Canadian Nuclear Laboratories (former AECL) focused mainly on the development of the electrochemical cell for CuCl/HCl electrolysis to generate hydrogen and hydrated cupric chloride. An essential part of the research involved development of advanced electrodes for the cathode. The cathodic half-cell reactions were studied in a three electrode electrochemical cell. The reference electrode of the half-cell was the standard calomel electrode (SCE), while the counter electrode was a graphite rod. Eventually, a patent was filled by CNL, in which multiple membrane cells were claimed as a new invention, as described in Ketter et al. (2015).

In this cell, cuprous chloride in hydrochloric acid is fed as anolyte. The cathode contains an electrocatalyst. Two or more cation exchange membranes are disposed in a manner in which the anolyte and catholyte are separated. The copper ions that may cross the membrane at the anode side are immediately removed or sequestrated at the center compartment. A flushing system is used at the middle compartment where the copper ions are removed with a stream of 1 to 12 M HCl solution. The system may result in a variant combination of anion and cation exchange membranes. Alternatively, the center compartment may be filled with a chelation or absorption agents or reticulated vitreous carbon for copper ions.

Fig. 4.79 shows the CNL process disclosed in Ketter et al. (2015). The process evolves hydrogen at cathode. The cathode half-reaction must be catalyzed and is described as follows:

$$2H^+ + 2e^- \rightarrow H_2(g) \tag{4.159}$$

The anolyte comprises chlorine ions from hydrochloric acid dissociation that occur as follows: $2HCl(aq) \rightarrow 2H^+(aq) + 2Cl^-(aq)$. The presence of chlorine ions in anolyte facilitates the following non-catalytic electrochemical half-reaction, as follows:

$$2CuCl(aq) + 2Cl^-(aq) \rightarrow CuCl_2 + 2e^- \tag{4.160}$$

Stolberg et al. (2008) at CNL, performed full-cell experiments that confirm that a current density higher than 1000 A/m² is achievable when 6 mol/L HCL anolyte is used with 1 mol/L CuCl molarity. The half-cell tests at AECL were performed with a catalyst-free graphite separator plate (GSP), coated GSP with Pt/XC-72R catalyst mixture, catalyst-free carbon fiber paper (CFP) type EC-TP1-060, platinized CFP, and coated CFP with Pt/XC-72R catalyst.

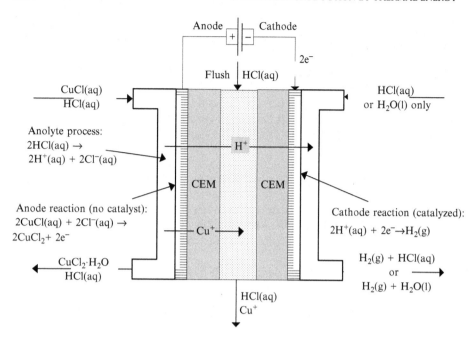

FIG. 4.79 Double membrane electrochemical process for Cu-Cl cycle. No solid copper can be formed at cathode. System developed by CNL, as described in Ketter et al. (2015).

The anolyte was stirred, and a solution of 1 mol/L CuCl/6 mol/L HCl was used. The results showed that a catalyst-free GSP performed the best with a saturation current density of 1200 A/m² at 0.5 V polarization, which is a 0.263 V anodic overpotential.

The tested electrodes for the cathode were fabricated by deposition of various platinized catalysts on a CFP type EC-TP1-060 Toray. The application of platinized catalysts was done by painting or direct spraying. The cathodic reaction was carried out in HCl (20% by weight). The following coatings were applied:

- Electrode #1: Pt/XC-72R catalyst mixture with Pt loading of 19% painted on graphite
- Electrode #2: Pt deposited on CFP type TGP-H-060 (Toray)
- Electrode #3: Pt/CV-72R catalyst mixture deposited on CFP type EC-TP1-060
- Electrode #4: Catalyst-free graphite
- Electrode #5: Catalyst-free CFP type EC-TP1-060

The polarization curves obtained at AECL with electrodes #1–#5 as cathodes are presented in Fig. 4.80. The best cathode is electrode #1, which shows 1000 A/m² at 0.33 V polarization versus SCE. Catalyst-free electrodes cannot be used for the cathode side reaction. Copper deposition on the cathode surface then occurs. AECL observed visually the copper deposits on the cathode after dismantling their single cell that had a cathode with 25 cm² area.

The half-cell process described by Eq. (4.152) does not require catalysts. Stolberg et al. (2008) determined the reversible potential of this reaction to be 0.237 V. With the aim of increasing the current density at the anode, Ranganathan and Easton (2010a) developed ceramic carbon electrodes (CCE) based on poly-aminopropyl siloxane (PAPS). The following main species are generated when cuprous chloride is dissolved in an aqueous solution of hydrochloric acid: $CuCl(aq)$, $CuCl_2^-(aq)$, $CuCl_3^{-2}(aq)$, and $CuCl_4^{-3}(aq)$. In addition, at higher HCl concentrations, the following species may be present: $Cu_2Cl^-(aq)$, $Cu_2Cl_2(aq)$, and $Cu_2Cl_3^-(aq)$. Therefore, the following concurrent reactions may occur at the anode:

$$CuCl_4^{-3}(aq) \rightarrow CuCl^+(aq) + 3Cl^- + e^- \text{ at 6 MHCl} \tag{4.161a}$$

$$CuCl_4^{-3}(aq) \rightarrow CuCl_2(aq) + 2Cl^-(aq) + e^- \text{ at 11 MHCl} \tag{4.161b}$$

Polarization curves for the anodic reaction were obtained by Ranganathan and Easton (2010a,b) with CCE electrodes as well as bare CFP electrodes. The results show the net improvement of CCE electrodes. Fig. 4.81 presents the polarization curves for double-sided coated electrodes (CCE) with a total exposed area (two sides) of 2 cm². The experiments were performed in a solution of 100 mmol/L CuCl. Data was obtained for 3 and 6 mol/L HCl in stirred and quiescent

FIG. 4.80 Polarization curves for cathode obtained at AECL with various electrodes. *Data from Suppiah, S., Stolberg, L., Boniface, H., Tan, G., McMahon, S., York, S., Zhang, W. (2009) Canadian nuclear hydrogen R&D programme: Development of the medium-temperature Cu-Cl cycle and contributions to the high-temperature sulphur-iodine cycle. Fourth international exchange meeting on nuclear hydrogen production. Oakbrook IL. USA. 13-16 April. Nuclear Energy Agency and Organisation for Economic Co-operation and Development, Development, NEA 6805.*

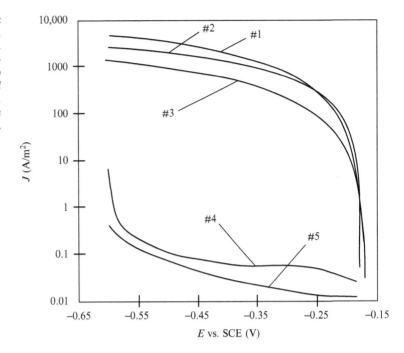

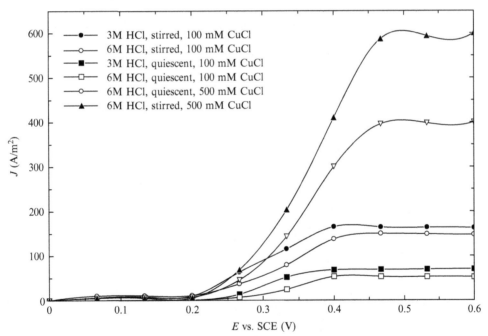

FIG. 4.81 Polarization curve for anodic half-reaction with a solution of 100 mmol/L CuCl using CCE electrodes prepared at UOIT. *Data from Ranganathan, S., Easton, E.B., 2010b. High performance ceramic carbon electrode-based anodes for use in the Cu–Cl thermochemical cycle for hydrogen production. Int. J. Hydrog. Energy 35, 1001–1007; cubic spline data smoothing is applied.*

solutions. The results show that the half-cell potential corresponding to the current saturation is approximately 0.4–0.5 V. The maximum current density is approximately 600 A/m² , whereas the corresponding overpotential for the half-cell reaction at the anode is ~0.2 V.

Regarding the cathodic reaction, Balashov et al. (2011) showed that it is not necessary that the catholyte have high concentration in HCl. These researchers reported encouraging experimental results obtained with water as the catholyte. The results were obtained in a full-cell water electrolysis system that used a proprietary membrane-electrode-assembly

(MEA) of which cathode materials were undisclosed. The MEA used by Balashov et al. (2011) were commercial types HYDRion 115 and HYDRion 117 produced by Electrochem, Inc.

Odukoya and Naterer (2011) modeled thermodynamically the cell for the electrochemical process P7. The concentration overpotential is calculated based on the Butler-Volmer and Tafel equations for a given current density (J) and under the assumption that the charge transfer coefficient a at the anode and cathode are similar. Therefore,

$$E_{conc} = \frac{RT}{2(1-a)F} \ln\left(\frac{J}{J_0}\right) \tag{4.162}$$

where J_0 is the exchange current density.

The ohmic overpotential is a summation of overpotentials due to effective electric charge transport through the electrolyte (r_e) and the electrical resistances of the anode (r_a) and cathode (r_c). These electrical resistances can be written per unit of surface area of the electrode; using area-specific resistances, the total ohmic overpotential is

$$E_{ohm} = J(R_e + R_a + R_c) \tag{4.163}$$

where R_e is the ohmic resistance of the PEM (electrolyte).

In the case of the CuCl/HCl electrolysis cell, one assumes that the electrolyte is a PEM. Therefore, the apparent electrical resistance of the electrolyte is obtained from the local electrical conductivity (σ_e) through a membrane via integration, as

$$R_e = \int_0^\delta \frac{dx}{\sigma_e(x)} \tag{4.164}$$

where

$$\sigma_e(x) = \sigma_{e0} \frac{c_{CuCl}(x)}{c_{H^+}} \tag{4.165}$$

Equation (4.165) was proposed in Odukoya and Naterer (2011) and assumed that σ_{e0} is the membrane conductivity for a molar concentration of cuprous chloride of unity. Moreover, the molar concentration of CuCl varies across the membrane as $c_{CuCl}(x)$ and the proton concentration $c_{CuCl}(x)$ was assumed constant. Determination of the concentration profile $c_{CuCl}(x)$ can be made based on Fick's law of diffusion that can be integrated over the membrane thickness, namely

$$c_{CuCl,a} = c_{CuCl,c} - \frac{n_d J}{DF}\delta \tag{4.166}$$

where subscripts a and c refer to the anode and cathode, respectively; D is the diffusion coefficient; δ is the membrane thickness; and n_d is the electro-osmotic drag coefficient.

Since the diffusive flux is proportional to the molar concentration, Ohm's law requires that the molar concentration increases linearly across the membrane. Therefore, from Eq. (4.158), one can obtain the equation

$$c_{CuCl}(x) = c_{CuCl,c}\left[1 + \left(\frac{x}{\delta} - 1\right)\frac{J}{J_L}\right] \tag{4.167}$$

where the limiting current density is introduced as follows

$$J_L = \frac{FD c_{CuCl,c}}{n_d \delta} \tag{4.168}$$

Equation (4.170) accounts for the fact that, when $j = j_L$, there is no presence of CuCl at the cathode electrolyte interface $c_{CuCl}(x = 0) = c_{CuCl,c} = 0$. Based on Eqs. (4.157), (4.167), and (4.168), one obtains for Eq. (4.164) the following form

$$R_e = \int_0^{\ddot{a}} \left[\left(\frac{c_{H^+}}{c_{CuCl,c}}\right) \bigg/ \left(1 + \left(\frac{x}{\delta} - 1\right)\frac{J}{J_L}\right)\right] dx \tag{4.169}$$

The activation overpotential is derived from Nernst equation as follows:

$$E_{act} = \frac{RT}{2aF} \times \ln\left(\frac{j}{j_0}\right) \tag{4.170}$$

The operating parameters assumed for the electrochemical cell modeling are given in Table 4.28. Fig. 4.82 shows a comparison between measured and predicted open circuit voltage (E_{ocv}) and activation overpotential (E_{act}).

The comparison shows that the magnitudes of the predicted overpotential in the new formulation has close agreement with experimental half-cell data reported previously by Stolberg et al. (2008) for cuprous chloride electrolysis. It can be observed that the activation potential shows only about a 6% difference between the experimental activation potential and the predicted results. Note that the comparison is made for a single operating condition of 298 K and current density is 2500 A/m^2.

Based on the parametric analysis shown in Fig. 4.83, the activation overpotential is much greater than the ohmic overpotential. The change in current density significantly affects the activation overpotential. This increase in overpotential is about 81%, when compared with its value at the lowest current density. The ohmic overpotential can be neglected at very high current densities, but it has a significant effect at low and intermediate current densities (0–7000 A/m^2). The ohmic resistance increases linearly with current density, as expected by the relationship of Ohm's Law.

Experimental tests with a commercially available electrolytic cell from Electrochem having a 25 cm^2 NAFION-117 (originally designed for water electrolysis) membrane (originally designed for water electrolysis) were performed at the University of Ontario Institute of Technology. The experimental setup is shown in a photograph in Fig. 4.84. The experiments were performed with a commercially available cell in which the anolyte was a CuCl + HCl solution with molarities varied as 0.5–1 M and from 6 to 10 M. The anolyte circuit was covered with aluminum foil since the CuCl is light-sensitive. The cell potential varied with the HCl concentration in anolyte, as shown in Fig. 4.85.

TABLE 4.28 Operating Parameters for Electrochemical Cell Modeling

Operating parameters	Value
Temperature of the cell, T (K)	373.15
Exchange current density, J_o (A/m^2)	0.7
Activation potential, η (mV)	0.515
Electrons transferred in the reaction, z	2
Charge transfer coefficient, a	0.5
Molar weight of mixture, M (kg/kmol)	0.17048
Universal Gas constant, R (J/molK)	8.314
Faraday's Constant, F (C/mol)	96,500
Concentration of CuCl (mol)	1.5
Concentration of HCl (mol)	6
Overpotential, E_0 (mv)	0.515

Source: Odukoya, A., Naterer, G.F., 2011. Electrochemical mass transfer irreversibility of cuprous chloride electrolysis for hydrogen production. Int. J. Hydrog. Energy 36, 11345–11352.

FIG. 4.82 Comparison between predicted and measured data. Predictions are based on modeling by Odukoya and Naterer (2011). Measured data is taken from Stolberg et al. (2008).

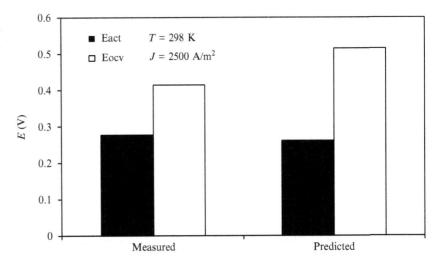

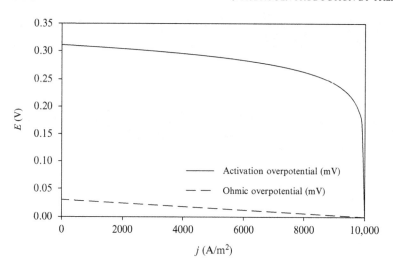

FIG. 4.83 Variation of ohmic and activation overpotentials with current density. *Data from Odukoya, A., Naterer, G.F., 2011. Electrochemical mass transfer irreversibility of cuprous chloride electrolysis for hydrogen production. Int. J. Hydrog. Energy 36, 11345–11352.*

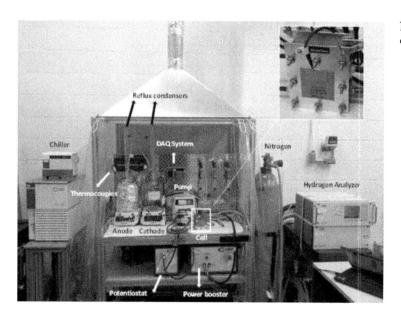

FIG. 4.84 Experiments at UOIT with an electrochemical cell setup as reported in Aghahosseini (2013).

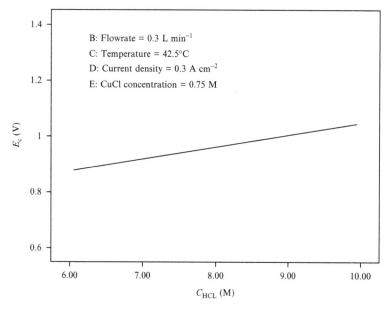

FIG. 4.85 The influence of HCl concentration in the anolyte on cell potential. *Data from Aghahosseini, S., 2013. System Integration and Optimization of Copper-Chlorine Thermochemical Cycle with Various Options for Hydrogen Production. PhD Thesis, University of Ontario Institute of Technology.*

In early development, the electrochemical cell assumed a cathodic reaction that generated metallic copper. This type of process was studied in past seminal work by Lewis et al. (2003), as follows:

$$CuCl(aq) + e^- \rightarrow Cu(s) + Cl^-(aq) \tag{4.171}$$

Electrowinning technology recommends the use of carbon electrodes at the cathode because copper precipitates as dendritic powder. The anolyte with aqueous cuprous chloride is fed at the bottom of the cathode compartment, and the product stream is extracted from the top side. Lewis et al. (2003) reported the formation of dendritic copper powder at the cathode with sizes from 3 to 100 μm. It appears that the development of copper-producing electrochemical cells for the copper-chlorine cycle have been set aside by most researchers because the alternative hydrogen-producing electrochemical process appears to be more competitive.

Another cell process will involve the use of an anion exchange membrane that will be resistant to copper crossover. The process is represented in Fig. 4.86. Chlorine ions cross the membrane and react non-catalytically with CuCl at anode to form cupric chloride.

The focus of researchers has been toward development of AEMs that are resistant to copper crossover, as reviewed in Naterer et al. (2011). However, in Cu-Cl electrolytes, both anions and cations ca be formed, and all can pose potential problems for a selective ion membrane. For example, the presence of chlorine ions may lead to formation of the following anionic species: $CuCl_2^-$, $CuCl_3^{2-}$, $[CuCl_3(H_2O)_3]^-$, $[CuCl_4(H_2O)_2]^{2-}$, $CuCl_4^{-3}$, Cu_2Cl^-, and $Cu_2Cl_3^-$.

4.5.5.2 Hydrolysis Reaction and Process

Hydrolysis reaction is not spontaneous and evolves at high temperature with required heat addition. Based on the assumption that the molar enthalpy and entropy of copper oxychloride ($CuO \cdot CuCl_2$) can be represented as a summation of enthalpies of cupric oxide (CuO) and cupric chloride ($CuCl_2$), the Gibbs free energy and enthalpy of the reaction at 673 K are calculated as ~40 kJ/mol H_2O and ~117 kJ/mol H_2O, respectively. Gibbs free energy of the reaction is positive; therefore, one may expect relatively low yields. In fact, hydrolysis is the most challenging reaction of the copper-chlorine cycle. There are several side reactions that may occur together with hydrolysis.

Table 4.29 gives the desired reaction and the possible side reactions of hydrolysis process. The main products in the hydrolysis reactor are copper oxychloride and hydrochloric acid. However, the temperature in the hydrolysis reactor is sufficiently high for copper oxychloride to partially decompose and possibly release gaseous oxygen and cuprous chloride.

Cuprous chloride may be present as solid, liquid, and/or gas. According to Ferrandon et al. (2008), if the temperature is lower, below the normal boiling point of CuCl, sublimation of solid CuCl may occur. For a temperature over

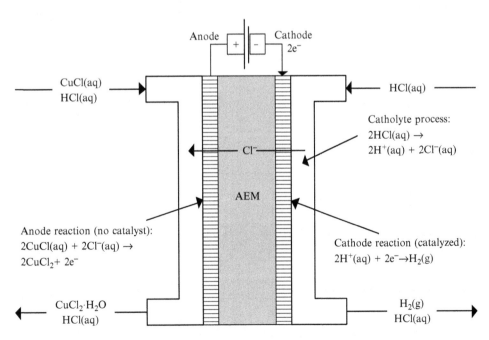

FIG. 4.86 Electrochemical cell with anion exchange membrane to avoid copper crossover.

TABLE 4.29 Desired Reaction and Possible Side Reactions in a $CuCl_2$ Hydrolysis Reactor

ID	Reaction	Remarks
1	$CuCl_2(s) + H_2O(g) \rightarrow CuO \cdot CuCl_2(s) + HCl(g)$	Desired reaction, hydrolysis
2	$CuO \cdot CuCl_2(s) \rightarrow 0.5O_2(g) + 2CuCl(s, l, g)$	Thermolysis of copper oxychloride
3	$CuCl_2(s) \rightarrow CuCl(s, l, g) + 0.5Cl_2(g)$	Thermolysis of $CuCl_2$ with CuCl formation
4	$3CuCl(g) \rightarrow Cu_3Cl_3(g)$	Formation of tricopper thrichloride
5	$Cl_2(g) + H_2O(g) \rightleftarrows 0.5O_2(g) + HCl(g)$	Reverse Deacon and Deacon reactions

the melting point, liquid and vapor CuCl are produced. The most important side reaction is the cupric chloride decomposition that generates cuprous chloride and chlorine.

It is also possible that tricopper trichloride (a gaseous compound, Cu_3Cl_3) is formed from gaseous cuprous chloride directly. Cuprous chloride formed in the hydrolysis reactor is not necessarily problematic because this is the desired product of copper oxychloride decomposition within the cycle. The main problem is the formation of molecular chlorine that must be recycled in the system. Formation of significant quantities of chlorine negatively impact the cycle of chemical reactions within the thermochemical cycle. Therefore, the process parameters and reactor must be selected and designed such that the chlorine yield is minimized or avoided. In addition, there may be traces of tricopper trichloride generated as side reactions; this must be avoided, too, because Cu_3Cl_3 is stable because it has a circular molecule.

Fig. 4.87 shows the calculated equilibrium yields as a function of temperature in a reaction vessel where 2 moles of cupric chloride and 1 mole of steam are supplied. It can be observed that at 573 K, there is no chlorine generation; however, $CuCl_2$ conversion is very low (below 5%). At approximately 773 K, the conversion reaches approximately 50%, but the chlorine yield is too high (approximately 0.5 moles per mole of $CuCl_2$). At 973 K, the chlorine production reaches an apparent maximum of 20%. Over 973 K, gaseous tricupric trichloride starts to be produced. This analysis demonstrates that molecular chlorine is the only relevant side product that occurs at temperatures of interest (623–723 K). If excess steam is used, then the chlorine generation decrease.

The extent of the hydrolysis reaction was analyzed by Daggupati et al. (2009). Accordingly, the extent of reaction can be expressed as

$$\xi = \sqrt{\frac{0.25 K_{eq} x_s^2}{K_{eq} + 4P}} \tag{4.172}$$

where P is the operating pressure and x_s is the molar ratio of steam versus cupric chloride supplied to the reaction.

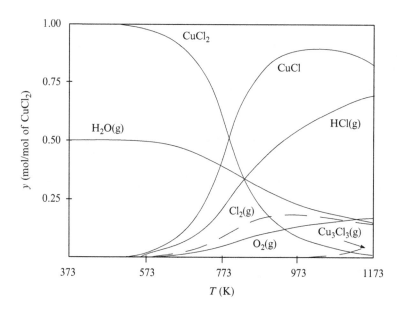

FIG. 4.87 Equilibrium yields of a $2CuCl_2 + H_2O$ reaction (dashed lines refer to the undesired side products). *Data from Lewis, M.A., Masin, J.G., O'Hare, P.A., 2009. Evaluation of alternative thermochemical cycles, Part I: the methodology. Int. J. Hydrog. Energy 34, 4115–4124.*

FIG. 4.88 Variation of the extent of reaction (ξ) for the hydrolysis reaction at various steam to CuCl$_2$ molar ratios (x_s). *Data from Daggupati V., Naterer G.F., Gabriel K., Gravelsins R., Wang Z. 2009. Equilibrium Conversion in Cu-Cl Cycle Multiphase Processes of Hydrogen Production. Thermochim. Acta 496, 117–123*

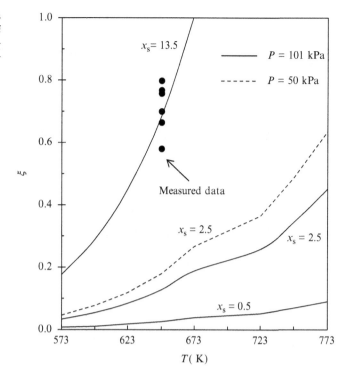

If the reaction is stoichiometric, then $x_s = 0.5$. The equilibrium constant is calculated from the Gibbs energy of the reaction at the specified temperature:

$$K_{eq} = -\Delta G^0(T)/(RT) \tag{4.173}$$

In Fig. 4.188 the variation of the extent of reaction with temperature is shown. Note that with excess steam of 13.5 more than stoichiometric (27 moles of steam per 2 moles of cupric chloride), the reaction goes to completion at 673 K; it has reasonable equilibrium yield at 650 K.

Various reactors were tested, such as packed bed, fluidized bed and spray reactors. The reactor sizes range from a small bench, operating with CuCl$_2$ samples of tens of grams, to larger engineering scale reactors at UOIT. Most of the experimental reactors were a batch type, and some were semi-batch (in which only the fluids are supplied and extracted continuously and the reacting solids remain in the reactor until the experiment ends) or flow reactors. In this section, the experimental research on reactor development is examined.

Based on the review from Naterer et al. (2013), the following types of reactors were proposed for hydrolysis processing: horizontal fixed bed reactor, vertical fixed bed reactor, vertical fluidized bed reactor and continuous spray reactor. The principle of vertical fluidized bed reactor for hydrolysis is illustrated in Fig. 4.89.

Vertical fluidized bed reactors were tested with 1.35 g CuCl$_2$ samples at ANL and 2250 g at UOIT. ANL also developed a bench-scale spray reactor that can operate with ca. 300 g cupric chloride di-hydrate CuCl$_2 \cdot 2$H$_2$O charge in a continuous flow system with solid products accumulated at the bottom of the reactor. Fig. 4.90 shows the principle of the ANL spray reactor for hydrolysis.

4.5.5.3 Copper-Chlorine Plant Development

The conceptual flowchart of a three-step copper-chlorine plant is shown in Fig. 4.91, based on ANL concept published in Ferrandon et al. (2008). In this three-step process, the following main reaction equipment are included: (i) an electrolytic cell with anion exchange membrane; (ii) a crystallizer to dehydrate CuCl$_2 \cdot n$H$_2$O, $n > 1$ to the form CuCl$_2 \cdot 2$H$_2$O; (iii) a spray reactor for hydrolysis; and (iv) an oxy-decomposer with bayonet heat exchangers. The electrolyzer operates at high pressure—24 bar—while the spray reactor operates in a slight vacuum or at 1 atm.

Water is fed into the anolyte buffer where there is a solution of hydrochloric acid, cuprous chloride and cupric chloride. This solution is pumped to 24 bar and heated to 373 K when the pressurized solution is still in the liquid phase. The anodic reaction occurs between cuprous chloride and chlorine ions, namely 2CuCl(aq) $+ 2$Cl$^-$(aq) $\rightarrow$ CuCl$_{2(aq)} + 2$e$^-$. Once formed, cupric chloride complexes with water molecules as CuCl$_2 \cdot n$H$_2$O, $n \geq 2$.

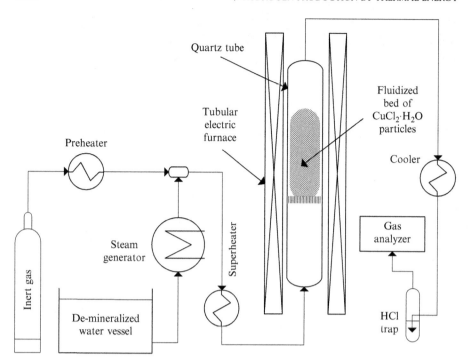

FIG. 4.89 Vertical fluidized bed reactor for hydrolysis experiments and auxiliary systems.

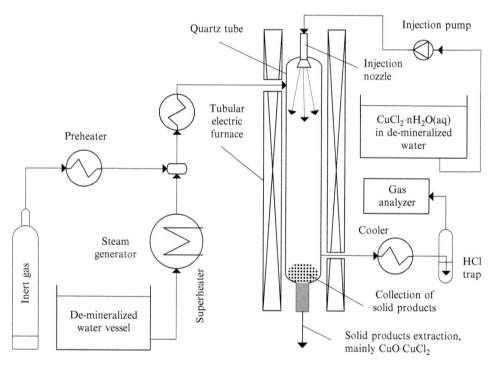

FIG. 4.90 Principle of a spray reactor system for $CuCl_2$ hydrolysis.

The anolyte is cooled down to ~303 K and passed to a crystallization unit where it partially dehydrates. Bayonet heat exchangers are used in the crystallization unit to remove heat. Next, cupric chloride in aqueous form is heated such that the aqueous solution can be concentrated in $CuCl_2 \cdot 2H_2O(aq)$. The solution is rapidly expanded in spraying nozzles that form micrometer droplets scattered in steam at 673 K where the hydrolysis reaction occurs with formation of copper oxychloride particles that deposit at the bottom of the reactor. Hydrochloric acid and steam are extracted as gases and cooled to 373 K.

A part of the liquid is further condensed and sub-cooled to ~303 K and further mixed with cooled cuprous chloride particles generated by the molten salt heat exchanger. The particles partially or totally dissolve in the acidic solution.

The copper oxychloride solids fall into the oxy-decomposer that is supplied with heat by bayonet heat exchangers sunk into a molten cuprous chloride bath. The decomposition reaction generates oxygen gas that is removed from the top of the vessel. The upper layer of molten salt is free of other solid impurities and passes over a baffle into a room from where it is extracted.

A molten salt heat exchanger is used for heat recovery and quenching of fine CuCl droplets. With a selected conveyor, CuCl is mixed with water, and hydrochloric acid is directed toward the anolyte reservoir. If the spray reactor and oxy-decomposer operate in a slight vacuum, a pump is needed to pressurize the liquid prior to the return in an anolyte buffer. A pump pressurizes the anolyte to 24 bar, the operating pressure of the electrolyzer and crystallization cells. Hot gases of HCl and steam at 373 K are directed toward the catholyte buffer where water condenses and HCl is dissolved. The hot catholyte is pumped to 24 bar and supplied at the cathode. Hydrochloric acid at high concentration generates protons and chlorine anions according to $HCl \rightarrow H^+ + Cl^-$. The chlorine anions cross the membrane while the protons reduce at the cathode forming hydrogen. A separation unit (not detailed) is used to separate and scrub hydrogen and to return the acidified liquid solution at lower pressure but high temperature to the catholyte buffer.

The calculated data for a copper-chlorine plant of 125,000 kg hydrogen per day was reported by Ferrandon et al. (2008). Accordingly, the plant required 210 MW thermal energy and ∼88 MW electric energy. If one assumes that electricity is generated with 40% efficiency, the total requirement of thermal energy is 430 MW; therefore, the energy efficiency is 29%. In order to generate the same amount of hydrogen with an alkaline electrolysis system of 65% energy efficiency, one needs ∼313 MW of electric power, or an equivalent of ∼780 MW electric energy, with an energy overall efficiency of 26%.

A scaled-up test loop of the copper-chlorine cycle is under development at UOIT for a projected production of 3 kg hydrogen per day, when all unit operations are connected and integrated together. The test loop consists of four separate main units: CuCl/HCl electrolyzer, dehydration reactor, hydrolysis reactor and oxy-decomposer. In addition, several auxiliary units necessary for system integration are under development. These include the heat recovery system for molten cuprous chloride and an apparatus for hydrochloric acid/water separation, among others.

A general layout of the scaled-up system is depicted in Fig. 4.92. Each major unit of the test bench is operational. Research studies have been conducted at UOIT on system integration of experimental unit operations of the Cu-Cl cycle, including integration of electrolysis, water separation and hydrolysis processes, as well as the integration of hydrolysis, oxygen and hydrogen generation processes.

The dominant parameter that influences the integration of the electrolysis and hydrolysis processes is the steam quantity required by a hydrolysis step. Another parameter that affects the processes of integration is the concentration of cupric chloride exiting the electrolytic cell. The output stream from the electrolytic cell has been simulated using the quaternary system of H_2O, $CuCl_2$, CuCl, HCl in the experiments with varied concentration of each constituent.

Crystallization and spray drying experiments have been performed to concentrate the $CuCl_2$ solution so as to accommodate the variations of electrolysis outputs and deliver $CuCl_2$ of different water contents to the hydrolysis reactor. Crystallization experiments of cupric chloride were also performed by varying the HCl concentration and temperature of the quaternary system. The $CuCl_2$ output from the crystallization and spray drying is delivered separately to the hydrolysis reactor to produce HCl liquid.

With respect to system integration of hydrolysis, oxygen and hydrogen generation processes, studies have also been conducted on the interactions between the Cu_2OCl_2 decomposition and $CuCl_2$ hydrolysis steps. The effects of completion of the hydrolysis reaction on the Cu_2OCl_2 decomposition was studied by varying the fraction of $CuCl_2$ in a mixture of $CuCl_2$ and Cu_2OCl_2 that are fed into the Cu_2OCl_2 decomposition reactor. The primary influences include the conversion of by-products such as Cl_2, partial evaporation of molten CuCl and heat requirements. The heat input to the oxygen production reactor was characterized by the enthalpy change of the Cu_2OCl_2 and $CuCl_2$ decomposition processes. A mixture of $CuCl_2$ and CuO was also used in the experiments to study the oxygen production reaction so as to accommodate different hydrolysis technologies that may reduce the excess steam and temperature requirements.

For separation of HCl and steam downstream of the hydrolysis reactor, the mixture of H_2O and HCl can be concentrated (individually or in combination) by distillation and extractive rectification performed near or below the azeotrope temperature (381 K, at 1 bar; 0.4 HCl mass fraction). Significant quantities of heat flow are required to condense the high temperature products, which then must be reheated for other processes. Thus the downstream design of the hydrolysis reactor can have a significant impact on the thermodynamic efficiency of the Cu-Cl cycle. Several techniques to condense the HCl/H_2O mixture are being analyzed and studied experimentally to determine and compare the effectiveness of each system.

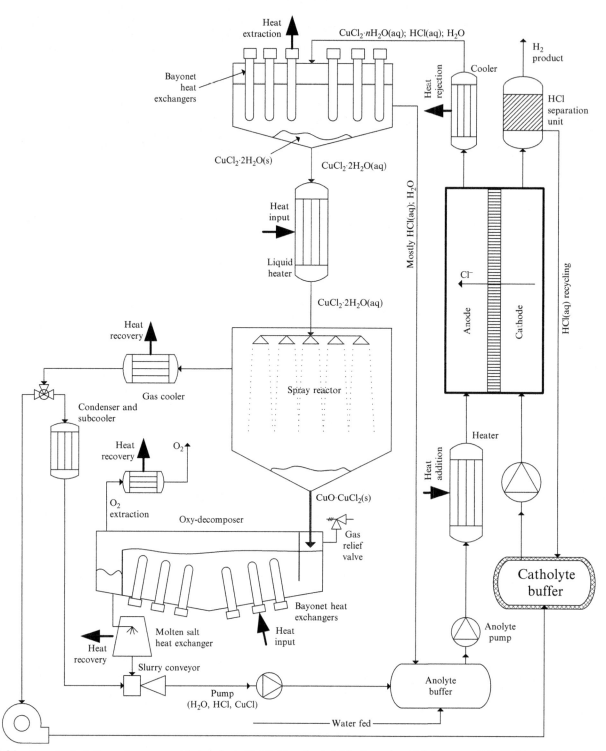

FIG. 4.91 Flowsheet of a three-step copper-chlorine plant. *Modified from Lewis, M.A., Ferrandon, M.S., Tatterson, D.F., Mathias, P., 2009b. Evaluation of alternative thermochemical cycles – Part III further development of the Cu-Cl cycle. Int. J. Hydrog. Energy 34, 4136–4145.*

Experimental work is also underway on the processing of solidified CuCl and entrained by-products produced from the oxygen production reactor. The processed CuCl will provide the feedstock to the downstream HCl/CuCl electrolyzer. The preparation of the ternary system of CuCl, HCl, and water at various concentrations was developed for the integration of oxygen and hydrogen production steps. It has been shown that the high conversion of $CuCl_2$ to Cu_2OCl_2 in the hydrolysis step has vital importance to reduce the challenges of the integration of hydrolysis, oxygen and hydrogen production steps.

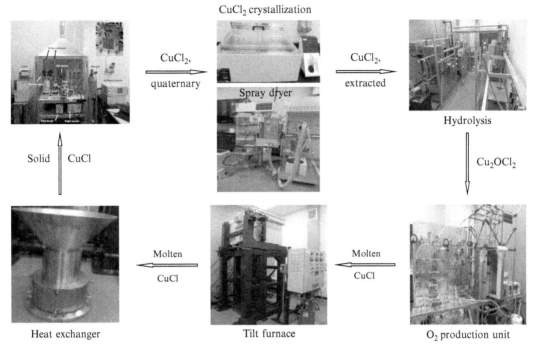

CuCl$_2$ crystallization

FIG. 4.92 System integration for scaled-up and integrated Cu-Cl cycle at UOIT.

4.6 GASIFICATION AND REFORMING FOR GREEN HYDROGEN PRODUCTION

When a solid fuel like coal or biomass or solid waste is converted into hydrogen gas or synthesis gas (hydrogen + carbon monoxide), the process is named gasification. When a fluid fuel (gas or liquid: alcohols, natural gas petroleum) is converted into synthesis gas, this process is denoted as reforming. If the primary source is a fossil fuel, then carbon sequestration can be applied to make the overall process green. Fig. 4.93 illustrates green gasification/reforming for hydrogen production.

4.6.1 Steam Reforming

The steam reforming of natural gas (NG), oil, and other hydrocarbons and the steam gasification of coal and other solid carbonaceous materials (e.g., biomass) can be expressed by the simplified net reaction

$$C_xH_y + xH_2O \xrightarrow{\text{High temperature heat}} \left(\frac{y}{2} + x\right)H_2 + xCO \tag{4.174}$$

Depending on the reaction kinetics and on the presence of impurities in the raw materials, other compounds may also be formed during the conversion. Examples of other possible reactions that can occur concurrently at various rates are written as

$$CH_4 + H_2O \rightarrow CO + 3H_2 - 206\,kJ/mol \tag{4.175a}$$

$$CH_4 + 2H_2O \rightarrow CO_2 + 4H_2 - 165\,kJ/mol \tag{4.175b}$$

$$C + H_2O \rightarrow CO + H_2 - 131\,kJ/mol \tag{4.175c}$$

$$C + H_2 \rightarrow CH_4 - 75\,kJ/mol \tag{4.175d}$$

$$C + CO_2 \rightarrow 2CO - 172\,kJ/mol \tag{4.175e}$$

Steam methane reforming proceeds with syngas production followed by a water gas shift reaction. The stoichiometric steam reforming of natural gas is conducted catalytically at approximately 1000 K according to the following reaction at equilibrium

$$CH_4 + H_2O \leftrightharpoons 3H_2 + CO \tag{4.176}$$

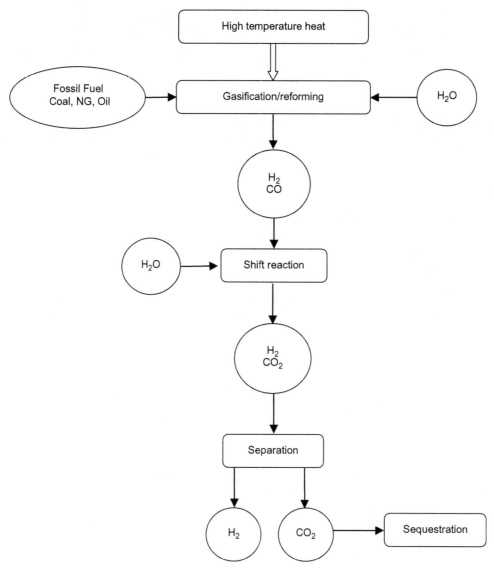

FIG. 4.93 Schematic diagram of green gasification/reforming process for hydrogen production.

Reaction (4.170) is endothermic with a heat of reaction of 206 kJ/mol. The reaction products from the reformer are then passed into a low temperature reactor where the so-called gas-shift reaction occurs exothermically at equilibrium. The heat of reaction is −41 kJ/mol. The reaction is conducted catalytically and its stoichiometry is described as follows:

$$CO + H_2O \rightleftharpoons H_2 + CO_2 \tag{4.177}$$

If a part of methane is combusted directly with oxygen to release sufficient heat to drive the endothermic reforming reaction, then an autothermal process is conducted. The overall autothermal process of steam methane reforming has the following stoichiometry that results from combining Eqs. (4.176) and (4.177), and considering a methane combustion process with oxygen to generate the 206 kJ/mol needed by the reforming, namely:

$$1.415CH_4 + 1.37H_2O + 0.63O_2 \rightarrow 4H_2 + 1.315CO_2 \tag{4.178}$$

In practice, the yield of hydrogen is 65% on average, which means that 1.54 more reactants are required to generate the stoichiometric amount of hydrogen from Eq. (4.178). This means that 0.54 moles of methane must be combusted with oxygen for 1 mol of hydrogen.

Biofuels can be converted to hydrogen by reformation. This method may be viewed as a green hydrogen production process, in the sense that it is carbon neutral. Reformation of methanol to hydrogen may occur according to the chemical equation

$$CH_3OH + H_2O \rightarrow 3H_2 + CO \tag{4.179}$$

which is then followed by the water-gas shift process according to Eq. (4.179).

Biomass-produced methanol can be reformed to hydrogen thermo-catalytically according to the following overall process as described in Fatsikostas et al. (2002), namely:

$$C_2H_5OH + H_2O \rightarrow 4H_2 + CO \tag{4.180}$$

Reaction (4.183) is endothermic with a heat of reaction of 239 kJ/mol. Nickel catalysts may be used at temperatures of approximately 900 K. Again, this reforming reaction is followed by a water-gas shift process to generate more hydrogen as given in Eq. (4.177).

The landfill gas or biogas also can be reformed to hydrogen via thermochemical methods, e.g., reforming. Biogas typically contains 60% to 70% methane and 30% to 40% carbon dioxide with traces of nitrogen, ammonia, hydrogen, and hydrogen sulfide. Since methane is the "active" constituent of biogas and landfill gas, their reformation is described with the same equations as that for steam-methane reforming.

The equilibrium constants for methane reforming and water gas shift process can be described by the following two equations:

$$K_{reform} = \frac{y_{e,H_2}^3 y_{e,CO}}{y_{e,CH_4} y_{e,H_2O}} \left(\frac{P}{P_0}\right)^2 \tag{4.181a}$$

$$K_{shift} = \frac{y_{e,H_2} y_{e,CO_2}}{y_{e,CO} y_{e,H_2O}} \tag{4.181b}$$

The equilibrium constants from Eq. (4.181) are given in the following regressed form with the constants given in Table 4.30:

$$\log K = a_0 + a_1 T + a_2 T^2 + a_3 T^3 + a_4 T^4 \tag{4.182}$$

Let us denote with n the number of moles of a species that participates to a methane reforming process according to the methane steam reforming and water-gas shift cascaded reactions. Denote with index "in" the species at inlet and "out" at outlet. Then based on the stoichiometry of steam methane reforming and water-gas shift reactions, the following equations can be written:

$$n_{out}^{CH_4} = n_{in}^{CH_4} - \xi_{reform} \tag{4.183a}$$

$$n_{out}^{CO} = n_{in}^{CO} + \xi_{reform} - \xi_{shift} \tag{4.183b}$$

$$n_{out}^{CO_2} = n_{in}^{CO_2} + \xi_{shift} \tag{4.183c}$$

Here ξ represents the extent of reaction.

A basic steam reforming process calculation implies the evaluation of the equilibrium constants according to Eq. (4.83) and a specified process temperature. Then Eqs. (4.181) and (4.183) are to be solved simultaneously to eventually correlate the molar species in the output stream to the species present in the input stream and the extent of reactions.

Fig. 4.94 shows a simplified flowsheet diagram of a steam methane reforming plant for hydrogen production. These types of plants implicitly include a carbon dioxide capture unit, typically a pressure swing absorption (PSA) system. The capture of carbon dioxide is required for purification of hydrogen from product stream.

TABLE 4.30 Values for Polynomials Coefficients of K_{eq} Regressions

Coefficient	Reforming	Shifting
a_0	−66.139	13.210
a_1	0.195	−3.915E−2
a_2	−2.252E−4	4.637E−5
a_3	1.241E−7	−2.547E−8
a_4	−2.631E−11	5.470E−12

Source: Fryda et al. (2008).

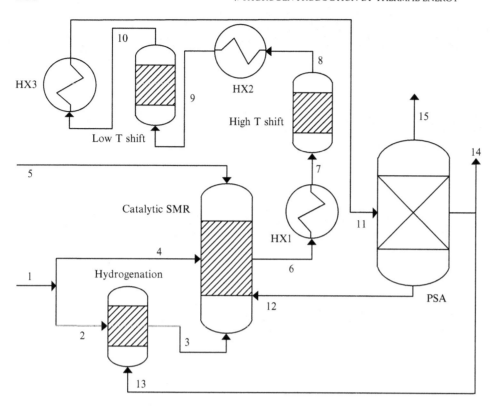

FIG. 4.94 Simplified flowsheet of a steam methane reforming plant for hydrogen production.

If the separated carbon dioxide gas is not expelled into the atmosphere, but rather is sequestrated, then a green hydrogen production process is obtained. In Fig. 4.94, stream #1 is the supply of methane (e.g., natural gas, biomethane sourced from biogas or landfill gas). Methane is fed into a hydrogenation reactor where it reacts catalytically with hydrogen taken from the plant output and fed into port #13. The hydrogenated methane is then fed at proper process parameters to the catalytic steam methane reformer (SMR) in #3 together with a main feedstock of methane in #4 and steam in #5. The product gas results at output port #6 and is cooled in HX1 and then supplied into the catalytic shift reactor at high temperature. After additional cooling, the stream is shifted again at lower temperature and cooled again in HX3. Pure hydrogen is produced by PSA and delivered at stream #14. Also, the carbon dioxide is extracted as pure stream #15.

Fig. 4.95 shows a typical pressure swing absorption system for carbon dioxide capture from product gas in a reforming plant. A typical solvent for carbon dioxide absorption is monoethanolamine (MEA), which operates at ~2.5 bar during the absorption phase and at standard pressure for the regeneration phase. The gas is cooled at approximately 45°C and pressurized at 2.5 bar using a blower as shown in the figure. These are the optimal temperature and pressure for the absorption process. At these conditions, most of the carbon dioxide is absorbed in liquid MEA, which is withdrawn at the bottom of the absorber. At the top of the absorber, the remaining hydrogen rich gas—having almost zero carbon dioxide content—is delivered as a final product.

The MEA rich in carbon dioxide is heated to about 110°C, which is the typical temperature for CO_2 release. Thereafter, the hot liquid can be expanded to standard pressure to recover some work, typically with the help of a Francis turbine. The rich liquid is injected into the regenerator, which is continuously heated so that it releases gaseous carbon dioxide, which is collected at the top of the regenerator. Due to the carbon dioxide release, the MEA solution is regenerated, and it can be pumped back to high pressures for a new absorption cycle.

Once the carbon dioxide is captured from a process, it has to be stored in some form. Storage of carbon dioxide is a sustainable alternative to emitting it into the atmosphere where it acts as GHG. The general term for carbon dioxide storage, when this is done with the purpose of avoiding emitting it into the atmosphere, is "sequestration." The storage can be short term or very long term (e.g., ocean deep sequestration has a time range of more than 100 years).

An option for carbon dioxide sequestration is the conversion into a useful product. This is typically done at fertilizer plants where a part of the generated CO_2 is used for urea fertilizer production, whereas the other part is released in the atmosphere. Urea results from a catalytic reaction of ammonia with carbon dioxide. Another useful fertilizer option is

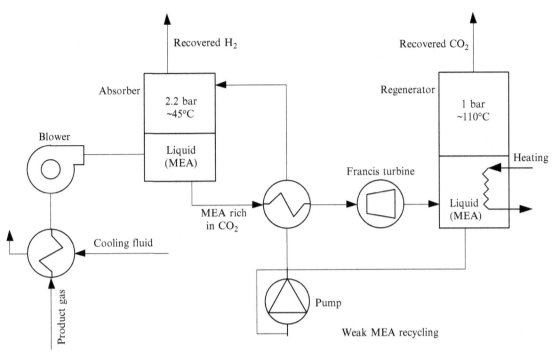

FIG. 4.95   Carbon dioxide capture system from product gas in a reforming plant.

the production of ammonium carbonate, which can be obtained in a spray reactor where ammonia water is sprayed through the CO_2 gas. The ammonia carbonation reaction occurs spontaneously, as follows:

$$CO_2(g) + NH_3(aq) + H_2O(l) \rightarrow NH_4HCO_3(s) \tag{4.184}$$

An advantage of the process given by Eq. (4.184) is that the ammonium carbonate precipitates as solid and can be easily separated. The reaction is exothermic and, therefore, allows for heat recovery to some degree.

4.6.2 Gasification

Gasification is the thermochemical conversion of either a solid (coal, coke, biomass, solid waste) or sometimes a heavy liquid (oil, tar, pitch) fuel into a synthesis gas (or syngas) composed primarily of H_2 and carbon monoxide (CO). Unlike combustion processes that produce only carbon dioxide (CO2) and water, gasification is a partial oxidation process that occurs in an oxygen-limited environment. The gasification process to generate hydrogen is typically coupled with the water-gas shift reaction—Eq. (4.169).

Biomass having moisture content lower than 35% can also be used as primary energy to extract hydrogen from steam. If the moisture content is too high, either the biomass must be dried before gasification or supercritical steam gasification can be applied. Supercritical steam gasification is a process that uses steam at supercritical pressure and temperature, and it is adapted to convert biomass to hydrogen regardless of its moisture content. Wood sawdust and sugar cane bagasse are among some general forms of biomass that can be used to produce hydrogen.

The biomass is introduced to a gasifier at an operating temperature range of 1000–1500 K. The global reaction of the biomass gasification process to produce hydrogen reads as

$$\alpha C_l H_m O_n + \gamma H_2O \xrightarrow{heat} aH_2 + bCO + cCO_2 + dCH_4 + eC + f\text{Tar} \tag{4.185}$$

where $\alpha C_l H_m O_n$ is the general chemical representation of the biomass.

The formation of tars is undesirable because of its negative effect on pipes, ducts and equipment by slugging and fouling. Tar formation can be diminished with proper control and by using various catalysts. The hydrogen yield from gasification of some biomasses are shown in Table 4.31 when gasification agent is steam. Table 4.32 describes the gasification agents in general.

Fig. 4.96 shows a temperature versus air-fuel ratio for thermochemical conversion of biomass, when air is the gasification agent. Depending on temperature and the AFR, the process used can be combustion, gasification, or pyrolysis.

TABLE 4.31   Indicative Parameters and H_2 Yields of Biomass Gasifiers

Biomass	T (°C)	Steam-biomass ratio	H_2 yield (%)
Pine and eucalyptus	880	0.8	41
Pine sawdust	750	0.5	40
Mixed sawdust	750	0.51	62.5
Mixed sawdust	800	4.7	57.4
Mixed sawdust	800	1.4	48.8
Mixed sawdust	800	1.1	46
General biomass	777	1.5	59

Source: Abuadala et al. (2010).

TABLE 4.32   Gasification Agents

Agent	Process temperature (°C)	Remarks
Oxygen	1000–1400	Syngas heating value is high: 10–15 MJ/m³. Oxygen handling is expensive and implies safety issues
Air	900–1100	Syngas has low heating value (4–6 MJ/m³) because it contains up to 60% N_2 and other contaminants like tars and hydrocarbons. It is the cheapest and most-used method
Steam	800–1200	The heating value of product gas is 8–10 MJ/m³ with approximately 45% H_2, 25% of CO and a large quantity of steam (~18%). Presence of steam creates problems with corrosion
Supercritical water	400–800	Operates at high pressures, approximately 30 MPa. This process can suppress the formation of tar and char. Drying of fuel (like biomass) can be avoided

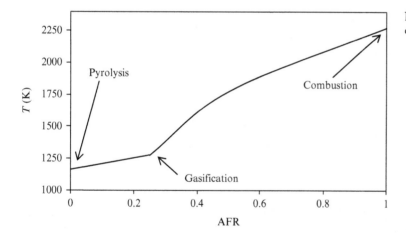

FIG. 4.96   Temperature versus air fuel ratio at thermochemical conversion of biomass.

According to the ultimate analysis, biomass consists of carbon, hydrogen, oxygen, nitrogen and minute traces of sulfur. In principle, one can categorize biomass sources as two kinds: energy crops and residual biomass materials. Here the residual biomass can comprise all sorts of waste woods such as from demolition, furniture factories (sawdust), building materials (sawdust, bark), residual fiber boards, straw residuals from agriculture, various residuals specific to forestry (tree branches, pruning, some trees), residuals from the food processing industry (kernels, seed shells, etc.), residuals from paper mills and any other kind of recoverable paper materials. Here are some examples of biomass:

- Wood: trees, tree stumps, dead trees, branches, wood residuals from forestry and wood processing industries
- Agricultural residues: straw, sugarcane fibre, rice hulls, animal wastes, dried dung
- Biodegradable wastes: from municipal and industrial sources
- Vegetable oils: palm oil, corn oil, peanut oil, soy oil, canola oil
- Energy crops: miscantus, sorghum, switchgrass, hemp, sugarcane, corn, poplar, eucalyptus, willow, aquatic plants.

Proximate analysis identifies the main categories of compounds in solid fuels such as biomass or coal by using certain specific methods. Volatile matter is the percentage of products that are volatile—that is, they can be released as gases during the pyrolysis process or by heating. Volatile matter refers to all other components except the water vapor that results from a mixture's evaporation. The loss of carbon dioxide from mineral embedded in coal structure is an example of volatile matter; other examples are hydrogen emanated by chloride minerals or of sulfur from pyrite. If the mass of volatiles, ash and moisture is extracted from the mass of coal, the remaining quantity is known as fixed carbon.

Table 4.33 gives the proximate analysis of some types of biomass, where AS = almond shell; WP = walnut pruning; RS = rice straw; WW = whole tree woodchips; OW = oak wood; SD = saw dust; SL = sludge; NP = non-recyclable waste paper; BB = beech bark; SG = switch grass; ST = straw. In addition to the proximate analysis, there is the so-called ultimate analysis that determines elemental composition on coal and that reports the weight fraction of carbon, hydrogen, sulfur, nitrogen, oxygen, and ash. The weight fraction of hydrogen and oxygen represents a special situation because these elements can be present in moisture (water, H_2O) and also in the chemical constituency of volatile mater fraction or fixed carbon fraction.

Table 4.34 presents the ultimate analysis data of 12 types of biomasses. The carbon content in biomass is high, ranging from ~35–50% by weight. The content of hydrogen in biomass is relatively low with typical fractions of 3–6% by weight. Table 4.35 presents the heating value and the specific chemical exergy of 15 types of biomasses. It is remarked that the exergy content of biomass sources is in the range of approximately 8–25 MJ/kg on wet basis.

The gross calorific value of biomass combustion can be calculated with the relation proposed by Van Loo and Koopejan (2008), according to the equation

$$GCV\,[\text{MJ/kg}] = 34.91X_C + 117.83X_H + 10.05X_S - 1.51X_N - 1.034X_O - 2.11X_{ash} \tag{4.186}$$

where $\mathcal{A}$ is the weight fraction of ash in a dry (moisture-free) biomass.

Szargut (2005) gives an equation for the chemical exergy of biomass (in which the contribution of the sulfur to the combustion process is neglected)

$$ex^{ch} = NCV\beta, \text{ where } \beta = 1.0347 + 0.014f_1 + 0.0968f_2 + 0.0493f_3 \tag{4.187}$$

TABLE 4.33 Typical Proximate Analysis of Some Types of Biomass

Constituent	AS	WP	RS	WW	OW	SD	SL	NP	BB	SG	ST
Moisture	8	12	6	7	6.5	34.9	8	6	8.4	11.9	14.4
Ash	24	3	19	3	0.3	0.7	35	1	7.1	4.5	7.5
Volatiles	54	71	65	78	73.0	55.1	52	85	67.5	70.8	64.3
Fixed Carbon	14	14	10	19	20.0	9.3	5	8	17.0	12.8	13.8

TABLE 4.34 Typical Ultimate Analysis of Some Types of Biomass

Constituent (wt. %)	AS	WP	RS	WW	OW	SD	SL	PW	NP	BB	SG	ST
C	36	48	38	51	47	40	36	48	49	48	43	40
H	4	4	3.5	3	6.5	6	4	6.7	5	6.0	6.1	5.6
N	1	0.6	0.5	0.35	0.3	0.5	6	0.10	0.1	0.7	0.7	1
S	0.05	0.03	0.06	0.05	0.2	0	1.1	~0.0	0.06	0.3	0.2	0.6
O	32	44	36	42	40	34.5	15	39.1	44	36	40	37.8
Ash	26.95	3.37	21.94	3.6	6	19	37.9	5.0	1.84	9	10	15

TABLE 4.35 Heating Values of Various Types of Biomasses

| Biomass | GCV (MJ/kg db) | Moisture (w%, w.b.) | NCV | | ex^{ch} |
			MJ/kg w.b.	MJ/dm³ w.b.	MJ/kg w.b.
Wood pellets	19.8	10	16.4	9.8	18–19
Pine wood	20.5	8	18.9	14	24.8
Woodchips	19.8	30–50	8.0–12.2	2.8–3.9	8–14
Grass	18.4	18	13.7	2.7	15–16
Cereals	18.7	15	14.5	2.5	16–17
Bark	20.2	50	8.2	2.6	9–10
Sawdust	19.8	50	8.0	1.9	8–9
Straw	18.7	15	14.5	1.7	16–17
Olive kernels	22.0	58	7.3	6.3	7–9
Almond shells	15.1	9	14.3	2.5	16.7
Walnut pruning	19	14.3	18.2	10.2	20.5
Rice straw	14.8	6	14.1	2.5	15.9
Whole tree woodchips	20.4	7.5	19.7	12	22.1
Sludge	15.4	8.6	14.5	2.5	15.9
Non-recyclable waste paper	18.8	6.3	17.7	10	20.1

Note: w%, percent by weight; d.b., dry basis; w.b., wet basis.

An alternative expression for factor β in Eq. (4.179) is proposed in Szargut and Styrylska (1964) for biomasses with low oxygen content, as

$$\beta(1 - 0.3035f_2) = 1.0412 + 0.216f_1 - 0.2499f_2(1 + 0.7884f_1 + 0.045f_4) \tag{4.188}$$

where factors f_i are defined as follows:

$$f_1 = \frac{X_H}{X_C}, f_2 = \frac{X_O}{X_C}, f_3 = \frac{X_S}{X_C}, f_4 = \frac{X_N}{X_C} \tag{4.189}$$

The net calorific value—NCV in Eq. (4.179)—can be calculated according to Van Loo and Koopejan (2008) as

$$NCV[\text{MJ/kg}] = GCV(1 - X_w) - 2.444X_w - 21.839X_H(1 - X_w) \tag{4.190}$$

where X_w is the fraction of moisture.

The following elementary reactions are also important in gasification:

$$C + O_2 \rightarrow CO_2, \Delta H^0 = -390 \text{kJ/mol} \tag{4.191a}$$

$$C + H_2O \rightarrow H_2 + CO, \Delta H^0 = 130 \text{kJ/mol} \tag{4.191b}$$

$$C + 2H_2 \leftrightarrows CH_4, \Delta H^0 = -71 \text{kJ/mol} \tag{4.191c}$$

In an autothermal gasification, the process is adiabatic. It is generally agreed that the adiabatic temperature of the gasifier substantially changes with the type of gasifying medium. Also the efficiency of air gasification is higher than steam gasification because the gasifier operates at a higher temperature.

The influence of operation factors, like temperature, and the equivalent ratio (ER) on the gasification performance process irreversibilities are assessed by considering steam and a steam-air mixture as gasifier mediums. The equivalent ratio is defined as the ratio of used oxygen for gasification of any material divided by the required oxygen for its stoichiometric combustion.

TABLE 4.36 Initial-Change-Equilibrium Chart for a Typical Gasification Process

Species	Initial	Change	Equilibrium
C	$n_{C,0}$	$-\xi_1 - \xi_2$	0
O_2	$n_{O_2,0}$	$-\xi_1$	0
CO	0	$\xi_2 - \xi_3$	$\xi_2 - \xi_3$
H_2O	$n_{H_2O,0}$	$-\xi_2 - \xi_3$	$n_{H_2O,0} - \xi_2 - \xi_3$
CO_2	0	$\xi_1 + \xi_3$	$\xi_1 + \xi_3$
H_2	0	$\xi_2 + \xi_3$	$\xi_1 + \xi_3$

Assume that all oxygen is consumed in the gasification process. Then the equilibrium constant in a gasification process involving H_2, CO, CO_2, H_2O as the main gas phase species can be approximated as follows:

$$K_{eq} = \frac{P_{H_2}P_{CO_2}}{P_{CO}P_{H_2O}} \tag{4.192}$$

Considering the elementary reaction at the gasification as just discussed, an initial-change-equilibrium chart is constructed for the general gasification process, as shown in Table 4.36. The extent of reaction for carbon oxidation is denoted here with ξ_1. For the water-gas-shift reaction the extent of reaction is ξ_2. The extent of reaction ξ_3 can be correlated to the equilibrium constant, the stoichiometry and the other two extents of reactions. The following intermediary coefficients are introduced for this correlation: $a_1 = 1 - K_{eq}$, $a_2 = \xi_1 + \xi_2 + K_{eq}n_{H_2O,0}$, and $a_3 = \xi_1\xi_2 - K_{eq}\xi_2(n_{H_2O,0} - \xi_2)$. The extent ξ_3 is then correlated as follows:

$$\xi_3 = \frac{-a_2 + \sqrt{a_2^2 - 4a_1a_3}}{2a_1} \tag{4.193}$$

The equilibrium concentrations in product gas result in

$$y_{CO} = \frac{\xi_2 - \xi_1}{n_{H_2O,0} + \xi_1 + \xi_2} \tag{4.194a}$$

$$y_{CO_2} = \frac{n_{H_2O,0} - \xi_2 - \xi_3}{n_{H_2O,0} + \xi_1 + \xi_2} \tag{4.194b}$$

$$y_{CO_2} = \frac{\xi_1 + \xi_3}{n_{H_2O,0} + \xi_1 + \xi_2} \tag{4.194c}$$

$$y_{H_2} = \frac{\xi_2 + \xi_3}{n_{H_2O,0} + \xi_1 + \xi_2} \tag{4.194d}$$

The exergy efficiency of the hydrogen production process through gasification can be generally defined as

$$\psi = \frac{\dot{n}_{H2}ex_{H_2}^{ch}}{\dot{Ex}_{feedstock} + \dot{W}_{aux}} \tag{4.195}$$

where $\dot{W}_{aux}$ represents any power (or even exergy rate) required by auxiliary processes.

Oxygen is beneficial for gasification. Oxi-gasification will use pure oxygen as oxidant rather than air with the advantage of obtaining a more efficient process with less-expensive separation operations. Operating at lower ER, smaller gasifier, reduction in NO_x emissions, and potential saving in the compression for the produced syngas are other advantages of using oxygen instead of air.

Fig. 4.97 shows the variation of exergy efficiency of two gasification processes with the equivalence ratio. In one of the processes, air is the oxidant, while in the other, it is pure oxygen. It can be remarked that the exergy efficiency of the oxi-gassification tends to be better than conventional gasification that uses air as oxidant. The figure is constructed from a case study by Aghahosseini (2013), but the results can be considered of general relevance.

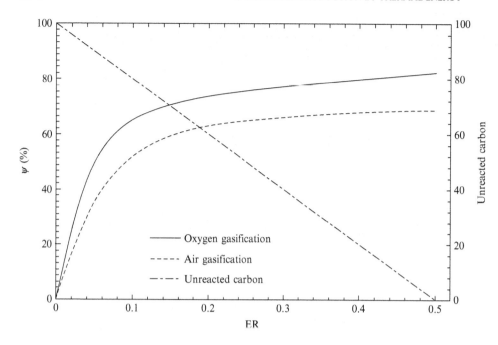

FIG. 4.97 Exergy efficiency and unreacted carbon versus equivalence ratio in a gasifier. *Data from Aghahos-seini, S., 2013. System Integration and Optimization of Copper-Chlorine Thermo-chemical Cycle with Various Options for Hydrogen Production. PhD Thesis, University of Ontario Institute of Technology.*

In general, gasification of biomass is conducted at higher temperature than gasification of coal. When biomass gasification is conducted at high temperature, the following additional reactions that may concurrently occur with the elementary reaction just discussed, should be considered:

$$CH_4 + \frac{3}{2}O_2 \rightarrow CO + 2H_2O \tag{4.196a}$$

$$CH_4 + \frac{1}{2}O_2 \rightarrow 2H_2 + CO \tag{4.196b}$$

$$CH_4 + CO_2 \rightleftarrows 2H_2 + CO_2 \tag{4.196c}$$

$$C_2H_4 + O_2 \rightarrow 2H_2 + 2CO \tag{4.196d}$$

$$C_2H_6 + O_2 \rightleftarrows 3H_2 + 2CO \tag{4.196e}$$

$$C_2H_4 + 2O_2 \rightarrow 2CO + 2H_2O \tag{4.196f}$$

$$C_2H_6 + \frac{5}{2}O_2 \rightarrow 2CO + 3H_2O \tag{4.196g}$$

$$C_3H_8 \rightarrow C_2H_4 + C_{H_4} \tag{4.196h}$$

The reaction rate of the most significant reaction in a biomass gasification can be modeled based on the Arhenius Eq. (4.30). Table 4.37 shows a compilation of rate constant correlation sourced from Abuadala (2010). The gasification process can be performed with or without a catalyst, depending on gasifier downstream use, and can take place in a fixed bed or fluidized bed gasifier and under atmospheric or super-atmospheric pressure. For cost-effective hydrogen production using this technology, large fuel resources are needed that require development of smaller, efficiently distributed gasification plants. The gasification process results in a continuous change in char composition, so its reactivity continuously varies.

Tar is an undesirable product from biomass gasification due to the various problems of fouling and slugging in the process equipment. There are hundreds of species in the tar sample, but in order to simplify the analysis, all the species are treated in a single lump.

Three methods are available to minimize tar formation: (i) proper design of a gasifier; (ii) proper control and operation; (iii) additives/catalysts. Tar is modeled as a benzene, C_6H_6. The modeling approach of a gasifier varies from homogeneous gas-gas, to heterogeneous gas-solid, from single to multiple regions, and from zero to three-dimensional modeling. Biomass is more reactive than coal, it pyrolyses very quickly, and its ash content is usually very low; therefore, in mathematical models, often ash can be neglected.

TABLE 4.37 Rate Constant Parameters for Elementary Reactions at Biomass Gasification

Reaction	Pre-exponential factor	Activation energy (kJ/mol)
$C + 0.5O_2 \rightarrow CO$	$\mathcal{A}(s^{-1}) = 6.67 \times 10^{-1}$	$E_a = 130,144$
$CO + 0.5O_2 \rightarrow CO_2$	$\mathcal{A}(m^{2.4}kmol^{-0.8}s^{-1}) = 4.8 \times 10^8$	$E_a = 16,000$
$H_2 + 0.5O_2 \rightarrow H_2O$	$\mathcal{A} = 4.9 \times 10^{10}$	$E_a = 21,500$
$CH_4 + H_2O \rightarrow 3H_2 + CO$	$\mathcal{A}(kmol\,m^{-3}s^{-1}) = 3.1$	$E_a = 122,010$
$CO + H_2O \rightarrow H_2 + CO_2,\ T > 1123\,K$	$\mathcal{A} = 2.7 \times 10^{-3}$	$E_a = 32,210$
$O + H_2O \rightarrow H_2 + CO_2,\ T < 1123\,K$	$\mathcal{A} = 10^6$	$E_a = 51,813$
$CH_4 + \dfrac{3}{2}O_2 \rightarrow 2H_2 + CO$	$\mathcal{A} = 7 \times 10^{11}$	$E_a = 245,647$

Source: Abuadala, A.G., 2010. Investigation of Sustainable Hydrogen Production From Steam Biomass Gasification. MSc Thesis, University of Ontario Institute of Technology.

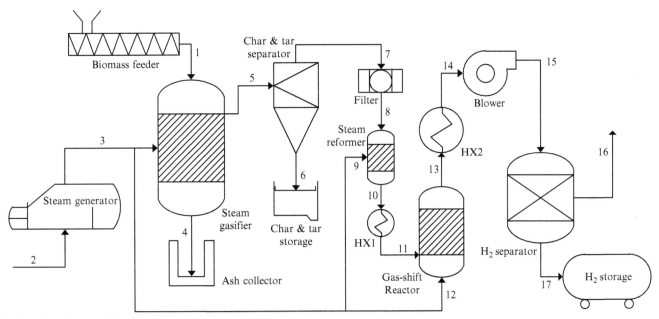

FIG. 4.98 Simplified flowchart of a biomass gasification plant.

A simplified flowchart of a biomass gasification plant is shown in Fig. 4.98. The steam gasifier requires a feed of biomass in #1 and process steam in #3 generated from water stream #2. In an autothermal gasification, a steam generator may not be required. Nevertheless, steam generation can be conducted in a separate unit, driven by biomass combustion. Some ash results from the gasification in #4. The product gas in #5 is separated from char and tar #6 and filtered as process #7–8. Furthermore, steam reforming reaction is conducted with additional steam in #9 that generates syngas in #10. The syngas is cooled in HX1 and fed to the gas-shift reactor in #11, where it reacts with additional steam provided at port #12 to generate a hydrogen—carbon dioxide rich gas. Since steam is the gasification agent, there will be only minor traces of other gases besides hydrogen and carbon dioxide in stream #13. Finally, process gas is conditioned and hydrogen gas is extracted as stream #17 and stored.

4.7 INTEGRATED SYSTEMS FOR GREEN-THERMAL HYDROGEN PRODUCTION

4.7.1 Nuclear-Based Systems

Nuclear energy is retrieved in the form of a high-temperature process heat that typically is transferred to a water stream to generate high-pressure superheated steam. The commercial nuclear reactors for power generation that are in operation today use to generate high-pressure superheated steam at approximately 580 K. Next-generation nuclear reactors are deemed to generate process heat at 1000 K, with a wide applicability to hydrogen and synthetic fuel production. In this section, some nuclear-based systems for hydrogen production are presented, based on a literature study. The main focus is on copper-chlorine-based nuclear hydrogen production.

4.7.1.1 Mark 10 Nuclear Hydrogen Production Plant

A conceptual flowsheet diagram for a hydrogen production plant based on the ISPRA Mark 10 thermochemical cycle connected to the high temperature test reactor (HTTR) of Japan Atomic Energy Agency was developed by Rosen (2008). The HTTR is a graphite-moderated helium gas-cooled reactor comprising long hexagonal fuel assembles. The HTTR is being operated successfully in Oarai, Japan, for a capacity of 30 MW. Rosen reasonably assumed that HTTR can be configured to supply 11.4 GWh power per day and 113 GWh high temperature heat at 1200 K.

Fig. 4.99 shows the black-box representation of the nuclear hydrogen production plant studied in Rosen (2008). Energy and exergy flows are indicated in percentages on the diagram. As seen, the energy efficiency of the integrated plant is 20.5%. The exergy efficiency is given by the ratio $19/76 = 25\%$. The ISPRA Mark 10 cycle is a purely thermochemical process including the following six main steps:

- hydrolysis

$$2H_2O(l) + I_2(s) + SO_2(g) + 4NH_3 \rightarrow 2NH_4I(s) + (NH_4)_2SO_4(s), \text{ at } 323 \text{ K} \qquad (4.197)$$

- hydrogen evolving reaction

$$2NH_4I(s) \rightarrow H_2(g) + 2NH_3(g) + I_2(g), \text{ at } 900 \text{ K} \qquad (4.198)$$

- oxygen evolving reaction

$$SO_3(g) \rightarrow 0.5O_2(g) + SO_2(g), \text{ at } 1123 \text{ K} \qquad (4.199)$$

- ammonia regeneration reaction

$$(NH_4)_2SO_4(s) + Na_2SO_4(s) \rightarrow 2NaHSO_4(s) + 2NH_3(g), \text{ at } 673 \text{ K} \qquad (4.200)$$

- steam regeneration reaction

$$2NaHSO_4(s) \rightarrow Na_2S_2O_7(s) + H_2O(g), \text{ at } 673 \text{ K} \qquad (4.201)$$

- sulfur trioxide regeneration reaction

$$Na_2S_2O_7(s) \rightarrow Na_2SO_4(s) + SO_3(g), \text{ at } 825 \text{ K} \qquad (4.202)$$

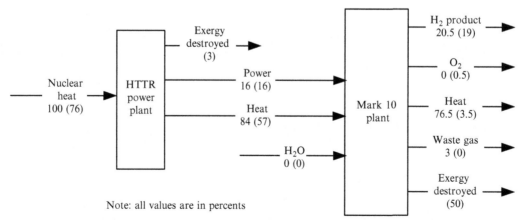

FIG. 4.99 Energy and exergy flows for a Mark 10 plant integrated with a HTTR. *Data from Rosen, M.A., 2008. Exergy analysis of hydrogen production by thermochemical water decomposition using the Ispra Mark-10 Cycle. Int. J. Hydrog. Energy 33, 6921–6933*

The process flowsheet diagram of the mark 10 process is shown in Fig. 4.100. The streams are indicated by numbers on the diagram. According to flowsheet, the plant has 23 components connected by 34 material streams. The flowsheet illustrates the complexity of the thermochemical plant, although some details are not shown, namely a heat exchanger network for internal heat recycling. There are a significant number of blowers and solid material conveyers; these consume power that must be included in thermodynamic analyses. Table 4.38 gives the composition of each material stream of the flowsheet. Table 4.39 gives the brief description of each component shown on the flowsheet.

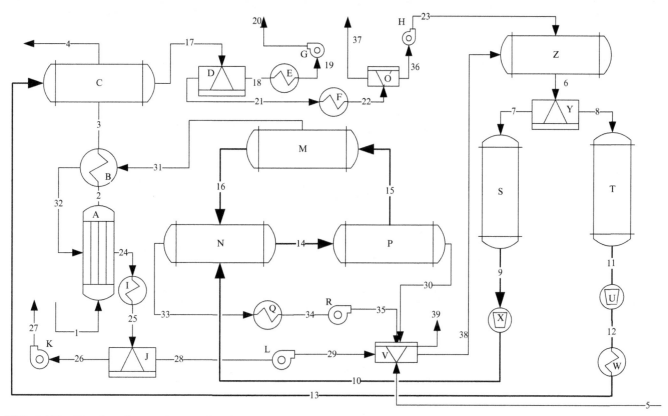

FIG. 4.100 Flowsheet diagram for a Mark 10 thermochemical cycle.

TABLE 4.38 Stream Definitions for Mark 10 Plant Shown in Fig. 4.100

Stream	Composition	Stream	Composition	Stream	Composition
1	Hot helium input	14	$Na_2S_2O_7(s)$ preheated	27	Oxygen, blown out
2	Helium	15	Heat transfer flux	28	$SO_2 + O_2$ cold
3	Helium	16	$2NaHSO_4(s)$ cooled	29	$SO_2 + O_2$ blown
4	Colder helium return	17	$H_2 + 2NH_3 + I_2$ hot	30	Steam, cooled
5	Liquid water input	18	Hydrogen purified, hot	31	$SO_3(g)$ colder
6	$2NH_4I(s) + (NH_4)_2SO_4(s)$	19	Cooled hydrogen	32	$SO_3(g)$ heated
7	$(NH_4)_2SO_4(s)$ cold	20	Pressurized hydrogen	33	$2NH_3(g)$ hot
8	$2NH_4I(s)$ cold	21	$2NH_3 + I_2 + H_2$ hot	34	$2NH_3(g)$ cooled
9	$(NH_4)_2SO_4(s)$ preheated	22	$2NH_3 + I_2 + H_2$ cooled	35	$2NH_3(g)$ blown
10	$(NH_4)_2SO_4(s)$ conveyed	23	$2NH_3 + I_2$ blown	36	$2NH_3 + I_2$ recovered
11	$2NH_4I(s)$ preheated	24	$0.5O_2(g) + SO_2(g)$ hot	37	Unrecoverable H_2
12	$2NH_4I(s)$ conveyed	25	$0.5O_2 + SO_2$ cooled	38	NH_3, H_2O, SO_2 recycled
13	$2NH_4I(s)$ hot	26	Oxygen, cold	39	Unrecoverable O_2

TABLE 4.39 Component List for Flowsheet Diagram from Fig. 4.100

S.	Component	S.	Component	S.	Component
A	Oxygen evolving reactor	J	Absorption separator for SO_2	S	$(NH_4)_2SO_4(s)$ preheater[a]
B	$SO_3(g)$ preheater	K	Oxygen blower	T	$2NH_4I(s)$ preheater[a]
C	Hydrogen evolving reactor	L	SO_2 blower	U	$2NH_4I(s)$ conveyer
D	Hydrogen separation reactor	M	Sulphur trioxide reactor	V	NH_3, H_2O, SO_2 recovery reactor
E	Hydrogen cooler[b]	N	Ammonia regeneration reactor	W	$2NH_4I(s)$ heater[a]
F	Cooler for $2NH_3(g) + I_2(g)$[b]	O	NH_3, I_2 recovery reactor	X	$(NH_4)_2SO_4(s)$ conveyer
G	Hydrogen compressor	P	Steam regeneration reactor	Z	Hydrolysis reactor
H	Blower $2NH_3(g) + I_2(g)$	Q	Ammonia cooler[b]		
I	Cooler for $0.5O_2(g) + SO_2(g)$[b]	R	Ammonia blower		

[a]Internal heat reuse.
[b]Heat recovery applied.

Within the thermochemical plant, water is supplied at state #5, and hydrogen is delivered at #20. Water must be supplied in excess, because neither all water can be converted nor all resulting hydrogen and oxygen can be recovered. The thermochemical plant also rejects heat at various points as it is not possible to recover and reuse all heat internally. The process simulation of this plant was performed with ASPEN Plus, and the results were reported in Rosen (2008).

Fig. 4.101 shows a black-box representation of the thermochemical plant and gives the energy and material flows as inputs and outputs. This figure summarizes the overall mass and energy balances for the thermochemical plant on the assumption that the plant is supplied with high temperature heat in the form of hot helium derived from HTTR and, in addition, electric power necessary to drive all auxiliary units.

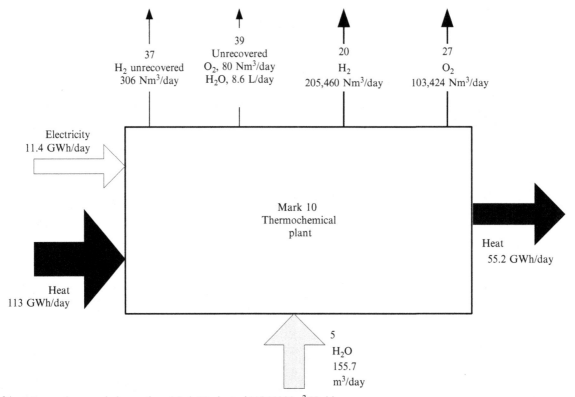

FIG. 4.101 Mass and energy balances for a Mark-10 plant of 205,000 Nm^3 H_2/day.

FIG. 4.102 Comparison of hydrogen production efficiencies as a function of the level of the process analysis.

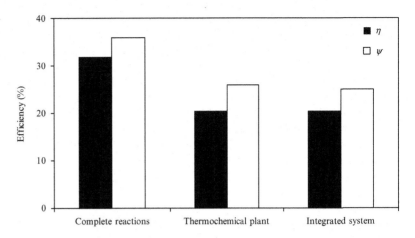

Based on the overall mass and energy balance, the energy and exergy flows can be calculated for the thermochemical plant itself. When the energy and exergy of each input/output flow is determined, the thermochemical plant is integrated with the nuclear plant that supplies both electrical and thermal energy to the water splitting process.

The level of process analysis will influence the predicted energy and exergy efficiency. A comparison of various efficiencies as a function of the level of the process analysis is presented in Fig. 4.102. For the first level of analysis—only the thermochemical cycle is studied, assuming complete reactions—the predicted efficiencies are highest. No electricity is found necessary to drive the processes since the system is idealized.

A flowsheet analysis of the thermochemical plant identifies the needs of auxiliary power. Consequently, the efficiencies drop sharply. Further integration with the power generation system does not reduce the energy efficiency. In the integrated system, all of the heat rejected by the power plant is recovered and reused. However, there is a 1% penalty on exergy efficiency because the integrated system adds more equipment and, therefore, more irreversibilities.

4.7.1.2 Nuclear-Based Sulfur-Iodine Plant

The very high temperature reactor (VHTR) offers a good opportunity for nuclear hydrogen production with a sulfur-iodine cycle in the near future. Directly linked combined cycles for power generation show the highest potential for efficient power generation in an integrated system for nuclear hydrogen and electricity production. Fig. 4.103 illustrates a simplified diagram of a combined cycle that uses a helium-based Brayton cycle (topping) and a double-reheat steam Rankine cycle (bottoming cycle) and a sulfur-iodine remote plant.

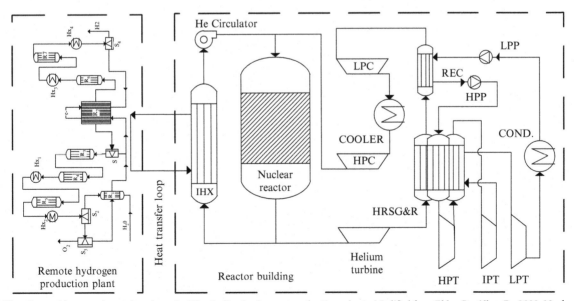

FIG. 4.103 General layout of a nuclear-based sulfur-iodine hydrogen production plant. *Modified from Elder, R., Allen, R., 2009. Nuclear hydrogen production: Coupling a very high/high temperature reactor to a hydrogen production plant. Prog. Nucl. Energy. 51 (2), 500–525.*

In this flowsheet, Brayton cycle has two stages of compression with intercooling. Heat is recovered from the expanded helium flow in the HRSG&R (heat recovery steam generator and reheater) to generate superheated steam for the high pressure turbine (HPT) and steam reheating prior to the intermediate-pressure turbine (IPT) and low pressure turbine (LPT). A recuperative heat exchanger is used to preheat water between the low-pressure pump (LPP) and high pressure pump (HPP), avoiding steam extraction from IPT or LPT.

As reported by Elder and Allen (2009), the efficiency of this cycle is in the range of 46–49% when the reactor outlet temperature varies from 1123 to 1223 K. This efficiency is the highest among the following cycles: direct Brayton cycle (not combined), combined cycles with indirect heat transfer and no intercooler, steam Rankine cycle with single reheat, steam Rankine cycle with no reheat, combined indirect Brayton-Rankine cycle and Brayton indirect cycle.

The hydrogen production plant is located remotely with respect to the nuclear reactor and power generation loop. A design challenge with linking the nuclear reactor to a downstream process such as hydrogen production is the possibility of tritium escaping in the primary loop of the coolant. Tritium is generated in the reactor core from side nuclear reactions involving lithium, helium and boron atoms, where helium is the coolant. In some instances, tritium atoms—which are radioactive—can escape from the primary to the secondary circuit through the intermediate heat exchanger wall and reach the secondary helium circuit. From there they contaminate the system radioactively by traversing the metallic wall of the Intermediate Heat Exchanger (IHX).

As reported by Elder and Allen (2009), in a VHTR, the release of tritium in the primary circuit is of order 60 µCi/s (Ci=Curie), whereas the tritium contamination of the hydrogen produced by a thermochemical plant connected to the IHX is of order 50 mCi per tonne of hydrogen product. This is far below the maximum admissible limits of contamination, but it must be considered thoroughly for design purposes. One potential design measure for preventing tritium migration to the secondary circuit is to pressurize the secondary circuit more than the primary circuit. The material selection of the IHX is also important; hastelloy XR is an option. The IHX must be also a compact design with a high overall heat transfer coefficient, capable of transferring hundreds of megawatts of thermal energy from the primary to the secondary side with a reduced temperature difference.

Table 4.40 shows operating parameters of the IHX of four different concepts proposed in past literature. Mizia (2008) mentions that a promising option for IHX is a printed circuit technology that consists of stacking steel plates via a diffusional bonding process. The plates have small flow channels obtained by mechanical pressing or chemical etching. The advantage of this concept is approximately six times greater compactness than a conventional tubular IHX for the same effective heat transfer rate.

Two types of printed circuit heat exchanger (PCHE) concepts exist, namely the Toshiba concept and the Heatric Corporation concept. The IHX is formed with PCHE modules linked with flow distribution and collection lines and is eventually embedded in an outer metallic shell. There are two IHX systems installed in series: one at the lower

TABLE 4.40 Main Design Parameters of IHX

Parameter			GA	JAEA	EURATOM	AREVA-NP
Capacity, $\dot{Q}$ (MW thermal)			600	170	613	608
LMTD (K)			24	154	48	50
Primary	Flow rate, $\dot{m}$ (kg/s)		321	324	210	240
	In	P (MPa)	7.03	5.0	5.5	5.5
		T (K)	1223	1223	1223	1123
	Out	P (MPa)	7.0	4.95	5.4	5.4
		T (K)	863	1123	660	623
Secondary	Flow rate, $\dot{m}$ (kg/s)		321	80	219	614
	In	P (MPa)	7.10	5.15	5	5.5
		T (K)	838	773	623	573
	Out	P (MPa)	7.07	5.0	4.9	5.4
		T (K)	1998	1073	1163	1073

Source: Eleder and Allen (2009).

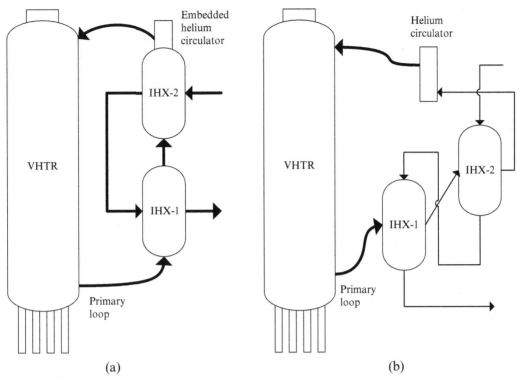

FIG. 4.104 Flow configuration for high temperature and low temperature IHX systems: (a) core-side configuration; (b) shell-side configuration.

temperature side and with an extended lifetime of 60 years; the other is a replaceable heat exchanger placed at the high temperature end. This heat exchanger has a shorter lifetime due to the higher operating temperature and more severe conditions.

According to Mizia (2008), two types of configurations—core-side and shell-side—are considered for development. A core-side configuration is shown in Fig. 4.104a, and the primary circuit has flow channels at the core side of the IHX. The hottest gas is circulated in the middle of the IHX, while the gas of the secondary loop comes from the outer site, which circulates through the shell. Inside the shell, there are PCHE modules. The low-temperature IHX (namely, IHX-2 in the figure) has an embedded helium circulator for the primary loop. The shell side configuration is presented in Fig. 4.104b. In this case, the hot fluid of the primary helium loop is circulated through the heat exchanger's shell side, and from there it is forced to flow through the PCHE module and exit from the opposite part of the shell side. In the core, the secondary helium flows at lower temperature, and it includes the shell distribution and collection systems for the secondary flow.

Atomic Energy Commissariat (CEA) of France supports developments of AREVA's VHTR via several research projects, including the design and testing of innovative concepts of crucial elements of helium circuits. As mentioned by Carre et al. (2010), a new concept of a compact IHX of high capacity (800 MW thermal energy) was recently developed.

The system is presented in a simplified diagram in Fig. 4.105. It shows the heat exchanger, primary isolation valve, secondary isolation valves and the helium circulator of the primary loop; all elements are part of a compact assembly, denoted as IHX.

Hot helium from the reactor enters the IHX assembly from the side, via a tube that turns up inside the vessel and directs the gas toward the top of the reactor, where it is placed with an isolation valve for the primary circuit. The hot gases from the top plenum are then forced downward and across a finned tube surface for enhanced heat transfer with helium from the secondary loop. At the bottom of the vessel, the cooled gases are pressurized by the helium circulator, which has a compact construction together with its motor.

The secondary helium enters at lower temperatures from the side of the IHX vessel, and it is fed to a distributor that diverts the flow in a number of vertical tubes with finned surfaces. Helium flows upward as indicated, and then it returns downward. At the lower side of the heat exchanger module, there is a collector that collects the flow from all parallel tubes and directs it toward the exit port. A secondary isolation valve is placed at the exit port. The IHX vessel is made of industrial-grade ferritic steel (9Cr1Mo) with a 7.3 m inner diameter and a height of 20 m, weight of approximately 1000 t and transferring 0.8 kW thermal energy per kg with a corresponding power density of approximately 0.94 MW/m^3.

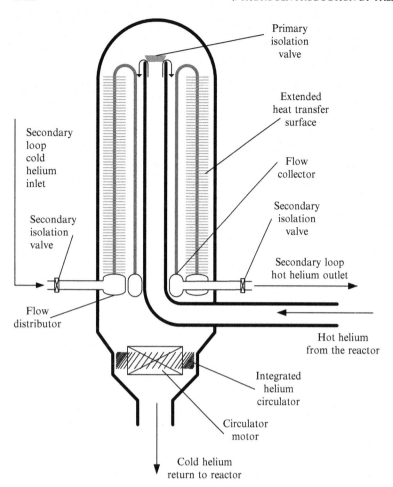

FIG. 4.105   IHX concept developed at AREVA-NP/CEA. *Modified from Carre, F., Yvon, P., Anzieu, P., Chauvin, N., Malo, J.-Y., 2010. Update of the French R&D strategy on gas-cooled reactors. Nucl. Eng. Des. 240, 2401–2408.*

The transfer of heat from the secondary loop to a hydrogen production process presents various challenges. In general—with high-temperature gas reactors—it implies the transfer of heat from flowing helium gas to a chemical reaction. One integration scheme investigated by General Atomics involved the use of IHX with a secondary helium loop to transfer high temperature heat at a remote location from the reactor where the S-I (sulfur-iodine) plant is located. The transmission distance may be a few hundred meters.

As described in Elder and Allen (2009), the GA concept, from the US-DOE NERI (Nuclear Energy Research Initiative), is designed around the modular helium reactor of 600 MW thermal energy output per unit. The integrated hydrogen production site will include four nuclear reactors, totaling 2.4 GW thermal power generation. The heat is transferred at 115 m from the reactor via a secondary helium loop that delivers heat at the consumption point at 1100 K to the S-I cycle. The return flow comes at 838 K at the IHX inlet. A schematic of the US-NERI integrated plant is presented in Fig. 4.106. The overall hydrogen generation efficiency of the process is 38% with respect to the LHV of hydrogen.

4.7.1.3 Nuclear-Based Hybrid Sulfur Plant

According to Elder and Allen (2009), a consortium involving the Shaw Group and Westinghouse conceptually developed an integrated plant for hydrogen production with a hybrid-sulfur process and PBMR reactor (pebble bed modular reactor). The configuration of the integrated plant is depicted in Fig. 4.107. For power generation, the plant uses a standard steam Rankine cycle that uses two heat sources: heat recovered from the hybrid sulfur plant and heat delivered from heat exchangers coupled in series with the IHX (helium) of the PBMR reactor, at the bottom side. One single module of the PMBR-based hybrid sulfur (HyS) plant comprises the VHTR, IHX, three helium loops and the HyS unit.

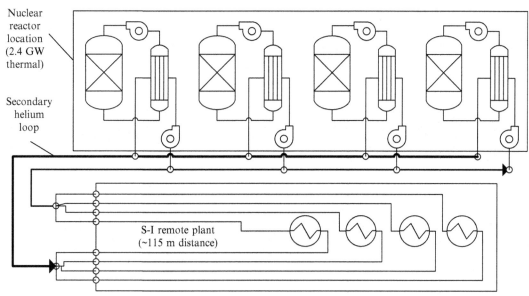

FIG. 4.106   Configuration of the integrated MHR/S-I plant proposed by DOE-NERI and GA.

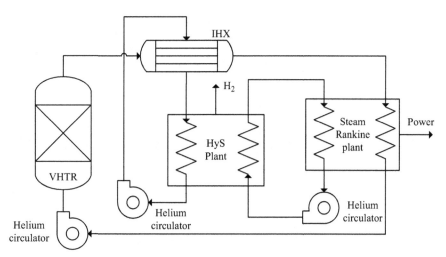

FIG. 4.107   Integrated plant with hybrid sulfur process and PBMR for hydrogen generation.

There are four linked PBMR/HyS modules on a steam Rankine cycle. The PBMR reactor unit generates 500 MW thermal while heating helium in the primary loop from 700 to 1223 K. In the IHX, the temperature of helium from the primary loop is reduced to 1020 K. The temperature of helium is reduced to ~700 K due to heat transfer to the Rankine cycle. From the 500 MW thermal output, approximately 195 MW are transferred to the HyS process via a secondary helium loop (see figure) with 1173 K temperature at the hot end and 970 K at the cold end.

Low grade heat (approximately 100 MW thermal) rejected by the HyS cycle is recovered and transferred to the power plant. As reported by Elder and Allen (2009), each HyS unit uses ~54 MW electrical energy provided by the power plant. The power plant unit generates ~ 600 MW electrical energy with an efficiency higher than 38%. Approximately 62% of the total power generated by the plant is delivered to the grid while the remainder is used to power the hydrogen production process (including its auxiliary equipment). The total hydrogen production rate of a full plant is ~20 Mt per day. In addition, 380 MW power output is generated. Note that the total heat input generated by the four reactors per plant is 2.4 GW.

4.7.1.4 Nuclear-Based Magnesium-Chlorine Plant

Ozcan (2015) proposed a nuclear-based magnesium-chlorine plant in which the nuclear reactor is a CANDU-SCWR (supercritical water reactor). The plant also includes a Linde-Hampson unit for production of liquefied/cryogenic hydrogen. The CANDU reactor also drives a steam Rankine power plant that generates the required power to drive the process. An intermediate heat exchanger is used to transfer heat to a secondary loop and thence to the magnesium-chlorine cycle.

Ozcan (2015) found that the overall energy efficiency of the plant for the base case is found to be 42.7%, where the exergy efficiency is 63.7%. The total electrical work requirement to run the cycle is found to be 156.46 MW to liquefy 1 kmole/s hydrogen. The total exergy destruction of the system is found to be 144.38 MW where the highest contribution is made by heat exchangers in the system. Fig. 4.108 shows the layout of the nuclear-based magnesium-chlorine plant for hydrogen production. As emphasized in the figure, there are three main functional units, namely the nuclear reactor with its high-pressure supercritical steam generator, the magnesium-chlorine plant connected to the nuclear-steam generator through a heat transfer line and the liquefaction unit.

The energy and exergy efficiencies of the main units and overall plant of nuclear hydrogen production with Mg-Cl cycle, as derived from Ozcan (2015), are represented in a bar chart format, as shown in Fig. 4.109. The best exergy efficiency is that of the hydrogen liquefaction plant, followed by the nuclear Rankine power plant and the magnesium-chlorine plant. The relative exergy destruction of the liquefaction plant is 25%, of the nuclear steam Rankine plant of 34% and of magnesium-chlorine cycle 41%.

4.7.1.5 Nuclear-Based Copper-Chlorine Cycle for Hydrogen Production

The copper-chlorine cycle was initially developed for the CANDU-SCWR. For safety reasons, indirect heat transfer methods with intermediate heat exchangers must be developed in order to link the SWCR to a Cu-Cl plant. The temperature levels of SCWR and the Cu-Cl cycle also have significant influence on the location of the heat extraction from an SCWR. The Cu-Cl cycle will influence the water flow arrangement of the SCWR.

Wang et al. (2010) proposed an integration scheme of a copper-chlorine thermochemical cycle with SWCR according to the flowchart in Fig. 4.110. The power plant is a steam Rankine cycle with a single reheat system that uses a

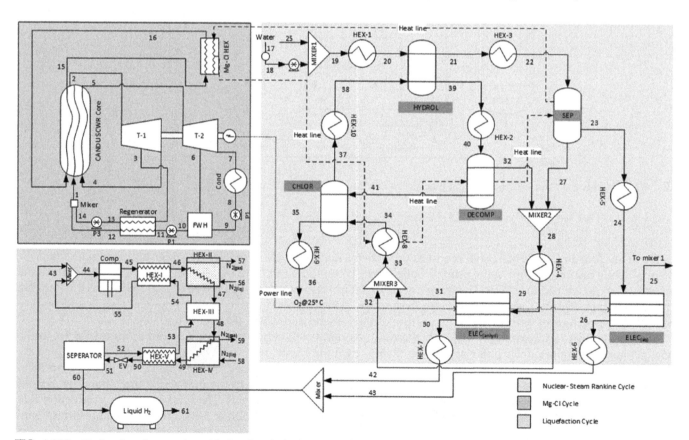

FIG. 4.108 Nuclear-based magnesium-chlorine plant for hydrogen production. *Modified from Ozcan, H., 2015. Experimental and Theoretical Investigations of Magnesium-Chlorine Cycle and Its Integrated Systems. PhD Thesis, University of Ontario Institute of Technology.*

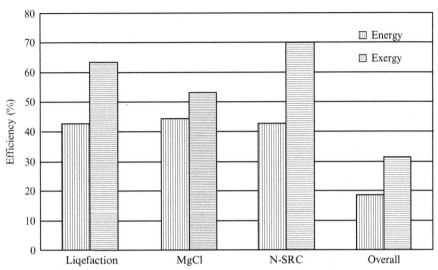

FIG. 4.109 Energy and exergy efficiencies of the main units and overall plant of nuclear hydrogen production with Mg-Cl cycle. *Data from Ozcan, H., 2015. Experimental and Theoretical Investigations of Magnesium-Chlorine Cycle and Its Integrated Systems. PhD Thesis, University of Ontario Institute of Technology.*

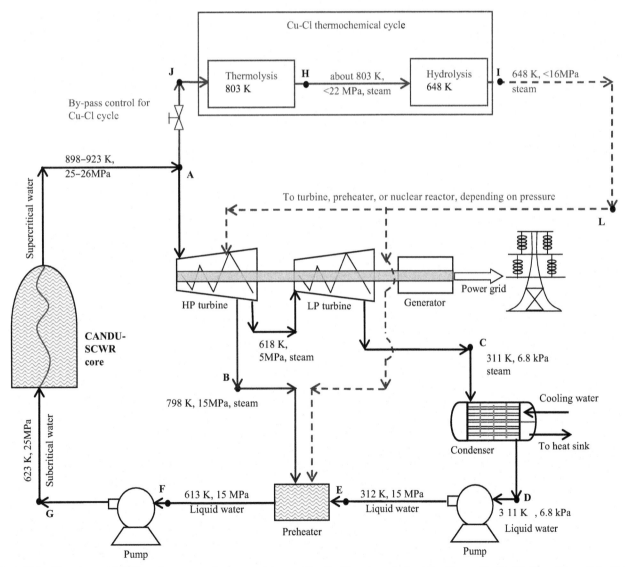

FIG. 4.110 Flowchart of SCWR coolant stream for heat exchange with a Cu-Cl cycle and reheat Rankine cycle. *Modified from Wang, Z.L., Naterer, G.F., Gabriel, K.S., 2010. Thermal integration of SWCR nuclear and thermochemical hydrogen plants. Second Canada-China Joint Workshop on Supercritical Water Cooled Reactor, Toronto ON, April 25–28.*

preheater and two turbines. In the proposed system, a part of the supercritical water stream at the outlet of the SCWR (state A on the diagram) is diverted via a by-pass control valve toward the copper-chlorine thermochemical plant.

This stream acts as a heat transfer fluid for two reactors of the copper-chlorine cycle: thermolysis and hydrolysis. First, heat is delivered to the thermolysis reactor where the endothermic chemical process is maintained at ~800 K. Subsequently, the hydrolysis reactor is used for the chemical process at ~640 K. After delivering heat to the copper-chlorine cycle, supercritical water converts into steam (state J in the figure) with the following parameters at the return point: ~16 MPa and 648 K. Steam is further used in the Rankine cycle either by injecting it at the high-pressure turbine or at the preheater.

A second linkage scheme between SCWR, Cu-Cl and Rankine plants is illustrated in Fig. 4.111. In this case, a double reheat Rankine plant is considered for more efficient power generation. Supercritical water at 26 MPa from the reactor outlet is diverted partially to a Cu-Cl plant and partially to the high pressure turbine. The stream of supercritical water is mainly used to provide reaction heat to the thermolysis and hydrolysis reactors.

However, an additional stream is diverted from the reactor to the copper-chlorine plant in the case of the flowchart in Fig. 4.111. The superheated steam generated in a reheater is split into two parts: one goes to the Cu-Cl plant and the other to the low pressure turbine. Steam at the low pressure turbine is assumed at 5 MPa. In order to use effectively the enthalpy of the superheated intermediate pressure steam, which is diverted to the hydrolysis reactor, a system comprising heat exchangers and a work recovery turbine can be devised such that eventually process steam is provided at

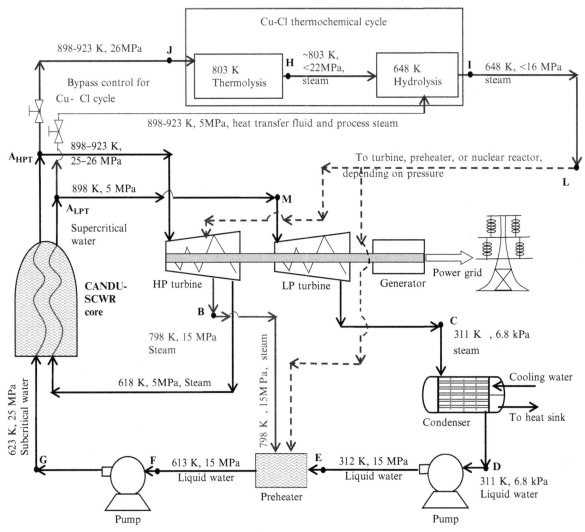

FIG. 4.111 Flowchart of SCWR coolant stream for heat exchange with the Cu-Cl cycle and a single reheat Rankine cycle. *Modified from Wang, Z.L., Naterer, G.F., Gabriel, K.S., 2010. Thermal integration of SWCR nuclear and thermochemical hydrogen plants. Second Canada-China Joint Workshop on Super-critical Water Cooled Reactor, Toronto ON, April 25–28.*

~650 K and ~0.1 MPa. In order to keep the water inventory in the Rankine and reactor circuit, the same amount as the consumed process steam is supplied as fresh water in the preheater vessel (state E in the figure).

A scheme similar to Fig. 4.111 was analyzed by Naidin et al. (2009) concluding that the single reheat scheme is advantageous with respect to no-reheat because it increases the power generation efficiency to >49%. An alternative way to integrate SWCR with a Rankine cycle and a thermochemical cycle is through chemical heat pumps. Chemical heat pumps perform cyclical chemical processes with a net effect of upgrading the temperature level from a heat source. In the context of nuclear driven thermochemical cycles, it is often required to match the temperature of the reactor coolant to that required by endothermic reactions of the thermochemical water splitting cycle.

Ozbilen (2013) developed a detailed flowsheet of a nuclear-based copper-chlorine plant for hydrogen and additional products generation. The assumed nuclear reactor is CANDU-SCWR which provides supercritical water at 25 MPa and 923 K. Water return is at 648 K. A part of the primary thermal energy is used to generate cooling and heating as by-products.

The concept of Ozbilen (2013) assumes that part of the generated hydrogen is directed to a fuel cell to produce power as a by-product. This system is shown in Fig. 4.112. The power, heating and cooling by-products may be utilized, for example, for local needs at the nuclear facility and to drive auxiliary power-consuming devices within the plant.

The plant (including detailed flowsheet of the copper-chlorine cycle) has been simulated with ASPEN Plus software. The oxychlorination reactor of the copper-chlorine cycle was simulated using the RGIBBS reactor model in ASPEN Plus based on 1 atm and 803 K and complete decomposition of copper oxychloride assumed reaction conditions. RGibbs is the only Aspen Plus block that deals with solid-liquid-gas phase equilibrium. A Gibbs free energy minimization was done to determine the product composition at which the Gibbs free energy of the products is at a minimum.

At more than 823 K, the kinetics of the oxidation reaction of cuprous chloride becomes favorable, which consumes oxygen and generates chlorine which is an undesired side product. Another possible and unwanted reaction generates cupric oxide and cupric chloride at temperatures above 673 K. The generated cupric chloride decomposes thermally

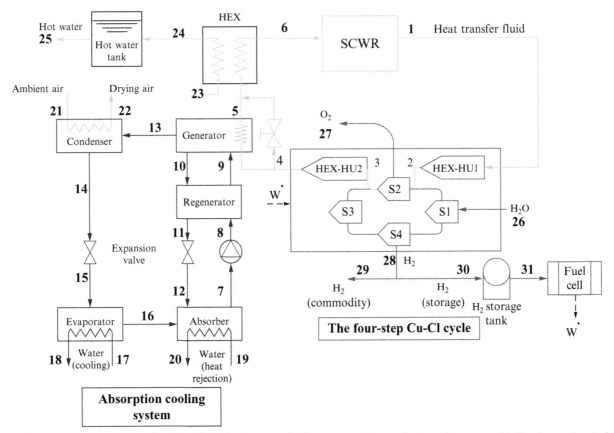

FIG. 4.112 Integrated system for CANDU-SCWR hydrogen and other commodities production with copper-chlorine thermochemical cycle. *Modified from Ozbilen, A.Z., 2013. Development, Analysis and Life Cycle Assessment of Integrated Systems for Hydrogen Production Based on the Copper-Chlorine Cycle. PhD Thesis, University of Ontario Institute of Technology.*

and produces chlorine, which is in stoichiometric amount with cupric oxides consumed in the chlorination process. If the operating pressure is 1 bar, then the release of oxygen at equilibrium becomes significant only after 700 K.

The simulation assumptions considered the reaction equilibrium and kinetics results obtained at UOIT in a scaled-up reactor setup, oxychloride at a scale 1000 times larger than prior proof-of-principle experiments. The schematic of the test reactor loop is shown in Fig. 4.113. The reactor consists of a vertical cylindrical vessel made of stainless steel 316 and hastelloy C. It contains a crucible fabricated in ceramics where the sample of copper oxychloride is placed. There is also a secondary containment to the sample that allows for channeling of the gases outside via a liquid seal that maintains a back pressure within the chamber slightly above atmospheric pressure. This method is used to prevent outside gases from entering the reactor during tests.

At equilibrium, the predictions show the conversion of the mixture of $CuO \cdot CuCl_2$ is between 50% and 80% at 800 K. The experimental decomposition of copper oxychloride resulted in conversions between 60% and 80% when operated at atmospheric pressure. Two reaction schemes are considered in the models developed, as reviewed in Naterer et al. (2013). These schemes are denoted as A and B. Scheme A, defined by the reaction sequence as follows:

$$CuO \cdot CuCl_2(s) \rightarrow CuO(s) + CuCl(s, l) + 0.5Cl_2(g) \tag{4.203a}$$

$$CuO(s) + 0.5Cl_2(g) \rightarrow CuCl(l) + 0.5O_2(g) \tag{4.203b}$$

Scheme B, defined by the reaction sequence:

$$CuO \cdot CuCl_2(s) \rightarrow CuO(s) + CuCl_2(s) \tag{4.203a}$$

$$CuCl_2(s) \rightarrow CuCl(l) + 0.5Cl_2(g) \tag{4.203b}$$

$$CuO(s) + 0.5Cl_2(g) \rightarrow CuCl(l) + 0.5O_2(g) \tag{4.203c}$$

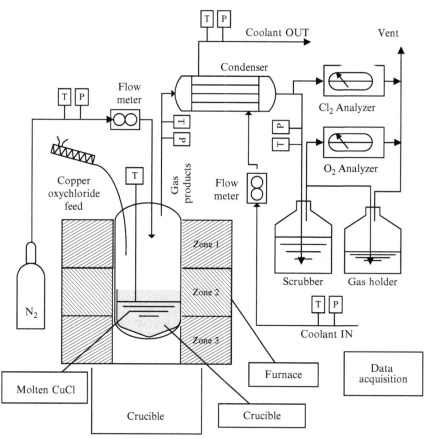

FIG. 4.113 Schematic of the scaled-up copper oxychloride reactor and its test loop at UOIT that validate the ASPEN process simulations. *Modified from Naterer, G.F., Dincer, I., Zamfirescu, C., 2013. Hydrogen Production from Nuclear Energy. Springer, New York.*

The decomposition rate falls close to the predictions of the reaction scheme A and B. In the analysis, the partial pressures of chlorine in the reactor rise up to four orders of magnitude larger than the partial pressure of CuCl vapor at temperatures between 723 and 833 K.

The experiments confirmed that once the temperature of 773 K is reached, the sample decomposes quickly with oxygen released at the peak rate of 0.006 mol/min. A total of 0.107 moles of oxygen were released with a yield of 89%. The yield was calculated by comparing the actual oxygen generated with the amount of oxygen that could have been generated if the reaction was completed to 100%. The entire decomposition of 0.236 mol of $CuO + CuCl_2$ occurred within 35 min of the initial 10 min, showing a marked oxygen evolution, while the balance of time showed the reaction rate slowly decaying to negligible oxygen release.

The four-step Cu-Cl plant flowsheet developed by Ozbilen (2013), specifically for integration into his concept of multi by-product nuclear-based hydrogen production plant is shown in Fig. 4.114. Four main reactors of the four-step Cu-Cl cycle are integrated using Aspen Plus. Heat exchangers, pumps, separators and mixers are also used wherever necessary. A novel heat exchanger network has been developed to recover heat internally and implement heat regeneration, using in this respect APEN Plus Energy Analyzer code.

A hierarchy block is also used for CuCl cooling, since molten CuCl undergoes two transformations: liquid-solid and solid-solid. Molten CuCl remains in a liquid phase down to 696 K, and it then changes phase to a solid β phase. With further cooling down of CuCl, another transformation at 685 K happens, and changes the solid β phase to another solid form referred to as (c solid). In this c-phase, CuCl has a different microstructure and different thermophysical

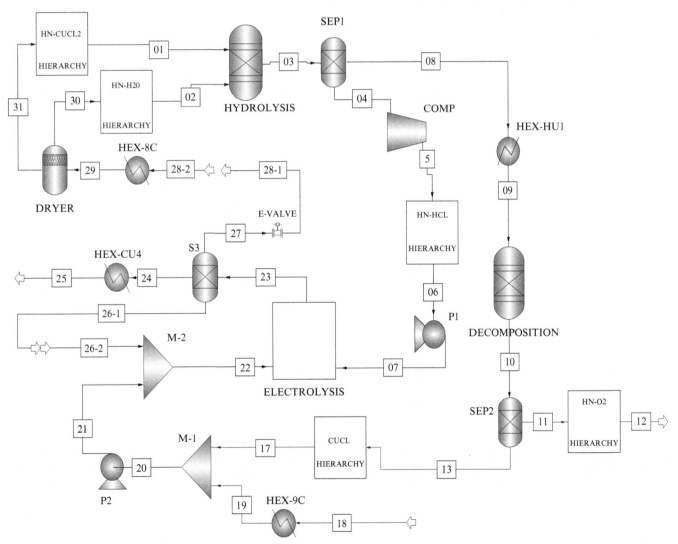

FIG. 4.114 ASPEN Plus simulation flowsheet of copper-chlorine cycle integrated within Ozbilen (2013) nuclear-based hydrogen production plant.

properties than in β phase. Figure 4.115 shows the CuCl phase transformation in a wider-range temperature vs heat transfer rate diagram transformations developed for the cycle flowsheet under the assumption of using stoichiometric reactors (Fig. 4.115).

A pinch analysis is developed for an integrated heat exchanger network to enable an effective heat recovery within the Cu-Cl cycle. The first step is to determine the optimum ΔTmin, minimum temperature difference between hot and cold streams. Fig. 4.116 shows the developed heat exchanger network. Abbreviation CU is used for the cold utility, and HU is used for the hot utility. The heat recovered by the hot streams is not only transferred to the cold streams but also to the dryer and the hydrolysis reactors.

According to the detailed simulations, the overall exergy efficiency of 35%, under the assumption of an effectiveness of 80–90% of all heat exchangers involved in the process. Because the system generates heating and cooling in addition to hydrogen, its energy efficiency is as high as 79%. In terms of exergy outputs, the system generates 90% hydrogen, 6% process cooling, 3% hot water, and 1% dried air.

4.7.2 Solar-Thermal-Based Systems

High temperature heat can be obtained from concentrated solar radiation. Therefore, thermochemical methods may be applied to generate hydrogen using the concentrated solar energy input. Several technologies are available to generate high temperature with concentrated solar radiation, including the central receiver plants with heliostat field, through-type concentrated solar systems and parabolic dish concentrators.

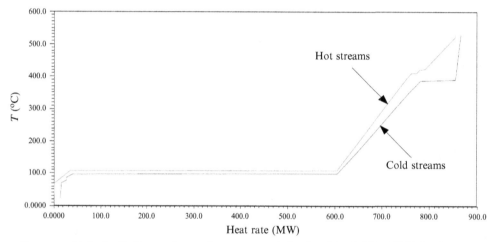

FIG. 4.115 Composite curves obtained with the new heat exchanger network developed by Ozbilen (2013) for the copper-chlorine cycle.

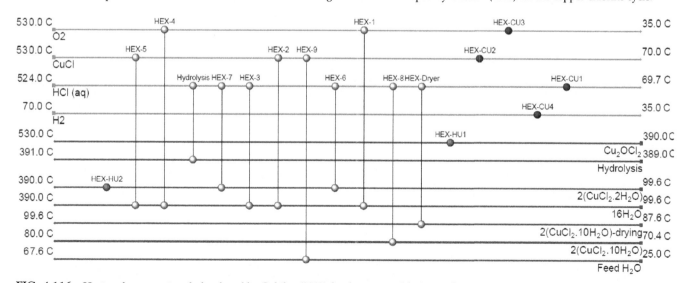

FIG. 4.116 Heat exchanger network developed by Ozbilen (2013) for the copper-chlorine cycle.

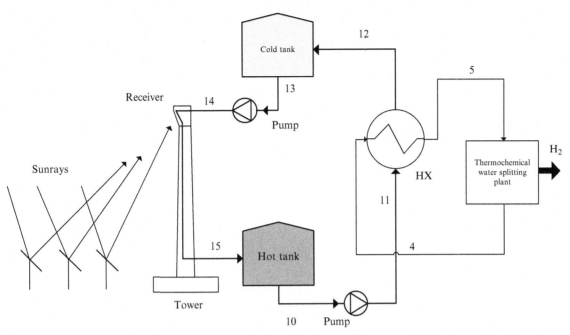

FIG. 4.117 Conceptual layout of a concentrated solar-thermal-hydrogen production plant integrated with a thermochemical water splitting cycle.

A layout of a hydrogen production plant that uses concentrated solar thermal radiation as energy input and a thermochemical water splitting cycle for hydrogen generation is illustrated in Fig. 4.117. A hot and a cold thermal storage tank will operate with a molten salt that maintains an average hot temperature during a 24 h cycle. The heat carried by the molten salt is transferred to a secondary circuit and carried to the thermochemical waters splitting cycle as shown in the figure.

4.7.2.1 Solar-Thermal-Based Magnesium-Chlorine Plant

Ozcan (2015) developed a solar-based hydrogen production plant that integrates the magnesium-chlorine cycle with a central receiver solar plant and a hydrogen compression station and a supercritical carbon dioxide cycle for power generation.

The solar data for the solar plant is taken for the Greater Toronto Area (GTA) by considering daily and annual data. Molten salt storage is considered for the system in order to run the system without source feed interruptions when the sun is out. The solar heat from the receiver is transferred to two consecutive heat exchangers to provide enough heat for the Mg-Cl and the supercritical CO_2(sCO_2) Brayton cycles. The integrated system is shown in Fig. 4.118.

According to the process simulation by Ozcan (2015), the total heat requirement for the Mg-Cl cycle is found to be 248.58 MW and total heat input for the solar gas turbine (GT) cycle is 698.9 MW. The solar intensity considered for the design point corresponds to the climacteric data for the month of July in Greater Toronto Area.

Energy load from the receiver is transferred to the Mg-Cl and the sCO_2-GT cycles' heat exchangers by splitting the energy stream. The total mass flow rate of the system is found to be 2644.2 kg/s, where 29.7% of this mass is transferred to supply heat for the Mg-Cl heat exchanger. High and low temperature energy storage range for the molten salt (60% $NANO_3$, 40% KNO_3) is 565 and 290°C, and this range is in perfect agreement with both GT and Mg-Cl cycles. The heliostat field layout was determined by Ozcan (2015) using NREL open software "System Advisory Model" (SAM), as shown in Fig. 4.119.

The supercritical CO_2 gas turbine cycle is a promising alternative to conventional gas turbine cycles with higher performance characteristics at lower maximum temperatures. A recuperative cycle is considered for internal heat recovery within the system. The T-s diagram with indicated state points is shown in Fig. 4.120. Effects of the pressure ratio, the approach temperature, and the turbine inlet temperature are investigated, where these parameters have slight to strong influence on the cycle performance. The low pressure side of the cycle is set to 74 bar, which is slightly above the critical pressure of CO_2. The high pressure side is not set as constant, but is instead decided by the pressure ratio.

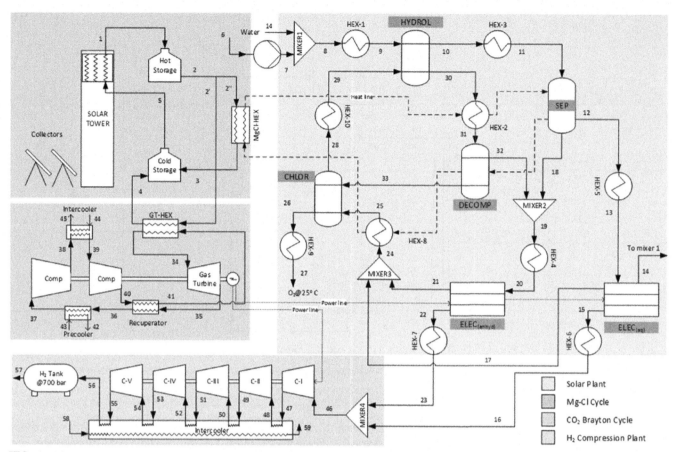

FIG. 4.118 Solar thermal-based magnesium-chlorine plant for hydrogen production. *Modified from Ozcan, H., 2015. Experimental and Theoretical Investigations of Magnesium-Chlorine Cycle and Its Integrated Systems. PhD Thesis, University of Ontario Institute of Technology.*

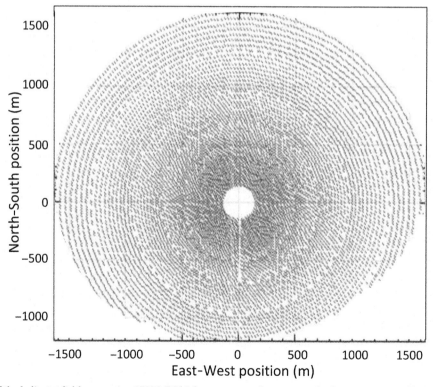

FIG. 4.119 Layout of the heliostat field generating 1500 MW high temperature heat to supply the magnesium-chlorine plant for hydrogen production. *Data from Ozcan, H., 2015. Experimental and Theoretical Investigations of Magnesium-Chlorine Cycle and Its Integrated Systems. PhD Thesis, University of Ontario Institute of Technology.*

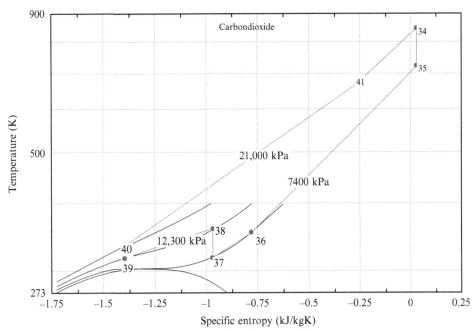

FIG. 4.120 Thermodynamic power cycle with supercritical CO$_2$ integrated within the magnesium-chlorine plant. *Data from Ozcan, H., 2015. Experimental and Theoretical Investigations of Magnesium-Chlorine Cycle and Its Integrated Systems. PhD Thesis, University of Ontario Institute of Technology.*

A slight decrease at both power consuming and producing devices are observed until a pressure ratio of 2, where a linear increase is then observed above 2. The mass flow rate of the cycle shows a significant decrease at higher pressure ratios, resulting in a more compact system. Even if a higher pressure ratio seems to be favorable for the system size, the system performance should also be analyzed at this range.

The results show that a higher pressure ratio enhances the system performances until 3.05, and a slight decrease is observed above this value. It should also be noted that the work consumption of compressors significantly increases at higher pressure ratios which would contribute to lower system performances. An optimum pressure ratio for both energy and exergy efficiencies is found to be 3.02. The internal heat recovery is one of the most crucial mechanisms within the GT cycle, where a reasonable assumption is made for the approach temperature.

Since the maximum temperature from the solar subsystem can be up to 565°C, a higher limit for maximum temperature is set as a constraint for the system. This maximum temperature can be higher than the mentioned value; however, it should be noted that molten salt is considered for the solar subsystem in order to store energy for night time use.

Although it is not as effective as pressure ratio and approach temperature, higher turbine inlet temperature slightly increases both energy and exergy efficiencies within the system. The base case model assumes maximum temperature at the heat exchanger exit. The modeled sCO$_2$-GT cycle consumes ~699 MW energy to produce the required power for the electrolysis of HCl in the Mg-Cl cycle and the compression of hydrogen by five-stage compressors.

The system energy and exergy efficiencies are found to be 46.5% and 60.9%, respectively. Use of a supercritical CO$_2$-GT system is superior to using conventional air gas turbine systems and shows the potential to perform at high performances at lower maximum temperature ranges. Even if this cycle shows relatively higher performance results, almost 208 MW energy is calculated as a total irreversibility within the cycle. A higher performing cycle with lower irreversibilities can be obtained by higher performing system components and proper selection and optimization of system parameters.

Energy and exergy efficiency comparisons of the system are illustrated in the bar chart diagram shown in Fig. 4.121. The unique product of the integrated system is hydrogen at 700 bar, where its energy and exergy contents are calculated as 250.28 and 255.82 MW, respectively. The total energy and exergy content of the energy input to the integrated system are calculated as 1535 and 1454 MW, respectively.

Energy and exergy efficiencies of the overall system are calculated as 16.31% and 17.6%, respectively. When the energy and exergy loads of the receiver are taken into account as the main inputs, energy and exergy efficiencies

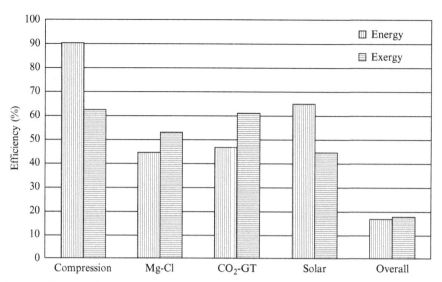

FIG. 4.121   Energy and exergy efficiency of the solar thermal magnesium-chlorine plant with hydrogen compression. *Data from Ozcan, H., 2015. Experimental and Theoretical Investigations of Magnesium-Chlorine Cycle and Its Integrated Systems. PhD Thesis, University of Ontario Institute of Technology.*

become 25.1% and 39.8%, respectively. Total exergy destruction within the system is found to be 1265 MW where the solar field contains almost 64% of the total irreversibility with a value of ~811 MW.

4.7.2.2 Solar Thermal-Based Copper-Chlorine Plant

Ozbilen (2013) developed an integrated solar-thermal-based copper-chlorine plant as shown in Fig. 4.122. Solar thermal energy, concentrated using heliostat solar tower, is the energy source. Molten salt is considered as a heat transfer fluid (HTF) to supply heat to the Cu-Cl cycle. Initially, heat is supplied to the copper oxychloride (Cu_2OCl_2) to increase the temperature from 390 to 530°C. Then the heat is transferred to the decomposition step (step 2) that has the highest temperature heat requirement (530°C) in the cycle. Finally, heat is supplied to the cupric chloride ($CuCl_2$). The heat requirement of hydrolysis and drying step is managed via thermal management within the cycle

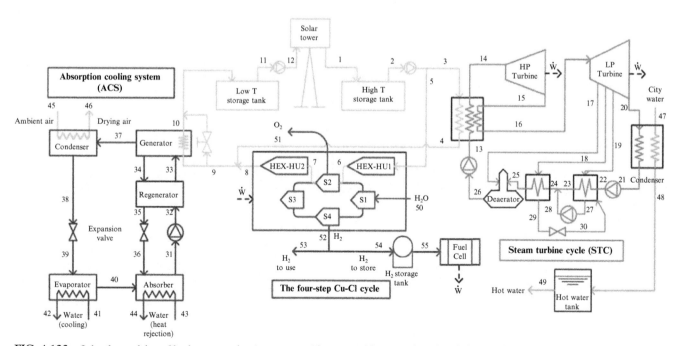

FIG. 4.122   Solar thermal–based hydrogen production system with copper-chlorine cycle and multi by-product generation. *Modified from Ozbilen, A.Z., 2013. Development, Analysis and Life Cycle Assessment of Integrated Systems for Hydrogen Production Based on the Copper-Chlorine Cycle. PhD Thesis, University of Ontario Institute of Technology.*

The temperature of the molten salt in a low temperature storage tank is higher than 250°C, which is about 30°C higher than the melting point of the molten salt. A hydrogen storage tank and fuel cell unit are also integrated into the Cu-Cl cycle for energy management. Energy management with a hydrogen storage option is promising, because hydrogen can be converted into electricity efficiently via fuel cells during peak hours. The system also includes a steam turbine cycle, which has a low pressure and a high pressure steam turbine, and a LiBr-H2O cooling cycle.

The energy efficiency of the overall system is 71.0%, whereas the exergy efficiency is 38.0%. However, if the heat released by the condenser of the steam turbine cycle is not utilized as hot water, the energy efficiency value would become 45.0% and the corresponding exergy efficiency 36.1%. The COP of the absorption cooling system is 77.0%, and the exergetic COP is 30.0%. The exergy efficiency of the Rankine cycle (STC) is greater than its energy efficiency, since exergy of output power is the power itself, even though exergy of input heat is lower than energy of the heat input.

Assuming that heat exchanger effectiveness within the system is superior to 80%, then the total exergy content in the outputs consists of 84% because of the generated hydrogen, 7% is electrical power generation, 4% is air drying service, 3% is hot water production and 2% is cooling generation.

Ozbilen (2013) developed an alternative to the solar-based copper-chlorine cycle system in which the solar energy resource is supplemented with a natural gas fuel combustion process. This system generates other by-products beside hydrogen, namely oxygen, hot water, air drying and cooling. The system flowchart is shown in Fig. 4.123. The main subunits of the system are the Cu-Cl cycle, a heliostat-based solar thermal tower with molten salt system for heat storage, a steam turbine cycle, a gas turbine cycle and a LiBr-H2O absorption cooling system. The gas turbine cycle comprises two compression stages with intercooler and a preheating air unit placed before the combustion chamber.

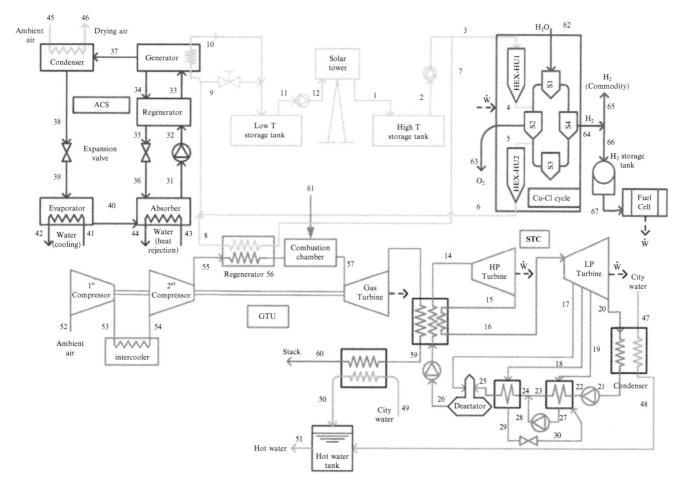

FIG. 4.123 Integrated hydrogen production system for thermochemical hydrogen production with copper-chlorine cycle using concentrated solar radiation and natural gas as energy supply. *Modified from Ozbilen, A.Z., 2013. Development, Analysis and Life Cycle Assessment of Integrated Systems for Hydrogen Production Based on the Copper-Chlorine Cycle. PhD Thesis, University of Ontario Institute of Technology.*

The overall energy efficiency of System III is calculated to be 75.9%, and the exergy efficiency is 34.1%. The exergy output carried by oxygen product is negligible, whereas the exergy carried by the generated hydrogen stream is 94% of all outputs. Also, hot water exergy is 2%, cooling service exergy is 2% and the exergy corresponding to the drying air service is 2%.

4.7.3 Biomass-Based Systems

A biomass gasification plant will need a certain amount of electric power to drive the auxiliary subsystems. One of the options is to integrate a solid oxide fuel cell for the auxiliary power generation. Abuadala (2010) proposed such a system, which is presented here.

The system aims to utilize the derived biomass steam gasification hydrogen (primary hydrogen) in producing power and to increase hydrogen yield by further processing of the other gasification by-products in steam reforming and water-gas-shift reactors. The main components of the system are as follows: gasifier, solid oxide fuel cell, compressors, turbine and heat exchangers. The system layout is shown in Fig. 4.124.

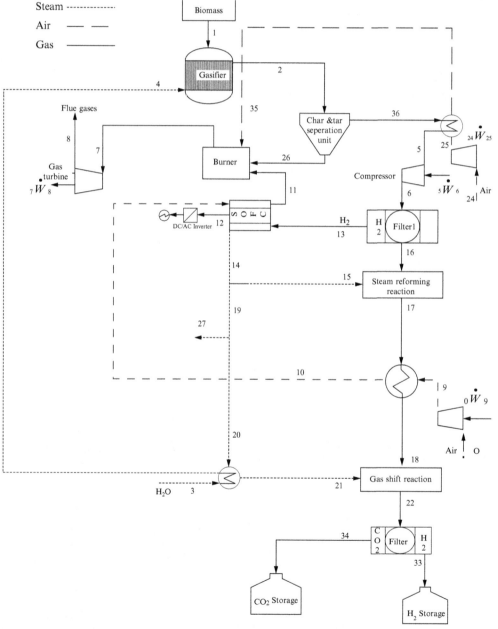

FIG. 4.124 Biomass-based hydrogen production system integrated with SOFC for auxiliary power generation. *Modified from Abuadala, A.G., 2010. Investigation of Sustainable Hydrogen Production From Steam Biomass Gasification. MSc Thesis, University of Ontario Institute of Technology.*

The produced gas is separated from the tar and char in the separation unit. The tar and char are sent to the burner to burn, where more energy is extracted. The gas is cooled to approximately 498 K. The cooling process is modeled by heat exchanger 36-5-25-35. The relative cool gas is compressed in the compressor 5-6. The gas is filtered to have pure hydrogen and the rest of the product gas. The pure hydrogen is known as primary hydrogen and is fed to the SOFC; the remaining product gas is further processed in gasifier bottoming reactors.

The SOFC is an external reforming SOFC. It operates at 1000 K and a pressure of 1.2 bar. The hydrogen from the filter enters the anode side of the SOFC through state point 13. Most of the primary hydrogen is oxidized to water. In the fuel cell, the hydrogen is converted into electricity and steam.

The unused hydrogen that leaves the anode and the cathode of gas are sent to the burner. In the burner, the unused hydrogen, char and tar are burned. The excess air required in the burning process is compressed in compressor 24-25 and is preheated by passing through the heat exchanger 36-5-24-25. The flue gas results from the burning are expanded in the turbine. In this system, the residual heat from the flue gas is assumed to be further unutilized.

The major part of the exergy destruction occurs in the SOFC stack followed by the turbine and the burner. Also, it is found that the total exergy destruction in the system components is at minimum when the gasification temperature is 1175 K, as shown in Fig. 4.125.

In the gasification temperature range, and for a given utilization factor and steam-biomass ratio, the overall exergy efficiency for electrical production from the system was based on exergy of biomass throughput versus gasification temperature, as shown in Fig. 4.126. The efficiency decreases from 56% to 49.4% in the studied gasification temperature range because of the decrease in the exergy efficiency of the turbine. From the exergy loss results, it was found that a major part of exergy destruction occurred in the SOFC. Also, its exergy destruction increased with the gasification temperature. Secondary hydrogen yield increased and, accordingly, its exergy increased, and thus its exergy efficiency increased from 22% to 32%.

Cohce (2010) developed and analyzed with ASPEN Plus software three biomass gasification flowsheets for hydrogen production. One of these systems is reviewed here. The system is designed for a biomass consisting of oil pal shells having 50% moisture content. The flowsheet diagram of the system is shown in Fig. 4.127.

In this system, the biomass feedstock is supplied at the dryer (DRY-REACT). The drying process of biomass is conducted in the heat exchanger (HX0) and the condenser (DRY-FLSH). The drier process reduces the amount of water in the biomass from 50% to 5.7%. It should be highlighted that this system's drier section is different than the other two system because, in this system, biomass and drier gas do not mix in the HE0, thus providing us with the means to use evaporated water from the biomass after the condensation process. The flow rate becomes 2000 tonne/day (dry basis) after the drier process and then goes into the decomposer (DECOMP).

The next step is implemented using decomposed dry biomass that splits into two parts: one part moves to the combustion (COMB1), where some amount of biomass (47%) and the leftover methane (CH4) and carbon monoxide (CO) from the PSA combust; and the second part (53%) moves into the splitter (SP2), where some amount of the biomass is split into two parts—one (25%) to the combustor (COMB2) and one to the gasifier (28%). In addition, in this simulation, the COMB1 and COMB2 provide heat for the gasifier.

Produced syngas passes through the cyclone (cyc1) to eliminate ash and other impurities, such as carbon (solid) and NOx. These impurities enter the combustion. The ash-free produced syngas goes through the HE2 and the HE3 to reach the reformer where the steam reforming takes place. After the reformer treatment, the syngas passes through the HE4 to enter the compressor (COMP). This process increases the pressure and temperature. The high pressure and high temperature syngas enter into the HE5 and HE6 to connect with the high temperature shift reactor (HTS) where shift reaction occurs.

The syngas passes the HX7 to reach to the pressure swing absorption system to purge the syngas from other unwanted gases in order to produce 99.9% purity hydrogen. Also, this current design includes a water tank mixer (MIXER1) to collect all the water and distribute it to the necessary areas to save energy and exergy, since adding a little water to the system contributes positively to the system's efficiency. In the same way this system has the turbine (TURB) to assess the extra steam to produce electricity in the system, it also increases the system's energy and exergy efficiency due to the heat recovery phenomena. The results will be discussed at the end.

The system assumptions for the plant were that the rate of dried biomass supplied at COMB1 is 41.5 t/h, the dried biomass at COMB2 is 19.5 t/h, the dry rate of biomass supplied to gasifier is 24.8 t/h, the steam input to the gasifier is 14.9 t/h, the derived methane within the biomass gasification and processing units is supplied internally to COMB1 at 8.96 t/h and the steam-to-biomass ratio is 0.6. Based on the simulations, the gasifier outlet temperature was 870°C, and temperature for combustor 1 and 2 outlet temperature was 1000°C.

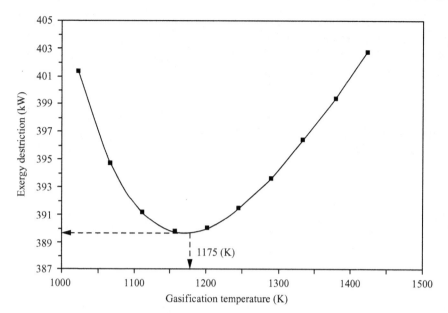

FIG. 4.125 Exergy destruction for a case study with the integrated biomass gasifier–SOFC system. *Data from Abuadala, A.G., 2010. Investigation of Sustainable Hydrogen Production From Steam Biomass Gasification. MSc Thesis, University of Ontario Institute of Technology.*

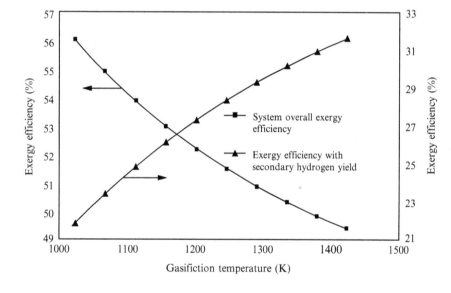

FIG. 4.126 Exergy efficiency of the system developed by Abuadala (2010) versus gasification temperature.

The main input and output streams of hydrogen production plant are given in Table 4.41. The gasifier outlet conditions are shown in Table 4.42. Hydrogen at outlet is of 99.9% purity and delivered at 120°C at 25 bar absolute pressure, with a flow rate of 3.7 t/h. The energy efficiency of the system results is 22%, whereas the exergy efficiency is 19%.

4.7.4 Clean Coal Process-Based Systems

Coal is the most abundant fossil fuel resource. The world energy supply depends on coal combustion, although the associated carbon dioxide emissions to the atmosphere by flue gas stacks contribute greatly to environmental pollution and global warming. Conventional combustion or coal gasification processes are not environmentally benign because of their large emissions of carbon dioxide. However, multiple technologies are known to make coal combustion or coal gasification cleaner. An extreme process is the total capture and sequestration of carbon dioxide. However, this process will not be economically competitive. Therefore, some alternatives are usually considered in which the process is at least cleaner than the conventional baseline.

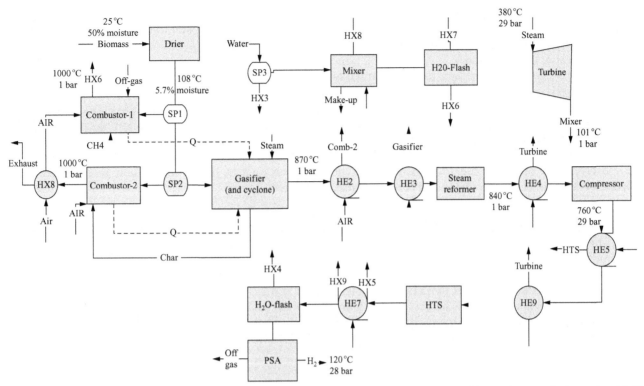

FIG. 4.127 Process flowsheet of a hydrogen production plant from palm oil shell gasification of 4000 t/day wet biomass feed capacity. *Modified from Cohce, M.K., 2010. Thermodynamic Performance Assessment of Three Biomass-Based Hydrogen Production Systems. MSc Thesis, University of Ontario Institute of Technology.*

TABLE 4.41 Main Input and Output Streams of Hydrogen Production Plant of Fig. 4.127

Input streams			Output streams		
Stream	$\dot{E}$(MJ/h)	$\dot{E}x$(MJ/h)	Stream	$\dot{E}$ (MJ/h)	$\dot{E}x$ (MJ/h)
Biomass (dry basis)	1957	2197	Hydrogen	450	439
Water	3.0	0.2	Exhaust gas	238	123

TABLE 4.42 Gasifier Outlet Conditions for Process Shown in Fig. 4.127

Component	H_2O	CO	CH_4	NH_3	H_2S	CO_2	C(s)	Ash	H_2
$\dot{m}$ (t/h)	4.6	25.5	0.03	0.08	0.13	5.97	0.001	1.14	2.98

Zamfirescu et al. (2012) proposed a new method to enhance the coal-fired boiler efficiency by system integration with a copper-chlorine cycle for water splitting and hydrogen generation. Instead of generating electricity only, in the new approach, the coal-fired power plant works synergistically with a copper-chlorine water splitting plant to ultimately produce electricity and hydrogen with reduced pollution.

A modified furnace concept with pulverized coal is introduced as a main component of an integrated system that couples a steam Rankine power plant with a copper-chlorine water splitting cycle to the same source of energy. A low pressure steam reheater is the first heat exchanger in the flue gas stream. The integrated system functions as follows:

- Heat of flue gases is transferred to the water preheater, boiler, superheater and reheaters of the steam Rankine plant; simultaneously, heat of the flue gases is transferred to a low pressure steam reheater that conveys thermal energy to the copper-chlorine plant. A new coal-fired furnace concept with a better temperature profile match between hot and cold streams is introduced.
- The copper-chlorine plant generates hydrogen and oxygen from water, and the hydrogen is stored in a compressed form, while oxygen is used to conduct an oxyfuel combustion process with pulverized coal. In order to fulfill the need of oxidant, additional oxygen is generated from the air using an air separation unit (ASU).
- The steam Rankine cycle generates power in which a part is supplied to the electrochemical reaction within the copper-chlorine cycle.
- Since oxyfuel combustion is applied, a very limited amount of oxygen is present in the stack gases, which is composed mainly of carbon dioxide, steam and oxygen. The stack gas is, therefore, partially recirculated and partially cooled to condense water and capture carbon dioxide.

Fig. 4.128 illustrates the coal-fired furnace concept that integrates the Rankine power plant with a Cu-Cl water splitting plant. Coal is pulverized at the lower part of the furnace where it devolatilizes and ignites in an oxidative atmosphere. The average temperature of gases at the lower part of the furnace (below point A in the figure) is typically approximately 650–700 K.

A calandria boiler system with pipes at the lower half of the furnace is installed. Heat transfer by radiation occurs between the hot gases and the pipes (with boiling water) placed at the channel periphery. In the first part of the combustion process, the volatiles are oxidized while the temperature of coal particles increases. Subsequently, carbon oxidizes, and the flue gas temperature increases further. This process occurs approximately between locations A and B in the figure. In point B, coal particles are completely consumed, and the flue gas temperature reaches its maximum value (~1200 K).

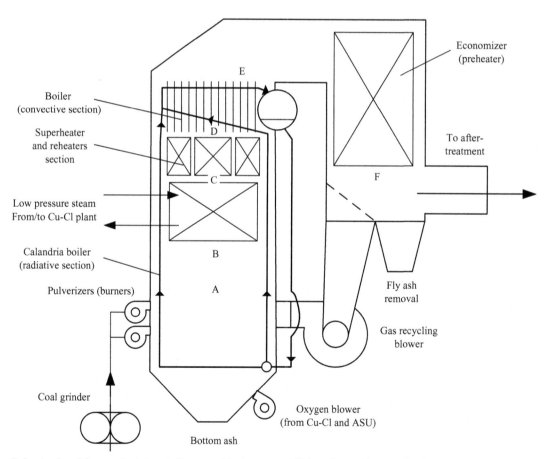

FIG. 4.128 Pulverized coal furnace for integrated copper-chlorine water splitting plant with steam Rankine power plant. *Modified from Zamfirescu, C., Dincer, I., Naterer, G.F., 2012. Comparing emissions from integrated Cu-Cl cycle with nuclear and coal power plants. Global Conference on Global Warming, Istanbul, July 8–12.*

Between states B and C (see the figure), hot flue gas exchanges heat by convection and radiation with low-pressure steam in a heat exchanger linked to the heat supply circuit for the copper-chlorine cycle. In this heat exchanger, superheated steam is reheated from 800 to 900 K. In the path of flue gases, two reheaters and the superheater of the steam Rankine power plant are installed accordingly. In point D, the flue gas temperature reaches about 950 K. Furthermore, heat is transferred to the last segment of the boiler (states D–E in the figure); note that state E at 820–850 K corresponds to the pinch point. It follows the economizer, after which the flue gas temperature decreases to about 550 K. A part of the combustion gases—comprising mainly carbon dioxide, steam and oxygen—is recirculated back to the combustion zone. The other part passes to the after-treatment section where gases are cleaned, particulate matter is extracted, further cooling water is condensed and carbon dioxide is captured.

Fig. 4.129 represents the integrated system of a coal-fired power plant with a copper-chlorine cycle for hydrogen production. The heat generated by the combustion process is transferred via multiple heat exchangers to the working fluid of the steam Rankine plant and to the copper-chlorine water splitting cycle. Radiative heat transfer exists at the bottom part of the furnace until the maximum flue gas temperature point, B, in Fig. 4.138 and Fig 4.130, between the hot gas and walls flanked with calandria pipes where forced convection boiling occurs.

This process is represented in the diagram from Fig. 4.129 by the "Calandria HX." Furthermore, hot flue gases in state B pass through the low pressure steam reheater and then are diverted to the superheating heat exchanger (SH) and the reheater heat exchangers (RH$_1$ and RH$_2$), connected in parallel at the hot stream side. Next the flue gas is directed to the boiler (D-E) and then the economizer (E-F). The lowest grade heat recovered at the economizer section (F-G) of the furnace is transferred to the dehydration process within the copper-chlorine cycle.

The steam generated in the low pressure superheater is transported to the copper-chlorine cycle for heating purposes; this is a secondary steam circuit. Steam for water splitting purposes is extracted from the low pressure turbine (LPT) of the power plant. In order to maintain the working fluid balance within the Rankine plant, fresh water is supplied to the direct contact heat exchanger (DHX) in the same amount as it is extracted as steam from the LPT. The component labels in Fig. 4.129 are as follows: COND—condenser, LPP—low pressure pump, LPHx—low pressure heat exchanger (preheater), DHX—direct contact heat exchanger (and deaerator), HPP—high pressure pump,

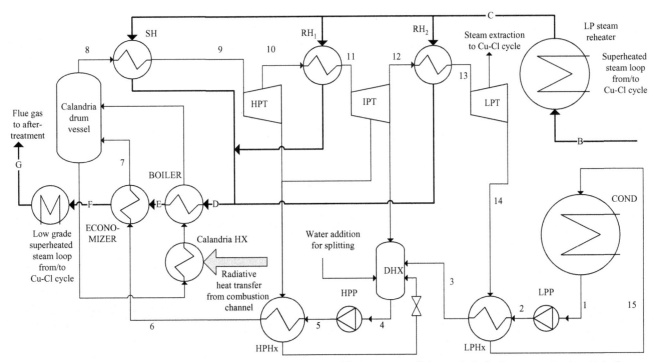

FIG. 4.129 Integrated system of coal-fired power plant with a copper-chlorine water splitting cycle for hydrogen production. Only the power plant and heat supply system to the Cu-Cl cycle are shown; bold lines represent a flue gas stream flow corresponding to Fig. 4.128. *Modified from Zamfirescu, C., Dincer, I., Naterer, G.F., 2012. Comparing emissions from integrated Cu-Cl cycle with nuclear and coal power plants. Global Conference on Global Warming, Istanbul, July 8–12.*

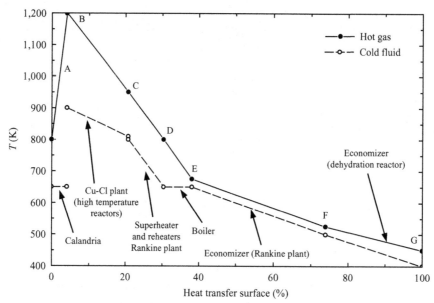

FIG. 4.130   Temperature versus heat transfer surface diagram for the modified coal-fired furnace. *Modified from Zamfirescu, C., Dincer, I., Naterer, G.F., 2012. Comparing emissions from integrated Cu-Cl cycle with nuclear and coal power plants. Global Conference on Global Warming, Istanbul, July 8–12.*

HPHx—high pressure heat exchanger (preheater), SH—superheater, RH—reheater, HPT—high pressure turbine, IPT—intermediate pressure turbine, LPT—low pressure turbine.

The temperature versus heat transfer surface diagram of the modified coal-fired furnace is determined as shown in Fig. 4.130. This diagram was obtained by determining heat transfer coefficients for each zone. Starting from the furnace bottom, the following heat transfer processes occur:

- Pulverized coal combustion and radiative heat transfer occur at the calandria tubes. As just mentioned, there is a first phase to this process of coal devolatilization (process before point A) followed by carbon combustion (A-B). The estimated heat transfer via thermal radiation is $200\,W/m^2K$, which corresponds as an order of magnitude to the general past studies in the literature.
- B-C—radiative and convective heat transfer to the low pressure steam, which transports heat to the high temperature reactors of the copper-chlorine plant.
- C-D—radiative and convective heat transfer to the reheaters and superheater of the steam Rankine plant.
- D-E—radiative and convective heat transfer to the boiler.
- E-F—convective heat transfer to the water preheater.
- F-G—convective heat transfer to low pressure steam to transport heat to the dehydration reactor of the copper-chlorine plant.

As indicated by the results shown in Fig. 4.131, the flow rate of coal is assumed to be 1 kmol/s; in the conditions considered here, it yields an exergy destruction in the furnace of 76 MW, and the furnace operates with 71% exergy efficiency. The exergy destruction in the Rankine plant is 37.6 MW, while its energy efficiency is 37%, and the exergy efficiency is 68%. The exergy efficiency of the copper-chlorine plant is estimated to be 79% provided that 50% of the heat is recovered inside the cycle and that the rejected heat has very low grade.

The last result is a comparative assessment of nuclear-based and coal-fired production of hydrogen and coal with an integrated system including the copper-chlorine water splitting cycle. The environmental impact of nuclear-based hydrogen production with a copper-chlorine water splitting cycle was studied in detail by Ozbilen et al. (2011). According to those past studies, it appears that the global warming potential of the system is on the order of $10^9\,kg\,CO_2$, for a plant with 125 t H_2 production per day. This estimate is based on a lifecycle assessment. The upper bound of carbon emissions of the system is slightly below 1 kg CO_2 per kg of H_2 produced.

This estimate considers the indirect emissions related to the nuclear power plant since no direct carbon dioxide pollution is produced at these facilities. In the plot of Fig. 4.132, comparisons are shown for the emissions of the reference coal gasification plant, the nuclear-based copper-chlorine integrated plant and the coal-fired integrated plant for hydrogen and power generation. Even without carbon capture, a coal-fired integrated plant is better from a pollution point of view than the coal gasification system. When carbon capture is applied, the coal-fired integrated system approaches (with respect to pollution mitigation) the nuclear-based system that is the most environmentally benign.

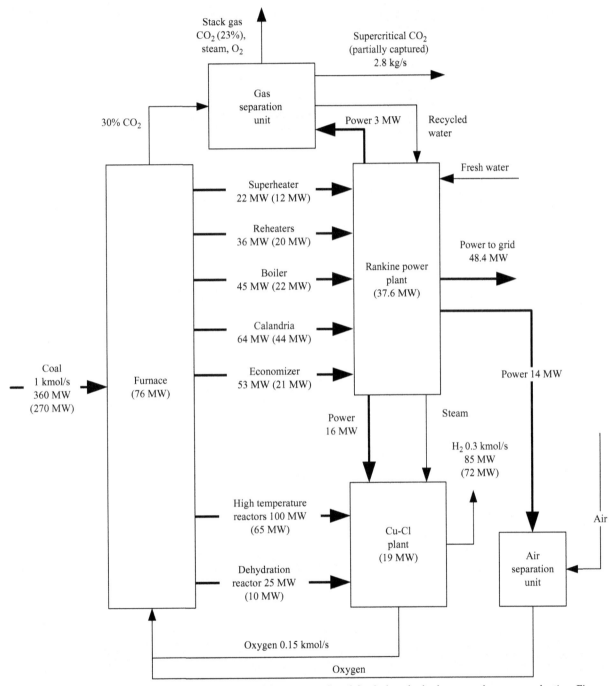

FIG. 4.131 Thermodynamic energy and exergy analysis of the integrated coal-fired plant for hydrogen and power production. Figures in the parentheses represent exergy. *Modified from Zamfirescu, C., Dincer, I., Naterer, G.F., 2012. Comparing emissions from integrated Cu-Cl cycle with nuclear and coal power plants. Global Conference on Global Warming, Istanbul, July 8–12.*

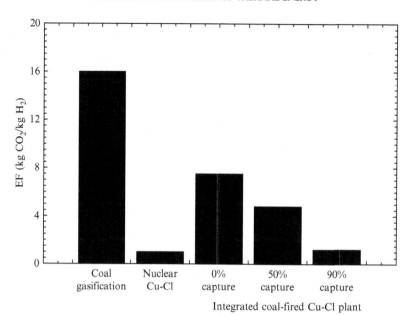

FIG. 4.132 Emissions comparison of two integrated hydrogen production systems based on copper-chlorine water splitting cycle (note: reference coal gasification system in the plot).

4.8 CONCLUDING REMARKS

In this chapter, the methods of hydrogen production by thermal energy are introduced and discussed. In the first section, the fundamentals of thermochemical hydrogen production are presented along with various important thermodynamic parameters and thermochemical reactions. Also, the main types of thermochemical processes for hydrogen production are introduced, including water thermolysis, thermochemical cycles and gasification and reforming of fuels. The chapter ends with a presentation of the some integrated systems to produce green hydrogen from thermal energy. These systems are based on nuclear energy, solar thermal energy, biomass energy and clean coal technology.

Nomenclature

a	chemical activity, $kmol/m^3$
A	area, m^2
$\mathcal{A}$	pre-exponential factor
c	molar concentration, $kmol/m^3$
D	diffusivity, m^2/s
$\dot{E}$	energy rate, kW
ex	specific chemical exergy, kJ/mol
$\dot{Ex}$	exergy rate, kW
f	activity coefficient
G	molar Gibbs free energy, kJ/kmol
H	molar enthalpy, kJ/kmol
HHV	higher heating value, kJ/mol
J	current density, A/m^2
k	reaction rate
$\mathcal{k}$	rate constant
K	equilibrium constant
l	length, m
$\dot{m}$	mass flow rate, kg/s
M	molar mass, kg/kmol
n	number of moles, kmol
$\dot{n}$	molar rate, kmol/s
P	pressure, kPa
$\dot{Q}$	heat rate, kW

r	mean pore radius, m
R	universal gas constant, kJ/kmolK
S	molar entropy, kJ/kmolK
t	time, s
T	temperature, K
V	volume, m^3
$\dot{V}$	volumetric flow rate, m^3/s
$\dot{W}$	power, W
x	length coordinate, m
y	molar fraction

Greek Letters

η	efficiency
μ	chemical potential, kJ/kmol
ψ	exergy efficiency
ξ	extend of reaction

Subscripts

0	reference state
a	activation
a	anode
abs	absorbed
act	activation
AE	alkaline electrolyzer
aux	auxiliary
b	backward
c	cathode
c	convection
cell	electrolysis cell
ci	cut in
co	cut out
conc	concentration
d	destroyed
d	diffusion
el	electrolysis
elch	electrochemical
eq	equilibrium
eq	equivalent
f	forward
f	fuel
g	gap
H	high temperature source
k	Knudsen diffusion
lim	limiting
loss	losses
m	advection (conduction)
oc	open circuit
oxi	oxidation
Ω	ohmic
P	products
pg	power generation
ph	photonic
pol	polarization
Q	reaction quotient
R	reactants
red	reduction
ref	reference
rev	reversible
RMC	root mean cube
sat	saturation
sc	short circuit
sep	separation

surr   surroundings
th   thermo-neutral
tot   total

Superscripts

0   standard state
ch   chemical
eff   effective

References

Abuadala, A.G., 2010. Investigation of Sustainable Hydrogen Production From Steam Biomass Gasification. MSc Thesis, University of Ontario Institute of Technology.

Abuadala, A., Dincer, I., Naterer, G.F., 2010. Exergy analysis of hydrogen production from biomass gasification. Int. J. Hydrog. Energy 35, 4981–4990.

Aghahosseini, S., 2013. System Integration and Optimization of Copper-Chlorine Thermochemical Cycle with Various Options for Hydrogen Production. PhD Thesis, University of Ontario Institute of Technology.

Balashov, V.N., Schatz, R.S., Chalkova, E., Akinfiev, N.N., Fedkin, M.V., Lvov, S.N., 2011. CuCl electrolysis for hydrogen production in Cu-Cl thermochemical cycle. J. Electrochem. Soc. 158, B266–B275.

Beghi, G.E., 1986. A decade of research on thermochemical hydrogen at the joint research centre, Ispra. Int. J. Hydrog. Energy 11, 761–771.

Bilgen, E., Bilgen, C., 1982. Solar hydrogen production using two-step thermochemical cycles. Int. J. Hydrog. Energy 7, 637–644.

Brown, L.C., Besenbruch, G.E., Schultz, K.R., Showalter, S.K., Marshall, A.C., Pickard, P.S., Funk, J.F., 2002. High Efficiency Generation of Hydrogen Fuels Using Thermochemical Cycles and Nuclear Power. General Atomics Report GA-A24326.

Bulan, A., Hansen, W., Gestermann, F., Grossholz, M., Pinter, H.D., 2006. U.S. Patent No. 7,128,824.

Carre, F., Yvon, P., Anzieu, P., Chauvin, N., Malo, J.-Y., 2010. Update of the French R&D strategy on gas-cooled reactors. Nucl. Eng. Des. 240, 2401–2408.

Carty, R.H., Mazumder, M.M., Schreiber, J.D., Pangborn, J.B., 1981. Thermochemical Production of Hydrogen. Final report 30517, vol. 1–4, Institute of Gas Technology, Chicago, IL.

Cohce, M.K., 2010. Thermodynamic Performance Assessment of Three Biomass-Based Hydrogen Production Systems. MSc Thesis, University of Ontario Institute of Technology.

Daggupati, V., Naterer, G.F., Gabriel, K., Gravelsins, R., Wang, Z., 2009. Equilibrium conversion in Cu-Cl cycle multiphase processes of hydrogen production. Thermochim. Acta 496, 117–123.

Dokyia, M., Kotera, Y., 1976. Hybrid cycle with electrolysis using Cu-Cl system. Int. J. Hydrog. Energy 1, 117–121.

Dokyia, M., Fukuda, K., Kameyama, T., Kotera, Y., Asakura, S., 1977. The study of thermochemical hydrogen preparation. (II) Electrochemical hybrid cycle using sulphur-iodine system. Denki Kagaku (Electrochemistry, Jpn) 45, 139–143.

Dokyia, M., Kameyama, T., Fukuda, K., 1979. Thermochemical hydrogen preparation – part V. A feasibility study of the sulphur iodine cycle. Int. J. Hydrog. Energy 4, 267–277.

Eames, D.J., Newman, J., 1995. Electrochemical conversion of anhydrous HCl to Cl_2 using a solid-polymer-electrolyte electrolysis cell. J. Electrochem. Soc. 142, 3619–3625.

Elder, R., Allen, R., 2009. Nuclear hydrogen production: Coupling a very high/high temperature reactor to a hydrogen production plant. Prog. Nucl. Energy 51 (2), 500–525.

Eom, H.C., Park, H., Yoon, H.S., 2010. Preparation of anhydrous magnesium chloride from ammonium magnesium chloride hexahydrate. Adv. Powder Technol. 21 (2), 125–130.

Ewan, B.C.R., Allen, R.W.K., 2005. Assessing the efficiency limits for hydrogen production by thermochemical cycles. In: AIChE Annual Meeting, Cincinnati October 30–November 4, Paper 210c.

Ferrandon, M.S., Lewis, M.A., Tatterson, D.F., Nankanic, R.V., Kumarc, M., Wedgewood, L.E., Nitsche, L.C., 2008. The hybrid CueCl thermochemical cycle. I. Conceptual process design and H_2A cost analysis. II. Limiting the formation of CuCl during hydrolysis. In: NHA Annual Hydrogen Conference, Sacramento Convention Center, CA; March 30-April 3.

Fatsikostas, A.N., Kondarides, D.I., Verykios, X.E., 2002. Production of hydrogen for fuel cells by reformation of biomass-derived ethanol. Catalysis Today 75, 145–155.

Fryda, L., Panopoulos, K.D., Karl, J., Kakaras, E., 2008. Exergetic analysis of solid oxide fuel cell and biomass gasification integration with heat pipes. Energy 33, 292–299.

Funk, J.E., 2011. Thermochemical hydrogen production: past and present. Int. J. Hydrog. Energy 26, 158–190.

Funk, J.E., Reinstorm, R.M., 1964. Final report energy depot electrolysis systems study. TID 20441 (EDR 3714), Vol. 2, Suppl. A.

Funk, J.E., Reinstorm, R.M., 1966. Energy requirement in the production of hydrogen from water. Ind. Eng. Chem. Process. Des. Dev. 5, 336–342.

Gorensek, M.G., Summers, W.A., 2009. Hybrid sulfur flowsheets using PEM electrolysis and a bayonet decomposition reactor. Int. J. Hydrog. Energy 34, 4097–4114.

Grimes, P.G., 1966. Energy depot fuel production and utilization. SAE Trans. 74, 65001.

Hesson, R.N., 1979. Kinetics of the Chlorination of Magnesium Oxide. Report no. IS-T-823. Ames Laboratory, IA, USA.

Huskinson, B., Rugolo, J., Mondal, S.K., Aziz, M.J., 2012. A high power density, high efficiency hydrogen–chlorine regenerative fuel cell with a low precious metal content catalyst. Energy Environ. Sci. 5, 8690–8698.

Ino, S., Ochiai, Y., 1961. Production of anhydrous $MgCl_2$ by chlorination. Gov. Chem. Ind. Res. Institute 56, 81–89.

Kasahara, S., Kubo, S., Hino, R., Onuki, K., Nomura, M., Nakao, S., 2007. Flow-sheet study of the thermochemical water splitting iodine-sulfur process for effective hydrogen production. Int. J. Hydrog. Energy 32, 489–496.

Kelley, K.K., 1945. Energy Requirements and Equilibria in the Dehydration, Hydrolysis, and Decomposition of Magnesium Chloride. Vol. 676. US Government Printing Office, Washington, DC.

Ketter, A., Stolberg, L., Li, H., Shkarupin, A., Suppiah, S., 2015. Electrolysis Cell With Multiple Membranes for CuCl/HCl Electrolysis in Hydrogen Production. US Patent 2015/0047988A1.

Klein, S.A., 2015. Engineering Equation Solver. F-Chart Software, Madison.

Kromer, M., Roth, K., Takata, R., Chin, P., 2011. Support for cost analyses on solar-driven high temperature thermochemical water-splitting cycles. Report DE-DT0000951, US Department of Energy.

Lamy, M., Ng, K.W., Harris, R., 2004. Mechanism of chlorination of MgO by HCl gas injection. Can. Metall. Q. 43, 555–560.

Lede, J., Lapicque, F., Villermaux, J., Gales, B., Ounalli, A., Baumard, J.F., Anthony, A.M., 1982. Production of hydrogen by direct thermal decomposition of water: preliminary investigations. Int. J. Hydrog. Energy 7, 939–950.

Lewis, M.A., Masin, J.G., 2009. The evaluation of alternative thermochemical cycles – part II: the down-selection process. Int. J. Hydrog. Energy 34, 4125–4135.

Lewis, M.A., Serban, M., Basco, J., 2003. Hydrogen production at 550°C using a low temperature thermochemical cycle. In: Proceedings of the OECD/NEA Meeting, Argonne National Laboratory.

Li, Y., Song, P., Xia, S., Li, W., Gao, S., 2006. Solubility prediction for the HCl–MgCl$_2$–H$_2$O system at 40°C and solubility equilibrium constant calculation for HCl·MgCl$_2$°7H$_2$O at 40°C. Calphad 30, 61–64.

McQuillan, B.W., Brown, L.C., Besenbruch, G.E., Tolman, R., Cramer, T., Russ, B.E., Vermillion, B.A., Earl, B., Hsieh, H.-T., Chen, Y., Kwan, K., Diver, R., Siegal, N., Weimer, A., Perkins, C., Lewandowski, A., 2010. High Efficiency Generation of Hydrogen Fuels Using Solar Thermal-Chemical Splitting of Water. General Atomics Project 3022.

Messner, G., 1974. Process and Apparatus for the Electrolysis of HCl Containing Solution with Graphite Electrodes Which Keep the Chloride and Hydrogen Gases Separate. US Patent 3,855,104.

Mizia, R.E., 2008. Next Generation Nuclear Plant Intermediate Heat Exchanger Acquisition Strategy. Idaho National Laboratory. Report INL/EXT-08-14054.

Naidin, M., Mokry, S., Baig, F., Gospodinov, Y., Zirn, U., Pioro, I., Naterer, G., 2009. Thermal design options for pressure channel SCWRs with cogeneration of hydrogen. J. Eng. Gas Turbines Power 131, 012901.

Nakamura, T., 1977. Hydrogen production from water utilising solar heat at high temperatures. Sol. Energy 19, 467–475.

Naterer, G.F., Dincer, I., Zamfirescu, C., 2013. Hydrogen Production from Nuclear Energy. Springer, New York.

Naterer, G.F., Suppiah, S., Stolberg, L., Lewis, M., Wang, Z., Daggupati, V., Gabriel, K., Dincer, I., Rosen, M.A., Spekkens, P., Lvov, L., Fowler, M., Tremaine, P., Mostaghimi, J., Easton, E.B., Trevani, L., Rizvi, G., Ikeda, B.M., Kaye, M.H., Lu, L., Pioro, I., Smith, W.R., Secnik, E., Jiang, J., Avsec, J., 2010. Canada's program on nuclear hydrogen production and the thermochemical Cu-Cl cycle. Int. J. Hydrog. Energy 35, 10905–10926.

Naterer, G.F., Suppiah, S., Stolberg, L., Lewis, M., Ferrandon, M., Wang, Z., Dincer, I., Gabriel, K., Rosen, M.A., Secnik, E., Easton, E.B., Trevani, L., Pioro, I., Tremaine, P., Lvov, S., Jiang, J., Rizvi, G., Ikeda, B.M., Luf, L., Kaye, M., Smith, W.R., Mostaghimi, J., Spekkens, P., Fowler, M., Avsec, J., 2011. Clean hydrogen production with the Cu-Cl cycle – progress of international consortium, I: experimental unit operations. Int. J. Hydrog. Energy 36, 15472–15485.

Nomura, M., Nakao, S., Okuda, H., Fujiwara, S., Kasahara, S., Ikenoya, K., Kubo, S., Onuki, K., 2004. Development of an electrochemical cell for efficient hydrogen production through the IS process. AICHE J. 50, 1991–1998.

Odukoya, A., Naterer, G.F., 2011. Electrochemical mass transfer irreversibility of cuprous chloride electrolysis for hydrogen production. Int. J. Hydrog. Energy 36, 11345–11352.

Ogawa M, Hino R, Inagaki Y, Kunitomi K, Onuki K, Takegami H (2009) Present status of HTGR and hydrogen production development at JAFA. Nuclear Hydrogen Production. Fourth Information Exchange Meeting Oakbrook, IL. Nuclear Energy Agency 6805

Ozcan, H., 2015. Experimental and Theoretical Investigations of Magnesium-Chlorine Cycle and Its Integrated Systems. PhD Thesis, University of Ontario Institute of Technology.

Ozbilen, A.Z., 2013. Development, Analysis and Life Cycle Assessment of Integrated Systems for Hydrogen Production Based on the Copper-Chlorine Cycle. PhD Thesis, University of Ontario Institute of Technology.

Ozbilen, A., DIncer, I., Rosen, M.A., 2011. Environmental evaluation of hydrogen production via thermochemical water splitting using the Cu-Cl cycle: a parametric study. Int. J. Hydrog. Energy 36, 9514–9528.

Perret, R., Chen, Y., Besenbruch, G., Diver, R., Weimer, A., Lewandowski, A., Miller, E., 2005. High temperature thermochemical solar hydrogen generation research. UNLV Research Foundation. Report to Department of Energy DE-FG36-03GO13062.

Ranganathan, S., Easton, E.B., 2010a. Ceramic carbon electrode-based anodes for use in the Cu-Cl thermochemical cycle. Int. J. Hydrog. Energy 35, 4871–4876.

Ranganathan, S., Easton, E.B., 2010b. High performance ceramic carbon electrode-based anodes for use in the Cu–Cl thermochemical cycle for hydrogen production. Int. J. Hydrog. Energy 35, 1001–1007.

Rosen, M.A., 2008. Exergy analysis of hydrogen production by thermochemical water decomposition using the Ispra Mark-10 Cycle. Int. J. Hydrog. Energy 33, 6921–6933.

Sato, S., 1979. Thermochemical hydrogen production. In: Otha, T. (Ed.), Solar-Hydrogen Energy Systems. Pergamon Press, New York.

Savage, R.L., Blank, L., Cady, T., Cox, K., Murray, R., Dee Williams, R., 1973. A hydrogen energy carrier. Systems Design Institute, NASA Grant NGT 44-005-114.

Sivasubramanian, P., Ramasamy, R.P., Freire, F.J., Holland, C.E., Weidner, J.W., 2007. Electrochemical hydrogen production from thermochemical cycles using a proton exchange membrane electrolyzer. Int. J. Hydrog. Energy 32, 463–468.

Stolberg, L., Boniface, H.A., McMahon, S., Suppiah, S., York, S., 2008. Electrolysis of the CuCl/HCl aqueous system for the production of nuclear hydrogen. In: Proceedings of the Forth International Topical Meeting on High Temperature Reactor Technology HTR-2008. September 28th–October 1st, Washington, DC.

Suppiah, S., Li, J., Sadhankar, R., Kutchcoskie, K.J., Lewis, M., 2005. Study of the hybrid Cu-Cl cycle for nuclear hydrogen production. In: Nuclear Hydrogen Production, Third Information Exchange Meeting, Oarai, Japan, October 5–7. Nuclear Energy Agency 6122.

Suppiah, S., Stolberg, L., Boniface, H., Tan, G., McMahon, S., York, S., Zhang, W. (2009) Canadian nuclear hydrogen R&D programme: Development of the medium-temperature Cu-Cl cycle and contributions to the high-temperature sulphur-iodine cycle. Fourth international exchange meeting on nuclear hydrogen production. Oakbrook IL. USA. 13-16 April. Nuclear Energy Agency and Organisation for Economic Co-operation and Development, NEA 6805.

Szargut, J., 2005. Exergy Method. Technical and Ecological Applications. Boston, WIT Press.

Szargut, J., Styrylska, T., 1964. Approximate evaluation of the exergy of fuels. Brennst.-Warme-Kraft 16, 589–596.

Van Loo, S., Koopejan, J., 2008. The Handbook of Biomass Combustion and Co-firing. Earthscan, Sterling, VA.

Wang, Z.L., Naterer, G.F., Gabriel, K., Gravelsins, R., Daggupati, V., 2009. Comparison of different copper-chlorine thermochemical cycles for hydrogen production. Int. J. Hydrog. Energy 34, 3267–3276.

Wang, Z.L., Naterer, G.F., Gabriel, K.S., 2010. Thermal integration of SWCR nuclear and thermochemical hydrogen plants. In: Second Canada-China Joint Workshop on Supercritical Water Cooled Reactor, Toronto, ON, April 25–28.

Wentorf, R.H., Hanneman, R.E., 1974. Thermochemical hydrogen generation. Science 185, 311–319.

Zamfirescu, C., Dincer, I., Naterer, G.F., 2012. Comparing emissions from Integrated Cu–Cl cycle with nuclear and coal power plants. In: Global Conference on Global Warming, Istanbul, July 8–12.

Zimmerman, W.H., Trainham, J.A., Law, C.G. Jr., Newman, J.S., 2001. Electrochemical Conversion of Anhydrous Hydrogen Halide to Halogen Gas Using an Electrochemical Cell. US Patent 6,203,675E1.

STUDY PROBLEMS

4.1 Categorize the methods for hydrogen production.

4.2 Calculate the reaction enthalpy and the Gibbs energy of water decomposition reaction at 25°C, 1000°C, and 2500°C and compare the results.

4.3 Make a definition of water thermolysis and explain the difference and similarities with water electrolysis.

4.4 Explain the concept of thermochemical water splitting.

4.5 Describe the sulfur-iodine cycle.

4.6 What is the difference between fuel reforming and gasification and how to illustrate these differences?

4.7 What are the envisaged hydrogen production methods coupled with nuclear reactors?

4.8 After making reasonable assumptions, estimate the energy and exergy of the five-step copper-chlorine hydrogen production cycle.

Hydrogen Production by Photonic Energy

5.1 INTRODUCTION

It is known to almost everyone that electromagnetic radiation surrounds everything on earth and space. This basic form of interaction does not require a material support to propagate; it can pass through a true vacuum and travel distances measured in light years without attenuation. Solar light, consisting of a spectrum of photons with a temperature around 6000 K, travels about 8 min before reaching earth. Life on earth depends on light and the photophysical and photochemical processes induced by light upon the earth's systems.

When photon interacts with matter, a multitude of photophysical processes may occur. Therefore, photons can be used to generate hydrogen out of water or other substances. There are known photophysical, photochemical and photoelectrochemical methods to generate hydrogen from water splitting. An interesting feature of photonic radiation is its ability to interact with matter across volumes rather than surfaces in transparent or semitransparent media. The possibility to transmit energy to every point of a volume is attractive for water-splitting processes.

Pure water, however, does not absorb radiation in the visible and near ultraviolet (NUV) ranges, so one cannot directly dissociate the water molecule using solar radiation at the terrestrial surface. This is why photocatalysis is necessary to promote chemical reactions stimulated by light in the range of visible spectrum, where solar radiation has a maximum energy. Photocatalysis is a method of acceleration of a chemical reaction driven by light. Presently, research of photocatalysis for hydrogen production is very intense worldwide.

Both homogeneous and heterogeneous photocatalysis systems are known that generate hydrogen from water under the exposure of photonic radiation (light). There are many approaches to hydrogen generation from water with homogeneous multicomponent photocatalytic systems, and some are based on supramolecular devices. Homogeneous catalysis, although seen as promising for future appears to be very impractical due to its low efficiency and extremely small production scale (generally less than micromoles per hour) and very expensive materials for catalysts (e.g., Ru, Ir, Rh, etc.). On the other hand, heterogeneous, dispersed particulate catalysts in aqueous electrolytes containing sulfides appear to be very promising for hydrogen production, showing the potential for large-scale practical applications of photocatalytic water-splitting reactors. A third approach is possible, although few past studies were reported, namely hybrid catalysis that implements both types of catalysis within a device. A fourth approach is a photoelectrochemical cell consisting of a modified electrolysis cell, which includes photoelectrodes instead of electrodes. This is, therefore, a heterogeneous photocatalytic system operating under an electric potential bias. A photoelectrochemical cell for water electrolysis is probably the most well known light-driven water-splitting device. Different versions are possible. ie. cell with photoanode, with photocathode or with both kinds of photoelectrodes.

Despite the substantial progress observed, none of the available photochemical or photoelectrochemical technology can operated under the whole range of solar light spectrum. Most of the photocatalysts can operate only in upper visible and NUV spectrum at wavelengths shorter than 400 nm where the energy carried by sunlight represents only 5% of the total spectrum. Therefore, integrated systems are required in which several technologies work together to harvest light and generate hydrogen from a wider solar energy spectrum.

In this chapter, hydrogen production systems driven by photonic energy are introduced. Fundamental aspects are studied first, including thermodynamic analysis of light and kinetic aspects of photochemical reactions. Further, the most relevant methods for photonic-based hydrogen production are presented. The chapter ends with a number of case studies presenting integrated systems for hydrogen production with solar light as energy input.

5.2 FUNDAMENTALS OF PHOTONIC HYDROGEN PRODUCTION

The water-splitting reaction is a multielectron process that requires a source of energy to meet the Gibbs free energy of the reaction, which is needed to rearrange the valance electrons to make the formation of H_2 and O_2 possible. This energy is equal to 2.458 eV to produce one molecule of hydrogen in standard conditions, representing the amount required to rearrange electrons under 1.229 eV of potential difference. For a full reaction to occur, two molecules of hydrogen are produced, as follows

$$2H_2O(l) \rightarrow 2H_2(g) + O_2(g) \tag{5.1}$$

Here, four valance electrons of two water molecules are dislocated, requiring an energy input of 4.915 eV. The required energy for the full reaction (4.915 eV) can be produced by one photon of ultraviolet light with a wavelength shorter than 252.3 nm, or by two photons in the visible spectrum with a wavelength shorter than 504.5 nm, or four infrared photons of 1.23 eV. Fig. 5.1 indicates the energy carried by photons as a function of their wavelength and the energy level needed to split the water molecule and generate one molecule of oxygen and two molecules of hydrogen. The most useful energy spectrum of the sunlight, is in the upper part of visible spectrum from 700 to 400 nm with an energy content representing 43% of the whole spectrum. In that region, two or three photons are ideally necessary to conduct the water-splitting process.

Hydrogen production using photons' energy can be classified by function of the nature of the photocatalysis process as homogeneous, heterogeneous and hybrid. Fig. 5.2 shows a classification of the devices for photonic-based hydrogen production. The systems with heterogeneous catalysis can be divided into two types: particulate photocatalyst systems and photoelectrochemical cells. In a particulate photocatalysis system, the photocatalysts are solid particles of below-micrometer size, which are churned mechanically inside a transparent reactor that is subjected to high-intensity illumination.

In a photoelectrochemical cell, the photocatalysts are coated on the photoelectrodes. In addition, a bias potential is applied to the photoelectrodes to facilitate a photoelectrochemical reaction that eventually generates hydrogen.

FIG. 5.1 Energy carried by photons versus the wavelength and free energy for water splitting.

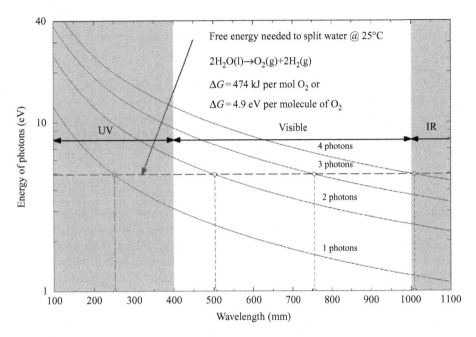

FIG. 5.2 Classification of photon-energy-based hydrogen production devices.

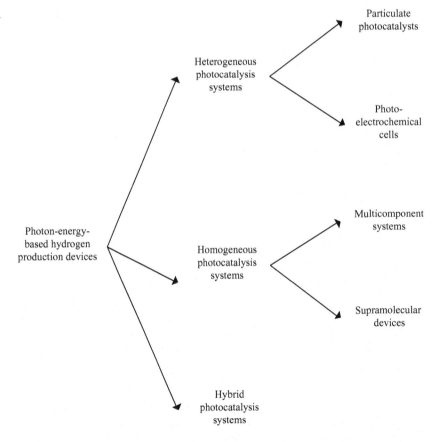

The homogeneous photocatalysis system usually consists of metallic-organic photocatalysts dissolved in a solution that is intensely illuminated.

Understanding and assessing photonic energy-driven hydrogen production processes requires some specific fundamental background that is reviewed herein. Here, the physics of photons and light is briefly discussed the exergy concept of light is introduced, and solar radiation resource is studied. Photophysics of light-matter interaction is discussed. Relevant photochemical processes and kinetic parameters are introduced. Some efficiency formulations for photonic energy-based hydrogen production system assessment are given.

5.2.1 Photons and Electromagnetic Radiation

Electromagnetic radiation is emitted or absorbed by any corpse in the universe that has a temperature superior to zero Kelvin. The electromagnetic radiation propagates through space as a wave, having a propagation speed in a vacuum equal to the universal speed of light constant. The electric and magnetic field oscillate in phase and are perpendicular to each other. The cross-product of $\mathbb{E} \times \mathbb{B}$ gives the direction of the wave propagation, where $\mathbb{E}$ and $\mathbb{B}$ are the intensity vectors of the electric and magnetic fields, respectively.

The electromagnetic radiation encountered on earth forms a very wide spectrum, which is categorized based on wavelength. The longer wavelengths form the long radio waves emitted at frequencies below 10 kHz and even 1 Hz and lower. The radio waves have wavelengths in the range from 1 to 2 km. Microwave wavelengths are on the order of cm. When the wavelength becomes shorter, the wave behavior of electromagnetic radiation is less observable. Note that photon is a package of electromagnetic radiation of high frequency having no rest mass and possessing a dual behavior of being a particle and a wave. At wavelengths shorter than a few mm, electromagnetic waves behave as quanta of energies, or photons. The infrared photons extend from micrometer to mm wavelengths. The visible spectrum is 400–700 nm. The ultraviolet spectrum extends in the range of 10 pm to 400 nm. The X-ray photons have wavelengths from 10 fm to a few pm. Gamma ray photons have wavelengths of pm scale and below.

Understanding the interaction of light with matter requires the introduction of the concept of photons. In fact, Einstein discovered the photon in relation to the theory of photoelectric effect. In the photoelectric effect, photons dislocate valence electrons in metal exposed to electromagnetic radiation. The metal electrons can be energized only if their energy is larger than a threshold value. In a metal, the electrons share a common energy level denoted as a conduction band (CB).

The interaction of photons with metals is one of the simplest to understand. More complicated is the interaction of photons with semiconductors and the generation of electron-hole pairs. In a semiconductor, two energy bands coexist, the valence and CB. When a photon has sufficient energy, then it can displace an electron from the valence to the CB. The interaction of photons with molecules and supra-molecules is an even more complex process. This will be discussed later in this chapter.

The process of photon-matter interaction involves energy dissipation in forms of vibrionic energy, phosphorescence, fluorescence, etc. This is why a photon interacts with matter transfering entropy. A well studied form of light-matter interaction is thermal radiation. As described in Bejan (1988), heat transfer through thermal radiation is intrinsically irreversible. The concept of blackbody radiation and the related theory emerged by 1860 with the works of Kirchhoff, Stefan, Boltzmann and Wien, who were the main contributors. This will be discussed next.

5.2.1.1 Blackbody Radiation

The blackbody model consists of an ideal photonic radiation absorber conceived in the form of a cavity with a pinhole. Light that enters through the pinhole is entrapped in the cavity and never escapes, but rather is thermalized (it dissipates thermally). Similarly, light is emitted from the cavity at the pinhole location, and with a blackbody spectrum that corresponds to the temperature of the cavity. Early discoveries showed that the emitted energy rate is proportional to the power fourth of the temperature. Spectroscopes were invented at that time, and light irradiance measured at every wavelength. It was observed that the spectral irradiance has a maximum that, according to the Wien law, depends on temperature only.

The schematics of the blackbody cavity receiver and the spectral distribution of the emissive power are shown in Fig. 5.3, as prepared with Engineering Equation Solver (Klein, 2015). As mentioned above, the blackbody cavity is conceptualized as a thermally insulated closed chamber with a pinhole. The chamber is at a uniform and constant temperature. Since the chamber is insulated, it only emits and receives radiation through the small pinhole.

In the figure, the displacement toward shorter wavelengths (ultraviolet) when temperature increases can be observed. This displacement is stated by the Wien law, which also shows that the emissive power of the blackbody does not depend on wavelength, but on the product λT. Experimentally, it has been determined that when I_λ is maximum, then the product $(\lambda T)_{max}$ is given as

$$(\lambda T)_{max} = \frac{c_2}{5} \tag{5.2}$$

where $c_2 = 0.014388$ mK is denoted as the second radiation constant.

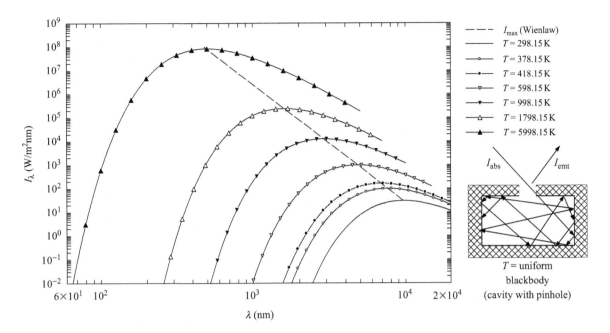

FIG. 5.3 Emission spectrum of blackbody radiation.

A form of Wien displacement law can be written as

$$I_{\lambda,b} = T^5 f(\lambda T) \tag{5.3}$$

where $I_{\lambda,b}$ is the spectral irradiance of the blackbody typically measured in W/m²nm, λ is the wavelength, T is the blackbody temperature and f is a function discussed subsequently.

The total emissive power of the blackbody is obtained by integration of the spectral emissivity from Eq. (5.3) for all spectrum of wavelengths, where $\lambda \in (0, \infty)$, namely

$$I_b = \int_0^\infty T^5 f(\lambda T) d\lambda = T^4 \int_0^\infty f(\lambda T) d(\lambda T) \tag{5.4}$$

Furthermore, Stefan-Boltzmann law gives the emissive power of the blackbody radiation in a vacuum as

$$I_b = \sigma T^4 \tag{5.5}$$

with the Stefan-Boltzmann constant $\sigma = 5.67 \times 10^{-8}$ W/m²K⁴.

Equating the two expressions for the emissive power of the blackbody radiation given by Eqs. (5.4) and (5.5), one obtains that for a blackbody at temperature T, the following holds

$$\int_0^\infty f(\lambda T) d(\lambda T) = \sigma \tag{5.6}$$

Eq. (5.6) suggests that electromagnetic (photonic) energy transfer can be quantized. Based on the hypothesis of quantization of the electromagnetic energy, Planck derived the form of the function $f(\lambda T)$ for blackbody spectral emissive power. This determines Planck's distribution of blackbody radiation, which is given by

$$I_{\lambda,b} = T^5 \frac{c_1 (\lambda T)^{-5}}{\exp(c_2/\lambda T) - 1} \tag{5.7}$$

where $c_1 = 3.7418 \times 10^{-16}$ Wm² is the first radiation constant.

Here, a new constant, Planck's constant, denoted with h, is defined as

$$h = \frac{c_1}{2\pi c} = 6.55 \times 10^{-34} \text{Js} \tag{5.8}$$

where c is the speed of light in vacuum constant.

Alternatively, the Planck constant can be expressed as

$$h = \frac{c_2 k_B}{c} \tag{5.9}$$

where k_B is the Boltzmann constant.

Moreover, with the help of Planck constant, the following expression can be found for Stefan-Boltzmann constant

$$\sigma = \frac{2\pi^5 k_B^4}{15 c^2 h^3} \tag{5.10}$$

Furthermore, the hypothesis of quantization of the electromagnetic energy leads to the determination of the energy content of the quanta of light, the photon. The energy of a photon is proportional to the reciprocal of the wavelength, whereas the proportionality factor is the Planck constant into the speed of light constant. Therefore, the energy of a photon depends only on its wavelength and is given as

$$E_\lambda = h\frac{c}{\lambda} \tag{5.11}$$

Based on the energy quanta of a photon given by Eq. (5.11), the spectral radiation of blackbody radiation becomes

$$I_{\lambda,b} = \dot{N}''_{\lambda,b} h\frac{c}{\lambda} \tag{5.12}$$

where $\dot{N}''_{\lambda,b}$ is the spectral photon distribution, that is, photon rate per unit of normal surface and wavelength (photons/m²nm).

Eq. (5.12) can be solved for spectral photon distribution to give

$$\dot{N}''_{\lambda,b} = \left(\frac{T^4}{hc}\right)\frac{c_1(\lambda T)^{-4}}{\exp(c_2/\lambda T) - 1} \tag{5.13}$$

As viewed at the quantic level, light (electromagnetic radiation) carries kinetic energy and momentum, both transferable as work. The kinetic energy content of a photon is written as

$$KE_\lambda = 0.5 h\frac{c}{\lambda} \tag{5.14}$$

Although at the microscopic level, the photon can transfer work, from the macroscopic viewpoint, the photon transfers heat when it interacts with matter. Henceforth, the blackbody radiation transfers entropy at the temperature of the blackbody. Furthermore, it can be argued that since the energy of a photon is quantized, so it must be the photonic entropy. Moreover, similarly as for energy, the entropy of a photon must depend only on its wavelength; photons with the same wavelength will carry the same quanta of entropy. This means that a spectral distribution of blackbody radiation entropy must exist, which is given as

$$\dot{S}''_{b,\lambda} = \dot{N}''_{\lambda,b} S_\lambda \tag{5.15}$$

where S_λ is a short notation of the entropy carried by a single photon.

The entropy flux of the blackbody radiation results from spectral integration of $\dot{S}''_{b,\lambda}$ as follows

$$\dot{S}''_b = \int_{\lambda=0}^{\lambda=\infty} \left(\frac{T^3}{hc}\right)\frac{c_1(\lambda T)^{-4}}{\exp(c_2/\lambda T) - 1} S_\lambda d(\lambda T) \tag{5.16}$$

Since the temperature of radiation is T and the intensity I_b, from Eq. (5.5), one obtains that entropy flux of the blackbody radiation is given as

$$\dot{S}''_b = \frac{I_b}{T} = \sigma T^3 \tag{5.17}$$

Combining Eqs. (5.16) and (5.17) the following equation results

$$\int_0^\infty \frac{(\lambda T)^{-4}}{\exp(c_2/\lambda T) - 1} S_\lambda d(\lambda T) = \frac{\sigma hc}{c_1} \tag{5.18}$$

The integral from Eq. (5.18) is a constant for any given blackbody temperature T. An analogy is claimed again between energy and entropy of the quanta of light. More specifically, since in Eq. (5.6) the integrand is a function of λT, so must be for the integrand from Eq. (5.18). If this hypothesis is valid, then S_λ must be a function of λT. However; however, since S_λ is a function of λ only as stated previously, it results that S_λ is a constant as the only remaining possibility.

Let us denote s_{ph} the entropy constant of a photon and replace S_λ with s_{ph} in Eq. (5.18) to emphasize that the entropy of a photon is a constant, not dependent on λ. It therefore comes out that the entropy constant of a photon is given as

$$s_{ph} = \frac{\sigma hc}{c_1 \int_0^\infty \frac{(\lambda T)^{-4}}{\exp(c_2/\lambda T) - 1} d(\lambda T)} \tag{5.19}$$

Chen et al. (2008) determined that the entropy constant of a photon is given as

$$s_{ph} = 2.69952 k_B = 3.7268 \times 10^{-23} \, \text{J/K} \tag{5.20}$$

5.2.1.2 Exergy of Photonic Radiation

If a photonic radiation interacts with a medium at a reference temperature T_0, then work can be produced in an amount corresponding to the exergy content of the radiation. Fig. 5.4 shows a model for light-matter interaction in which the photons will pump electrons generating work potential. The system is equivalent with a reversible heat engine configuration in which the radiation at a temperature T_{rad} is provided at the source side, while the heat sink is at the reference environment temperature. The exergy of the photonic radiation will depend on Carnot factor. The work potential due to this light-matter interaction can be retrieved in many ways; for example, it can drive a photochemical reaction to generate hydrogen.

With the use of Carnot factor, the exergy of a radiation at a temperature T_{rad} is given as

$$\dot{E}x'' = \left(1 - \frac{T_0}{T_{rad}}\right) I \tag{5.21}$$

where I is the normal irradiance given in W/m^2.

Consider a given spectral distribution of a nonblackbody light radiation defined by the photon flux $\dot{N}_\lambda''$. Then, the spectral irradiance of the nonblackbody radiation is

$$I_\lambda = \frac{hc}{\lambda} \dot{N}_\lambda'' \tag{5.22}$$

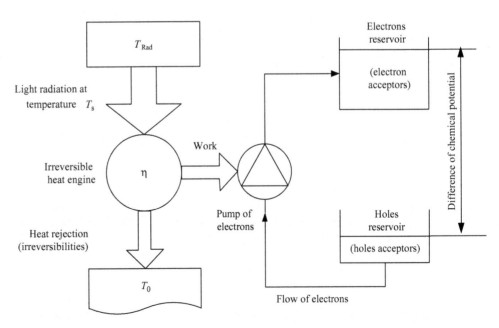

FIG. 5.4 Interaction of photonic radiation with matter at a reference temperature T_0. The radiation can deliver its exergy content as work.

and the normal irradiance becomes

$$I = \int_0^\infty I_\lambda d\lambda = \frac{hc}{\lambda} \int_0^\infty \dot{N}_\lambda'' d\lambda \tag{5.23}$$

The spectral entropy flux carried by this radiation becomes

$$\dot{S}_\lambda'' = s_{ph} \dot{N}_\lambda'' \tag{5.24}$$

Therefore, the entropy flux of the radiation becomes

$$\dot{S}'' = s_{ph} \int_0^\infty \dot{N}_\lambda'' d\lambda \tag{5.25}$$

In the same time, the entropy flux can be expressed based on the radiation temperature T_{rad} and irradiance I, as

$$\dot{S}'' = \frac{I}{T_{rad}} = \frac{1}{T_{rad}} \int_0^\infty \frac{hc}{\lambda} \dot{N}_\lambda'' d\lambda \tag{5.26}$$

Equating the two expressions for the entropy flux, namely Eqs. (5.25) and (5.26) one can solve for the temperature of the nonblackbody radiation of a given spectrum, as follows

$$T_{rad} = \frac{hc}{s_{ph}} \frac{\int_0^\infty \frac{\dot{N}_\lambda''}{\lambda} d\lambda}{\int_0^\infty \dot{N}_\lambda'' d\lambda} \tag{5.27}$$

Alternatively, Eq. (5.27) can be expressed based on the spectral irradiance of the nonblackbody radiation as

$$T_{rad} = \frac{hcI}{s_{ph} \int_0^\infty \lambda I_\lambda d\lambda} \tag{5.28}$$

Of particular importance on earth is the exergy of the solar radiation. It is clear from Eqs. (5.21) and (5.28) that this exergy depends only on the spectral irradiance and the reference temperature T_0. The solar light spectral irradiance at the earth's surface varies depending on the geographic location, local conditions and the astronomic time. ASTM AM 1.5 (2008) is a reference standard for solar spectrum adopted by the American Society for Testing and Materials (ASTM).

Fig. 5.5 shows the ASTM AM 1.5 spectrum giving four spectral irradiation components: extraterrestrial $I_{\lambda,etr}$, global on a 37 degree tilt southfacing surface $I_{\lambda,gt}$, direct normal and circumsolar $I_{\lambda,dc}$ and diffuse radiation $I_{\lambda,diff}$. This reference spectrum corresponds to a zenith angle $\theta_z = 48.2$ degree (solar zenith angle is between the local vertical and the direction of the sun).

The extraterrestrial radiation intensity (measured in W/m^2) corresponds to the light intensity at the upper edge of the atmosphere; it is obtained by the integration of the extraterrestrial spectral irradiance over the all-spectrum of wavelength. According to Duffie and Beckman (2013), the extraterrestrial solar radiation varies according to the day of the year n_{day} (counted from the 1st of January) according to

$$I_{ext} = \int_0^\infty I_{\lambda,ext} d\lambda = I_{sc} \left[1 + 0.033 \cos\left(\frac{360}{365} n_{day} \right) \right] \tag{5.29}$$

where I_{sc} is commonly called *solar constant*.

The solar constant represents the annual average extraterrestrial solar radiation intensity, whose value is $I_{sc} = 1367$ W/m^2. Note that the extraterrestrial radiation varies along the year, because it depends on the astronomical distance between the sun and earth. This distance is maximum at aphelion (on the 3rd of July) with 152×10^6 km and a minimum at perihelion (on the 3rd of January) with 147×10^6 km.

The most important parameter that influences both the intensity of solar radiation at earth's surface and the spectrum is the air mass. The air mass is defined as the ratio between the path length of sunrays through the atmosphere and the effective atmosphere thickness at local zenith. Air mass depends on the zenith angle, the day of the year and the geographical latitude. At sea level, when the sun is at zenith, then air mass is $AM = 1$, while if sun is at horizon,

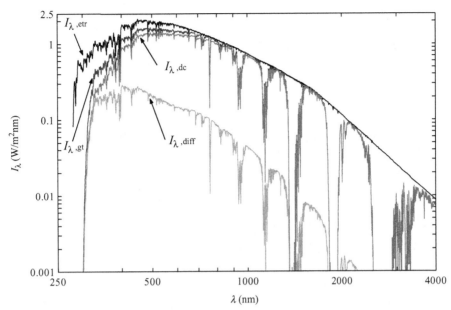

FIG. 5.5 The ASTM 1.5 reference spectrum for solar irradiation. *Data from ASTM, 2008. G173-03 Reference Spectra Derived from SMARTS v. 2.9.2. Accessed from* http://rredc.nrel.gov/solar/spectra/am1.5/ASTMG173/ASTMG173.html.

then $AM = 38.2$. Several types of spectral distributions of solar light can be obtained as a function of air mass as follows:

- AM0—the extraterrestrial spectrum at the edge of the atmosphere
- AM1—the extraterrestrial spectrum with sun at the zenith (equator, sunbelt regions)
- AM1.1—obtained when zenith angle is ~25 degree and applicable to tropical regions
- AM1.5—this is the most widely adopted case ($\theta_s = 48.2$ degree discussed above)
- AM2—for zenith angle of 60 degree for high latitudes (>45 degree)
- AM3—for zenith angle of 70 degree for even higher latitudes (>60 degree)
- AM38.2—is the air mass spectrum corresponding to horizon ($\theta_z = 90$ degree)

The air spectra can be predicted according to the methodology adopted by the National Renewable Energy Laboratory (NREL), which is based on the paper of Gueymard et al. (2002) and can be calculated with the help of software SMARTS described in Gueymard (1995). Besides the air mass, the solar spectrum depends on the water content and ozone in the atmosphere, as well as on turbidity, aerosol types and concentration, cloudiness and haziness and optical thickness of the atmosphere. In Fig. 5.6, the most important AM spectra are presented in detail. Note that the data in this graph are calculated with NREL SMARTS software (Gueymard, 1995).

The direct radiation is obtained directly from the solar disk, which is viewed at an angle of 5 degree. The direct radiation has two subcomponents: the direct beam that is the radiation of the solar disc itself viewed at an angle of 0.53 degree, and the circumsolar radiation that refers to the ring around the sun, which covers the angular region from 0.53 degree until 5 degree. The diffuse component, which is the light coming indirectly to the observer, is more intense when air mass is higher. In general, the diffuse radiation is only 10% of the global radiation, the rest being the direct and circumsolar components.

The AM spectra from Fig. 5.6 shift very slightly toward shorter wavelengths when the air mass increases. Also the intensity of the radiation decreases with the air mass. For AM1, the global normal radiation is ~1040 W/m²; for AM 1.1, it is ~1015 W/m²; for AM1.5, it is 1000 W/m²; for AM2, it is 840 W/m²; for AM3, it is ~715 W/m²; while for AM38, it is only 20 W/m². An important parameter for practical solar radiation estimation is the global horizontal intensity, which can be calculated with the following equation from Sing and Tiwari (2005)

$$\frac{I_{gh}}{f_{gd}I_{ext}\cos\theta_z} = \exp\left[-\left(AM\left(4.5\times10^{-4}AM^2 - 9.67\times10^{-3}AM + 0.108\right)f_{ch} + f_{at}\right)\right] \qquad (5.30)$$

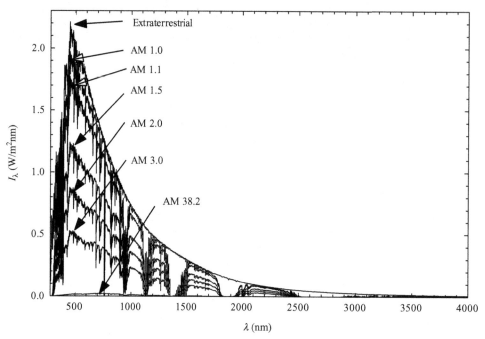

FIG. 5.6   Global spectral radiation on horizontal surface as a function of air mass.

Here, the extraterrestrial solar radiation I_{ext} is given in Eq. (5.29) in function of the day of the year. The zenith angle varies depending on the hour and geographical location. The factor f_{ch} represents the cloudiness and haziness of the atmosphere. This factor is dimensionless, and takes values in the range of 2–6, depending on the geographical location on earth's surface. The term f_{at} represents the direct radiation depletion factor due to compounded effect of aerosol level and thermal disturbances in the atmosphere. The range of practical values for this factor is 0.2–0.5. The factor f_{gd} represents the ratio between the intensity of global radiation and direct radiation. This factor can be correlated with the number of sunshine hours. From Sing and Tiwari (2005), for five hours of sunshine, the daily diffuse radiation is 75% of the global radiation, and for 9 h of sunshine, it reduces only 25%. All factors f_{ch}, f_{at} and f_{gd} are dimensionless.

5.2.2 Photophysical and Photochemical Processes

Light-matter interaction is a physical process in which electromagnetic radiation of high-enough frequency (or short-enough wavelength) displaces electrons and induces elementary physical processes and energy dissipative processes in substances comprising atomic structures. There are many physical processes induced by light, either light-absorbing or light-emitting or other processes. All those processes, which do not end with chemical reactions, are denoted as photophysical processes. Briefly said, light-matter interaction is photophysical.

Table 5.1 gives the duration time ranges of main photophysical processes. Photon absorption by matter at rest is the fastest among all known processes from the universe. The duration of this process δt can be found based on the

TABLE 5.1   Duration of Photophysical Processes

Process	Duration
Absorption	10^{-17}–10^{-15} s
Vibrational relaxation	10^{-13}–10^{-12} s
Fluorescence	10^{-9}–10^{-7} s
S-T intersystem crossing	10^{-12}–10^{-6} s
T-S intersystem crossing	10^{-9}–10^{1} s
Phosphorescence	10^{-6}–10^{-3} s
Internal conversion	10^{-12}–10^{-6} s

Heisenberg principle, which states that $\delta t \geq \lambda/(4\pi c)$, where λ is the wavelength and c is the speed of light in a vacuum. This time is as short as the order of 10^{-17} s. The photon absorption is a complex process in which electrons displace from a ground state of energy to an excited state. The process depends on the actual structure of the substance, whether it is a metal, a semiconductor crystal, or a molecular system of other various kinds.

In a molecular system, the absorption can occur with the passage of the electron from an orbital to another more energetic orbital. Typically, this transition is from a singlet to a singlet configuration. Molecular orbital (MO) structures are described by MO theory. Within the molecule, the electrons occupy certain orbitals around nuclei that form a molecule, which can be predicted as a linear superposition of all possible electronic orbitals on all the atoms of the molecule. The electrons do not appertain to an atom individually, but rather they are spread across the entire molecule. Because electrons are electric charges, the nature of energy that moves electrons must be electromagnetic. This energy can be categorized as two kinds: vibrational (thermal) and radiative.

The atomic orbital wave function Ψ_a describes the position of each of the nuclei and electrons. According to the molecular orbital theory, the wave function of the molecule becomes

$$\Psi = \sum c_i \Psi_{a,i} \tag{5.31}$$

where the coefficients c_i are determined by application of variational mathematics to the Schrödinger equation.

Once the wave-function is determined, the relative position of the nuclei and molecular orbitals is known. Fig. 5.7 illustrates a relative scale of the calculated molecular orbitals for some basic atoms and molecules, as obtained using Discovery Studio 3.5. The wave function changes from one that corresponds to the "ground state" of the molecule, where electrons are in the lowest energy molecular orbitals. With incident radiation absorbed, the most probable process is a jump of the valence electron(s) to the highest energy orbitals. The electrons being on the highest occupied molecular orbital (HOMO) can interact with photons, and they do so, as they are displaced to the lowest unoccupied molecular orbital (LUMO). In other words, during the light-matter interaction, HOMO loses one electron and LUMO receives one. The electron-transition process is extremely fast, as it occurs in femtoseconds (10^{-15} s).

The second process given in Table 5.1 is the vibrational relaxation. This is the process through which the molecule transfers vibrational energy to the surrounding medium or it transfers the energy to itself by intramolecular vibrational redistribution "rearranging" of the energy levels. Fluorescence is the spontaneous emission of photonic radiation by relaxation of the energy level with retention of spin multiplicity.

The singlet-triplet (S-T) intersystem crossing is a process of changing the spin multiplicity of the molecule from singlet to triplet. This occurs isoenergetically and without radiation emission. The process is immediately followed by vibrational relaxation. The triplet-singlet (T-S) intersystem crossing is similar to the S-T process, but the transition is from a triplet to singlet state and is immediately followed by vibrational relaxation.

Phosphorescence is the spontaneous radiation emission of an excited molecule while relaxing to the ground state with a change of spin multiplicity. The phosphorescence occurs typically at system transition from the first triplet state (T_1) to the ground singlet state (S_0). Internal conversion is an isoenergetic process occurring without radiation emission between electronic states with the same spin multiplicity, ie, from S_2 to S_1. The longest photophysical process is phosphorescence, occurring by relaxation from triplet to singlet states. This process is very important, because it

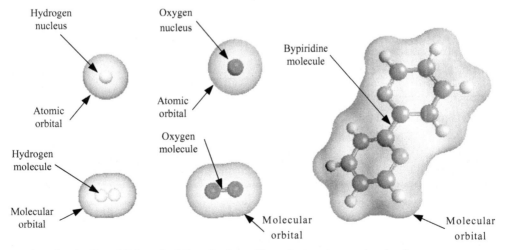

FIG. 5.7   Representation of molecular orbitals and relative atomic positions of some atoms and molecules.

may increase the probability that the energy captured by light absorption is transferred to a molecular system to induce a desired photochemical reaction.

The processes that occur during light-matter interaction can be described with the help of the Jablonski diagram as shown in Fig. 5.8. The diagram indicates the energy level of the singlet ground state (S_0), the higher level (excited) singlet states (S_1, S_2) and the triplet state (T_1). The following processes occur:

- process 1–2: photon absorption
- process 2–3: vibrational relaxation
- process 3–4: fluorescence with a S_1–S_0 transition
- process 3–5: intersystem crossing and the vibrational relaxation
- process 5–6: molecular relaxation to the ground state S_0
- process 5–7: phosphorescence
- process 1–8: high-energy photonic absorption with an electron jump in the vibronic energy level of the second excited state (S_2)
- process 8–9: vibrational relaxation to S_2
- process 9–10: internal conversion
- process 10–11: vibration relaxation to S_1
- process 11–12: fluorescence
- process 11–5: S-T intersystem crossing

Normally, the ground state of a molecule is a shell singlet state, as this state is the most stable for most molecules. In a singlet state, the magnetic fields of constituent atoms have opposing magnetic spins. The singlet state is denoted, for example, by $^1O_2$, referring to a singlet oxygen molecule. When the magnetic spins of the constituent atoms are aligned, the molecular configuration is denoted with a triplet, e.g., $^3O_2$. Some rules apply when photons interact with molecules, as follows:

- Kasha's rule: the most probable excited state is the very first superior to the ground state
- Stark-Einstein rule: the energy of the absorbed photon must match the difference between the ground state and the excited state
- A single electronic transition is possible per photon absorption

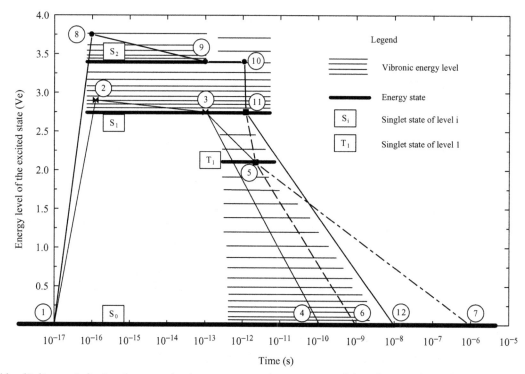

FIG. 5.8 Jablonski diagram indicating the energy levels, energy conversion processes and their duration during light-matter interaction (example corresponding approximately to the molecular complex Ru(bpy)$_3$$^{2+}$, with bpy = bipyridine).

The quantum efficiency of any process, either photophysical or photochemical, is defined as the number of events (x) per number of photons absorbed (n_a). It signifies the probability that a molecular system undergoes the desired process caused by light-matter interaction. According to the definition, the quantum efficiency is

$$\phi(\lambda) = \frac{n_x}{n_a} \tag{5.32}$$

Photophysical processes can induce photochemical reactions. For a photochemical reaction to occur, an appropriate quantity of energy must be transmitted to the orbiting electrons such that they can "jump" temporally to a higher-energy orbital and then recombine, forming the desired products.

The excited states are very reactive and as a consequence, chemical reactions can be induced by photophysical processes. The most relevant photochemical processes will be discussed next. Such processes can be seen as alternative paths for hydrogen production from light, either starting from water or from other hydrogen-containing materials.

There are four main classes of photochemical reaction pathways, proposed by Förster, which can be explained with the help of the notion of a *potential energy surface*. As the name implies, a potential energy surface is a region (surface) where the potential energy described by a molecular wave-function is constant, or in other words, the energy level is preserved. For example, to the ground state S_0, it corresponds to a certain potential energy surface and so it does for any excited state S_i, T_i.

If the distance between atoms is larger, or the molecular orbitals are far from the atoms, the potential energy level is higher. Sometimes, a lower- and a higher-energy surface are very close to each other, or even intersect, which facilitates the occurrence of chemical reactions. "Moving along" a potential energy surface refers to changes in a molecular configuration that do not affect the potential energy; however, this process may require energy or it may release energy (other than potential), depending on the case.

One important class of photochemical reaction pathway is illustrated in Fig. 5.9. This pathway is known as diabatic photoreaction pathway. Assume that certain reactants (molecules or atoms) are brought closer to interact; thus their wave functions couple, and in general, one models the wave function of this system through the product $\Psi_R = \Pi\Psi_i$, where the index "i" refers to molecule and R to reactants. The products must obey the same wave-function in their ground state, since the only difference between the products and reactants in the molecular system is a rearrangement of bonds and orbitals. Thus, $\Psi_P = \Psi_R$. Typically, products and reactants are lying on the same singlet potential energy surface, as indicated in Fig. 5.9 with S_0.

Between reactants and products, however, there is a vibrational energy barrier indicated on the figure with ΔE. The usual chemical path is to provide the vibrational energy thermally to the reactant's molecular system such that the system moves along the potential energy surface following the direction 1–4–5 and generating the products in a stable form. Note that both products and reactants are stable, because they lie on energy minima. The photochemical pathway is different, as it involves certain photophysical processes first. Once excited by the photonic radiation, the

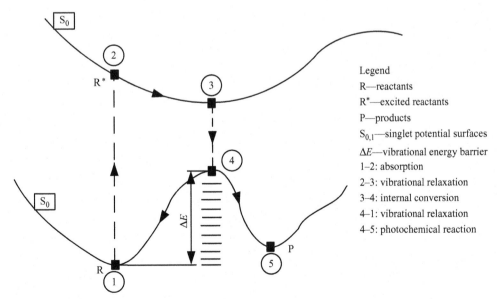

FIG. 5.9 Illustration of the diabatic photoreaction pathway.

molecular system formed by the reactants absorbs light and jumps to the excited state R*, typically on the first superior singlet potential surface (process 1–2). In the diabatic photoreaction pathway, the potential surface on S_1 presents a valley (see Fig. 5.9), which forces vibrational relaxation to the minimum in state 3. Close to 3, the ground level potential surface shows a maxima, thus the internal conversion process 3–4 is facilitated. From the maxima 4, two processes are almost equally probable, namely, process 4–1, which is a vibrational relaxation to the ground state 1, or the photoreaction process 4–5, which reconfigures the molecular assembly to form the products, which are chemical species other than R.

Note that only one subprocess of the diabatic pathway is photochemical—the process 4–5—while the others are photophysical processes. Three other pathways are possible, as mentioned above; these are introduced briefly as follows.

- *Adiabatic reaction*: the absorption process excites the reactant's molecular assembly to the vibrational level of the immediate superior singled state; the system has enough energy to jump over the energy barrier of the excited potential surface S_1 and to form products in the excited phase according to the reaction R*→P*; the excited products relax to a ground state, releasing energy in the form of radiation or vibration.
- *Hot ground state reaction*: in this case, after the absorption, the R* relaxes immediately, but has enough energy to jump over the energy barrier and form the products evolving along the S_0 potential surface.
- *Reaction via intermediate*: the reactants in the excited state evolve toward a minimum and form an excited intermediate EI*, which has a long life.

Water has a molecule with an important dipole moment. When water dissociates, one of the possible reactions is $H_2O \rightarrow H^+ + OH^-$. Assume that this process occurs in a vacuum, and the resulting ions are separated apart at enough distance to split, say 100 molecular diameters, which is about 200 pm. The Coulomb energy needed to perform this work is calculated based on Coulomb's law according to

$$E_{Coulomb} = \frac{e^2}{4\pi\epsilon_0\epsilon_r l} \tag{5.33}$$

where e is the elementary electric charge, ϵ_0 is the electric permittivity of vacuum, ϵ_r is the relative permittivity of a vacuum ($\epsilon_r = 1$), and l is the distance between the released ions.

The Coulomb energy becomes in this case 29 eV; if instead of a vacuum, the molecular assembly is dissolved in water, the relative permittivity is 78 and the Coulomb energy becomes 0.37 eV. This numerical example indicates that it is favorable to operate photochemical reactions in solution.

Assume that water is present in a liquid solution and it is found in the vicinity of a light-absorbing molecule. Consider the modalities of energy transfer between the light-absorbing molecule and the molecule of water. In general terms, if one denotes with R the reactants, and D the energy donor (which is the photosensitizer (PS)), the energy transfer mechanism between D and R can be described with: $D^* + R \rightarrow D + R^*$. In typical photochemical equations, in this reaction, the reactants are denoted with A (from acceptor, since they are energy acceptors). Two basic mechanisms of energy transfer between D* and R are the most relevant for water photodissociation: Förster mechanism (electrostatic energy transfer) and Dexter mechanism (collisional energy transfer).

In the Förster mechanism, the energy transfer is explained by the interaction of a wave function of the two molecular complexes D* and R, which exchange energy. Subsequently, the Förster mechanism is described based on Klan and Wirz (2009). The coupling of wave functions can be written as

$$\Psi_f = \Psi_i \tag{5.34}$$

where $\Psi_i = \Psi_{D^*}\Psi_A$ and $\Psi_f = \Psi_D\Psi_{A^*}$. The two molecules must be at a close distance such that this interaction can occur.

The distance must not be very small, however, because the Förster energy transfer mechanism occurs only through resonance of dipole moments. This means that the wave functions must not overlap. The dipole resonance process is known as FRET (Förster resonance energy transfer).

Triplet-triplet energy transfer does not occur by this mechanism; the only allowed interaction is singlet-singlet. The reaction for a Förster mechanism is

$$^1D^* + {}^1R \rightarrow {}^1D + {}^1R^* \tag{5.35}$$

The process is possible if the distance between molecules is smaller than 10 nm. Note that the Förster energy is isoenergetic. The excited state energy of $^1D^*$ is completely transferred to $^1R^*$; however, it is possible that the received energy by R falls in the vibronic level of its S_1 state, in which $^1R^*$ dissipates energy irreversibly through a vibrational relaxation process. The rate of energy transfer by FRET depends on the rate of decay of $^1D^*$ in the absence of $^1R^*$, namely $k_{D^*}^0$ and on the distance between molecules l, according to the Förster relation

$$k_{FRET} = k_{D^*}^0 \left(\frac{l_0}{l}\right)^6 \tag{5.36}$$

where l_0 denotes the critical transfer distance, which depends on the luminescence quantum yield of the excited state $^1D^*$ and the overlap between the emission spectrum of D and the absorption spectrum of A.

The overlap integral is expressed as

$$J = \int_\lambda I_{D^*,\lambda} \epsilon_R(\lambda) \lambda^4 d\lambda \tag{5.37}$$

where $\epsilon_R(\lambda)$ is the molar absorption coefficient of R and λ is the wavelength.

The Förster equation for the critical length is

$$l_0 = a \left(\phi_{D^*} J / n^4\right)^{1/6} (nm) \tag{5.38}$$

where n is the refraction index of the solution and a is a constant that depends on the dipole moment's orientation of the molecules D and R. It has typical values of 0.023–0.027.

The average distance between the molecules dissolved in solution is calculated based on the equivalent diffusion coefficient $D_{eq} = D_D + D_R$, where the diffusion coefficient of each molecule is given by the Stockes-Einstein equation

$$D = \frac{k_B T}{6\pi\mu r} \tag{5.39}$$

where k_B is the Boltzmann constant, T is temperature, μ is the dynamic viscosity of the solution and r is the molecule radius.

Thus the average distance between D and R in the solution can be approximated with

$$l = \sqrt{2Dk_D} \tag{5.40}$$

where k_D is the rate of decay of $^1D^*$ in the absence of R.

The reaction rate of internal conversion and intersystem crossing processes are related to the energy jump $\Delta E = \tilde{\nu} hc$, where $\tilde{\nu}$ is the wave number (μm^{-1}). These rates can be estimated as follows

$$\log_{10} k = a - 2\tilde{\nu} \tag{5.41}$$

where a is a parameter that takes values of 12 for internal conversion and 7 for slow internal crossing or 10 for fast internal crossing.

Based on the Förster model and the relative concentration of the acceptor molecules R, denoted with $\tilde{c}$ is given as

$$k_D = \frac{2k_D^0}{\sqrt{\tilde{C} + 4} - 2\tilde{C}} \tag{5.42}$$

The relative concentration is defined with respect to the critical concentration of species R, up to which the Förster mechanism is possible; this concentration is denoted here with C_0 and $\tilde{C} = C/C_0$. If the concentration is higher than C_0, then the molecules are so close to each other that their wave functions overlap physically and the Förster mechanism is replaced by a Dexter mechanism for energy transfer. If there is no dissolved species R, then this yields $k_D = k_D^0$ (initial rate). If the concentration of R is at maximum, then the value $k_{D,max} = 8.5k_D^0$. The critical concentration is very low, on the order of 10^{-25} mol/dm^3. The critical concentration is defined as a function of critical length according to the following equation

$$C_0 = 4.473 \times 10^{-25} l_0^{-1} \tag{5.43}$$

where l_0 is expressed in nm, and C_0 results in mol/dm^3.

The Dexter mechanism requires the molecule to collide such that their wave functions interact in a manner in which they can exchange electrons. This mechanism is possible from triplet states according to

$$^3D^* + {}^3R \rightarrow {}^3D + {}^3R^* \tag{5.44}$$

The LUMO electron of $^3D^*$ transfers to the LUMO of $^3R^*$, while in the same time, the HOMO electron of ^3R jumps to the HOMO of ^3D. The overlap between molecular orbitals of the two molecular counterparts must be high, so the distance between molecules becomes close to the sum of the van der Walls radii. The Dexter mechanism is very important for photosensitization, because most of the PSs are excited triplet states (some examples will be given in the subsequent sections).

There are some models in past literature that correlate the rate constant of processes with the energy difference between the triplet state of the acceptor and donor, ie, $\Delta E = {}^3E(R) - {}^3ED$. A practical formula for the Dexter constant rate, denoted with k_{et}, where "et" stands for electron transfer, is given as

$$k_{et} = k_d \exp\left(-\frac{\Delta E}{RT}\right) \tag{5.45}$$

where k_d is the constant rate of the diffusion process of the dissolved molecules, which can be approximated based on the solvent viscosity with $k_d = 8RT/3\mu$.

5.3 SYSTEMS WITH HOMOGENEOUS PHOTOCATALYSIS

5.3.1 Homogeneous Photocatalysis Processes

Homogeneous photocatalysis for hydrogen production is typically conducted in aqueous or nonaqueous (liquid) solutions. The process involves at least one catalyst dissolved in the liquid phase (as known, homogeneous catalysis implies that the catalyst is of the same phase as the reaction participants/products). Homogeneous photocatalysis is based on the interaction of complex molecular structures and water in a solution.

Several molecular complexes participate in the process, performing different functions such as photosensitization, charge separation, charge transfer, electron acceptance or donation and catalysis. Fig. 5.10 shows a conceptual description of a hydrogen-evolving process (HEP) with homogeneous photocatalysis.

The homogeneous photocatalysis process shown in Fig. 5.10 can be described as follows:

- The events initiate with process 1, which is the photon absorption of the PS; thus the PS enters into an excited state according to $PS + h\nu \rightarrow PS^*$.
- For hydrogen-evolving reactions, the PS interacts with the catalyst and then transfers an electron according to process 2: $PS^* + ACC \rightarrow PS^+ + ACC^-$, where ACC stands for "active catalytic center."
- The PS becomes reactive and recovers an electron from the electron donor dissolved in the solution, according to process 3, namely $PS^+ + ED \rightarrow PS + ED^+$. At this time, the PS is reactivated, while the electron donor remains stable in the solution. The active center catalyzes water decomposition by facilitating electron transfer; for a complete reaction, four catalytic cycles are needed, thus the PS must absorb photons four times.
- The catalytic process 4 evolves according to $4H_2O + 4ACC^- \rightarrow 2H_2 + 4OH^- + 4ACC$.

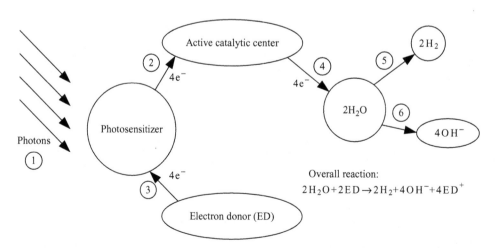

FIG. 5.10   Conceptual description of a homogeneous photocatalysis system evolving hydrogen from water splitting.

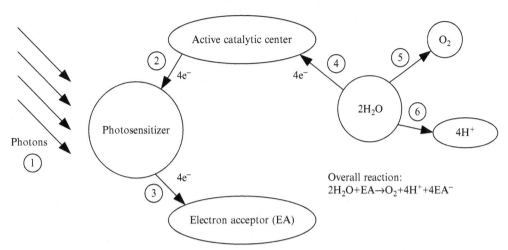

FIG. 5.11 Conceptual description of a homogeneous photocatalysis system evolving oxygen from water splitting.

In many cases, oxygen-evolving reactions are conducted through homogeneous photocatalysis to generate hydrogen and hydroxyl from water. One such process is conceptually described in Fig. 5.11. The oxygen-evolving process (OEP) transfers the electrons in an opposite direction, as it requires an electron acceptor instead of an electron donor; its overall reaction is $2H_2O + EA \rightarrow O_2 + 4H^+ + 4EA^-$. The OEP is important for producing a complete water-splitting cycle that eventually couples an OEP to a HEP.

A typical PS is the ruthenium tris-2,2'-bipyridyne, namely $[Ru(bpy)_3]^{2+}$, which embeds a ruthenium atom in three bipyridine molecules. It can act both as a reductant and an oxidant. Its molecular structure is shown in Fig. 5.12. The precursory of this PS is the bipyridine, a key chemical compound derived from pyridine C_5H_5N, which is an aromatic heterocyclic organic compound. Bipyridine, with a chemical formula of $(C_5H_4N)_2$, comprises two pyridine rings. It can be linked to a ruthenium metal active center to have the role of PS. The Ru PS "dislocates" electrons each time when the active center is hit by photons of appropriate energy. The ruthenium tris-2,2'-bipyridyne molecule has the ability to store electronic energy in an excited orbital around the Ru atom. It absorbs photons at 450 nm (2.75 eV) and generates an excited singlet state $^1[Ru(bpy)_3]2+*$ which rapidly loses 0.65 eV energy by an intersystem crossing from a singlet to triplet according to the following reaction

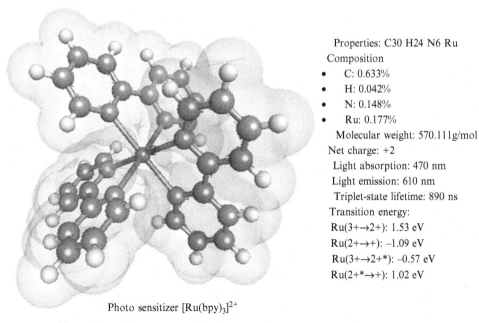

Properties: C30 H24 N6 Ru

Composition

- C: 0.633%
- H: 0.042%
- N: 0.148%
- Ru: 0.177%

Molecular weight: 570.111g/mol

Net charge: +2

Light absorption: 470 nm

Light emission: 610 nm

Triplet-state lifetime: 890 ns

Transition energy:

Ru(3+→2+): 1.53 eV

Ru(2+→+): −1.09 eV

Ru(3+→2+*): −0.57 eV

Ru(2+*→+): 1.02 eV

Photo sensitizer $[Ru(bpy)_3]^{2+}$

FIG. 5.12 Ruthenium tris-2,2'-bipyridyne photosensitizer.

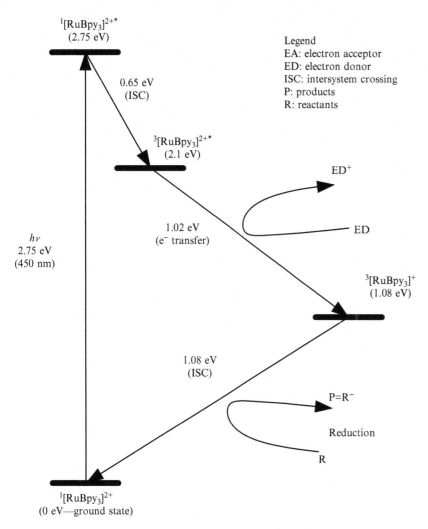

Legend
EA: electron acceptor
ED: electron donor
ISC: intersystem crossing
P: products
R: reactants

$$^1\left[Ru\left(bpy\right)_3\right]^{2+*} \rightarrow {}^3\left[Ru\left(bpy\right)_3\right]^{2+*} \tag{5.46}$$

where bpy means 2,2'-bipyridine and * means excites state.

The excited state can induce both water oxidation or water reduction reactions, as it is very long-lived at 890 ns. Fig. 5.13 shows the reductive quenching cycle of ruthenium tris-2,2'-bipyridyne photosensitizer. In the reductive pathway, an electron donor is used to give an electron to the excited triplet state and form the reductive catalytic center $[Ru(bpy)_3]^+$.

Fig. 5.14 shows the required Gibbs energy to oxidize water with $[Ru(bpy)_3]^{3+}$ and what is needed to reduce water with $[Ru(bpy)_3]^+$ as a function of pH. Reductive quenching is favored at high pH levels, while oxidative quenching is favored at low levels. Thus, the dissociation of a water molecule with $[Ru(bpy)_3]^{2+}$ is possible in both pathways. In practice, it is important to have systems that not only split water, but also separate the product streams. If produced oxygen and hydrogen are confined in the same reactor space, it may create an explosive danger.

Fig. 5.15 shows the oxidative cycle of the ruthenium tris-2,2'-bipyridyne photosensitizer. In this cycle, an electron acceptor is used to transfer an electron from the excited triplet $^3[Ru(bpy)_3]2+*$ to form $[Ru(bpy)_3]^{3+}$ at a lower energy level. This compound is very reactive and can catalyze the desired oxidation reaction, which leads to the formation of the singlet ground state, according to the following reaction

$$^3\left[Ru\left(bpy\right)_3\right]^{3+} + R \rightarrow {}^3\left[Ru\left(bpy\right)_3\right]^{2+} + R^+ \tag{5.47}$$

FIG. 5.14 Reductive and oxidative pathways for water dissociation with $[RuBpy_3]^{2+}$.

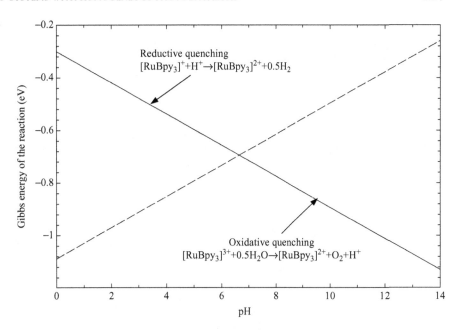

Many electron donors and acceptors have been identified, as reported in the open literature, and tested for oxygen- or hydrogen-evolving reactions with homogeneous catalysis systems. Some known electron acceptors are $K_2S_2O_8$, $AgNO_3$, $[Co(NH_3)_5Cl]Cl_2$, $Na_2S_2O_8$, $S_2O_8^{2-}$ and $[Co(NH_3)_5Cl]^{2+}$. After donating electrons, they transform irreversibly in other compounds. An example is persulfate, which acts according to the following reaction

$$2\left[Ru(bpy)_3\right]^{2+} + S_2O_8^{2-} \rightarrow 2\left[Ru(bpy)_3\right]^{3+} + 2SO_4^{2-} \tag{5.48}$$

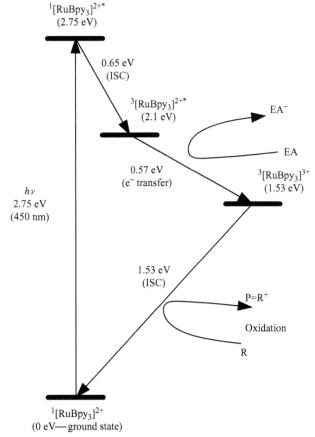

FIG. 5.15 Reductive quenching cycles of ruthenium tris-2,2′-bipyridyne photosensitizer.

The peroxodisulfate ($S_2O_8^{2-}$) is consumed as it transforms in sulfate (SO_4^{2-}). Once excited to the state $\left[Ru(bpy)_3\right]^{3+}$, the photosensitizer is ready to give electrons to a water oxidation process. Christensen et al. (1985) used colloidal RuO_2 as a catalyst as follows

$$4\left[Ru(bpy)_3\right]^{3+} + 2H_2O \xrightarrow{RuO_2} 4\left[Ru(bpy)_3\right]^{2+} + O_2 + 4H^+ \tag{5.49}$$

In the reaction from Eq. (5.47), the photosensitizer takes one electron from the RuO_2 catalyst at each photocatalytic cycle. Then, during four catalytic cycles, it transfers +4e "holes" to water and becomes able to split two water molecules.

Duan et al. (2010) developed an oxygen evolution process with homogeneous catalysis for which they reported a generation rate of 80 $\mu dm^3/h$ of oxygen at a turnover rate of 0.23 s^{-1} and a turnover number (TON) of 550. The system operates in pH 7 solution and uses the following complex as PS: $[Ru(bpy)_2(dcb)]^{2+}$, where dcb = 4,4'-dicarboxylethyl. Once excited by the photonic radiation, the PS releases electrons to the electron acceptor, which in this case was $S_2O_8^{2-}$. Once it donates electrons, the electron acceptor transforms irreversibly, according to $S_2O_8^{2-} \rightarrow 2SO_4^{2-}$, and it remains in the solution. The activated PS delivers electrons to the catalyst $[RuL(pic)_3]$ (pic = 4-picoline, H_2L = 2,6-pyridinecarboxylic acid). When activated, this generates one oxygen molecule per four catalytic cycles. Table 5.2 gives a description of the water oxidation process by Duan et al. (2010).

There are also many available electron acceptors, e.g., N,N-dimethylaniline (DMA). Among the PSs, the most widely used is ruthenium tris-2,2'-bipyridyne, for example $[Ru(bpy)_2(dcb)]^{2+}$, where dcb = 4,4'-dicarboxylethyl. Past experiments with multicomponent photocatalysis systems reported production rates on the order of tens of $\mu mol/h$ obtained in laboratory glassware arrangements.

The multicomponent systems for homogeneous photocatalysis present an intrinsic disadvantage that the molecular components that perform various functions like photosensitization, charge separation, electron transfer and catalysis are not interconnected. Thus the quantum efficiency of the process is reduced, since many processes compete with each other. A recent approach aiming to obtain better performance of water dissociation is to develop supramolecular systems that perform multiple functions. In such systems, the PS is connected to the active center with bridging ligands and charge transfer paths. Thus, once excited by light, the photosensitizer releases an electron that travels through a "designed" path and achieves higher probability at the active center.

Fig. 5.16 illustrates a tested configuration of multicomponent supramolecular systems for water dissociation. The system from Fig. 5.16a shows an approach to water splitting at the molecular level. This system comprises the photosensitizer, electron donor, electron acceptor, oxidation catalyst and reduction catalyst. When PS is excited by light, its LUMO electron is transferred to the electron acceptor via a bridging (molecular) ligand. The electron passes further to the catalytic center where a proton reduction process occurs. The PS, depleted by one electron, relaxes to a ground state by taking an electron from the electron donor, which further takes an electron from the active catalytic center for oxidation. Oxidation of water occurs in generating molecular oxygen and protons. Note that four photonic excitation cycles are needed for a complete photochemical reaction for water splitting.

The system in Fig. 5.16b represents a "triad," comprising three subassemblies: the oxidative catalytic center, the photosensitizer and the electron acceptor. Many systems of this kind were developed in recent years. The aim of such systems is to maintain long-lived charge separation, such that the desired catalytic reaction can occur. These systems can accumulate multiple electrons at the active center by successive photoexcitations. The system in Fig. 5.16c represents a multielectron photoreduction device for water splitting. The system comprises two PSs connected through

TABLE 5.2   Characteristics of a Water Oxidation System with Homogeneous Photocatalysis

Parameter	Description and conditions	Remarks
PS	Bi(2,2'-bipyridil)Ruthenium-II(dcb); $\left[Ru(bpy)_2(dcb)\right]^{2+}$	Concentration 5.0×10^{-4} mol/dm^3
Catalyst	Tris(4-picoline)Ruthenium-L; $[RuL(pic)_3]$	Concentration 5.0×10^{-5} mo/dm^3
Electron acceptor	Dissolved $Na_2S_2O_8$	Concentration 9.0×10^{-3} mol/dm^3
Illumination	Filtered light with 400nm < λ < 780nm @ 0.3 W/cm^2	Power density ~900 W/dm^3
pH adjuster	Phosphate/acetonitrile (10/1.1 per volume)	pH 7.2
Performance	At turnover number 550 and turnover frequency 0.23 s^{-1}	O_2 generation rate 2 $\mu mol/dm^3s$

From Duan, L., Xu, Y., Gorlov, M., Tong, L., Andersson, S., Sun, L., 2010. Chemical and photochemical water oxidation catalyzed by mononuclear ruthenium complexes with a negatively charged tridentate ligand. Chem. Eur. J. 16, 4659–4668.

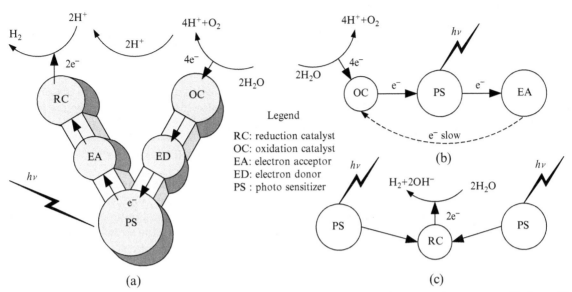

FIG. 5.16 Configurations of multicomponent molecular system for photocatalysis: (a) molecular water-splitting system; (b) photooxidation triad; (c) multielectron photoreduction device.

bridging ligands to the active center. Upon interaction with light, both PSs dislocate electrons that travel toward the catalytic center. The PS is based on ruthenium bipyridyne, while the catalytic center is a rhodium atom. According to the review by Vayssieres (2010), the longest lived system is based on ruthenium-tris-bypiridine as a PS and a catalytic center with two magnesium atoms; its recorded charge separation time has been 0.6 ms.

In a seminal paper, Campagna et al. (1991) reported a similar triad catalyst of the structure Ru—Ru—Ru. Being a triad, the catalyst performs three functions: PS, electron transfer and catalysis. The catalyst structure is similar as that general structure explained previously in relation to Fig. 5.16c. The catalyst is capable of capturing two excited electrons at its active center. This catalyst structure includes two ruthenium centers as PSs and a ruthenium center as electron collector and catalysis active center. The chemical formula of this supramolecular photocatalyst is {Ru[(dpp)Ru (bpy)$_2$]$_2$Cl$_2$}(PF$_6$)$_4$, with dpp = 2,3-bis(2-pyridile) pyrazine, bpy = 2,2′-bipbyridine. The electrons energized by the PSs are sent through the ligands to the electron collection center where the catalysis process occurs.

A slightly modified version of the photocatalyst by Campagna et al. (1991) that also preserves the same way of operation is the triad [Ru(bpy)$_2$(dpp)]$_2$RuCl$_2$(PF$_6$)$_4$, reported in Brauns et al. (1997). Replacing the ruthenium active center in the triad Ru—Ru—Ru with a rhodium atom to obtain a triad such as Ru—Rh—Ru has been pursued by Swavey and Brewer (2002), who developed the photocatalyst [Ru(bpy)$_2$(dpp)]$_2$RhCl$_2$(PF$_6$)$_5$. This photocatalyst, comprehensively studied at Virginia Tech as published in Elvington and Brewer (2006) has been eventually reported in Brewer and Elvington (2006) to evolve hydrogen from water dissolved in an organic solvent.

Further studies report in Rangan et al. (2009) demonstrated that replacement of the chlorine halogen with bromine leads to a similar photocatalyst [Ru(bpy)$_2$(dpp)]$_2$RhBr$_2$(PF$_6$)$_5$ which performs better, generating hydrogen more effectively. The photocatalyst is shown to generate hydrogen from water under light illumination with a wavelength shorter than about 520 nm; but better at 470 nm. The catalyst dissolves in acetonitrile or dimethylformamide (DMF). There is a need of an electron donor in the solution. Typically, a DMA (dimethylaniline) is used as an electron donor source.

White et al. (2012) further developed a variant of photocatalyst reported previously in Rangan et al. (2009) in which the bpy groups are replaced by Ph$_2$phen = 4,7-diphenyl-1,10-phenanthroline. This supramolecule with the structure [Ru(Ph$_2$phen)$_2$(dpp)]$_2$RhBr$_2$(PF$_6$)$_5$ shows very high water splitting performance with a production of 8 ml of hydrogen per hour using DMA when exposed to a monochromatic light source of 470 nm for 5 h.

5.3.2 Case Study: A Supramolecular Photocatalyst

Let us consider here an illustrative example of a supramolecular photocatalysis for hydrogen evolution from water with a homogeneous photocatalysis process. The selected photocatalyst is a metal-organic supramolecule with the chemical structure given as follows: {[Ru(bpy)$_2$ (dpp)]$_2$RhBr$_2$}$^{5+}$. This catalyst has been developed at Virginia Tech as reported in White et al. (2012) and further studied at University of Ontario Institute of Technology as reported in

Zamfirescu et al. (2011, 2012) and Roberts (2013). Photocatalyst structure and system, photocatalyst preparation, molecular charge transfer modeling and hydrogen photocatalyst performance are described.

5.3.2.1 Photocatalyst and Process Description

The photocatalyst $\{[Ru(bpy)_2(dpp)]_2RhBr_2\}^{5+}$ is a supramolecular ion with a charge +5 that is dissolved in the acetonitrile organic solvent in which water and DMA electron donors are also dissolved. The catalyst is prepared in the form $[Ru(bpy)_2(dpp)]_2RhBr_2(PF_6)_5$. When dissolved in acetonitrile the following dissociation occurs:

$$\left[Ru(bpy)_2(dpp)\right]_2RhBr_2(PF_6)_5 \rightarrow \left\{\left[Ru(bpy)_2(dpp)\right]_2RhBr_2\right\}^{5+} + 5PF_6^- \tag{5.50}$$

Being a complex molecule with an octahedral geometry of the central rhodium atom, the coordination chemistry establishes that the supramolecule can have more than one possible configuration, that is, stereoisomers of types *cis* and *trans*. The type of isomer is important, as it affects the absorption spectra, the catalytic cycle and the ability of the catalyst to work effectively in splitting the water molecule. For the *cis* type isomers, the atoms of interest are adjacent to each other on the central atom. In *trans* type isomers, the atoms of interest are opposite each other with the atom between them. Some *cis* isomers of the catalyst were investigated as reported in White et al. (2012). A *trans* isomer is reported in Roberts (2013), who analyzed six possible configuration of the dissolved catalyst and reported that both the *cis* and *trans* variants have two arrangements, one with the V shape of both dpp groups facing the same way as noted with ∧•∧, and one with them reversed, represented with the notation ∧•∨ (Fig. 5.17).

Fig. 5.18 shows a *cis*-∧•∧ configuration of the nonactivated photocatalyst dissolved in acetonitrile. Here, they are calculated and represented with Discovery Studio 3.5 software. The nonactivated photocatalyst will have the two bromine atoms bonded to the molecule. Fig. 5.19 shows a *trans*-∧•∨ configuration of the catalyst. Other nonactivated photocatalyst configurations are possible as follows: *cis*-∧•∨, *trans*-∧•∧ as calculated and represented with Discovery Studio 3.5 software.

The catalyst activation supposes that phonons of sufficient energy will reach the photosensitizing centers of the catalyst. The reach of a photon will displace an electron which moves to the rhodium center. Although there are two PSs, it is unlikely that both are activated at the same time and that two electrons reach the rhodium center at almost the same time. If this happens, the two bromine atoms will be simultaneously released. Once released, the catalyst activates and bromine may remain in the solution as an ionic species or even form a dissolved gas; however, the most probable event appears to be the release of a single bromine, when the catalyst absorbs an energetic photon.

Four configurations were found possible with the help of Discovery Studio software by Roberts (2013). These are again the *cis* and *trans* isomers with ∧•∧ and ∧•∨ arrangements. Fig. 5.20 shows as an example the nonactivated photocatalyst with a bromine atom removed as described by the formula $\{[Ru(bpy)_2(dpp)]_2RhBr\}^{5+}$ of a ∧•∧ configuration, as calculated and presented with Discovery Studio 3.5 software. The valence electron of the catalyst-bromine bond remains with the catalyst MO.

Once a second photon will be absorbed, then the other bromine atom will be removed. The configuration of the molecule will change and the photocatalyst fully activates; however, the configuration of the activated photocatalyst will depend on the actual intermediary configuration of the molecule having a single bromine bonded. Nevertheless, as previously commented in Roberts (2013), two possible planar (*trans*) configurations are stable and most probable, one being of the arrangement ∧•∧ and the other of the arrangement ∧•∨ (*cis* configurations are less probable). Fig. 5.21 shows the activated photocatalyst of configuration *trans*-∧•∧, as calculated and presented with Discovery Studio 3.5 software. The activated catalyst has the general formula $\{[Ru(bpy)_2(dpp)]_2Rh\}^{5+}$ and is obtained according to the following photochemical reaction

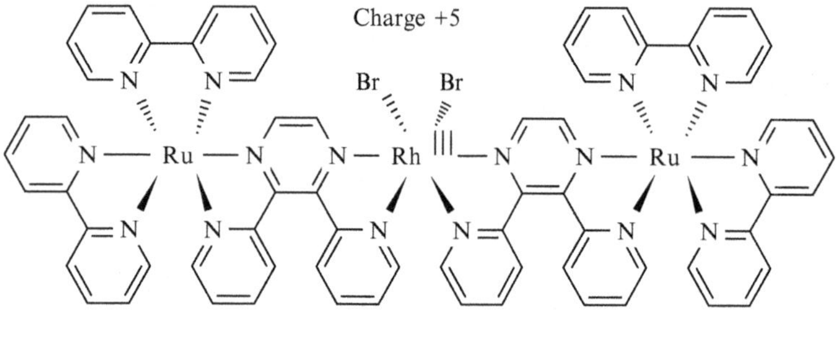

FIG. 5.17 Structure of the photocatalyst supramolecule dissolved in acetonitrile.

FIG. 5.18 Nonactivated photocatalyst of *cis-∧•∧* configuration with a bromine atom removed dissolved in acetonitrile.

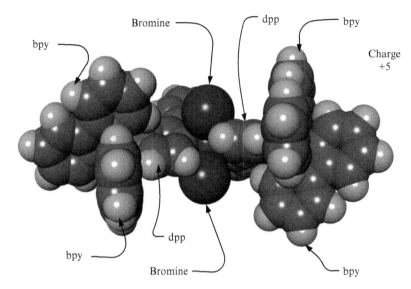

FIG. 5.19 Nonactivated photocatalyst of *trans-∧•∨* configuration dissolved in acetonitrile.

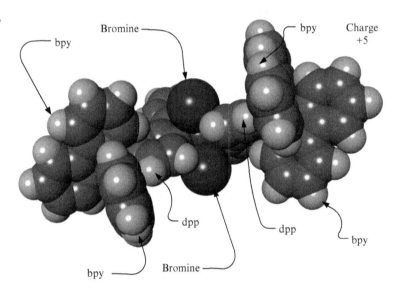

FIG. 5.20 Nonactivated photocatalyst of *trans-∧•∨* configuration with a bromine atom removed dissolved in acetonitrile.

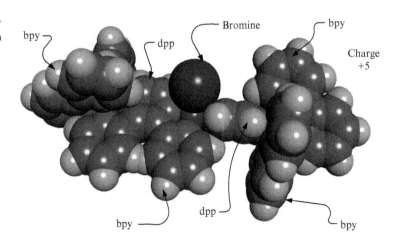

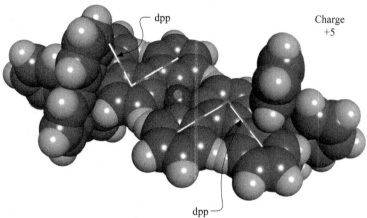

FIG. 5.21   Activated photocatalyst $\left\{\left[\text{Ru}(\text{bpy})_2(\text{dpp})\right]_2\text{Rh}\right\}^{5+}$ of *trans-*∧•∧ configuration dissolved in acetonitrile.

$$\left\{\left[\text{Ru}(\text{bpy})_2(\text{dpp})\right]_2\text{RhBr}_2\right\}^{5+} \overset{h\nu}{\rightarrow} \left\{\left[\text{Ru}(\text{bpy})_2(\text{dpp})\right]_2\text{Rh}\right\}^{5+} + \text{Br}_2(\text{disolved}) \qquad (5.51)$$

The other configuration of the activated catalyst is *trans-*∧•∨ {[Ru(bpy)₂(dpp)]₂Rh}⁵⁺, which is represented in Fig. 5.22. It appears from a comparison of Figs. 5.21 and 5.22 that the planar structure *trans-*∧•∧ confers more room to accommodate water at the active center. This *trans-*∧•∧ molecule has a calculated volume of 991.1 Å³, as calculated with Discovery Studio 3.5. The molecule is represented in detail in Fig. 5.23.

As mentioned in Roberts (2013), the activated catalyst has the ability to bond molecules to the active center. This is a competing mechanism among species dissolved in the solution, with those having a higher bond energy being more active in bonding at the active center. For the discussed system, the bonding energy of involved species are in the following order: bpy > CH₃CN > H₂O > OH⁻ > Br⁻. In this hierarchy, the results show that bromine breaks the easiest with the active center. The activated catalyst can bond water instead. If this happens, the water molecule attaches to the active center as shown in Fig. 5.24. Two water molecules were placed in the vicinity of the catalyst under a 50% steric, 50% electrostatic force assumption and a Dreiding-like field. Their quasistable position is determined by the software (Discovery Studio) in the near vicinity of the rhodium center at about 3 Å distance; the molecules are at the same side of the rhodium center, in enough proximity to each other for the catalytic reaction to occur. The molecules will eventually attach at the two free bonds of the rhodium center at the same side. If they attach at the opposite side of the planar molecular structure, the configuration shown in Fig. 5.25 is formed accordingly, as shown through Discovery Studio 3.5 software.

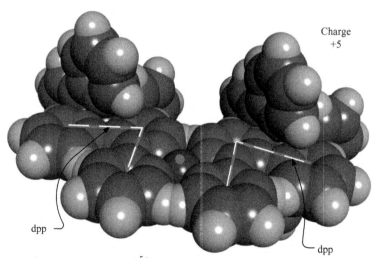

FIG. 5.22   Activated photocatalyst $\left\{\left[\text{Ru}(\text{bpy})_2(\text{dpp})\right]_2\text{Rh}\right\}^{5+}$ of *trans-*∧•∨ configuration dissolved in acetonitrile.

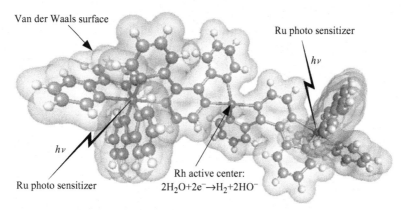

FIG. 5.23 Detailed representation of the activated $\left\{[\text{Ru(bpy)}_2\,(\text{dpp})]_2\text{Rh}\right\}^{5+}$ of *trans-*∧•∧ photocatalyst.

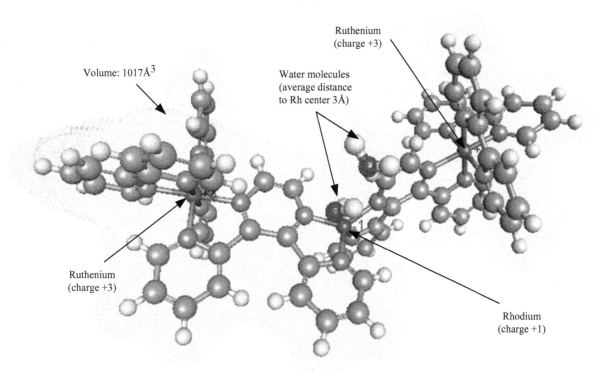

FIG. 5.24 Calculated structure of the catalyst with two water molecules in the vicinity of the active center in acetonitrile.

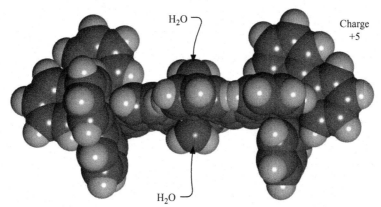

FIG. 5.25 The *trans-*∧•∧ $\left\{[\text{Ru(bpy)}_2(\text{dpp})]_2\text{Rh(H}_2\text{O)}\right\}^{5+}$ supramolecule with bonded water.

Once the catalyst of the configuration reported in Fig. 5.25 (calculated and presented with Discovery Studio 3.5 software) absorbs water, the water-splitting reaction may occur, as the two attached water molecules will receive electrons and release a molecule of hydrogen. It is possible that the hydroxyls remain bonded at the active center or they may be released into the solution. The release of hydroxyl will occur in the excited state similarly with the first release of bromine at the first-time activation of the catalyst. If hydroxyls remain bonded after water dissociation, the configuration shown in Fig. 5.26, as calculated and presented with Discovery Studio 3.5 software, is formed, which assumes that the two free bonds of the catalyst are on the same side. In this case, the catalysis probability is enhanced with respect to having water molecules bonded at opposite sides.

Once a molecule of hydrogen is formed, the hydroxyls will remain attached until the next excitation. The hydroxyls can also be displaced by new water molecules having a stronger bonding energy.

Acetonitrile can displace water from the active center and bond itself to it as it has a higher bond energy. If this happens, the quantum efficiency of hydrogen production is diminished. Experiments reported in White et al. (2012) confirm that if acetonitrile is replaced with another solvent, e.g., DMF, then better hydrogen production efficiency is obtained.

The next phase of the catalytic cycle is the electron donation. The DMA will donate electrons. Fig. 5.26 shows a representation of the quenching process of the catalyst with attached hydroxyls at the active center and the presence of DMA molecules in the vicinity. The catalytic cycle is summarized in Fig. 5.27. In summary, there are four important energy states of the supramolecular device (SMD) photocatalyst during its operation, as follows:

- ^{1}GS—singlet ground state of the SMD
- ^{1}MLCT—singlet excited state of SMD produced by absorption of two photons at Ru centers, according to ^{1}GS + 2$h\nu \rightarrow$ ^{1}MLCT; after the absorption of photons, a charge transfer occurs between ruthenium atoms (metals) and the pyridine-based ligands of the molecule; this transfer process and the associated state is denoted metal-to-ligand charge transfer (MLCT)
- ^{3}MLCT—triplet excited state that occurs through intersystem crossing and associated dissipated energy, according to ^{1}MLCT $\rightarrow$ ^{3}MLCT + ΔE_{IC}^{diss}

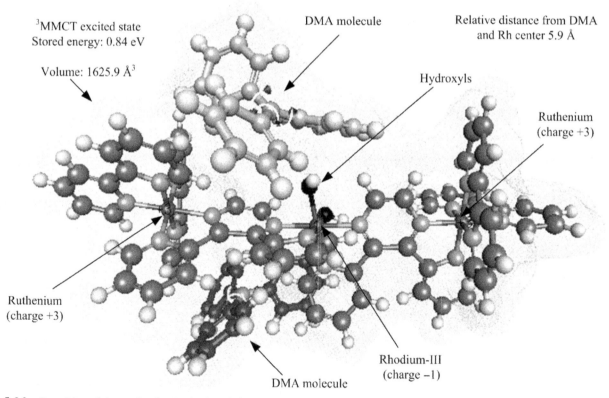

FIG. 5.26 Quenching of the catalyst having hydroxyls bonded at the active center with DMA in acetonitrile.

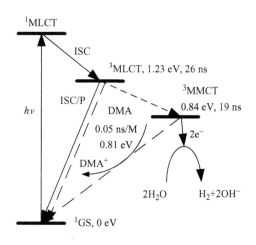

Molecular device properties:

$C_{68}H_{52}N_{16}Ru_2Rh$

C:0.584, H:0.037, N:0.16, Rh:0.074, Ru: 0.145

1399.17 kg/kmol, Charge +5

Van der Waals volume: 982.08 Å^3

Van der Waals area: 1208.3 Å^2

λ_{max}: abs=520 nm, em=760 nm

Note: eV given versus standard calomel electrode

Notations:

1GS: ground state, singlet

1MLCT: singlet metal-to-ligand charge transfer

ISC/P: intersystem crossing/phosphorescence

^{3}MLCT: triplet metal-to-ligand charge transfer

^{3}MMCT: triplet metal-to-metal charge transfer

DMA: *N,N*-dimethylaniline

FIG. 5.27 State diagram for the hydrogen-evolving photocatalytic process.

- ^{3}MMCT—triplet excited state that occurs by ligand-to-metal charge transfer according to ^{3}MLCT $\rightarrow$ ^{3}MMCT + ΔE_{ET}^{diss}; the energy is dissipated during this process due to irreversible intrasystem charge transfer that occurs during this transformation; the overall process of metal-to-ligand and ligand-to-metal electron transfer is denoted as metal-to-metal charge transfer (MMCT).

5.3.2.2 Catalyst Preparation

A supramolecular catalyst must be prepared based on a building-block approach. The preparation will start with the precursors available as off-the-shelf chemicals. Serroni et al. (2002) established a building-block procedure applied to this class of supramolecular catalysts synthesis. Roberts (2013) at the University of Ontario Institute of Technology applied this method to produce the [Ru(bpy)$_2$(dpp)]$_2$RhBr$_2$(PF$_6$)$_5$ photocatalyst. The process is done in three stages.

In stage one, 153.1 mg of the off-the-shelf catalyst precursor [(bpy)$_2$Ru(dpp)](PF$_6$)$_2$ is added in a 45 ml de-aerated solution of 2:1 (v/v) of ethanol-water created from 30 ml 95% ethanol and 15 ml distilled water. Further, 30.6 mg of rhodium bromide dihydrate (RuBr$_3$·H$_2$O) is added to the solution, forming a molar ratio of 1:2 with the precursor [(bpy)$_2$Ru(dpp)](PF$_6$)$_2$. The solution is heated at reflux at 95°C for 1 h and then cooled at room temperature.

Stages two and three are intended to separate and extract the photocatalyst from the reaction mixture. The building-block approach of catalyst construction is illustrated as shown in Fig. 5.28. In stage two, the solution containing the catalyst obtained in stage one is added drop by drop to a concentrated solution of potassium hexafluorophosphate (KPF$_6$). This concentrated solution is obtained by dissolving KPF$_6$ in 100 ml water. The drop-by-drop addition induces precipitation. Vacuum filtration is applied to separate the dark red precipitate from the solution. The precipitate is washed with diethyl ether and then dried using vacuum desiccation.

FIG. 5.28 Building the photocatalyst from two precursory blocks $\left[(bpy)_2Ru(dpp)\right]^{2+}$ and RuBr$_3$. *Modified from Roberts, R.R., 2013. Experimental Investigation of a Catalyst under Various Solar Light Conditions for Hydrogen Production, MSc Thesis. University of Ontario Institute of Technology.*

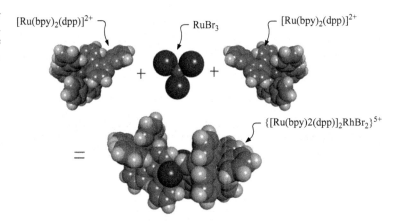

In stage three, the filtered deep red precipitate is dissolved in a minimum volume of acetonitrile to obtain a concentrated solution. This solution is further rotary-evaporated to dryness. The resulting product is again dissolved in acetonitrile and added to diethyl ether to induce precipitation again. Next, vacuum filtration is applied to extract the dark precipitate. The extracted precipitate is then dried using the vacuum desiccation to form a deep red solid.

5.3.2.3 Test Reactor for Homogeneous Photocatalysis Research

Photocatalysis experiments of hydrogen evolution require illumination of the solution under controlled conditions and while metering the gas emanation. The typical research in this area is done in closed flask under constant volume conditions. The gas evolution will increase the pressure above the solution. Once pressure is metered, the number of moles generated in the gas phase can be deduced. Of course, the batch reaction can be tested for a limited time as the reaction is slowed down due to pressure build into the gas space.

Another approach is to conduct a photocatalysis experiment in a semi-batch reactor, able to extract and meter continuously the gaseous emanation and thence to conduct the reaction at constant pressure. In both types of approaches, constant volume or constant pressure, the solution must be mixed to stimulate the reaction and gas separation.

A photocatalysis reactor that can operate both at constant volume and constant pressure has been reported in Zamfirescu et al. (2011), which has been developed at the University of Ontario Institute of Technology to study the previously discussed reaction. The custom-made photoreactor consists of a 110 ml cylindrical borosilicate glass that is 2.8 mm thick. As indicated in Fig. 5.29, the reactor vessel includes a magnetic bar for stirring and is cooled with air via forced convection. The illumination is made with 20 blue LEDs of 470 nm placed uniformly around the reactor in four vertical rows of five LEDs. A photograph of the photocatalysis reactor is shown in Fig. 5.30.

The produced gas is collected from the upper side and drawn automatically in a cylinder with a piston actuated by a servomotor. A pressure sensor is installed, through which the displacement of the piston is controlled, such that the pressure remains constant. A digital caliper is fixed to the piston rod to monitor its movement (with <0.01 mm accuracy) and thus determine the stroke and the collected gas volume.

The LEDs are blue, each emitting at 470 nm peak band with a bandwidth of 25 nm and minimum/maximum wavelengths of 460/490 nm, respectively. There are in total 20 LEDs consuming an electric power of around 100 W and producing light at 700 lm (lumen). The LEDs are placed in four banks symmetrically arranged around the glass cylinder, each having five LEDs. The distance between the light source spot and the glass surface is ~29 mm. Reflectors with a 25-degree view angle are used for each LED. The distance between two LEDs on a bank is ~22.2 mm.

All systems are made with parts that are easy to dismantle and clean. The glass is effectively sealed with EPDM (ethylene-propylene-diene monomer) rubber rings. The system allows flexibility in experiments. At least three ways to conduct hydrogen evolution tests are envisaged as follows:

- water photolysis—light is applied and the evolved is captured in the cylinder at constant pressure
- water electrolysis—electricity is applied under the absence of light and the composition and quantity of the resulting gas (hydrogen/oxygen mixture) is measured
- hybrid process of water electrolysis and photolysis—both electricity and light are applied

5.3.2.4 Modeling of the Photocatalytic Process

The photocatalytic process can be modeled by accounting for charge transfer processes and photochemical processes that affect the overall quantum efficiency. Here, the past work of Zamfirescu et al. (2012) in which the molecular charge transfer and quantum efficiency of the homogeneous supramolecular catalyst (SMC) $\{[Ru(bpy)_2(dpp)]_2RhBr_2\}^{5+}$ is briefly reviewed.

A series of processes compete after the initial photon absorption to influence the reaction kinetics. One first process is the absorption of a photon. The process duration of photon absorption can be approximated by its wavelength divided by the speed of light

$$\tau_{^1MLCT} \cong \lambda/c \tag{5.52}$$

For a photon with a wavelength of 470 nm, the absorption duration is about 2.5×10^{-15} s; this is the time of ^1MLCT formation. After absorption, a process of vibrational relaxation occurs, followed by intersystem crossing and formation of the triplet state ^3MLCT. Zamfirescu et al. (2012) approximated the time constant for the formation of the triplet MLCT state as

$$\tau_{^3MLCT} \cong 10^{0.002\lambda - 7} \tag{5.53}$$

where λ is given in nm; for a photon with 470 nm, a singlet-triplet process occurs in 8.7×10^{-7} s.

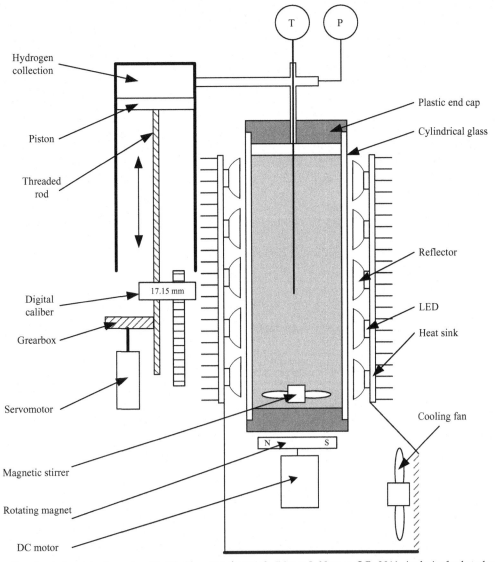

FIG. 5.29 Schematics of a photocatalysis reactor. *Modified from Zamfirescu, C., Dinçer, I., Naterer, G.F., 2011. Analysis of a photochemical water splitting reactor with supramolecular catalysts and a proton exchange membrane. Int. J. Hydrog. Energy 36, 11273–11281.*

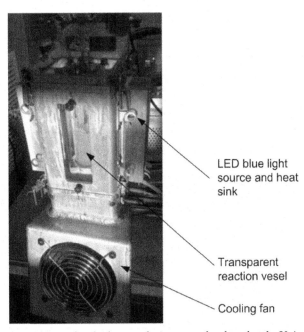

FIG. 5.30 Photograph of the constant pressure semi-batch photocatalysis reactor developed at the University of Ontario Institute of Technology.

The formation of the ^{3}MMCT state occurs by electron transfer from ruthenium to rhodium metals; the duration of this process is estimated to $\tau_{MMCT} \cong 0.192 \times 10^{-7}$ s. Competitively with the formation of the ^{3}MMCT, the ^{3}MLCT can relax to a ground state via radiative (phosphorescence) and nonradiative (vibrational) processes. The lifetime of the ^{3}MLCT state after which it relaxes to the ground state (GS) is estimated to $\tau_{GS,1} = 26$ ns. The ^{3}MMCT can relax to a ground state also via radiative and nonradiative processes. The time constant of the ^{3}MMCT to ground state via triplet-singlet intersystem crossing and phosphorescence is estimated to $\tau_{GS,2} \cong 19$ ns.

These previously mentioned time constants can be further used to estimate the quantum efficiencies of relevant subprocesses. The quantum efficiency of catalyst activation via ^{3}MLCT quenching is given as

$$\Phi_{SMC}^{MLCT} = \frac{k_{d,ED}}{\sum k_{d,i} + k_{GS,1} + k_{MMCT}} \tag{5.54}$$

where $k_{d,ED}$ is the rate of diffusion of the electron donor and $\sum k_{d,i}$ is the rate of diffusion of all relevant species present in the solution.

The quantum efficiency of ^{3}MMCT formation is determined similarly as

$$\Phi_{MMCT} = \frac{k_{MMCT}}{\sum k_{d,i} + k_{GS,1} + k_{MMCT}} \tag{5.55}$$

The quantum efficiency of catalyst activation via ^{3}MMCT quenching is given as

$$\Phi_{SMC}^{MMCT} = \frac{k_{d,ED}}{\sum k_{d,i} + k_{GS,2}} \tag{5.56}$$

The catalyst can be activated via two concurrent routes, both being initiated at the ^{3}MLCT state; they are direct quenching, or indirect quenching via the ^{3}MMCT. The overall quantum efficiency of catalyst activation (considering both routes) is therefore

$$\Phi_{SMC} = \Phi_{SMC}^{MLCT} + \Phi_{MMCT}\Phi_{SMC}^{MMCT} \tag{5.57}$$

Once the catalyst is activated, water reduction can occur according to the following reaction during which the photocatalyst is regenerated: $^3CAT + 2H_2O \rightarrow H_2 + \,^1GS$. This reaction is diffusion-controlled, because it requires the diffusive encounter of the photocatalyst and two water molecules. Water diffusion to the active center competes with diffusion of other molecules present in the solution (including the solvent). The quantum efficiency of the catalytic water reduction process becomes

$$\Phi_{RED} = \frac{k_{d,H_2O}}{\sum k_{d,i} + 1/\tau_{GS}} \tag{5.58}$$

where $k_{d,i}$ is the rate of diffusion of species i determined as

$$k_{d,i} = \frac{8}{3}\left(\frac{RT}{\mu}\right)C_i \tag{5.59}$$

where R is the universal gas constant, T is temperature in Kelvin, μ is the dynamic viscosity of the solvent and C_i is the molar concentration of the species i.

Finally, the quantum efficiency of hydrogen production is calculated by

$$\Phi_{H_2} = \Phi_{CAT}\Phi_{RED} \tag{5.60}$$

The knowledge of quantum efficiencies of all relevant subprocesses determines an energy-time diagram of the overall process. If several events compete at a certain moment, and each event can occur with a reaction rate k_j, then the time constant of the process of interest "i" is

$$\tau_i = \left(\Phi_i \sum_j \frac{1}{k_j}\right)^{-1} \tag{5.61}$$

where Φ_i is the quantum efficiency of the process "i."

FIG. 5.31 Energy-time diagram of the photocatalytic process with SMD.

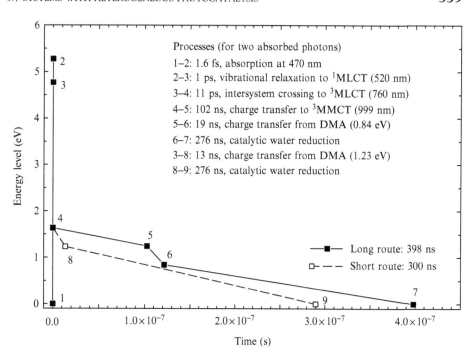

The predicted energy-time diagram of the photocatalytic process, as determined in Zamfirescu et al. (2012) based on Eqs. (5.51)–(5.60) is shown in Fig. 5.31. The energy of the SMC is represented in the diagram with reference to the ground state energy. One clearly observes in this diagram the two parallel photocatalytic routes initiated at the ^{3}MLCT state. The short route duration is 300 ns and its occurrence probability (given by quantum efficiency) is 7.1%. The long route takes 398 ns to form a molecule of hydrogen with a probability of 6.13%. Thus the weighted average duration of the catalytic cycle is 340 ns.

5.4 SYSTEMS WITH HETEROGENEOUS PHOTOCATALYSIS

5.4.1 Heterogeneous Photocatalysis Processes

In heterogeneous catalysis process, the photocatalyst is in typically solid phase and the water dissociation occurs in the liquid phase. Other substances that can derive hydrogen can be used instead of water, but the heterogeneous photocatalyst will be in a different phase than the reaction mixture. In this section, particulate heterogeneous photocatalysts made of semiconductor materials are discussed. One of the reasons to use semiconductors is their small bandgap between the valence and CBs.

Both photosensitization and photocatalytic behavior are observed with some semiconductors (like TiO_2). With the incident of light, the catalytic sites for water-splitting reaction are activated, photons are absorbed by semiconductors, which dislocate electrons from the valance bond and move them to CB. As a result, redox reactions are aided by the photocatalyst. It is important to note that the photocatalyst is not consumed in the reaction, but it is used to reduce the activation energy.

Fig. 5.32 illustrates the solar light-driven, water-splitting process on heterogeneous photocatalysts schematically. As can be seen from Fig. 5.32, the CB is separated from the valence band (VB) by a bandgap. The energy of the irradiated light must be larger than the bandgap in order to generate electrons and holes in the CBs and VBs, respectively.

These electrons and holes cause redox reactions similar to electrolysis: electrons reduce the water molecules to form hydrogen and holes oxidize the water molecules to form oxygen. Therefore, width of the bandgap and energy levels of the CBs and VBs strongly affect the performance of a photocatalyst. Since the bottom level of the photocatalyst has to be more negative than the redox potential of H^+/H_2 and the top level of the VB has to be more positive than the redox potential of O_2/H_2O, the bandgap should be wider than 1.23 eV. Furthermore, the photocatalyst must be stable in aqueous solutions under photoirradiation.

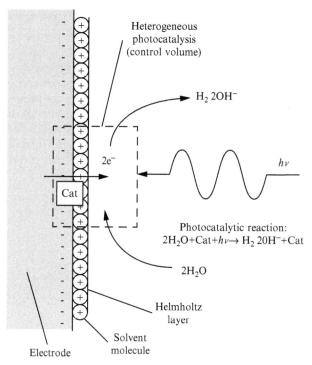

FIG. 5.32   Heterogeneous photocatalysis mechanism for water reduction and oxidation.

Novel semiconductor materials to develop efficient heterogeneous photocatalysts have been extensively studied in the literature. One of the challenges of the photocatalytic water splitting is preventing backward reactions. One important issue is to keep the photogenerated electrons and holes separated. Crystallinity and surface properties of photocatalysts strongly affect the electron-hole generation and separation processes.

In general, photocatalysts with high crystallinity have higher photocatalytic activities. Photocatalysts with less surface defects (increased crystallinity) could prevent electron-hole recombination sites, and this can increase the lifetime and mobility. Higher photocatalytic activity can also be obtained by reducing the particle size of a photocatalyst, because the diffusion length for photogenerated electron-hole pairs can be shortened. Also, the surface area of the photocatalyst grain strongly affects the number of active sites.

Oxide-based semiconductors are extensively used as heterogeneous photocatalysts since they are reported to be stable against photocorrosion. It is quite challenging to develop an oxide semiconductor photocatalyst that has both a sufficiently negative CB for hydrogen production and a sufficiently narrow bandgap (ie, <3.0 eV) for visible light absorption, because of the highly positive VB (at ca. +3.0 V vs. NHE) formed by the O 2p orbital.

Most visible-light-responsive oxide photocatalysts cannot produce hydrogen from water due to their CBs being too low for water reduction. Among the photocatalytic systems that have been reported to be active for overall water splitting (simultaneous generation of O_2 and H_2), most of them only absorb ultraviolet light ($\lambda \ll 400$ nm), which occupies only 3–5% of solar energy, causing a low efficiency of energy conversion.

Although some nonoxide (e.g., sulfide-based) semiconductors possess appropriate band levels for water splitting under visible light, they are generally unstable and readily become deactivated through photocorrosion or self-oxidation, rather than evolving O_2. For example, cadmium sulfide (CdS) has appropriate band levels for water reduction and oxidation as well as a narrow bandgap that permits visible light absorption; however, CdS is not stable in its water oxidation reaction to form O_2, because the S^{2-} anion is more susceptible to oxidation than water, causing the CdS catalyst itself to be oxidized and degraded.

Because of its advantages like stability, noncorrosiveness, environmentally friendly nature, and cost-effectiveness, Titania (titanium dioxide, TiO_2) has been a widely used photocatalyst for photocatalytic water-splitting reactions. One of the most important advantages of TiO_2 is its appropriate energy levels to initiate the water-splitting reaction: the CB of TiO_2 is more negative than the reduction energy level of water $\left(E_{H^+/H_2} = 0 \text{ V}\right)$, while the VB is more positive than the oxidation energy level of water $\left(E_{O_2/H_2O} = +1.23 \text{ V}\right)$.

Despite its many advantages, however, TiO_2 has a quite low photocatalytic water-splitting efficiency under solar light. One of the reasons of this low efficiency is that the photogenerated electrons in the CB of TiO_2 may recombine with the VB holes quickly to release energy in the form of unproductive heat or photons.

Another reason is the large positive Gibbs free energy ($\Delta G = 237$ kJ/mol) of the decomposition of water into H_2 and O_2, which makes the backward reaction (recombination of H_2 and O_2 into water) easy to proceed. Last, but not least, the bandgap of TiO_2 is about 3.2 eV, and as a result, only UV light can be utilized to activate the photocatalyst. Since, as mentioned above, UV light only accounts for ~4% of solar energy, the inability to utilize most of the solar light limits the efficiency of TiO_2 in solar photocatalytic hydrogen production.

Many tantalates have been reported to be highly active photocatalysts since the mid-1990s; however, many tantalates have wide bandgaps of 4.0–4.5 eV and they generally only have relatively high activity for overall water splitting under UV irradiation. WO_3 functions as a stable photocatalyst for O_2 evolution under visible light in the presence of an appropriate electron acceptor; the bottom of the CB of the material is higher than the potential for water reduction. As a result, WO_3 cannot reduce H^+ to H_2.

In the first phase of a heterogeneous photocatalysis process, a photon is absorbed, transferring energy to the semi-conductor. This energy is greater than the bandgap energy of the photocatalyst. As a result, photoexcited electron-hole pairs are formed as a second phase of the overall process. These electron-hole pairs migrate to the surface without recombination. Then, in a third phase of photocatalysis, electrons and holes are used to produce H_2 and O_2, respectively. Although the first two steps strongly depend on structural and electronic properties of the photocatalyst, the third step usually requires the use of a cocatalyst.

Usually, cocatalysts are noble metals (e.g., Pt, Rh) or transition-metal oxides (e.g., NiO_x, RuO_2). They are loaded onto the base catalyst as a dispersion of nanoparticles (typically <50 nm in size) to (i) extract photogenerated electrons and holes from the photocatalyst; (ii) generate and/or introduce active sites; and (iii) reduce the activation energy for product gas evolution. Consequently, the overall efficiency of a photocatalytic system depends on the type/quality of the loaded cocatalyst.

To date, Pt, Pd, Ru, Rh, Cu, Au, and Ni have been applied as a cocatalyst for photocatalytic overall water splitting. Among them, Pt is known to be a promising cocatalyst for the reduction of protons to produce hydrogen molecules; however, a Pt cocatalyst does not necessarily deliver the highest photocatalytic activity among other alternatives (e.g., Ru and Rh).

The efficiency of a photocatalytic system is affected by several factors, such as band-gap potential and working mechanism of the photocatalyst and the loaded cocatalyst. When designing an efficient cocatalyst for a photocatalytic system, the processes ((i), (ii), and (iii)) should be taken into account. The structural characteristics and essential catalytic properties of a cocatalyst for H_2 (or O_2) evolution are important factors affecting the overall photocatalytic system's performance.

Another important requirement of the cocatalyst is to be inactive for water formation from H_2 and O_2. In order to avoid the backward reaction, transition-metal oxides that do not exhibit activity for water formation from H_2 and O_2 are usually applied as cocatalysts for overall water splitting. Naturally, there are some exceptional cases in which noble metals have been effectively utilized as cocatalysts for photocatalytic water splitting.

Despite the fact that some photocatalysts have been reported to exhibit high activity without the presence of a cocatalyst, a cocatalyst usually enhances the overall efficiency of the reaction. Thus, developing a cocatalyst that efficiently promotes photocatalytic water splitting is an important research focus. Pt cocatalysts are widely used in the literature due to their shown catalytic activities for hydrogen production under solar light like irradiation.

The uphill nature of the photocatalytic reaction generally makes the overall water-splitting reaction difficult to achieve. Especially due to rapid recombination of photogenerated CB electrons and VB holes, it is difficult to achieve water splitting for hydrogen production using photocatalysts in pure water. Therefore, photocatalytic activities of a compound for water reduction or oxidation reactions are usually studied in the presence of electron donors, electrolytes- (e.g., methanol, Na_2SO_3, Na_2S, KI, etc.). It should be noted that electrolytes do not undergo reduction or oxidation by CB electrons and VB holes like a sacrificial reagent would; however, when added to the reaction solution, they have a significant effect on the performance of photocatalytic water-splitting reaction.

When the photocatalytic reaction is conducted in the presence of an electron donor, photogenerated holes in the VB irreversibly oxidize the electron donor instead of H_2O, thus facilitating water reduction by CB electrons if the bottom of the CB of the photocatalyst is located above the water reduction potential. On the contrary, in the presence of an electron acceptor, photogenerated electrons in the CB irreversibly reduce electron acceptors instead of H^+. As a result, water oxidation by VB holes is promoted if the top of the VB of the photocatalyst is more positive than the water oxidation potential. It is important to note that the ability of a photocatalyst to both reduce and oxidize water separately does not guarantee the capability to achieve overall water splitting without sacrificial reagents.

Adding electron donors or sacrificial reagents to react with the photogenerated VB holes is an effective measure to enhance the electron-hole separation, resulting in higher quantum efficiency; however, the drawback of this technique is the need to continuously add electron donors in order to sustain the reaction, since they will be consumed during photocatalytic reaction. An attractive solution to this issue is to identify waste materials from industry that can be recovered and used as electron donors in photocatalytic systems. One option is to use aqueous wastes containing sulfide/sulfite that can well act as electron donors. Many sulfurous waste streams can be recovered from petroleum and chemical industries. Other types of pollutant streams can be used for the same purpose.

Here, we exemplify the heterogeneous photocatalysis process with a sulfite/sulfide-based photocatalytic system that is very common, as reviewed in Zamfirescu et al. (2013). Fig. 5.33 shows one probable process pathway for heterogeneous photocatalysis.

Consider a control volume around a catalyst particle. Once a photon is absorbed, an excited state is formed, provided that the photon energy is higher than the bandgap. Then an electron will be displaced from the VB to the CB and catalytic and hole-scavenging processes can both be triggered. Because formation of a hydrogen molecule is a two-electron process, two subsequent catalytic half-cycles of one-electron transfer involving hydronium

$$H_3O^+ + e^- \rightarrow H\cdot + H_2O \tag{5.62}$$

are required for the complete process.

One can describe the photocatalytic process in a simplified manner with the help of Fig. 5.33, as follows

- Two light quanta of enough energy displace two electrons in CB
- The electrons contribute to the catalytic process to eventually generate hydrogen from water, according to the reaction: $2H_2O + 2e^- \rightarrow H_2 + 2OH^-$
- A hole-scavenging process that reactivates the catalyst takes place by transferring two electrons from a sulfide ion, according to the sulfur generation reaction: $S^{2-} \rightarrow S + 2e^-$
- The presence of sulfite in a solution and of sulfur make probable the following spontaneous reaction that generate thiosulfate $S + SO_3^{2-} \rightarrow S_2O_3^{2-}$. This is in fact a desired process for avoiding sulfur deposition to the catalysts, a fact that may cause inactivation.

Note that the overall reaction for the photocatalytic system described in Fig. 5.33 and denoted here as the "main process route" occurs according to the following reaction

$$2H_2O + S^{2-} + SO_3^{2-} \xrightarrow{2h\nu} H_2 + 2OH^- + S_2O_3^{2-} \tag{5.63}$$

Fig. 5.34 presents the alternative route of the photochemical process that becomes possible during the system operation when the concentration of hydroxyls in the electrolyte is high enough. In this case, the process occurs as follows

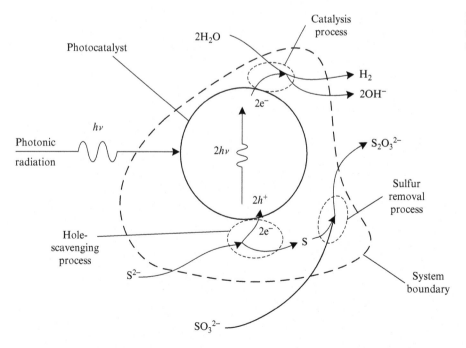

FIG. 5.33 Representation of a heterogeneous photocatalysis process with sulfite/sulfide.

FIG. 5.34 Representation of an alternative photocatalysis process with sulfite/sulfide.

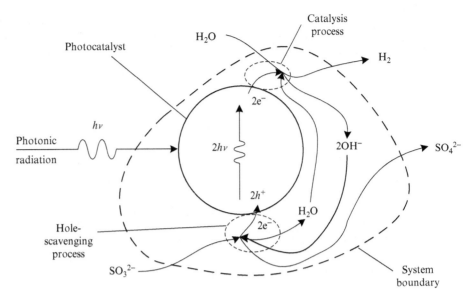

- Two light quanta of enough energy to displace two electrons in CB
- The electrons contribute to the catalytic process to eventually generate hydrogen from water according to the reaction $2H_2O + 2e^- \rightarrow H_2 + 2OH^-$
- The hole-scavenging process is triggered by the presence of hydroxyls in the vicinity of the catalysts, which makes possible the following reaction to occur: $2OH^- + SO_3^{2-} \rightarrow H_2O + SO_4^{2-} + 2e^-$. Occurrence of this reaction is beneficial for preventing the photocorrosion of the catalyst and hence its passivation.

The overall process of the alternative route is described thus by the following chemical reaction

$$H_2O + SO_3^{2-} \overset{2h\nu}{\rightarrow} H_2 + SO_4^{2-} \tag{5.64}$$

Other concurrent reactions exist, such as the formation of disulfide from photocatalytically produced sulfur and sulfide ($S + S^{2-} \rightarrow S_2^{2-}$) and thiosulfate formation from disulfide and sulfite ($S_2^{2-} + SO_3^{2-} \rightarrow S_2O_3^{2-} + S^{2-}$). These reactions are favored by the presence of sulfite in the electrolyte and are hence beneficial due to the fact that impeding the deposition of sulfur helps impede accumulation of disulfide, which degrades the transparency of electrolyte (changing the color to dark yellow).

The concurrent reactions do not change the overall process described by the main and alternative route as mentioned above; however, the actual photocatalysis process can be represented by a combination of main and alternative routes and depends on pH conditions and the concentration ratio of sulfite and sulfide ions in the electrolyte. In an actual reactor with continuous and steady operation that maintains the pH, the oxidation rate of hole scavengers S^{2-} and SO_3^{2-} tend to become equal. Hence, it is possible to set the reactor operational parameter such that the rates of main and alternative photocatalytic routes are equal. Therefore, in this case, the overall catalytic process can be described in an approximate manner by the following overall reaction

$$3H_2O + S^{2-} + 2SO_3^{2-} \overset{4h\nu}{\rightarrow} 2H_2 + 2OH^- + S_2O_3^{2-} + SO_4^{2-} \tag{5.65}$$

5.4.2 Heterogeneous Photocatalysts

Photocatalysts can be assessed based on some specific parameters. The TON of the catalyst quantifies its ability to act upon more than one set of reactant molecules. It is defined as the number of reactants that a catalyst acts upon before it degrades. The *turnover frequency* (TOF) is another important parameter that represents the number of reactants that a photocatalyst can convert per unit of time; thus, it is measured in s^{-1}.

An alternative way of define the TOF is as the rate of hydrogen formation expressed as the molar rate of hydrogen production per mass unit of catalyst, denoted with $\dot{n}_m$ and measured in $\mu mol/h\text{-}g_{cat}$ (micromoles of hydrogen production per hour per gram catalyst).

A parameter somehow similar to the TOF is the mass specific molar rate of hydrogen production (μmol/h-g_{cat}), a second criterion is used to evaluate the hydrogen production rates of selected photocatalysts; this is the area-specific molar hydrogen production rate.

The unit is μmol/h-m_{cat}^2, which represents micromoles of hydrogen production per hour per unit for a specific area (m^2) of the photocatalyst. This amount is calculated by dividing the μmol/h-g_{cat} data mentioned above by Brunauer-Emmett-Teller (BET) surface area (m^2/g_{cat}). The BET surface area is defined in association with the sorption isotherm introduced by Brunauer et al. (1938), also known as the BET isotherm. This isotherm provides an effective way to predict the local adsorption rate with respect to other water-binding mechanisms.

BET analysis provides precise specific surface area evaluation of materials by nitrogen multilayer adsorption measured as a function of relative pressure using a fully automated analyzer. The technique encompasses external area and pore area evaluations to determine the total specific surface area in m^2/g, yielding important information in studying the effects of surface porosity and particle size in many applications.

The photocatalytic performance of a photocatalyst strongly depends on its electronic band structure and band-gap energy, E_g. For an efficient photocatalyst, the band-gap energy should be smaller than 3 eV to extend the light absorption into visible region to efficiently utilize the solar energy. Apart from the maximum band-gap requirement, the minimum bandgap of semiconductor photocatalysts for water splitting should be 1.23 eV, and effective photocatalysts have been shown to exhibit bandgaps larger than 2 eV.

Table 5.3 gives a number of heterogeneous photocatalysis systems as compiled by Acar et al. (2013). The photocatalytic systems are given by specifying the chemical structure of the photocatalyst, the required cocatalyst and the electrolyte. In the table, the energy gap of the photocatalyst is given, where available.

Fig. 5.35 shows a comparison of the TOF of some relevant heterogeneous photocatalysts reviewed in the previous work by Acar et al. (2013). The turnover frequency is expressed in terms of μmol hydrogen produced per hour and grams of catalyst. They show that TiO_2-C-362 has the highest performance in terms of μmol/h-g_{cat}. while Au-CdS has the second highest hydrogen production rate per grams of photocatalyst. Following these two, In(OH)$_y$S:Ag-Zn, GO-TiO_2, $K_4Nb_6O_{17}$/CdS and CdS/Ta_2O_5 have similar μmol/h-g_{cat} performances.

The hydrogen production rates per specific area performances of the selected photocatalysts are slightly different than that of a per-gram catalyst. In the latter case, the highest rates are observed with g-C_3N_4-$SrTiO_3$:Rh, $Ag_{0.03}Mn_{0.40}Cd_{0.60}S$ and $(CuAg)_{0.15}In_{0.3}Zn_{1.4}S_2$, as presented in Fig. 5.36.

Considering technical, cost, health and environmental impact, and the performance criteria of representative catalysts from literature, Acar et al. (2013) identified a list of the top-five catalysts. Those are given and compared as shown in Fig. 5.37. The ideal case is represented as the outer shell. Closer rankings to this ideal case represent a better performing photocatalyst. The trade-off between technical performance and cost and environmental-health impact can be seen in Fig. 5.37 as well. Any potential improvement on photocatalytic performance (e.g., via band-gap engineering) while keeping the cost and environmental-health impact minimal could be a major accomplishment in the photocatalyst development phase of hydrogen production.

5.4.3 Case Study: An Engineered Heterogeneous Photocatalyst

In this case study, a photochemical system uses an engineered photocatalyst of solid-solution type with general formula $Cd_{1-x}Zn_xS$ ($x = 0.21$), which is presented based on a previous study by Zamfirescu et al. (2013). The system uses sulfur ions as electron donors and generates hydrogen from water under illumination with photons of wavelengths shorter than 500 nm. The considered system functions as previously explained in relation to Figs. 5.33 and 5.34.

5.4.3.1 Heterogeneous Photocatalyst Development

A photocatalyst of solid-solution type has been prepared by sequential precipitation of Cd^{2+} and Zn^{2+} ions with sulfide (sourced from Na_2S) in an aqueous solution, followed by hydrothermal crystallization for 72 h at 200°C. After cooling, the photocatalyst sample has been filtered and washed with distilled water and dried under a vacuum at 80°C for 3 h. The general formula of the catalyst is $Cd_{1-x}Zn_xS$. Furthermore, a palladium cocatalyst has been deposited by using $PdCl_2$ as a palladium source. The added cocatalyst has been an amount of 0.2% by weight PdS.

The VB potential of PdS is approximately +1.34 V versus NHE, which is less positive than that of CdS (+1.5 V). This implies that the holes transfer from $Cd_{0.79}Zn_{0.21}S$ to PdS is a favorable process, and that the PdS could act as an oxidation cocatalyst. Platinum is commonly used as a reduction cocatalyst. Hydrogen-producing photocatalytic systems with PdS/CdS-based catalysts and aqueous electrolyte with sulfite only or with sulfite/sulfide electron donors do not necessarily require platinum cocatalyst because during the photocatalysis process, a metallic palladium is formed on the surface of the CdS particle.

TABLE 5.3 Heterogeneous Photocatalysis for Hydrogen Production

Photocatalyst	Co	Electrolyte	E_g (eV)
$(Ru/SrTiO_3:Rh)-(BiVO_4)-FeCl_3$	None	$K_2SO_3 + Na_2S$	N/a
$F-TiO_2$	Pt	BPA	N/a
$ZnIn_2S_4$	Pt	$Na_2SO_3 + Na_2S$	2.59
$Bi_{1.5}Zn_{0.99}Cu_{0.01}Ta_{1.5}O_7$	None	$Na_2SO_3 + Na_2S$	2.6
$Ag_{0.03}Mn_{0.40}Cd_{0.60}S$	Pt	$Na_2SO_3 + Na_2S$	2.59
$CaTa_{0.8}Zr_{0.2}O_{2.2}N_{0.8}$	Pt	HCOOH	2.59
$Cd_{0.8}Zn_{0.2}S/S_{15}$	None	$Na_2SO_3 + Na_2S$	2.23
$K_2La_2Ti_3O_{10}$	CdS	Na_2S	3.5
$K_2Ti_4O_9$	CdS	$Na_2SO_3 + Na_2S + KOH$	2.4
$SrTiO_3:Ni/Ta$	Pt	CH_3OH	N/a
$0.93\ TiO_2$-$0.07\ ZrO_2$	Cu	CH_3OH	3.25
$SrTiO_3:Ni/Ta/La$	Pt	CH_3OH	N/a
Cr/N doped $SrTiO_3$	Pt	CH_3OH	2.39
$In(OH)_yS:Ag-Zn\ Zn(Ag)/In = 0.04$	None	$Na_2SO_3 + Na_2S$	1.65
TiO_2/SnO_2	None	CH_3OH	2.9
$Rh/Cr_2O_3/GaZn$ oxide	None	CH_3OH	2.6
ZnO/ZnS	None	Glycerol	3.4
$Cd_{0.4}Zn_{0.6}S$	NiS	$Na_2SO_3 + Na_2S$	2.4
TiO_2-C-362	None	CH_3OH	N/a
CdS	Pt	$(NH_4)_2SO_3$	2.4
N-doped $In_2Ga_2ZnO_7$	Pt	CH_3OH	2.5
5% Fe–4% Ni/TiO_2	None	C_2H_6O	2.41
$GO-TiO_2$	Pt	TEOA	N/a
$Au-TiO_2$-AC	None	EDTA	3.2
TiO_2-SiO_2	None	DEA	3.26
TiO_2-ZnO	None	C_2H_6O	3.06
$2Au-TiO_2$	None	Ascorbic acid	3.28
$K_4Nb_6O_{17}/CdS$	None	$Na_2SO_3 + Na_2S + KOH$	3.1
TiO_2-NiS	None	Lactic acid	N/a
$g-C_3N_4-SrTiO_3:Rh$ (0.3 mol%)	Pt	CH_3OH	3.2
Au-CdS	None	$Na_2SO_3 + Na_2S$	2.4
25 wt% $PbS/K_2Ti_4O_9$	None	$Na_2SO_3 + Na_2S + KOH$	N/a
5 wt% In_2O_3/Ta_2O_5	Pt	CH_3OH	2.8
CdS/Ta_2O_5	None	Lactic acid	2.4
$Cd_{0.1}Zn_{0.9}S$	Cu	$Na_2SO_3 + Na_2S$	2.78
Pt-PdS-CdS	None	$Na_2SO_3 + Na_2S$	2.4
CdS/Ti	None	TEA	N/a
$K_{1.025}Sr_2Nb_{2.9875}Cr_{0.0125}O_{10}$	None	CH_3OH	3.5
7.5% Bi-doped $NaTaO_3$	Pt	CH_3OH	2.64
$(CuAg)_{0.15}In_{0.3}Zn_{1.4}S_2$	Ru	KI	1.9

Co = cocatalyst.

From Acar, C., Dinçer, I., Zamfirescu, C., 2013. A review on selected heterogeneous photocatalysts for hydrogen production from water. Int. J. Energy Res. 38, 1903–1920.

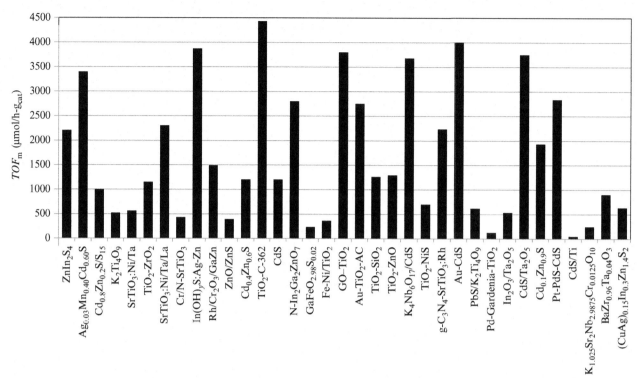

FIG. 5.35 Turnover frequency TOF_m in µmol/h-g_{cat} of selected heterogeneous particulate photocatalysts. *Data from Acar, C., Dinçer, I., Zamfirescu, C., 2013. A review on selected heterogeneous photocatalysts for hydrogen production from water. Int. J. Energy Res. 38, 1903–1920.*

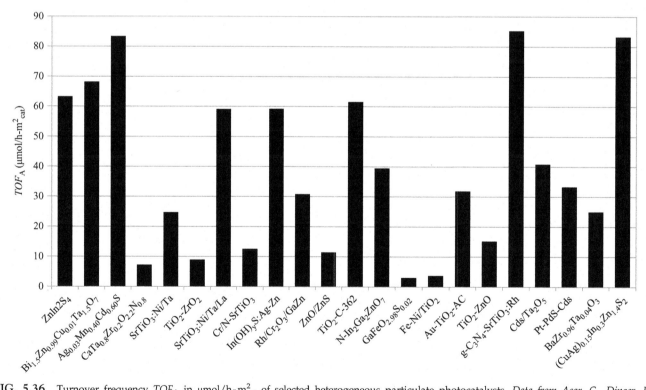

FIG. 5.36 Turnover frequency TOF_A in µmol/h-m^2_{cat} of selected heterogeneous particulate photocatalysts. *Data from Acar, C., Dinçer, I., Zamfirescu, C., 2013. A review on selected heterogeneous photocatalysts for hydrogen production from water. Int. J. Energy Res. 38, 1903–1920.*

FIG. 5.37 Comparison chart of selected heterogeneous photocatalysts for hydrogen generation. *Data from Acar, C., Dinçer, I., Zamfirescu, C., 2013. A review on selected heterogeneous photocatalysts for hydrogen production from water. Int. J. Energy Res. 38, 1903–1920.*

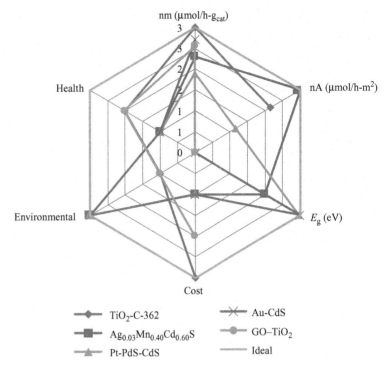

The metallic palladium can act as a reduction cocatalyst for hydrogen generation, hence substituting for the need for platinum. The formation of metallic palladium is explained by the fact that during photocatalysis, the Pd(II) ions are reduced to Pd-metallic. The elimination of platinum can lead to simpler recovery and recycling processes of metals from the passivized photocatalysts and, as well, a better cost effectiveness of the platinum-free photocatalyst.

Palladium has a price of 2.5 times lower than platinum. The use of palladium as a cocatalyst represents a cheaper solution without essential loss of performance. The cocatalyst is necessary to increase the stability of the photocatalyst and its active lifetime. It is widely known that by coating the catalyst with small amounts of platinum group metals (e.g., Pt, Pd) acting as cocatalyst, a protective layer is obtained.

The photocatalyst has been characterized by multiple methods. XRD patterns of the sample were confirmed by X-ray powder diffraction using an X'Pert PRO MPD PANalytical diffractometer with CuK_α ($\lambda = 1.5406\,Å$) radiation. The ultraviolet-visible diffuse reflectance spectra (DRS) were measured on a Lambda 950 Perkin-Elmer spectrometer with an integrating sphere. The nanocrystal's morphology was investigated using an FEI Inspect S scanning electron microscope, and elemental composition was determined using energy dispersive X-ray (EDX) spectroscopy.

The catalyst powder is of yellowish color. The morphology results are presented in Fig. 5.38, which shows a scanning electron microscope (SEM) image of the photocatalyst. The catalyst has a hexagonal-bipyramidal morphology with angles of 120 degree. The maximum sizes of particle was found to be 450–500 nm; the average particle size can be estimated to 250 nm with an average volume of catalyst grain of $V_{cat} \cong 0.008\,\mu m^3$.

The X-ray diffraction spectrum of the power of photocatalyst is shown in Fig. 5.39. The spectra confirm the crystalline structure of the catalyst as $Cd_{1-x}Zn_xS$ with hexagonal morphology. Preferential crystal growth is given by a diffraction maximum, which appears at $2\theta = 26.957$ degree as shown by the catalyst XRD spectra (Fig. 5.39). This peak is close to the Bragg reflection situated at $2\theta = 26.914$ degree (JCPDS card no. 00-040-0835) in hexagonal $Cd_{0.805}Zn_{0.195}S$ solid solution. A plane with Miller indices (002) is corresponding to this reflection.

The elemental composition of the photocatalyst has been determined qualitatively and quantitatively by EDX analysis, as shown in Fig. 5.40. The atomic ratio of Zn:(Zn + Cd) is found in about 0.21 which corresponds to an elemental structure of the photocatalyst of $Cd_{0.79}Zn_{0.21}S$. The sulfur content is confirmed according to the atomic ratio S:(Cd + Zn) = 0.98 (which is very close to one). The existence of oxygen atoms in elemental composition observed in the cassette of Fig. 5.40 can be attributed to chemical bonds of the type $Cd_{1-x}Zn_xS_{(bulk)}$-Cd-OH that may exist at catalyst particle surface or due to adsorbed water molecules.

Oxygen atoms are derived from water during the catalyst synthesis process, which is done hydrothermally. The XRD patterns shown above in Fig. 5.39 confirm that there are no diffraction peaks corresponding to any crystalline

FIG. 5.38 Catalyst morphology observed by scanning electron microscope.

Image zoom
(observe hexagonal morphology with
angles of 120°)

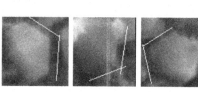

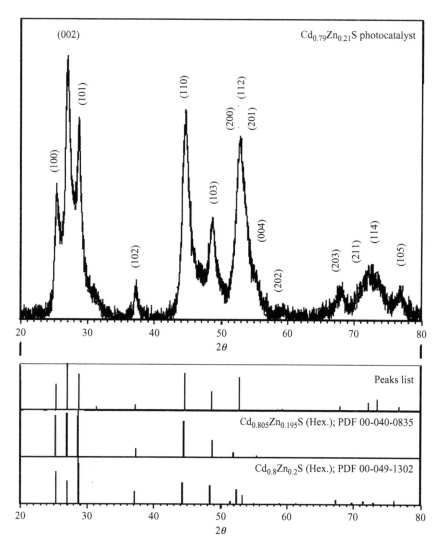

FIG. 5.39 X-ray diffraction spectra of the $Cd_{0.79}Zn_{0.21}S$ photocatalyst and two JCPDS cards for materials with similar composition. *Data from Zamfirescu, C., Dinçer, I., Naterer, G.F., Banica, R., 2013. Quantum efficiency modeling and system scaling-up analysis of water splitting with $Cd_{1-x}Zn_xS$ solid-solution photocatalyst. Chem. Eng. Sci. 97, 235–255.*

FIG. 5.40 The EDX spectrum of the photocatalyst and the global quantitative analysis. *Data from Zamfirescu, C., Dinçer, I., Naterer, G.F., Banica, R., 2013. Quantum efficiency modeling and system scaling-up analysis of water splitting with Cd$_{1-x}$Zn$_x$S solid-solution photocatalyst. Chem. Eng. Sci. 97, 235–255.*

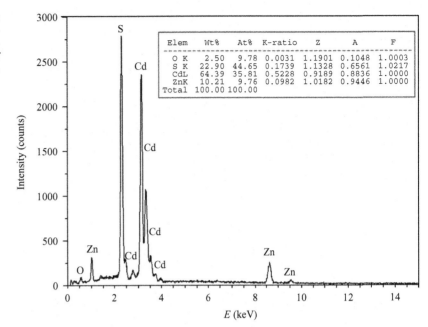

Elem	Wt%	At%	K-ratio	Z	A	F
O K	2.50	9.78	0.0031	1.1901	0.1048	1.0003
S K	22.90	44.65	0.1739	1.1328	0.6561	1.0217
CdL	64.39	35.81	0.5228	0.9189	0.8836	1.0000
ZnK	10.21	9.76	0.0982	1.0182	0.9446	1.0000
Total	100.00	100.00				

phase comprising oxygen. Diffuse spectral reflectance analysis has been measured in ultraviolet and visible ranges. Two reflectance spectra of the catalysts were obtained for unloaded and for loaded catalysts, respectively. The loaded catalyst comprises 0.2% by weight PdS cocatalyst, while the unloaded one has no palladium. The reflectance spectra are presented comparatively in Fig. 5.41. From the figure, it clearly results that the catalyst can absorb well at wavelength shorter than ~500 nm, whereas for longer wavelengths, the reflectivity of catalysts is higher. This is a first indication that the bandgap energy corresponds to a wavelength of ~500 nm. Furthermore, for lower photon energy, the loaded catalyst shows a reflectance of 40%, while that of the unloaded catalyst is about double.

PdS is an n-type semiconductor with a bandgap of 1.60 eV (775 nm). Hence, PdS absorbs very intensely in visible spectrum. According to Ferrer et al. (2007), the absorption coefficient is higher than 10^5 cm^{-1} for wavelengths shorter than 620 nm. Therefore, when loaded with PdS, the engineered photocatalyst Cd$_{0.79}$Zn$_{0.21}$S will have a reduced reflectance with respect to the unloaded catalyst at wavelengths shorter than the cut-off (which is beneficial, because it leads to an electron density increase in CB).

FIG. 5.41 Diffuse reflectance spectra of unloaded and loaded catalyst.

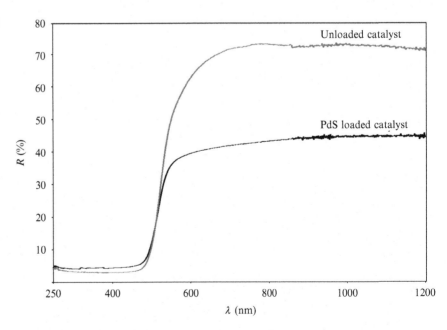

On one hand, in order to reduce catalyst reflectance, the number of PdS cocatalyst clusters loaded on $Cd_{1-x}Zn_xS$ surface must be large enough such that the additional electron-holes generated by PdS will separate before recombination. On the other hand, adding too much PdS is detrimental, because the PdS cannot catalyze water reduction. The excess of PdS disfavors light absorption in the visible spectrum by $Cd_{1-x}Zn_xS$ with the consequence of a reduction of electron density in the CB and reduced catalyst productivity. Furthermore, for cost effectiveness of the loaded catalyst, only minor amounts of palladium must be used.

The band-gap energy can be calculated using the Kubelka-Munk theory, which provides a correlation between reflectance and absorbance of weakly absorbing samples. The Kubelka-Munk function (f_{K-M}) is proportional with the Kubelka-Munk absorption coefficient (α_{K-M}) and inversely proportional with the scattering coefficient (s) of the sample. The Kubelka-Munk function can be calculated based on reflectance spectrum of the infinite thick sample (R_∞) extracted from Fig. 5.41, using the following equation

$$f_{K-M} = \frac{(1-R_\infty)^2}{2R_\infty} \tag{5.66}$$

The absorption coefficient is related to the Kubelka-Munk function and scattering coefficient s as follows

$$\alpha_{K-M} = s\, f_{K-M} \tag{5.67}$$

For direct band-gap semiconductors, the band-gap energy E_g is related to the absorption coefficient α_{K-M}, through the following equation

$$\alpha_{K-M} h\nu = \text{const.}\,(h\nu - E_g)^{1/2} \tag{5.68}$$

where $h\nu$ is the photon energy and const. is a proportionality constant.

Because both CdS and ZnS display direct transitions, the band-gap energy results from the graph representing the variation of $(\alpha h\nu)^2$ versus $h\nu$. Based on the above discussion, the quantity $(\alpha h\nu)^2$ is given by $(s\, f_{K-M} h\nu)^2$. It is reasonable to assume that the scattering coefficient has a constant value.

The graph from Fig. 5.42, which plots the variation of $(f_{K-M} h\nu)^2$ against $h\nu$, is obtained. Its linear portion is extrapolated until the abscissa is intersected (meaning that the absorption coefficient is zero and reflectance is 100%) as indicated in figure. It is thus confirmed that the bandgap of the photocatalyst is 2.48 eV, which corresponds to 500 nm. This value is slightly superior to a CdS catalyst with 2.42 eV and inferior of a ZnS catalyst with 3.8 eV due to the replacement of ~20% of cadmium ions with zinc ions.

Once a catalyst has been fabricated and characterized, constant pressure photocatalysis experiments were performed under monochromatic light illumination at 470 nm wavelength for which no other reaction participant except the catalyst can absorb light. Fig. 5.43 is a photograph of the catalyst in a still electrolyte solution, deposited on the bottom of an Erlenmeyer flask.

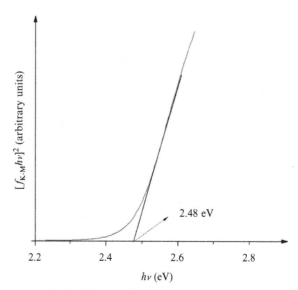

FIG. 5.42 Determination of band-gap energy of loaded photocatalyst.

FIG. 5.43 The prepared $Cd_{1-x}Zn_xS$ solid-solution photocatalyst sediment at the bottom of a flask with electrolyte.

The LEDs are controlled by regulating the current intensity. Three values of current intensity were used during experiments, which correspond to three levels of photon rates of ~8, 16 and 24 micromoles photons per second. The spectral emissivity of LEDs is presented in the plot from Fig. 5.44, where the catalyst reflectance spectrum is also superimposed to demonstrate that photon energy was superior to band-gap energy. Simple analysis of this graph demonstrates that more than ~95% of the emitted photons energy can contribute to catalyst excitation.

Under an illumination, intensity corresponding to 16 micromoles photons per second, the total generation after 3 h of experiment, is higher than 10.76 mmol. Two subsequent runs of 3 h each were also conducted at constant pressures using photon rates of 8 micromoles per second and 24 micromoles per second, respectively. The quantity of gas extracted was 196 and 297 ml, respectively. These quantities correspond to metered hydrogen gas evolutions of 8.15 and 12.35 mmol, respectively.

FIG. 5.44 Emission spectrum of LED superimposed on reflectance spectrum of photocatalyst. *Data from Zamfirescu, C., Dinçer, I., Naterer, G.F., Banica, R., 2013. Quantum efficiency modeling and system scaling-up analysis of water splitting with $Cd_{1-x}Zn_xS$ solid-solution photocatalyst. Chem. Eng. Sci. 97, 235–255.*

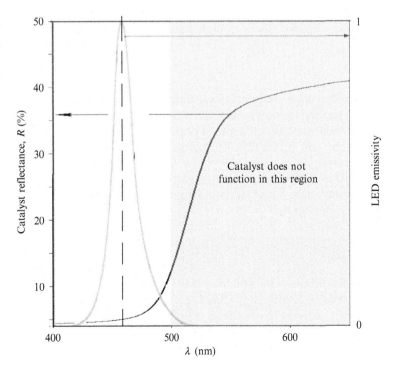

5.4.3.2 Modeling of the Photocatalytic Process

The model formulation for the overall photocatalysis process must consider three types of submodels. Firstly, and most importantly, one has to formulate a model of quantum efficiency that expresses the link between the rate of photon occurrence at the catalytic site and the rate of molecular hydrogen production. Secondly, a model to determine the percentage of incident photons that actually reach at the catalytic site must be proposed. Thirdly, mass transport processes at a system boundary must be modeled and boundary conditions must be specified.

Quantum efficiency has two components: probability of successful photon absorption (Φ_{abs}), and probability of sucessful catalysis (Φ_{cat}). Denote k_{H_2} as the quantum rate of hydrogen formation (at a catalytic site) and $\sum k_i$ the sum of the rates of all possible events triggered by absorption of photons. With these notations, the probability of catalysis is expressed as follows

$$\Phi_{cat} = \frac{k_{H_2}}{\sum k_i} \tag{5.69}$$

where the quantum rate of hydrogen occurrence represents the reciprocal of the sum of the time constants for all favorable events that lead to the generation of one molecule of hydrogen.

These events are specific to each type of photocatalytic system and include mainly the time constants for: excited state formation (τ^*), electron donor encounter (τ_{ED}), electron transfer (τ_{et}), water molecule encounter (τ_{H_2O}), hydrogen atom formation and their molecular desorption and diffusion far from the catalytic site (τ_f). The following algebraic expression is derived in Zamfirescu et al. (2013) for the quantum efficiency

$$\Phi = \left[\left(1 + f_{bg}(1 - f_{cat})\tilde{k}_{ph} \right) \left(1 + \tilde{k}_{GS} + \tilde{k}_{enc} \left(1 + \tilde{C}_b + \tilde{C}_d \right) \right) \right]^{-1} \tag{5.70}$$

The factor f_{bg} in Eq. (5.70) is defined as follows. Assume that photons occurring at the catalytic site have a spectral distribution such that only the fraction $f_{bg} \in [0, 1]$ represents photons with energy higher than the bandgap of the photocatalyst. The factor f_{bg} will be equal to unity if the radiation is monochromatic and at sufficient energy. Using the factor f_{bg}, one can express the rate of high-energy photons as the product $f_{bg}\dot{N}_{ph}$, where $\dot{N}_{ph}$ was defined above as rate of photon occurrence at the catalytic site.

The factor f_{cat} in Eq. (5.70) is defined assuming that the photon rate occurring in the vicinity of single monodisperse photocatalyst grain is influenced by the volumetric fraction of photocatalyst inside the reactor. If one denotes $f_{cat} \in (0, 1)$ the volumetric fraction occupied by the amount of catalysts inside the reactor, then the molar rate of photons that are scattered by the reactor is $f_{cat}\dot{n}_{ph}$. Hence, the molar rate of photons occurring inside the reactor is given by $(1 - f_{cat})\dot{n}_{ph}$. The factor $\tilde{k}_{ph}$ in Eq. (5.70) is important for quantifying the ability of the catalyst to absorb photons. The factor is introduced as the rate ratio of photon occurrence at the catalytic site and of excited state decay, and expressed as follows

$$\Phi \tilde{k}_{ph} = \frac{\dot{n}_{ph}}{k_{esd}} \frac{N_A}{N_{cat}} \tag{5.71}$$

If excited state decay rate is essentially much higher than the rate of photon occurrence, then the probability of photon absorption becomes ~100%; the dimensionless parameter $\tilde{k}_{ph}$ is in this case zero. Otherwise, if $\tilde{k}_{ph}$ becomes essentially higher such that the rate of photon occurrence at the catalytic site is of the same order of the rate of excited state decay, then the probability of photon absorption decreases substantially; more photons escape and fewer are used in the process.

The dimensionless factor $\tilde{k}_{GS}$ is introduced by $\tilde{k}_{GS} = k_{GS}/k_{H_2}$, where k_{GS} is the quantum rate of catalyst relaxation to the ground state n and k_{H_2} is the quantum rate of hydrogen generation. The dimensionless parameter is introduced ($\tilde{k}_{enc}$) according to

$$\tilde{k}_{enc} = \frac{8RT}{3\mu k_{H_2}} \mathcal{C}_{H_2} \tag{5.72}$$

For facilitating a parametric study, it is useful to introduce the dimensionless factor representing the concentration ratio between bubbles and dissolved hydrogen $\tilde{C}_b = \mathcal{C}_b/\mathcal{C}_{H_2}$ and the concentration ratio between other dissolved species and dissolved hydrogen $\tilde{C}_d = \sum \mathcal{C}_d/\mathcal{C}_{H_2}$. Here, the symbols $\mathcal{C}_{H_2}$, $\mathcal{C}_b$ and $\mathcal{C}_d$ denote the concentrations of hydrogen dissolved in solution, of hydrogen present in dispersed phase (bubbles) and other dissolved species concentrations, respectively. Furthermore, based on quantum efficiency as given by Eq. (5.70), the following algebraic expression can be derived for the molar rate of hydrogen generation

$$\dot{n}_{H_2,gen} = \frac{(1-f_{cat})\dot{n}_{ph}}{2\left(1+f_{bg}\tilde{k}_{ph}(1-f_{cat})\right)\left(1+\tilde{k}_{GS}+\tilde{k}_{enc}\left(1+\tilde{C}_b+\tilde{C}_d\right)\right)} \tag{5.73}$$

If hydrogen bubbles are present in a solution, they rise and reach the liquid-gas interface where they bounce with high frequency ($\sim$1 kHz) for a few oscillations until the rupture of the liquid film occurs such that the bubble bursts by creating a weak shock wave and gas jet. The bursting of submillimeter bubbles at the surface of pure water occurred at a time constant of $\tau_b \cong 4$ ms. The amount of gas carried by a single bubble depends on the bubble's size and the pressure inside the bubble. According to the well known Laplace equation, the pressure of the gas enclosed in a bubble depends on the pressure in the medium surrounding the bubble and bubble diameter

$$P_b = P_{amb} + 4\sigma/d_b \tag{5.74}$$

where σ is the surface tension of the liquid, d_b is bubble diameter, and P_{amb} is the gas pressure in the space above the liquid.

On the other hand, the bubble diameter is influenced by the pressure of the medium surrounding the bubble. In general, the bubble diameter at departure is proportional to the reciprocal of pressure. There are many studies in past literature regarding the bubble diameter after its initial formation (departure bubble diameter). As discussed by Zamfirescu et al. (2013) for water show, that for pressures of 1–20 bars, the bubble diameter decreases from $\sim$0.3 to 0.03 mm for which the pressure difference $P_b - P_{amb}$ is negligible.

The number of moles carried by a bubble is proportional with its volume (assumed spherical) and the molar-specific volume at the temperature and pressure of operation. If the concentration of hydrogen above the liquid is smaller than the concentration of hydrogen dissolved in the liquid, then a mass-transfer process occurs at the liquid-gas interface by diffusion. Thus, in this case, the molar rate of hydrogen evolved by diffusion process is dependent on the superficial diffusion coefficient D, free surface area A, thickness of mass transport boundary layer δ, the concentration difference between dissolved hydrogen C_{H_2} and hydrogen concentration in the atmosphere above electrolyte C_{amb}

$$\dot{n}_d = DA\frac{C_{H_2} - C_{amb}}{\delta}, C_{H_2} > C_{amb} \tag{5.75}$$

In general, there may be four cases of hydrogen transport across the free surface of the electrolyte as follows:

- dissolved hydrogen is at saturation and hydrogen concentration above the liquid is smaller than concentration in the liquid, in which case hydrogen crosses the interface both by diffusion and bubble-bursting process
- dissolved hydrogen is at saturation, but the concentration of hydrogen inside the liquid is not higher than the concentration of hydrogen above the liquid, in which case only bubble bursting is possible at the interface
- the concentration of hydrogen inside the liquid is smaller than the saturation limit, but higher than the concentration of hydrogen above the liquid; thus, bubbles cannot occur and hydrogen crosses the interface by diffusion only
- the hydrogen concentration in liquid is below the saturation limit and lower than the concentration of hydrogen above the liquid, in which case there cannot be any mass transfer of hydrogen across the interface, from liquid toward the gas above liquid.

The bubble concentration is proportional to pressure. If pressure increases (e.g., due to gas accumulation in a constant volume), then the bubble concentration in the solution must increase; however, the quantum efficiency is inversely proportional with bubble concentration. An increase of bubble concentration leads to a reduction of hydrogen production rate, although the photon rate remains constant.

Assume that the photochemical reactor operates in batch mode, in which case the control volume for reactor modeling comprises the electrolyte and a gaseous space above the liquid. The volume of electrolyte is assumed constant and no chemical enters the reactor. During the photocatalysis process, the reactants deplete while products are formed.

The molar concentration of hydrogen in the reactor as dissolved phase either is constant in time, if saturation is achieved, or, it increases in time proportionally with the difference between generated hydrogen rate and the mass transfer rate of hydrogen across the liquid-gas interface. This statement can be written in the form of the following differential equation

$$\frac{dC_{H_2}}{dt} = \begin{cases} \frac{1}{V_{elec}}\left(\dot{n}_{H_2,gen} - \dot{n}_{H_2,extr}\right), if\, C_{H_2} < C_{H_2},max \\ 0, if\, C_{H_2} = C_{H_2},max \end{cases} \tag{5.76}$$

where V_{elec} represents the volume of the electrolyte in dm^3.

The molar hydrogen concentration in dispersed phase can have a change only if bubbles exist in the reactor. This is of course valid in the situation when $C_{H_2} = C_{H_2},max$ and more hydrogen is generated. Consequently, the following differential equation can be written for hydrogen concentration in dispersed phase

$$\frac{dC_{b}}{dt} = \begin{cases} \frac{1}{V_{elec}}\left(\dot{n}_{H_2,gen} - \dot{n}_{H_2,extr}\right), if\, C_{H_2} = C_{H_2},max \\ 0, if\, C_{H_2} < C_{H_2},max \end{cases} \tag{5.77}$$

The major competing process with the chain of events that led to the formation of a molecule of hydrogen is the relaxation of the catalyst to the ground state. For the case of semiconductor suspension, once an electron is displaced from the VB to the CB, a tendency of electron-hole recombination occurs. The recombination process is favored by the presence of crystal defects within the semiconductor at its surface. Thus, for impeding a rapid recombination process, it is necessary to produce a crystal free of defects.

The time constant of catalyst activation (for the semiconductor) has the same order of magnitude of electron movement from valence to CBs, which is $\tau_* \cong 10^{-10}$ s. The time constant for sulfide encounter depends on the rate of molecular diffusion, $k_d = 8/3\mathcal{R}T/\mu \cong 7.5 \times 10^6$ m^3s^{-1}/mol. Typical concentrations of sulfide (also used in the experimental work as detailed subsequently) are $0.2{-}0.4$ mol/dm^3; thus the encounter rate of the electron donor has the order of 2×10^6 s^{-1} with a corresponding time constant of $\tau_{ED} \cong 4.5 \times 10^{-7}$ s. The typical range of intermolecular electron transfer in solution is $\tau_{et} = 10^{-9} - 10^{-7}$ s.

The predictions of pressure increase during constant volume operation of the photoreactor are compared to the measurements as indicated in Fig. 5.45. The increase of light intensity leads to higher gas generation and thus to faster increase of pressure of the enclosed gas. This trend is predicted correctly by the model. Moreover, the pressure increase is slightly concave for the hour of recordings, as indicated both by the measurements and the predictions. There is a bias between predicted and metered pressures at high photon rate. It is speculated that this bias is due to consumption of electron donors for that particular run.

The model can take into account the changes of electron donor concentration by the value of the time constant of hydrogen formation, τ_{H_2}, which must increase when donors are consumed. In the previous section is mentioned that

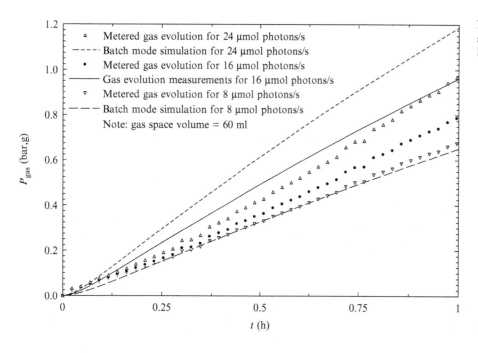

FIG. 5.45 Predicted and measured pressure increase at photocatalytic hydrogen evolution during constant volume operation.

FIG. 5.46 Variation of quantum efficiency and amounts of generated and extracted hydrogen for the initial phase of batch simulations at constant pressure.

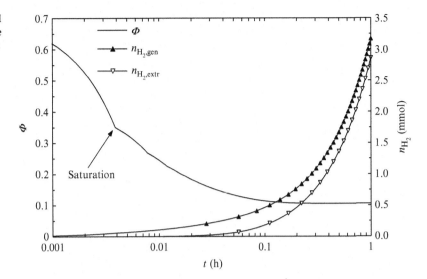

the value for τ_{H_2} has taking constant under the assumption that the electron donor concentration does not change significantly during experiments. This situation is relevant especially for short-time batch-flow reactors and for continuous-flow reactors operating steadily.

In addition, it may be observed from Fig. 5.46 that quantum efficiency decreases in time from $\sim$0.62 to 0.1 after 1 h, and then it stabilizes. Recall that for this case, the previously calculated average quantum efficiency was 0.127 (see above). The quantum efficiency curve has an inflexion point at approximately the moment when bubbles occur in the electrolyte. The hydrogen generation curve ($n_{H_2,gen}$) shows a higher rate of hydrogen production than it shows the curve, which indicates the cumulative hydrogen extraction from the electrolyte ($n_{H_2,extr}$). At the end of the first hour of operation, the generation and extraction curves become parallel, which demonstrate that generation and extraction rate of hydrogen are quasi-equal, therefore the process is steady.

In Fig. 5.47, the temporal variations of quantum efficiency for low- and high-photon rate based on numerical simulation with the model are illustrated. The predicted quantum efficiency is higher when the photon rate is lower, although the production of hydrogen increases when light intensity increases.

Let us consider now a reactor with continuous operation at a steady state. The amount of catalyst is also an important design parameter that should not be wrongly chosen. It is convenient to express the catalyst amount with the help of volumetric fraction f_{cat}, introduced previously. The number of catalytic sites can be approximated from f_{cat} provided that the average volume of the grain V_{cat} and the electrolyte volume V_{elect} are specified, according to

FIG. 5.47 Variation of quantum efficiency for constant pressure simulations at lower and higher light intensities.

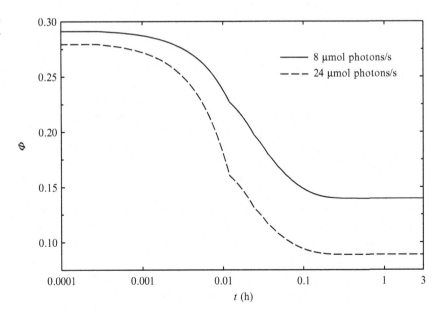

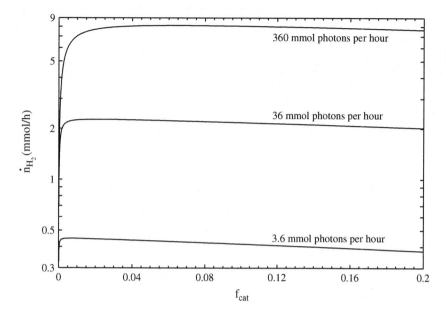

FIG. 5.48 Rate of hydrogen production in function of catalyst fraction and three values of photon rates under monochromatic radiation, $f_{bg} = 1$ in Eq. (5.72).

$$N_{cat} = f_{cat} \frac{V_{elect}}{V_{cat}} \tag{5.78}$$

As seen from the graphs presented in Fig. 5.48, the rate of hydrogen production has a maximum when plotted against the volumetric fraction of catalyst in the electrolyte; however, the optimum of catalyst fraction tends to be robust, namely, if chosen correctly, the catalyst amount influences little the hydrogen production rate. Moreover, if the catalyst is too little or too much, the production rate is clearly influenced. The influence of catalyst amount is higher when the photon rate is enhanced. An increase of photon rate with an order of 100 (from 3.6 to 360 mmol/h) displaces the optimal volumetric fraction of catalyst grains from ~1% to 7%.

5.5 PHOTOELECTROCHEMICAL CELLS

5.5.1 Photoelectrochemical Process Description

A method to enhance the efficiency of water electrolysis is by application of photoelectrodes. Typically a photoelectrode consists of a conductive material doped with photocatalysts. These can be noble metals or even more ordinary metals like iron or compounds of these. Since the electrode is a conductive material, its VB overlays with its CB. Thus electrons can be provided to the active catalytic sites where water electrolysis is performed. The photoelectrodes are exposed to solar radiation via transparent windows.

Photophysical processes like photon absorption occur at the photocatalyst site. Thus, the active site of the photocatalyst is "activated" via two channels: directly by incident light; and through an electrical potential bias created by an external electrical supply connected at the electrode. When affected by light, the active semiconductor of the photoelectrodes absorbs photons that dislocate electrons from the VB and move them in the CB. Consequently, redox reactions are facilitated at the photocatalyst. In order to enhance the redox potentials, a bias voltage is applied at the electrode via an external power supply. Fig. 5.49A explains the process of heterogeneous photocatalysis that conducts a water reduction reaction. For a water oxidation process in an electrochemical cell, the processes are illustrated as shown in Fig. 5.49A.

In order for the photocatalytic process to occur in a photoelectrochemical (PEC) cell, the required energy levels must be satisfied; this means that the photon energy must be higher than the difference between conduction and VBs $h\nu > E_C - E_V$. In addition, the energy level of the reduction reaction must be below that of the CB (of the p-type semiconductor); for oxidation reactions, the level of the VB must fall below the level of the energy level of the oxidation reaction.

Some typical photoelectrolysis cell configurations are suggested schematically as shown in Fig. 5.50. The electrodes are made commonly from various kinds of semiconductors, possibly doped or coated with other compounds that have a role of photocatalysts or PSs. Table 5.4 gives energy levels of some heterogeneous photocatalysts for photo-electrodes.

FIG. 5.49 Reductive (a) and oxidative (b) process in a photoelectrochemical (PEC) cell.

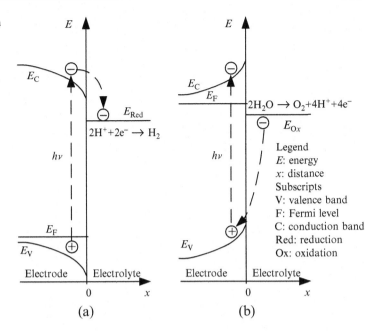

The cell in Fig. 5.50a comprises a photoanode immersed in an electrolyte exposed to light through a transparent window. The potential generated at the photoanode is sufficient to drive the electrolysis with the selected semiconductor and counter electrode (of metallic kind). The oxygen-evolving reaction occurs at the anode, while at the cathode, protons and electrons combine to evolve hydrogen. A proton exchange membrane and an external electric circuit are used to transport protons and electrons.

The second configuration is illustrated in Fig. 5.50b. It comprises, in addition to the previously mentioned elements, a PV array and battery. This system is a PV-assisted photoelectrochemical cell. The semiconductor selected at the photoanode does not generate enough potential to drive the reaction; in this case, an electric bias is generated by the external voltage. The system can operate during nights as a regular electrolyzer. During days, it is driven directly with light. The third system presented here in Fig. 5.50c combines a photoanode and a photocathode, which generate sufficient potential to drive the complete electrolysis reaction. Both electrodes are exposed to the light in this case. The anode is always an n-type semiconductor, while the cathode is of p-type.

5.5.2 Dye-Sensitized Tandem Photoelectrolysis Cell

One of the major challenges with photoelectrochemical cells is the bandgap of photoelectrodes, such that the incident light is better harvested. In general, photoanodes absorb light in other parts of the spectrum than photocathodes.

Semi-transparent electrodes were developed to allow a sandwich type configuration of the cell such that the light transmitted by one of the photoelectrodes is absorbed by the other. The total active area exposed to the incident light is reduced in this way. An additional approach for better efficiency is to coat the electrodes with dyes containing PSs, which help reduce the energy gap and thus utilize light from a broader spectrum.

TABLE 5.4 Energy Levels of Heterogeneous Photocatalysts

Parameter	SrTiO$_3$	TiO$_2$	GaP	Remarks
Type	n	n	p	Type of semiconductor
$E_C - E_V$	3.2 eV	3.2 eV	2.3 eV	Difference between conduction and valence band energy levels
$E_{V,s} - E_V$	0.3 eV	0.2 eV	−0.6 eV	Valance band energy at the interface with respect to bulk
$E_R - E_V$	1.9 eV	2.0 eV	−1.3 eV	Energy level at redox reaction with respect to valence band energy

Note: for n-type photocatalysts, the redox reaction is water oxidation $2H_2O \rightarrow O_2 + 4H^+ + 4e^-$; for p-type catalysts, the redox reaction is proton reduction $2H^+ + 2e^- \rightarrow H_2$.

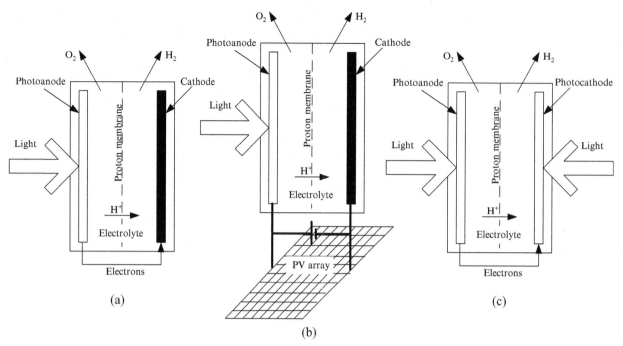

FIG. 5.50 Photoelectrochemical cells for water electrolysis; (a) cell with photoanode and metallic cathode, (b) PV-assisted cell, (c) combined photoanode/photocathode cell.

Dye-sensitized tandem electrolysis cells are hybrid devices comprising photoelectrodes and PSs. The functioning of the tandem cell is described in Fig. 5.51, according to the description given in Grätzel (2001). The cell has two compartments, one comprising an aqueous electrolyte, and the other an iodine-based electrolyte. Light enters through transparent windows in both compartments. The first compartment is where photoelectrolysis is performed. Light enters in this compartment and crosses the electrolyte, which is quasitransparent to solar radiation. Then light is absorbed partially by an extremely thin photoanode based on wolfram trioxide (WO_3), which is a semiconductor. This electrode is transparent to the lower visible range and infrared.

The transmitted light falls on a second photoelectrode, which is sandwich-like and comprises a semiconductor layer in titanium dioxide (TiO_2) as a second layer dye coating. The dye contains ruthenium-bipyridyne based PSs. This dye absorbs light in mostly the lower visible spectrum. Through the opposite window, incident light passes and crosses a thin layer of platinum that has the role of counter electrode. It also crosses the iodine-based electrolyte and eventually it is absorbed by the dye. This tandem cell (a system comprising two cells that work jointly) generates ultimately hydrogen and oxygen from water supplied at the bottom, as shown in the figure.

The electron flow and the relevant processes for the dye-sensitized photoelectrochemical cell are described in Fig. 5.52, which can be analyzed in conjunction to Fig. 5.51. The process occurring in state 1 is water oxidation (1.23 eV). The process 1–2 represents the electron transfer from the reaction sites, through the Helmholtz layer to the photoanode surface.

The anodic reaction is $2H_2O \rightarrow O_2 + 4H^+ + 4e^-$. Under the influence of incident light, the electron jumps from the VB to the CB of the semiconductor, process 2–3. The absorbed energy by the semiconductor is 2.6 eV, or 477 nm (upper visible range). Next the electron drops on the Fermi level (process 3–4) and it is transferred through an external circuit (process 4–5) to the platinum counter-electrode in the other cell. The generated electrical charge crosses the electrolyte at the Fermi level (process 5–6). The cathode reaction is

$$I_3^- + 2e^- \rightarrow 3I^- \tag{5.79}$$

while the anode reaction (at the dye surface) is

$$2PS^+ + 3I^- \rightarrow 2PS + I_3^- \tag{5.80}$$

where PS stands for photosensitizer.

The PS is excited by light to jump over the CB level of the TiO_2 semiconductor (process 6–7). The magnitude of the jump is 1.6 eV, which corresponds to photon absorption at 775 nm (lower visible range). This process is

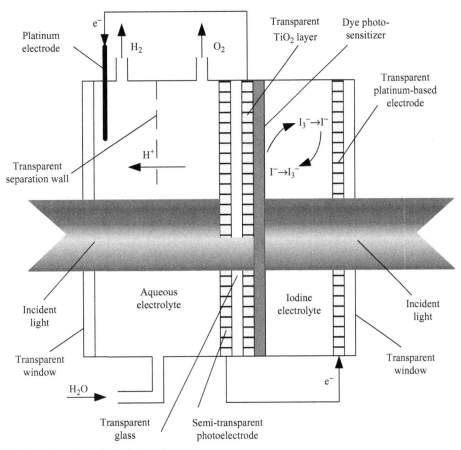

FIG. 5.51 Dye-sensitized tandem photoelectrolysis cell.

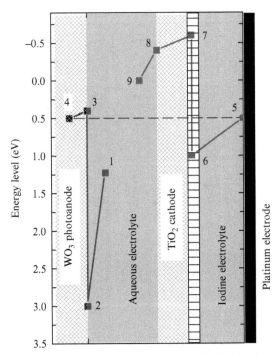

FIG. 5.52 Approximation of energy levels of processes within a tandem dye-sensitized photoelectrochemical cell for water splitting.

written as $PS + h\nu \rightarrow PS^*$. The excited PS relaxes by dropping one electron in the CB of titanium dioxide, namely $PS^* \rightarrow PS^+ + e^-$. The electron energy level drops to reach the CB of the semiconductor according to the process 7–8 (see Fig. 5.52). The electron is transferred next to the electrolyte through a Helmholtz layer (process 8–9) and loses energy to reach the level of a hydrogen-evolving reaction, which occurs in (9), according to $2H^+ + 2e^- \rightarrow H_2$. The photonic energy efficiency of the cell can be calculated as $1.23/(2.6 + 1.6) = 29\%$.

5.5.3 Efficiency Definitions

In Fig. 5.53, four examples of photoelectrochemical cell configurations and their relevant energy levels are schematically shown. The cell from Fig. 5.53A consists of a $SrTiO_3$ photoanode and a metallic cathode. The states and processes indicated on the figure are explained as follows: state #1 is the water oxidation level; state #2 is the VB level within the semiconductor. Process 1–2 represents the energy loss by electron transfer between the electrolyte and semiconductor. Process 3–4 is the photon absorption, followed by electron displacement from the VB to a CB; for process 3–4, the electron diffuses into the bulk semiconductor, where the CB is lower than at the surface.

Process 4–5 is the electron transport via external circuit (at a CB level). In process 5–6, the electron reaches the Fermi level within the electrolyte, process 6–7 compensates the ohmic losses at the metallic cathode, and state #7 represents the energy level of a proton reduction reaction. The difference in energy levels between states #1 and #7 is 1.23 eV, which corresponds to the Gibbs free energy for splitting one water molecule.

The photonic energy efficiency of the PEC system, expressed in terms of energy used to split the water molecule by the energy of the absorbed photon, is written as

$$\eta_{ph} = \frac{E_0}{E_{ph}^{abs}} \tag{5.81}$$

where $E_0 = 1.23\,eV$, and E_{ph}^{abs} is the energy of the absorbed photon.

The energy of the absorbed photon must be higher than the difference between conduction and VBs of the semiconductor, $E_{ph}^{abs} \geq E_C - E_V$. For system Fig. 5.53a, photonic energy efficiency becomes $\eta_{ph} = 1.23/3.2 = 38\%$.

Another way to define the energy efficiency of the system is by dividing the energy embedded in the produced hydrogen, namely $\dot{E}^{H_2} = \dot{n}HHV$, where molar $\dot{n}$ is the rate of hydrogen production and HHV is the molar-based higher heating value of hydrogen. The energy consumed to generate the hydrogen is the energy of the photons' incident on the photoelectrode, given by $\dot{E}_{ph}^{inc} = I_{T0}A$, where I_{T0} is the solar irradiation on a tilted surface and A is the area of the electrode. Thus, the energy efficiency is introduced as

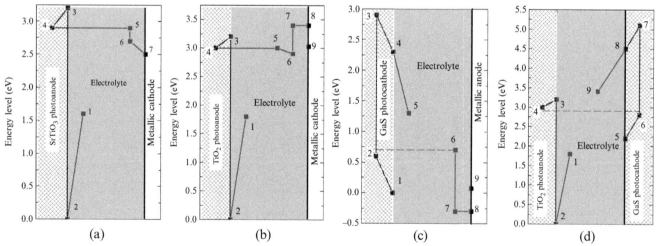

FIG. 5.53 Approximation of energy levels for several photoelectrochemical cell arrangements for hydrogen production; (a) $SrTiO_3$ photoanode; (b) TiO_2 photoanode and external 0.5 V source bias; (c) GaS photocathode and 1 V source bias; (d) coupled photoanode/photocathode cell with TiO_2 and GaS photoelectrodes.

$$\eta_{\text{syst}} = \frac{\dot{n}\text{HHV}}{I_{T0}A} \tag{5.82}$$

An additional way to quantify the efficiency of the system is by exergy. In this case, one defines the exergy efficiency as the ratio of the exergy embedded chemically in the produced hydrogen and the exergy of the incident light. This can be written in the following manner

$$\psi_{\text{syst}} = \frac{\dot{n}ex_{\text{H}_2}^{\text{ch}}}{I_{T0}A(1 - T_0/T_{\text{rad}})} \tag{5.83}$$

where chemical exergy of hydrogen is 236.12 kJ/mol or 2.447 eV per molecule, T_0 is the reference temperature and T_{rad} is the temperature of the photonic radiation.

The system from Fig. 5.53b uses titanium dioxide as a photoelectrode. This system requires PV or battery assistance to compensate for the low electromotive voltage generated by the system. The process 1–2 represents water oxidation, which is followed by photon absorption (3.2 eV), a fall of energy level corresponding to the difference between the semiconductor surface and the bulk (process 3–4), electron transport to the cathode (4–5), fall of the Fermi level (5–6), voltage bias (6–7), and ohmic losses at the metallic cathode.

The system from Fig. 5.53c uses a photocathode instead of a photoanode. This system uses a GaP semiconductor of p-type to evolve hydrogen at its surface, while oxygen evolves at a metallic anode. The system requires 1 V bias generated by an external power source.

The system from Fig. 5.52d, represents a coupled photoelectrode system comprising a TiO_2 photoanode and a GaS photocathode. In order to operate, the photoanode requires 3.2 eV of photonic energy, which can be obtained from light with wavelengths shorter than 387.5 nm (top edge of the visible spectrum). The photocathode requires 2.3 eV of light with a wavelength shorter than 539.1 nm (in the middle of the visible range).

Note that the efficiency of the single photoelectrode cell with $SrTiO_3$ is 1.2% and that of a coupled cell (Fig. 5.52D) is 5.9%. The corresponding exergy efficiencies are calculated based on a quality factor of 0.94% and 4.6% for the single and coupled cells, respectively.

An estimation of the energy efficiency of the tandem cell can be determined as follows. The energy portion from incident radiation carrying photons with a wavelength shorter than 477 nm (which can be absorbed by the photoanode) is 13%. The portion of photons with wavelengths shorter than 774 nm (which can be absorbed by the dye) represents 53%. Thus the photons in the range 477–744 nm carry 40% from incident radiation. This means that the light energy that crosses the photoanode is sufficient to excite the dye. Therefore, only 13% from incident solar radiation can be used by the system. It results in an energy efficiency of 7.3%, while the exergy efficiency is 5.7%. Regarding dye-sensitized tandem photoelectrolysis cell, Vayssieres (2010) states that their lab-scale practical implementations have demonstrated 3% efficiency.

5.5.4 Case Study 1: Development of a Novel Polymeric Membrane PEC

This section presents the experimental and theoretical work reported in the previous study by Casallas et al. (2016) on a newly developed polymeric membrane-based photoelectrochemical cell. The PEC is based on a newly developed membrane photoelectrode assembly (MPEA) with electrodeposited dual-band active photocatalytic islands.

The polymeric membrane photoelectrochemical cell is similar to a proton exchange membrane electrolysis cell that has a transparent window at the cathodic side, and the cathode is modified to a photocathode. The structure of the cell and its operation are presented as shown in Fig. 5.54. A membrane electrode assembly is fabricated such that the cathode compartment has deposited semiconductor photocatalysts, able to displace electrons when displaced by photons. Water is supplied to the anode where it releases oxygen and forms protons in the form of hydronium.

The polymeric cation exchange membrane is hydrated and crossed by the hydronium ions carrying the protons formed at the anode. Eventually, the protons that reach the cathode will combine with electrons to generate hydrogen. Both reactions are catalyzed, but especially the cathodic reaction requires platinum group materials for catalysis. The photocathode is realized by electrodeposition of CuO/Cu_2O photocatalyst islands (bandgap 1.2/2.137 eV) over the cathodic side surface of the polymeric membrane electrode assemble. The composite membrane that is formed thereto is referred as MPEA.

The test cell is shown in Fig. 5.55. An original mechanical system is devised to provide the necessary electrical contacts, the necessary electrical insulation between the anode and cathode and the necessary sealing between the anolyte and catholyte. There are no bipolar plates in the current construction, but rather the electrical contacts are made as

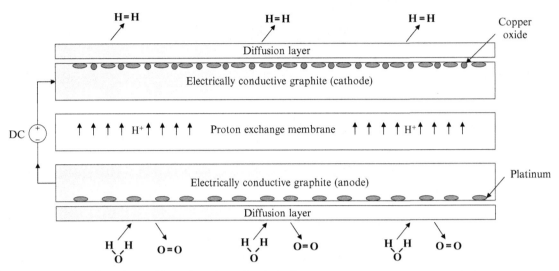

FIG. 5.54   Schematics of a polymeric membrane photoelectrochemical cell.

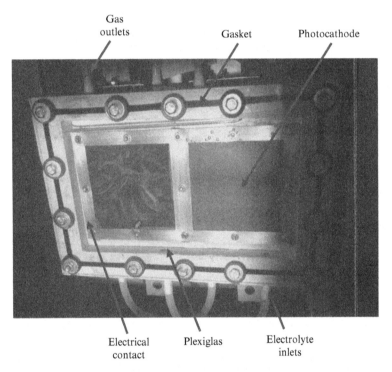

FIG. 5.55   Test PEC with membrane photocathode assembly, having two membranes of 100 cm^2.

frames in SS 316 for both the anode and cathode. The anode side contact is to be coated with titanium. The sealing between the anolyte and catholyte is assured by a nitrile gasket of 1/16 inch thickness, which is compressed 20% to obtain a firm electrical contact at the electrodes.

The frame of the photoelectrochemical cell is made of ABS material and 3D printed. Nitrile gaskets of 3 mm are used to seal the front and back glassing. Regarding the flow configuration, the anolyte and catholyte flows are in cocurrent crossing flow. The feed is in liquid phase; however, at the cathode, hydrogen evolves and is carried as a two-phase flow toward the upper side exit ports. In order to facilitate the draining of the gas phase (bubbles), the upper side of the cathode compartment (the roof) is tapered at 2 degree.

Fig. 5.56 shows an exploded view of the polymeric membrane PEC design. The cell has been specifically designed to drive multiple types of hydrogen-evolution reactions, with simple reconfiguration. It allows for simple replacement of membranes. If pure water is supplied, then the process will be a water photoelectrolysis that generates hydrogen and

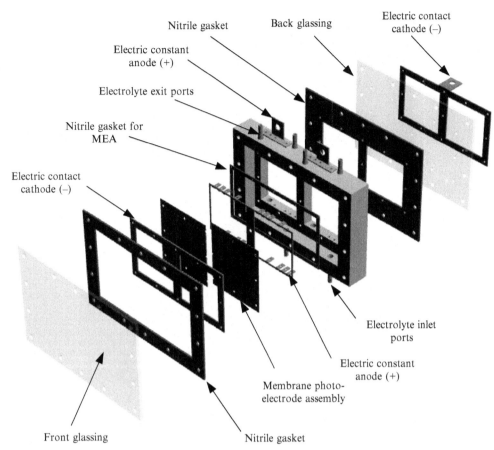

FIG. 5.56 Exploded view showing the polymeric membrane PEC design.

oxygen. This is the case reported in the current section. Further sections will report the use of the cell for a photochlor-alkali process and a copper-chlorine process.

The membrane preparation starts from a commercially available membrane electrode assembly (MEA) of $100\,cm^2$ that comprises a NAFION® membrane coated at both faces with porous carbon to form electrodes. Platinum-black catalysts are dispersed into the cathode layer. An electrodeposition setup is then constructed to deposit CuO/Cu$_2$O photocatalyst islands on the cathode surface.

The potentiostatic method of electrodeposition has been pursued. The electrodeposition procedure followed is outlined in Siripala et al. (1996). An amount of 700 mL of electrolyte of 0.1 M sodium acetate, anhydrous, 99% from Alfa Aesar, and 1.6×10^{-2} M copper(II) acetate 99% pure, anhydrous, from Acros Organic has been prepared.

Two membranes were electrodeposited, denoted as MPEA 1 and MPEA 2. The voltages applied were 50 mV for MPEA 1 and 250 mV for MPEA 2 with respect to the reference electrode and the membrane being set to negative as the working electrode. The counter electrode was a graphite rod with platinum wire coiled around it with a projected area of $10\,cm^2$. For the electrodeposition, a silver chloride reference electrode was used, which provided a constant reference point for measurement and for the potentiostat. A stirrer was also used to mix the electrolyte during the procedures.

Fig. 5.57 shows the electrodeposition bath having the membrane at the bottom with the cathode side facing upwards. When the MPEA 1 has been electrodeposited, the counterelectrode rod has been maintained in a fixed position. For MPEA 2 deposition, the counter electrode has been moved in parallel sections so that the membrane surface has been covered more uniformly. An optical-digital microscope image with $8 \times$ magnification has been obtained for the membrane before and after deposition. The images are shown in Fig. 5.58.

The color in Fig. 5.58a corresponds to islands of electrodeposited copper oxide, which visually confirms the procedure. The electric charge (in Coulomb) for each electrodeposition has been metered and the resulted layer thicknesses of photocatalyst islands calculated. This thickness varied between a few nanometers to 1 μm.

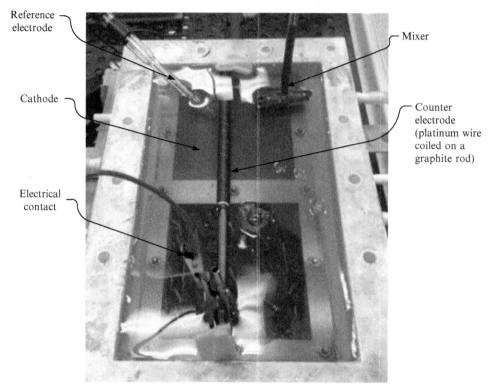

FIG. 5.57 Setup for photocatalysts electrodeposition to fabricate the PEC photocathode.

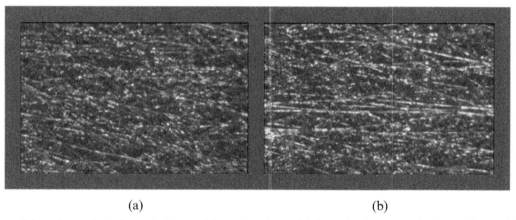

(a)   (b)

FIG. 5.58 Successful membrane electrodeposition layer on MPEA 1 membrane: (a) shows the electrodeposited copper layer, while (b) shows the original surface.

A TriSol Solar Simulator Class AAA was used to provide a stable source of light approximating the solar spectrum. This device was capable of illuminating an area larger than the reactor. During experiments consisting of hydrogen measurement, this was measured using a volume displacement method, which consists of a glass U tube filled with water, where both inlets are at the two top ends. When hydrogen enters one end, it displaces an equal amount of water in volume at the other end, which is at atmospheric pressure.

The cell operation has also been investigated by linear sweep voltammetry under dark and illuminated conditions. Initially, linear sweep voltammetry tests are performed in dark conditions with the original membrane electrodes assemblies before the photocatalyst electrodeposition. For MEA 2, one reference cell test has been conducted under constant voltage and with hydrogen production rates measured.

MEA 1 and MEA 2 under their original form without photocatalysts did not respond to light. Photosensitizers obtained from the first procedure on MPEA 1 showed a response to light with an increase in reaction rate and current density when exposed to light. Constant voltage experiments under light at constant voltage on MPEA 2 and MEA 2 show that under steady conditions, the reaction rate of the hydrogen production in MPEA 2 is doubled with respect to MEA 2 under light. The photosensitizers present in MPEA 2 were likely responsible for the increase in performance.

The results of the experiments show there is a significant light response both instantaneously and under steady conditions. Evidence is encountered to support that there is a photocurrent generated at the photocathode. By applying an external voltage based on a photovoltaic source, the process uses light as the single source, while also improving the evolution rate and effectiveness of the system.

The exergy destruction is found to be 69.7% when using concentrated solar light under a case study, which would render the light-based hydrogen production greater than the hydrogen produced from the applied current, assuming all the photonic current is used in the reaction.

5.5.5 Case Study 2: Photoelectrochemical Cell for Chloralkali Process

The electrochemical chloralkali process has been introduced in Chapter 3 as a method for the simultaneous production of hydrogen, chlorine and sodium hydroxide. A photoelectrochemical cell has been developed as reported in Ghosh (2014). The cell is the same one described in the previous section. In order to conduct the chloralkali process, the cell is fed with a saturated solution of NaCl. The membrane chloralkali process has been modified by adding a photocathode. The electrode, anolyte and catholyte reactions for the process are given in Table 5.5.

The flowsheet of the test system for the new chloralkali photoelectrochemical cell is shown in Fig. 5.59. A saturated aqueous solution of sodium chloride (NaCl) and water (H_2O) is prepared in electrolyte feed tank B.1. The saturated brine forms the anolyte and is fed to the anode side of the reactor. Likewise, a low concentration solution of sodium hydroxide (NaOH) in water is prepared in electrolyte feed tank B.2 to form the catholyte. The electrolytes in feed tanks B.1 and B.2 are heated to the process temperature using hot plates I.1 and I.2, respectively, and also stirred using magnetic stirrer bars. Before entering the reactor, the electrolytes pass through stop valves and then through the peristaltic pumps (C.1 and C.2), where the flow rates are externally connected and controlled by a designated computer.

After passing through the reactor, the depleted aqueous solution of sodium chloride is dumped to the electrolyte dump tank D.1 for latter analysis. Here the gas (mostly chlorine) is allowed to escape from the top of the dump tank and fed to the gas analyzer (E.1), where the concentration of chlorine gas is measured. On the other hand, an aqueous solution of sodium hydroxide, along with the gases produced, are fed to the electrolyte dump tank (D.2).

In the dump tank D.2, hydrogen gas is separated from the liquid and is allowed to escape from the top of the dump tank and fed to the hydrogen gas analyzer (E.2), where the concentration of hydrogen gas is measured. The dotted lines in Fig. 5.59 represent the nitrogen gas pipe. Nitrogen is purged through the reactor prior, as well as at the conclusion of the experiment, to keep the PEC reactor away from contact with air (especially oxygen) at all times. Fig. 5.60 shows a photograph of the experimental setup developed at the Clean Energy Research Laboratory of the University of Ontario Institute of Technology, where the chloralkali PEC is exposed to artificial solar radiation generated with TriSol Solar Simulator Class AAA.

The following equation is derived in Ghosh (2014) to express the molar flow rate of generated hydrogen with the cell, under the irradiance with the photonic radiation of known spectral irradiance I_λ, namely

$$\dot{n}_{H_2} = \frac{\Phi_{PEC}\eta_{opt}A}{2N_Ahc} \int_{\lambda_g}^{\infty} (1 - R_{\lambda,PEC})\lambda I_\lambda d\lambda \qquad (5.84)$$

TABLE 5.5   Electrode, Anolyte and Catholyte Reactions for the Chloralkali PEC

Reaction	Chemical equation
Anolyte	$NaCl(s) \rightarrow Na^+(aq) + Cl^-(aq), E^0 = 1.36\,V$
Anode	$2Cl^-(aq) \rightarrow Cl_2(g) + 2e^-$
Photocathode	$2H_2O(l) + 2e^- \rightarrow H_2(g) + 2OH^-(aq), E^0 = -0.8277\,V$
Catholyte	$Na^+(aq) + OH^-(aq) \rightarrow NaOH(aq)$
Cell	$2NaCl(s) + 2H_2O(l) \rightarrow 2NaOH(aq) + Cl_2(g) + H_2(g), E^0 = 2.187\,V$

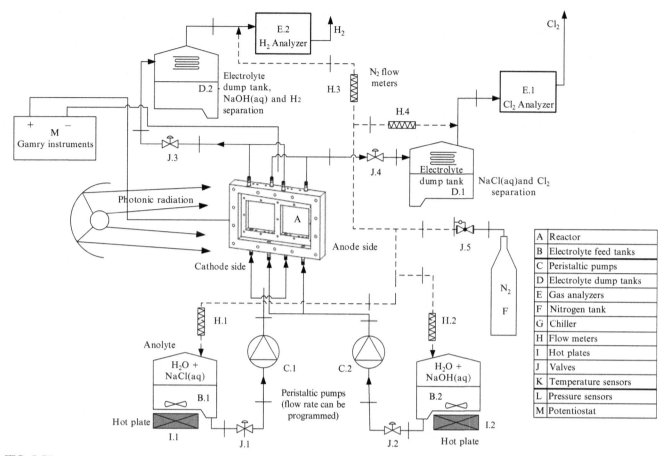

FIG. 5.59 Flowsheet of the test system for the new chloralkali photoelectrochemical cell. *Modified from Ghosh, S., 2014. Experimental Investigation of a New Light-Based Hydrogen Production System. MSc Thesis, University of Ontario Institute of Technology.*

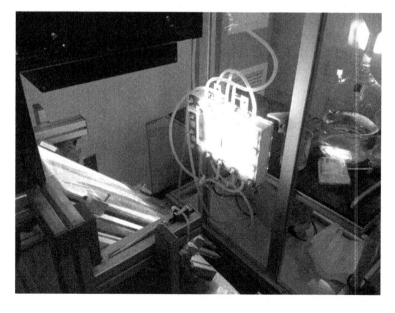

FIG. 5.60 The chloralkali PEC exposed to photonic radiation under the solar simulator. *Courtesy of Ghosh, S., 2014. Experimental Investigation of a New Light-Based Hydrogen Production System. MSc Thesis, University of Ontario Institute of Technology.*

TABLE 5.6   Design Point Conditions of the Test Chloralkali PEC With MPEA

State	$\dot{m}$ (g/s)	h (kJ/kg)	s (kJ/kg K)	ex^{ch} (kJ/kg)
Saturated NaCl brine input	10	−1939	7.847	80.67
Depleted NaCl brine output	7.319	−3954	1.089	60
Chlorine gas extraction	1.635	1.18E−7	3.146	1743
Water input	10	−13,422	10.48	49.96
NaOH solution output	10.94	−3669	9.436	200.6
Hydrogen gas extraction	5.36E−2	−1.3E−6	64.82	147,486

where Φ_{PEC} is the quantum efficiency of the chloralkali PEC process, η_{opt} is the optical efficiency of the setup, A is the active area of the photocatode, $R_{\lambda,PEC}$ is the spectral reflectance of the cell window at the photocathode side and λ_g is the bandgap wavelength of the photocatalyst.

In Eq. (5.84), the quantum efficiency is related to the level of bias voltage applied externally to the cell. For a practical cell, the applied voltage can be assumed to be $V_{PEC} = 1.55$ V. The design point conditions of the test chloralkali PEC with MPEA are given in Table 5.6. The ambient temperature is kept constant at 25°C and the operating temperature is varied accordingly.

Ghosh (2014) made a series of parametric studies with the photoelectrochemical chloralkali cell, which are summarized next. Hydrogen production is enhanced by a higher cell temperature. The prime reason for this is the fact that increasing the temperature increases the conductivities of the membrane and the electrolyte solutions, which implies more H_2 for the same input energy. A higher temperature also means a more derived (high concentration) NaCl. With the steady-state assumptions, this means that more Na^+ and Cl^- is produced, hence more OH^- are neutralized and more H_2 is produced. Increasing the temperature decreases the input work rate. This is mainly because for higher temperatures, the voltage losses decrease (especially the voltage drop across the catholyte and anolyte solutions), which means for the same current, the required cell voltage is less, which decreases the input work rate.

Fig. 5.61 shows that increasing the operating temperature increases the required current. As mentioned previously, increasing the temperature increases the concentration of NaCl in the input brine solution, which increases the H_2 production. In accordance with Faraday's law, the input current required is directly proportional to the hydrogen produced. Hence, higher current is needed to produce the increased H_2. Fig. 5.61 also shows that increasing the operating temperature decreases the input voltage. As discussed previously, increasing the temperature decreases the voltage losses. Hence, the input voltage decreases for the same current.

FIG. 5.61   Effect of operating temperature on input current and chloralkali PEC voltage. *Data from Ghosh, S., 2014. Experimental Investigation of a New Light-Based Hydrogen Production System. MSc Thesis, University of Ontario Institute of Technology.*

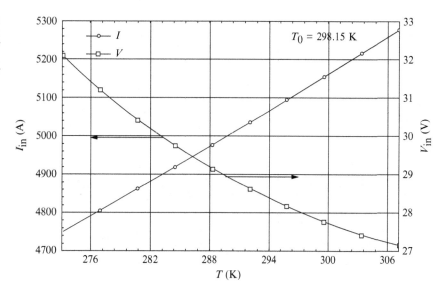

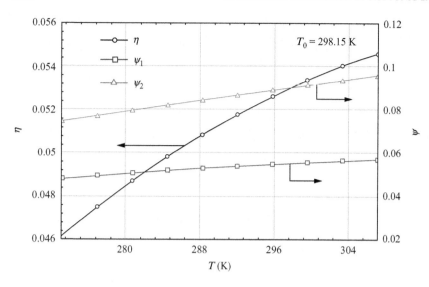

FIG. 5.62 Effect of operating temperature on energy and exergy efficiencies of chloralkali PEC. *Data from Ghosh, S., 2014. Experimental Investigation of a New Light-Based Hydrogen Production System. MSc Thesis, University of Ontario Institute of Technology.*

The exergy efficiency of the PEC can be defined based on hydrogen generation only. This efficiency will be equal to the exergy in the generated hydrogen divided by the exergy input to the cell. This hydrogen-only efficiency is referred here as ψ_1. When chlorine and sodium hydroxide are taken into account as useful outputs, then the exergy efficiency ψ_2 is defined as the ratio of the sum of exergies' outputs (hydrogen, chlorine, sodium hydroxide) to the total exergy input into the process.

Fig. 5.62 shows the effect of varying operating temperatures on energy and exergy efficiencies of the PEC system. As mentioned earlier, increasing the operating temperature decreases the losses in the PEC system. So in general, by increasing the operating temperatures, the system becomes more efficient both energetically and exergetically. It is important to note that there is a significant increase in ψ_2 (around 2%) as opposed to ψ_1 (around 1%).

The results are as yet preliminary, but they show that as the current increases, the cell voltage increases sharply, after which the voltage reaches saturation; however, the experimental results don't reach saturation. This is because the voltage losses are much higher and the current was only measured up to 2.8 A.

5.6 HYBRID PHOTOCATALYSIS SYSTEMS

Heterogeneous photocatalysis systems with nanoparticulate photocatalysts can be hybridized by the application of electrodes for an enhanced production of photonic hydrogen. These types of reactors were previously reviewed in Minggu et al. (2010). In these systems, two photochemical reactors are joined together to generate gaseous outputs in different manifolds. The configuration of hybridized systems is shown in Fig. 5.63.

The first reactor is a hydrogen-evolving photochemical reactor, requiring an electrolyte that comprises electron acceptors. A photoanode is also placed in this reactor. The incident photons will induce a water oxidation process and an electron acceptor will reduce $M^+ \rightarrow M$. Once passing into the next reactor occurs, the conditions for

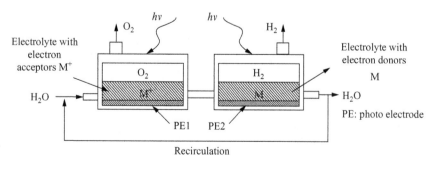

FIG. 5.63 Schematic of a dual-cell hybrid photocatalysis system. *Modified from Minggu, L.J., Daud, W.R.W., Kassim, M.B., 2010. An overview of photocells and photoreactors for photoelectrochemical water splitting. Int. J. Hydrog. Energy 35, 5233–5244.*

hydrogen-evolving reactions are set. The induced light will excite the photocatalysts and the photocathode. The electron donors will give electrons to the electrolyte from which hydrogen gas is generated.

Several system variants may exist, e.g., with electrodes polarized or not, with different or the same materials for electron donors and acceptors in both reactors, with particulate photocatalysts stirred in the solution or without, with membrane separators or without. Here, we will review some relevant systems with photocatalysis as devised and tested in the Clean Energy Research Laboratory at the University of Ontario Institute of Technology.

5.6.1 Hydrogen-Oxygen Generating Process

Baniasadi et al. (2013) developed a dual-cell hybrid photocatalytic system for enhanced hydrogen evolution, as shown in Fig. 5.64. The system is a kind of photoelectrochemical cell, having an anion exchange membrane and a specific catholyte. A sulfur-based photochemical system is used as the catholyte with stirred ZnS nanoparticulate photocatalysts and dissolved electron donors.

The photocatalysts were pure-grade 99.99% ZnS 325 mesh nanoparticulate from Alpha Aesar with used sodium sulfide electron donors given as 3% w/v in an aqueous solution. The aqueous solution is alkalinized to pH 13.2 using sodium hydroxide in both the photoreactor and dual-cell. The reactions occurring in the cathode compartment will be of two kinds. First is the water reduction that is catalyzed heterogeneously at the photocatalysts surface and at the electrode surface. The two reactions are

$$2H_2O(l) + 2e^- \xrightarrow{\text{electrode}} H_2(g) + 2OH^-(aq) \tag{5.85a}$$

$$2H_2O(l) + 2e^- \xrightarrow{h\nu} H_2(g) + 2OH^-(aq) \tag{5.85b}$$

Sodium sulfide dissociates in catholyte, and the following reaction occurs both at the catalyst and electrode surfaces

$$2S^{2-}(aq) \rightarrow S_2^{2-}(aq) + 2e^- \tag{5.86}$$

The obtained polarization curve of the cell is shown in Fig. 5.65. Since the delivered electrons to the reaction site are supplied by both photons and power supply, hydrogen evolution takes place in a hybrid manner. Both hydrogen and oxygen reactions initiate with higher rate constants, but the production rate becomes steady after almost 20 min of operation and the production rate decreases to an almost constant value as shown in Fig. 5.66.

FIG. 5.64 Dual-cell hybrid photoelectrochemical reactor with anion exchange membrane. *Modified from Baniasadi, E., Dincer, I., Naterer, G.F., 2013. Measured effects of light intensity and catalyst concentration on photocatalytic hydrogen and oxygen production with zinc sulfide suspensions. Int. J. Hydrog. Energy 38, 9158–9168.*

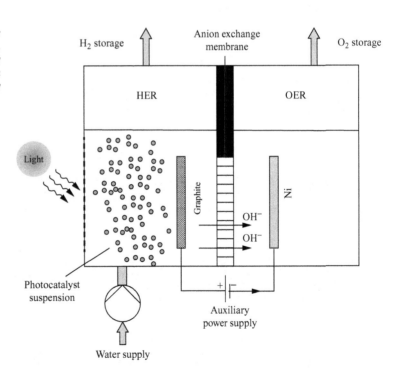

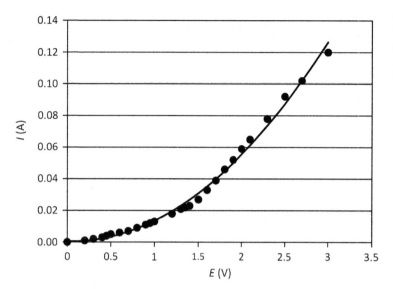

FIG. 5.65 Experimental polarization curve with the hybrid PEC shown in Fig. 5.64. *Data from Baniasadi, E., Dincer, I., Naterer, G.F., 2013. Measured effects of light intensity and catalyst concentration on photocatalytic hydrogen and oxygen production with zinc sulfide suspensions. Int. J. Hydrog. Energy 38, 9158–9168.*

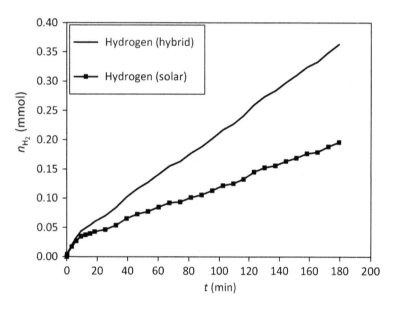

FIG. 5.66 Generated hydrogen versus time with and without the application of an external bias at 1000 W/m² irradiance with 1.5% w/v ZnS photocatalyst. *Data from Baniasadi, E., Dincer, I., Naterer, G.F., 2013. Measured effects of light intensity and catalyst concentration on photocatalytic hydrogen and oxygen production with zinc sulfide suspensions. Int. J. Hydrog. Energy 38, 9158–9168.*

5.6.2 Hybrid Photoelectrochemical Chloralkali Process

A hybrid photoelectrochemical process has been investigated using a dual-cell setup, and a hydroxyl selective membrane has been reported in Rabbani et al. (2014). This system produces hydrogen, chlorine and caustic soda in an aqueous solution. Electricity is applied to the electrodes, while at the same time, light is exposed at the cathode. ZnS/CdS nanopowder photocatalysts are stirred within the catholyte consisting of an aqueous solution of Na_2S-Na_2SO_3.

While hydrogen forms both at the cathode and the surface of the nanopowder photocatalysts, bubbles are generated and collected in a container above the solution during a constant-volume process. The hydroxyls cross the membrane and combine with sodium ion in the anode compartment, while chlorine dissociates electrochemically and is also collected. Tests under light exposure and in dark show the positive influence of the light on the reaction rate, which is thenceforth enhanced. Nevertheless, valorization of chlorine gas and caustic soda will diminish the return on investment and lower the hydrogen production cost.

The cell is shown in Fig. 5.67 and a photograph of the experimental setup is shown in Fig. 5.68. The experiments were performed under high-energy photonic radiation (upper visible spectrum) generated with a mercury lamp. The results are presented in Fig. 5.69.

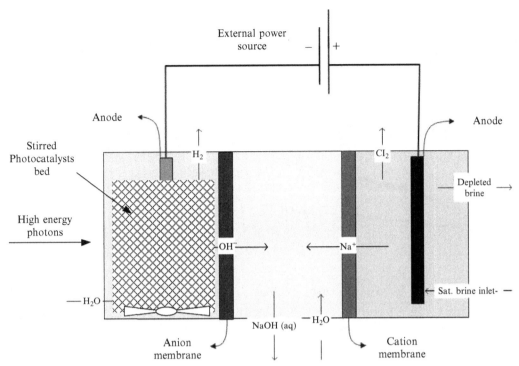

FIG. 5.67 Hybrid photoelectrochemical reactor for chloralkali process. *Modified from Rabbani, M., 2013. Design, Analysis and Optimization of a Novel Photo-Electrochemical Hydrogen Production System. PhD Thesis, University of Ontario Institute of Technology.*

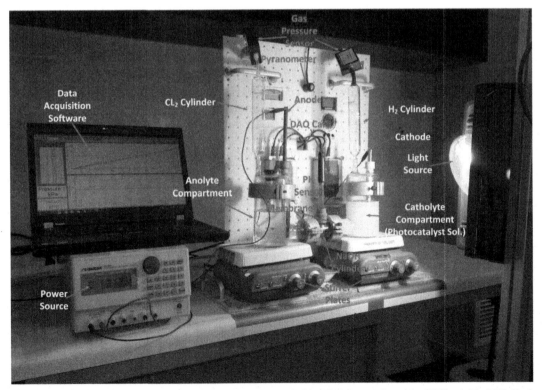

FIG. 5.68 Photograph of the experimental cell setup photoelectrochemical chloralkali setup. *Courtesy of Rabbani, M., 2013. Design, Analysis and Optimization of a Novel Photo-Electrochemical Hydrogen Production System. PhD Thesis, University of Ontario Institute of Technology.*

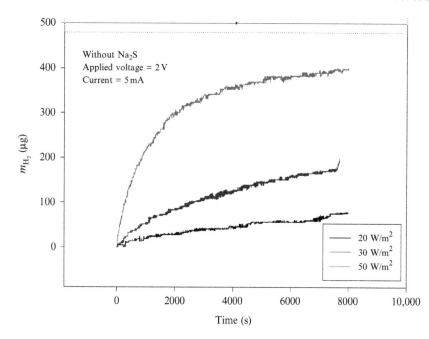

FIG. 5.69 Experiments under high-energy photon flux (mercury lamp) with no hole scavenger used and 2 g/425 ml ZnS photocatalyst loading. *Data from Rabbani, M., 2013. Design, Analysis and Optimization of a Novel Photo-Electrochemical Hydrogen Production System. PhD Thesis, University of Ontario Institute of Technology.*

In the results presented in Fig. 5.69, sodium sulfide (which is a hole scavenger material and the only consumable in the system), is not added to the catholyte solution. The supply of electrons is only provided by a cylindrical nickel electrode. At lower intensities, the amount of hydrogen produced is almost linear. After ~20 min at higher intensity levels, however, the rate of hydrogen production is set to a steady-state constant value. During these experiments, neither chlorine nor sodium hydroxide was produced. The production of hydrogen, however, confirms that the hole scavenger can be replaced with an electrode in a photoelectrochemical process. Eliminating the hole scavenger means there are no consumables in the system except water, which is available in enormous amounts.

5.6.3 Hybrid Photoelectrochemical Process for Cu-Cl Cycle

A hybrid dual-cell test system has been developed to demonstrate the coupling between a hydrogen evolution photochemical process and another selected proton-generation half reaction. In this system, an aqueous solution of HCl and CuCl containing nanopowder TiO_2 catalysts is prepared at the anode. The catholyte consists of an aqueous solution of Na_2S-Na_2SO_3 in which nanopowder ZnS/CdS photocatalysts are stirred.

The cell process is shown in Fig. 5.70. The specific electrodes are immersed in catholyte and anolyte. Each one or both reaction vessels are exposed to concentrated solar radiation under the solar simulator. Voltage is applied to the electrodes. The experiments done in dark and illumination conditions indicate the positive and effective action of light, which essentially augments the production rate.

While the cathodic reaction is a photocatalyzed water-reduction process using ZnS nanopowder photocatalysts, Na_2S hole scavengers and a polarized cathode, the anodic reaction is

$$2CuCl(aq) + 2HCl(aq) \rightarrow 2CuCl_2(aq) + 2H^+(aq) + 2e^- \tag{5.87}$$

Zamfirescu et al. (2012) originally proposed a similar process as shown in Fig. 5.70, except using a different photocatalysis system with homogeneous photocatalysis dissolved in catholyte. Ratlamwala (2013) further developed this system with heterogeneous catalysts at both the anolyte and catholyte compartments. The photograph of the dual-cell setup is shown in Fig. 5.71. In one system variant reported in Ratlamwala and Dinçer (2014), only the catholyte compartment includes photocatalytic systems and is illuminated, whereas the anode compartment is dark.

The catholyte is an aqueous solution of NaOH (33 g/l) with dissolved Na_2S (220 g/l). Nanoparticulate ZnS photocatalysts were stirred continuously at the cathode during the experiments. The cation exchange membrane (CEM) is NAFION 115, obtained from Ion Power Inc., having a thickness of 127 μm and 0.1 S/cm conductivity. The anolyte is an aqueous solution HCl with dissolved CuCl in a range of concentrations from 33 to 66 g/l. Tests are also performed with a titanium dioxide nanopowder photocatalyst stirred in the anolyte and with a TiO_2 photoanode prepared by screen printing on conductive glass.

FIG. 5.70 Hydrogen-generating photoelectrochemical process for the copper-chlorine cycle.

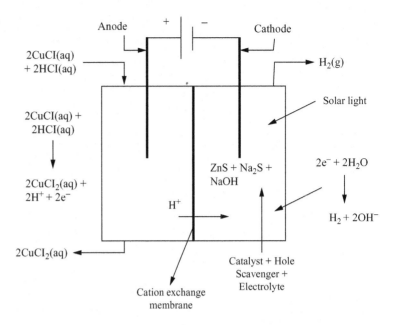

FIG. 5.71 Dual-cell hybrid system for H_2 generation within Cu-Cl water-splitting cycle. *Courtesy of Ratlamwala, T.A.H., 2013. Design, Analysis and Experimental Investigation of Cu-Cl Based Integrated Systems. PhD Thesis, University of Ontario Institute of Technology.*

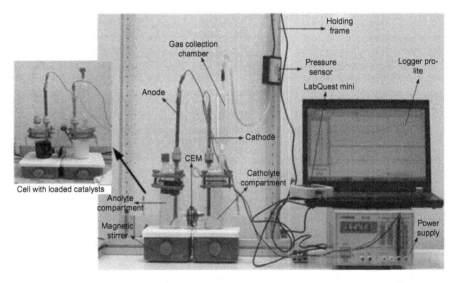

Fig. 5.72 shows the process flow diagram of the experimental setup with the experiment performed under artificial light produced with an OAI Trisol simulator. The analysis of variance was applied to the experimental data to correlate the hydrogen production rate with various parameters, such as light intensity, voltage, CuCl concentration in anolyte and particulate photocatalyst concentration in catolyte. The generated hydrogen is collected in a glass collection chamber, which is a recipient of constant volume. The pressure in the vessel is monitored.

The number of moles of gas within the collection chamber is calculated based on the ideal gas law, knowing the pressure, volume and temperature. Based on this type of calculation, the production rate of hydrogen can be derived. Fig. 5.73 shows the variation of hydrogen production rate with the applied voltage. For the data from Fig. 5.73, the CuCl concentration is fixed at 53 g/l. The increase of light intensity with 25% leads to a 12.5% increase of productivity (on average). This confirms the beneficial effect of direct light application in photoelectrolysis.

Further confirmation in this respect is provided by the results from Fig. 5.73, which shows an increase of hydrogen production when light intensity is increased. For the test cell, most of the overpotential is due to the anode losses. The anodic overpotential accounted for more than 60% of cell voltage.

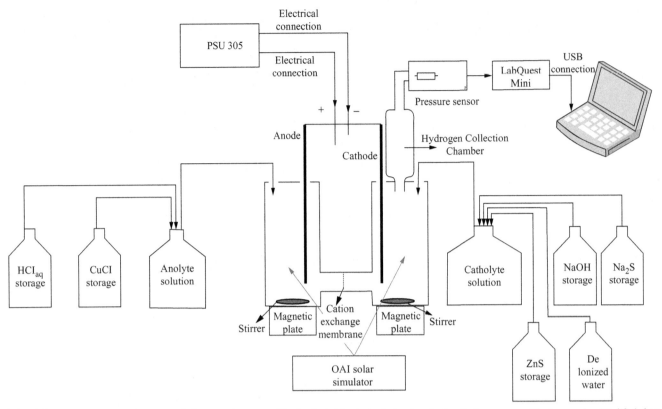

FIG. 5.72 Process flow diagram of the experiments with the hybrid photoelectrochemical cell for copper-chlorine cycle. *Modified from Ratlamwala, T.A.H., Dinçer, I., 2014. Experimental study of a hybrid photocatalytic hydrogen production reactor for Cu-Cl cycle. Int. J. Hydrog. Energy 39, 20744–20753.*

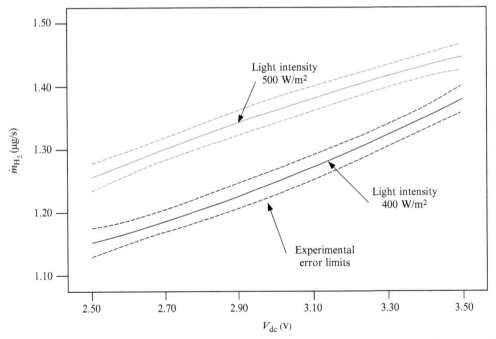

FIG. 5.73 Effect of light intensity on photoelectrolytic cell for $CuCl_2$ and H_2 production ($CuCl$ concentration 53 g/l. *Data from Ratlamwala, T.A.H., 2013. Design, Analysis and Experimental Investigation of Cu-Cl Based Integrated Systems. PhD Thesis, University of Ontario Institute of Technology.*

When no voltage is applied from an external source to the electrodes, the experimental determinations show the hydrogen generation rate is ~0.3 μg/s. This reflects the fact that in hybrid photoelectrochemical operation, about one-fourth of hydrogen production has been due to the photocatalytic effect, while the rest of production is due to electrochemical effect. Fig. 5.74 shows more results on the influence of solar light on hydrogen production. As seen, the increase of hydrogen production rate is nonlinear and is favored by higher intensities.

The photocatalytic effect has been increased to 32% when titanium dioxide nanoparticle were added in the anolyte and illuminated, as they act as photocatalysts. This can be observed in Fig. 5.75, where the hydrogen production amount is plotted against the time for two situations: with and without particulate photocatalysts at the anode. If photocatalysts are used at the anode, then the reaction rate clearly increases.

Fig. 5.76 shows the variation of cell energy and exergy efficiencies with the current density when the cell operation is hybrid. The average value energy efficiency of the cell is 5%. The average exergy efficiency of the cell is 4.5%. The

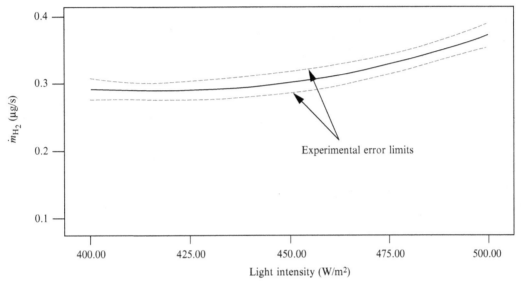

FIG. 5.74 Variation of hydrogen production rate with light intensity when CuCl concentration at anode is set to 50 g/l, and concentration of photocatalyst at cathode is set to 20 g/l. *Data from Ratlamwala, T.A.H., 2013. Design, Analysis and Experimental Investigation of Cu-Cl Based Integrated Systems. PhD Thesis, University of Ontario Institute of Technology.*

FIG. 5.75 Influence of titanium dioxide photocatalyst (at anode) on hydrogen production. *Data from Ratlamwala, T.A.H., 2013. Design, Analysis and Experimental Investigation of Cu-Cl Based Integrated Systems. PhD Thesis, University of Ontario Institute of Technology.*

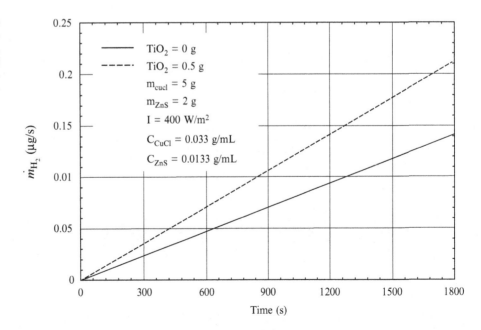

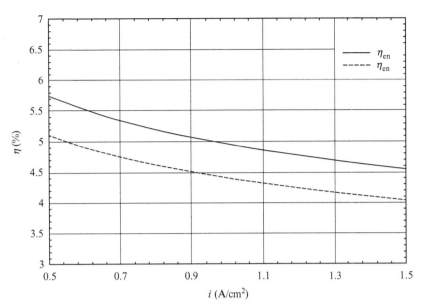

FIG. 5.76 Influence of current density on energy and exergy efficiency of the photoelectrochemical cell. *Data from Ratlamwala, T.A.H., 2013. Design, Analysis and Experimental Investigation of Cu-Cl Based Integrated Systems. PhD Thesis, University of Ontario Institute of Technology.*

maximum determined energy efficiency of the cell is 5.55%. When current density increases, the efficiencies tend to decrease.

The minimum distance between the electrodes has been set to 10 mm for the experiments, while the maximum distance was 60 mm. When the distance is decreased to a minimum, the productivity of the cell will increase by ~7%. There is clear potential of efficiency increase if the dual cell configuration is replaced with a planar MEA type of cell. In dual-cell, the distance between the electrodes is rather high.

5.7 INTEGRATED PHOTONIC ENERGY-BASED HYDROGEN PRODUCTION SYSTEM

5.7.1 Solar-Through Photocatalytic Reactor Integrated with PV-Electrolysis

Higher rates of hydrogen generation can be obtained with higher photon flux incident on a photochemical reactor, because an increased TOF can be obtained. The increase of photon flux can be obtained by light concentration. Since more hydrogen is generated per unit of volume of reactor, the integration of photochemical reactors with solar concentrators appears to be a good design choice in designing better photonic-based hydrogen production systems.

An increased TOF means a higher production rate for a smaller catalyst quantity, and therefore, a reduced investment cost in expensive material from platinum group (included in catalyst production). On the contrary, there will be an additional cost of silver and glass required for the solar concentrator; however, it appears that an investment in a solar concentrator will balance competitively the cost reduction for the catalyst.

One more aspect must be thoroughly considered, namely, the required bandgap energy of the photocatalyst. Due to the bandgap energy limitation, only part of the solar radiation photons are useful for a photoreaction. In general, most of the spectrum is lost, since in general, photocatalysts work better starting from blue to higher-energy colors.

Here, a solar-through photocatalytic system integrated with PV-electrolysis is presented as a practical solution for better conversion of sunlight photons to hydrogen extracted from water. A small-scale solar proof-of-concept purposed through-type concentrator has been constructed and integrated with a PV-electrolysis system at the Clean Energy Research Laboratory of the University of Ontario Institute of Technology. The system has been tested under a solar simulator providing 26 μmol useful photons per second.

Solar through-concentrators are typical devices that can concentrate sunlight on a tube placed at a focal point. The tube is parallel with the through. These types of concentrators can be constructed at very large scale according to the current technology at capacities of 200 MW harvested light radiation. These systems were demonstrated for power generation, e.g., nine power-generating stations totaling 354 MW were built in the 1980s in the Mojave Desert of California. Very relevant to the through-type concentrator integration with photocatalytic system on a large scale

FIG. 5.77 Principle diagram of continuous-flow photocatalytic reactor.

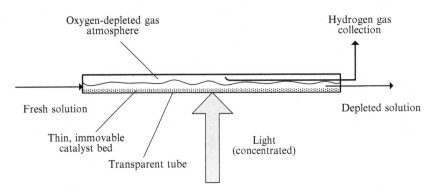

is the fact that well established technology do exist to construct glass air evacuated tubes as solar receivers are installed at the focal point.

The lab-scale system incorporates: (i) a photocatalytic reactor operating with a heterogeneous catalyst; (ii) a linear PV array placed behind the photoreactor; (iii) a small electrolysis cell driven by the PV array; (iv) a water coil to generate warm water from the residual solar heat; (v) a peristaltic pump to feed water to the heating coil; and (vi) a constant-temperature water bath.

The principle of a novel flow reactor is presented schematically in Fig. 5.77. It comprises a thin, immovable catalyst bed entrapped in a glass tube reactor with continuous flow operation. Concentrated radiation is provided at the tube bottom. A fresh liquid solution for catalysis is provided at the left-side end of the tube. Once the photocatalysis progresses, hydrogen evolves in the free space above the liquid level. The gas is collected from the free space, while the depleted solution is extracted (quasi)continuously from the right end of the reactor.

The size of the reactor has been designed to fit the OAI Trisol solar light simulator of the lab, which generates light according to the solar spectrum on a surface of 20×20 cm. A solar through concentrator has been devised from second-surface mirror strips shimmed along a parabolic line such that light is focused along a line at the focal point. A laser pen has been used to align the mirrors, which were glued on two parallel parabolic supports spaced 15 cm apart. The photoreactor, which is made of a 15 mm o.d., 10-inch long glass tube, is placed at the focal point. The photoreactor tube has an instrument purging port for nitrogen (or argon), a fresh solution injection port, a depleted solution extraction port and a hydrogen (gas) extraction port (Fig. 5.78).

Once the concentrated light falls on the photochemical reactor tube, part of the photon flux is absorbed, another part is reflected and the other is scattered. A PV array has been attached after the reactor in order to harvest the remaining portion of the solar radiation, which is transmitted through in a scattered manner. The PV array is made of four PV modules connected in a series. Because concentrated radiation converts partially into heat, a planar heat exchanger has been custom-made by bending a 4 mm outside diameter glass tube in the form of a U-shape. The coil has been attached to the active face of the PV array. In the tube coil so-made water is circulated with a flow rate finely controlled by an adjustable peristaltic pump. The photograph in Fig. 5.79 shows the arrangement around the tubular reactor with the PV array and cooling coil attachment. The coil is in good thermal contact with the blackish-color surface of the PV array, which warms up.

The lab-scale proof-of-concept setup is shown in Fig. 5.80. The system also comprises a fresh electrolyte supply system made of a syringe attached to one of the reactor tube sides with the help of a rubber stopper. Similarly, a syringe is installed in the other part to extract the depleted electrolyte. The ports for inlet and outlet are always submerged in the liquid electrolyte and never connect directly with the gas space. A gas intake port is made in the rubber stopper through which nitrogen-purging gas is supplied. A gas outlet port is attached at the other side to extract (sample) product gas for hydrogen concentration measurement. Experiments were also performed with the nitrogen-purging

FIG. 5.78 The cylindrical tube photochemical reactor illuminated with concentrated radiation produced with the OAI trisol solar spectrum simulator.

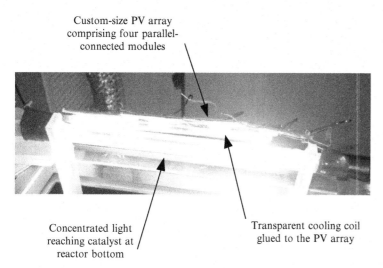

Custom-size PV array
comprising four parallel-
connected modules

Concentrated light
reaching catalyst at
reactor bottom

Transparent cooling coil
glued to the PV array

FIG. 5.79 Photograph showing the attachment of the transparent cooling coil and the PV array to the photochemical reactor exposed to concentrated radiation.

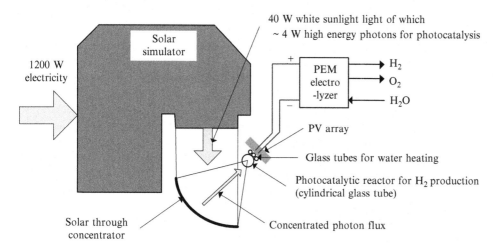

Solar
simulator

40 W white sunlight light of which
~ 4 W high energy photons for photocatalysis

1200 W
electricity

PEM
electro
-lyzer

H_2

O_2

H_2O

PV array

Glass tubes for water heating

Photocatalytic reactor for H_2 production
(cylindrical glass tube)

Solar through
concentrator

Concentrated photon flux

FIG. 5.80 Lab-scale proof-of-concept photochemical reactor setup integrated with PV-electrolysis and heat recovery system for hot water generation.

gas switched off. In this case, the outlet port has been connected to a syringe pump that accurately metered the emanated volume while maintaining the pressure constant in the sealed reactor enclosure.

The setup comprises a water circuit, which supplies water to the coil and returns it in a tank. Only the outlet (warm) water temperature is measured. The PV array discharges on an off-the-shelf miniature PEM electrolysis cell to generate hydrogen as a different manifold. Fig. 5.81 shows a photograph of the setup during an experimental run.

An engineered cadmium-zinc-sulfate photocatalyst loaded with palladium cocatalyst has been prepared. The electrolyte consists of an aqueous solution of SO_3^{2-}. This class of reactions has been discussed previously in this chapter. The reaction products consist mainly of hydrogen gas and an aqueous solution slightly more basic that the fresh electrolyte, since hydroxyls are generated during the process. The product solution will have dissolved SO_4^{2-} and $S_2O_3^{2-}$. A black box description of the photocatalytic process is shown in Fig. 5.82.

The preliminary results shown in Fig. 5.83 in the form of an energy balance are promising. Practical lab tests showed a 10% increase of hydrogen production efficiency with respect to PV electrolysis, due to the simultaneous use of photocatalysis. Potentially, this figure can be well improved. In addition, optical and heat losses can be dramatically reduced in a next generation of reactor design.

5.7.2 Integrated Photocatalytic Systems with Heliostat Field-Central Receiver Concentrators

Many variants of the central receiver system integrated with photochemical reactions were conceptualized in the specialized literature. A heliostat field will concentrate light radiation atop a solar tower. The concentrated light is

FIG. 5.81 Photograph of the solar-through-based photocatalytic reactor setup.

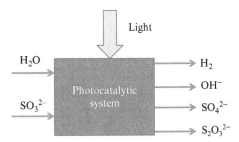

FIG. 5.82 Black box simplified description of the conducted photocatalytic process.

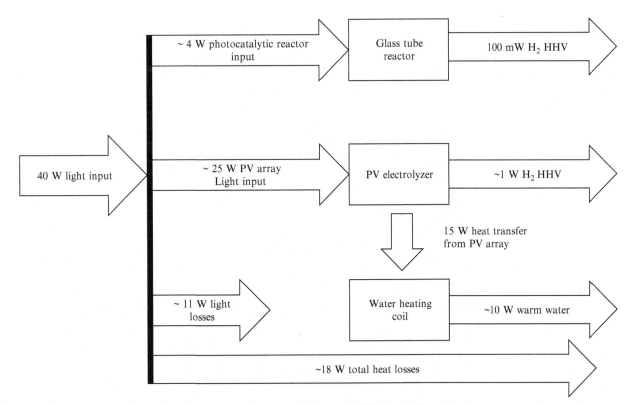

FIG. 5.83 Energy balance of the tested photochemical reactor with integrated PV-electrolysis and hot water cogeneration system.

processed (e.g., through spectral splitting) and eventually a highly intense radiation of the right spectrum is diverted toward a multitude of photoreactors for hydrogen production. Several variants of this type of system for green hydrogen production are introduced in this section.

5.7.2.1 Lab-Scale Proof-of-Concept System with Miniature Solar Tower

A continuous-flow photocatalysis system with concentrated radiation and PV integration is developed and tested by Shamim (2013) at the Clean Energy Research Laboratory of the University of Ontario Institute of Technology. The system is further detailed in Shamim et al. (2014).

A cavity receiver has been devised that receives concentrated radiation (10 suns) through this aperture. In the centerline, a glass reaction vessel is placed for photocatalysis, whereas at the walls, multiple PV cells are installed. The concentrated radiation is focused on the reaction vessel, but this is partially scattered by the semiconductor nanoparticulate photocatalysts.

A sufficient part of the scattered light, especially the lower-energy spectrum, is effectively used by the PV cells to generate power. The overall setup includes an electrolyte supply that is continuously fed to the photoreactor at an optimized rate. The liquid electrolyte carrying hydrogen bubbles and dissolved hydrogen are transferred into a dumping vessel in which a vacuum is maintained automatically such that gas separates and is metered. The electricity generated by the PV cells covers well the requirement for catalyst stirring and electrolyte pumping. This reactor successfully demonstrated hydrogen production at ml/h scale.

Heliostats are designed by cutting small 30 mm square pieces from a back-surface silver mirror. For each individual heliostat, an adjustable stand is designed using small brass hinges. The hinge is compressed in a vice to make it rigid, so that it may be adjusted to focus the reflected light beam in the vertical direction. Then the hinge is attached to the heliostat base using a single screw that allows adjustment of the light beam in the horizontal direction. Heliostats are mounted on the base using a heat-resistant adhesive. Once the heliostats are calibrated, an adhesive is applied to each stand to fix the position. The base of the heliostat field is covered by nonreflective material in order to absorb the incident light that is not reflected by the heliostats. Fig. 5.84 shows the constructed miniature heliostat field covering a square area of 20×20 cm^2.

The cavity receiver is built as shown in Fig. 5.85. It comprises nine PV modules fixed inside the cavity at the walls. The solar cells need to be installed in a manner that does not change the status of the photocatalytic process as the primary process. The cells, therefore, must not interfere with light incident on the photocatalytic reaction vessel, but should harness the radiation scattered by the photocatalytic process. The size of the PV cells is an important factor to consider, since a smaller size provides greater flexibility in fitting the greatest number of cells within the constraints of the existing reactor design.

The cavity receiver increases the efficiency of a system on the premise of trapping the photons inside the cavity. Since all exit routes are closed off for photons, the incoming photons from the aperture have to be used by any one of the three processes occurring simultaneously. Failing that, they are used for direct thermal usage for either

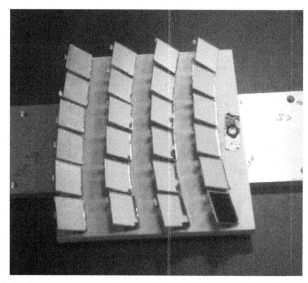

FIG. 5.84   Custom-made miniature heliostat field for the integrated system, reported in Shamim (2013).

FIG. 5.85 Custom-made cavity receiver with installed PV module. The photocatalytic reactor location is at the center. *Courtesy of Shamim, R., 2013. Experimental and Theoretical Investigation of a New Integrated Solar Tower System for Photocatalytic Hydrogen and Power Production. MSc Thesis, University of Ontario Institute of Technology.*

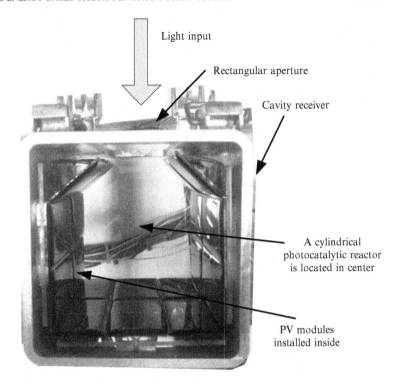

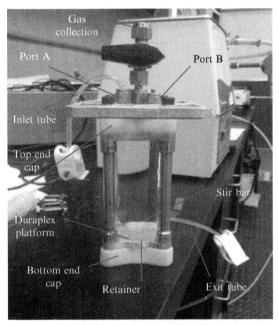

FIG. 5.86 Photograph of the reaction vessel. *Courtesy of Shamim, R., 2013. Experimental and Theoretical Investigation of a New Integrated Solar Tower System for Photocatalytic Hydrogen and Power Production. MSc Thesis, University of Ontario Institute of Technology.*

hot air or hot water. In this case, the top of the reactor housing is not closed and provides an escape route for the photon, but it is not desirable to close it off due to a need for air flow, which is blown from below.

The reaction vessel is shown in Fig. 5.86. The reactor is removable and dismountable. For this reason, the existing configuration of steel tubing that connects the reaction vessel to the piston cylinder is modified and a valve is attached that can be closed off to allow the detachment of the reaction vessel from the main reactor body.

The process flow diagram of a photocatalytic process is shown in Fig. 5.87. The fresh solution is stored in a sealed compressible bag in order to maintain the reservoir pressure. The design is constrained by the availability of a single pump. Flow from the fresh reservoir is driven by a pressure differential and controlled by the restrictor. The reservoir is at atmospheric pressure. The reactor is operated at a pressure below atmosphere. The fresh solution enters the reaction vessel at the top. The used solution is extracted from the bottom, through a retainer. The vacuum pump maintains the pressure in the waste reservoir. Keeping the pressure constant at all points minimizes variations in flow. The flow rate is calculated by taking weight measurements at regular intervals.

The micro flow pump is a major component of the flow system. A peristaltic pump is used to prevent the electrolyte from coming into contact with any pump materials, thereby preventing corrosive effects of sulfite ions. The pump is digitally controlled and is able to provide a maximum flow rate of 500 ml/h between equal pressure ends. A general view of the integrated lab-scale test system is shown in Fig. 5.88.

As a main result, the experimental tests showed that as the concentration of the electrolyte is decreased, the yield at the end of the time period chosen for the study and also the initial reaction rate decrease. Fig. 5.89 compares the production of hydrogen with respect to time for three different concentrations. At the highest concentration of 0.3 M, the production is almost linear for the time period. This is because the production rate of hydrogen is small, and due to the large amount of electron donor present in the solution, the change in the solution composition is not large enough to decrease the reaction rate. When the initial concentration of the solution is decreased, the parabolic shape of the plot is more visible.

Considering the first couple minutes of the reaction to be linear, based on the experimental results, the initial reaction rates can be calculated to be 0.0019, 0.0015, and 0.0011 mmol/min for the concentrations of 0.3, 0.03, and 0.003 M, respectively.

Increasing the concentration of the solution is only beneficial up to about 0.1 M. Beyond this point on the plot, the initial rate of the reaction does not increase steeply, and may not be enough to justify the increased complexity of a system of higher concentration. For the batch product, the productive reaction time is at higher concentrations longer,

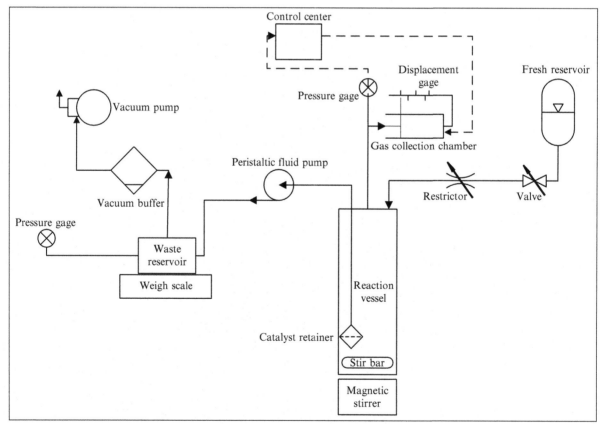

FIG. 5.87 Process flow diagram of the experimental setup with integrated photocatalysis reactor. *Modified from Shamim, R., 2013. Experimental and Theoretical Investigation of a New Integrated Solar Tower System for Photocatalytic Hydrogen and Power Production. MSc Thesis, University of Ontario Institute of Technology.*

FIG. 5.88 Overall view of the integrated photocatalysis system. *Courtesy of Shamim, R., 2013. Experimental and Theoretical Investigation of a New Integrated Solar Tower System for Photocatalytic Hydrogen and Power Production. MSc Thesis, University of Ontario Institute of Technology.*

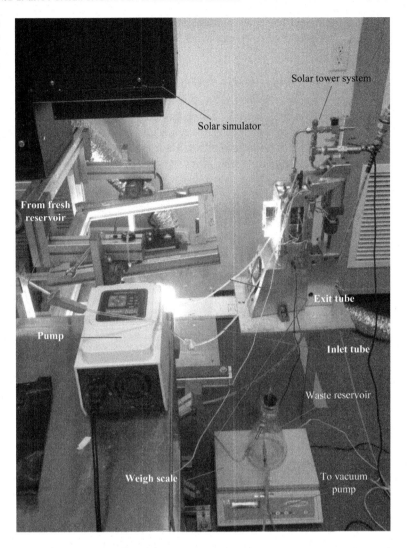

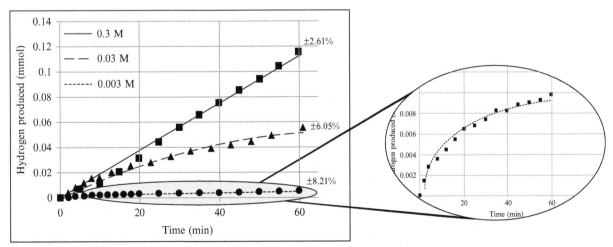

FIG. 5.89 Photocatalytic hydrogen production comparison at different electrolyte concentrations. *Data from Shamim, R., 2013. Experimental and Theoretical Investigation of a New Integrated Solar Tower System for Photocatalytic Hydrogen and Power Production. MSc Thesis, University of Ontario Institute of Technology.*

and so is the final hydrogen quantity achieved after 60 min, but from the point of view of a continuous flow process, the initial reaction rate is more important than the batch process product.

5.7.2.2 Large-Scale Concept System of Solar-Tower-Based Photocatalysis

A large-scale solar-tower-based photocatalytic system integrated with concentrated PV power generation and ORC (organic Rankine cycle) power generation is developed by Shamim (2013). A very similar system that integrates alkaline electrolyzer is reported in Zamfirescu and Dinçer (2014), whereas in Shamim (2013) a large-scale PEM electrolysis plant is integrated. The process flow diagram is shown in Fig. 5.90. In this system, the concentrated radiation is split into three useful spectra.

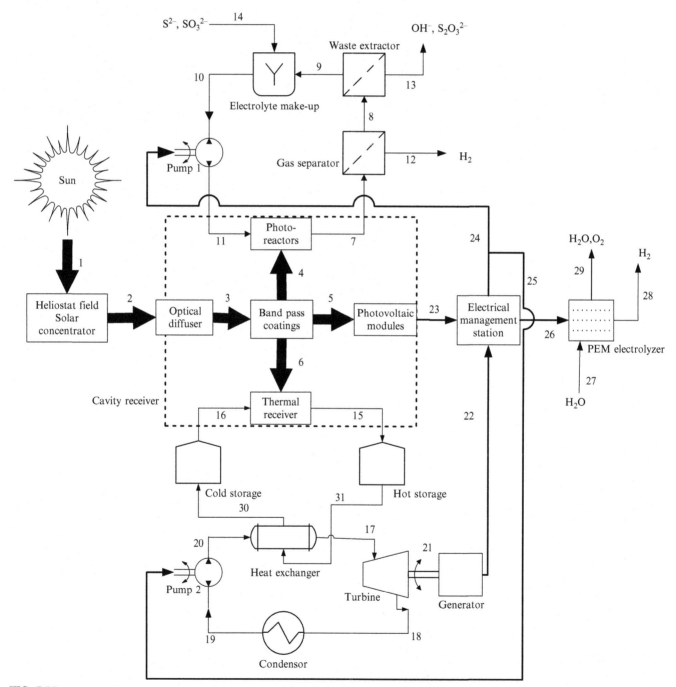

FIG. 5.90 Integrated solar-tower-based system with photocatalysis, electrolysis, ORC and concentrated PV. *Modified from Shamim, R., 2013. Experimental and Theoretical Investigation of a New Integrated Solar Tower System for Photocatalytic Hydrogen and Power Production. MSc Thesis, University of Ontario Institute of Technology.*

The heat source at the ORC is relatively low grade, as recovered from the cavity receiver. For a chosen day in August, the total hydrogen produced by the system is 52 tonne. The photocatalytic process and photovoltaic process both operate intermittently according to the solar radiation. The heat engine is sized so as to operate continuously for 24 h, through utilization of a thermal energy storage system. The energy efficiency varies from 34% to 40% and the exergy efficiency varies from 27% to 31% for changing solar conditions throughout the day.

The previously mentioned system reported in Zamfirescu and Dinçer (2014) has been designed for 599 MWh concentrated radiation on its aperture. For creating this energy, a heliostat field of 5526 acre is required, where the total installed reflective surface is 913,289 m². The layout of the heliostat field is described as shown in Fig. 5.91.

The optimized heliostat field has been designed for the area of Calgary. The energy efficiency of the heliostat field for each hour of system operation during the year is shown in Fig. 5.92. The annual average of solar field efficiency is 0.622 ± 0.18.

From the results, it appears that if hydrogen is produced by conversion of thermal energy into electricity with Rankine cycle, followed by water electrolysis, then the production is 28.7 tonne per day, whereas with the integrated system, it is at 41.4 tonne/day (annual averages). Whence, from a production of 15 kt H_2/year a revenue of $60 million from hydrogen sale is expected with $3/kg. This covers an investment of ~$400 mill. in ~7 years and gives a net benefit of $1100 mill. at the end of 25 years. If sulfur is valorized, then 3.5 kg sulfur per kg H_2 is generated. Sulfur price per kg is roughly at par with hydrogen price.

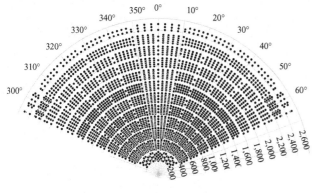

Heliostat field layout

Heliostat diameter: 20 m; Heliostats number: 2997
Land: 5526 acre; Max. radius: 2553 m; Span angle: 130°
Reflective area: 913,289 m²; Tower height: 255 m
Focal point aperture: 29 m width × 25 m height

Number of heliostats on each sector

r (m)	0°	10.9°	21.8°	32.7°	43.6°	54.5°
289.9	7	7	7	7	7	7
486.6	20	20	20	20	20	20
683.4	24	24	24	24	24	24
880.2	26	26	26	26	26	26
1077	27	27	27	27	27	27
1274	27	27	27	27	27	27
1471	27	27	27	27	27	27
1667	26	26	26	26	26	26
1864	26	26	26	26	26	26
2061	26	26	26	26	26	26
2258	25	25	25	25	25	25
2455	12	12	12	12	12	9

Note: r = radius to sector center; angle = span angle of sector; the field is symmetric

FIG. 5.91 Optimized layout of heliostat field at Calgary for 500 MWh light intensity on aperture. *Modified from Zamfirescu, C., Dinçer, I., 2014. Assessment of a new integrated solar energy system for hydrogen production. Sol. Energy 107, 700–713.*

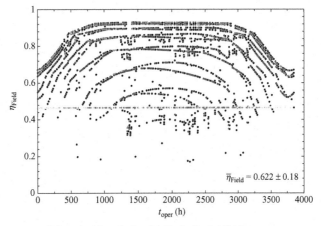

FIG. 5.92 Annual variation of the compounded optical loss factor for the heliostat field.

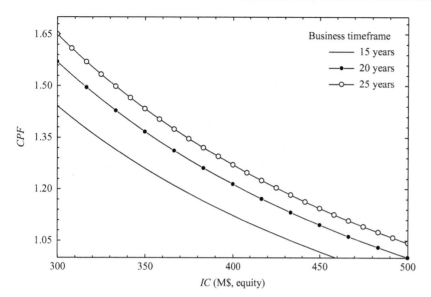

FIG. 5.93 Capital productivity factor (CPF) in function of invested capital (*IC*).

Based on system efficiency and insolation conditions, one determines from the above presented results that 52 kt of product (hydrogen and pure sulfur) is generated per year. Thence, the annual levelized income is of M$85.8 and the levelized operation and maintenance costs are M$8.58. For a 25-year timeframe, three cases are considered for an equity investment capital: M$300, M$400 and M$500. The following generated capital is obtained for each case, respectively: M$194.9, M$104.9 and M$ 21.1. The capital productivity factor is therefore 1.650, 1.262 and 1.042, respectively.

A plot is shown in Fig. 5.93 to illustrate an economic parametric study in which the investment capital is varied from M$300 to M$500 and three business timeframes are considered: 15, 20 and 25 years. From the study, it results that if the business is conducted for 15 years only, then the equity investment must be inferior to ~M$425 for profitability. In the case of a 25-year business timeframe, the equity investment up to M$500 appears to be well justified. In a more comprehensive economic analysis, the justifiable equity investments determined here can be compared with a predicted capital and operation and maintenance cost for the system.

Nevertheless, for a first approximation, judging based on previous experience with central receiver systems, a capital investment in the range of M$300 to M$500 in a 500 MW integrated system (comprising a solar field, tower, PV arrays, photoreactors, power plant, electrolyzer, chemical storage and auxiliary subsystems) appears plausible.

5.7.3 Integrated Chloralkali Photoelectrochemical Cell with Water Desalination

Here, a conceptual large-scale integrated system for the chloralkali photochemical process is presented as developed by Ghosh (2014). A block illustration is presented in Fig. 5.94 that describes the large-scale conceptual proposed system. The solar radiation harvested by the heliostat field (#1) is concentrated on the aperture of a hot mirror, placed at the top of the solar tower. The hot mirror, acting as spectral splitter, consists of a selected dielectric coating deposited on the face of a borosilicate glass. The hot mirror diverts (by reflection) the long wave spectrum portion (#4) of the concentrated radiation (#2) toward a volumetric solar receiver placed at the tower foot.

The transmitted radiation (#3) is dispersed and partially scattered into a large-size cavity receiver at the back of the hot mirror. Inside the cavity receiver are PV arrays and photoreactors. Special coatings are applied to the PV arrays and photoreactors. Owing to the coating, the PV arrays reflect back the photons of high-energy spectrum while absorbing only in the range of 500–800 nm. The photoreactors consist of cylindrical tubes of glass, covered with high band pass coating, which reflects wavelengths longer than 500 nm.

The optical arrangement inside the enclosure is such that it acts as a spectral splitter, which divides the incoming light in (#3) into a high-energy spectrum (#5) and middle spectrum (#6). Ocean/salty water (#8) is utilized by the desalination system (#9), along with the volumetric solar receiver placed at the tower foot. The desalination process produces fresh water (#14) and saturated salt water as a byproduct. The photoelectrochemical reactor system consumes high-energy spectrum light (#5), electricity from PV arrays (#7) and saturated brine (#10). From these aqueous wastes,

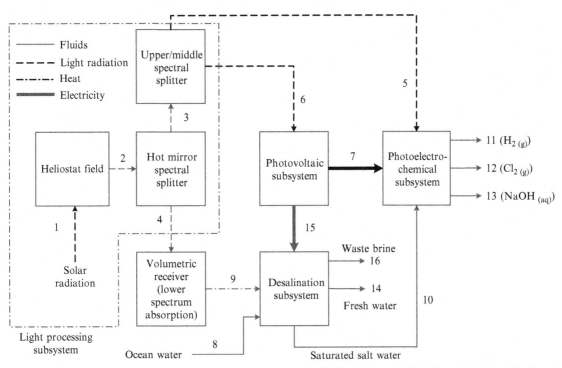

FIG. 5.94 Process flowsheet of an integrated photoelectrochemical process with desalination. *Modified from Ghosh, S., 2014. Experimental Investigation of a New Light-Based Hydrogen Production System. MSc Thesis, University of Ontario Institute of Technology.*

the photoelectrochemical system generates three valuable products: hydrogen (#11), chlorine (#12) and sodium hydroxide (#13).

The desalination subsystem is detailed in Fig. 5.95. The purpose of the desalination subsystem integrated within the solar hydrogen production plant is twofold:

- To generate a concentrated NaCl solution for the photoelectrochemical chloralkali process that extracts three valuable chemicals from it (H_2, Cl_2, NaOH)
- To generate fresh water as an additional marketable by-product

The desalination subsystem has two functional units, namely, the desalination unit (based on reverse osmosis) and the brine concentration unit (based essentially on thermal distillation) as shown in the figure. Ocean water is assumed to be at atmospheric temperature and pressure with a salinity of $\omega_8 = 1.55$ g/kg. For the thermodynamic analyses, the dead state of sea water is assumed to be $\omega_0 = 35$ g/kg salinity.

The pie chart from Fig. 5.96 shows the exergy destructions within the light-processing subsystem. It can be seen that the heliostat field accounts for 91% of the exergy destruction within the light-processing subsystem, followed by 5% and 4% for the upper/middle spectral splitter and hot mirror spectral splitters, respectively. The total exergy destruction corresponds to 182.71 W per square meter of heliostat.

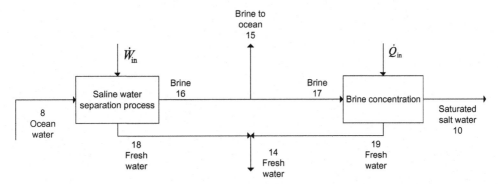

FIG. 5.95 Process flow diagram for the desalination subsystem. *Modified from Ghosh, S., 2014. Experimental Investigation of a New Light-Based Hydrogen Production System. MSc Thesis, University of Ontario Institute of Technology.*

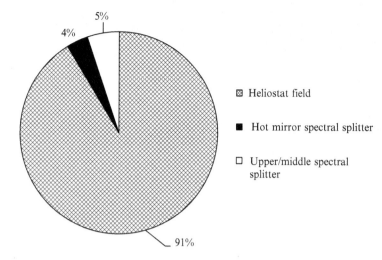

FIG. 5.96 Exergy destructions within light-processing subsystems. *Data from Ghosh, S., 2014. Experimental Investigation of a New Light-Based Hydrogen Production System. MSc Thesis, University of Ontario Institute of Technology.*

- ⊠ Heliostat field
- ■ Hot mirror spectral splitter
- ☐ Upper/middle spectral splitter

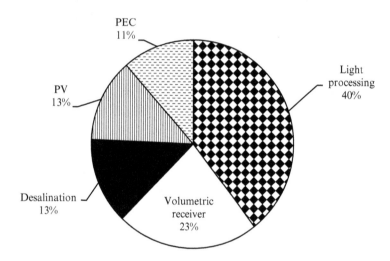

FIG. 5.97 Exergy destructions of the integrated system for solar hydrogen production. *Data from Ghosh, S., 2014. Experimental Investigation of a New Light-Based Hydrogen Production System. MSc Thesis, University of Ontario Institute of Technology.*

The exergy destructions for the integrated hydrogen production system are determined and reported as shown in Fig. 5.97. The total exergy destruction is 462.75 W per square meter of heliostat area. It is clear that the light-processing system accounts for the maximum of the exergy destructions (40%), followed by the volumetric receiver (23%), while the desalination system, PV subsystem and the PEC account for equal amounts of exergy destructions.

The annual production for the square meter of heliostat is 2.8 kg of hydrogen, 47.59 kg of chlorine, 112 kg of caustic soda and 50 kg of water. Assuming that the market price of products are $1/kg for hydrogen, $1/kg for chlorine, $0.5/kg for caustic soda and $0.1/kg of water, then the total income is $61 with respect to 1 square meter of heliostat.

In order to account for the operation and maintenance, the levelized price of products is given as $66 per square meter and 1 year has been assumed. In these conditions, the capital productivity factor has been determined for three time spans, namely 10 years, 25 years and 40 years. As shown in Fig. 5.98, the investment is profitable for a specific invested capital (in $/m^2 of heliostat) of up to 400 $/m^2, provided that the business time span is equal or superior to 25 years.

For a 10-year timespan, the business becomes nonprofitable if the specific invested capital is superior to $290/m^2. More importantly, the selling price of hydrogen can be kept as low as $1/kg, because the revenue from the additional products makes the business favorable. This is really relevant in the context of producing sustainable hydrogen, because the product price becomes competitive with conventional technologies.

FIG. 5.98 Capital productivity factor as a function of the specific invested capital given in dollars per square meter of heliostat field. *Data from Ghosh, S., 2014. Experimental Investigation of a New Light-Based Hydrogen Production System. MSc Thesis, University of Ontario Institute of Technology.*

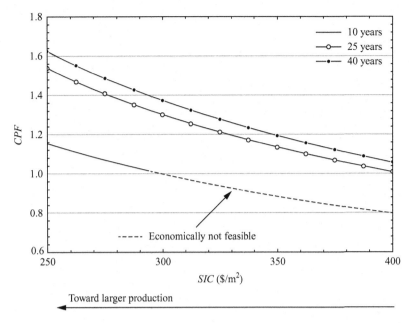

Toward larger production

5.8 CONCLUDING REMARKS

In this chapter, the methods of hydrogen production by photonic energy are introduced and discussed. The chapter parallels Chapter 4 in the fact that the focus is now on photochemical reaction rather than the thermochemical reactions. The fundamentals of photochemical hydrogen production are presented, and significant thermodynamic parameters and photochemical reactions are introduced for discussion. Also, the main types of photochemical processes for hydrogen production are introduced.

The chapter ends with a presentation of some integrated systems to produce green hydrogen from photonic energy. The systems do not emit harmful chemicals or GHG during operation. By converting the potential of solar energy and hydrogen-containing materials such as water, sea water and industrial waste water into commercially viable products, the approached systems also increase the energy and exergy efficiencies. Due to the possibility of waste recovery and multiproduct generation, the presented integrated system will lower hydrogen production costs.

Nomenclature

A	area (m^2)
c	speed of light constant (m/s)
c_λ	photon constant (m/s)
e	electron charge (C)
$\dot{E}$	energy rate (W)
$\dot{Ex}$	exergy rate (W)
F	factor
FF	filling factor
h	Planck's constant
J	current density (A/m^2)
$\hbar$	extinction coefficient
k	Boltzmann constant
n	refraction index
N	number of elements
$\dot{N}$	photon rate (s^{-1})
N_A	Avogadro's number
P	pressure (kPa)
R	reflectance
S	entropy (J/K)
$\dot{S}$	entropy rate (W/K)

V	voltage (V)
t	time (h)
T	temperature (K)

Greek Letters

η	energy efficiency
ψ	exergy efficiency

Subscripts

abs	absorbed
bkw	backward
cat	catalysis
dni	direct normal spectral irradiation
fwd	forward
h	heliostats
λ	wavelength
loss	energy losses
oc	open circuit
oper	operation
PC	photocatalysis
ph	photon
PV	photovoltaic
refl	reflective surface
sc	short circuit
TH	thermal
year	year

References

Acar, C., Dinçer, I., Zamfirescu, C., 2013. A review on selected heterogeneous photocatalysts for hydrogen production from water. Int. J. Energy Res. 38, 1903–1920.

ASTM, 2008. G173-03 Reference Spectra Derived from SMARTS v. 2.9.2. Accessed from http://rredc.nrel.gov/solar/spectra/am1.5/ASTMG173/ASTMG173.html.

Baniasadi, E., Dincer, I., Naterer, G.F., 2013. Measured effects of light intensity and catalyst concentration on photocatalytic hydrogen and oxygen production with zinc sulfide suspensions. Int. J. Hydrog. Energy 38, 9158–9168.

Bejan, A., 1988. Advanced Engineering Thermodynamics. John Wiley and Sons, New York.

Brauns, E., Jones, S.W., Clark, J.A., Molnar, S.M., Kawnishi, Y., Brewer, K.J., 1997. Electrochemical, spectroscopic, and spectroelectrochemical properties of synthetically useful supramolecular light absorbers with mixed polyazine bridging ligands. Inorg. Chem. 36, 2861–2867.

Brewer, K.J., Elvington, M., 2006. Supramolecular Complexes as Photocatalysts for the Production of Hydrogen from Water. US Patent 7,122,172 B2.

Brunauer, S., Emmett, P.H., Teller, E., 1938. Adsorption of gases in multimolecular layers. J. Am. Chem. Soc. 60, 309–319.

Campagna, S., Denti, G., Serroni, S., Ciano, M., Balzani, V., 1991. Hexanuclear homo- and heterobridged ruthenium polypyridine complexes: syntheses, absorption spectra, luminescence properties, and electrochemical behavior. Inorg. Chem. 30, 3728–3732.

Casallas, C., Dinçer, I., Zamfirescu, C., 2016. Experimental investigation and analysis of a novel hydrogen production cell with polymeric membrane photocathode. Int. J. Hydrog. Energy 41, 7968–7975

Christensen, P.A., Erbs, W., Harriman, A., 1985. Photo-oxidation of water in non-sacrificial systems. J. Chem. Soc. Faraday Trans. 81, 575–580.

Chen, Z.S., Mo, S.P., Hu, P., 2008. Recent progress in thermodynamics of radiation—exergy of radiation, effective temperature of photon and entropy constant of photon. Sci. China Ser. E: Technol. Sci. 51 (8), 1096–1109.

Duan, L., Xu, Y., Gorlov, M., Tong, L., Andersson, S., Sun, L., 2010. Chemical and photochemical water oxidation catalyzed by mononuclear ruthenium complexes with a negatively charged tridentate ligand. Chem. Eur. J. 16, 4659–4668.

Duffie, J.A., Beckman, W.A., 2013. Solar Engineering of Thermal Processes, fourth ed. John Wiley and Sons, Hoboken, NJ.

Elvington, M., Brewer, K.J., 2006. Photoinitiated electron collection at a metal in a rhodium-centered mixed-metal supramolecular complex. Inorg. Chem. 45, 5242–5244.

Ferrer, I.J., Díaz-Chao, P., Pascual, A., Sánchez, C., 2007. An investigation on palladium sulfide (PdS) thin films as a photovoltaic material. Thin Solid Films 515, 5783–5786.

Ghosh, S., 2014. Experimental Investigation of a New Light-Based Hydrogen Production System. MSc Thesis, University of Ontario Institute of Technology.

Grätzel, M., 2001. Photoelectrochemical cells. Nature 414, 338–344.

Gueymard, C., 1995. SMARTS2, simple model of the atmospheric radiative transfer of sunshine: algorithms and performance assessment. Report FSEC-PF-270-95, Florida Solar Energy Center, Cocoa, FL.

Gueymard, C.A., Myers, D., Emery, K., 2002. Proposed reference irradiance spectra for solar energy systems testing. Sol. Energy 73, 443–467.

Klan, P., Wirz, J., 2009. Photochemistry of Organic Compounds: From Concepts to Practice. Wiley-Blackwell, Chichester, West Sussex, UK.

Klein, S.A., 2015. Engineering Equation Solver. F-Chart Software.

Minggu, L.J., Daud, W.R.W., Kassim, M.B., 2010. An overview of photocells and photoreactors for photoelectrochemical water splitting. Int. J. Hydrog. Energy 35, 5233–5244.

Rabbani, M., 2013. Design, Analysis and Optimization of a Novel Photo-Electrochemical Hydrogen Production System. PhD Thesis, University of Ontario Institute of Technology.

Rabbani, M., Dinçer, I., Naterer, G.F., 2014. Experimental investigation of processing parameters and effects on chloralkali products in an electrolysis based chloralkali reactor. Chem. Eng. Process. Process Intensif. 82, 9–18.

Rangan, K., Arachchige, S.M., Brown, J.R., Brewer, K.J., 2009. Solar energy conversion using photochemical molecular devices: photocatalytic hydrogen production from water using mixed-metal supramolecular charges. J. Energy Environ. Sci. 2, 410–419.

Ratlamwala, T.A.H., 2013. Design, Analysis and Experimental Investigation of Cu-Cl Based Integrated Systems. PhD Thesis, University of Ontario Institute of Technology.

Ratlamwala, T.A.H., Dinçer, I., 2014. Experimental study of a hybrid photocatalytic hydrogen production reactor for Cu-Cl cycle. Int. J. Hydrog. Energy 39, 20744–20753.

Roberts, R.R., 2013. Experimental Investigation of a Catalyst under Various Solar Light Conditions for Hydrogen Production. MSc Thesis, University of Ontario Institute of Technology.

Serroni, S., Campagna, S., Puntoriero, F., Juris, A., Denti, G., Balzani, V., Venturi, M., 2002. A luminescent decanuclear ruthenium(II) polypyridine complex: a convergent approach to a dendritic structure employing the "complexes as metals/complexes as ligands" synthesis strategy. Inorg. Synth. 33, 10–18.

Shamim, R., 2013. Experimental and Theoretical Investigation of a New Integrated Solar Tower System for Photocatalytic Hydrogen and Power Production. MSc Thesis, University of Ontario Institute of Technology.

Shamim, R.O., Dinçer, I., Naterer, G.F., Zamfirescu, C., 2014. Experimental investigation of a solar tower based photocatalytic hydrogen production system. Int. J. Hydrog. Energy 39, 5546–5556.

Sing, H.N., Tiwari, G.N., 2005. Evaluation of cloudiness/haziness factor for composite climate. Energy 20, 1589–1601.

Siripala, W., Perera, L.D.R.D., De Silva, K.T.L., Jayanetti, J.K.D.S., Dharmadasa, I.M., 1996. Study of annealing effects of cuprous oxide grown by electrodeposition technique. Sol. Energy Mater. Sol. Cells 44, 251–260.

Swavey, S., Brewer, K.J., 2002. Visible light induced photocleavage of DNA by a mixed-metal supramolecular complex: $[\{(bpy)_2Ru(dpp)\}_2RhCl_2]^{5+}$. Inorg. Chem. 41, 6196–6298.

Vayssieres, L., 2010. Solar Hydrogen and Nanotechnology. John Wiley and Sons, New York.

White, T.A., Knoll, J.D., Arachchige, S.M., Brewer, K.J., 2012. A series of supramolecular complexes for solar energy conversion via water reduction to produce hydrogen: an excited state kinetic analysis of Ru(II), Rh(III), Ru(II) photoinitiated electron collectors. Materials 5, 27–46.

Zamfirescu, C., Dinçer, I., 2014. Assessment of a new integrated solar energy system for hydrogen production. Sol. Energy 107, 700–713.

Zamfirescu, C., Dinçer, I., Naterer, G.F., 2011. Analysis of a photochemical water splitting reactor with supramolecular catalysts and a proton exchange membrane. Int. J. Hydrog. Energy 36, 11273–11281.

Zamfirescu, C., Dinçer, I., Naterer, G.F., 2012. Molecular charge transfer simulation and quantum efficiency of a photochemical reactor for solar hydrogen production. Int. J. Hydrog. Energy 37, 9537–9549.

Zamfirescu, C., Dinçer, I., Naterer, G.F., Banica, R., 2013. Quantum efficiency modeling and system scaling-up analysis of water splitting with $Cd_{1-x}Zn_xS$ solid-solution photocatalyst. Chem. Eng. Sci. 97, 235–255.

STUDY PROBLEMS

5.1 Explain the main principle of photonic water-splitting process.

5.2 Classify the photonic hydrogen production methods and describe each with an illustration.

5.3 Describe blackbody radiation.

5.4 Illustrate how photonic radiation interacts with the reference temperature.

5.5 Sketch the Jablonski diagram and analyze it thermodynamically.

5.6 Illustrate and describe the diabatic photoreaction pathways.

5.7 Conceptually describe and analyze homogeneous photocatalytic systems evolving hydrogen and oxygen water-splitting processes and compare them.

5.8 Illustrate the reductive quenching cycles and describe them.

5.9 Describe a supramolecular photocatalyst and use it in photocatalytic hydrogen production.

5.10 Sketch a state diagram for the hydrogen-evolving photocatalytic process.

5.11 Describe and illustrate the heterogeneous photocatalysis mechanism.

5.12 Provide a comparative assessment of some common heterogeneous photocatalysts for hydrogen production and discuss their pros and cons.

5.13 Describe and illustrate a photoelectrochemical process for hydrogen production.

5.14 Discuss the potential types for photoelectrochemical cells.

5.15 Define all possible efficiencies for photocatalytic hydrogen production.

5.16 Define all possible efficiencies for photoelectrochemical hydrogen production.

5.17 Explain how polymeric membranes are used in photoelectrochemical cells.

5.18 Describe the chloralkali process and how it can be incorporated into a photoelectrochemical cell.

5.19 Analyze and assess the integrated system shown in Fig. 5.90 thermodynamically through energy and exergy methods. Make conceptually correct assumptions and consider the data provided in the system-related section.

5.20 Analyze mathematically the optical loss factor for a heliostat field.

Hydrogen Production by Biochemical Energy

6.1 INTRODUCTION

Biochemical energy is manipulated by living organisms to construct their needed materials. This type of energy is generally stored in glucose, sucrose, cellulose, carbohydrates, glucose, and proteins. The photosynthetic mechanisms allow plants and microorganisms to produce those fuels. In order to work, biological cells have specific mechanisms to synthesize enzymes. Those are complex supramolecular structures that possess metallic, active centers to catalyze specific chemical reactions.

One of the key processes in living systems is the metabolism. In order to work consistently, the chemical reactions in living systems are organized in specific pathways or metabolic sequences. A large fraction of the material in a cell are protein chains that are encoded in the structural genes that direct reactions' pathways. Heat is generated during metabolism as a dissipation process associated to the chemical reaction fueled by high-energy chemicals. The resulting materials from the synthesis process are highly organized and possess low entropy. The cell are also able to break down those compounds when needed, in a process called catabolism.

Hydrogen can be generated from a variety of biochemical substrates; however, efficiency of any biochemical hydrogen generation process is generally low. For example, dark fermentation may be one of the biochemical hydrogen production options that can generate 2–3 moles of hydrogen from a mole of glucose substrate (Hallenbeck and Benemann, 2002). When light is used as energy input for biochemical energy conversion, the process is known as biophotolysis.

A hydrogen-producing enzyme is strictly necessary to generate hydrogen gas from a substrate. In this catalytic process, metabolic factors are limiting. The most important enzymes for biochemical hydrogen are the nitrogenase and hydrogenase. Two hydrogenases are most often encountered, NiFe-hydrogenase, and Fe-hydrogenase. The hydrogenases are most effective for hydrogen generation. Clostridium pasteurianum produce hydrogenases shown at a 6000 s^{-1} hydrogen generation turnover frequency, whereas *Desulfovibrio* spp shows 9000 s^{-1}, respectively.

Green algae and cyanobacteria are known microorganisms that generate hydrogen in some conditions when exposed to photonic radiation. Melis (2002) reviewed photobiochemical hydrogen production from green algae, *Chlamydomonas reinhardtii*, which spontaneously induces a hydrogenase catalytic pathway of electron transport in the chloroplast, which then generate hydrogen photosynthetically. Kava-Cordeiro and Vargas (2015) described a model for genetic modification of microalgae for hydrogen production.

An indirect pathway to generate biochemical hydrogen is through biogas and biofuel reforming. Well established technology does exist to generate biogas and biofuels, such as methanol, ethanol, and biodiesel. These fuels are valuable products in and of themselves, but they can also be used to generate green hydrogen. In this respect, biogas reforming can be applied. Also biomethanol and bioethanol can be reformed to hydrogen.

In this chapter, hydrogen production by biochemical energy is revised. In the first part of the chapter, the main biochemical and photobiochemical processes for hydrogen production are presented. The main types of reactors and systems are presented next. The chapter continues with some case studies of water photolysis and other biochemical hydrogen production methods given as illustrative example. Biofuels and biogas generation is presented. Some lifecycle assessments (LCAs) of biogas production and other green pathways of biochemical green hydrogen generation are provided and discussed for various operating conditions.

6.2 BIOCHEMICAL PROCESSES

Biochemical hydrogen production will depend first on the type of substrate, namely, the type of microorganisms and food. Some microorganisms can grow incredibly fast in proper conditions. For example, the acetic acid bacteria is able to oxidize ethanol from beer and wine to take energy, $CH_3CH_2OH + O_2 \rightarrow CH_3COOH + H_2O$. Bacteria can grow incredibly fast because it's small. The order of magnitude of the surface-to-volume ratio of a bacterium is $10^6 \, m^{-1}$, or that of an ameba is $10^4 \, m^{-1}$. Some bacteria are autotrophic (self-nourishing), being able to synthetize all their cell constituents only starting from inorganic compounds of sulfur, nitrogen, water and carbon dioxide. The chemoautotroph bacteria can take energy from hydrogen oxidation or from H_2S oxidation for H_2SO_4. Many bacteria are chemoheterotrophic, taking energy from oxidizing organic compounds. The photoautotrophs take energy from light. The anaerobic bacteria produce fermentation to create sugars in absence of oxygen. Nevertheless, all living structures are interrelated and form hierarchies. Fig. 6.1, in this regard, shows the universal phylogenic tree.

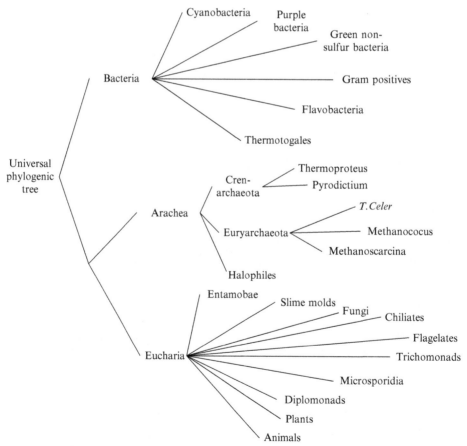

FIG. 6.1   The universal phylogenetic tree for classification.

One of the key oxidative processes in living cells is carried out by a biological agent denoted as nicotinamide adenine dinucleotide (NAD^+) with an oxidation-free energy of pH 7 of 22.1 kJ/mol. The oxidation reaction is

$$Acetate^- + 4NAD^+ + 2H_2O \rightarrow 4NADH + 2CO_2 + 3H^+ \tag{6.1}$$

where NADH is the reduced NAD^+.

Reaction from Eq. (6.1) is slightly endothermic; however, the NADH is further oxidized in mitochondria with a large release of ready-to-use energy, as follows

$$4NADH + 4H^+ + 2O_2 \rightarrow 4H_2O + 4NAD^+, \Delta G(pH7) = -876.1\frac{kJ}{mol} \tag{6.2}$$

Another key energy-releasing reaction is the hydrolysis of adenosine triphosphate (ATP). In a simplified version, the ATP hydrolysis can be written as follows:

$$ATP \rightarrow ADP + Pi \tag{6.3}$$

where ADP is adenosine diphosphate and Pi is inorganic phosphate (HPO_4^{2-}).

Around 1.5 mol of hydrogen can be generated per mole of sucrose based on a series of processes catalyzed by hydrogenase and nitrogenase enzymes, according to the following prototypical reaction

$$C_{12}H_{22}O_{11} + H_2O \rightarrow CH_3COOH + 4H_2(g) \tag{6.4}$$

Table 6.1 gives the Gibbs free energy of formation and NAD^+ oxidation of some biochemical compounds. In the case of aerobic fermentation (which essentially occurs in the presence of oxygen), typically 7 mol of hydrogen can be

TABLE 6.1 Gibbs Free Energy of Formation and NAD^+ Oxidation of Some Biochemical Compounds

Compound	ΔG_f^0 (kJ/mol)	ΔG_{ox}^0 (NAD^+, pH 7)
Acetaldehyde, C_3H_4O	−139.7	−28.3
Acetic acid, $C_2H_4O_2$	−369.4	9.3
Acetate$^-$	−369.2	22.1
Acetylene, C_2H_2	209.2	−140.0
Creatine, $C_4H_9O_2N_3$	−264.3	−80.8
Dihydroxyacetone-P	−1293.3	−144.2
Ethanol, C_2H_6O	−181.5	−4.6
Formaldehyde, CH_2O	−130.5	−63.0
Formic acid, CH_2O_2	−356.1	−56.5
Fructose, $C_6H_{12}O_6$	−915.4	−245.7
α-D-Glucose, $C_6H_{12}O_6$	−917.2	−243.8
Glycerol, $C_3H_8O_3$	−488.5	−110.2
α-Lactose, $C_{12}H_{22}O_{11}$	−1515.2	−569.7
Methane, CH_4	−50.8	58.2
Methanol, CH_4O	−175.2	−36.4
Ribose-5P	−1599.9	−224.6
Sucrose, $C_{12}H_{22}O_{11}$	−1551.8	−533.1
Urea, CH_4ON_2	−203.8	−7.8
Uric acid, $C_5H_4O_3N_4$	−356.9	−118.1

Source: Metzler, D., 2003. Biochemistry. The Chemical Reactions of Living Cells. Elsevier, New York.

generated per mole of glucose. According to Das and Veziroglu (2008), the energy efficiency of hydrogen production by aerobic digestion can reach remarkably high values for such substrates, e.g., 28% when molase is the substrate.

Nitrogenase, which is one of the most complex enzymes, performs ammonia synthesis in a very intelligent way, by not breaking dinitrogen directly, but rather bond by bond. Nitrogenase consists in two protein clusters: one that has an electron acceptor active site based on iron (Fe) and molybdenum (Mo) having the stoichiometry $MoFe_7S_9N$ (in some alternative versions of nitrogenase, the active center is based on Fe and V); and the second having an iron-sulfur center that hydrolyzes ATP to obtain energy and provide the active site of the first cluster with electrons.

Nitrogenase uses directly gaseous nitrogen (N_2), which is "captured" by organisms from the atmosphere by specific respiration mechanisms. Hydrogen is not used by nitrogenase in a molecular form. Rather, hydrogen is used in the form of protons produced by H_2 ionization at the electron acceptor active site and electrons provided by the electron donor active site. Intensive efforts were dedicated in recent years to clarify the ammonia production cycle by nitrogenase. The main findings are summarized, e.g., in papers by Hinnemann and Nørskov (2006) and Kästner and Blöchl (2007). The synthesis process at the active site is presented in Fig. 6.2. Ammonia synthesis consists of 14 steps in which various intermediate compounds are formed at the active site. All intermediates are recycled, and globally, from one nitrogen molecule and six hydrogen molecules, two NH_3 molecules are produced per cycle.

Nitrogenase is produced by a number of microbes that live in symbiosis with root nodules of legumes and plants such as alfalfa, clover or peas. There are also free-living microbes that produce nitrogenase, most of them being anaerobic (e.g., *Clostridium*, *Klebsiella pneumoniae*, *Bacillus polymyxa*, *Bacillus macerans*, *Escherichia intermedia*, *Rhodobacter sphaeroides*, *Rhodopseudomonas palustris*, *Rhodobacter capsulatus*) and others aerobics (e.g., *Azotobacter vinelandii*, *Anabaena cylindrical*, *Nostoc commune*). Hydrogen can be easily extracted from biochemically produced ammonia, e.g., by thermos-catalytic cracking. This establishes a potential pathway to biochemically green hydrogen.

Hydrogen can be produced through dark fermentation. The substrate consists of waste biomass streams or microalgae or higher-value biomass. There is a high technological potential with metabolic engineering of dark fermentation microorganisms for redirecting the electron flux toward hydrogen production. Pyruvate's metabolism appears to be the most promising for dark fermentation with hydrogen production. Fig. 6.3 shows a simplified schematic representation of the dark fermentation process for hydrogen production.

In this mechanism, bacteria grows on a substrate degraded by oxidation. If the environment is maintained anoxic, then the electrons generated from oxidation will reduce protons. Glucose is initially converted to pyruvate and then pyruvate is oxidized to form acetyl coenzyme A (CoA) with generation of ATP through a reaction pathway described as follows

$$Pyruvate + CoA + 2Fd(ox) \rightarrow Acetyl - CoA + 2Fd(red) + CO_2 \tag{6.5}$$

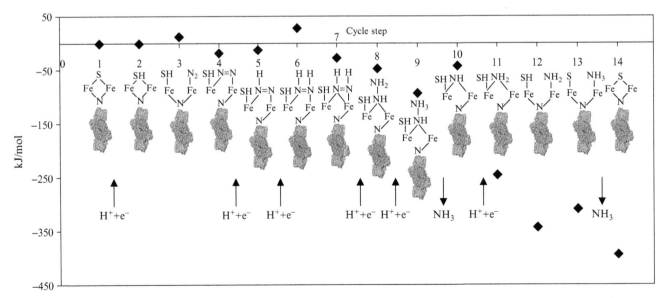

FIG. 6.2   The 14 steps nitrogenase biochemical cycle. *Data from Kästner, J., Blöchl, P.E., 2007. Ammonia production at the FeMo cofactor of nitrogenise: results from density functional theory. J. Am. Chem. Soc. 129, 2998–3006.*

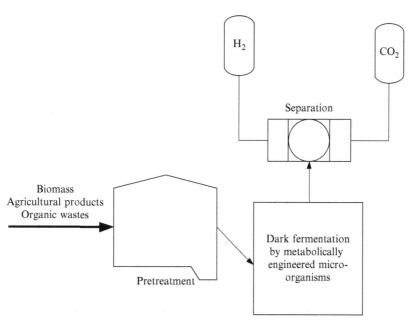

FIG. 6.3 Simplified schematic of a dark fermentation process.

where Fd is ferredoxin and CoA stands for coenzyme A, referring to β-mercaptoethylamina + ADP with 3′-phosphate group + pantothenic acid.

The reduced protein ferredoxin reduces protons and generates hydrogen as

$$Fd(red) + 2H^+ \rightarrow H_2 + Fd(ox) \tag{6.6}$$

One major group of hydrogen production processes is biophotolysis. There are two subclass methods: direct biophotolysis and indirect photolysis. In this process, the photosynthetic apparatus provides electrons to a low potential reductant able to reduce the hydrogenase. Of course, the energetic photons are produced by the absorbed photons. The substrate is water and the enzyme is ferredoxin-hydrogenase system generated with green algae *C. reinhardtii*. Direct photolysis systems require oxygen absorbers to continuously remove oxygen from the reaction mixture; otherwise, hydrogenase will stop working.

As mentioned earlier in Hallenbeck and Benemann (2002), direct photolysis with heterocystous cyanobacteria can potentially confer more advantages than green algae for direct photolysis. This is because heterocystous cyanobacteria possess two different types of cells that separate spatially the hydrogen and oxygen evolutions. A review of the process is given in Benemann and Weare (1974); however, it appears that the low efficiency of the heterocystous cyanobacteria process makes the system nonpractical (Figs. 6.4–6.5).

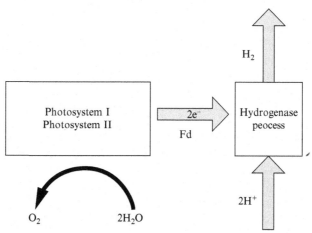

FIG. 6.4 Simplified representation of direct photolysis process for hydrogen production. *Modified from Hallenbeck, P.C., Benemann, J.R., 2002. Biological hydrogen production; fundamentals and limiting processes. Int. J. Hydrog. Energy 27, 1185–1193.*

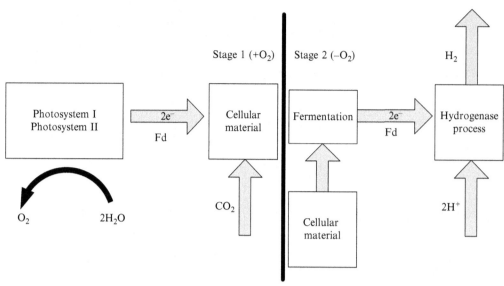

FIG. 6.5 Simplified representation of indirect photolysis for hydrogen production. *Modified from Hallenbeck, P.C., Benemann, J.R., 2002. Biological hydrogen production; fundamentals and limiting processes. Int. J. Hydrog. Energy 27, 1185–1193.*

Note that indirect photolysis is conceived as a method to separate hydrogen- and oxygen-evolving processes in a photolysis system. Therefore, the process takes place in two stages. Stage I is aerobic and evolves oxygen while consuming carbon dioxide. Stage II is anaerobic, and evolves hydrogen and carbon dioxide. Both stages may require light as energy input. The aerobic stage I process is described as

$$12H_2O + 6CO_2 + h\nu \rightarrow C_6H_{12}O_6 + 6O_2 \tag{6.7}$$

The stage II is anaerobic and described as

$$C_6H_{12}O_6 + 6H_2O + h\nu \rightarrow 12H_2 + 6CO_2 \tag{6.8}$$

As mentioned in Kava-Cordeiro and Vargas (2015), in indirect photolysis, the microalgae are deprived from sulfur nutrients to increase the hydrogen production with the photosystem II. The biochemical process needs to be conducted by first feeding the microalgae and allowing them to grow sufficiently. The biomass will grow with the air supply in stage I. Once air supply is cut in stage II, hydrogen production is obtained. A bioreactor should then operate in a steady periodic regimen to generate hydrogen continuously.

A well stirred bioreactor with steady periodic operation can be semiempirically modeled, as suggested in Kava-Cordeiro and Vargas (2015), by correlating the mass fraction of algae y_a with the growth rate, as

$$\frac{dy_a}{dt} = y_a \left(\mu_a - 1.069 \times 10^{-6} \right) \tag{6.9}$$

where μ_a is the growth rate of algae in s^{-1}.

A large-scale reactor is reported in Vargas et al. (2014) for a microalgae-derived hydrogen process with indirect photolysis. The reactor has $2 \times 5\,m$ and is $8\,m$ tall. A photograph of the bioreactor is shown in Fig. 6.6. The reactor consists of transparent horizontal tubes through which green algae is circulated. The system energy efficiency for hydrogen production as reported in Kava-Cordeiro and Vargas (2015) is better than 6% with perspective to more than 13% as predicted due to genetic modification of the microorganisms.

Photofermentation is a hybrid biophotolysis system based on photosynthetic bacteria that are able to simultaneously use the energy from a substrate and incident photons. Photosynthetic bacteria lack the photosystem II to synthesize a nitrogenase system and evolve hydrogen, as illustrated in the diagram from Fig. 6.7. Because the photosystem II misses, the photosynthetic bacteria is not able to generate oxygen; therefore, the hydrogen production is not inhibited. Das and Veziroglu (2008) mention Rhodospirillaceae photoheterotrophic bacteria, which can grow in dark using CO to generate ATP with a simultaneous release of hydrogen and carbon dioxide. Purple nonsulfur bacteria is known to operate based on the scheme from Fig. 6.7 and to evolve hydrogen from the substrate when nitrogen is deficient. The reactor needs to be operated in an anaerobic mode. The process can be described as (Fig. 6.8)

FIG. 6.6 Photograph of a bioreactor for hydrogen production with indirect photolysis. *Courtesy of J.V.C. Vargas, Federal University of Parana, Brazil.*

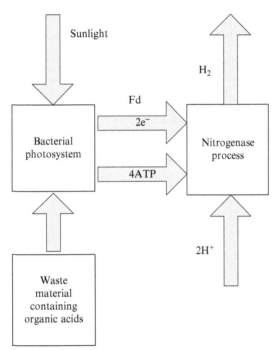

FIG. 6.7 Photofermentation process for biochemical hydrogen production. *Modified from Hallenbeck, P.C., Benemann, J.R., 2002. Biological hydrogen production; fundamentals and limiting processes. Int. J. Hydrog. Energy 27, 1185–1193.*

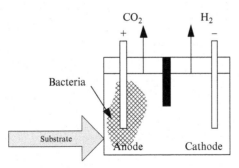

FIG. 6.8 Schematic of a microbial electrolysis cell.

$$CH_3COOH + 8e^- + 16ATP \xrightarrow{h\nu} 4H_2 + 16ADP + 16Pi \qquad (6.10)$$

Microbial electrolysis cells that catalyze the half reactions with the help of enzymes were illustrated and elaborated in Das and Veziroglu (2008) with *Geobacter, Shewanella* sps., and *Rhodoferax ferrireducens*. The anodic reaction is written as

$$CH_3COOH + 2H_2O \rightarrow 8H^+ + 8e^- + 2CO_2 \qquad (6.11)$$

Other biofuels are considered concomitantly with biohydrogen. Methane produced through anaerobic digestion of microalgae biomass after lipid extraction is discussed in Spolaore et al. (2006). Microalgae-derived biodiesel has been studied in Chisti (2007). Furthermore, ethanol derived from microalgae is reported in Harun et al. (2011).

6.3 INTEGRATED SYSTEM FOR GREEN BIOCHEMICAL HYDROGEN PRODUCTION

6.3.1 Biogas Facility Integrated With Natural Gas Reforming

The system presented here integrates a biochemical process and a thermochemical process. Biogas is generated in a digestion process (biochemically). The methane-rich biogas is purified by removing CO_2 while passing through an absorption unit. Then natural gas steam reforming is applied to the purified biomethane, which is then converted to hydrogen. The main biochemical processes during anaerobic digestion in a biogas reactor can be described as

- Ethanol formation from the organic substrate

$$C_6H_{12}O_6 \rightarrow 2C_2H_5OH + 2CO_2 \qquad (6.12)$$

- Acid acetic formation from ethanol decomposition

$$2C_2H_5OH + CO_2 \rightarrow CH_4 + CH_3COOH \qquad (6.13)$$

- Acetic acid decomposition to release methane

$$CH_3COOH \rightarrow CH_4 + CO_2 \qquad (6.14)$$

- Carbon dioxide hydrogenation to release methane (Fig. 6.9)

$$CO_2 + 4H_2 \rightarrow CH_4 + 2H_2O \qquad (6.15)$$

A LCA case study is presented here as an illustrative example with a biogas production facility from biowastes. Biowaste contains biogenous household waste, yard waste and food waste. The technology of the biogas production plant is thermophile, single-stage digestion with post-composting. In this LCA analysis, following processes are included:

- Data on the environmental exchanges due to biowaste pretreatment (inclusively the disposal of contaminants) biowaste digestion and post-composting of digested matter
- Emissions to soil due to the use of press water and digested matter as a fertilizer in agriculture are recorded
- Spreading of the fertilizer as well as transport from biowaste plant to farms are taken into account

The multioutput process "biowaste to anaerobic digestion" delivers three co-products/services: biogas, disposal of biowaste and application of digested matter as a fertilizer in agriculture. Allocation for operation expenditures and emissions to air occurring at the biowaste treatment plant has been performed by taking into account the revenues of a plant with a yearly treatment capacity of 10,000 tons of biowaste, respectively.

The above exchanges are allocated as follows: 31% biogas production and 69% waste disposal. For heavy metals and emissions of trace elements to soil, a practical allocation has been performed: 50% of the latter are assigned to the disposal service, and the remaining 50% are allocated to the application of press water and digested matter in agriculture. Biowaste input is assumed to have a share of dry matter of 40%, and an organic share of dry matter of 70% with a carbon content of 53%. The share of carbon decomposition during digestion is 76%. Biowaste treatment fulfills three

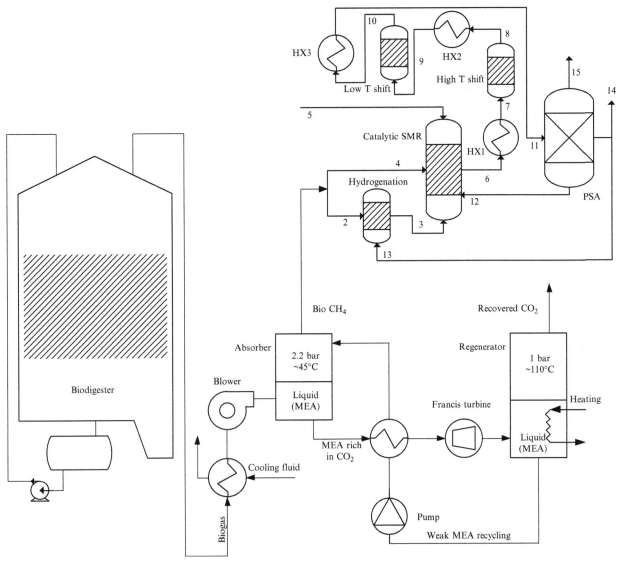

FIG. 6.9 Biomass facility integrated with biogas reforming to generate green hydrogen.

functions: biowaste disposal, biogas production and fertilizer production. Transport processes are only accounted for with household biowastes.

The LCA is performed with a specialized software that implements the CML 2001 and Eco-indicator 99 methods. The impact categories considered for the CML 2001 method are global warming and human toxicity. The impact categories considered for Eco-indicator 99 method are: human health, ecosystem quality and resources. Fig. 6.10 shows the calculated results of global warming potential (GWP) related to biogas production. Since 1 m^3 of biogas comprises about 0.5 m^3 of biomethane, and for each mole of methane, two moles of hydrogen can be reformed, Fig. 6.10 shows that 0.11 kg CO_2 eq. are emitted per cubic meter of biohydrogen produced when biomethane is reformed due to methane leaks in the atmosphere: 0.085 kg for the emitted carbon dioxide and 0.051 kg for N_2O for 1 m^3 of biohydrogen. The total emissions add up to 0.247 kg CO_2 equivalent per m^3 of biohydrogen (Fig. 6.11).

The impact on human health due to human toxicity is measured in kg-equivalent of 1,4-dichlorbenzene equivalent, which refers to toxicity produced in air, freshwater, seawater or land. The total toxicity is 0.0204 kg dcb per cubic meter of biohydrogen. Most of the toxicity is due to polycyclic aromatic hydrocarbons, benzene and arsenic emissions.

The ozone depletion potential (ODP) is determined in the form of kg-CFC 11 emission equivalent. The substances that contribute to ozone depletion are bromofluorohalons, HCFC-22 and other substances. These emissions will occur during lifecycle operations. Fig. 6.12 shows the obtained ODP for LCA of the integrated biogas-reforming plant. The total ODP is 9.95E-9 kg-CFC 11 equivalent.

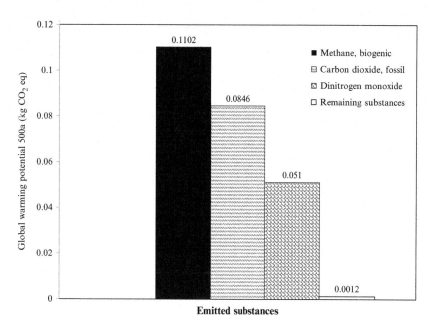

FIG. 6.10 Global warming potential corresponding to 1 m³ of biohydrogen production from biogas reformation.

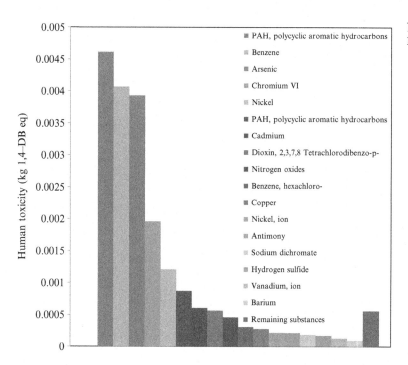

FIG. 6.11 Human toxicity impact factor for 1 m³ of biohydrogen from biogas reforming.

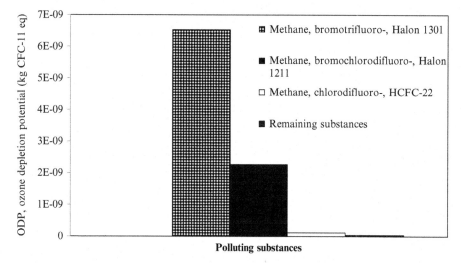

FIG. 6.12 Ozone depletion potential for 1 m³ of biohydrogen produced by biogas reforming.

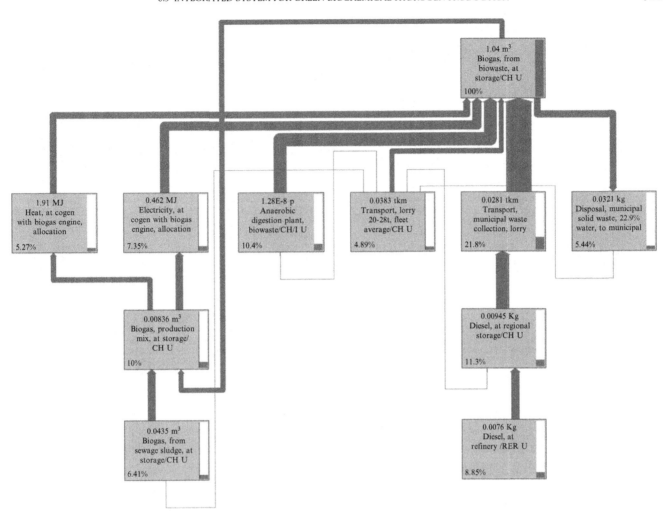

FIG. 6.13 Single-score Eco-indicator 99 contribution of lifecycle units.

Eco-indicator 99 is a single-score compounded indicator that is able to quantify the impact of the process units to overall lifecycle. Fig. 6.13 shows the Eco-indicator 99.

In the LCA, many detailed aspects are considered, such as the use of conventional fuels for material transportation. Diesel fuel usage for regional transportation of biomass is accounted for in the LCA of biohydrogen production. Diesel fuel impacts 11.3 on the Eco-indicator 99. The total impact due to transportation including waste collection, lorry operation and diesel-related emissions for refinery operation represent 41.95% environmental impact. The CML 2001 impact categories are shown in Fig. 6.14 for the ODP (left) and GWP (right).

6.3.2 Conceptual System Integration of Microbial Electrolysis With PV-Arrays

In this section, an integration concept of a microbial electrolysis with PV-electricity is presented. The system consists of a hybrid electrolysis cell comprising a microbial anode and an inorganic photocathode. A Fresnel-lens-type solar concentrator is used to split the solar spectrum into two subspectra using a dielectric mirror. The high-energy spectrum is directed toward the photocathode, and the lower-energy spectrum is directed toward PV array. The power generated by the PV array is applied to the electrolysis cell.

The concept is presented in Fig. 6.15. Better energy harvesting of solar photons is obtained when spectral splitting is applied. It is noteworthy, though, that the dielectric mirror devices for spectral splitting are rather expensive. Therefore, the use of a simple acrylic-made Fresnel lens will produce a concentrated radiation on a small-area dielectric mirror. Now, the photons with wavelengths of approximately 550 nm or shorter can be directed onto a photocathode, whereas the longer wavelength photons will reach the PV array. The microbial electrolysis cell will be supplied with energy from at least three sources: the PV array, the electrons excited at photocathode, the biochemical energy of the microbial substrate.

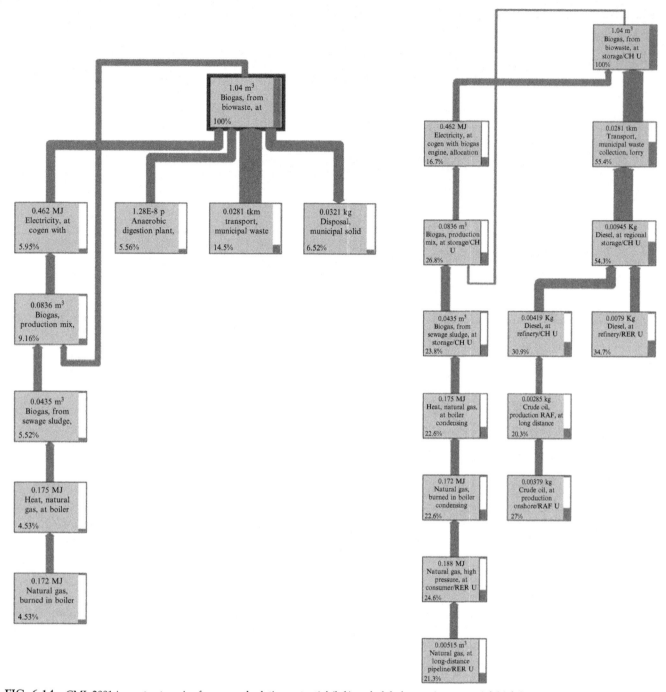

FIG. 6.14 CML 2001 impact categories for ozone depletion potential (left) and global warming potential (right).

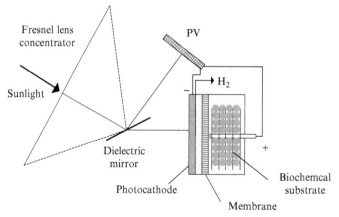

FIG. 6.15 Solar spectral splitter integrated with a microbial electrolysis cell and PV array.

TABLE 6.2 Hydrogen Production Rate and Yield at 0.6 V Applied Voltage

Substrate	y_{H_2}	Rate (day^{-1})
Glucose	8.55	1.23
Cellulose	8.20	0.11
Acetic acid	3.65	1.10
Butyric acid	8.01	0.45
Lactic acid	5.45	1.04
Propionic acid	6.25	0.72
Valeric acid	8.77	0.14

Source: Regan, J.M., Yan, H., 2014 Bioelectrochemical systems for indirect biohydrogen production. In: Zanoni, D., De Philippis, R. (Eds.), Microbial Bioenergy: Hydrogen Production. Advances in Photosynthesis and Respiration, vol. 38. Springer, Netherlands (Chapter 10).

This type of hybrid cell is usually denoted as bioelectrochemical system (BES) or bioelectrochemical cell as referred by Regan and Yan (2014). It appears that the electrode potential of those systems is lower than 0.25 V. Hydrogen is not a product of the biological metabolism, but rather is derived from a microbial-induced current. This means that at the cathode, typical methods may be applied from electrochemistry. One option is to use a photocathode, which itself will provide electrons in an excited state. Virtually any biodegradable electron donor can be used.

The following microorganisms have been found active for electrolysis cell anode, being able to donate electrons to the anode, namely, *Comamonas denitrificans*, *Ochrobactrum anthropi*, *Enterobacter cloacae*, although the rate of hydrogen production with the device has been 3.5 times smaller than for a dark-fermentation process, Regan and Yan (2014).

The substrate influences the hydrogen production rate. In Table 6.2, the yield and hydrogen production rate are given for a fixed-cell polarization of 0.6 V. The production rate is given in $m^3/day/m^3$ of reactor. Other products of a dark-fermentation process can be used downstream as a feedstock substrate for a microbial electrolysis cell and made to generate hydrogen under small applied voltage.

For some microbial cell configuration, the membrane is not needed. Consider acetate in standard biological conditions of pH 7, T_0 and P_0 as a hydrogen-evolving substrate with the following anodic decomposition reaction, emanating already hydrogen

$$CH_3COO^- + 4H_2O \rightarrow 2HCO_3^- + H^+ + 4H_2 \tag{6.16}$$

The Gibbs free energy reaction from Eq. (6.16) is $\Delta G^0 = 104.6$ kJ/mol. The reversible cell potential of the cell is written as

$$E_{rev} = \frac{\Delta G^0}{zF} = -0.114\,V \tag{6.17}$$

where the number of charges is $z = 8$, since the half reaction is written as

$$CH_3COO^- + 4H_2O \rightarrow 2HCO_3^- + 9H^+ + 8e^- \tag{6.18}$$

The reversible potential at anode becomes

$$E_{rev,a} = E_{rev,a}^0 - \frac{RT}{8F} \ln\left(\frac{[CH_3COO^-]}{[HCO_3^-]^2[H^+]^9}\right) \tag{6.19}$$

with $E_{rev,a}^0 = 0.187$ V

For standard condition $E_{rev,a} = -0.279\,V$ and for the cathodic proton-reduction reaction

$$E_{rev,c} = -\frac{RT}{2F} \ln\left(\frac{P_{H2}}{[H^+]^2}\right) \tag{6.20}$$

Regan and Yan (2014) showed that the electrical energy input required by the cell represents only about 10% of what a typical electrolysis cell will require.

6.3.3 Integrated Bioethanol-Based Systems

In this system, an integrated bioethanol-based system for hydrogen production is discussed. The system integrates a bioethanol plant with an ethanol thermochemical reforming plant. Bioethanol-derived hydrogen can be used directly in solid oxide fuel cell applications, since those are insensitive to poisoning with traces of carbon monoxide. Bioethanol is produced by fermentation of biomass material, which can be described according to the equation

$$C_6H_{12}O_6 \xrightarrow{\text{Fermentation}} 2CH_3CH_2OH + 2CO_2 \tag{6.21}$$

Many agricultural products and wastes such as sugar cane, potatoes, switchgrass, corn, starch-rich and lignocellulosic material can be converted to ethanol by fermentation. The diagram shown in Fig. 6.16 describes the production process of bioethanol from lignocellulosic materials. Two main processes are required: fermentation and simultaneous saccharification and fermentation. The lignocellulosic biomass comprises cellulose, hemicellulose and lignin, which are complex polymeric carbohydrates, which, for conversion to ethanol, require a pretreatment process to transform those polymers into fermentable sugars. The yeast bacteria *Saccharomyces cerevisiae* is used to convert sugars to ethanol. The pretreatment implies hydrolysis hemicellulose and lignin and an enzymatic cellulose treatment with specific pentose fermentation yeasts, such as *Candida shehatae* or *Pichia stipitis*.

The flowsheet diagram of the integrated plant is shown in Fig. 6.17. Bioethanol is produced in the first stage according to a biochemical conversion pathway. In the second stage, bioethanol is reformed with steam to generate pure hydrogen, following a thermochemical conversion process. The biochemical pretreating of biomass feedstock is the first phase of the first-stage process. The lignocellulosic matrix is broken in, sufficiently producing enough amorphous cellulose for the enzymatic fermentation process. Sulfuric acid and some additional enzymes are used in this process. The resulting products are liquids and solid materials comprising hydrolyzed hemicellulose, lignin, soluble pentose, xylose, and glucose.

The process materials after pretreatment are a slurry of liquids and solids. The liquids are separated, as a pentose-rich stream. This is treated first by ionic exchange and then by fermentation. The solids are a cellulose and hexose-rich stream. The fermenter unit will enzymatically ferment the feedstock consisting of a complex mixture of cellulose, hexose, and prefermented pentose, and finally generate liquid ethanol from the enzymatic process. A complex distillation unit is required to extract the desired fraction of ethanol, as shown in the flowsheet from Fig. 6.17. The resulting pure ethanol is now mixed with water under pressure and heated up to a high temperature for the reformation process.

The ethanol reformation process involves a series of chemical reactions that depend on the ethanol/water fraction, operating conditions and catalysts. If sufficient steam supply does exist at proper operating conditions, then the reforming process can be described by the following chemical equation

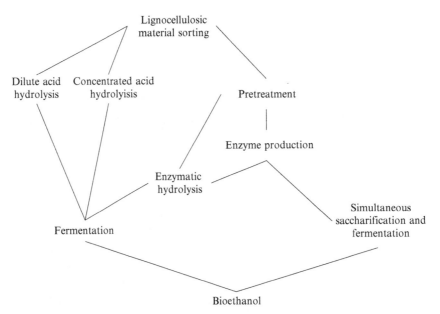

FIG. 6.16   Bioethanol production process from lignocellulosic materials.

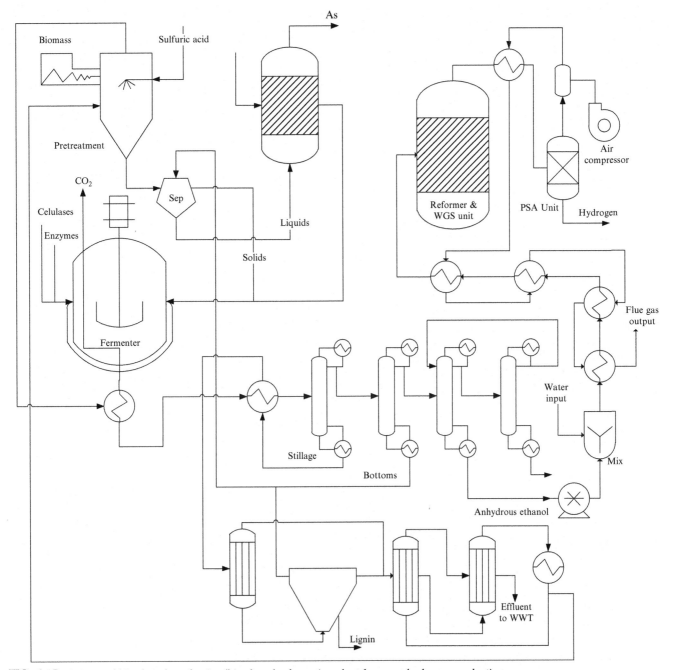

FIG. 6.17 Integrated bioethanol production/bioethanol reformation plant for green hydrogen production.

$$C_2H_5OH + 3H_2O \rightarrow 6H_2 + 2CO_2 \qquad (6.22)$$

However, the actual operation condition and re finiteness of the reformer size will not allow for the ideal conversion, so undesirable CO and CH$_4$ by products are formed

$$C_2H_5OH + H_2O \rightarrow 4H_2 + 2CO \qquad (6.23a)$$

$$C_2H_5OH + 2H_2 \rightarrow 4CH_4 + H_2O \qquad (6.23b)$$

The following ethanol dehydrogenation reaction-forming acetaldehyde may occur

$$C_2H_5OH \rightarrow H_2 + C_2H_4O \qquad (6.24)$$

The acetaldehyde may decompose as

$$C_2H_4O \rightarrow CH_4 + CO \qquad (6.25)$$

or be reformed with steam as

$$C_2H_4O + H_2O \rightarrow 3H_2 + 2CO \qquad (6.26)$$

Further reaction pathways are defined through the decomposition of ethanol as

$$C_2H_5OH \rightarrow H_2 + CO + CH_4 \qquad (6.27a)$$

$$2C_2H_5OH \rightarrow 3H_2 + C_3H_6O + CO \qquad (6.27b)$$

$$C_2H_5OH \rightarrow \frac{1}{2}CO_2 + \frac{3}{2}CH_4 \qquad (6.27c)$$

$$C_2H_5OH \rightarrow C_2H_4 + H_2O \qquad (6.27d)$$

Other side reactions and recombinations are written as

$$CO + 3H_2 \rightarrow CH_4 + H_2O \qquad (6.28a)$$

$$CO_2 + 4H_2 \rightarrow CH_4 + 2H_2O \qquad (6.28b)$$

$$CH_4 \rightarrow 2H_2 + C \qquad (6.28c)$$

$$2CO \rightarrow CO_2 + C \qquad (6.28d)$$

The catalysts selection is very important in optimization of ethanol reforming. As mentioned in Ni et al. (2007), nickel supported on yttrium oxide or lanthanum oxide or aluminum oxide are good candidates for ethanol steam reforming. The reforming process ends up with the water-gas shift reaction $CO + H_2O \rightarrow H_2 + CO_2$. Hydrogen is then formed at 773 K by partial oxidation of ethanol, which will lead to an autothermal process as follows

$$C_2H_5OH + 2H_2O + \frac{1}{2}O_2 \rightarrow 5H_2 + 2CO_2 \qquad (6.29)$$

with an exothermic heat of reaction, namely −50 kJ/mol.

6.4 CONCLUDING REMARKS

In this chapter, the methods of hydrogen production by biochemical energy and biophotochemical energy are introduced, classified and discussed. The basic pathways of biochemical hydrogen production are presented through the processes, either driven by a biochemical energy supply or by photonic energy manipulated by living cells. It is emphasized that the action of enzymes is necessary in any biochemical processes. Two enzymes are introduced as they are related to hydrogen production, namely nitrogenase and hydrogenase.

The chapter ends with a presentation of some integrated systems to produce green hydrogen from biochemical energy either light-assisted or based on dark fermentation. Most of the integrated systems assume generation of an intermediate biofuel based on common methods that produce biogas or bioethanol. The biofuel is then reformed with steam to generate a green hydrogen in an integrated manner.

Nomenclature

G	molar Gibbs free energy, kJ/mol
F	frataday number, C/mol
P	pressure, kPa
R	universal gas constant, J/kmolK
t	time, s
T	temperature, K
y	molar fraction
z	number of charges

Greek Letters

μ	growth rate

Subscripts

a	algae
a	anode
c	cathode
f	formation
ox	oxidation
rev	reversible

Superscripts

0	reference state

Acronyms

ADP	adenosine diphosphate
ATP	adenosine triphosphate
CoA	coenzyme A
Fd	ferredoxin
LCA	lifecycle assessment
NAD^+	nicotinamide adenine dinucleotide
NADH	reduced nicotinamide adenine dinucleotide
Pi	inorganic phosphate (HPO_4^{2-})

References

Benemann, J.R., Weare, N.M., 1974. Hydrogen evolution by nitrogen-fixing Anabaena cylindrical cultures. Science 184, 174–175.

Chisti, Y., 2007. Biodiesel from microalgae. Biotechnol. Adv. 25, 294–306.

Kava-Cordeiro, V., Vargas, J.C.V., 2015. Microalgae derived hydrogen production enhancement via genetic modification. In: Sixth International Conference on Hydrogen Production, May 3–6, UOIT, Oshawa, Ontario, Canada.

Das, D., Veziroglu, T.N., 2008. Advances in biological hydrogen production. Int. J. Hydrog. Energy 33, 6046–6057.

Hallenbeck, P.C., Benemann, J.R., 2002. Biological hydrogen production; fundamentals and limiting processes. Int. J. Hydrog. Energy 27, 1185–1193.

Harun, R., Jason, W.S.Y., Cherrington, T., Danquah, M.K., 2011. Exploring alkaline pre-treatment of microalgal biomass for bioethanol production. Appl. Energy 88, 3464–3467.

Hinnemann, B., Nørskov, J.K., 2006. Catalysis by enzymes: the biological ammonia synthesis. Top. Catal. 37, 55–70.

Kästner, J., Blöchl, P.E., 2007. Ammonia production at the FeMo cofactor of nitrogenise: results from density functional theory. J. Am. Chem. Soc. 129, 2998–3006.

Melis, A., 2002. Green alga hydrogen production: progress, challenges and prospects. Int. J. Hydrog. Energy 27, 1217–1228.

Ni, M., Leung, D.Y.C., Leung, M.K.H., 2007. A review on reforming bio-ethanol for hydrogen production. Int. J. Hydrog. Energy 32, 3238–3247.

Regan, J.M., Yan, H., 2014. Bioelectrochemical systems for indirect biohydrogen production. In: Zanoni, D., De Philippis, R. (Eds.), Microbial Bioenergy: Hydrogen Production. In: Advances in Photosynthesis and Respiration, vol. 38. Springer, Netherlands (Chapter 10).

Spolaore, P., Joannis-Cassan, C., Duran, E., Isambert, A., 2006. Commercial applications of microalgae. J. Biosci. Bioeng. 101, 87–96.

Vargas, J.V.C., Mariano, A.B., Corrêa, D.O., Ordonez, J.C., 2014. The microalgae derived hydrogen process in compact photobioreactors. Int. J. Hydrog. Energy 39, 9588–9598.

STUDY PROBLEMS

6.1 How many moles of glucose substrate is required to generate one mole of hydrogen in dark fermentation?

6.2 Define the notion of turnover frequency of an enzyme. What is the turnover frequency of hydrogenase in *Desulfovibrio* spp?

6.3 Describe the photo-induced hydrogenase process in *C. reinhardtii*?

6.4 For what reason is indirect photocatalysis an advantageous process?

6.5 What is the role of nicotinamide adenine dinucleotide in biochemical processes?

6.6 Explain dark fermentation according to Fig. 6.3?

6.7 What is the advantage of indirect photolysis with respect to direct photolysis?

6.8 What type of microorganisms can be used for indirect photolysis?

6.9 Explain the difference between direct photolysis and photofermentation?

6.10 Explain the operation of a microbial electrolysis cell?

Other Hydrogen Production Methods

7.1 INTRODUCTION

A number of atypical hydrogen production methods are discussed in this chapter based on a literature survey of past publications. The focus of the chapter is on older or relatively known methods, rather than newer and less-known ones, which are covered in Chapter 8. Some of these older methods may be interest in newer green hydrogen production applications.

An example is the Yokohama Mark 2A cycle proposed by Yokohama National University of Japan. This system is not photochemical (Chapter 5) and not thermochemical (Chapter 4), but rather a hybrid photo-thermochemical process. It was developed in the 1970s and was demonstrated with a lab-scale proof-of-concept prototype. The oxygen evolution step of this cycle was found to be inefficient; therefore, the researchers proceeded to a change one of the intermediate reagents and eventually obtained the first photo-electro-thermochemical water splitting cycle, now named Yokohama Mark 6.

Radiolysis is another process that is encountered in some limited situations. Water can be split into hydrogen and oxygen radiolytically, although with less efficiency, when exposed to high intensity nuclear radiation, especially alpha, beta, and gamma radiation. At the nuclear reactor, water management requires special precautions related to the issue about radiolytic water decomposition, which generates (undesired) hydrogen in the loop. Radio-thermochemical cycles were proposed in the past to split water, as reviewed herein.

Another interesting area is coal hydrogasification based on process heat derived from nuclear reactors. This process was studied intensively in Germany in the 1970s and 1980s. Though the process has true potential for hydrogen and synthetic fuel production on a large scale, research on the process was abandoned, because Germany began focusing on renewable energy. A revival of interest in this process can easily be observed in recent publications. Similarly, nuclear-based natural gas reforming for hydrogen production is of high interest for future development.

The use of concentrated solar radiation as a high temperature heat source for conventional and synthetic fuels reforming to hydrogen is viewed as a potential method for better harvesting of the solar energy. Thermocatalytic reactors were developed in the form of cavity volumetric solar receivers to reform natural gas or synthetic methane, methanol, propane, ammonia, and other fuels.

Molten alkali salts have long been known as water solvents, and only but relatively recently were applied to high temperature water electrolysis. In this system, water is the solute at temperatures of 773 K or more. The hydroxyls in the molten salt system present an excellent ion conduction medium for water electrolysis. Some NaOH and KOH eutectics were devised for effective water electrolysis, as is reviewed in this chapter. The advantage of high temperature operations relates to better kinetics and more favorable thermodynamics.

The chapter ends with a section discussing the paradigm of green hydrogen from ammonia. It appears beneficial in converting green hydrogen or even non-green hydrogen into ammonia to store (even seasonally) and to use as a commodity and hydrogen source. Hydrogen can be derived easily from ammonia with thermos-catalytic methods.

A hydrogen-ammonia economy integrates perfectly with the natural oxygen, nitrogen and water cycle, whereas, according to the hydrogen economy paradigm, a perfect integration with natural oxygen and water cycles is sought. Therefore, ammonia economy is viewed as a perfect transition phase to a completely established hydrogen economy. An electrochemical method of simultaneously producing hydrogen and NH_3 along with a membrane thermocatalytic reactor for NH_3 conversion to hydrogen has future high scaling-up potential, as reviewed in the chapter.

7.2 PHOTO-THERMOCHEMICAL WATER SPLITTING

Water splitting through a sequence of thermochemical and photochemical reaction was proposed in the past as a method for hydrogen generation. The cycle Yokohama Mark 2A, previously mentioned in Section 7.1, was probably the first hybrid photo-electro-thermochemical water splitting cycle proposed. This cycle comprised four intermediate reagents fully recyclable, namely I_2, HI, $FeSO_4$, and $Fe(OH)SO_4$. The process diagram of the cycle is shown in a simplified manner in Fig. 7.1.

The starting process is photo-hydrolysis in which an aqueous solution of iodine and ferrous sulfate is exposed to light radiation. The iodine in aqueous solution dissociates, forming two ionic species, both of charge −1, namely I^- (iodide) and I_3^- (triodide).

In an aqueous solution, iodine dissociates according to the following reaction at equilibrium:

$$I_2(aq) + I^-(aq) \rightleftarrows I_3^-(aq) \tag{7.1}$$

The dissolved ferrous sulfide dissociates in water, as follows:

$$FeSO_4(aq) \rightarrow Fe^{2+}(aq) + SO_4^{2-}(aq) \tag{7.2}$$

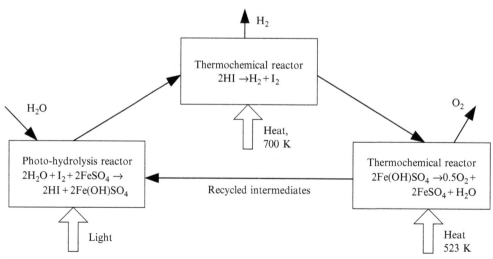

FIG. 7.1   Simplified process diagram of Yokohama Mark 2A photo-electro-thermochemical cycle.

The aqueous solutions of triiodide and ferric ion are photosensitive, promoting the following photochemical reaction:

$$2Fe^{2+}(aq) + I_3^-(aq) \xrightarrow{h\nu} 2Fe^{3+}(aq) + 3I^-(aq) \tag{7.3}$$

The process described by Eqs. (7.1)–(7.3) constitutes the photochemical step of the cycle, which can be expressed as an overall reaction, as follows:

$$2H_2O + I_2 + 2FeSO_4 \xrightarrow{h\nu} 2HI + 2Fe(OH)SO_4 \tag{7.4}$$

The hydrogen iodide can be separated as gas from the aqueous solution. Once separated, hydrogen iodide can be thermally decomposed at 700 K to produce hydrogen gas as flows:

$$2HI(g) \rightarrow H_2(g) + I_2(g), \text{Thermochemical}, \; \Delta H = 158.3 \, kJ/mol \tag{7.5}$$

Iodine can be easily separated from hydrogen and reused in the photochemical process described previously by Eq. (7.4). Iron hydroxide sulfate that resulted as an output from the photochemical process can be decomposed thermally at 523 K to generate oxygen and regenerate ferrous sulfate as follows:

$$Fe(OH)SO_4 \rightarrow 0.5O_2 + 2FeSO_4 + H_2O \tag{7.6}$$

The experiments done with the Yokohama Mark 2A cycle showed that the oxygen evolution process is relatively inefficient. Eventually the cycle was modified into a Yokohama Mark 5 version, which uses an electrochemical hydrogen evolution process and an electrochemical oxygen evolution reaction. The Yokohama Mark 5 cycle version is presented in a simplified manner as shown in Fig. 7.2.

The cycle "Yokohama Mark 5" is a peculiar one because it comprises two electrochemical steps. This cycle uses $Fe_2(SO_4)_3$ as an intermediary reagent instead of $Fe(OH)SO_4$. With this change, the photochemical process remains essentially the same as for Yokohama Mark 2A.

One problem with this cycle design is the choice of the photochemical process. The process does not require photocatalysts, which is advantageous. However, it requires light at wavelengths shorter than 350 nm, which are very scarce in the solar spectrum. The photochemical step of the cycle is as follows:

$$H_2SO_4 + I_2 + 2FeSO_4 \xrightarrow{h\nu} 2HI + 2Fe_2(SO_4)_3, \text{Photochemical}, \; \lambda \leq 350 \, nm \tag{7.7}$$

The hydrogen production step is performed electrochemically as follows:

$$2HI \rightarrow H_2 + I_2, \text{Electrochemical}, \Delta G = 109.7 \frac{kJ}{mol} \tag{7.8}$$

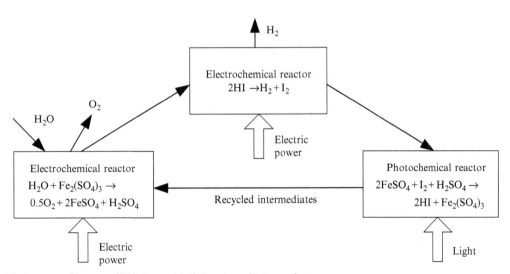

FIG. 7.2 Simplified process diagram of Yokohama Mark 5 water splitting cycle.

The oxygen production step has also been performed electrochemically for the Yokohama Mark 5 cycle. This process is as follows:

$$Fe(OH)SO_4 \rightarrow 0.5O_2 + 2FeSO_4 + H_2O, \text{ Electrochemical, } \Delta G = 82.5 \text{ kJ/mol} \tag{7.9}$$

Because it has two electrochemical steps, with a high consumption of power, and has the ability to use only a scarce spectrum of solar radiation, Yokohama Mark 5 cannot compete with water electrolysis as a baseline case in terms of hydrogen production. Therefore, this cycle was abandoned in favor of a more promising one, the Yokohama Mark 6, which will be discussed latter in this chapter.

A four-step photo-thermochemical process called the "photolytic sulfur ammonia cycle" was discussed by Perret (2011). This process includes a photolysis reaction (that is a photochemical reaction of hydrolysis) which evolves hydrogen at 353 K, and a thermal decomposition reaction with oxygen evolution at 1143 K. There are two thermochemical reactions that recycle the intermediate reagents, one at a low temperature of 393 K and the other at 773 K. The simplified cycle diagram is presented schematically in Fig. 7.3.

The photochemical hydrolysis reaction requires an expensive photo-catalyst made in noble metals based on cadmium sulfide and a platinum-palladium-ruthenium complex. The photo-catalyst requires light at a wavelength shorter than 520 nm. The photo-reactor may require a source of green/blue light, which can be derived from solar radiation with adequate spectral splitters. The photochemical reaction is described based on the following equation:

$$H_2O + (NH_4)_2SO_3 \rightarrow H_2 + (NH_4)_2SO_4, \text{ Photochemical, } \lambda < 520 \text{ nm} \tag{7.10}$$

The thermochemical reactions are three in total, as follows:

$$H_2O + SO_2 + 2NH_3 \rightarrow (NH_4)_2SO_3, 393 \text{ K} \tag{7.11a}$$

$$ZnSO_4 \rightarrow 0.5O_2 + ZnO + SO_2, 1143 \text{ K} \tag{7.11b}$$

$$(NH_4)_2SO_4 + ZnO \rightarrow 2NH_3 + ZnSO_4 + H_2O, 773 \text{ K} \tag{7.11c}$$

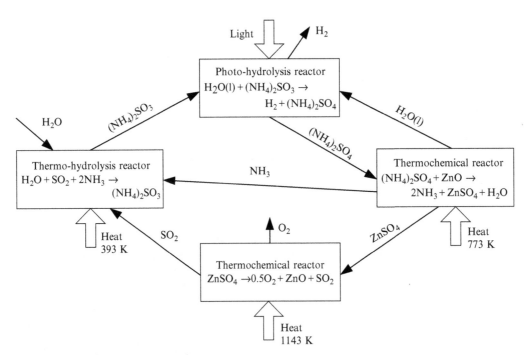

FIG. 7.3 Photo-thermochemical sulfur-ammonia cycle.

7.3 PHOTO-ELECTRO-THERMOCHEMICAL WATER SPLITTING

A possible scheme of a three-step photo-electro-thermochemical water splitting cycle is shown in the block diagram in Fig. 7.4. A compound generically denoted here with AB is hydrolyzed thermochemically by the addition of high temperature heat. The flowing reaction will produce AH_2 and an oxide BO in the following representative process:

$$H_2O(l) + AB \rightarrow AH_2 + BO \tag{7.12}$$

The compound AH_2 is thereafter decomposed photo-catalytically as follows:

$$AH_2 \rightarrow H_2 + A \tag{7.13}$$

The remaining process or intermediate reagent regeneration is generally thermodynamically unfavorable; therefore, it is "forced" electrochemically, as follows:

$$BO + A \rightarrow 0.5O_2 + AB \tag{7.14}$$

Provided that all reactions are complete, a chemical loop is closed with the only consumed material being water and with produced chemicals being hydrogen and oxygen. The consumed energy has three sources: high temperature heat, light radiation, and electric power.

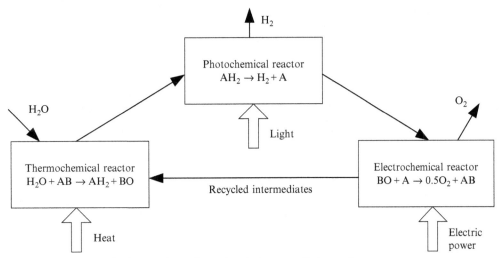

FIG. 7.4 Simplified process diagram of a thermo-photo-electrochemical water splitting cycle.

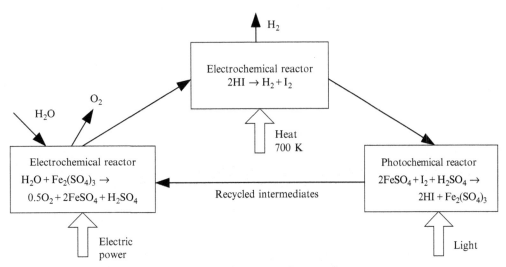

FIG. 7.5 Simplified process flow diagram of the Yokohama Mark 6 water splitting cycle.

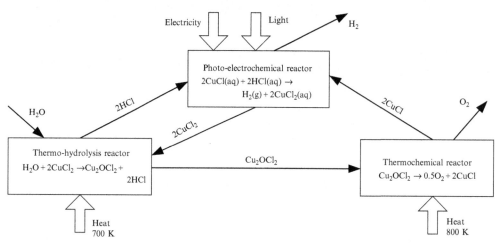

FIG. 7.6   Simplified process flow diagram for the photo-electro-thermochemical copper-chlorine water splitting cycle.

In the Yokohama Mark 6 implementation, the hydrogen evolving reaction is performed thermochemically at 700 K. This cycle is represented in the simplified flowsheet diagram from Fig. 7.5. This reaction is given previously in Eq. (7.5). The photochemical step of the cycle is described according to Eq. (7.7). The third and last step of the cycle is the electrochemical process described according to Eq. (7.8). The proved efficiency of the Yokohama Mark 6 cycle is 30%, but the cycle keeps the same deficiency of the lack of a reliable/sustainable light source of a wavelength shorter than 350 nm.

Zamfirescu et al. (2012) proposed a photo-electro-chemical method to replace the electrochemical step within the copper-chlorine cycle. This process was further confirmed experimentally by Ratlamwala and Dincer (2014), as previously presented in Chapter 5. The cycle integrated with the photo-electro-chemical process was introduced by Naterer et al. (2013).

A simplified process flow diagram of the thermo-photo-electro-chemical copper-chlorine cycle is shown in Fig. 7.6. The cycle has three reaction steps. Here the photo-electro-chemical process evolves hydrogen as described in the following equation:

$$2CuCl(aq) + HCl(aq) \rightarrow 2CuCl_2(aq) + H_2(g) \tag{7.15}$$

The cycle continues with two thermochemical steps, discussed previously in Chapter 4. There is a thermolysis reaction at 700 K, described as follows:

$$H_2O + 2CuCl_2(s) \rightarrow CuOCl_2(s) + 2HCl(g) \tag{7.16}$$

The closing reaction evolves hydrogen in a thermolysis process of copper oxychloride:

$$CuOCl_2(s) \rightarrow 0.5O_2(g) + 2CuCl(l) \tag{7.17}$$

7.4 RADIO-THERMOCHEMICAL WATER SPLITTING

High energy nuclear radiation, especially gamma and alpha, can split water molecules and generate hydrogen. Although less developed than other hydrogen production processes, radio-thermochemical water splitting processes may present some potential for further consideration because of their unique potential to directly use nuclear radiation to promote chemical reactions—either at a nuclear reactor site, or at nuclear fuel processing plants or spent fuel processing facilities.

Von Federsdorff (1974) proposed a radio-thermochemical water splitting cycle based on a specific radiochemical gas-phase reaction occurring at 500 K. This reaction is as follows:

$$CO_2(g) \xrightarrow{\text{Radiolysis}} 0.5O_2(g) + CO(g) \tag{7.18}$$

When carbon dioxide is subjected to high energy nuclear radiation at 500 K, several processes may occur, among which the one described by Eq. (7.18) is very probable. Other possible radiolysis processes are as follows:

$$CO_2(g) \xrightarrow{\text{Radiolysis}} C + 2O \tag{7.19a}$$

$$CO_2(g) \xrightarrow{\text{Radiolysis}} CO + O \tag{7.19b}$$

Von Federsdorff (1974) proposed to conduct reaction (7.18) in a specially designed nuclear reactor. As suggested, the separation of oxygen gas from carbon monoxide requires an additional chemical reaction in which mercury is used to form mercury oxide MgO. Thereafter, the mercury oxide is decomposed thermally to release oxygen while mercury is condensed. Other CO and O_2 separation methods to consider may be based on physical processes such as those using absorption columns.

Dicarbon monoxide (C_2O) can be thermochemically formed as a side product of radiolytic carbon dioxide decomposition, as

$$CO + C \rightarrow C_2O \tag{7.19}$$

and another side product that may further form thermochemically is tricarbon dioxide, (C_3O_2)

$$C_2O + CO \rightarrow C_3O_2 \tag{7.20}$$

Both the dicarbon monoxide and tricarbon dioxide are stable carbonaceous deposits that form a buildup of scale on the reactor walls. The closure reaction for the radio-thermochemical cycle can be a carbon monoxide hydrolysis conducted thermochemically at 573 K, with hydrogen evolution. Hydrogen must then be separated from carbon dioxide gas according to established technology (e.g., pressure swing absorption). This closure reaction is written as follows:

$$H_2O(g) + CO(g) \rightarrow H_2(g) + CO_2(g) \tag{7.21}$$

The original method proposed in Von Federsdorff (1974) for cycle closure involved two intermediate chemical reagents, namely FeO(s) and Fe_3O_4(s). This closure proceeds in two reaction steps as follows:

$$H_2O(g) + 3FeO(s) \rightarrow H_2(g) + Fe_3O_4(s) \tag{7.22a}$$

$$Fe_3O_4(s) + CO(g) \rightarrow 3FeO(s) + CO_2(g) \tag{7.22b}$$

The process described by Eq. (7.22a) is a high temperature hydrolysis conducted at 673 K. The reaction (7.22b) is endothermic and needs to be conducted at 1000 K.

7.5 COAL HYDROGASIFICATION FOR HYDROGEN PRODUCTION

Past research in the 1970s and 1980s reported hydrogasification of coal as a safe method to generate hydrogen from sustainable heat sources, mainly from nuclear energy. A pioneering study of nuclear coal gasification was published by Schrader (1975). In a hydrogasification process, coal was combined with hydrogen to produce synthetic natural gas with an exothermic reaction. When hydrogen is allowed to flow slowly through a fixed bed of coal particles, a hydrogasification process occurs exothermically in which methane is formed from carbon, according to the following spontaneous reaction:

$$C(s) + 2H_2(g) \rightarrow CH_4(g) \tag{7.23}$$

A nuclear-based coal hydrogasification process for hydrogen production functions as described in the diagram shown in Fig. 7.7. The plant comprises five main units of operation: nuclear reactor, power plant, steam generator, hydro-gasifier, and steam-methane reformer. Hydrogasification of coal to generate methane evolves at a high rate because hydrogen is very reactive with coal. Because the reaction is exothermic, there is less need for coal preheating. The nuclear reactor produces heat for the steam generator. Superheated steam is reacted with methane in an allothermal steam methane reforming unit in which the required high temperature thermal energy is supplied by heat transfer from the nuclear reactor. The temperature level at which the reforming process must be heated is approximately 1000 K. This implies the utilization of the very high temperature reactor (VHTR) of the next generation of nuclear reactors. Therefore, in this process, neither coal nor methane is oxidized, and the resulting CO_2 emissions are minimized.

An integrated coal hydrogasification plant to produce hydrogen is shown in Fig. 7.7. The Generation IV nuclear reactor concept VHTR is used as the high temperature heat source. In the primary loop of the reactor, helium coolant

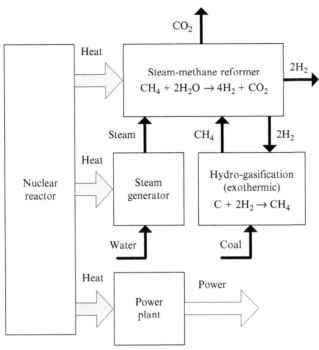

FIG. 7.7 Simplified representation of a nuclear-based coal hydrogasification plant for hydrogen production.

is circulated. Heat is transferred to a secondary loop through which the gasification agent consisting of steam and hydrogen mixture is circulated. The gasification agent is heated up to 1173 K. The helium circuit also transfers heat to the steam generator of a Rankine cycle to produce power.

The hottest gasification agent passes through a heat exchanger system installed within the methane steam reformer unit of operations. The overall process of methane steam reforming can be described compactly, based on the following endothermic reaction:

$$CH_4(g) + 2H_2O(g) \rightarrow 4H_2(g) + CO_2(g) \tag{7.24}$$

The gasification agent is thereafter injected at 1100 K at the bottom of the gasifier. Fig. 7.8 shows the schematic of an integrated hydrogen production plant based on hydrogasitifation of coal using heat derived from a very high temperature reactor.In the gasifier, dried coal particles are fed from above. Coal is milled and pulverized in a fluidized bed dryer. Steam from the Rankine cycle is extracted to be used for the gasification process. This steam is sourced from supply water that is fed to the Rankine cycle pump. A pilot plant based on hydrogasification of coal was demonstrated at the "Rheinische Braunkholenwerke AG" in Germany, and the results were reported in Schrader et al. (1975). The process involved helium preheated at temperature levels that simulate the situation in an actual VHTR. The pilot plant was devised for 400 kg raw lignite per hour or 150 kg hard coal per hour. It generated 700 m^3 per hour of raw gas at 4–10 MPa. Based on the experimental demonstration, some full-scale conceptual designs were provided. It was shown that a VHTR of 3 GW thermal can process 2.5 Mt/h lignite and generate 410,000 Nm^3/h of methane and 180 MW electrical power. It also yields 330 t/h residual char. The system efficiency of methane production is estimated at 72%.

Therefore, accounting for the reaction heat for steam methane reforming, one may expect a hydrogen generation efficiency of more than 60% for the integrated plant of hydrogen production via coal hydrogasification. It appears, therefore, that this type of process has significant potential for large-scale hydrogen production.

7.6 NUCLEAR-BASED NATURAL GAS REFORMING FOR HYDROGEN PRODUCTION

Process heat derived from the next generation of nuclear reactors can be at a sufficient temperature level to drive a steam-methane reaction process. If this is done, then less carbon dioxide emissions are obtained because no methane combustion is required to sustain the process. According to Verfondern (2007), at least 30% more hydrogen is

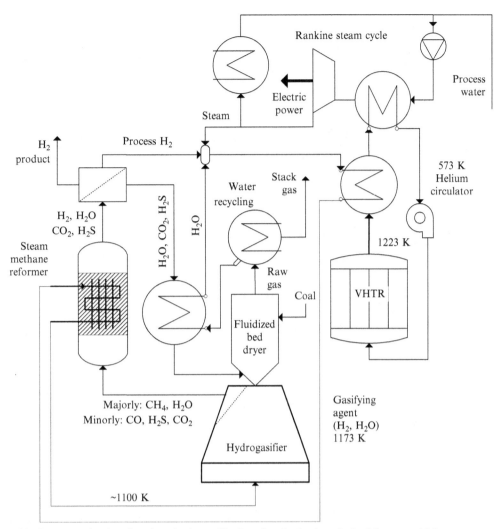

FIG. 7.8 Integrated hydrogen production plant from hydrogasification of coal using heat derived from very high temperature reactor (VHTR) of Generation IV concept.

generated for the same amount of CO_2 emission if the heat is derived from a nuclear reactor rather than from combustion of an additional quantity of natural gas.

The most technologically advanced developments of nuclear-based natural gas reforming were developed in Japan and Germany where specific research programs were initiated three to four decades ago. The one best suited for application to steam methane reforming technology is the VHTR. This reactor generates a sufficiently high process temperature, which allows for good yield of the thermocatalytic steam-methane reformation process, namely at a temperature above 900 K.

In a VHTR plant, the steam can be heated directly in the IHV (intermediate heat exchanger) and supplied at elevated temperatures to a reformer. Alternatively, a helium loop can be used to deliver heat both to a steam generator and to a steam-methane reformation reactor. The most straightforward approach is a system that does not require any modification of the reactor and intermediate heat exchanger (IHX). Schematically, this system is depicted in Fig. 7.9.

The VHTR facility generates hot helium at approximately 1100 K, which is directed toward the steam-methane reformer. A possible design of the reformer is shown in the figure. The hot helium heats the catalytic bed as it is circulated through a long coil. The colder helium is used to transfer heat to a heat recovery steam generator (HRSG) and recirculated through the IHX of the VHTR. Water is fed into the system (in excess of the stoichiometric amount) and preheated using heat recovered from expelled gases and product hydrogen gas, after which superheated steam is produced. Preheated compressed natural gas (CNG) and superheated steam are fed to the reactor and passed through the hot catalytic bed where successive reactions occur (steam-methane reaction and water-gas shift reaction). A hydrogen selective membrane (palladium-based) can be used to extract high purity hydrogen in the vicinity

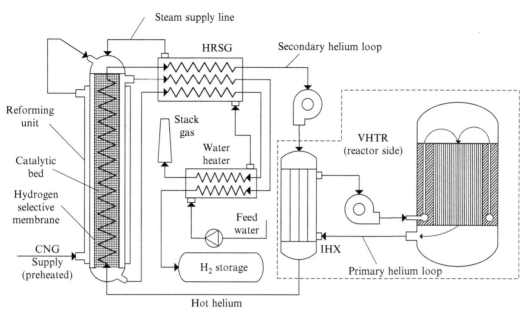

FIG. 7.9 System integration of steam methane reforming with VHTR.

of the reaction sites. The membrane can have a tubular geometry, and hydrogen is collected at the outer shell. The remaining (unreacted) gases are collected at the bottom and are used for heat recovery, and then further expelled.

A detailed description of past German research on nuclear-based natural gas reforming was reported in Verfondern (2007). A helium loop—heated electrically such that it simulates a VHTR system—was used as a heat source for the reforming process during long-term experiments at the Jülich Research Centre in Germany. A bayonet type reformer system—named EVA—was developed with hot helium supplied at the outer shell and reformation processes in the core tube. The heating capacity was 10 MW with generation of helium at 4 MPa and 1223 K.

The steam reformer was placed in sequence with a steam generator along the helium circuit. Heat is supplied to the steam reformer as helium cools down from 1223 to 923 K, while heat is supplied to the steam generator when helium cools down from 923 to 623 K. The process gas is supplied to the reactor in a proportion of 1:4 methane-to-steam at a temperature of 500 K. Because of the heat addition, the gas temperature after leaving the catalytic bed becomes 1073 K. The heat from the product gas (comprising H_2, CO_2, CO, and CH_4) is recovered in the range of 1073–723 K. The reaction rate is expressed as 1000 Nm^3 of hydrogen per m^3 of catalyst per hour. The heat transfer coefficient at the process side was reported at approximately 1000 W/m^2K, while at the hot helium side, it was ~500 W/m^2K.

From the past German research reported by Verfondern (2007), it results that, in order to generate 100,000 Nm^3/h hydrogen using steam reforming of natural gas, there is a need for nuclear reactor heat of 84.82 MW. The recommended reforming unit design comprises a cylindrical shell housing of 3.8 m diameter, which accommodates about 300 tubes of 17 m length filled with catalysts deposited on Rasching rings. The flow rate of helium that transfers heat from the nuclear reactor facility must be around 75 kg/s.

7.7 SOLAR FUEL REFORMING FOR HYDROGEN PRODUCTION

Solar energy is a widely available commodity with a vast capacity that to this point, with the exception of biological products, has been sparsely harvested. Going forward, solar energy is expected to take over an increasing portion of the energy market; however, effectively using solar energy presents many challenges. Solar energy may be used to provide high temperature heat to thermochemical reactors to reform fuels to hydrogen. One possibility at hand is methane reformation, as shown previously in Eq. (7.24). Another attractive possibility is methanol reforming, which is conducted as follows:

$$CH_3OH(g) + H_2O(g) \rightarrow 3H_2(g) + CO_2(g) \tag{7.25}$$

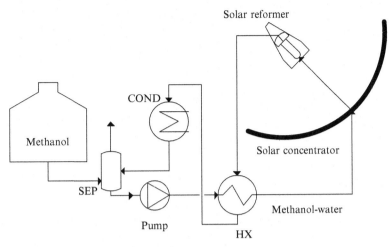

FIG. 7.10 Solar concentrator system with fuel reforming to hydrogen.

The heat reaction in Eq. (7.25) is 50.7 kJ/mol. This reaction can be conducted at a range of 450–600 K, which is one of the lowest temperature level requirements for a reforming process. A similar process is propane reforming with a reaction heat of 37.4 kJ/mol.

The conceptual design of a solar-heated methanol reforming system is shown in Fig. 7.10. Methanol from the tank is mixed with water and then pressurized; then heated and directed toward the reformer unit placed in the focal point of a solar concentrator. In the reformer, methanol is partially decomposed to form hydrogen. The product gases are returned and cooled and then condensed to recover water and separate hydrogen gas. A heating flux of 1 kW/m^2 is required at the solar receiver for a small-scale facility for a 5 kW unit, as reported in Liu et al. (2009).

The reforming reactor essentially consists of a small tube in which a catalyst is installed in a packed bed or fixed bet arrangement. Liu et al. (2009) reported excellent results with an off-the-shelf catalyst based on Cu/ZnO/Al$_2$O$_3$. The process must be set for an optimal water-to-methanol molar ratio 1.6–1.8.

The energy efficiency of the system reported in Liu et al. (2009) was an average of 50% for light irradiance in the range of 500–800 W/m^2. Of course, fuels other than methanol can be reformed using the concentrated solar energy input in a system, as that just discussed. Fig. 7.11 shows a possible alternative system for fuel reforming that integrates a Rankine cycle in addition to previous components. A Rankine cycle helps for better recovery of the temperatures levels in the streams. The additional work generated with the Rankine cycle will find local usefulness (e.g., running auxiliary systems). A seen, the flow scheme assumes that the exit stream from the reactor is split into two streams, which can facilitate system tuning in which the temperatures in points 5 and 7 are about the same, such that the least amount of exergy is destroyed.

The Rankine cycle has intrinsic thermodynamic irreversibility because of the pinch point. So, an alternative system that eliminates the Rankine cycle is shown in Fig. 7.12, where the products are generated in a pressurized state at

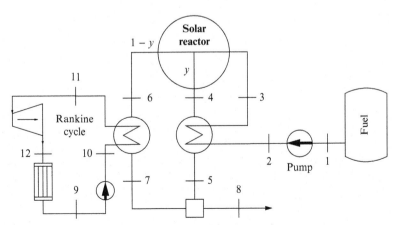

FIG. 7.11 Integrated system with concentrated solar fuel reforming and Rankine cycle.

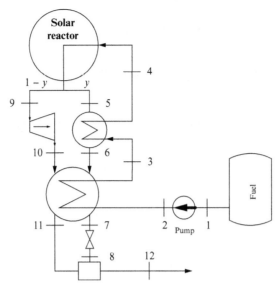

FIG. 7.12   Integrated solar concentrator system with fuel reforming and turbine expansion.

reactor exit and where a fraction $(1 - y)$ of them are expanded, according to process 9–10. The system confers an excellent match of the temperature profiles.

By using a direct expansion cycle such as that shown in Fig. 7.12, pinch point issues can be avoided; furthermore, points of adjustability are worked into the cycle and allow for a wider range of operation that is less sensitive to fluctuations in conditions.

Urea is one possible material that can be utilized as fuel that can be reformed to generate hydrogen. Concentrated solar radiation at 400–600 K will be suitable to drive the processes. Urea melts at 406 K, and with more heat addition, it decomposes releasing ammonia gas and isocyanic acid (also gas), according to the following pyrolysis reaction:

$$CO(NH_2)_2(l) \rightarrow NH_3(g) + HNCO(g) \qquad (7.26)$$

Fig. 7.13 shows an integrated system with concentrated solar energy that pyrolyses and reforms urea to generate hydrogen. Urea prills are typically used as fertilizer. In the system shown in the figure, they are removed from the tanker and squeezed into a tubular reactor by a helical auger. The pyrolysis reaction proceeds into the first part of the tube indicated by 3 and continues until reaching 4. Process 1–2 is the urea prills preheating process until at the melting temperature. Process 2–3 is a constant temperature melting process of urea.

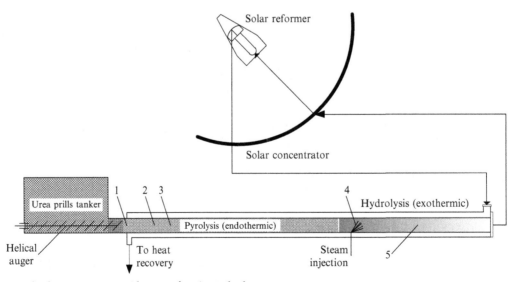

FIG. 7.13   Integrated solar concentrator with urea reforming to hydrogen.

After reaching 473 K, the reaction system is hydrolyzed with superheated steam, and an exothermic reaction of isocyanic acid hydrolysis occurs, as follows:

$$HNCO(g) + H_2O(g) \rightarrow NH_2(g) + CO_2(g) \tag{7.27}$$

Furthermore, ammonia is decomposed thermocatalytically in the reactor placed in the solar receiver. The resulting gases comprise hydrogen, nitrogen carbon dioxide and steam. The system assumes that part of the stream is recovered and recycles. Hydrogen can be easily separated at the reactor setting (at focal point) on a palladium membrane; this technology will be discussed later in the chapter.

7.8 ELECTROLYSIS IN MOLTEN ALKALI HYDROXIDES FOR HYDROGEN PRODUCTION

Typical alkaline electrolysis is done in an aqueous sodium or potassium hydroxide solution. If the solution temperature is increased, then electrolysis process is benefited by better ion mobility and a reduced requirement of Gibbs free energy for the reactions. If the temperature is increased above the normal boiling point of water, then the alkaline electrolyzer must be pressurized so that water will not boil. However, with alkali hydroxides, some special cases occur that provide the opportunity to develop an electrolysis process in molten phases at high temperature.

In order to introduce the electrolysis process in molten alkali hydroxide, let us review some theoretical aspects of liquid solutions. In a liquid solution, two phases coexist at any time, the solvent and the solute. Depending on the actual type of solution (e.g., zeotropic, azeotropic) the vapor pressure can depend or not on the concentration. The vapor pressure is higher at a higher temperature with saturated solutions. Furthermore, the solubility may increase substantially with the temperature, as is often the case. Because of this effect, at even higher temperatures, the high solubility of the solute causes a reduction the vapor pressure.

Because of this intriguing process, if the aqueous solution of alkali hydroxide is saturated, then water will never boil out of it. The process is nicely explained in Fletcher (2001) as the basis of high temperature electrolysis in molten alkali hydroxides. At a high temperature, water (steam) can be dissolved in molten alkali hydroxides (NaOH, KOH), which plays the role of an electrolyte, and be dissociated in such a molten bath. One great advantage of alkali hydroxides is that they are not volatile. They remain in molten phase and offer only a medium full of hydroxyls that are sufficiently mobile to facilitate the electrolysis process.

The melting point of NaOH is 591 K and the normal boiling point is 1827 K. The saturation pressure of sodium hydroxide are determined by the equation given in Fletcher (2001):

$$P_{sat} = 10^{5.113 - \frac{966}{T}} \tag{7.28}$$

When a saturated sodium hydroxide + water solution at 1 bar is heated, the molar fraction of NaOH must be increased to maintain the saturation state. Water will not boil if the saturation state is maintained. Once the normal melting point of sodium hydroxide is reached at 591 K, then the molten hydroxide becomes the solvent while water is the solute.

Licht et al. (2003) provided the diagram shown in Fig. 7.14 in which the measured cell potential of water electrolysis in a H_2O/NaOH system are plotted against the temperature. The cell potential starts to reduce as temperature increases. At a point, the cell voltage drops below the reversible cell potential of water electrolysis at 1 bar. Licht et al. (2003) explain this process as being due to a hydrogen-oxygen recombination that is stimulated in the cell in the calculated operating conditions. Because a part of products recombine, the cell potential drops and the production of species decrease.

If the gases are not allowed to recombine, then the cell potential is higher, but the production rate is also higher. For runs at 100 mA/cm^2, the actual cell potentials measured by Licht et al. (2003) are shown in Fig. 7.15. It can be observed that the average cell overpotential is approximately 0.45 V. Since, in an industrial alkaline electrolyzer, the overpotential is typically 0.6 V, it appears that the operation in molten salt is technically sound. In addition, there is a net decrease of cell potential because of the high temperature operation.

The recombination of hydrogen and oxygen in an alkaline cell can be exploited as a means of optimizing an integrated system that comprises photovoltaic cells and electrolyzer without the essential electronics for charge regulation. The optimum design point for the cell should fall close to the optimum design point for the electrolyzer. This principle is illustrated in Fig. 7.16. The cell design can be adjusted to allow for some, for more or for no recombination of oxygen and hydrogen gases within the molten NaOH.

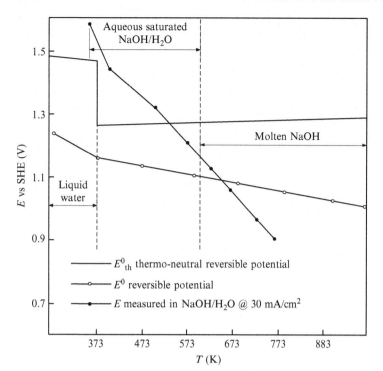

FIG. 7.14 Reversible cell potential of water dissociation $(H_2O \rightarrow H_2 + 0.5O_2)$ and the measured cell potential of water electrolysis in molten sodium hydroxide for a range of temperatures and 1 bar. *Data from Licht et al. (2003).*

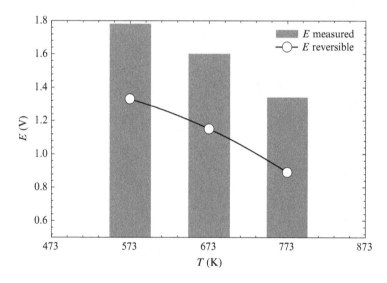

FIG. 7.15 Reversible cell potential and actual cell potential in a molten NaOH electrolysis cell with separated electrodes operating at 100 mA/cm² and 1 bar. *Data from Licht et al. (2003).*

Three cases are shown in the figure. Case 1, shown with long dash lines, assumes that no gas recombination is allowed, in which case, the production rates with the electrolysis cell will be at their maximum. However, the *I-V* characteristics of the PV cell and the electrolysis cell intersect at a current value that is too low. This means that PV cannot give sufficient voltage to the electrolysis cell, which will produce too little hydrogen.

In case 2, shown with continuous line, some recombination is allowed within the cell. For this reason, the cell voltage decreases for the same current. The cell will not produce hydrogen at the maximum that it could. However, in these conditions, the *I-V* characteristics of the PV cell and electrolysis cell intersect at high current, actually right at the maximum power point of the PV cell. This means that the cell transmits the maximum power that it can to the electrolysis process, which is very beneficial. Furthermore, the cell produces more hydrogen case 1 since the current is much higher.

In case 3, shown with short dash lines, more recombination is allowed from the cell design. This means that the cell has an even lower voltage requirement. However, in this condition, the *I-V* curve of the PV cell and of the electrolysis

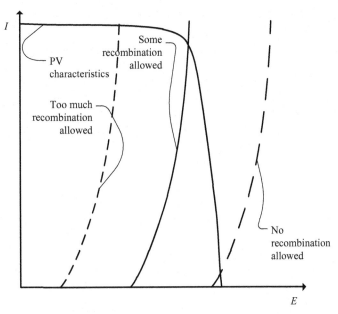

FIG. 7.16 Design optimization principle of a molten NaOH electrolysis system connected to a PV cell with given characteristics.

cell intersect at a current that is too high, more than the optimum point of the PV. Therefore, less power is transferred to the electrolysis cell with respect to what the PV could do. Furthermore, too much recombination implies a low hydrogen production rate. This case is well below optimal and generates a low hydrogen rate.

The optimum case is obviously case 2. More aspects can be explored if 2 or 3 PV cells or 1–2 electrolysis cells are connected in series. Licht et al. (2003) showed a situation where an optimum coupling of three PV cells with two molten NaOH electrolysis cells were found. The design combinations are compared in an *I-V* diagram, as shown in Fig. 7.17. It can be easily deduced from the diagram that the best design uses three PV cells and two electrolysis cells in series. No gas recombination is allowed for the optimum design. These results were obtained by Licht et al. (2003) when working with a molten NaOH cell at 773 K and the PV cell type was HECO 335 Sunpower with 1.561 cm^2 illuminated at 50 suns. The molten electrolyte was prepared from NaOH granules in an inert argon gas atmosphere. The granules were heated by steam injection until the cell reached the desired operation temperature. The cathode was a platinum gauze of 200 mesh 100 cm^2.

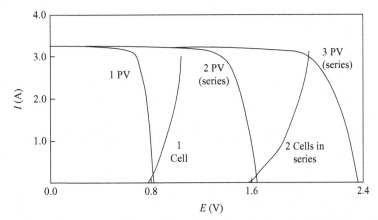

FIG. 7.17 Illustrative example of design optimization strategy when integrating 1 or 2 molten NaOH electrolysis cells with 1–3 concentrated PV cells. *Data from Licht et al. (2003).*

7.9 GREEN HYDROGEN FROM AMMONIA

Green hydrogen can be produced from green ammonia. Ammonia is a major commodity, with much of its value being in the fertilizer industry. It is also used extensively in refrigeration on an industrial scale. Ammonia is much used as a NO_x reduction agent in power plants and on vehicles (in which case, ammonia is carried in the form of urea solutions). Ammonia is a major chemical in many chemical processes and pharmaceutics.

At the same time, as pointed out in Zamfirescu and Dincer (2008, 2009), ammonia is a fuel and a source of hydrogen. It appears that ammonia will play a major role together with methanol in a transition period toward a fully implemented hydrogen economy. In this respect, ammonia is a more compact and convenient method to store and distribute hydrogen.

Hydrogen economy integrates well with water and oxygen cycles in nature. Hydrogen-ammonia economy integrates perfectly with water, oxygen, and nitrogen cycles in nature. Fig. 7.18 shows this integration. At the production point, hydrogen is made of water, and the ammonia is made from nitrogen and hydrogen. Ammonia is stored and distributed. At the utilization point, hydrogen is produced on demand with release of nitrogen back into the air. When hydrogen is combusted with oxygen from air, water is again formed and released into nature.

It appears, therefore, that the hydrogen-ammonia system will conserve perfectly the natural balances, while conveying useful work as extracted and converted from sustainable forms of energies. In this section, the role of ammonia as a hydrogen source and the related technology of green ammonia production and of ammonia conversion to hydrogen are reviewed.

7.9.1 Ammonia Synthesis

Ammonia is manufactured worldwide from nitrogen and hydrogen in Haber-Bosh synthesis loops. In this process, the activation energy of this reaction—which is equivalent to that needed for breaking the triple covalent bond of nitrogen molecule ($N{\equiv}N \rightarrow 2N^{3+}+6e^-$), namely 460 kJ/mol—appears to be very high.

The Haber-Bosh process was invented at the beginning of the 20th century to combine hydrogen with nitrogen thermocatalytically. The principle of this process is based on increasing the temperature such that the nitrogen molecule receives enough energy to be cracked. The catalyst breaks the nitrogen bonds at the surface. If the temperature is not high enough, nitrogen atoms remain strongly bound at the surface and "poison" the catalyst, which is, therefore, not able to perform a new catalytic cycle. However, the forward reaction is facilitated by low temperatures and high pressures. Since the reaction temperature cannot be set low (because of catalysis poisoning), the operating pressure must be extremely high. Typically, the operating temperature and pressure are 600°C and 100–250 bar, respectively, for a 25–35% conversion. For this reason, the loop consumes rather high energy. Licht et al. (2014) appreciated the fact that the ammonia synthesis process is responsible for 2% of the world's energy consumption.

A simplified flowsheet of the Haber-Bosh conversion loop is shown in Fig. 7.19. Make-up gas consisting of hydrogen and nitrogen is provided as input and compressed up to an intermediate pressure. The make-up is combined with unreacted gases returned from the loop and compressed further up to the conversion pressure. The feed is

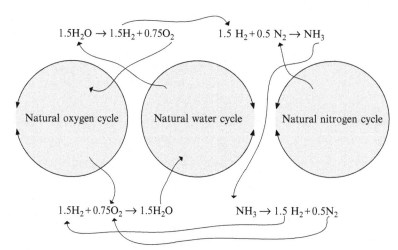

FIG. 7.18   Integration of hydrogen-ammonia economy in natural cycles.

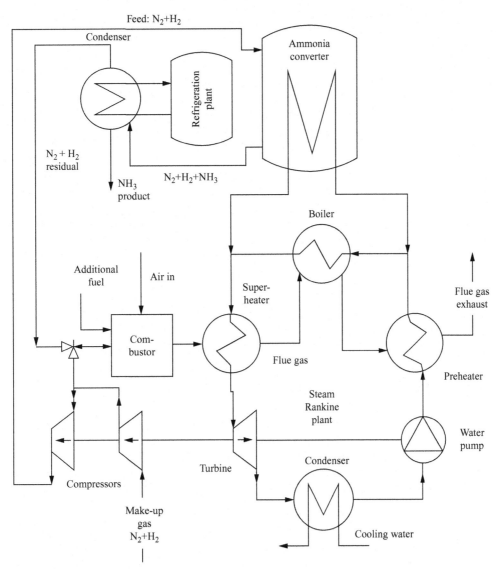

FIG. 7.19 Simplified flowsheet of a Haber-Bosh ammonia synthesis plant.

directed toward the catalytic converter, which contains mainly iron-based catalysts. The resulting gases containing converted ammonia product enter in the ammonia separator that operates at the intermediate pressure. There ammonia is separated by condensation and collected as liquid from the bottom of the separator. A refrigeration plant based on ammonia is used to cool, condensate and separate the product. The remaining gases, containing mainly unreacted nitrogen and hydrogen are partly recycled (recompressed together with the make-up gas) and partly used in a combustor to produce process heat. Additional fuel may be used also. The flue gases and the heat of the exothermic ammonia reaction are used to generate steam in a Rankine cycle that drives a turbine.

For better efficiency, power conversion is avoided by using the turbine to directly drive the compressors and the water pump, whole-mounted on the same shaft. Pressure is the parameter that controls the ammonia conversion. For a typical case at 200 bar, ammonia conversion is ~15% and increases up to 25% at 400 bar. Two types of catalytic ammonia converters are commonly used. The first is cooled internally with a coil running through the catalyst bed; the second divides the chemical reactor in modules, and after each module, the products are cooled in separate heat exchangers.

Because of the importance of the heat generated at ammonia synthesis (ie, 2.7 GJ per tonne of ammonia), great emphasis is placed on heat recovery. High pressure steam further expanded in turbines (as just explained) is seen as the most beneficial solution for heat recovery. In principle, high pressure steam at ~125 bar can be generated. Advanced ammonia plants produce ~1.5 of high pressure steam per tonne of ammonia representing 90% recovery of ammonia formation enthalpy. Other gas-handling operations (e.g., hydrogen separation from returning stream, purge gas management, nitrogen separation, and hydrogen production) consume some amount of primary energy and degrade the synthesis loop efficiency.

The efficiency of ammonia production from a primary thermal energy source, through the Haber-Bosch process, varies between 37% and 65%. If hydrogen is derived from a sustainable source (meaning that no CO_2 emission could be associated with the hydrogen production), then the greenhouse gases (GHGs) equivalent to the energy needed to run the plans are on the order of 0.4 tCO_2/tNH_3; see Rafiqul et al. (2005). Typical CO_2 emissions are 2.2 tCO_2/tNH_3 if hydrogen is produced from natural gas, and 16.2 tCO_2/tNH_3 if coal is the primary source; the minim possible value for GHG emission with today's technology is 1.6 tCO_2/tNH_3.

An electrochemical method coproducing hydrogen and ammonia from water and air was recently proposed by Licht et al. (2014). This process can be scaled up and appears to have high potential for industrial adoption in the near future. The system uses specific iron-based catalysts dissipated in the molten electrolyte. The experimental proof-of-concept by Licht et al. (2014) produced ammonia at rates in the range of 2.9×10^{-9} to 6.7×10^{-9} mol s^{-1} cm^{-2} with efficiencies up to 35% and coproduces hydrogen at 30% efficiency. The cell uses a KOH-NaOH eutectic mixture with Nano-iron oxide catalysts suspended in the mixture working as low as 200°C in atmospheric pressure.

The overall reaction that occurs in the hydrogen-ammonia synthesis cell at a reversible cell potential of 1.18 V is as follows:

$$0.5N_2(g) + 16.5H_2O(g) \rightarrow NH_3(g) + 15H_2(g) + 8.25O_2(g) \tag{7.29}$$

The half-reaction at the cathode with the reference of 1 mole of ammonia is as follows:

$$0.5N_2 + 33H_2O + 33e^- \rightarrow NH_3(g) + 15H_2(g) + 33OH^- \tag{7.30}$$

The anode half-reaction is found to be

$$16.5H_2O + 33OH^- \rightarrow 8.25O_2(g) + 33H_2O + 33e^- \tag{7.31}$$

A cell based on reactions (7.30) and (7.31) was developed recently at University of Ontario Institute of technology as reported in Casallas and Dincer (2016). The reactor shown in Fig. 7.20 is an improved version of the system proposed by Licht et al. (2014), which operated under similar conditions. The design is modified to include a lid allowing for replacement of materials inside the reactor, study of materials and corrosion, and retesting under modified conditions. The reactor is designed to operate under high temperatures between 200 and 500°C. The catalyst for the reaction is

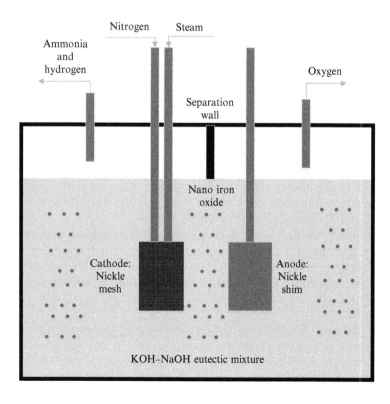

FIG. 7.20 Molten salt electrochemical reactor representation for hydrogen-ammonia production. *Modified from Casallas, C., Dincer, I., 2016. An integrated system for solar based electrochemical synthesis of ammonia and electrolysis of hydrogen. Int. J. Hydrog. Energy (submitted).*

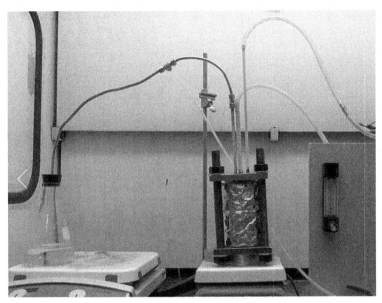

FIG. 7.21 Experimental setup for electrochemical hydrogen-ammonia production in molten salt, tested at University of Ontario Institute of Technology.

Nano-iron oxide that is suspended in the molten salt mixture. The reactor is also designed so that it is possible to implement a mixing mechanism to keep the Nano particles suspended instead of settling at the bottom. The design also incorporated a custom electric resistance heating design to heat reactants to 200°C prior to entering the reactor through alumina ceramic tubes.

The experimental setup is shown in Fig. 7.21. The reactor consists of the main reactor structure with two independent lids that have a groove to fit the alumina crucible. In between them is a gasket that creates a seal. The metal rods connecting the two lids are oversized for stability and support. The electrodes consist of the mesh and shim for the cathode and anode, respectively. The alumina tubes act as mechanical supports for the electrodes as well as inlet and outlet mechanisms. The mesh is selected to optimize the area of reaction, and nickel wire is used as support and to deliver electrical contacts inside the reactor. Steam is let in from the left flask, while nitrogen is delivered and controlled from a tank regulator and a flow meter.

The experiment lasted approximately 1000 s at a constant voltage of 1.44 V. Previous to applying the voltage, the system had been supplied steam for approximately 5 min and nitrogen was supplied at the same time as the voltage. The control of the steam was set to maximum supply and nitrogen to minimum.

The sensor used to measure the ammonia is an Arduino interphase device with a MQ-135 gas sensor that was calibrated and tested for ammonia gas prior to the experimentation. The ammonia was initialy detected when the sudden peaks in current were seen. This was not a steady transition because the rate is highly dependant on the location and distribution of the catalyst in the electrolyte. When the ammonia rate increased, the current steadily decreased from an average of 500 to about 450 mA. After reaching a peak, it quickly diminished. This meant there was an obstruction in the reaction, possibly a change in temperature because of the cold reactants. The increase in electric current suggests that the rate of hydrogen electrolysis increased.

Table 7.1 shows how the nitrogen flow rate affected the ammonia rate. The nitrogen rates were recorded according to the approximate ammonia concentration seen after a reasonable steady value was reached. It is evident that

TABLE 7.1 Nitrogen Flow Rate and Ammonia Concentration

Nitrogen flow rate (mL/min)	PPM NH$_3$
60	47
100	20
140	106

Modified from Casallas, C., Dincer, I., 2016. An integrated system for solar based electrochemical synthesis of ammonia and electrolysis of hydrogen. Int. J. Hydrog. Energy (submitted).

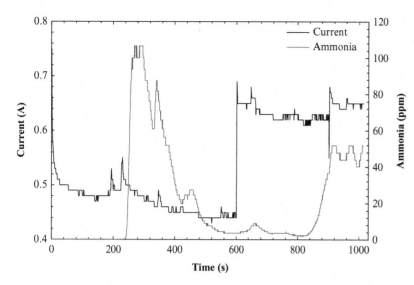

FIG. 7.22 Current and ammonia cocentration variations at 1.44 V. *Modified from Casallas, C., Dincer, I., 2016. An integrated system for solar based electrochemical synthesis of ammonia and electrolysis of hydrogen. Int. J. Hydrog. Energy (submitted).*

ammonia rate increases with increasing nitrogen flow rate. This is shown in the table with the flow increasing four times its initial value, while the rate increases six times its initial value.

The current density was found to be the same at a value approximating 25 mA/cm² validating the data reported by Licht et al. (2014). The steam was controlled by use of a simple valve. The valve was closed halfway, cutting off approximately half of the supply by outletting the rest to the environment. This resulted in an immediate increase of ammonia production. The impact of the valve was not measurable because of the lack of instruments.

In the experimental run reported as shown in Fig. 7.22, the voltage was increased to 1.7 V. At this point, the nitrogen flow rate was also increased, and the steam flow rate was decreased. As seen in Fig. 7.23, the production steadily increased because of the higher voltage and nitrogen supply. These results also conclude that the steam supply was excessive in the first experiment.

The second experiment maintained the same voltage but with an increasing nitrogen supply rate. The nitrogen was constantly increased in small increments that continued to increase the ammonia concentration from the previous experiments. It was a general increasing trend from the initial conditions, meaning that nitrogen is one of the most important parameters that directly and greatly affects the ammonia. This is reasonable because the reaction is limited to the amount of nitrogen supplied. Steam was initially over supplied with approximately 3.3 L/min. Fluctuations in current were observed throughout the experiment. This is perhaps because of the variation in reaction rates. Looking at the reactions carefully, it is evident that there are different half-reactions at each electrode and that these can limit each

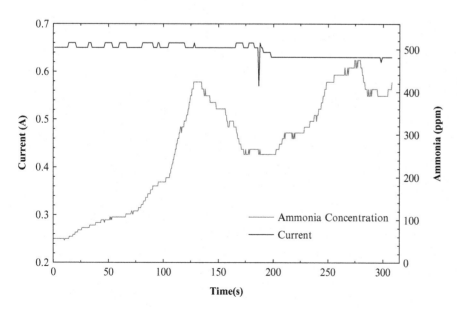

FIG. 7.23 Current and ammonia cocentration variations at 1.44 V with an increased nitrogen flow rate. *Modified from Casallas, C., Dincer, I., 2016. An integrated system for solar based electrochemical synthesis of ammonia and electrolysis of hydrogen. Int. J. Hydrog. Energy (submitted).*

FIG. 7.24 Current and ammonia cocentration variations at 1.70 V. *Modified from Casallas, C., Dincer, I., 2016. An integrated system for solar based electrochemical synthesis of ammonia and electrolysis of hydrogen. Int. J. Hydrog. Energy (submitted).*

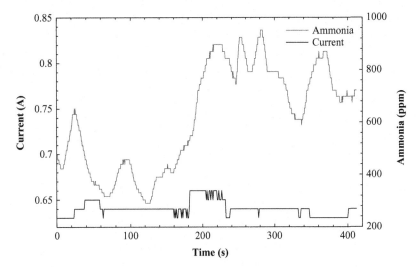

other if there is a lack of reactants. These half-reactions may be constantly changing. They affect each other when they increase or decrease because they have a common reactant. This explains the changes in current, and is further verified through the constantly changing ammonia production rate. There was no steady state point reached with the ammonia production.

In the third experiment, the voltage was set at 1.7 V. There was a peak reached at about 950 ppm of ammonia, at which point the reaction began to decrease, as shown in Fig. 7.24. This point was marked by a measurement of 220 ml/min of nitrogen. The current increased much higher, which is evidence of electrolysis being concurrently driven because of the large supply of water. This was likely caused by the settling of the Nano-iron oxide which was the catalyst, or another obstacle that stopped the production of ammonia. When the setup was taken apart, it was observed that the cathode was corroded. This may have been part of the reason the production stopped, since there was no longer a nickel surface for the reaction.

Table 7.2 shows the second set of flow rates to concentration measurements taken during the third experiment; this was an attempt to show the most optimum flow rates for further experiments. The nitrogen flow rate affected the concentration much more than any other parameter. This was due to the water content held by the mixture that would balance any changes in water flow. The reaction continued at high currents after the third experiment, but the ammonia was no longer detected because of the settling of the Nano particles, the corrosion later observed at the cathode mesh, and the temperature variances because of the flow of cold reactants. The cold reactants may have affected the eutectic mixture.

7.9.2 Ammonia Storage and Distribution

Because of the major interest of industrial ammonia as fertilizer, large-capacity seasonal storage tanks are developed. Ammonia demand peaks during summers when it must be spread on the agricultural fields. Ammonia is produced throughout the year, and the winter's production is stored for the summer season. Tanks with a capacity of 15,000–60,000 m³ were constructed before the 1970s. Ammonia is stored in refrigerated state at ambient pressure and its normal boiling point, which is −33°C. The tanks are of cylindrical construction with a 38–52 m inner diameter, and 18–32 m useful heights were built. In order to compensate for the heat penetrations, the entire construction is well

TABLE 7.2 Nitrogen Flow Rate and Ammonia Concentration

Nitrogen flow rate (mL/min)	PPM NH$_3$
140	391
180	500
220	815

Modified from Casallas, C., Dincer, I., 2016. An integrated system for solar based electrochemical synthesis of ammonia and electrolysis of hydrogen. Int. J. Hydrog. Energy (submitted).

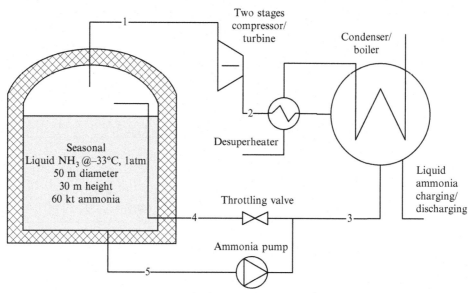

FIG. 7.25 Seasonal ammonia storage system in refrigerated state at −33°C.

insulated (double-wall technology is used), and compressors are employed to remove the heat penetration by creating a refrigeration plant in which the tank plays the role of evaporator.

Fig. 7.25 presents a typical seasonal ammonia storage system. Cold vapors in 1 are aspired and compressed with a two-stage compression station up to state 2 that corresponds to a condensation temperature for the winter season. The liquid condensate at ambient temperature in 3 is throttled and returned as a cold two-phase vapor-liquid mixture into the tank. The circulated ammonia flow rate must be such that it compensate for the effect of heat penetration from outside and the associated ammonia evaporation. In practice, 0.1% from stored liquid evaporates per day. The latent heat of ammonia at 1 bar is 1370 kJ/kg; thus about 1.4 kJ of cooling must be provided per kg of liquid ammonia, each day. During winter, ammonia refrigeration plants operating at −33°C evaporation and +15°C condensation temperature with a condensation can achieve COP = 2.5; therefore, the corresponding compressor shaft energy is 0.6 kJ/kg every day, or 110 kJ/kg per 6 months of storage. The total energy for running a 60 kt tank is, therefore, 6.6 GJ per storage season (season = 6 months winter-to-summer storage period).

Charging of ammonia into the tank is normally done by liquid at high pressure, ie, at condenser level. During charging, the liquid is expanded to 1 bar, and therefore the cooling must compensate only for the fraction of generated vapors that must be condensed. Vapor fraction in winter conditions during ammonia filling is ~15%; therefore, about 9 kt vapors must be condensed for the 60 kt tank, or 4.9 GJ shaft energy must be provided for complete filling. In total, the needed shaft energy to drive refrigeration is ~11.5 GJ per season.

The liquid in the tank is subcooled because it is subjected to a hydrostatic overpressure $P = \rho g \Delta z$, where Δz is the mean height of the liquid. The overpressure is estimated in these conditions to be about 1 bar, and the estimated specific exergy is 19 kJ/kg or 1.1 GJ per 60 kt. In principle, 50% of this exergy can be recovered through a heat engine operating between the ambient and low temperature of the cold storage; the recovered exergy represents ~5% from the energy spent to fill the tank and keep it refrigerated for a whole season.

Ammonia is stored in smaller quantities in tanks made from regular carbon steel, designed for ~20 bar operating pressure where ammonia is kept liquid at ambient temperature. A rule of thumb is that at least 3 t of ammonia can be stored per tonne of steel. Therefore the tank weight is about one-fourth from the ammonia mass. Various sizes of cylinders are available in industry. The size of pressure ammonia tankers is limited for practical reasons to about 300 t.

There is considerable experience with distributing ammonia by trucks, barges, ships and rail. With road transport, the typical cisterns have 45 k litres capacity, while rail cars have ~130 k litres. Ocean ships transport ammonia in low temperature storage tanks in capacities up to 50 kt. In regard to pipeline transportation, it is easy to derive that distribution energy efficiency is 93% with respect to HHV at an energy density of 14 GJ/m^3.

Storage of ammonia can be made in chemical forms, such as metal amines or ammonia boranes, which are recently developed as physical-chemical reversible methods—see e.g., Heldebrant et al. (2008) and Christensen et al. (2005). In their proposed technology, ammonia is adsorbed in a porous metal amine complex—e.g., hexaaminemagnesium chloride, $Mg(NH_3)_6Cl_2$, which is done by passing NH_3 over an anhydrous magnesium chloride ($MgCl_2$) powder at room temperature. The absorption and desorption of ammonia in and from $MgCl_2$ is completely reversible. The metal amine can be shaped in the desired form and can store $0.09\ kgH_2/kg$ and $100\ kgH_2/m^3$.

7.9.3 Hydrogen Generation from Ammonia

Ammonia can be cracked thermocatalytically to obtain hydrogen, according to the following endothermic reaction with a standard heat of reaction of 30.1 kJ per mole of H_2:

$$\frac{2}{3}NH_3 \rightarrow H_2(g) + \frac{1}{3}N_2 \tag{7.32}$$

The activation energy of ammonia decomposition reaction is 196E3 kJ/mol, and the pre-exponential factor is 2.098E6. Here the required enthalpy represents 10.6% of higher heating value or 12.5% of lower heating value of the produced hydrogen. Ammonia cracking reaction does not need catalysis to be performed at high temperatures (e.g., more than1000 K); however, at lower temperatures, the reaction rate is too low for practical applications as hydrogen generation for energy conversion. Nevertheless, at 400°C, the equilibrium conversion of NH_3 is very high at 99.1%, and at about 430°C, almost all ammonia is converted to hydrogen at equilibrium, under atmospheric pressure conditions.

A large array of catalysts are applicable to ammonia decomposition (e.g., Fe, Ni, Pt, Ir, Pd, Rh), but ruthenium (Ru) appears to be the best one when supported on carbon nanotubes generating hydrogen for more than 60 kW equivalent power per kg of catalyst (Yin et al., 2004). Over Ruthenium catalysts, at temperatures lower than ~300°C, recombination of nitrogen atoms is rate-limiting, while at temperatures higher than 550°C, the cleavage of ammonia's N—H bond is rate-limiting. However, the activation energy is higher at lower temperatures (180 kJ/mol), and it lowers at higher temperatures (21 kJ/mol). The best temperature range for ammonia decomposition over Ruthenium catalysts may be 350–525°C, which suggests that flue gases from hydrogen internal combustion engines, or other hot exhausts from combustion processes, or electrochemical power conversion in high temperature fuel cells can be used to drive ammonia decomposition.

Fig. 7.26 presents three possible reactor configurations for ammonia decomposition. The direct products of decomposition consist of hydrogen and nitrogen and traces of unreacted ammonia. For pure hydrogen generation, membranes technology can be applied either in the same reactor or separately. The reactor shown in Fig. 7.26A is the simplest one and does not separate the products in the output stream. It consists of a simple tube (which can be coiled) filled with the catalytic bed and heated from the outside with flue gases. The reaction occurs at surface and cannot go

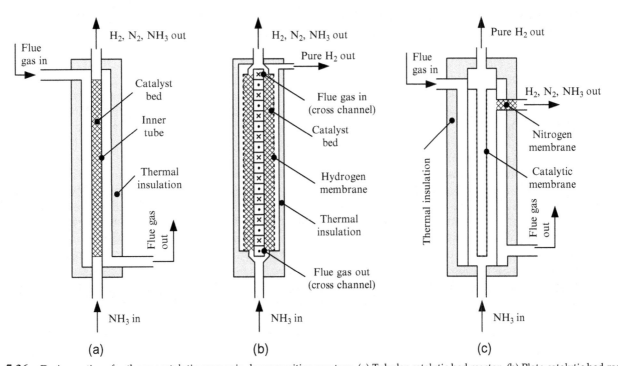

FIG. 7.26 Design options for thermocatalytic ammonia decomposition reactors. (a) Tubular catalytic bed reactor. (b) Plate catalytic bed reactor with H_2 selective membrane. (c) Catalytic membrane reactor.

beyond the chemical equilibrium conversion at the temperature of operation. Some old trials reported by Grimes (1966) to produce this kind of reactor were based on iron catalysis and achieved, for operation at 900°C, a production of 1.3 kW power equivalent of hydrogen (with respect to HHV) for 1 liter of reactor plus auxiliary heat exchangers. Works by Hacker and Kordesch (2003) describe a tubular reactor based on Ni-Ru catalyst that produced hydrogen equivalent (HHV) with 60 kW at 600°C and 240 kW at 800°C per liter of reactor.

Fig. 7.26B suggests the construction of a plate-type catalytic reactor with an integrated hydrogen selective membrane. The reactor is heated with flue gases circulated in cross-flow with the reactants-products streams. Ammonia is fed at the bottom and passes over the catalytic bed where the disassociation reaction occurs. The catalytic bed is surrounded by a hydrogen selective membrane that allows only pure hydrogen to pass through. Palladium-based membranes are the most efficient known for hydrogen separation. The reactor produces a pure stream of hydrogen and a stream of residuals, containing mainly nitrogen and traces of hydrogen and ammonia. Reactors of this kind were described by several researchers and tested up to the present data in the laboratory. Garcia-Garcia et al. (2008) used Ru based catalysis and Pd membrane and obtained ~20% conversion enhancement with respect to conventional (tubular) catalytic reactor; the conversion obtained at 350°C has been 95%. Ganley et al. (2004) showed that hydrogen production up to a 170 kW H_2 HHV equivalent is possible for 1 liter of reactor.

The third decomposition reactor, shown in Fig. 7.26C, has a catalytic membrane and has been proposed by Skodra et al. (2006). The catalysts used were based on Ni on alumina support. In this approach, the hydrogen selective membrane is doped with ammonia-cracking catalysts to form a catalytic membrane. The testing conditions were 500–800°C, 2–10 bar and 0.5–1 s residence time, which are consistent with the situations specific to vehicle propulsion. At 550°C, the conversion has been 85% for 2 bar pressure and 20% or 30% at 10 bar. In Fig. 7.26C, it is suggested that better product separation could be achieved if a nitrogen selective membrane is placed at the outlet port of unreacted gases. Separating simultaneously the nitrogen and hydrogen products represents a way to shift the reaction equilibrium toward the right. By extracting nitrogen from the reactor, the recombinative nitrogen effect can be avoided, and higher reaction rates could be achieved.

A continuous flow membrane catalytic reactor for hydrogen generation from ammonia cracking—referred to also as decomposition and separation unit—was developed by Dincer and Zamfirescu (2009). An experimental proof-of-concept prototype was developed at University of Ontario Institute of Technology by Nieminen (2011). The reactor has a loosely packed bed of Ni_2O_3 catalyst powder and a palladium catalytic membrane. The experimental study examined the effects of operating temperature, operating pressure and residence time on hydrogen production rates, overall system efficiency, ammonia conversion efficiency, transient ammonia concentration and ammonia decomposition rate.

The best results were obtained at 20.0 bar, 500.0°C and a residence time of approximately 3.965 seconds. The maximum hydrogen production was approximately 287.0 mL/min. This resulted in a maximum energy ratio of were 6.77:1 and an ammonia conversion of 95.64%.

The use of palladium membrane is crucial in this type of reactor. Very thin palladium is attached to a tubular porous steel. The phenomenon of hydrogen permeation in bulk palladium membranes comprises several steps in series: molecular transport from the bulk gas to the gas layer adjacent to the surface, disassociative adsorption of hydrogen onto the membrane surface, transition of atomic hydrogen from the surface into the bulk of the palladium membrane, Fickian diffusion of atomic hydrogen through the palladium membrane, transition of atomic hydrogen from the opposite membrane surface, recombinative desorption of hydrogen from the membrane surface and molecular gas transport away from the membrane surface and into the free stream.

As a hydrogen selective membrane, palladium and related alloys exhibit great temperature stability from 300.0 to 600.0°C. Furthermore, they show the highest selectivity to hydrogen and highest hydrogen flux compared to other types of membranes. The only material limitations on palladium and related alloys mentioned by the researcher were damage because of α–β phase transition and potential poisoning by hydrogen sulfide, hydrochloric acid, and carbon monoxide. Poisoning of the membrane surfaces had been mentioned to reduce the hydrogen flux by upward of 20.0%. There will be an optimum trans-membrane pressure drop in the reactor that maximizes the hydrogen recovery, as reported by Nieminen (2011) and as shown in Table 7.3.

A schematic of the main experimental setup is illustrated in Fig. 7.27. The operating temperature and pressure ranges are between 400.0 and 500°C and 5.0–20.0 bar, respectively. Anhydrous ammonia is transferred from the pressurized tank into a 1.0-inch-diameter stainless steel tube that is 18.0 inches long. It is capped with a 1.0-inch stainless steel end cap on one end, and is reduced on the other end to be fitted with a Swagelok PGI series process pressure gauge and input/output Swagelok 4PT plug valves. It was placed into a 2.0-inch-diameter black pipe that is 20.0 inches long.

Reactor pressure is controlled by a Swagelok KPR series forward pressure regulator. Flow into the reactor is measured by an Omega FMA-1812 electronic flow meter that is interfaced with a National Instruments NI-cDAQ9174 data

TABLE 7.3 Maximum Hydrogen Recovery and Optimum Trans-Membrane Pressure Drop

Temperature (K)	Maximum hydrogen recovery (%)	Maximum trans-membrane pressure drop (bar)
673.0	92.0	4.964 bar @ 33.67% draw
773.0	92.5	5.248 bar @ 33.67% draw

Modified from Nieminen, J.L., 2011. Experimental Investigation of Hydrogen Production from Decomposition of Ammonia. MSc Thesis, University of Ontario Institute of Technology.

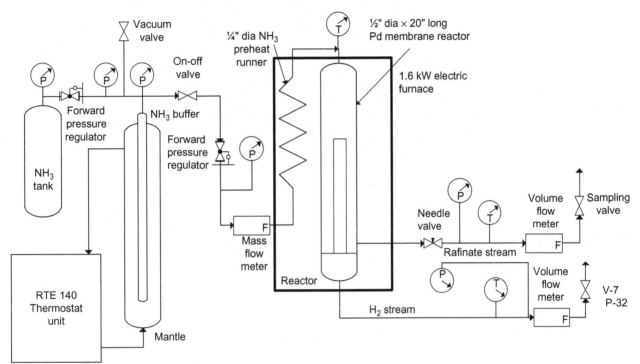

FIG. 7.27 Schematic of main experimental setup for H_2 production from NH_3 decomposition. *Modified from Nieminen, J.L., 2011. Experimental Investigation of Hydrogen Production from Decomposition of Ammonia. MSc Thesis, University of Ontario Institute of Technology.*

acquisition system. Flow out of the reactor is controlled by Swagelok SS-4MG needle valve. Heat is supplied to the reactor by an electrical resistance furnace that can be connected in either series or parallel for a power consumption of 0.3 or 1.2 kW, respectively. The electrical resistance furnace is controlled by an Omega CN7523 P.I.D controller that is passed temperature measurements from a 20.0-inch K-type thermocouple placed down the center of the furnace. The electrical resistance furnace was designed in-house and was built specifically for this experiment. It consists of two banks that contain a semicircular array of 13 vertical wires embedded 0.250″ deep in dense pourable ceramic that are 2.625″ × 5.25″ × 18.0″. The banks are then attached to their frames, which are then attached via hinges and filled with Rockwool insulation. The entire assembly is then wrapped with approximately 1.0″ of Rockwool insulation and piping insulation. All joints are taped with aluminum tape.

Because of the hydrogen selective nature of a palladium membrane, there are two outlet streams; they are called the hydrogen and raffinate streams. Sampling ports are on each stream to allow for withdrawing samples of fixed volumes and/or to accommodate for the potential vacuuming of the entire system. An additional two K-type thermocouples are installed in both product streams to measure and record their temperatures. Two Omega FL-5000 series 65.0 mm glass tube flow meters are installed to measure the flow rates in each product stream. Fig. 7.28 shows a photograph of the experimental setup.

The experimental conditions and results with the reactor are given in Table 7.4. Fig. 7.29 shows the recoded hydrogen production rates at 5 bar and 673 K. Fig. 7.30 shows the recoded hydrogen production rates at 20 bar and 773 K. A positive temperature and pressure dependence of ammonia decomposition over a Ni_2O_3 catalyst is observed.

FIG. 7.28   Photograph of the 300 ml/min hydrogen-generating membrane thermocatalytic reactor test setup developed at University of Ontario Institute of Technology.

TABLE 7.4   Hydrogen Production Rate of Various Test Conditions With Membrane Catalytic Reactor

Run #	Temperature (°C)	Pressure (bar)	NH₃ flow rate (mL/min)	Residence time (s)	Max H₂ output (std mL/min)
1, 2	400.0, 500.0	5.0	12.0	3.748	75.0
3, 4	400.0, 500.0	10.0	5.06	7.058	75.0
5, 6	400.0, 500.0	20.0	2.25	15.86	75.0
7, 8	400.0, 500.0	5.0	24.0	1.784	150.0
9, 10	400.0, 500.0	10.0	10.12	3.529	150.0
11, 12	400.0, 500.0	20.0	4.50	7.929	150.0
13, 14	400.0, 500.0	5.0	48.0	0.937	300.0
15, 16	400.0, 500.0	10.0	20.24	1.764	300.0
17, 18	400.0, 500.0	20.0	9.0	3.965	300.0

Modified from Nieminen, J.L., 2011. Experimental Investigation of Hydrogen Production from Decomposition of Ammonia. MSc Thesis, University of Ontario Institute of Technology.

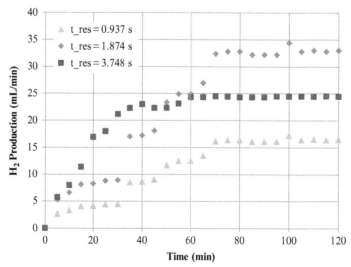

FIG. 7.29   Recorded hydrogen production rates at 5 bar and 673 K. *Modified from Nieminen, J.L., 2011. Experimental Investigation of Hydrogen Production from Decomposition of Ammonia. MSc Thesis, University of Ontario Institute of Technology.*

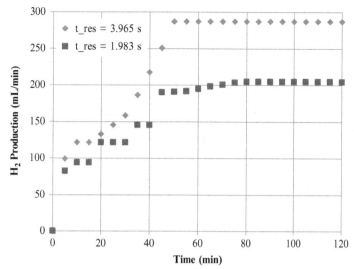

FIG. 7.30 Recorded hydrogen production rates at 20 bar and 773 K. *Modified from Nieminen, J.L., 2011. Experimental Investigation of Hydrogen Production from Decomposition of Ammonia. MSc Thesis, University of Ontario Institute of Technology.*

The effects of increasing the operating pressure were determined to be threefold. Firstly, a higher pressure increases the ammonia density and so reduces the volumetric flow rate. Because of this, a higher residence time was achieved, and more hydrogen was produced for a given mass flow rate. Secondly, because of the higher hydrogen partial pressure in the reactor associated with the higher operating pressure, there was a greater trans-membrane pressure difference. This increase in trans-membrane pressure difference was responsible for effectively generating a larger driving force that increased the flux of the hydrogen through the membrane.

At 773 K and with a residence time in the neighborhood of 3.6 s, it was found that a quadrupling of reactor pressure caused an increase in hydrogen production by a factor of approximately 12.0. Moreover, the author concluded that the reason for such seemingly low hydrogen production at 673 K was due to the catalyst not being particularly active.

Changes in catalyst surface morphology and composition, because of being exposed to reacting ammonia at high temperatures (400.0–500.0°C) and pressures (5.0–20.0 bar), are shown comparatively in Figs. 7.31 and 7.32 for 773 K. Fig. 7.31 shows the SEM image of an unexposed Ni_2O_3 catalyst, whereas Fig. 7.32 shows the exposed catalyst structure.

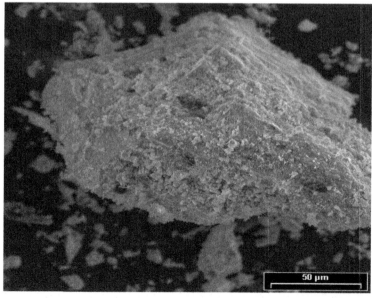

FIG. 7.31 SEM image of an unexposed Ni_2O_3 catalyst particle obtained at University of Ontario Institute of Technology.

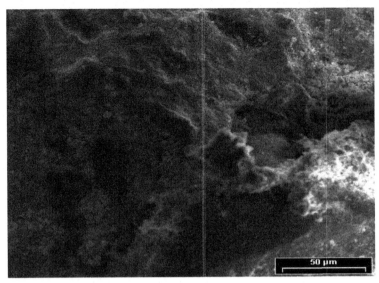

FIG. 7.32 SEM image of an exposed Ni_2O_3 catalyst particle obtained at University of Ontario Institute of Technology.

7.10 CONCLUDING REMARKS

In this chapter, the methods of hydrogen production through some hybrid water splitting cycles such as photo-thermochemical, photo-electro-thermochemical and radio-thermochemical are introduced, categorized and discussed. Some nuclear-based hydrogasification of coal and nuclear-based natural gas reforming as two potential methods are also introduced for very large-scale hydrogen production. Solar fuel reforming for hydrogen production is also covered with potential application at small and intermediate scales of production. Some fuels envisioned for this process are natural gas or synthetic natural gas and methanol. Other less common fuel is urea, which is also a fertilizer. Urea is a source of hydrogen, too, as it can be converted thermochemically. Water electrolysis in molten alkali hydroxides is studied as a method to conduct electrolysis at high temperature. If nitrogen is supplied to the molten electrolyte, then ammonia molecule is cogenerated together with hydrogen. The chapter ends with a focus on a hydrogen-ammonia paradigm by discussing ammonia synthesis, storage, and conversion to hydrogen.

Nomenclature

E voltage, V
g gravity constant, m/s^2
I current intensity, A
P pressure, kPa
T temperature, K
z elevation, m

Greek Letters

ρ density, kg/m^3

References

Casallas, C., Dincer, I., 2016. An integrated system for solar based electrochemical synthesis of ammonia and electrolysis of hydrogen. Int. J. Hydrog. Energy (submitted).

Christensen, C.H., Sørensen, R.Z., Johannessen, T., Quaade, U.J., Honkala, K., Elmøe, T.D., Køhlera, R., Nørskov, J.K., 2005. Metal ammine complexes for hydrogen storage. J. Mater. Chem. 15, 4106–4108.

Dincer, I., Zamfirescu, C., 2009. Methods and Apparatus for Using Ammonia as Sustainable Fuel, Refrigerant and NO$_x$ Reduction Agent. US Patent 61/064,133.

Fletcher, E.A., 2001. Some considerations on the electrolysis of water from sodium hydroxide solutions. J. Sol. Energy Eng. 123, 143–146.

Ganley, J.C., Seebauer, E.G., Masel, R.I., 2004. Development of a microreactor for production of hydrogen from ammonia. J. Power Sources 137, 53–61.

Garcia-Garcia, F.R., Ma, Y.H., Rodrigues-Ramos, I., Guerrero-Ruiz, A., 2008. High purity hydrogen production by low temperature catalytic ammonia decomposition in a multifunctional membrane reactor. Catal. Commun. 9, 482–486.

Grimes, P.G., 1966. Energy Deport Fuel and Utilization. Transaction of the Society of Automotive Engineers, paper #650051.

Hacker, V., Kordesch, K., 2003. Ammonia crackers. In: Handbook of Fuel Cells – Fundamentals, Technology and Applications. John Wiley and Sons, Chichester.

Heldebrant, D.J., Karkamkar, A., Linehan, J.C., Autrey, T., 2008. Synthesis of ammonia borane for hydrogen storage applications. Energy Environ. Sci. 1, 156–160.

Licht, S., Halperin, L., Kalina, M., Zidman, M., Halperin, N., 2003. Electrochemical potential tuned solar water splitting. Chem. Commun. 24, 3006–3007.

Licht, S., Cui, B., Wang, B., Li, F.-F., Lau, J., Liu, S., 2014. Ammonia synthesis by nitrogen and steam electrolysis in molten hydroxide suspensions on nanoscale Fe_2O_3. Science 345, 637–640.

Liu, Q., Hong, H., Yuan, J., Jin, H., Cai, R., 2009. Experimental investigation of hydrogen production integrated methanol steam reforming with middle-temperature solar thermal energy. Appl. Energy 86, 155–162.

Naterer, G.F., Dincer, I., Zamfirescu, C., 2013. Hydrogen Production from Nuclear Energy. Springer, New York.

Nieminen, J.L., 2011. Experimental Investigation of Hydrogen Production from Decomposition of Ammonia. MSc Thesis, University of Ontario Institute of Technology.

Perret, R., 2011. Solar Thermochemical Hydrogen Production. Thermochemical Cycle Selection and Investment Priority. Sandia National Laboratory, report 3622.

Ratlamwala, T.A.H., Dincer, I., 2014. Experimental study of a hybrid photocatalytic hydrogen production reactor for Cu-Cl cycle. Int. J. Hydrog. Energy 39, 20744–20753.

Schrader, L., Strauss, W., Teggers, H., 1975. The application of nuclear process heat for hydro-gasification of coal. Nucl. Eng. Des. 34, 51–57.

Verfondern, K., 2007. Nuclear energy for hydrogen production. Energy Technology Series, Vol. 58. Research Center Jülich, Germany.

Von Federsdorff, C.G., 1974. Non-Fossil Fuel Process for Production of Hydrogen and Oxygen. US Patent 3,802,993.

Zamfirescu, C., Dincer, I., 2008. Using ammonia as a sustainable fuel. J. Power Sources 185, 459–465.

Zamfirescu, C., Dincer, I., 2009. Ammonia as a green fuel and hydrogen source for vehicular applications. Fuel Process. Technol. 90, 729–737.

Zamfirescu, C., Naterer, G.F., Dincer, I., 2012. Photo-electro-chemical chlorination of cuprous chloride with hydrochloric acid for hydrogen production. Int. J. Hydrog. Energy 37, 9529–9536.

STUDY PROBLEMS

7.1 Explain the structural difference between iodide and triodide?

7.2 Define energy and exergy efficiency formulations of a photo-thermochemical water splitting cycle?

7.3 Compare qualitatively the Yokohama Mark 5 cycle and the sulfur-ammonia cycle. Which of the two can make more use of solar radiation?

7.4 What is the technical advantage of the hydrogasification of coal with respect to steam coal gasification?

7.5 Why is a hydrolysis phase required for urea conversion to hydrogen?

7.6 Why doesn't water boil from a saturated alkali salt solution heated at over 100°C?

7.7 What would be the advantages of water electrolysis in molten alkali salts at high temperature as compared to solid oxide high temperature electrolysis technologies?

7.8 What are the advantages of electrochemical ammonia synthesis?

7.9 Why can hydrogen-ammonia economy be beneficial?

7.10 What is the benefit of using hydrogen selective membrane in a thermos-catalytic hydrogen production reactor from ammonia decomposition?

Novel Systems and Applications of Hydrogen Production

8.1 INTRODUCTION

In the book, the primary focus is placed on sustainable hydrogen production and applications. Moreover, in this final chapter of the book, a further step is taken to go beyond what has been presented on commonly known sustainable hydrogen production methods in the previous chapters and to discuss the recent progress made on some novel and newly developed hydrogen production processes, methods, techniques, technologies, and so on; and to investigate the future of hydrogen economy technologies and trends. A detailed literature survey, including patents, research articles, company reports, and more is conducted to determine the new and original hydrogen production options from various universities, research institutions, and companies. There are numerous hydrogen production methods ranging from biological to photoelectrochemical options. Various pathways through which the four kinds of energies drive hydrogen production can be obtained from energy sources. Electrical and thermal energy can be derived from renewable energies, such as solar, wind, geothermal, tidal, wave, ocean thermal, hydro, and biomass; from nuclear energy; and from recovered energy.

8.2 FOSSIL AND BIOFUELS BASED NOVEL HYDROGEN PRODUCTION OPTIONS

Among the fuels with high energy density, such as methanol, ethanol, and propane, methanol steam reforming (MSR) has the obvious advantages of low reaction temperature, good miscibility with water, high hydrogen concentration and low carbon monoxide concentration in the reformate gas. Utilization of coal with environmentally benign methods yields low greenhouse gas emissions and is, therefore, more acceptable than other means. In this section, new processes for MSR, coal gasification and other hydrocarbon utilization options are discussed.

8.2.1 Methanol Steam Reforming by an Innovative Microchannel Catalyst Support

To improve the energy conversion of a microchannel reactor, an innovative microchannel catalyst support with a microporous surface was proposed and fabricated by a layered powder sintering and dissolution method (LPSDM) in a study by Mei et al. (2016), as illustrated in Fig. 8.1. The catalyst adhesion strength was analyzed by an ultrasonic vibration test and results showed the catalyst had a 7 wt% loss on the catalyst support with a microporous surface and a 16 wt% loss on the nonporous surface. Subsequently, the coated catalyst support was applied to MSR for hydrogen production. The results showed that the microchannel catalyst supports with microporous and nonporous

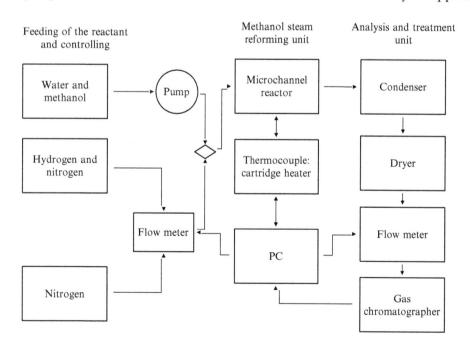

Feeding of the reactant and controlling

Methanol steam reforming unit

Analysis and treatment unit

FIG. 8.1 Schematic diagram of the performance testing system for hydrogen production. *Modified from Mei et al. (2016).*

surfaces, respectively, had a hydrogen production rate of 18.07 and 9.65 mL/min at 573 K under the inlet flow rate of 30 mL/min. Compared with the catalyst support with a nonporous surface, the catalyst support with a microporous surface showed better catalyst adhesion and obvious advantages in hydrogen production rate and methanol conversion.

Scanning electron microscopy (SEM) and LSCM results of fabricated microchannel catalyst support with porous surface show micropores with diameters of 100–150 mm and depth of 50–80 mm distributed uniformly on the surface of microchannels. The micropores on microchannel surfaces are helpful in increasing the surface area and residence time for the catalytic reaction and in facilitating the heat and mass transfer. After the microchannel catalyst supports are coated with $CuO/ZnO/Al_2O_3$ catalyst by a wash-coating method, an ultrasonic vibration test is carried out to analyze the adhesion strength of the catalyst, and results indicate that a porous surface structure can improve adhesion strength significantly. Subsequently, the microchannel catalyst supports with porous and nonporous surfaces coated with catalyst are applied to MSR for hydrogen production. When the reaction temperature and inlet flow rate are 573 K and 30 mL/min, respectively, the microchannel catalyst support with porous surface has a hydrogen production rate of 18.07 mL/min and 55.9% methanol conversion, while the catalyst support with nonporous surface has a hydrogen production rate of only 9.65 mL/min and 31.3% methanol conversion. The results showed that the hydrogen production rate and methanol conversion of MSR by microchannel catalyst support with a porous surface are obviously higher than those by the catalyst support with a nonporous surface (Mei et al., 2016).

8.2.2 Reaction-Integrated Novel Gasification Process

A new process for hydrogen production called HyPr-RING was suggested by Lin et al. (2001). This new method, where high pressure steam is thermochemically decomposed in the presence of carbon to CO_2 with Ca-based sorbents in a single reactor, produces hydrogen with high yield and a small release of CO_2. In order to develop this new hydrogen production process, thermodynamic analysis and an experimental study on the characteristics of the coal gasification/steam/sorbent mixture was systematically carried out. The main characteristics of CO_2 adsorption at elevated temperatures and pressure through Ca were also examined. Thermodynamic analysis of the gaseous products of the equilibrium components in the production of hydrogen from a $coal/CaO/H_2O$ mixture to elevated temperatures concluded that a pressure higher than 3 MPa was needed to achieve a substantial sorption of CO_2 with the Ca-based sorbent and H_2 production with high yield. For the experiments with gasification using a batch autoclave, pipe reactor and a continuous reactor, it was confirmed that the theoretical prediction of temperature and pressure for adsorption of CO_2 are suitable. For efficient adsorption of CO_2 multiple cycles using sorbents on the basis of Ca at elevated temperature and pressure in the vapor phase, hydration of the calcined adsorbent was found to enhance the reactivity of the sorbent and to maintain the strength of the sorbents. Based on the fundamental studies, a small-scale unit with a capacity of 50 kg coal/day was designed and built to demonstrate the process as a schematic diagram, as illustrated in Fig. 8.2.

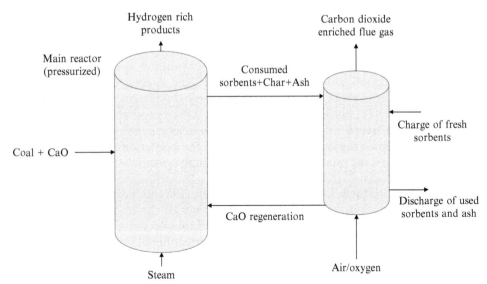

FIG. 8.2 Hydrogen production by reaction integrated novel gasification process. *Modified from Lin, S., Suzuki, Y., Hatano, H., Oya, M., Harada, M., 2001. Innovative hydrogen production by reaction integrated novel gasification process (HyPr-RING). J. South. Afr. Inst. Min. Metall. 101 (1), 53–59.*

Here are the main chemical reactions taking place within the system:

Steam gasification (partially):

$$C + H_2O \rightarrow H_2 + CO$$
$$CO + H_2O \rightarrow H_2 + CO_2$$

In situ CO_2 sorption with sorbents:

$$CO_2 + CaO \rightarrow CaCO_3$$

Overall reaction:

$$C + 2H_2O + CaO \rightarrow 2H_2 + CaCO_3$$

The use of natural gas and oil with current technology can produce and supply enough hydrogen to meet the needs of the refineries and chemical industry. But to meet the expectations of future demands for hydrogen needed for the hydrogen-fueled transportation sector, the cost of production, delivery and storage of hydrogen from coal or natural gas must be reduced. In particular, coal is an important substance as a primary energy source in the coming decades because of its huge reserves and geographically uniform distribution.

For a sustainable and stable supply of hydrogen, the chemical potential contained in coal should be converted into clean and efficient energy source—, such as hydrogen—and is used with a minimal loss of exergy. The HyPr-RING project is one of Japan's calls for the development of advanced technology to produce hydrogen from coal. HyPr-RING is a novel hydrogen production from advanced reactions of steam and, coal, and from increased sorption of CO_2, using lime. Many of the fundamental characteristics of the chemical reactions associated with HyPr-RING have been clarified, but technology research is needed to create a robust design and for commercialization of this new process of gasification (Lin et al., 2001).

8.2.3 Novel Oxidative Hydrogen Production From Methanol

Methanol as a hydrogen carrier is attractive because it can be stored as a liquid at ambient temperature, and it is an abundant chemical commodity. Several different methods are used for methanol, including methanol decomposition, steam reforming (SMR), partial oxidation of methanol (POM) produced in the presence of hydrogen and the combination of steam and oxidation reforming (CMR). Combustion is done through three channels, including volume combustion, combustion impregnated substrate, and the so-called second wave-dipping combustion process for preparing a catalyst composition containing copper, zinc, zirconium, and palladium oxide. By X-ray diffraction (XRD), X-ray photoelectron spectroscopy (XPS) and specific surface area (BET) nitrous oxide decomposition methods of these catalysts were characterized and evaluated for their activity and selectivity for the POM of methanol. Novel secondary combustion wave methods showed excellent activity with palladium metal compared to conventional combustion synthesis mode. Palladium was also shown to significantly reduce the bulk CuO reduction temperature.

The combustion synthesis-based approach shows promise for a methanol reforming synthesis catalyst. Three composite multifunctional catalysts containing copper, zinc, zirconium, and palladium oxide are used to synthesize based on the combustion method. These catalysts were tested with a part of hydrogen from methanol oxidation activity and characteristics of BET, N_2O decomposition, XRD, XPS, temperature-programmed reduction (TPR), and SEM. Active and selective catalysts for the POM by a new combustion synthesis technique include the volume of the combustion synthesis, impregnated support combustion, and the preparation of the second wave of impregnation.

It was noted that the catalysts do not contain a high selectivity of the hydrogen palladium, yet this slight increase in the yield of hydrogen requires significantly higher temperatures. This effect is less desirable because of the increased formation of by-products, formation of carbon monoxide and reducion of the service life time of the catalyst. Decreased production of hydrogen-free palladium catalysts may also be associated with the reactor access points and changes in the oxidation state of copper. Normally, the maximum selectivity to hydrogen and carbon dioxide is greater than 90% of the active catalysts. The catalysts produced by conventional techniques were very sensitive to the fuel/oxidizer ratio in the preparation. A list of conversion and selectivity data for selected catalysts at maximum hydrogen selectivity is shown in Table 8.1. The materials synthesized with a fuel-rich mixture had a lower catalyst surface and activity with a high carbon content and a larger crystal size. Fresh fuel-rich products also contain monoclinic ZrO_2 and thin

TABLE 8.1 Conversion and Selectivity Data for Selected Catalysts

Catalyst	CH$_3$OH (% conversion)	H$_2$ (% selectivity)	CO$_2$ (% selectivity)	CO (ppm)
7/Cu/3Zn/0Zr/0Pd-VCS-0.5	95	96	98	1000
7/Cu/3Zn/0Zr/3Pd-VCS-0.5	92	94	93	2000
7/Cu/3Zn/1Zr/3Pd-VCS-0.5	94	97	94	1700
7/Cu/3Zn/1Zr/0Pd-VCS-0.5	95	95	99	800
6/Cu/2Zn/3Zr/3Pd-VCS-3.0	74	67	83	3500
7/Cu/3Zn/1Zr/1Pd-ISC-Al	79	70	80	4800
7/Cu/3Zn/1Zr/3Pd-ISC-Zr	87	85	92	2100
7/Cu/3Zn/0Zr/3Pd-SWI-0.5	94	94	94	2600
7/Cu/3Zn/1Zr/3Pd-SWI-0.5	91	96	96	1000

From: Schuyten, S., Dinka, P., Mukasyan, A.S., Wolf, E., 2008. A novel combustion synthesis preparation of CuO/ZnO/ZrO$_2$/Pd for oxidative hydrogen production from methanol. Catal. Lett. 121 (3), 189–198.

preparations contained only distributed tetragonal ZrO$_2$. It was concluded that addition of the palladium catalyst significantly improves overall catalytic activity (Schuyten et al., 2008).

8.2.4 Chemical-Looping Reforming and Steam Reforming with CO$_2$ Capture by Chemical-Looping Combustion

Chemical-cycle combustion is an innovative combustion technology that can be used for CO$_2$ capture in energy processes in which direct contact between fuel and combustion air do not occur. Instead, a solid carrier is available that performs the task of oxygen for the air/fuel ratio. Chemical-cycle combustion takes place in two separate reactors, as indicated in Fig. 8.3. The oxygen carrier nuclear fuel is reduced by the oxidation of fuel. The reactor is air-oxidized by an air oxygen carrier. Chemical cycle reforming is similar to chemical combustion cycle, but complete oxidation of the fuel is prevented by the low air/fuel ratio. Accordingly, chemical cycle reforming can be described as a process for the POM of the hydrocarbon fuel with the undiluted use of chemically cyclic oxygen. This is a significant advantage compared to conventional technology because the need for expensive and energy intensive air separation is eliminated.

Two new approaches for the production of hydrogen, namely reforming of natural gas and the use of chemical looping, are presented. Chemical-looping cycle is the process of POM of fossil fuels, in which a solid oxygen carrier is used as a source of concentrated oxygen. Steam regeneration cycle with CO$_2$ capture recalls chemical vapor combustion conventionally, but the oven containing reformer tubes are replaced with a chemical combustion cycle. It was found that both variants have the potential to achieve efficiencies of about 80%, except for the heat loss, but include

FIG. 8.3 Schematic description of chemical-looping combustion and chemical-looping reforming. *Modified from Rydén, M., Lyngfelt, A., Mattisson, T., 2006. Two novel approaches for hydrogen production: chemical-looping reforming and steam reforming with carbon dioxide capture by chemical-looping combustion. Paper Presented at the Proceedings of the 16th World Hydrogen Energy Conference, Lyon, France.*

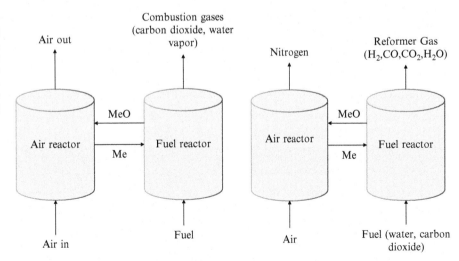

CO_2 capture and compression. In addition, the appropriateness of the chemical cycle for fuel reformation was proven in a fixed bed reactor, and in a continuous flow reactor comprising two interconnected fluidized chemical-looping combustion units of operation. This is a new approach to oxidation of the fuel.

First, oxygen enters the fuel with a solid oxygen carrier, the metal oxide. Then flameless oxidation occurs at moderate temperatures (800°C or higher), and there is no mixing of air and fuel. It can be concluded that the chemical-looping cycle is a promising method for the production of synthesis gas and H_2. This concept was proven in a continuous reactor. If a chemical loop pressurized reformer integrates with a gas turbine, it is possible to obtain an overall efficiency of about 81% for the production of H_2 from CO_2 absorption. The chemical-looping cycle also seems like an attractive method for the production of synthesis gas. Steam reforming with CO_2 capture and chemical combustion cycle also seem well suited to large-scale production of high-purity H_2 with CO_2 capture. With the concept of using conventional techniques under medium temperature and pressures, it may be easier to implement the chemical-looping cycle with an achievable overall efficiency of about 80%, respectively.

As indicated in Fig. 8.4, air is compressed (AC) before entering the air reactor (AR). The fuel is mixed with some steam before entering the fuel reactor (FR) to suppress carbon formation. The outlet from the FR is cooled and extra steam is added before high-temperature shift (HTS) and low-temperature shift (LTS). H_2O is removed from the shifted gas in a condenser (COND) before CO_2 is captured by absorption with methyldiethanolamine (MDEA) solvent. CO_2 for sequestration is obtained by regenerating the MDEA solvent in a stripper column operating at low pressure. If desired, some of the H_2 produced could be burned in a separate combustor to increase the power output of the gas turbine (GT).

The concept is similar to conventional steam reforming, but the furnace is replaced by chemical combustion cycles. The AR and FR operates at atmospheric pressure, as shown in Fig. 8.5. Steam reforming (SR) occurs at a high pressure in the converter tubes filled with catalyst disposed within the reactor fuel. Gas Reformer treatment occurs in the water-gas shift reactor. Water is removed from the process stream while cooling the condenser (COND) before entering the pressure swing adsorption (PSA). The H_2 that is produced is delivered at high pressure, but a small part is needed for cleaning and regeneration of the adsorbent. The waste gas is supplied at low pressure and used as fuel and the fluidizing gas in the reactor fuel and chemical cycles (Rydén et al., 2006).

8.2.5 Hydrogen and Oxygen Cogeneration Using Sorption-Enhanced Reforming Process

Hydrogen and oxygen are significant industrial gases. The oxygen is also widely used in the metallurgical industry, chemical industry etc. CO_2 can be detached at the site of the reaction gas mixture as it is formed at the time of adding a

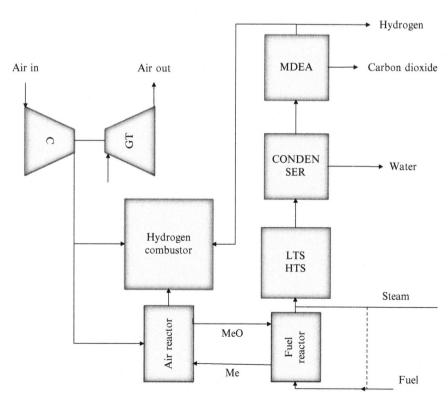

FIG. 8.4 Schematic process description of H_2 production with CO_2 capture by chemical-looping reforming at elevated pressure. *Modified from Rydén, M., Lyngfelt, A., Mattisson, T., 2006. Two novel approaches for hydrogen production: chemical-looping reforming and steam reforming with carbon dioxide capture by chemical-looping combustion. Paper Presented at the Proceedings of the 16th World Hydrogen Energy Conference, Lyon, France.*

FIG. 8.5 Schematic description H$_2$ production by steam reforming with CO$_2$ capture by chemical-looping combustion. *Modified from Rydén, M., Lyngfelt, A., Mattisson, T., 2006. Two novel approaches for hydrogen production: chemical-looping reforming and steam reforming with carbon dioxide capture by chemical-looping combustion. Paper Presented at the Proceedings of the 16th World Hydrogen Energy Conference, Lyon, France.*

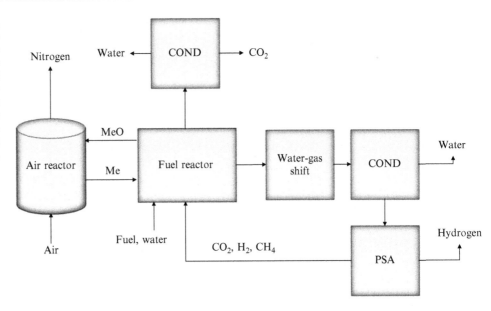

CO$_2$ sorbent reforming catalyst. Normal thermodynamic equilibrium reforming and water-gas shift reactions differ and high-purity hydrogen can be achieved in one step without the need for a shift reaction and purification steps. Sorption-enhanced steam methane reforming (SMR) simplifies the whole SMR process in a two-step reaction and sorbent regeneration, which has the potential of decreasing the cost of producing hydrogen. Sorption-enhanced SMR process is noticeably thermally neutral and does not require extra fuel and costly heat exchangers to offer energy for the reforming reaction. Additional energy is required to regenerate the sorbent waste CO$_2$, but a decline of 20–25% of energy consumption is estimated. The hydrogen and oxygen coproduction process comprises four subsystems: sorption-enhanced SMR process for producing hydrogen, a cryogenic distillation process at atmospheric pressure to produce oxygen, the process of the gas turbine to generate electricity, and the heat recovery steam generation (HRSG), as shown in Fig. 8.6.

The sorption-enhanced reforming process contains mainly a reformer, a regenerator and other units such as heat exchangers, cyclones, and turbines. The core of the cryogenic air distillation process is the dual distillation columns coupled by a combination of the reboiler/condenser. A multiple flow heat exchanger is another component that is essential for the heat integration process. The gas turbine produces power for supplying compressors and other apparatuses that use electricity. The feed to the combustion chamber in the process includes fuel gas and air. Gas fuel is derived from waste gas from the purification of hydrogen sorption-enhanced reforming. Extra fuel to generate additional power is required for the overall process. The flow of nitrogen waste from the distillation of air, which is rich in oxygen, is diversified with fresh air to supply oxygen to the combustion chamber. The generation of steam for the recovery of heat comprises a series of heat exchangers, pumps, and turbines. Liquid water is heated to produce steam at high temperature by exhaust gases from the gas turbine through the heat exchanger train. Part of the steam is used in gas turbines to produce power, and the remaining steam is supplied to the steam reforming process. Steam needed for the improved sorption SMR process is produced in the HRSG. HRSG takes heat from high-temperature exhaust gas from the gas turbine process. The fuel required to regenerate the CO$_2$ absorbent in the flue gas originates from the purification of H$_2$ in the sorption-enhanced SMR. The power consumption of each compressor and electrical facility is generated by gas and steam turbines. The overall process is balanced in power and does not need external power, as clearly outlined elsewhere (Peng, 2003).

8.3 WATER DECOMPOSITION-BASED NOVEL HYDROGEN PRODUCTION OPTIONS

Water is one of the most abundant materials on earth. Hence, producing hydrogen from water is a sustainable method if adequate processes are selected. Water electrolysis is one of the most basic methods to generate almost pure hydrogen and is driven by the movement of electrons that are continuously circulating through an external circuit. Electrolysis requires catalysts for enhanced reaction rate and current density. Note that platinum is typically applied

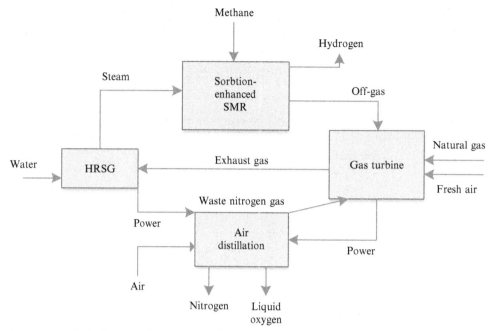

FIG. 8.6 Schematic diagram of the hydrogen and oxygen coproduction process. *Modified from Peng, Z., 2003. A Novel Hydrogen and Oxygen Generation System. Faculty of the Louisiana State University and Agricultural and Mechanical College in Partial Fulfillment of the Requirements for the Degree of Master of Science in Chemical Engineering, In The Gorden A. and Mary Cain Department of Chemical Engineering by Zhiyong Peng BE, Tianjin University.*

on the electrodes to achieve a heterogeneous catalysis process. Instead, homogeneous catalysts can be dissolved/deferred in the electrolyte. Homogeneous catalysts are treated as a favorable option for the future because they are less costly and have a high turnover rate.

8.3.1 Decoupled Catalytic Hydrogen Evolution from a Molecular Metal Oxide Redox Mediator

The new method proposed by Rausch et al. (2014) allows greater amounts of hydrogen to be formed at atmospheric pressure using lower power compared to characteristic of those generated by green energy resources. It also resolves essential safety matters that have so far restricted the usage of intermittent renewable energy for hydrogen generation. The process utilizes a liquid that allows the hydrogen to be locked up in a liquid-based inorganic fuel. By using a liquid sponge known as a redox mediator that can soak up electrons and acid, it is plausible to make a system where hydrogen can be formed in a detached chamber without any extra energy input after the water electrolysis occurs. The link between the rate of water oxidation and hydrogen generation has been overcome, permitting hydrogen to be released from the water 30 times faster than the leading PEM electrolysis process on a per milligram of catalyst basis.

As shown in Fig. 8.7, the water is oxidized to give oxygen gas, protons, and electrons. The electrons travel through an external circuit, and the protons diffuse through a semipermeable membrane separating the chambers. An electron-coupled proton buffer (ECPB) in another compartment is reduced by electrons while balancing the charge to protons,

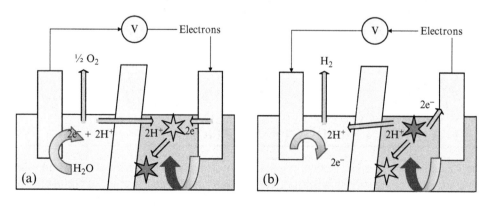

FIG. 8.7 Decoupled catalytic hydrogen evolution from a molecular metal oxide redox mediator in water splitting (A) water oxidization and (B) reoxidation of the electron-coupled proton buffer. *Modified from Symes and Cronin (2013).*

turning from yellow to dark blue. Reoxidation of the ECPB releases protons, which travel via the membrane to the other electrode where it combines with electrons detached from ECPB to produce hydrogen gas. ECPB returns to its original color yellow. In respect to receiving a stream of hydrogen from water splitting, which is substantially free of oxygen and sufficiently pure for use in industrial processes or systems for fuel cells in which O_2 and H_2 are individual points in the space currently important. I, in the conventional electrochemical cells, OER (the oxygen-evolving reaction) and HER (the hydrogen-evolving reaction) are closely interlinked. Therefore, one of the half a reactions can occur only if the other half-reaction occurs at a considerable rate.

Natural photosynthetic systems that are capable of splitting water to separate production of H_2 and O_2 equivalent in time and space. In addition, some thermochemical cycles based on heavy reaction conditions can lead to H_2 and O_2 in water at different points on a large scale. They present an alternative approach to water splitting, whereby the electrons and protons generated during the oxidation of water to O_2 are taken up reversibly by an ECPB, rather than being used directly to produce H_2. After reoxidation, ECPB discharges these protons and electrons to produce hydrogen. Therefore, ECPB acts as a reversible electron-proton donor/acceptor with a redox couple, an energetic intermediate position between the OER and HER. The complete result of utilizing an ECPB is that H_2 and O_2 can be attained from the electrolysis of water at entirely different times; that is, the OER is entirely decoupled from the HER to generate O_2 essentially free of H_2, and H_2 essentially free of O_2. In this way, two smaller energy inputs are used to split water to give H_2 and O_2 at different times, as opposed to a single energy input that produces H_2 and O_2 simultaneously (Rausch et al., 2014):

$$ECPB + H_2O \rightarrow \left[ECPB_{(two\ electron\ reduced)}\right]\left[H^+\right]_2 + \frac{1}{2}O_2$$

$$\left[ECPB_{(two\ electron\ reduced)}\right]\left[H^+\right]_2 \rightarrow H_2 + ECPB$$

8.3.2 Low-Temperature, Manganese Oxide-Based, Thermochemical Water-Splitting Cycle

The thermochemical cycles are recognized, as presented previously, as heat using (thermally driven) cycles to disassociate water into hydrogen and oxygen under the temperatures below 1000°C without causing any potential issues on the environment. They are apparently desirable as they can convert heat into chemical energy in the form of hydrogen. In this section, the focus is placed on the manganese oxide-based, thermochemical water-splitting cycle. As reported by Xu et al. (2012), the manganese-based thermochemical cycle with a maximum operating temperature of 850°C is fully recyclable and does not involve any toxic or corrosive components. The defined thermochemical cycle uses the redox reaction of Mn(II)/Mn(III) oxides. Shuttling of Na^+ inside and outside of oxides of manganese in hydrogen-oxygen evolution step, respectively, provides the main driving force and allows the thermodynamic cycle to be closed at a temperature less than 1000°C. Hydrogen and oxygen are completely reproducible at least five cycles. In this method, instead of using heated oxidized metal to drive the oxygen from water, they used magnesium oxide. The reactions are facilitated by shuttling ions in and out of it. The use of sodium decreases the temperature needed for the reaction to 850°C, versus the 1000°C needed in other cases. As mentioned by Xu et al. (2012), the technology is still in its early stage and needs further development to work properly and be sustainable.

The four-reaction cycle begins with a manganese oxide and sodium carbonate, and is a totally closed arrangement, as illustrated in Fig. 8.8. The water that enters the system in the second step becomes entirely transformed into hydrogen and oxygen during each cycle. That is important since it implies that none of the hydrogen or oxygen has vanished, and the cycle can run multiple times, splitting water into the two gases. In the mentioned system, the researchers ran their recently formed cycle five times to demonstrate reproducibility (Xu et al., 2012).The thermochemical cycle comprises four key steps: (i) heat treatment of a physical mixture of Na_2CO_3 and Mn_3O_4 producing MnO, CO_2, and α-$NaMnO_2$ at 850°C; (ii) oxidation of MnO in the presence of Na_2CO_3 with water to form H_2, CO_2, and α-$NaMnO_2$ at 850°C; (iii) extraction of Na^+ from $NaMnO_2$ suspended in aqueous solution in the presence of bubbled CO_2 at 80°C; and (iv) recovery of Mn_3O_4 thermal reduction of sodium ion extracted solids produced in step (iii) at 850°C. The overall reaction is a stoichiometric splitting water into hydrogen and oxygen, without the by-product. The combination and extraction of Na^+ into and out of the manganese oxides are the important steps in dropping the temperature compulsory for both the hydrogen evolution and the thermal reduction steps.

The highest temperature essential for Mn-based cycle is close to the temperatures used with other low-temperature multistep water-splitting cycles. Both hydrogen and oxygen are carried out in stages of 850°C in a Mn-based cycle, a similar temperature for the higher temperature stage system in the piloted sulfur iodine. A Mn-based system does not

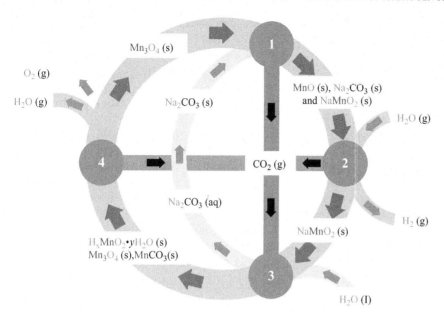

FIG. 8.8 Schematic representation of the low-temperature, Mn-based thermochemical cycle. *Adapted from Xu, B., Bhawe, Y., Davis, M.E., 2012. Low-temperature, manganese oxide-based, thermochemical water splitting cycle. Proc. Natl. Acad. Sci. 109 (24), 9260–9264.*

imply any corrosive materials, and nearly all the previously reported multistage low-temperature cycles of the thermal water splitting have toxic and/or corrosive intermediates in one or more of the stages of the cycle. In particular, in each step of the sulfur iodine cycle, there is at least one of the following chemicals: SO_2, H_2SO_4, I_2, and HI, which present significant environmental and engineering problems (Xu et al., 2012).

8.3.3 Aluminum-Based Hydrogen Production Using Sodium Hydroxide as a Catalyst

It is generally well known in the field of chemistry that aluminum reacts with water to generate hydrogen and heat spontaneously. The reaction between aluminum and water is not maintainable at room temperature because of the protective oxide layer on the metal-forming surface. Therefore, the use of aluminum as a fuel for generating hydrogen gas to heat requires that the protective layer is removed continuously and effectively and that the reaction is maintained at an elevated temperature.

A number of hydrogen generators are commercially available for use with fuel or as a source of heat batteries for internal combustion engines, for example. These generators are typically complex and risky for unqualified personnel to operate. These hydrogen generators are planned to be used by scientists and other professionals working in laboratory conditions. A process for producing hydrogen according to the present patented invention comprises reacting aluminum with water in the existence of sodium hydroxide as a catalyst:

$$2Al + 3H_2O \xrightarrow[\text{catalyst(NaOH)}]{} Al_2O_3 + 3H_2$$

The device for implementing the method uses pressure and temperature of the reaction to control the degree of immersion of a fuel cartridge and therefore in water to control the force and duration of the reaction. The method and apparatus of the present invention provide a convenient process and a secure device for use by the public for generating heat, light and hydrogen gas in blackout situations or places where electricity is not available. In addition, the method and apparatus of the present patent utilize readily available aluminum waste in household garbage and metal-working shops, to promote conservation, recycling and energy. Further details about the patent are provided by Andersen and Andersen (2003).

8.3.4 Tungsten Disulfide as a New Hydrogen Evolution Catalyst for Water Decomposition

Properties of hydrogen evolution and corrosion resistance of WS_2 make it a good candidate as an active catalyst component for photonic hydrogen by photocatalytic decomposition of water. Tungsten disulfide has been studied extensively as a catalyst in hydrogenation and hydrodesulphurization of organic compounds. In the study by Sobczynski et al. (1988), both unsupported and silica or alumina-supported WS_2 were prepared. Promotion WS_2 with

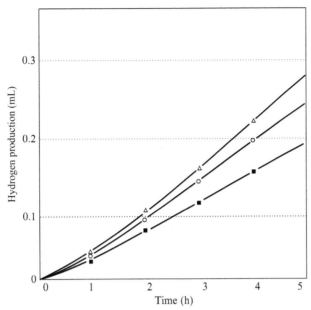

FIG. 8.9 Hydrogen production from water-methanol-KOH using various WS$_2$/SiO$_2$ powders with ($\triangle$) WO$_3$/SiO$_2$ strengthened at 400°C in air for 3 h, reaction with H$_2$S at 300°C for 2 h; (○) WO$_3$/SiO$_2$ strengthened at 400°C in air for 3 h, reaction with H$_2$S at 300°C for 3 h; and (■) WO$_3$/SiO$_2$ dried at 110°C, reaction with H$_2$S at 300°C for 3 h. *Modified from Sobczynski, A., Yildiz, A., Bard, A.J., Campion, A., et al., 1988. Tungsten disulfide: a novel hydrogen evolution catalyst for water decomposition. J. Phys. Chem. 92 (8), 2311–2315.*

cobalt and nickel sulfides improves the catalytic activity of the active center. Silica support prepared tungsten disulfide by reaction WO$_3$/SiO$_2$ with H$_2$S at 300°C hexagonal WS$_2$ was confirmed by XRD studies. The hydrogen yield property of WS$_2$/SiO$_2$ was defined by catalytic and photocatalytic tests in the presence of fluorescein or CdS as sensitizers and electrochemical measurements. They were compared with platinum-supported on silica. Although a graphite electrode was immersed in a suspension of Pt/SiO$_2$ evolved hydrogen at an applied potential more positive than a WS$_2$/SiO$_2$ suspension, silica-supported tungsten disulfide was a more active and stable catalyst than Pt/SiO$_2$ for the catalytic production of hydrogen, as well as in the dark and under illumination of visible light with cadmium sulfide as a sensitizer. The results of hydrogen evolution from water-methanol-KOH with different WS$_2$/SiO$_2$ powders is shown in Fig. 8.9.

The tungsten disulfide with high-specific surface area was prepared by deposition on a silica backing. Hexagonal WS$_2$ was confirmed by XRD. The Auger electron spectroscopy (AES) and XPS studies indicated that WS$_2$ included minor amounts of O$_2^-$ ions, which were created from the origin WO$_3$. Electrochemical studies showed that, although tungsten disulfide silica support did not show any photoresponse, it had good properties of hydrogen evolution. Compared to Pt/SiO$_2$, WS$_2$/SiO$_2$, it showed better release of hydrogen activity and stability in a solution of V^{2+} in 1 M H$_2$SO$_4$ and a slurry of spent CdS/SiO$_2$ in water-methanol 0.1 M KOH. The activity of the particulate WS$_2$/SiO$_2$-CdS/SiO$_2$ system in water-methanol-0.1 M KOH was steady within 7 days of visible light. The extensive initiation period in the formation of hydrogen observed with WS$_2$/SiO$_2$ can be linked to the hydrogen adsorption WS$_2$ and/or reaction with atomic hydrogen surface W^{4+} ions (Sobczynski et al., 1988).

8.3.5 Membrane-Less Electrolyzer for Hydrogen Production

An ion conductive membrane is regarded as an important component in the electrolytic water. Since its initiation in 1960, it has become the standard Nafion proton conducting membrane in engineering and academic institutions. Despite the high cost and limited lifetime and limitation to strongly acidic pH, researchers have used the material mainly because of its excellent stability and ionic conductivity. Membrane-less electrolysis arrangements can solve these problems and lead to even more effective tools due to the high conductivity of the liquid electrolyte.

The investigation conducted by Hashemi et al. (2015) is recognized as a first study in the area and clearly introduces a membrane-less device that can operate continuously and robustly with different kinds of catalysts and electrolytes across the pH scale, simultaneously producing a flow of hydrogen gas with an oxygen content much lower than the

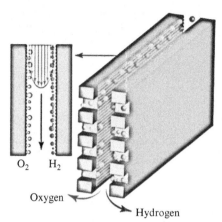

FIG. 8.10   Schematic diagram of the membrane-less electrolysis. *Adapted from Hashemi, S.M., Modestino, M.A., Psaltis, D., 2015. A membrane-less electrolyzer for hydrogen production across the pH scale. Energy Environ. Sci. 8 (7), 2003–2009.*

required safety limit. The improvement of deployable water splitting strategies is delayed by the price of electricity and the absence of steady ion-conducting membranes that can function across the pH scale, execute low ionic resistances and avoid product mingling. The use of a membrane-less method, as shown in Fig. 8.10, has proven that the electrolyzer can function with a lower resistance than the benchmark membrane-based electrolyzers using virtually any electrolyte. The method separates the product gases by controlling the delicate balance between the hydrodynamic forces on the device. The devices presented may divide water at current densities more than 300 mA/cm^2, with more than 42% energy conversion efficiency and with the crossover of the hydrogen gas in the oxidation side as low as 0.4%, which led to a non-flammable and uninterrupted flow of hydrogen fuel.

Note that two parallel plates are covered with hydrogen and oxygen evolution catalysts, respectively, and are detached by less than a few hundred micrometers. The electrolyte streams between the catalyst plates and the produced gases intersect close to the matching catalyst surface due to the Segre-Silberberg influence. Each of the gaseous product streams was collected in special places. Stacks of these horizontal planes can be used for higher bandwidth. The two gas streams have no convective mixing due to the forces that push toward the adjacent walls of lifting. The only mixing mechanism is diffusion of dissolved gases, which is minimal at high flow rates.

Hashemi et al. (2015) confirmed that the membrane-less electrolyzer is capable of producing hydrogen flow that stays incombustible and stable over the pH scale. By comparing the ohmic resistance of the apparatus with the devices based on Nafion membrane, it is clear that their device shows potential to exceed expectations about the performance of their water-splitting apparatus based on ion-conducting membranes for separation. Though a self-contained electrode pair can generate only a restricted amount of fuel, it can be scaled up on multistack panels for improved throughput or for the application of huge area electrodes. Because the only element to be kept small is the inter-electrode distance, further research is aimed at developing high-throughput devices, which require a large area for the flat electrodes used as the sidewalls of narrow channels of the electrolyte. In addition, the electrolyzer platform can be used in reverse as a fuel cell electrolyte with two streams, each rich in H$_2$ and O$_2$, allowing the generation of electricity. The researchers stated that the design of the electrolytic membrane-less production can facilitate mass production, especially using high resolution 3D printers or molding.

8.3.6 Nanosilicon: Splitting Water without Light, Heat, or Electricity

Common methods for splitting water include electrolysis, thermolysis, photochemical and combinations. Water can also be divided using chemicals that can be oxidized with water, such as aluminum or silicon. The reactions of silicon-water are still not developed enough to generate hydrogen on demand, because the process is slow and can be self-limiting, with the formation of oxide. Nevertheless, silicon can ideally release two moles H$_2$ per mole of silicon, plentifully and safely; it has high energy density and does not release carbon dioxide. The limited literature on the topic indicates that silicon is an encouraging material for the production of hydrogen from water since it is stable and transportation is easy.

FIG. 8.11 Schematic showing hydrogen generation from silicon nanoparticles. *Adapted from Erogbogbo, F., Lin, T., Tucciarone, P.M., LaJoie, K.M., Lai, L., Prasad, P.N., Patki, G.D., 2013. On-demand hydrogen generation using nanosilicon: splitting water without light, heat, or electricity. Nano Lett. 13 (2), 451–456.*

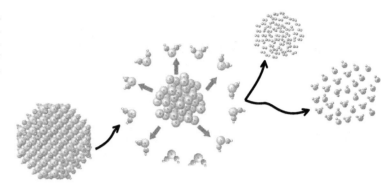

$$Si(cr) + 4H_2O(l) \rightarrow Si(OH)_4(aq) + 2H_2(g)$$

$$Si(cr) + 2H_2O(l) \rightarrow SiO_2(s) + 2H_2(g)$$

It was shown that nanoscale silicon that is about 10 nm in diameter reacts with water to produce hydrogen 1000 times quicker than bulk silicon, 100 times faster than formerly reported Si assemblies, and six times faster than opposing metal preparations. The rate of production of H_2 using a 10 nm Si is 150 times that attained when using 100 nm particles, sharply higher than the projected influence of increased surface to volume ratio. This may be attributed to changes in the process dynamics at the nanoscale level which evolves from an from anisotropic etching of greater silicon toward an efficient isotropic etching of 10 nm silicon. These results indicate that nanosilicon can provide a practical approach to produce hydrogen on demand without the addition of heat, light, or electricity. The schematic diagram shown in Fig. 8.11 by Erogbogbo et al. (2013) implies that the production of hydrogen from microsilicon nanoparticles occurs much faster than expected from extrapolation of the rate obtained using larger particles.

Silicon nanoparticles can have practical applications to generate hydrogen on demand, based on their increased activity compared to other air-stable hydrogen-generating materials such as aluminum and zinc. The integration of nanosilicon technology with the appropriate cartridge may provide "just add water" hydrogen on-demand technology to facilitate the adoption of hydrogen fuel cells to power portable applications. Nevertheless, scalable and energy-efficient processes for the production of nanoparticles should be implemented to empower the use of H_2 generation based on silicon beyond niche applications. Since laser pyrolysis was demonstrated in kg/h scale, it may be one of the candidate processes (Erogbogbo et al., 2013).

8.3.7 Hydrogen Production from Sunlight and Rainwater

The production of hydrogen involving a combination of electrolysis, photocatalysis and photovoltaic modules was suggested by Chua (2015), as illustrated in Fig. 8.12. This new approach significantly increases the rate of hydrogen production using relatively less energy compared to other conventional methods. The idea of this system lies in the components of a new generation of high-performance photocatalysts that harness sunlight and solar heat for hydrogen production. Water is pumped through a reactor tube where inexpensive photocatalyst powder is added to the electrolysis process of water splitting. The generated hydrogen can be directed to the storage system, and oxygen can

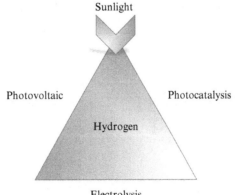

FIG. 8.12 Schematic diagram of sunlight- and rainwater-based hydrogen production system including three various technologies.

be released in the air or stored for other applications. Only one or two membranes are required for hydrogen production to achieve a high purity of about 99.8%, which is seen in many manufacturing industries as environmentally friendly. Moreover, such a system can also provide additional shielding from sunlight on the roofs of houses, resulting in a cooler room due to the reduction of solar heat. The rooftop system comprises three reactor tubes, a novel hybrid hydrogen filtration system and two solar panels. Each reactor tube can produce about 35–40 L/h hydrogen. With three reactors, more than 100–120 L/h hydrogen can be produced. The system collects rainwater to split. Using clean and renewable sources to produce hydrogen—namely, rainwater and sunlight—the system can be easily and inexpensive installed on rooftops.

8.4 SOLAR-BASED NOVEL HYDROGEN PRODUCTION OPTIONS

Solar radiation is in essence a source of photonic energy. It's understood that the photonic energy is proportional to the frequency of the radiation. Photons can be used to dislocate electrons through their interaction with matter. While electrons are dislocated, the obtained electrical charge can be used to manipulate the valence electrons of chemical species in order to conduct chemical reactions photocatalytically. Photoelectrolysis includes the application of a heterogeneous photocatalysts at one electrode that is open to solar irradiation. Furthermore, the electrolysis cell is provided with electricity at the electrodes. Because of the action of photonic irradiation, the necessary electrical energy is decreased. Photoelectrochemical cell (PEC) is a new application of photoelectrolysis that contains photosensitive semiconductors immersed in an electrolyte and counter electrodes.

8.4.1 Water Photolysis via Perovskite Photovoltaics

Although sunlight-driven splitting of water is a promising path to sustainable production of hydrogen fuel, the full implementation is hindered by costly photovoltaic and photoelectrochemical apparatus. In a study by Luo et al. (2014), a highly efficient and inexpensive water-splitting cell was described that integrates a perovskite tandem solar cell with a bifunctional and inexpensive catalyst. The catalyst electrode and NiFe-layered double hydroxide showed high activity against both the oxygen and hydrogen evolution reactions in alkaline electrolyte. The arrangement of these two made the splitting water photocurrent density about $10 \, \text{mA}/\text{cm}^2$, resulting in a solar to hydrogen efficiency of 12.3%. Presently, perovskite instability restricts the lifespan of cells.

Using the inexpensive perovskite solar cell and bifunctional water-splitting catalyst, the complete water-splitting cell was constructed. Perovskite solar cells were placed side by side and connected by wires to electrodes, and embedded catalysts and simulated solar radiation provided the energy to split water. A layout of the energy diagram is shown in Fig. 8.13, in which two perovskites solar cells are connected in series in a tandem cell to split water. The researchers built a device that uses electricity and catalytic materials to make hydrogen and oxygen from water.

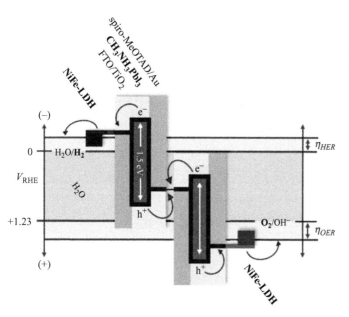

FIG. 8.13 Perovskite tandem cell combined with NiFe DLH/Ni foam electrodes for hydrogen production via splitting of water. *Adapted from Luo, J., Im, J.-H., Mayer, M.T., Schreier, M., Nazeeruddin, M.K., Park, N.-G., David Tilley, S., Fan, H.J., Grätzel, M., 2014. Water photolysis at 12.3% efficiency via perovskite photovoltaics and Earth-abundant catalysts. Science 345 (6204), 1593–1596.*

The proposed device was accepted as remarkable because it met three of the four criteria required for a practical device: high efficiency, low cost and the use of abundant materials, which leads to large-scale construction of the system. The device used moderately high-voltage solar panels to produce the electricity needed, along with low-cost novel catalytic materials based on nickel and iron for two electrodes that produce hydrogen and oxygen separately. It was shown that nickel hydroxide is a favorable catalyst, and the addition of iron can improve it. Iron nickel hydroxide was added to form a multilayer structure, and the catalyst was put on porous nickel foam to raise the area through which the reaction took place by accelerating them. A solar water splitter retains and uses 12.3% of the energy in sunlight to generate hydrogen. Considering the solar cell power conversion efficiency of about 16%, the efficiency is quite suitable for hydrogen storage (Luo et al., 2014).

8.4.2 Biomass Valorization and Hydrogen Production in a PEC

A PEC can directly use the photogenerated electron-hole pairs in semiconductor electrodes for fuel production, although nature makes it through photosynthesis. In a typical PEC, fuel is formed by reduction reactions at the cathode, which consume photoexcited electrons. Water can be reduced to produce H_2 and CO_2 can be reduced to produce a carbon-based fuel such as methanol and methane. In order to complete the cycle, the oxidation reaction takes place at the anode and consumes photogenerated holes. In general, the oxidation of water to produce O_2 is used as an anode reaction, which is environmentally safe and does not require additional species in the electrolyte. However, water oxidation is not a kinetically favored reaction, and its product, O_2, is not of significant value. Therefore, to anodic reactions, which have more favorable kinetics and, can generate value added fuels would be useful to improve the overall effectiveness and usefulness of the PEC.

For the anodic reaction, Cha and Choi (2015) used oxidation of 5-hydroxymethylfurfural (HMF) to 2,5 furandicarboxylic acid (FDCA). HMF is a crucial intermediate in the transformation of biomass, which can be derived from cellulose; FDCA is a significant molecule for the formation of polymers. Accordingly, the oxidation of HMF in FDCA has received considerable research attention. Much of the early work exploring the conversion of HMF in FDCA used aerobic oxidation with the use of heterogeneous catalysts. However, an alternative approach is electrochemical oxidation, which is caused by the electrochemical oxidation potential applied to the electrode to exclude the use of O_2 or other chemical oxidants. The researchers developed an effective method for the electrochemical oxidation of HMF to FDCA at room temperature and atmospheric pressure using water as a source of oxygen and 2,2,6,6-tetramethylpiperidine-1-oxyl (TEMPO) as a mediator, as shown in Fig. 8.14. Then this oxidation reaction was used in the anode reaction of PEC, which produced hydrogen on the cathode. It is reported that the electrochemical oxidation of HMF to FDCA using 2,2,6,6-tetramethylpiperidine 1-oxide (TEMPO) as a mediator and the catalyst can achieve almost 100% yield.

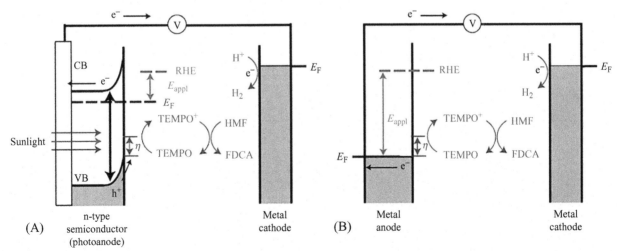

FIG. 8.14 (A) Photoelectrochemical TEMPO-mediated HMF oxidation (B) Electrochemical TEMPO-mediated HMF oxidation. *Adapted from Cha, H.G., Choi, K.-S., 2015. Combined biomass valorization and hydrogen production in a photoelectrochemical cell. Nat. Chem. 7 (4), 328–333.*

The PEC is proven to oxidize HMF as an anodic reaction on the expense of solar energy input. This PEC, n-type $BiVO_4$ was used as photoanode to produce and discrete the electron-hole pairs. The photoexcited electrons were transferred to the cathode of Pt on the products H_2, whereas the photogenerated holes were used at the $BiVO_4$ surface for HMF oxidation. The technical and economic advantages and the effectiveness of TEMPO-mediated electrochemical and the photoelectrochemical oxidation of HMF and its feasibility was shown in the study. Depending on these promising results, more varied biomass conversion and organic synthesis reactions can be examined as anode reactions for PECs (Cha and Choi, 2015).

8.4.3 Splitting Water with an Isothermal Redox Cycle

Solar thermal-driven water-splitting cycles are recognized as an appropriate means of producing hydrogen gas from water and sunlight. In this regard, an isothermal redox cycle was proposed by Muhich et al. (2013) as a two-step, metal oxide-based cycle to produce H_2 by a consecutive high-temperature reduction and water reoxidation of a metal oxide. The temperature changes between the reduction and oxidation steps (which are seen as essential for these cycles) reduced the cycle's total efficiency owing to thermal and time losses that occur during regular heating and cooling of the metal oxide. A study by Muhich et al. (2013) showed that these temperature swings are needless and that isothermal water splitting at 1350°C using the hercynite cycle shows H_2 generation capacity three and 12 times that of hertsinita and cerium oxide, respectively, by per mass of active material when decreased at 1350°C and oxidized at 1000°C. The proposed technique is simply a huge solar-thermal-sunlight-concentrated system on a giant central tower by a bulky array of mirrors, which increases the temperature of the tower to about 1371°C. The heat is then transmitted into a reactor with chemical mixtures known as metal oxides. As the metal oxide compounds' temperature increases, oxygen atoms are released, which alters the compounds' material conformation, causing the freshly formed compounds to search for new oxygen atoms. Adding steam to the system, as in Fig. 8.15—which could be created by boiling water in the reactor with the concentrated sunlight radiated to the tower—would cause oxygen from the water molecules to stick to the surface of the metal oxide, freeing up hydrogen molecules for gathering as hydrogen gas.

The closed system operation of the researchers' solar-thermal water splitting includes the water splitting reactor that takes in solar heat and splits water and a fuel cell that generates work in the form of electricity by recombining the water, as illustrated in Fig. 8.15. The main difference between this method and previous methods is that this method makes it feasible to conduct two chemical reactions at the same temperature. Although there are no working models, conventional theory holds that making hydrogen through the metal oxide method necessitates heating the reactor to a high temperature to remove oxygen; then cooling it to a low temperature before inserting steam to reoxidize the compound in order to discharge hydrogen gas for collection. The quantity of hydrogen formed for fuel cells or for storage is totally reliant on the amount of metal oxide that is made up of a mixture of iron, cobalt, aluminum, and oxygen and the amount of steam. One of the designs projected is to construct reactor tubes roughly a foot in diameter and several feet

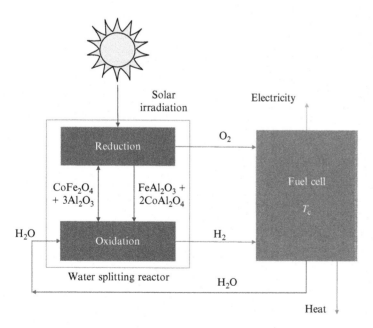

FIG. 8.15 Closed system operation of solar-thermal water splitting. *Modified from Muhich, C.L., Evanko, B.W., Weston, K.C., Lichty, P., Liang, X., Martinek, J., Musgrave, C.B., Weimer A.W., 2013. Efficient generation of H_2 by splitting water with an isothermal redox cycle. Science 341 (6145), 540-542.*

long, fill them with the metal oxide material and stack them on top of each other. A working system to generate a noteworthy amount of hydrogen gas would need a number of the tall towers to collect concentrated sunlight from several acres of mirrors surrounding each tower.

Hercynite cycle active materials do water splitting in two isothermal redox cycles, as evidenced by the results of isothermal water splitting at 1350°C. Generation of H_2 under isothermal water splitting conditions has been demonstrated with the initial O_2 plateau increased due to incomplete removal of the H_2O system. After all the steam was cleared from the reactor, the O_2 peak happened as the materials reduced, followed by the normal exponential decline as the reaction went to conclusion. The isothermal water splitting at 1350°C produced considerably more O_2 and H_2 than 1350°/1000°C STWS and to some extent more O_2 and H_2 than 1500°/1200°C through the cycle. The isothermal water splitting was a higher rate of H_2 production, and its lack of irreversible heat losses related to the change in temperature between the oxidation and reduction steps (which was required for the cycle) and gave isothermal water splitting a higher theoretic performance than TSWS under the circumstances in that study (Muhich et al., 2013).

8.4.4 WSe₂ 2-D Thin Films for Photoelectrochemical Hydrogen Production

Various materials have been considered for use in the direct solar conversion technologies in hydrogen, but 2D materials have been recently recognized as an encouraging candidate. In general, these materials include the class of graphene and have unusual electronic features. However, utilizing the used amount of solar energy requires large areas of solar panels, and it is difficult and expensive to fabricate thin film's 2D material to such an extent and maintain good performance. Therefore, the research focused on one of the best 2D materials for solar water splitting called tungsten diselenide.

The studies have shown that this material has a high efficiency for converting solar energy directly into fuel hydrogen and is also highly stable. Uniform dispersion of the material needed to be achieved, though. Therefore, the material was mixed with the tungsten disulfide powder with a liquid solvent, using sonic vibrations to exfoliate it into thin, 2D flakes, and then distinctive chemicals were added for balance of the mixture. To form high-quality, thin films, Yu et al. (2015) injected the tungsten diselenide ink at the boundary between two liquids that do not mix, as shown in Fig. 8.16. The selected immiscible liquids were oil and water. A rolling pin that made high quality 2D flakes with minimal adhesion at the interface between the immiscible liquids a had been used. Fluids were sensibly removed after that, and the thin film was moved to a bendable plastic support, which is much cheaper than standard solar modules.

The thin film was produced and tested and found to be better than films made of the same material but that used other comparable methods. In this proof-of-concept step, direct conversion efficiency of solar to hydrogen of about 1% is an important improvement compared to thin films produced by other methods, and with substantial potential for higher efficiency in the future (Yu et al., 2015).

8.4.5 Novel Nanoscale Semiconductor Photocatalyst for Solar Hydrogen Production

Hydrogen sulfide acidic gas (H_2S) is a pollutant of the environment due to its toxic, smelly and corrosive nature. Every year, millions of tons of H_2S are carried out worldwide by oil refineries and natural gas production, which is

FIG. 8.16 Schematic of injection of tungsten diselenide. *Adapted from Yu, X.Y., Prevot, M.S., Guijarro, N., Sivula, K., 2015. Self-assembled 2D WSe₂ thin films for photoelectrochemical hydrogen production. Nat. Commun. 6, 7596.*

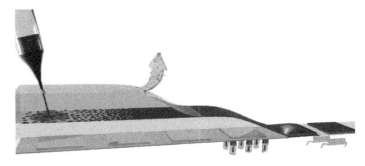

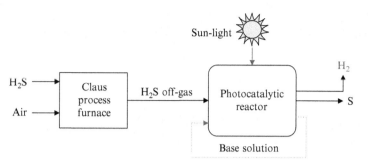

FIG. 8.17 Photocatalytic solar H_2 production from hydrogen sulfide (H_2S) decomposition. *Modified from Baeg, J.O., 2008. A novel nanoscale semiconductor photocatalyst for solar hydrogen production. Solar alternative Energy SPIE Newsroom.*

projected to increase in the future. Decomposition of H_2S has, therefore, attracted attention because of the environmental problems it causes. At a level higher than the threshold limits of 10 ppm, its harmful impacts on lives are observed. Although H_2 can also be obtained by the photocatalytic splitting of water at water's decomposition to H_2 and O_2, higher efficiencies for visible light photocatalyst still are not achieved. H_2S decomposition requires far less energy ($\Delta G° = 33.44$ kJ/mol) compared to the energy required to split water ($\Delta G° = 237.19$ kJ/mol). These combined considerations show that the decomposition of H_2S may be the perfect way to produce H_2. In addition, H_2S degradation with concomitant formation of H_2 offers products for sustainable energy and abatement of environmental pollution.

In the Claus process, H_2S is separated from the host gas stream using amine extraction. Then it is fed to the Claus unit, where it is transformed into two steps. In the first thermal step, H_2S is partly oxidized with air. This is done in a reaction furnace at high temperatures (1000–1400°C). Sulfur is made, but a little H_2S leftovers are unreacted, and some SO_2 is made. In the second catalytic step, the remaining H_2S is reacted with the SO_2 at lower temperatures (about 200–350°C) over a catalyst to make more sulfur, as it is illustrated photocatalytically in Fig. 8.17.

The original draft of the new generation of photocatalysts metal mixtures (in block D of the periodic table), or D0 or D10 configuration or by combining D0 and D10 ions, were developed by employing computational quantum-mechanics-based tools. The involvement of electronic energy levels of the constituent ions in single or composite metal oxide photocatalysts were examined to get the suitable density of states in valence and conduction bands, and appropriate band positions for reduction and oxidation. The effect of chemical modification—for example, nitridation or sulfurization—the provisions of the bands and the highest occupied and lowest unoccupied molecular orbitals were studied in the design process. In quantum-mechanically designed photocatalysts were synthesized using innovative methods. Chemical modification processes such as nitridation, sulfurization, and injection of boron were considered for converting UV photocatalysts, so they were stimulated by visible light. Individual or composite metal-oxide or nonoxide photocatalysts in the form of oxynitrides, oxychalcogenides, oxycarbides, oxyborides, oxyphosphides, and oxysilicides can be arranged (Baeg, 2008).

8.4.6 Photoelectrochemical Hydrogen Production Using Novel Carbon-Based Material

The most important prerequisite of a photocatalyst is its ability to generate electron-hole pairs, collecting a significant part of the available spectrum of solar radiation. Over the past four decades, there has been considerable effort to develop the visible light photocatalyst to split water. These efforts have led to significant progress in understanding the appliances involved, but there are still some problems in the development of visible-light responsives, photostables and cost-effective materials for water photoelectrolysis. However, the large bandgap of these materials allows photoconversion of only UV radiation, which is about 4% of the solar energy spectrum. Several successful efforts have been made to reduce the gap, but to increase the efficiency of solar photochemical generation of hydrogen remains challenging.

Recently, carbon-based materials received much attention as photocatalysts, since these materials are abundant, renewable, and cheap. Carbon-based photochemical materials are ideal and unique in that they absorb energy in the range of visible light, not UV range. Many materials based on carbon are made in efficient photocatalysts by doping them with several extremely electronegative elements and metals. Sharma et al. (2014) reported about a novel, renewable supply-based phosphorus, nitrogen co-doped carbon material (PNDC) arranged by a microwave-assisted technique for photocatalytic splitting of water and hydrogen generation. The synthesis of PNDC includes microwave predecessor by mingling suitable amounts of tannin, melamine, and hexamine.

In their study, it was established that a new photo-active carbon-based material can be used in photoelectrochemical hydrogen production. This material can be adapted to control the bandgap and charge transport properties. The band

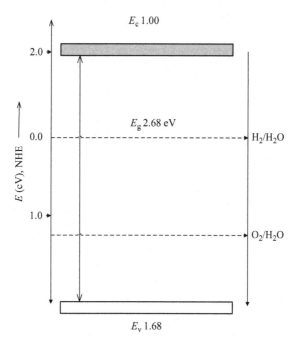

FIG. 8.18 Schematic illustration of the electronic band structure of PNDC. *Modified from Sharma, R., Arnoult, K., Ramasahayam, S.K., Azam, S., Hicks, Z., Shaikh, A., Viswanathan, T., 2014. Photo-electrochemical hydrogen production using novel carbon based material. Paper Presented at the Industry Applications Society Annual Meeting, 5–9 October 2014. IEEE.*

structure of PNDC, shown in Fig. 8.18, represents good prospects for visible light-mediated photocatalysis, which leads to the formation of H_2. Photoelectrochemical response of PNDC as defined in that study is comparable to or better than some oxide semiconductors that have been studied for the photoelectrochemical generation of hydrogen. This can be attributed to a combination of factors, such as the availability of N individuals who contribute photocatalysis, increasing the absorption of visible light by reducing the gap width, and with high surface area and thus increased contact area of the electrode-electrolyte to provide maximum optical absorption and effective transmission charge (Sharma et al., 2014).

8.5 BIOMASS AND BIOLOGICAL-BASED NOVEL HYDROGEN PRODUCTION OPTIONS

Among all the novel processes, biological hydrogen production has two key benefits over the standard methods: it produces less greenhouse gas and couples the metabolic activity of hydrogen producing microorganisms with the concurrent removal of human-derived wastes rich in organics, such as local and food industry wastewaters.

8.5.1 Hydrogen Production from Biomass by in Vitro Metabolic Engineering

The most common agricultural residue in the US is corn stover consisting of leaves and stalks. It consists of about 36% cellulose and 23% xylan whose main component is hemicellulose, which together accounts for about 92% of the total weight of sugars biomass. To release the free glucose and xylose for in vitro production of hydrogen, two entities were used for pretreatment of biomass: cellulose solvent, and organic solvent-based fractionation of lignocellulosic (COSLIF) and dilute acid pretreatment. Pretreated biomass as a result of both procedures were then hydrolyzed using a commercial cellulase/hemicellulase mixture to maximize the production of C6 and C5 sugars. In waste corn stalks, researchers used an enzyme process to break stover into hydrogen and carbon dioxide (Rollin et al., 2015). They used a specially designed set of genetic algorithms to help evaluate each enzymatic process that converts corn stalks into hydrogen and carbon dioxide. They also confirmed the ability of this method to use both sugars found in plant material—glucose and xylose—simultaneously, thereby accelerating the rate at which hydrogen can be produced. By using artificial enzymatic pathways, as shown in Fig. 8.19, it is predicted that, with the increase in the usual limit

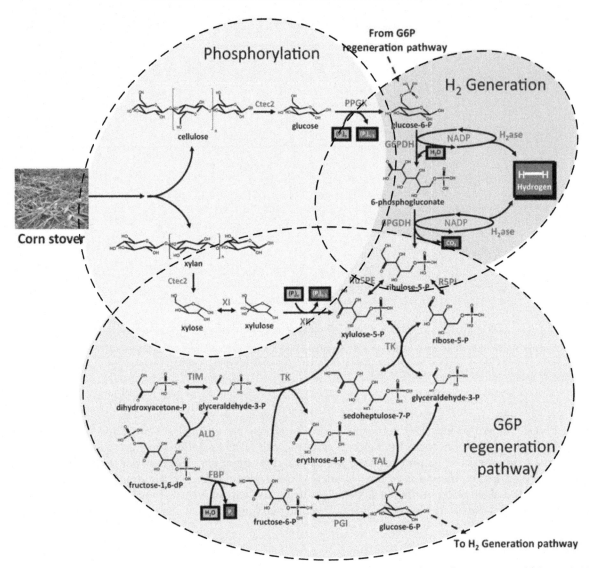

FIG. 8.19 Pathway depicting the enzymatic conversion of biomass to hydrogen and CO_2. *Adapted from Rollin, J.A., Martin del Campo, J., Myung, S., Sun, F., You, C., Bakovic, A., Castro, R., Chandrayan, S.K., Wu, C.-H., Adams, M.W.W., Senger, R.S., Zhang, Y.-H.P., 2015. High-yield hydrogen production from biomass by in vitro metabolic engineering: Mixed sugars coutilization and kinetic modeling. Proc. Natl. Acad. Sci. 112 (16), 4964–4969.*

of hydrogen-producing microorganisms and the avoidance of complex sugar regulatory issues, hydrogen production can be easily achieved in the network of hydrogen filling stations.

The resulting hydrogen is also clean enough to be an ideal candidate for use in hydrogen fuel cells, such as those used in the future hydrogen vehicle fuel cell. Rollin et al. (2015) showed complete conversion of glucose and xylose from plant biomass in H_2 and CO_2 based in vitro enzymatic synthetic way. Glucose and xylose were simultaneously converted to H_2 with a yield of two H_2 per carbon, the maximum possible yield.

Factors of nonlinear kinetic models were prepared with experimental data using genetic algorithm, and global sensitivity analysis was used to classify enzymes that have the greatest effect on the reaction rate and yield. After optimization of enzyme load using the model, the H_2 volumetric productivity was increased three times to 32 mmol H_2/L h. The performance was further increased to 54 mmol H_2/L h by increasing the reaction temperature, substrate and enzyme concentrations 67 times from baseline studies of the method. It is predicted that enzymatic hydrogen production rates could be even more improved by increasing the reaction temperature to 80°C or higher, by means of substrate channeling synthetic metabolons and through finding and using more highly active enzymes and biocatalytic modules. Overall, the production of hydrogen from local biomass is a promising way to achieve world production of green energy (Rollin et al., 2015).

8.5.2 Biological Hydrogen Production from Starch Wastewater

The starch processing industry consumes a lot of water, resulting in a huge amount of industrial wastewater. The wastewater is discharged into main sewage networks without treatment, causing serious pollution problems. Chemical oxygen demand (COD) levels of starch waste water range from 10 to 30 g/L, and it certainly puts significant pressure on the environment, resulting in high costs in terms of waste disposal. Fortunately, starch wastewater is biodegradable, rich organic matter, and its temperature is relatively high (35–40°C). In addition, starch wastewater contains a high percentage of carbohydrates, fiber, protein, and nutrients that are important, energy-rich sources that could possibly be converted into a variety of useful products.

The bioconversion process is a favorable technique to recover beneficial sources from starch waste water, particularly for more-valued products such as microbial biomass protein and biopesticides. Nevertheless, end users feel free to use microbial biomass protein because of its taste, despite the fact that it is high in nucleic acids and digests slowly. The high cost of making them and the practical obstacles to large-scale application biopesticides also limit their use. Therefore, it is necessary to find promising sustainable approaches to treatment and simultaneous conversion of starch wastewater in renewable energy in the form of H_2, as proposed by Nasr et al. (2013).

Fig. 8.20 shows the up-flow anaerobic staged reactor (UASR) used in their study. The displacement reactor was 28 L. The dimensions of the reactor was 19.5 cm long, 19.5 cm wide and 75 cm in height. The reactor was made of Plexiglas material with a pyramidal shape at the bottom. The reactor was equipped with sloping walls to increase the contact time between the H_2 and the bacteria that produced flows substrate. Five sampling ports were distributed on the height of the reactor in order to evaluate intermediates, pH and concentration of residual COD. The volume of gas was measured by the gas meter each day. Wastewater was constantly stirred to prevent deposition of coarse particulates, and the temperature was maintained at 35°C with a thermostatic controlled heater. Starch wastewater was used as the sole substrate during the study. The substrate was diluted with tap water to attain COD concentrations in a range of 5–30 g/L (Nasr et al., 2013).

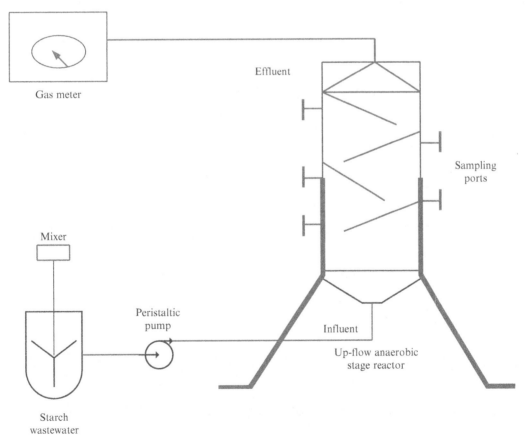

FIG. 8.20   Schematic diagram of an up-flow anaerobic staged reactor (UASR). *Modified from Nasr, M., Tawfik, A., Ookawara, S., Suzuki, M., 2013. Biological hydrogen production from starch wastewater using a novel up-flow anaerobic staged reactor. BioResources 8 (4), 4951–4968.*

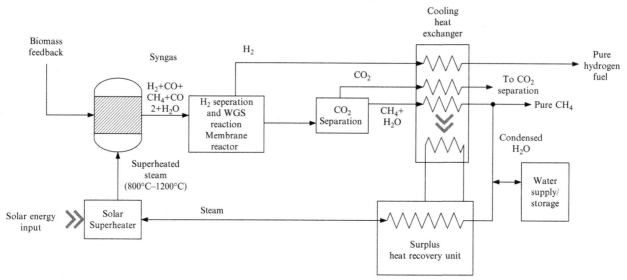

FIG. 8.21 Schematic of the integrated biomass to energy concept system. *Modified from Chung, J.N., 2014. A theoretical study of two novel concept systems for maximum thermal-chemical conversion of biomass to hydrogen. Front. Energy Res. 1, 12.*

8.5.3 Thermal-Chemical Conversion of Biomass to Hydrogen

An air-free process for steam gasification at high temperature has the potential to revolutionize converting biomass to energy technologies and to support the development of a new dimension with systems of high efficiency and minimum consumption of water that do not adversely affect the environment and climate.

As proposed by Chung (2014), the concept for this system is based on a thermochemical process of high-temperature steam gasification of lignocellulose biomass and municipal solid waste. The main objective of the system is to develop the scientific, engineering and technological solutions for converting lignocellulosic biomass and agricultural, forestry and municipal waste into clean energy and to minimize water consumption and the negative impact of energy production on the environment. Production of superheated steam is done by hydrogen combustion using secondary hydrogen, In the second concept, solar energy is used to produce steam. A membrane reactor performs reaction conversion of hydrogen separation and water gas involved in both systems for a pure hydrogen and CO_2 absorption. Based on achieving the maximum rate of hydrogen production, the hydrogen recycling ratio is about 20% for a hydrogen combustion steam-heating system.

The proposed integrated system based on solar heating is shown in Fig. 8.21. Here a concentrated solar heating unit replaces the hydrogen combustion chamber for superheated steam to facilitate gasification. The device is used to concentrate solar energy to heat steam up to the desired temperature gasification. Water vapor is generated from the condensed water vapor, resulting in the gas generator with the addition of an external source.

The overall rate of hydrogen production is a direct reflection of a synthesis gas composition, as shown in Fig. 8.22. For example, at 100 kg/h of biomass consumption rate, the maximum rate of production of H_2 is about 16.8 kg/h. Thermochemical conversion by air-free, high-temperature, supercritical steam at about 1000°C provides the necessary technology to convert biomass into synthesis gas, mainly H_2 and CO, which can be further converted into pure hydrogen or catalytically converted into liquid hydrocarbon fuels such as biodiesel or green fuel and chemicals. Several models of thermochemical reaction showed that the synthesis gas composition is approximately the same temperature when steam is 1000°C. The steam gasification process runs at the atmospheric pressure; therefore, no elevated pressure is required (Chung, 2014).

8.5.4 Biological Hydrogen Production from Corn Syrup Waste

Hafez et al. (2009) conducted a comprehensive study on biological hydrogen production from corn syrup waste. They recognized that converting organic waste into hydrogen is attractive from both pollution control and energy recovery points of view. However, few studies have been conducted to produce hydrogen from wastewater through the real problems associated with inhibition and microbial changes. The maximum specific growth rate for mixed cultures of bacteria to produce hydrogen, $0.333\ \text{h}^{-1}$, corresponds to the minimum retention time of solids (SRT_{min}) of 3 h;

FIG. 8.22 Net hydrogen generation rate for concept system II as a function of STBR. *Data from Chung, J.N., 2014. A theoretical study of two novel concept systems for maximum thermal-chemical conversion of biomass to hydrogen. Front. Energy Res. 1, 12.*

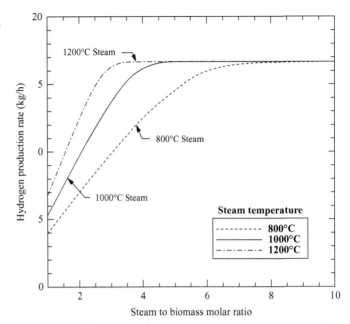

thus CSTRs exploited for hydrogen production are characterized by a hydraulic retention time (HRT) of 3–8 h. However, high levels of dilution resulted in a marked reduction of biomass in a reactor cell through leaching and serious system failure follows. The fill and draw reactors used for the production of hydrogen consistently suffered from unreliable hydrogen making and methane production. In order to overcome issues related to biomass washout in hydrogen reactors, a decoupling of SRT from HRT in hydrogen bioreactors was completed principally by means of biofilms in numerous media, including synthetic plastic media and treated anaerobic granular sludge, activated carbon, expanded clay and loofah sponge, glass beads and membranes. The difficulties with the progress of methanogenic biofilms on media bearing badly affected the stability of the process, which is crucial for sustainable hydrogen generation.

An experimental setup for the biohydrogenerator was constructed as shown in Fig. 8.23. The overall results of experiments done by Hafez et al. (2009) show that the biological production of hydrogen from corn syrup using heat pretreated fermented in anaerobic sediment can be achieved. The rate of hydrogen production is a function of the speed of the organic load; it increased from 10 to 34 L H_2/L d with increasing OLR from 26 to 81 g COD (chemical oxygen demand)/L d. No inhibition of hydrogen production was observed at loads reaching 81 g COD/L d. The largest yield of hydrogen achieved during the experimental period was 3.2 mol H_2/mol hexose, which corresponded to 430 ml H_2/gCOD and secondary glucose hydrogen conversion efficiency to 73%. The relatively high yield of hydrogen was tested with high molar ratio of acetate butyrate and mass balance of the EPC. The contribution of

FIG. 8.23 Experimental setup for the biohydrogenerator. *Modified from Hafez et al. (2009).*

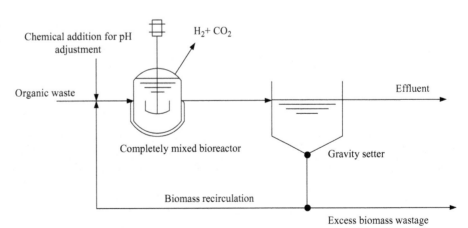

the high hydrogen generating acetate way improved from 68% of the total hydrogen at 26 gCOD/L·d to 78% at 81 gCOD/L d. Also, the fraction of non-hydrogen making bacteria in the biomass reactor diminished from 20% at 26 gCOD/L d to only 12% at 81 g COD/L d. The decoupling of SRT from HRT in bio hydrogen generation systems simplified by the better sludge settling features of hydrogen producers, evaluated in this work, validated the promise of using a gravity settler after a CSTR to preserve high biomass retention in the system and decline biomass washout, thus refining hydrogen yield and sustainability of hydrogen production, rendering the system as a competitor for biological hydrogen production from waste (Hafez et al., 2009).

8.6 OTHER NOVEL HYDROGEN PRODUCTION OPTIONS

For particular metal reactants that can induce hydrogen-evolving chemical reactions, aluminum and its alloys are accepted as one of the most appropriate metals applicable for upcoming hydrogen generation, and there is a trend to use them as an energy material, particularly in recent years. Moreover, the metal utilization is recognized as an effective, user-friendly and safe method for both hydrogen generation and energy storage. In this section, recent developments in hydrogen production and storage based on aluminum, metal hydrides, Ca-Cu thermochemical cycle and radio wave energy options are presented in detail.

8.6.1 Hydrogen Production Using Hydrolysis of Alane and Activated Aluminum

Although aluminum hydride (AlH_3) is not a novel material, it has been notably considered in the last few years as a hydrogen storage material for fuel cell uses because of its high volumetric and gravimetric hydrogen volumes as well as its promising desorption kinetics. There are many polymorphs of AlH_3; nevertheless, the most studied one has been α polymorph owing to its stability. A main disadvantage to the use of this material is that rehydrogenation of aluminum necessitates in excess of 105 bar H_2 and is presently unfeasible for automotive applications. The hydrolysis of Al as well as polymerically or oxide stabilized α-AlH_3 will readily occur in the presence of promoter additives (NaOH, $NaAlH_4$, and NaH) to rapidly release large quantities of H_2 at room temperature even when a stoichiometric volume of water is added to the composite:

$$2Al + 6H_2O \rightarrow 2Al(OH)_3 + 3H_2$$

The promoter additives provide an alkaline environment to remove the oxide layer on Al or polymeric coating for α-AlN in order to allow the reaction to propagate. The use of $NaAlH_4$ and NaH as promoter additives is far greater than the alkaline hydroxide promoter additives (NaOH) usually used. This is due to the large amount of heat released upon their hydrolysis as well as their inherent H_2 content.

The system is ideal for operation in environments where an external water supply is readily available. The α-AlH_3: promoter additives composites can provide a reliable H_2 supply for mobile applications. It is one aspect of at least one of the present embodiments to provide a process of generating hydrogen comprising the steps of: providing a source of AlH_3; heating the AlH_3 to a temperature of at least about 150°C and more preferably to a temperature of between 185°C and 200°C, thereby forming a reaction product consisting of a fast hydrogen gas and activated aluminum; adding a promoter to the activated aluminum reaction product, the promoter being selected from the group consisting of $NaAlH_4$, NaH, and NaOH to provide a composite material; and adding a stoichiometric amount of water to the composite material, thereby generating a second hydrogen gas source. Further details can be found in the patent of Zidan et al. (2014).

8.6.2 Novel Hydrogen Generator and Storage Based on Metal Hydrides

A compact unit was proposed by Rangel et al. (2009) as a new electrochemical system that combines hydrogen production, storage and compression in one device. It has a relatively low cost and high efficiency than classical electrolyzers do, and it provides an opportunity for cost-effective integration of renewable energy sources.

During the electrochemical charging, hydrogen was absorbed in metal hydride and corresponding oxygen was transported from the system. Conversely, in the case of unloading, released hydrogen was brought to the H_2 storage pressure 15 bar. In their system, production and storage of hydrogen were located in one compartment referred to as a hydrogen generator. In this case, the maximum pressure H_2 delivered depended on the mechanical design of the system. The developed hydrogen generator/storage as an electrochemical system consisted of six cell assemblies in a

FIG. 8.24 Schematic drawing of the hydrogen generator prototype integrating hydrogen generation and storage in a novel compact electrochemical system based on metal hydrides. *Modified from Rangel, C.M., Fernandes, V.R., Slavkov, Y., Bozukov, L., 2009. Novel hydrogen generator/storage based on metal hydrides. Int. J. Hydrog. Energy 34 (10), 4587–4591.*

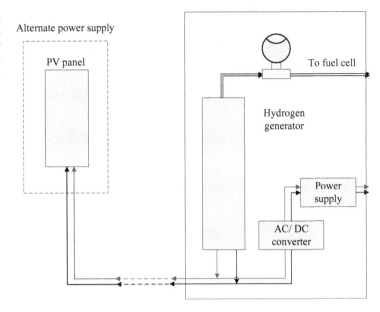

35 wt% KOH solution used as the electrolyte. Metal hydride working electrodes obtained using $LaNi_{4.3}Co_{0.4}Al_{0.3}$ alloy and counter electrodes made of nickel plates were set in a block of stainless steel. A schematic representation of the developed prototype device that shows the electrochemical reactor and means for connecting to a power source is shown in Fig. 8.24.

Water electrolysis, identified as a secondary process during the charging procedure, brought down the efficiency in the last 25% of the charging time. Improving the efficiency of the process was achieved by reducing the resistance between the electrodes of the cell assembly. An integrated system comprising 1.2 kW photovoltaic modules and 1 kW proton exchange membrane fuel cell is discussed in Rangel et al. (2009) and shown to achieve increased operational efficiency at a reduced cost of investment.

8.6.3 Hydrogen Production Based on a Ca-Cu Chemical Cycle

Despite the low specific emissions characteristic of a natural gas combined cycle (NGCC), CO_2 capture in these systems is a necessary condition to achieve the ambitious targets for climate change mitigation. The work conducted by Martínez et al. (2013) proposed the integration into a NGCC of a novel H_2 production process based on a double chemical Ca-Cu loop. This H_2 production process was based on the Sorption-Enhanced Reforming (SER) process including a CaO-based solid as a high-temperature CO_2 sorbent. The addition of a second CuO/Cu loop provided the energy required to carry out sorbent regeneration. The coupling between this process and a combined cycle is discussed in their paper, and an efficiency assessment of the whole plant was accomplished. Ca-Cu looping performance is likely to be improved if fewer exigent conditions are chosen, which will add to the inherent advantages of a cheaper reactor design; feasible solid materials; and fewer process units that will make the Ca-Cu looping process emerge as a potential pre-combustion CO_2 capture technology.

The first step of the process A in Fig. 8.25 consists of producing H_2 from CH_4 and steam via the SER in the presence of a CaO-based sorbent, a Ni-based reforming catalyst and a Cu-supported material. The use of high temperatures would allow higher CH_4 conversions in reforming reaction, but CO_2 capture efficiency would decrease. However, as the overall reaction during SER step is almost thermally neutral, H_2 production would not be significantly affected by temperature. The optimal temperature to carry out the SER must be around 650°C to maximize H_2 yield and promote carbonation reaction. The reaction rates of reforming and carbonation reactions were considered to be fast enough to approach equilibrium when the operation performed at the operating conditions normally employed in a SER process. The H_2-rich product obtained in this step needs to be cooled down to around 300°C before being sent

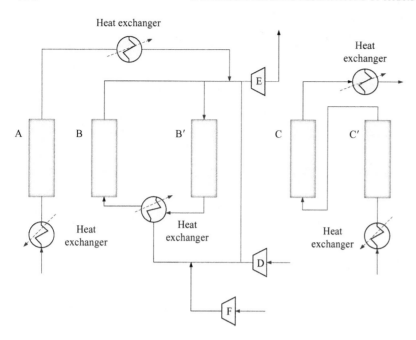

FIG. 8.25 General scheme of the H_2 production process based on the Ca-Cu chemical loop integrated with a combined cycle (D and E: compressor and expander, respectively; F: auxiliary air compressor). *Modified from Martinez et al. (2013).*

to the combustion chamber of the gas turbine. This fuel temperature is considered reasonable in order not to have operational issues in handling hot-H_2 fuel in the gas turbine, and because small thermodynamic benefits would result from further heating beyond this value.

$$CH_4 + H_2O + 206 \, kJ/mol \leftrightarrow CO + 3H_2$$

$$CO + H_2O \leftrightarrow CO_2 + H_2 + 41 \, kJ/mol$$

$$CaO + CO_2 \leftrightarrow CaCO_3 + 178.8 \, kJ/mol$$

Martinez et al. (2013) confirmed that once the CaO in the fixed bed is depleted, the next step of the process (B) starts. In this step, diluted air is fed into the solid bed to oxidize Cu and produce the CuO necessary for subsequent sorbent regeneration (stage C). The design of an integrated combined cycle with a novel H_2 production process based on a double Ca-Cu chemical loop with inherent CO_2 capture was developed in their study.

For the operating conditions chosen to maximize H_2 yield and reduce CO_2 emissions, an efficiency penalty of around 9.5% points was assessed with respect to a NGCC without capture, while having 12% of the emissions of this reference plant. Compared with other pre-combustion technologies focused on power production assessed on literature, similar efficiency penalties and CO_2 capture rates derive from this integrated layout. However, inherent advantages of the Ca-Cu looping process such as cheaper reactor design, fewer operation units or feasible materials, make this novel H_2 production process emerge as a potential pre-combustion CO_2 capture technology (Martínez et al., 2013).

8.6.4 Hydrogen Production Using Radiowave and Microwave Energy

The microwaves of radiowaves can also be utilized for hydrogen production as was proposed by Capistrano et al. (2015). In their research, it was expressed that exposure of methane and/or steam to radiowave (RW) and microwave (MW) energy using copper and rhodium catalysts can produce hydrogen gas. The feasible pathways of chemical reactions leading to its production are given as follows:

$$CH_4 + 2H_22 + MW/RW \, energy + catalyst \rightarrow CO_2 + 4H_2$$

$$4H_2O + MW/RW + catalyst \rightarrow 2O_2 + 4H_2$$

The generation of hydrogen gas was true for both sources of energy whether separately or in combination. The amount of hydrogen produced using radiowave energy was, however, greater by more than 53% as compared to the output of hydrogen using microwave energy. Microwave at 1145 W power using 18.3 g rhodium catalyst yielded an average of 15,000 ppm hydrogen. Using radiowave energy, the same quantity of rhodium catalyst yielded a very

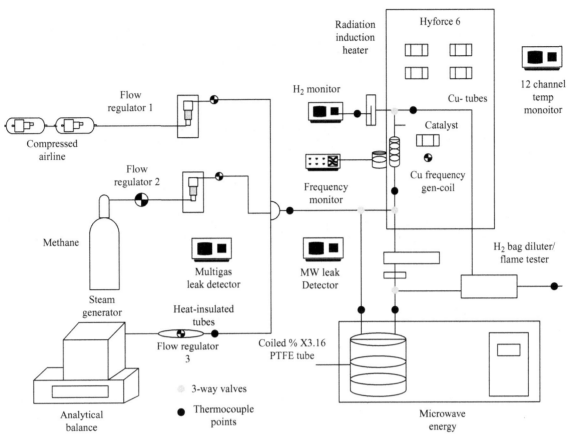

FIG. 8.26 Experimental setup of the radiowave energy-based hydrogen production. *Modified from Capistrano et al. (2015).*

high amount of 542,151 ppm hydrogen. Using 153.3 g copper catalyst, radiowave produced an average amount of 30,820 ppm hydrogen gas. The optimum conditions for these reactions to move forward and yield the highest amount of hydrogen gas were defined as the following: methane average gas flow rates of 1.0 mL/s; average steam flow rates of 5 g/s; radiowave coil glow time of 1 min and greater; a time range of 4–5 min at a power of 500–600 W for microwave; 475 kHz for a radiowave generator; and a coil glow time of 1 min and higher. Methane and steam when subjected to microwave and/or radiowave energy in the presence of catalyst(s) produced hydrogen gas.

The process of subjecting methane gas and/or steam to these two forms of energy followed the pathways in the experimental rig shown in Fig. 8.26. Reagent grade methane gas from liquid air and steam from distilled water through a steam generator quantified precisely by analytical balance were used. Flow meters monitored the gases passing through catalysts (copper and rhodium) subjected to induction heater and microwave energy. Various conditions such as flowrates, temperature, microwave power, gas combinations and radiowave coil glow time were adjusted to observe the effects on hydrogen gas production. The amount of hydrogen gas produced was measured precisely using a gas analyzer. They concluded that hydrogen gas can be produced from methane and/or steam using both microwave and radiowave energy sources. Microwave gave peak production of 15,080 ppm. A higher generation of 210,664 ppm was produced by using radiowave. Using a suppression method by varying reactants, conditions and radiowave generator parameters, the optimum production of hydrogen was recorded at 542,151 ppm. The potential likelihood, however, of the reaction leading to the optimum production of hydrogen was primarily, as follows:

$$2H_2O + microwave/radiowave\ energy + Rh/Cu \rightarrow O_2 + 4H_2$$

This reaction was not seen in the literature, but the reactants were known, and the products were identified by a hydrogen gas analyzer and was validated to support the Le Chatelier principle of reactions. It was also highlighted that the addition of a small quantity of methane at 1 mL/s, augmented the production of hydrogen tremendously. The optimum conditions determined for both microwave and radiowave induction energy that produce maximum hydrogen output were namely: Rhodium superior to copper catalyst; steam flow rate of 5 g/s and low methane flow rates of 1 mL/s; radiowave coil glow time of one minute or more at frequency of 473 kHz; and a catalyst temperature range of 475–508°C (Capistrano et al., 2015).

8.7 CONCLUDING REMARKS

This chapter is presented to discuss the recent progress made on some novel and newly developed hydrogen production processes, methods, techniques, technologies, etc., in order to reflect potential options and future directions on hydrogen production technologies. The materials covered include numerous hydrogen production methods, ranging from biological to photoelectrochemical options. Various pathways through which the four kinds of energies drive hydrogen production can be obtained from energy sources. The electrical and thermal energy can be derived from renewable energies such as solar, wind, geothermal, tidal, wave, ocean thermal, hydro, and biomass or from nuclear energy, or from recovered energy.

Acronyms

AES	Auger electron spectroscopy
AR	air reactor
BET	specific surface area
CMR	combination of steam and oxidation reforming
COD	chemical oxygen demand
COND	condenser
ECPB	electron-coupled proton buffer
FDCA	furandicarboxylic acid
FR	fuel reactor
HER	hydrogen-evolving reaction
HMF	hydroxymethylfurfural
HRSG	heat recovery steam generation
HRT	hydraulic retention time
HTS	high-temperature shift
ITWS	isothermal water splitting
LPSDM	layered powder sintering and dissolution method
LTS	low-temperature shift
MDEA	methyldiethanolamine
MSR	methanol steam reforming
MW	microwave
NGCC	natural gas combined cycle
OER	oxygen-evolving reaction
PEC	photoelectrochemical cell
PEM	proton exchange membrane
PNDC	nitrogen co-doped carbon material
PSA	pressure swing adsorption
RW	radiowave
SEM	scanning electron microscopy
SMR	steam methane reforming
SRT	retention time of solids
STWS	solar-thermal water splitting
UASR	up-flow anaerobic staged reactor
UV	ultraviolet
XPS	X-ray photoelectron spectroscopy
XRD	X-ray diffraction
$\Delta G°$	Gibbs free energy

References

Andersen, E.R., Andersen, E.J., 2003. Method for Producing Hydrogen. U.S. Patent No. 6,506,360 B1. U.S. Patent and Trademark Office, Washington, DC.

Baeg, J.O., 2008. A novel nanoscale semiconductor photocatalyst for solar hydrogen production. Solar Alternative Energy SPIE Newsroom.

Capistrano, R.F., 2015. Novel techniques of generating huge amount of hydrogen gas using microwave and radiowave energy. Int. J. Technol. Enhancements Emerg. Eng. Res. 3 (7), 113–116.

Cha, H.G., Choi, K.-S., 2015. Combined biomass valorization and hydrogen production in a photoelectrochemical cell. Nat. Chem. 7 (4), 328–333.

Chua, K.J., 2015. Novel Hydrogen Production from Sunlight and Rain-water. National University of Singapore, Department of Mechanical Engineering. Accessed from http://www.ae-innovation.org/CMS/uploadfile/2014/0429/Energy%20Study%20Poster%20Solar%20Hydrogen-NUS.pdf.

Chung, J.N., 2014. A theoretical study of two novel concept systems for maximum thermal-chemical conversion of biomass to hydrogen. Front. Energy Res. 1, 12.

Erogbogbo, F., Lin, T., Tucciarone, P.M., LaJoie, K.M., Lai, L., Patki, G.D., Prasad, P.N., Swihart, M.T., 2013. On-demand hydrogen generation using nanosilicon: splitting water without light, heat, or electricity. Nano Lett. 13 (2), 451–456.

Hafez, H., Nakhla, G., Naggar, H.E., 2009. Biological hydrogen production from corn-syrup waste using a novel system. Energies 2 (2), 445–455.

Hashemi, S.M., Modestino, M.A., Psaltis, D., 2015. A membrane-less electrolyzer for hydrogen production across the pH scale. Energy Environ. Sci. 8 (7), 2003–2009.

Lin, S., Suzuki, Y., Hatano, H., Oya, M., Harada, M., 2001. Innovative hydrogen production by reaction integrated novel gasification process (HyPr-RING). J. South. Afr. Inst. Min. Metall. 101 (1), 53–59.

Luo, J., Im, J.-H., Mayer, M.T., Schreier, M., Nazeeruddin, M.K., Park, N.-G., David Tilley, S., Fan, H.J., Grätzel, M., 2014. Water photolysis at 12.3% efficiency via perovskite photovoltaics and Earth-abundant catalysts. Science 345 (6204), 1593–1596.

Martínez, I., Murillo, R., Grasa, G., Fernández, J.R., Abanades, J.C., 2013. Design of a hydrogen production process for power generation based on a Ca-Cu chemical loop. Energy Procedia 37, 626–634.

Mei, D., Feng, Y., Qian, M., Chen, Z., 2016. An innovative micro-channel catalyst support with a micro-porous surface for hydrogen production via methanol steam reforming. Int. J. Hydrog. Energy.

Muhich, C.L., Evanko, B.W., Weston, K.C., Lichty, P., Liang, X., Martinek, J., Musgrave, C.B., Weimer, A.W., 2013. Efficient generation of H_2 by splitting water with an isothermal redox cycle. Science 341 (6145), 540.

Nasr, M., Tawfik, A., Ookawara, S., Suzuki, M., 2013. Biological hydrogen production from starch wastewater using a novel up-flow anaerobic staged reactor. BioResources 8 (4), 4951–4968.

Peng, Z., 2003. A Novel Hydrogen and Oxygen Generation System. Faculty of the Louisiana State University and Agricultural and Mechanical College in Partial Fulfillment of the Requirements for the Degree of Master of Science in Chemical Engineering, In The Gorden A. and Mary Cain Department of Chemical Engineering by Zhiyong Peng BE, Tianjin University.

Rangel, C.M., Fernandes, V.R., Slavkov, Y., Bozukov, L., 2009. Novel hydrogen generator/storage based on metal hydrides. Int. J. Hydrog. Energy 34 (10), 4587–4591.

Rausch, B., Symes, M.D., Chisholm, G., Cronin, L., 2014. Decoupled catalytic hydrogen evolution from a molecular metal oxide redox mediator in water splitting. Science 345 (6202), 1326–1330.

Rollin, J.A., Martin del Campo, J., Myung, S., Sun, F., You, C., Bakovic, A., Castro, R., Chandrayan, S.K., Wu, C.-H., Adams, M.W.W., Senger, R.S., Zhang, Y.-H.P., 2015. High-yield hydrogen production from biomass by in vitro metabolic engineering: Mixed sugars coutilization and kinetic modeling. Proc. Natl. Acad. Sci. 112 (16), 4964–4969.

Rydén, M., Lyngfelt, A., Mattisson, T., 2006. Two novel approaches for hydrogen production: chemical-looping reforming and steam reforming with carbon dioxide capture by chemical-looping combustion. In: Paper Presented at the Proceedings of the 16th World Hydrogen Energy Conference, Lyon, France.

Schuyten, S., Dinka, P., Mukasyan, A.S., Wolf, E., 2008. A novel combustion synthesis preparation of $CuO/ZnO/ZrO_2/Pd$ for oxidative hydrogen production from methanol. Catal. Lett. 121 (3), 189–198.

Sharma, R., Arnoult, K., Ramasahayam, S.K., Azam, S., Hicks, Z., Shaikh, A., Viswanathan, T., 2014. Photo-electrochemical hydrogen production using novel carbon based material. In: Paper Presented at the Industry Applications Society Annual Meeting, 5–9 October 2014. IEEE.

Sobczynski, A., Yildiz, A., Bard, A.J., Campion, A., et al., 1988. Tungsten disulfide: a novel hydrogen evolution catalyst for water decomposition. J. Phys. Chem. 92 (8), 2311–2315.

Xu, B., Bhawe, Y., Davis, M.E., 2012. Low-temperature, manganese oxide-based, thermochemical water splitting cycle. Proc. Natl. Acad. Sci. 109 (24), 9260–9264.

Yu, X.Y., Prevot, M.S., Guijarro, N., Sivula, K., 2015. Self-assembled 2D WSe_2 thin films for photoelectrochemical hydrogen production. Nat. Commun. 6, 7596.

Zidan, R., Teprovich, J.A., Motyka, T., 2014. Two Step Novel Hydrogen System Using Additives to Enhance Hydrogen Release from the Hydrolysis of Alane and Activated Aluminum. US Patent No. 9,199,844 B2. U.S. Patent and Trademark Office, Washington, DC.

STUDY PROBLEMS

8.1 What is the critical role of micro-channel catalyst in hydrogen production from methanol steam reforming?

8.2 What is the contribution of CaO in the reaction integrated gasification process?

8.3 What does chemical-cycle combustion stand for?

8.4 What is the significant advantage of chemical cycle reforming compared to conventional technology?

8.5 What is the difference between sorption-enhanced reforming process and standard steam methane reforming process?

8.6 What are the specific properties of electron-coupled proton buffer (ECPB)?

8.7 Explain the main steps of manganese oxide-based thermochemical water splitting cycle.

8.8 What is the catalyst used in aluminum-based hydrogen production?

8.9 How does membrane-less electrolyzer work?

8.10 Express the methodology of splitting water using nanosilicon.

8.11 What is the contribution of perovskite to photovoltaics efficiency?

8.12 Express the main differences between solar thermal water splitting (STWS) and isothermal water splitting (ITWS).

8.13 In which aspect does photocatalytic reactor contribute to Claus process?

8.14 Explain the steps of biological hydrogen production from starch wastewater and corn syrup waste.

8.15 What are the key stages of Ca-Cu chemical cycle based hydrogen production?

8.16 Does radiowave or microwave energy yield higher hydrogen production?

A

Conversion Factors

Quantity	SI to ESU conversion	ESU to SI conversion
Area	$1 m^2 = 10.764 ft^2$ $= 1550.0 in.^2$	$1 ft^2 = 0.00929 m^2$ $1 in.^2 = 6.452 \times 10^{-4} m^2$
Density	$1 kg/m^3 = 0.06243 lb_m/ft^3$	$1 lb_m/ft^3 = 16.018 kg/m^3$ $1 slug/ft^3 = 515.379 kg/m^3$
Diffusivity	$1 m^2/s = 10.7639 ft^2/s$	$1 ft^2/s = 0.0929 m^2/s$
Dynamic viscosity ($kg/m \, s = N \, s/m^2$)	$1 kg/ms = 0.672 lb_m/fts$ $= 2419.1 lb_m/fth$	$1 lb_m/ft \, s = 1.4881 kg/m \, s$ $1 lb_m/ft \, h = 4.133 \times 10^{-4} kg/m \, s$ $1 centipoise (cP) = 10^{-2} poise$ $= 1 \times 10^{-3} kg/ms$
Energy	$1 J = 9.4787 \times 10^{-4} Btu$ $1 J/kg = 4.2995 \times 10^{-4} Btu/lb_m$	$1 Btu = 1055.056 J$ $1 cal = 4.1868 J$ $1 lb_f \, ft = 1.3558 J$ $1 hp \, h = 2.685 \times 10^6 J$ $1 Btu/lb_m = 2326 J/kg$
Force	$1 N = 0.22481 lb_f$	$1 lb_f = 4.448 N$ $1 pdl = 0.1382 N$
Heat flux	$1 W/m^2 = 0.3171 Btu/h \, ft^2$	$1 Btu/h \, ft^2 = 3.1525 W/m^2$ $1 kcal/h \, m^2 = 1.163 W/m^2$ $1 cal/s \, cm^2 = 41,870.0 W/m^2$
Heat transfer coefficient	$1 W/m^2 \, K = 0.1761 Btu/h \, ft^2 \, °F$	$1 Btu/h \, ft^2 \, °F = 5.678 W/m^2 \, K$ $1 kcal/h \, m^2 \, °C = 1.163 W/m^2 \, K$ $1 cal/s \, m^2 \, °C = 41870.0 W/m^2 \, K$
Heat transfer rate	$1 W = 3.4123 Btu/h$	$1 Btu/h = 0.2931 W$
Kinematic viscosity	$1 m^2/s = 10.7639 ft^2/s$ $= 1 \times 10^4 stokes$	$1 ft^2/s = 0.0929 m^2/s$ $1 ft^2/h = 2.581 \times 10^{-5} m^2/s$ $1 stoke = 1 cm^2/s$
Length	$1 m = 3.28084 ft$ $1 m = 39.3701 in.$	$1 ft = 0.3048 m$ $1 in. = 25.4 mm$
Mass	$1 kg = 2.2046 lb_m$ $1 ton (metric) = 1000 kg$ $1 grain = 6.47989 \times 10^{-5} kg$	$1 lb_m = 0.4536 kg$ $1 slug = 14.594 kg$
Mass flow rate	$1 kg/s = 7936.6 lb_m/h$ $= 2.2046 lb_m/s$	$1 lb_m/h = 0.000126 kg/s$ $1 lb_m/s = 0.4536 kg/s$
Power	$1 W = 1 J/s = 3.4123 Btu/h$ $= 0.737562 lb_f \, ft/s$ $1 hp (metric) \, 0.735499 kW$ $1 ton \, of \, refrig. = 3.51685 kW$	$1 Btu/h = 0.2931 W$ $1 Btu/s = 1055.1 W$ $1 lb_f \, ft/s = 1.3558 W$ $1 hp^{UK} = 745.7 W$

Continued

Quantity	SI to ESU conversion	ESU to SI conversion
Pressure	$1\,\text{Pa} = 0.020886\,\text{lb}_f/\text{ft}^2$ $= 1.4504 \times 10^{-4}\,\text{lb}_f/\text{in.}^2$ $= 4.015 \times 10^{-3}$ in water $= 2.953 \times 10^{-4}$ in Hg	$1\,\text{lb}_f/\text{ft}^2 = 47.88\,\text{Pa}$ $1\,\text{lb}_f/\text{in.}^2 = 1\,\text{psi} = 6894.8\,\text{Pa}$ 1 stand. atm. $= 1.0133 \times 10^5\,\text{Pa}$ $1\,\text{bar} = 1 \times 10^5\,\text{Pa}$
Specific heat	$1\,\text{J/kg K} = 2.3886 \times 10^{-4}\,\text{Btu/lb}_m\,{}^\circ\text{F}$	$1\,\text{Btu/lb}_m\,{}^\circ\text{F} = 4187\,\text{J/kg K}$
Surface tension	$1\,\text{N/m} = 0.06852\,\text{lb}_f/\text{ft}$	$1\,\text{lb}_f/\text{ft} = 14.594\,\text{N/m}$ $1\,\text{dyn/cm} = 1 \times 10^{-3}\,\text{N/m}$
Temperature	$T(\text{K}) = T({}^\circ\text{C}) + 273.15$ $= T({}^\circ\text{R})/1.8$ $= [T({}^\circ\text{F}) + 459.67]/1.8$ $T({}^\circ\text{C}) = [T({}^\circ\text{F}) - 32]/1.8$	$T({}^\circ\text{R}) = 1.8T(\text{K})$ $= T({}^\circ\text{F}) + 459.67$ $= 1.8T({}^\circ\text{C}) + 32$ $= 1.8[T(\text{K}) - 273.15] + 32$
Temperature difference	$1\,\text{K} = 1\,{}^\circ\text{C} = 1.8\,{}^\circ\text{R} = 1.8\,{}^\circ\text{F}$	$1\,{}^\circ\text{R} = 1\,{}^\circ\text{F} = 1\,\text{K}/1.8 = 1\,{}^\circ\text{C}/1.8$
Thermal conductivity	$1\,\text{W/m K} = 0.57782\,\text{Btu/h ft }{}^\circ\text{F}$	$1\,\text{Btu/h ft }{}^\circ\text{F} = 1.731\,\text{W/m K}$ $1\,\text{kcal/h m }{}^\circ\text{C} = 1.163\,\text{W/m K}$ $1\,\text{cal/s cm }{}^\circ\text{C} = 418.7\,\text{W/m K}$
Thermal resistance	$1\,\text{K/W} = 0.52750\,{}^\circ\text{F h/Btu}$	$1\,{}^\circ\text{F h/Btu} = 1.8958\,\text{K/W}$
Velocity	$1\,\text{m/s} = 3.2808\,\text{ft/s}$ $1\,\text{km/s} = 0.62137\,\text{mi/h}$	$1\,\text{ft/s} = 0.3048\,\text{m/s}$ $1\,\text{ft/min} = 5.08 \times 10^{-3}\,\text{m/s}$
Volume	$1\,\text{m}^3 = 35.3134\,\text{ft}^3$ $1\,\text{L} = 1\,\text{dm}^3 = 0.001\,\text{m}^3$	$1\,\text{ft}^3 = 0.02832\,\text{m}^3$ $1\,\text{in.}^3 = 1.6387 \times 10^{-5}\,\text{m}^3$ $1\,\text{gal}^{US} = 0.003785\,\text{m}^3$ $1\,\text{gal}^{UK} = 0.004546\,\text{m}^3$
Volumetric flow rate	$1\,\text{m}^3/\text{s} = 35.3134\,\text{ft}^3/\text{s}$ $= 1.2713 \times 10^5\,\text{ft}^3/\text{h}$	$1\,\text{ft}^3/\text{s} = 2.8317 \times 10^{-2}\,\text{m}^3/\text{s}$ $1\,\text{ft}^3/\text{min} = 4.72 \times 10^{-4}\,\text{m}^3/\text{s}$ $1\,\text{ft}^3/\text{h} = 7.8658 \times 10^{-6}\,\text{m}^3/\text{s}$ $1\,\text{gal}^{US}/\text{min} = 6.309 \times 10^{-5}\,\text{m}^3/\text{s}$

Conversion factors under the column heading "Quantity" are given alphabetically.

Index

Note: Page numbers followed by *f* indicate figures and *t* indicate tables.

Printed in the United States
By Bookmasters